The CSE Manual

NINTH EDITION

The Council of Science Editors (CSE) is an international membership organization for editorial professionals publishing in the sciences. CSE's purpose is to serve its approximately 800 members in the scientific, scientific publishing, and information science communities by fostering networking, education, discussion, and exchange. CSE's aim is to be an authoritative resource on current and emerging issues in the communication of scientific information.

CSE Style Manual Task Force: Kaitlyn Aman Ramm; Janaynne do Amaral, PhD; Emmanuel A. Ameh, MBBS, FWACS, FACS; Jessica S. Ancker, MPH, PhD; Stanley N.C. Anyanwu, MBBS, FMCS, FWACS, FACS, FICS; Kenneth April; Elizabeth A. Arnold, ELS; Patricia K. Baskin, MS; Kevin R. Brown; Michelle Cathers, MLS; Jae Hwa Chang, MA, ELS; Soo-Hee Chang, MS, ELS; Dana M. Compton; Craig Damlo, MSc; Sherri Damlo, MS, ELS; Heather DiAngelis, MA; Simona Fernandes, MSc, ELS; Tina L. Fleischer; Peter J. Freeman, PhD; Beth E. Hazen, PhD; Sun Huh, MD, PhD; Andrew Hunt, PhD; Jacob Kendall-Taylor; Raymond Lambert, MLS, ELS; Trevor Lane, DPhil, PGCELT; Thomas A. Lang, MA; Jessica LaPointe; Ellen Lazarus, MD, ELS; Iris Y. Lo; Rachel J. Lowery; Audrey Daniel Lusher; Amy McPherson; Stephanie Mowat, MA; Beva Nall-Langdon; Kelly Newton; Joanna Odrowaz; Peter J. Olson, ELS; Heather Poirier, MA; Andrea Rahkola, ELS; Dara R. Rochlin; David M. Schultz, PhD, FAMS, FRMetS, FGS, SFHEA; Brit Stamey; Julie Steffen, CAE; Jessica L. Striley; Kristen Swendsrud; Caitlyn Trautwein; and Mary Warner, CAE

CSE Style Manual Advisory Group: Dana M. Compton, CSE Board of Directors' liaison; Mary E. Laur, executive editor at The University of Chicago Press; Russell Harper, freelance editor and writer; Mollie McFee, senior editorial associate at The University of Chicago Press; and Michael E. Fitzgerald, CAE, project manager for *The CSE Manual*.

CSE is grateful to The University of Chicago Press for its extensive contributions to the production of this style manual.

Other CSE publications include the following:

CSE's Recommendations for Promoting Integrity in Scientific Journal Publications (continuously updated recommendations paper) at https://www.councilscienceeditors.org/recommendations-for-promoting-integrity-in-scientific-journal-publications
Science Editor (quarterly publication) at https://www.csescienceeditor.org/

Direct inquiries to:

Council of Science Editors
CSE@CouncilScienceEditors.org
www.councilscienceeditors.org

THE
CSE
MANUAL

Scientific Style and Format for Authors, Editors, and Publishers

NINTH EDITION
Council of Science Editors
STYLE MANUAL TASK FORCE

Published by the Council of Science Editors in cooperation
with The University of Chicago Press
The University of Chicago Press   Chicago and London

The University of Chicago Press, Chicago 60637
The University of Chicago Press, Ltd, London
© 2006, 2014, 2024 by the Council of Science Editors
All rights reserved. No part of this book may be used or reproduced in any manner whatsoever without written permission, except in the case of brief quotations published in critical articles and reviews. For more information, contact The University of Chicago Press, 1427 E 60th St, Chicago, IL 60637.
Published 2024
Printed in the United States of America

33 32 31 30 29 28 27 26 25 24   1 2 3 4 5

ISBN-13: 978-0-226-68394-2 (cloth)
DOI: https://doi.org/10.7208/cse9

Library of Congress Cataloging-in-Publication Data

Names: Council of Science Editors. Style Manual Task Force, author.
Title: The CSE manual : scientific style and format for authors, editors, and publishers / Council of Science Editors Style Manual Task Force.
Other titles: Scientific style and format | Council of Science Editors manual
Description: Ninth edition. | Chicago : Published by the Council of Science Editors in cooperation with The University of Chicago Press, 2024. | Previously published under title: Scientific style and format. | Includes bibliographical references and index.
Identifiers: LCCN 2023039383 | ISBN 9780226683942 (cloth)
Subjects: LCSH: Technical writing—Handbooks, manuals, etc.
Classification: LCC T11 .C675 2024 | DDC 808.06/6—dc23/eng/20230901
LC record available at https://lccn.loc.gov/2023039383

♾ This paper meets the requirements of ANSI/NISO Z39.48-1992 (R2009) (Permanence of Paper).

CONTENTS

PREFACE

ABOUT THIS EDITION

"What's *SSF*? Never heard of it. I use the CSE manual." Despite being used for 30 years and 3 editions, *Scientific Style and Format* never caught on as the main title for the style manual produced by the Council of Science Editors (CSE). Nor did the manual's edition-specific initialisms, *SSF6*, *SSF7*, and *SSF8*. Instead, survey results and anecdotal evidence indicate that CSE members, researchers, and others refer to the style guide as simply "the CSE manual."

Consequently, for the ninth edition, the CSE Board of Directors retitled the style manual to align with the prevailing preference of those who use it, as well as to emphasize that the manual is the council's official style guide. By transposing elements from the manual's main title and subtitle, the new title, *The CSE Manual: Scientific Style and Format for Authors, Editors, and Publishers*, echoes the titles used for the manual's third, fourth, and fifth editions. In essence, to quote electrical engineer Steven Magee, "Sometimes you make more progress by going backwards."

Reviving one of the manual's previous titles is not the only way that the ninth edition is building on previous editions.[1-8] Here are some of the others.

Commitment to Diversity, Equity, and Inclusivity

The entire team that revised *The CSE Manual* sought to ensure that its contents reflect CSE's commitment to promoting diversity, equity, and inclusivity (DEI) in scientific publishing. The following changes were made to the ninth edition with the understanding that societal standards will evolve and the guidelines in this edition reflect the revision team's advice at the time of publication.

The DEI guidelines in the ninth edition mirror the CSE Board of Directors' 2020 policy statement on capitalizing racial and ethnic designations.[9] When race or ethnicity is pertinent to research manuscripts, the ninth edition specifies capitalizing the designations of "Black," "White," and "Indigenous," and it introduces "Latinx" as a gender-neutral alternative to "Latino" and "Latina." These updates are detailed in Section 7.4, "Inclusive Language," and Section 8.3, "Human Groups." To ensure that these guidelines reflect CSE's commitment to DEI, both sections were peer reviewed by the 2021–2023 chairs of CSE's Diversity, Equity, Inclusion, and Accessibility Committee: Otito Frances Iwuchukwu, PharmD, MA, PhD, FCP, CPTD, and Leonard Jack Jr, PhD, MSc.

Section 7.4 now also offers advice on when using the "singular they" is justified, as when referring to individuals who identify as nonbinary and to those whose gender is unknown or undisclosed for confidentiality or study-masking purposes. The section also stresses that using plural antecedents with plural pronouns is usually preferable to using singular antecedents with the "singular they."

Another concern addressed in Section 7.4 is avoiding stereotypical language, such as by substituting common-gender terms like "meteorologist" for gender-specific terms like "weatherman" and by using person-first descriptions, such as "patients with diabetes mellitus," instead of depersonalizing terms, such as "diabetics."

In addition, for the ninth edition, Section 7.11 was renamed from "Difficulties for Authors for Whom English Is a Second Language" to "Scientific English for Multilingual Authors." This section was refocused to address the challenges that authors face when writing scientific English if they are familiar with academic traditions that differ from Western scientific traditions. For example, Section 7.11 describes the differences between digressive writing and direct scientific English.

To make the manual less focused on male scientists in the United States conducting medical research, many examples throughout the manual were revised to highlight the contributions of female scientists, researchers from around the world, and investigators from a wide variety of scientific disciplines. Such examples in the ninth edition include Chinese paleontologist Meemann Chang, PhD; Norwegian psychologist and neuroscientist May-Britt Moser, PhD; French explorer and botanist Jeanne Baret; and Canadian physician and chemist Maud Leonora Menten, MD, PhD.

Furthermore, the ninth edition was revised and peer reviewed by a wide variety of science editors from 6 continents. By recruiting an international team to revise *The CSE Manual*, CSE not only enhanced the manual's relevance for English-language publications around the world, but the council also demonstrated its commitment to strengthen its relationships with such organizations as the African Journal Partnership Program, the Brazilian Association of Science Editors, the Korean Council of Science Editors, and the Asian Council of Science Editors.

Countries represented by chapter editors and peer reviewers:

Australia	Brazil
Canada	China, specifically Hong Kong
India	Nigeria
South Korea	Switzerland
United Kingdom	United States

The diverse team of contributors to the ninth edition includes editors in chief, publication directors, managing editors, manuscript editors, freelance editors, and even a space systems engineer. The chapter editors and reviewers included 18 with academic doctoral degrees and another 18 with master's degrees. Six hold the physician degree of either MD or MBBS, 20 are certified as editors in the life sciences by the Board of Editors in the Life Sciences, and 1 is a doctor of jurisprudence. Additionally, 4 are former CSE presidents, and 2 are former editors of CSE's publication *Science Editor*.

Finally, the online version of the ninth edition complies with the international standards of the World Wide Web Consortium's Web Content Accessibility Guidelines (WCAG) 2.1, Level AA, including the guidelines for using alternative text for all content presented as art. The manual's online version is available at https://www.csemanual.org.

Enhanced Digital Guidelines

To address the perpetually evolving electronic environment in scientific writing and publishing, a chapter titled "Digital Standards of Scholarly Journal Publishing" was added to *The CSE Manual*. Written by Sun Huh, MD, PhD, while he served as the president of both the Korean Council of Science Editors and the Korean Association of Medical Journal Editors, the new Chapter 33 covers topics ranging from the basic elements of journal home pages and the variety of Crossref services to journal metrics and artificial intelligence in scholarly publishing.

Additionally, content throughout the rest of the manual has been updated to reflect the growing prominence of electronic publishing compared with print publishing. For example, Chapter 32, "Proof Correction," was reorganized so that it covers annotating PDF proofs before it addresses marking paper proofs.

Overhauled Recommendations on References

Building on the eighth edition's switch to using the citation–sequence format for references, Chapter 29, "References," has undergone a major overhaul for *The CSE Manual*'s ninth edition. The requirements for end references were revised to promote conciseness and efficiency. In the process, the chapter was substantially shortened by eliminating redundant, obsolete, and other unnecessary content.

One major change in the ninth edition is to the rule on the number of authors to list in end references to documents with 6 or more authors. Under the revised recommendation, only the first author should be named before using "et al," instead of the first 5 authors. This change brings CSE's reference style in line with that of prominent scientific journals, most notably the American Association for the Advancement of Science's journal *Science* and the *Proceedings of the National Academy of Sciences of the United States of America*.

To further promote conciseness, the revised chapter recommends removing access dates from the end references to most online sources because those dates are no longer as relevant as they were in the early days of online publishing. Under the updated guidelines, access dates are recommended only for online sources for which the date of publication, copyright, or revision cannot be determined. Also removed from end references is the place of publication for book publishers, which has become less relevant because research works are easy to find online and because many book publishers have multiple locations.

The ninth edition recommends digital object identifiers (DOIs) as the preferred uniform resource locators in end references for online documents, when DOIs are available.

The revisions to Chapter 29 also include a new section on referencing journal preprints.

Expanded Guidance on Tables and Figures

To bolster the manual's guidance for visually representing data in scientific manuscripts, Chapter 30, "Tables, Figures, and Indexes," features 12 new figures that illustrate designs for graphs and 6 revised figures that illustrate how tables should be designed.

American versus British Conventions

The ninth edition takes care to distinguish between American conventions and British conventions, but the manual does not specify which is preferred. Instead, the ninth edition recommends that publications select one convention and apply it consistently throughout the publications.

Among the most apparent differences between American and British conventions is the use of quotation marks. Where American convention uses double quotation marks, British convention uses single quotation marks, and vice versa. In many circumstances, British convention places periods and commas outside closing quotation marks, while American convention never does.

Whereas the eighth edition recommended a hybrid between American and British conventions for quotation marks, the ninth edition recommends that publications pick one convention or the other (see Sections 5.3.4, "Quotation Marks," and 10.2.1, "Run-in Quotations"). However, to demonstrate how to follow a convention consistently, the ninth edition adheres to the American convention throughout the rest of the manual.

Real Examples

The CSE Manual was enhanced by removing examples that had been fabricated and those that were from general literature and replacing them with real examples from scientific literature. As a consequence, examples in the ninth edition use the names of known scientists and recognized scientific institutions. For instance, where the eighth edition illustrated how to separate family name suffixes from surnames by using as an example Franklin D. Roosevelt Jr, a son of the 32nd US president, the ninth edition uses pharmacologist and biochemist Earl Wilbur Sutherland Jr, MD, and hyperbaric medicine researcher Edmund Converse Peirce II, MD.

Sensible Expectations for Authors

Some guidelines in *The CSE Manual* have limited value to authors, even though they are significant to manuscript editors and copy editors. As chapter editor and former CSE president Thomas A. Lang, MA, stressed throughout the process of developing the ninth edition, "Telling authors how to create an en and an em dash or the point sizes of rules in a table is a waste of time. A good copy editor would flag those issues." Consequently, the ninth edition advises authors on such issues. For example, Section 5.3.5.1, "En Dash," advises, "Authors can rely on manuscript editors to spot any hyphens that should be converted to en dashes."

Nevertheless, for those authors interested in using correct punctuation and symbols, the ninth edition provides Unicodes when recommending characters not on keyboards (eg, Unicode 2032 for the prime symbol and Unicode 2009 for a thin space).

Refreshed Online Edition

The online version of the ninth edition includes a fresh design and new accessibility features. Corrections and updates will be made to this version as necessary. The online version is available at https://csemanual.org.

General Revisions

Each time *The CSE Manual* is revised, all existing chapters are reviewed for content that should be added, updated, clarified, or deleted, and the manual is revised accordingly. One of the primary goals for the ninth edition was to make the manual easier to read and comprehend. The chapter editors accomplished this by refining the language and creating a more coherent manual.

Other significant updates made to the ninth edition include the following:

- Cross-references to other sections in the manual were updated to include the section titles, not just the section numbers. This update provides users with clues about the content provided in the cross-references. In addition, as with the eighth edition, cross-references in the online version of the ninth edition are hyperlinked to the corresponding sections so that users can quickly and easily navigate to those sections.
- Guidance on avoiding sentence fragments was enhanced in Section 5.3.3.4, "Incorrect Uses."
- The rationale for using quotation marks instead of italics for words as words and phrases as phrases is clarified in Section 5.3.4.1, "Double Quotation Marks." Because *The CSE Manual* relies heavily on words as words and phrases as phrases, this convention is used far more extensively throughout the manual than is common in most publications.
- Section 9.3.1.1, "Multiterm Titles," clarifies that the rules on title-style capitalization dictate capitalizing or lowercasing words based on their parts of speech, not on word length.
- Section 11.2, "Punctuation and Typography," now calls for using commas before and after initialisms for degrees and other credentials used after surnames.
- Section 11.5, "Abbreviations of Common Latin Terms," recommends eliminating periods for standard Latin abbreviations such as "eg" and "ie" to bring these abbreviations into adherence with *The CSE Manual*'s general recommendation to eliminate periods in abbreviations whenever reasonable.
- Section 26.2, "Capitalization and Lowercase," specifies that the term "Earth" should be capitalized in nearly all cases. The lowercased "earth" should be reserved for referring to soil.
- The index has been revised and expanded.

Scientific Scope

As with previous editions of this style manual, the ninth edition focuses primarily on scientific nomenclature, symbols, and other style issues. The principles governing these guidelines are presented in sufficient detail to make clear the rationale for them. The manual does not provide comprehensive lists of scientific terms and conventions for all scientific disciplines, nor does the manual outline the fundamental concepts of all these disciplines. Authoritative documents are available in many disciplines to guide authors to understand these larger issues. References to many such resources are provided in the cited and additional references at the end of relevant chapters.

Organization

The overall organization of the ninth edition follows that of the eighth edition. The manual is divided into the following 4 parts:

- Part 1, "Publishing Fundamentals," serves as an overview of essential issues related to scientific publishing, including editorial policies and copyright.
- Part 2, "General Style Conventions," outlines guidelines for style issues that are common to general and scientific publishing.
- Part 3, "Special Scientific Conventions," focuses on style issues specific to a variety of scientific disciplines.
- Part 4, "Technical Elements of Publications," recommends formats applicable in scientific journals, books, and other media, including formats for references, tables, figures, and indexes.

Part 3, whose content is the most specific to scientific publishing, is organized according to a rising scale of dimensions. The part starts with chapters on the fundamental units of matter; proceeds up through chapters on chemical and cellular components, microorganisms, and macroorganisms; and finishes with chapters on the planet Earth and the rest of the universe.

Within each chapter, the section numbering is hierarchical. Primary sections are those that are numbered by the chapter number, a decimal point, and only one other number (eg, Section 13.2, "Time"). Secondary sections have numbers with 2 decimal points (eg, Section 7.3.2, "Active versus Passive Voice"), and tertiary sections have numbers with 3 decimal points (eg, Section 25.11.2.1, "Salinity of Seawater"). Each additional lower section has one additional decimal point and number.

Throughout the manual, primary and lower sections are all labeled as "sections," instead of designating them as "sections," "subsections," "subsubsections," and beyond.

In the online version of CSE's ninth edition, each primary section and its accompanying lower sections constitute a single web page. Because web pages are the main means of dividing the manual's content online, initialisms and other abbreviations are generally introduced at first use of the spelled-out terms within each primary section. Once introduced in a primary section, abbreviations are used throughout that section and its lower sections in both the online and print editions. Abbreviations are reintroduced with each new primary section, even when primary sections are adjacent to each other.

Qualification on Examples

Although great care was taken to ensure that the examples in the ninth edition adhere to the style guidelines within *The CSE Manual*, users are advised to give preference to the manual's written guidelines over its examples should they disagree.

RECOMMENDATIONS FOR FUTURE EDITIONS

This manual reflects the ongoing, collective knowledge of CSE members and others in the scientific publishing community. Given the dynamic nature of scientific fields and publishing practices, some topics will continue to evolve beyond what is covered in the ninth edition. For example, as online and other digital platforms evolve, *The CSE Manual* will need to focus even more on electronic publishing. Additionally, artificial intelligence is likely to play an ever-greater role in writing, processing, and publishing manuscripts. Future editions may even have to offer style guidance for metaverse-based platforms, complete with avatars.

Future editions of *The CSE Manual* will also benefit from updating examples with ones that cite new research and researchers from around the world in a wide variety of scientific disciplines.

Future editions should explore whether diversity, equity, and inclusivity can be promoted by modifying end references to include the given names of authors and other creators instead of initials for first and middle names. Current reference practices obscure the contributions of female researchers because end references do not provide these researchers' first names.

Another opportunity to enhance cited and additional references in future editions is to include more entries from CSE's quarterly publication, *Science Editor*, which can be accessed at https://www.csescienceeditor.org/. Doing so will strengthen CSE's connection to *The CSE Manual*.

Because monitoring the scientific literature for new style conventions, nomenclature, and notations is a huge task, CSE urges scientific journals, societies, committees, working groups, individual scientists, publishing professionals, and others to share with CSE recommendations for improving scientific style and format. Such recommendations can range from proposing additional topics to cover in the manual to pointing out style discrepancies in the ninth edition. The CSE Style Manual Task Force will review those suggestions in consultation with subject-matter experts and decide which recommendations to incorporate into future editions.

Send suggestions to CSE via email to CSE@CouncilScienceEditors.org.

CITED REFERENCES

1. Conference of Biological Editors, Committee on Form and Style. Style manual for biological journals. American Institute of Biological Sciences; 1960.

2. Conference of Biological Editors, Committee on Form and Style. Style manual for biological journals. 2nd ed. American Institute of Biological Sciences; 1964.

3. Council of Biology Editors, Committee on Form and Style. CBE style manual. 3rd ed. American Institute of Biological Sciences; 1972.

4. Council of Biology Editors, CBE Style Manual Committee. Council of Biology Editors style manual: a guide for authors, editors, and publishers in the biological sciences. 4th ed. Council of Biology Editors; 1978.

5. Council of Biology Editors, Style Manual Committee. CBE style manual: a guide for authors, editors, and publishers in the biological sciences. 5th ed, revised and expanded. Council of Biology Editors; 1983.

6. Council of Biology Editors, Style Manual Committee. Scientific style and format: the CBE manual for authors, editors, and publishers. 6th ed. Cambridge University Press; 1994.

7. Council of Science Editors, Style Manual Committee. Scientific style and format: the CSE manual for authors, editors, and publishers. 7th ed. Council of Science Editors; 2006.

8. Council of Science Editors, Style Manual Subcommittee. Scientific style and format: the CSE manual for authors, editors, and publishers. 8th ed. Council of Science Editors in cooperation with The University of Chicago Press; 2014. Also available at https://www.scientificstyleandformat.org/

9. Council of Science Editors, Board of Directors. Scientific Style and Format update: capitalize racial and ethnic group designations. Council of Science Editors; 2020. https://doi.org/10.36591/SE-D-4303-97

ACKNOWLEDGMENTS

All 9 editions of the Council of Science Editors (CSE) manual were made possible by volunteer science editors. During the course of more than 60 years, those volunteers documented style and format to enable fellow science editors to apply to publishing a level of discipline, precision, and integrity akin to that which scientists apply to their research.

Just as science evolves, so does scientific publishing. The volunteers who labored over the ninth edition of *The CSE Manual: Scientific Style and Format for Authors, Editors, and Publishers* revised the manual so that it will remain alive and relevant as an authoritative reference for writing and editing documents in various scientific disciplines.

CSE owes enormous gratitude to the many individuals who contributed to this edition of *The CSE Manual*, especially the volunteer chapter editors and peer reviewers. In ways both great and small, the contributors named below made a significant difference that should benefit all editors and authors who use *The CSE Manual*.

Special thanks are owed to the 8 chapter editors from the eighth edition who returned for the ninth: Emmanuel A. Ameh, MBBS, FWACS, FACS; Dana M. Compton; Tina L. Fleischer; Beth E. Hazen, PhD; Raymond Lambert, MLS, ELS; Thomas A. Lang, MA; Beva Nall-Langdon; and Mary Warner, CAE. Also worthy of special praise is Sun Huh, MD, PhD, the only contributor who drafted a new chapter for the ninth edition.

If anyone who helped prepare this manual was omitted from the following list, it was unintentional, and CSE sincerely apologizes for the error.

PROJECT MANAGER FOR *THE CSE MANUAL*
Michael E. Fitzgerald, CAE

PRODUCTION PROJECT MANAGER FOR *THE CSE MANUAL*
Peter J. Olson, ELS

ADVISORY GROUP
Dana M. Compton, CSE Board of Directors' liaison

Mary E. Laur, executive editor at The University of Chicago Press

Russell Harper, freelance editor and writer

Mollie McFee, senior editorial associate at The University of Chicago Press

Michael E. Fitzgerald, CAE, project manager for *The CSE Manual*

STYLE MANUAL TASK FORCE

The Style Manual Task Force for *The CSE Manual*'s ninth edition consists of experts throughout the world of scientific publishing who volunteered to revise the manual's chapters. Most worked in teams of 2 to 4. The chapter editors' credentials and the chapters these editors revised are listed below.

Kaitlyn Aman Ramm
Senior digital multimedia and graphics coordinator for the *Neurology* journals at the American Academy of Neurology
Chapter 25

Janaynne do Amaral, PhD
Postdoctoral research associate at the School of Information Sciences at the University of Illinois Urbana-Champaign in the United States and translations editor for CSE's quarterly publication, *Science Editor*
Chapters 1 and 27

Emmanuel A. Ameh, MBBS, FWACS, FACS
Editor in chief of the *Journal of West African College of Surgeons*, assistant editor in chief of the *World Journal of Surgical Infections*, founding editor in chief of the *Nigerian Journal of Surgical Research*, former editor in chief of the *Annals of African Medicine* and the *African Journal of Pediatric Surgery*, and professor of pediatric surgery and chief consultant pediatric surgeon at the National Hospital in Abuja, Nigeria
Chapter 2

Jessica S. Ancker, MPH, PhD
Professor and vice chair for educational affairs in the Department of Biomedical Informatics at the Vanderbilt University Medical Center in Nashville, Tennessee, United States
Chapters 12 and 30

Stanley N.C. Anyanwu, MBBS, FMCS, FWACS, FACS, FICS
Professor of surgery (surgical oncology) and director of the Institute of Oncology at Nnamdi Azikiwe University, Nnewi campus, in Nigeria
Chapter 13

Kenneth April
Senior manager of production in the Publications Department at the American Society of Hematology
Chapter 14

Elizabeth A. Arnold, ELS
Manager of manuscript editing at the Journal of Neurosurgery Publishing Group
Chapters 6 and 11

Patricia K. Baskin, MS
Executive editor of the *Neurology* journals at the American Academy of Neurology; CSE's 2016–2017 president; and 2012–2014 editor of CSE's quarterly publication, *Science Editor*
Chapter 21

Kevin R. Brown
Senior manuscript editor at the JAMA Network
Chapter 32

Michelle Cathers, MLS
Production director at the American Pharmacists Association
Chapter 32

Jae Hwa Chang, MA, ELS
Manuscript editor for the South Korean manuscript-editing company InfoLumi
Chapter 27

Soo-Hee Chang, MS, ELS
Manuscript editor for the South Korean manuscript-editing company InfoLumi
Chapter 24

Dana M. Compton
Managing director and publisher at the American Society of Civil Engineers and CSE's 2019–2020 president
Chapters 4 and 10

Craig Damlo, MSc
Senior manager in human space flight for space systems development at Blue Origin
Chapter 26

Sherri Damlo, MS, ELS
Managing editor at Damlo Edits, former editorial board member and social media editor for CSE's quarterly publication, *Science Editor*, and board member of the Space Ecology Workshop
Chapter 26

Heather DiAngelis, MA
Associate director of publications for the Transportation Research Board at the National Academies of Sciences, Engineering, and Medicine in the United States
Chapters 4 and 10

Simona Fernandes, MSc, ELS

Freelance editor and writer in science, technology, engineering, and math in Toronto, Ontario, Canada, and 2021 CSE Scholarship recipient

Chapters 27 and 31

Tina L. Fleischer

Quality engineer at Aries Systems Corporation

Chapter 19

Peter J. Freeman, PhD

Lecturer in health care sciences in the Division of Informatics, Imaging and Data Science at the Faculty of Biology, Medicine and Health of The University of Manchester in the United Kingdom

Chapter 21

Beth E. Hazen, PhD

Principal at Willows End scientific editing and writing

Chapters 22 and 23

Sun Huh, MD, PhD

Professor at Hallym University in South Korea, 2020–2022 president of the Korean Council of Science Editors, and 2020–2023 president of the Korean Association of Medical Journal Editors

Chapter 33

Andrew Hunt, PhD

Adjunct associate professor at the Queensland University of Technology in Brisbane, Australia, and freelance editor of scientific manuscripts

Chapter 1

Jacob Kendall-Taylor

Director of editorial systems at the JAMA Network

Chapter 13

Raymond Lambert, MLS, ELS

Senior managing editor for journals at Duke University Press

Chapters 12 and 17

Trevor Lane, DPhil, PGCELT

Publishing consultant, Hong Kong, and 2022–2027 trustee and 2016–2025 council member of the Committee on Publication Ethics

Chapters 5 and 8

Thomas A. Lang, MA

Principal, Tom Lang Communications and Training International, and CSE's 2001–2002 president

Chapters 2, 12, and 30

Jessica LaPointe

Managing copy editor at the American Meteorological Society

Chapter 29

Ellen Lazarus, MD, ELS

Freelance copy editor at Lazarus Editing and Consulting, LLC, and volunteer copy editor for CSE's quarterly publication, *Science Editor*

Chapter 24

Iris Y. Lo

Deputy managing editor at the JAMA Network

Chapter 29

Rachel J. Lowery

Journal editor for *Operative Neurosurgery* at the Congress of Neurological Surgeons

Chapter 6

Audrey Daniel Lusher

Independent contractor at ADL Editorial

Chapters 11 and 31

Amy McPherson

Director of publications at the Botanical Society of America and managing editor of the *American Journal of Botany*

Chapters 22 and 23

Stephanie Mowat, MA

Manuscript editor at the *Canadian Medical Association Journal*

Chapter 24

Beva Nall-Langdon

Editor at Biotext Scientific Writing and Editing

Chapters 7 and 20

Kelly Newton

Director of scientific publications at the Society for Neuroscience

Chapters 3 and 29

Joanna Odrowaz

Editorial consultant, technical writer, and writing coach in science and medicine in Toronto, Ontario, Canada

Chapter 5

Peter J. Olson, ELS

Freelance manuscript editing coordinator at the JAMA Network

Chapter 29

Heather Poirier, MA

Medical writer II with Technical Resources International

Chapter 9

Andrea Rahkola, ELS
Production editor of the *Neurology* journals at the American Academy of Neurology
Chapter 9

Dara R. Rochlin
Owner of Dara Rochlin Book Doctor and freelance copy editor for CSE's quarterly publication, *Science Editor*
Chapters 5 and 8

David M. Schultz, PhD, FAMS, FRMetS, FGS, SFHEA
Editor in chief of *Monthly Weather Review* from 2008 to 2022 and professor of synoptic meteorology at The University of Manchester in the United Kingdom
Chapters 25 and 26

Brit Stamey
Associate director of content management at J&J Editorial and 2021–2023 treasurer for the Coalition for Diversity and Inclusion in Scholarly Communications
Chapter 8

Julie Steffen, CAE
Chief publishing officer at the American Astronomical Society
Chapter 26

Jessica L. Striley
Journals production manager of Neurosurgery Publications at the Congress of Neurological Surgeons
Chapter 2

Kristen Swendsrud
Production coordinator for *Neurology* at the American Academy of Neurology
Chapter 28

Caitlyn Trautwein
Journal editor of *Neurosurgery* at the Congress of Neurological Surgeons
Chapter 7

Mary Warner, CAE
Vice president for publishing at the American Pharmacists Association
Chapters 15, 16, 17, and 32

PEER REVIEWERS

As part of CSE's commitment to following best practices in scientific publishing, each chapter of the ninth edition was peer reviewed by 1 to 3 experts in scientific publishing. Listed below are the credentials of these volunteer peer reviewers and the chapters they reviewed.

Terry Anderson, ELS, owner of Alan Edits, LLC (Chapter 8)

Tony Alves, senior vice president for product management at HighWire Press (Chapter 33)

Elizabeth A. Bales, MA, senior manuscript editor at *The New England Journal of Medicine* (Chapter 6)

James W. Bales, PhD, associate director of the Massachusetts Institute of Technology's Edgerton Center in the United States (Chapter 15)

Stacy L. Christiansen, MA, managing editor of *JAMA* and chair of the *AMA Manual of Style* (Chapter 27)

Heather DiAngelis, MA, associate director of publications for the Transportation Research Board at the National Academies of Sciences, Engineering, and Medicine in the United States (Chapter 1)

Maryann P. DiEdwardo, MA, EdD, independent writing coach, writer, and editor, as well as adjunct professor of computer science and engineering at Lehigh University in Bethlehem, Pennsylvania, United States, and adjunct professor of English at the University of Maryland Global Campus (Chapter 4)

Jo Ann M. Eliason, MA, ELS (D), Brookmere Biomedical Writing and Editing, LLC (Chapter 3)

Leah Ellman-Stortz, MS, soil and crop science conservationist (Chapter 25)

Kay Endriss, ELS, MS, NBCT, statistician and former teacher at the Winston-Salem/Forsyth County Schools in North Carolina in the United States (Chapters 12 and 13)

Barbara Meyers Ford, retired consultant and 1997–1998 CSE president (Chapter 1)

Catherine Forrest, ELS, associate director for publications production at the American Society of Clinical Oncology (Chapters 9 and 32)

Barbara Gastel, MD, MPH, ELS (H), professor of veterinary integrative biosciences and of humanities in medicine and director of the Master of Science Program in Science and Technology Journalism at Texas A&M University in the United States, as well as the 1998–2010 editor of CSE's quarterly publication, *Science Editor* (Chapter 7)

Sindhoora Bhargavi Gopala Reddy, PhD, ELS, scientific editor at Philip Morris International in Neuchâtel, Switzerland (Chapter 20)

Tamara Hanna, PhD, director of new product innovation at ACS Publications at the American Chemical Society (Chapter 17)

Russell Harper, freelance editor and writer (Chapter 29)

Jan Higgins, PhD, ELS, managing editor of JID Innovations (Chapter 21)

Kristin S. Inman, PhD, ELS, science editor and contractor for *Environmental Health Perspectives* at the National Institute of Environmental Health Sciences in the United States (Chapter 3)

Renata Iskander, MSc, doctoral degree candidate at McGill University in Montreal, Quebec, Canada (Chapter 6)

Otito Frances Iwuchukwu, PharmD, MA, PhD, FCP, CPTD, associate professor of pharmaceutical sciences at the Fairleigh Dickinson University School of Pharmacy in Florham Park, New Jersey, United States, and 2021–2023 co-chair of CSE's Diversity, Equity, Inclusion, and Accessibility Committee (Chapter 20, Section 7.4, and Section 8.3)

Leonard Jack Jr, PhD, MSc, editor in chief of the journal *Preventing Chronic Disease* at the US Centers for Disease Control and Prevention and 2021–2023 co-chair of CSE's Diversity, Equity, Inclusion, and Accessibility Committee (Section 7.4 and Section 8.3)

Jill Jackson, managing editor and publishing administrator of the *Annals of Internal Medicine* and co-chair of CSE's Editorial Policy Committee (Chapter 2)

Yateendra Joshi, freelance copy editor and trainer (Chapter 30)

Bill Kasdorf, principal at Kasdorf & Associates, LLC, and founding partner of Publishing Technology Partners (Section 1.4)

Holly Koppel, ACB, CL, senior managing editor for peer review for the American Society of Civil Engineers' journals (Chapter 28)

Thomas A. Lang, MA, principal, Tom Lang Communications and Training International, and CSE's 2001–2002 president (Chapter 2)

Kristin Mitchell, assistant managing editor of *Arthritis Care & Research*, an official journal of the American College of Rheumatology (Chapter 31)

Natalie Ngo, managing editor of the *Clinical Journal of the American Society of Nephrology* (Chapter 5)

Joanna Odrowaz, editorial consultant, technical writer, and writing coach in science and medicine in Toronto, Ontario, Canada (Chapter 7)

Karla D. Passalacqua, PhD, MWC, ELS, lead— medical writing and education at Henry Ford Health in Detroit, Michigan, United States (Chapters 22 and 23)

Katie Petronella, editing operations lead for the American Chemical Society's Journals Publishing Group (Chapter 16)

R. Michelle Sauer Gehring, PhD, ELS, senior research scientist in the Department of Advanced Cardiopulmonary Therapies and Transplantation at The University of Texas Health Science Center at Houston in the United States (Chapter 11)

Michele Springer, CMPP, deputy director of medical editing at Caudex, an IPG Health company (Chapter 14)

Jürgen Stohner, PhD, FRSC, professor of physical chemistry at the School of Life Sciences and Facility Management at the Zürich University of Applied Sciences in Switzerland (Chapters 18 and 19)

Havah Strickler, senior medical copy editor at Shoreland, Inc (Chapter 11)

William S. Strong, JD, author of *The Copyright Book: A Practical Guide* and a partner at the law firm Kotin, Crabtree & Strong, LLP, in Boston, Massachusetts, United States (Chapter 3)

Ann Tennier, ELS, managing editor at The University of Texas Southwestern Medical Center in Dallas in the United States (Chapter 10)

Nadja Torres, MD, MPH, board-certified ophthalmologist at the Howerton Eye Clinic in Austin, Texas, United States (Chapter 24)

Unending thanks go to Mary Laur, executive editor at The University of Chicago Press, for her constant support, guidance, and patience throughout this adventure. Much gratitude also goes to Laur's team of Russell Harper and Mollie McFee. Throughout the development process, Harper lent his expertise from revising *The Chicago Manual of Style*. In addition, Harper's peer review of *The CSE Manual*'s Chapter 29, "References," enhanced the outstanding work that the chapter's editors did in revising and simplifying the manual's rules on in-text and end references. To McFee goes the credit for managing all of the new and reused artwork in the ninth edition.

Among the others at The University of Chicago Press to whom CSE owes a vote of thanks are Jenni Fry, managing editor, Christine Schwab, assistant managing editor, and Jenny Ringblom, associate marketing director for reference works. Fry and Schwab shepherded the manual through copyediting and production, while Ringblom focused on promoting the manual to authors and editors in the science community.

A great debt is owed to Dana M. Compton, who as the CSE Board of Directors' liaison guided this project through the Board and represented CSE's interests to The University of Chicago Press. Finally, all those who served on the CSE Board of Directors during the years that the ninth edition was being revised should be lauded for their steadfast support of this style manual.

Michael E. Fitzgerald, CAE
Project manager for *The CSE Manual*

1

Publishing Fundamentals

1 Elements of Scientific Publication

Editors: Andrew Hunt, PhD, and Janaynne do Amaral, PhD

1.1 TYPES OF SCIENTIFIC PUBLICATIONS

Publishing results is an integral part of the scientific research process. Scientific research, developments, and debates are published in myriad forms, the most common of which are scholarly books and journals. Scholarly books are usually written by one author or a team of authors, they focus in depth on a central theme, and they employ a consistent format. By comparison, scholarly journals publish relatively short articles written by a wide variety of authors on different topics within the journals' scientific disciplines. Articles in the same journal can have various structures and formats, including original research reports, literature reviews, methodological articles, and opinion pieces.

This chapter focuses on the fundamentals of journal articles, as does Chapter 27, "Journal Style and Format." For the basics of scholarly books, see Chapter 28, "Books and Other Monographs."

The scope of content and the types of articles that a scientific journal publishes are determined by the scientists and other content experts serving as the journal's editorial board, including the journal's editor in chief and associate and assistant editors. The journal will usually provide this information on its website in detailed instructions for authors. Besides describing desired content, these instructions detail the journal's target audience, the formatting requirements for each type of article the journal publishes, and the journal's submission process. Authors can use this information to target their submissions to journals likely to publish their research. In addition, by adhering to a journal's author instructions, authors ensure their work will not be rejected simply because it is outside the journal's scope or because their article was not organized in the manner specified for that type of article. For additional information on journal articles, see Section 27.7.5, "Types of Scientific Articles."

1.1.1 Original Research Reports

In scholarly journals, original research reports are the primary article type for publishing novel scientific discoveries and insights. Original research reports are the heart of a journal, and they are expected to report on significant contributions to the knowledge base within the journal's scientific discipline. In addition to thoroughly reporting original research, these articles, along with supplementary materials, provide information necessary for such purposes as replicating the research and reviewing pertinent literature before conducting new research. These articles, therefore, need to be clear and comprehensive so that others can build on the research.

Original research reports take many forms. For example, in the medical and pharmacological fields, randomized, double-blind, placebo-controlled studies are the hallmark of original research reports. However, when research fields cannot meet the requirements of randomized controlled trials, they rely on such alternatives as observational, qualitative, and quasi-experimental studies.

The specific research questions being addressed and the logistical, ethical, and regulatory requirements of conducting a study determine the most appropriate study design. In their original research report, researchers must weigh the strengths and limitations of their study's design and qualify the corresponding level of evidence generated by their research.

The expectations for different approaches to scientific research are outlined in guidelines and checklists to ensure standardized reporting of results based on research methodology. For medical research, many health care journals and editorial groups, including the Council of Science Editors, support the Consolidated Standards of Reporting Trials (CONSORT).[1] Other guidelines include those developed by the National Institute of Neurological Disorders and Stroke[2] in the United States and the guidelines for Animal Research: Reporting of *In Vivo* Experiments (ARRIVE).[3] Observational and qualitative research practices are addressed by the Strengthening the Reporting of Observational Studies in Epidemiology (STROBE) initiative[4] and in the Standards for Reporting Qualitative Research (SRQR) guidelines,[5] respectively. Outside the field of medicine, physical and social science disciplines have developed publication standards for their research journals.

1.1.2 Review Articles

Researchers perform literature reviews to inform the research community of the current state of knowledge regarding a specific topic. In doing so, researchers synthesize multiple original research reports to reach evidence-based conclusions, make recommendations, and propose new hypotheses for scientific inquiry.

Among the most valuable review articles are systematic reviews. For these reviews, researchers define a specific research question and conduct a thorough and structured search of the scientific literature for all articles addressing the topic. The results from various studies are aggregated, and the quality of the evidence and the limitations of the studies are objectively evaluated. A systematic review may be based on a meta-analysis, a statistical technique that combines quantitative results from independent studies. The Preferred Reporting Items for Systematic Reviews and Meta-analyses 2020 (PRISMA 2020) statement[6] aims to standardize the way in which systematic reviews and meta-analyses are conducted and reported.

When evidence on a specific research question is too limited for a systematic review, researchers may perform a narrative review. In this type of review, researchers broadly discuss a topic of inquiry and bring together diverse and disparate sources of information to generate new ideas and establish research priorities. Another value of narrative reviews is that they can summarize original research reports and other scientific reports into easily understood recommendations for public policy.

1.1.3 Methodological Articles

A key element of high-quality scientific research is that it be reproducible by other researchers and institutions following the same methodology. To make that possible, methodological articles describe new or modified protocols for conducting research so that standardized methods can be implemented across studies and effective comparisons can be made to the results of those studies.

1.1.4 Opinion Pieces

Researchers can voice their perspectives to the academic community through opinion pieces such as editorials and letters to the editor. In journal editorials, subject-matter experts provide their perspectives on or interpret the results of research reported in the same or earlier issues. Letters to the editor allow readers to challenge, support, or otherwise respond to recently published articles.

1.1.5 Other Article Types

Among other types of scholarly articles are case studies, theses, dissertations, and preprints. A case study is an in-depth analysis based on observations on a specific situation, such as a patient with an unusual medical condition, pathogens affecting a specific community, or unexpected chemicals found in a wastewater treatment plant. When such a study provides observations on several cases with the same condition, the study is called a case series.

Upon completing their graduate research, candidates for master's and doctoral degrees compile their work into theses and dissertations. The candidates' academic institutions often make these documents available online for the wider academic community to access.

Preprints are draft research papers that authors make available online before they are peer reviewed. The authors do so with the expectation that they will receive wide-ranging feedback that they can use to improve the papers before they are formally peer reviewed by journals. As preprint servers became ubiquitous, many journals allowed this practice, provided that the journals would hold the rights to publish the version of record and that permission and appropriate citation would be required to reproduce an article once the version of record is published. However, not all scholarly journals are supportive of this practice. Some journals will not publish manuscripts that have been posted as preprints because those journals consider these works as already published.

1.2 COMPONENTS OF SCIENTIFIC PUBLICATIONS

The basic components of an original research report are well established: Introduction, Methods, Results, and Discussion (IMRAD) (see also Section 1.2.2, "Text"). But just as journals specify the scope of their content and the types of articles they publish, journals may specify how each type of article should be organized. Although most scholarly journals follow the basic IMRAD structure for original research reports (eg, *The New England Journal of Medicine* and the *Brazilian Journal of Medical and Biological Research*), some journals vary the IMRAD structure based on their subject matter. For example, *Nature* and *Cell* modify the IMRAD structure by placing the Methods at the end of articles. In addition, journals may specify other components for article types other than original research reports.

The basic components of each article type form an important part of a journal's instructions for authors and peer reviewers. These guidelines impose a consistent structure within each type of article in a given journal, which benefits authors, other researchers, and peer reviewers.

After an article is submitted to a journal, the journal's peer-review process will assess not only the scientific quality of the work but also the authors' adherence to the journal's instructions.[7] Peer reviewers may identify errors in an original research report, such as in the study's methods or analysis techniques, which in turn raise questions about the validity of the results.[8]

Peer review is used by journals to validate the research described in submitted manuscripts and to help editors decide whether to publish the manuscripts.[8] The process involves the authors who wrote the manuscripts; peer reviewers, who provide expert opinions on and constructive criticism of the manuscripts; and the editors, who decide whether manuscripts should be accepted, rejected, or returned to the authors for revisions.[7]

For more on peer review, see Section 27.3.1.1, "Overall Process."

1.2.1 Abstract

As noted in Section 27.7.1.4, "Abstract," both the National Information Standards Organization (NISO)[9] in the United States and the International Organization for Standardization (ISO)[10] stipulate that an abstract should be published with every journal article, essay, and discussion. An abstract helps readers to decide whether an article would be of interest to them. Therefore, an abstract should reflect the article's content as closely as possible within the length specified by the journal.

Most journals specify how abstracts are to be organized. For example, medical journals tend to follow the guidelines for abstracts detailed in the *AMA Manual of Style: A Guide for Authors and Editors*,[11] while chemistry journals may adhere to the American Chemical Society's *ACS Guide to Scholarly Communication*.[12] Standardized structured abstracts have become the norm for original research reports and other research articles. To ensure that they capture all pertinent information from the articles, structured abstracts have formal headings that typically parallel the IMRAD structure.

Beyond benefiting readers, abstracts expand the reach of research when they are included in abstracting and indexing databases such as the US National Library of Medicine's PubMed, EBSCO Information Services, BIOSIS Previews, and Elsevier's Scopus Preview. By making abstracts easily searchable and accessible, abstracting services increase articles' exposure and citations, benefiting journals and authors while speeding scientific discovery.

1.2.2 Text

The text of an article should follow the outline specified in the instructions for authors of the journal to which the authors intend to submit their article. Research articles and systematic reviews usually adhere to some variation of the IMRAD structure followed by a reference section. The organization of other types of articles is much more variable. For example, the IMRAD structure is not applicable to qualitative research, case reports, and reviews that are not systematic.[13]

The next 4 sections outline the requirements of the IMRAD structure.

1.2.2.1 Introduction

The Introduction must contain relevant background information on the research topic, identify gaps in current knowledge, provide a sound justification for conducting the research, state the hypothesis and aims of the study, and cite relevant literature.[14] Some journals request authors to report the principal conclusion of their studies in the Introduction.[14] However, the International Committee of Medical Journal Editors (ICMJE) recommends that authors avoid including data or conclusions in the Introduction.[15]

The Introduction must be concise, and most of it should be written in the present tense and in the active voice[14] (see also Section 7.3.1.1, "Verb Tenses in the Introduction Section"). Verbose and overly complex sentences should be avoided.[14]

1.2.2.2 Methods

The Methods should describe in detail the steps taken to conduct the research. This section is particularly important because it allows readers to assess whether the study results are pertinent to their own research and to identify flaws that may influence the validity of the outcomes. The Methods must include sufficient detail to allow other researchers to repeat the study. Like the Introduction, the Methods should be written in direct and precise language, but unlike the Introduction, the Methods should be written in the past tense.[16]

The following is a general checklist of information to include in a research article's Methods section:

(1) A detailed description of the experimental procedure used in the study and, if applicable, the procedure's potential hazards or a reference to a previously published report that thoroughly describes the same procedure and its potential hazards[12]
(2) The types of tests performed during the study, including statistical tests[17]
(3) The apparatuses used, including model numbers and manufacturers for specialized equipment
(4) Unambiguous identification of the nonbiological materials used (eg, chemicals), including the sources of such materials[12]
(5) For biological studies, unambiguous identification of genus, species, and strain; the source of any organisms used (eg, cell line or animal stock); and the age, sex, weight, and condition of organisms (for information on stable identifiers for species and genes, see Section 21.1.2, "Existing Standards and Unambiguous Identifiers")
(6) For clinical studies, pertinent details about human subjects, including methods of recruitment and relevant physical characteristics
(7) For studies involving fieldwork, the specific locations of study sites, with appropriately detailed maps[18,19]

Specific journals' instructions for authors may provide more detailed checklists for the Methods. An example is the instructions for the *Journal of Parasitology*,[20] which specifies what should be included in this section and how it should be written. Furthermore, some journals, such as *BMC Medical Research Methodology*,[21] publish checklists for specific types of studies, such as clinical trials.

1.2.2.3 Results

The Results should typically consist of 3 components: text, tables, and figures.[16] The relationship among these components is complementary in that each does not duplicate information in the other 2 components.[16]

In research involving humans, the Results section generally begins by describing the subjects and outlining whether and how they were divided into groups based on different study protocols. This information helps readers to evaluate whether the study's sample meets the research criteria described in the Methods section and to compare the study's sample with what readers encounter in their work.[16]

The Results should follow a logical sequence in which the main or most important findings are described first.[15] Like Methods, Results should be written in the past tense. Data and other information should be presented concisely, which often can be achieved with tables and figures (see Chapter 30, "Tables, Figures, and Indexes"). Data sets should be reported only if their collection techniques are described in the Methods.[16] If space is limited, as is often the case in printed publications, journals may make supplemental tables and figures available online. Depending on their online capabilities, journals also may post videos and other data formats to provide additional information about published studies.

1.2.2.4 Discussion

Although the least structured section in research articles, the Discussion typically covers the main results, comparisons with earlier studies, the studies' strengths and limitations, the generalizability and real-world implications of the results, and the authors' conclusions.[22] These components usually follow that order, but they are not necessarily identified by subheadings. Regardless of whether subheadings are used in the Discussion, each component is highlighted by a topic sentence at the beginning of a paragraph.

The ICMJE recommends beginning the Discussion by briefly presenting the main findings and offering possible explanations for those findings.[15] Other ICMJE recommendations are to avoid replicating data and other information described in the Introduction or Results and to refrain from making statements or drawing conclusions that are not supported by the data.

Because the Discussion can combine conclusions, results, and suggestions, tense can vary from present to past to future (see also Section 7.3.1.4, "Verb Tenses in the Discussion Section").

1.2.3 Other Requirements

Depending on the subject of an article and the journal in which it will be published, other elements may be required. For example, biomedical journals that follow the ICMJE's recommendations require that clinical trials be registered in at least one acceptable trial registry before patient enrollment begins.[23] Many journals require researchers to follow ethical standards and guidelines. For research involving humans, institutional review boards and human research ethics committees at the institutions conducting the research must review and approve the studies' ethics protocols. In addition, researchers must obtain voluntary and informed consent from human subjects before they participate in research, according to principles of the ethical conduct outlined in such documents as the World Medical Association's Declaration of Helsinki.[24]

For experimental studies involving animals, journals typically require a statement confirming that the researchers adhered to ethical guidelines for the care and treatment of animals under study, such as the guidelines developed by the American Psychological Association.[25] The fields of engineering and physical sciences also consider the importance of ethical conduct in their research methodologies and practices,[26,27] and the emerging field of data science is grappling with myriad ethical concerns arising from using big data.[28]

In addition, journals typically require authors to reveal the sources of financial and other support for their research and to submit statements of potential conflicts of interest to help editors and readers assess whether the research may be biased (see Section 2.5, "Competing Interests"). Also commonly required are signed copyright release forms or license agreements, which typically state that authors agree to transfer all or some intellectual property rights for their work to the publisher on acceptance of the manuscript (see Section 3.1.3.3, "Transfer of Ownership"). These agreements also ask authors to confirm that the work is their own and that they have or will obtain permission to reproduce any elements owned by other sources.

1.3 AUTHORSHIP OF SCIENTIFIC PUBLICATIONS

Authorship is a fundamental element of scientific publications. Yet what constitutes a substantial enough contribution to a scientific work to qualify contributors to be recognized as authors remains controversial. Being named as an author of a study is a highly sought-after privilege because it enhances a researcher's track record of contributing to their field, and it is pivotal for obtaining academic promotion and securing grant funding.

1.3.1 Authorship Criteria

Broadly, all named authors must have substantially contributed to the research being reported. Although the precise criteria vary as to what constitutes "substantial contribution," many academic journals have adopted the recommendation of the International Committee of Medical Journal Editors (ICMJE)[29] that authors meet all 4 of the following key criteria:

(1) Contribute to the study design or to the collection, analysis, or interpretation of the data
(2) Draft or critically revise the manuscript
(3) Approve the final version for publication
(4) Be accountable for all aspects of the research

Many journals have standardized the language used in reporting the activities of authors and other contributors by embracing the Contributor Roles Taxonomy, better known as CRediT, which has been adopted by the National Information Standards Organization in the United States.[30] Although CRediT does not define authorship, it provides decriptions of 14 roles that various contributors play in scientific research, ranging from conceptualization and funding acquisition to formal analysis and writing.

1.3.2 Author Order

The order in which authors' names appear on a published article can be as highly contentious as determining which contributors qualify as authors. The position of first author is usually the most prized because it is highly visible to readers of the work and citations to it. Accordingly, the first author is frequently the researcher who made the greatest contribution to the research, playing a pivotal role in all aspects from conception to publication. However, some journals require that the corresponding author be listed first, even if another author contributed the most to the research. If a journal does not list the corresponding author first, the journal may designate that author by another means, such as placing an asterisk after the author's name.

In some cases, the position of last author is reserved for the principal investigator, senior researcher, or supervisor. Regardless of how the first and last authors are selected, other authors are typically named in order of relative contribution to the work.

An option for studies with a vast number of authors is to order names alphabetically.

1.3.3 Acknowledgments

Scientific papers often include an acknowledgments section in which people who made significant contributions below the level of authorship are named and thanked for their roles. As noted in Section 1.3.1, "Authorship Criteria," CRediT has standardized the language used in describing the activities of such contribtors, as well as the activities of authors.[30]

The acknowledgments section is also used to disclose funding and other financial matters.

1.4 STRUCTURE AND XML TAGGING

One of the major advantages of ensuring that scientific articles follow established formats is that data can be compared across studies. Such comparisons and other analyses of the scientific literature usually begin by searching electronic databases, a process made easier when publications use structured tagging.

Extensible Markup Language (XML)[31] is a metalanguage that uses express tag vocabularies to mark up documents in a consistent and well-defined way. XML tags facilitate processing documents by different systems and converting documents into other formats for editorial, production, and dissemination purposes.[32] Ideally, the tags are based on a standard vocabulary, but sometimes, they are user defined.

XML tags specify the structure of a document rather than its appearance. They identify various components of each document and the relationships among them. The components may be standard elements in the document, such as the title and the author byline, or they may be parts of the document's content, such as the organism under investigation and the statistical tests performed. Although not required in XML tagging, the role of each element can be defined in a formal model called a "schema," such as a "document type definition" (DTD).[31,33] XML-tagged documents that use DTDs or other schemas can be systematically searched and analyzed for content regardless of the terminology used in the documents themselves.

The scientific community has been taking advantage of XML tagging in a variety of ways. Discipline-specific tagging systems have been developed (eg, Mathematical Markup Language,[34] better known by the abbreviation "MathML"). Research groups have defined DTDs and other schemas to facilitate viewing, sharing, and merging data from different laboratories. Publishers and publisher groups have developed DTDs and XML tag sets.

Among the more prominent DTDs is the Journal Article Tag Suite[35] (JATS), which the National Center for Biotechnology Information in the United States developed for journal articles submitted to PubMed Central, the US National Library of Medicine's free full-text archive of biomedical and life sciences journal literature. JATS has an extension for books called the Book Interchange Tag Suite[36] (BITS). Together, JATS and BITS provide publishers, hosting platforms, researchers, and others with access to data throughout most of scholarly publishing.

For more tagging, see Section 31.5, "Markup Languages" and Section 33.3.4, "Journal Article Tag Suite Extensible Markup Language (JATS XML)."

CITED REFERENCES

1. Schulz KF, Altman DG, Moher D, CONSORT Group. CONSORT 2010 statement: updated guidelines for reporting parallel group randomised trials. BMC Med. 2010;8:18. https://doi.org/10.1186/1741-7015-8-18

2. National Institutes of Health (US), National Institute of Neurological Disorders and Stroke. Clinical study quality control/quality assurance checklist. National Institute of Neurological Disorders and Stroke (US); [accessed 2023 Jan 20]. https://www.ninds.nih.gov/sites/default/files/migrate-documents/study _checklist_508C.pdf

3. Percie du Sert N et al. Reporting animal research: explanation and elaboration for the ARRIVE guidelines 2.0. PLOS Biol. 18(7):e3000411. 2020. https://doi.org/10.1371/journal.pbio.3000411

4. von Elm E et al. The Strengthening the Reporting of Observational Studies in Epidemiology (STROBE) statement: guidelines for reporting observational studies. Lancet. 2007;370(9596):1453–1457. https://doi .org/10.1016/S0140-6736(07)61602-X

5. O'Brien BC, Harris IB, Beckman TJ, Reed DA, Cook DA. Standards for reporting qualitative research: a synthesis of recommendations. Acad Med. 2014;89(9):1245–1251. https://doi.org/10.1097/acm .0000000000000388

6. Page MJ et al. The PRISMA 2020 statement: an updated guideline for reporting systematic reviews. BMJ. 2021;372:n71. https://doi.org/10.1136/bmj.n71

7. Peer review: the nuts and bolts: a guide for early career researchers. Sense about Science; 2012. https:// senseaboutscience.org/wp-content/uploads/2016/09/peer-review-the-nuts-and-bolts.pdf

8. How to peer review. Springer Nature; [accessed 2023 Jan 20]. https://www.springernature.com/in /authors/campaigns/how-to-peer-review

9. National Information Standards Organization (US). ANSI/NISO Z39.14-1997 (R2015) guidelines for abstracts. NISO Press; 2015. http://www.niso.org/publications/ansiniso-z3914-1997-r2015-guidelines -abstracts

10. International Organization for Standardization. Documentation—abstracts for publications and documentation (ISO 214:1976). International Organization for Standardization; 1976 [reviewed and confirmed 2020]. http://www.iso.org/iso/home/store/catalogue_tc/catalogue_detail.htm?csnumber=4084

11. JAMA Network Editors. AMA manual of style: a guide for authors and editors. 11th ed. Oxford University Press; 2020. Also available at https://www.amamanualofstyle.com

12. Banik G, Baysinger G, Kamat P, Pienta, N, American Chemical Society. The ACS Guide to Scholarly Communication. American Chemical Society; 2020. https://pubs.acs.org/doi/book/10.1021/acsguide

13. Batmanabane G. The IMRAD structure. In: Sahni P, Aggarwal R, editors. Reporting and publishing research in the biomedical sciences. Springer; 2017 Dec 13. p 1–4. https://link.springer.com/chapter/10 .1007/978-981-10-7062-4_1

14. Ghoshal UC. The introduction section. In: Sahni P, Aggarwal R, editors. Reporting and publishing research in the biomedical sciences. Springer; 2017 Dec 13. p 5–8. https://link.springer.com/chapter/10 .1007/978-981-10-7062-4_2

15. International Committee of Medical Journal Editors. Preparing a manuscript for submission to a medical journal. International Committee of Medical Journal Editors; 2019. http://www.icmje.org /recommendations/browse/manuscript-preparation/preparing-for-submission.html#one

16. Aggarwal A. The methods section. In: Sahni P, Aggarwal R, editors. Reporting and publishing research in the biomedical sciences. Springer; 2018. p 9–20. https://link.springer.com/chapter/10.1007/978-981-10-7062-4_3

17. Lang TA, Secic M. How to report statistics in medicine: annotated guidelines for authors, editors, and reviewers. 2nd ed. American College of Physicians; 2006.

18. Hansen WR, editor. Suggestions to authors of the reports of the United States Geological Survey. 7th ed. United States Geological Survey; 1991. 289 p. https://doi.org/10.3133/7000088

19. Bates RL, Adkins-Heljeson MD, Buchanan RC, editors. Geowriting: a guide to writing, editing, and printing in Earth science. 5th rev ed. American Geological Institute; 2004.

20. Clopton, RE. Journal of Parasitology policies and instructions to authors. J Parasitol. 2018:104(4):441–450. https://doi.org/10.1645/18-93

21. Bornhöft G et al. Checklist for the qualitative evaluation of clinical studies with particular focus on external validity and model validity. BMC Med Res Methodol. 2006;(6):56. https://doi.org/10.1186/1471-2288-6-56

22. Fletcher RH, Fletcher SW. The discussion section. In: Sahni P, Aggarwal R, editors. Reporting and publishing research in the biomedical sciences. Springer; 2017 Dec 13. p 39–48. https://link.springer.com/chapter/10.1007/978-981-10-7062-4_5

23. International Committee of Medical Journal Editors. Uniform requirements for manuscripts submitted to biomedical journals: writing and editing for biomedical publication. International Committee of Medical Journal Editors. [updated 2008]. https://www.icmje.org/recommendations/archives/2008_urm.pdf

24. World Health Organization. World Medical Association Declaration of Helsinki: ethical principles for medical research involving human subjects. Bulletin of the World Health Organization, 2001;79(4):373–374. https://apps.who.int/iris/bitstream/handle/10665/268312/PMC2566407.pdf?sequence=1&isAllowed=

25. American Psychological Association, Committee on Animal Research and Ethics. Guidelines for ethical conduct in the care and use of animals. American Psychological Association; 2022. http://www.apa.org/science/leadership/care/guidelines.aspx

26. Deb D, Dey R, Balas VE. Ethics in engineering research. In: Engineering research methodology: a practical insight for researchers. Intelligent Systems Reference Library, vol 153. Springer; 2018 Dec 15. p 43–48. https://doi.org/10.1007/978-981-13-2947-0_5

27. American Physical Society. APS statement: guidelines on ethics; 2019. https://www.aps.org/policy/statements/guidlinesethics.cfm

28. Martens, D. Data science ethics: concepts, techniques, and cautionary tales. Oxford University Press; 2022.

29. International Committee of Medical Journal Editors. Recommendations for the conduct, reporting, editing, and publication of scholarly work in medical journals: defining the role of authors and contributors. International Committee of Medical Journal Editors; [updated 2022 May]. https://www.icmje.org/recommendations/

30. National Information Standards Organization. Contributor Roles Taxonomy (CRediT). National Information Standards Organization; [accessed 2023 Jan 20]. http://credit.niso.org/

31. Extensible Markup Language (XML). World Wide Web Consortium; [updated 2016]. https://www.w3.org/XML/

32. Kasdorf B. XML and PDF: why we need both. An introduction to the two core technologies for publishing. JP News J Publ. 2003;(2):1,3–14.

33. Bryan M. An introduction to the Extensible Markup Language (XML). Bull Am Soc Inf Sci. 1998;25(1):11–14. https://doi.org/10.1002/bult.104

34. Mathematical Markup Language (MathML) version 3.0. 2nd ed. World Wide Web Consortium. 2014. https://www.w3.org/TR/MathML3/

35. Journal publishing tag set. National Library of Medicine (US), National Center for Biotechnology Information; [updated 2021]. https://jats.nlm.nih.gov/publishing/1.3/

36. Book interchange tag set: JATS extension. National Library of Medicine (US), National Center for Biotechnology Information. https://jats.nlm.nih.gov/extensions/bits/

ADDITIONAL REFERENCES

Council of Science Editors, Editorial Policy Committee. CSE's recommendations for promoting integrity in scientific journal publications. Council of Science Editors. https://www.councilscienceeditors.org/recommendations-for-promoting-integrity-in-scientific-journal-publications

Malički M et al. Systematic review and meta-analyses of studies analysing instructions to authors from 1987 to 2017. Nat Commun. 2021;12:5840. https://doi.org/10.1038/s41467-021-26027-y

2   Publication Policies and Practices

Editors: Jessica L. Striley; Emmanuel A. Ameh, MBBS, FWACS, FACS; and Thomas A. Lang, MA

2.1 SOURCES AND SCOPE

This chapter is drawn largely from the Council of Science Editors' recommendations paper, which is titled "CSE's Recommendations for Promoting Integrity in Scientific Journal Publications."[1] Additional detailed statements on editorial policy in scientific publishing are available from the International Committee of Medical Journal Editors[2] (ICMJE), the World Association of Medical Editors[3] (WAME), the Committee on Publication Ethics[4] (COPE), and the EQUATOR Network[5] (Enhancing the QUAlity and Transparency Of health Research). Another useful resource is the *AMA Manual of Style.*[6]

Many professional societies have guidelines on ethical publication that are specific to their disciplines. Some of those guidelines are cited in the Additional References at the end of this chapter.

When unethical behavior, such as scientific or publishing misconduct, is uncovered during the editorial review of manuscripts or after their publication, editors, editorial boards, and journal editorial offices need to respond appropriately. Most scientific publications published by professional societies rely on the publications' ethical guidelines and their editors' experience to address unethical behavior. Journal staff may also consult ICMJE and COPE guidelines, as well as relevant laws. When authors are found to have behaved unethically, journal editors may reject the authors' manuscripts if they have yet to be published or withdraw them if they have been published. In addition, editors may levy such sanctions as banning the authors from submitting other manuscripts to the editors' journals, and editors may share their concerns with the authors' institutions.

Most research institutions and universities, in turn, have internal mechanisms to address research and publishing misconduct. In some countries, independent bodies arbitrate or investigate research and publishing misconduct. For example, in the United States, some forms of research misconduct fall under the jurisdiction of the Office of Research Integrity of the Public Health Service[7] and the Office of Inspector General of the National Science Foundation.[8] In the United Kingdom, the Research Integrity Office,[9] which is an independent nonprofit organization, provides expert and objective guidance for addressing unethical conduct. In India, the University Grants Commission[10] and the Council of Scientific & Industrial Research[11] provide some guidance on research misconduct.

To ensure that high ethical standards are upheld in scientific publishing, all parties should work together with trust and respect in such ways as the following[1]:

- Authors and other contributors should be responsible for conducting their research in a way that advances or confirms scientific knowledge and that can be replicated by others.
- Institutions should create and encourage an ethical research environment, and they should be responsible for the conduct of their researchers.
- Journals should be responsible for the conduct of their editors and for ensuring that manuscripts are appropriately evaluated before publishing them.
- Peer reviewers should offer objective assessments on the scientific value and quality of reported research.

Detailed information on the relationships among these parties can be found in the CSE recommendations paper[1] and on the websites of the ICMJE[2] and COPE.[12]

2.2 RESPONSIBILITY TO PUBLISH RESEARCH RESULTS

Publishing research in peer-reviewed scientific journals has long been the standard for formally presenting new findings to fellow scientists and the public (see Table 2.1 for general responsibilities). Publication serves as both the final stage of research and the first stage of deliberating its importance.

In scientific publishing, concerns have arisen about such issues as determining

Table 2.1. General Responsibilities of Those Submitting and Publishing Research

Authors	• Document and publish research and results. • Conduct research according to the ethical guidelines of sponsoring institutions and funders before submitting manuscripts.	• Adhere to publication and ethical policies of the target journal. • In the cover letter, explain any variances from policies. • Declare sources of funding and potential conflicts of interest, as well as intended republication of copyrighted material.
Journal editors	• Uphold the principle that the highest priority in publishing a scientific journal is to advance science. • Adhere to established ethical and publishing standards. • Develop internal and external editorial ethics policies. • Publish only articles of interest to the journal's readers. • Responsibly assess and solicit manuscripts. • Review submitted manuscripts consistently, fairly, and promptly. • Review manuscripts for ethical violations.	• Recuse yourself from handling manuscripts in cases of competing interests. • Resist pressures that could compromise the review process. • Publish notices of any misconduct that occurred during the publishing process. • Maintain editorial independence with "firewalls" between the journal's editorial and administrative functions and between the editorial functions and the journal's sponsoring organization.
Peer reviewers	• Be familiar with current research and other developments in the field. • Understand and adhere to each journal's editorial and ethical policies. • Commit to accepting manuscripts that advance the field, including those reporting negative results. • Review only manuscripts you are qualified to evaluate, can review objectively, and can complete promptly, maintaining confidentiality.	• Maintain the confidentiality of the review process. • Communicate only with the commissioning editor about reviewed manuscripts unless instructed otherwise. • Do not disclose or use any knowledge gained from the review until after the article is published. • Report ethical concerns to the commissioning editor only. • Do not decline to review a manuscript because of ethical concerns without reporting the reason.

appropriate avenues for publication (eg, "primary" versus "gray" versus "predatory" literature), suppressing findings for illicit or unethical purposes,[13] and withholding research from publication for other reasons (eg, authors' or journals' lack of interest in publishing nonstatistically significant results).[14]

2.2.1 Primary, Gray, Predatory, and Counterfeit Literature

Scientific literature can be divided into 4 broad categories, which differ roughly by source and quality.

(1) "Primary literature" is research published in established, peer-reviewed print and electronic journals or books that are indexed, archived, and publicly available online or from libraries. In general, primary literature consists of the most rigorous and best research, and it constitutes the core of science.

(2) "Gray literature" is research that is not formally published in sources such as books and journals. It includes many types of documents, such as internal government reports, dissertations, white papers, conference abstracts, and in-house corporate reports. Gray literature may contain proprietary or confidential information. It is usually not readily

available through conventional sources, and it is less likely to be indexed than is primary literature.[15]

A major concern with gray literature is that it is likely to contain a great deal of sound, relevant research that is being withheld for proprietary or inappropriate reasons, such as suppressing findings that would reflect badly on a product or its manufacturer, avoiding compromising a pending patent, concealing previous research that did not adhere to current ethical standards, and avoiding the time and expense of preparing the research for publication. In addition, gray literature varies in quality, and it is more likely than primary literature to report negative results.[15] Consequently, when gray literature is included in systematic reviews and meta-analyses, it may strengthen or dilute research findings, depending on the quality of the gray literature.

(3) The main goal of "predatory" journals is making profits, not providing legitimate publication services. Because many predatory journals have reputations for being deceptive or even fraudulent, the research published in these journals is almost universally considered either to be tainted or to be of low quality.

Predatory journals typically prey on inexperienced authors and exploit the open-access, author-pay publication model.[16] These journals characteristically recruit articles with aggressive email campaigns that promise authors rapid publication and low article-processing charges. Such journals may have websites that look like those of legitimate journals, but these websites may sport fictitious editorial boards, false citation metrics, and misleading descriptions of the services they provide (eg, peer review, typesetting, and indexing). These journals often succeed in recruiting manuscripts because authors believe they are submitting their work to legitimate journals.[17]

Predatory journals may publish every manuscript they receive, or they may publish none, shutting down when their scam comes to light.[18]

(4) "Counterfeit" literature reports "systematically fabricated data following a recurring template."[19,20] These counterfeit manuscripts are created and sold by "paper mills" to so-called authors who generally need to publish for professional reasons but who are unwilling or unable to conduct legitimate research. Sometimes, counterfeit articles report fictitious data, and other times, they are "hijacked" articles, which are legitimate, published articles that have been heavily modified to avoid detection by plagiarism software. Such articles are often difficult to detect, and they are usually identified only when similar texts or figures are published in 2 or more journals. Paper mills often represent themselves as language-editing services for authors who are non-native English speakers, especially researchers in medical fields.[21]

2.2.2 Suppressing Research Findings

Although suppressing research findings may be legitimate in some instances, extended delay or total suppression prevents science from advancing and, at the very least, can lead to duplicate and unnecessary research (ie, "research waste").[22] Research findings may be suppressed to protect intellectual property and other proprietary interests. In its most egregious form, research suppression occurs because the findings would act against the financial or other interests of the authors, their employers, or funders. Global terrorism and "bad actor" nations have led to controversy over publishing articles on "dual-use" technologies (ie, research that has both peaceful and military applications). The scientific community is questioning whether, where, and when such articles should be published or even whether they should be suppressed to prevent their use by terrorists. Journal editors in the United States and the United Kingdom developed recommendations for balancing scientific advancement with national security.[13] These recommendations have been published in the *Proceedings of the National Academy of Sciences*, *Science*, *Nature*, and other journals.

2.2.3 Data Sharing

Attention has increasingly been focused on sharing data, which has been partially facilitated by technology that enables large data sets to be readily accessed and shared. The International Committee of Medical Journal Editors (ICMJE), for example, has stipulated since 2018 that clinical trials need to submit data-sharing plans when those trials are registered with recognized public clinical trial registries.[23]

During the pandemic caused by the coronavirus disease of 2019 (COVID-19), it became apparent just how important rapid, transparent, and international data sharing is to protecting humanity. Throughout the pandemic, nearly every peer-reviewed medical journal wanted to quickly publish data regarding the disease.

As early as 2003, the US National Research Council's Committee on Responsibilities of Authorship in the Biological Sciences concluded that all academic, government, and commercial investigators are obligated "not only [to] release data and materials to enable others to verify or replicate published findings (as journals already implicitly or explicitly require) but also to provide [research findings] in a form on which other scientists can build with further research."[24] Underpinning this conclusion is the understanding that all members of the scientific community derive benefit from open communication and are obligated to openly communicate their data. To this end, the National Research Council's responsibilities committee developed guiding principles and explicit recommendations for making research data and materials available to the scientific community and the public.

Several online public databases have been created into which researchers can place large data sets so that they can be shared with others. Examples include clinical trial registration websites (eg, ClinicalTrials.gov[22] and the European Union Drug Regulating Authorities Clinical Trials database,[25] better known as EudraCT) and databases offering high-resolution structural data and nucleotide sequences (eg, the Protein Data Bank[26] of the Research Collaboratory for Structural Bioinformatics and the databases of the member organizations of the International Nucleotide Database Collaboration[27]).

For their part, many scientific disciplines have identified areas in which data sharing is essential for driving science forward. The peer-reviewed journals and other publications in these disciplines should list those essential areas in their instructions for authors. Those publications should specify whether authors are encouraged or mandated to share their data, as well as recommend appropriate databases in which to deposit their data. As another means of broadly disseminating scientific findings, many governmental, commercial, and philanthropic organizations mandate that publications based on research they fund be deposited in public databases. Two prominent public databases are PubMed Central,[28] sponsored by the National Library of Medicine in the United States, and Europe PMC,[29] which is supported by 27 biomedical and life sciences research funders. Authors are responsible for submitting their scholarly works to PubMed

Central and Europe PMC, but in many cases, journals automatically deposit articles on their authors' behalf.

Many institutions have established institutional repositories where staff can post their publications. Often, individual authors can also "self-archive" their publications. In both cases, the institution's and publisher's guidelines should be followed. Two important issues that must be considered are copyright protections and at what stage manuscripts can be posted (eg, before peer review, after acceptance, or after publication).

2.2.4 Sharing Negative Findings

Another critical aspect of data sharing is the need for researchers and journals to publish negative findings, which show that hypotheses are not supported by research. Negative findings can provide valuable information that may reduce research waste by keeping researchers from pursuing unproductive lines of investigation and unnecessarily repeating unproductive experiments.

Unfortunately, negative findings are often viewed as uninteresting, and articles that report such findings are unlikely to be cited. Thus, many authors do not submit manuscripts with negative findings, and journals are less likely to publish such manuscripts. An early challenger to this thinking is the *Journal of Negative Results in Biomedicine*,[30] which was launched in 2002. Since then, several journals focused on publishing negative results have emerged in various fields, including *PLOS ONE's* "Missing Pieces" collection,[31] *The All Results Journals*,[32] the *Journal of Articles in Support of the Null Hypothesis*,[33] and the *Journal of Pharmaceutical Negative Results*.[34]

To encourage researchers to write and submit manuscripts on negative findings, the editors of other journals could take such steps as creating sections within their journals that focus on negative results. Of equal importance is depositing negative data in public online databases (see Section 2.2.3, "Data Sharing").

2.2.5 Authors' Responsibilities

The first obligation of authors is to document their results with the goal of publishing them. In meeting this obligation, authors should comply with the ethical guidelines of their institutions and the publishing requirements stipulated by their research funding before submitting their manuscripts to peer-reviewed journals and other publications. Authors should also become familiar with the editorial policies, instructions for authors, and disclosure requirements of the publications to which they intend to submit their manuscripts. If full compliance with a journal's expectations is not possible, authors should explain the exceptions in detail in a cover letter so that the journal's editor can decide whether to grant allowances.

2.2.6 Editors' Responsibilities

Journal editors are responsible for curating submissions and publishing content that adheres to ethical and pragmatic publishing responsibilities, as well as meets the needs of their targeted readers. In addition to assessing submitted manuscripts, editors can ethically solicit manuscripts by using such tools as "calls for submissions," which encourage authors to submit manuscripts on specific topics. Editors also can encourage readers to submit narrative review articles and editorials.

As the gatekeepers for the review process, editors should scrutinize manuscripts for ethical violations and report any they uncover. However, many journals are not able to investigate or adjudicate claims of misconduct.

Publication-related misconduct can erode a journal's credibility, especially if the journal ignores the misconduct. When published articles have been tainted by misconduct, editors should publish notices to inform readers of that misconduct. The World Association of Medical Editors (WAME) and the Committee on Publication Ethics (COPE) have both established recommendations for reporting scientific misconduct.[35,36]

Journal editors should develop editorial policies to guide authors, reviewers, editors, and journal staff through the range of ethical issues and pragmatic details related to journal publishing. In cases in which journals are published by professional societies, these policies may need to be developed in conjunction with the societies' governing bodies. The decision algorithms developed by COPE[4] are among the resources that editors can use to develop such policies.

Editors must review all manuscripts consistently and fairly within equitable time-lines. Editors must resist external and internal pressures that could infringe on the integrity of the review process. Professional societies and other publishers should protect editors when they resist such pressure by granting their journals editorial independence and by establishing a "firewall" between editorial and administrative functions. When editors are concerned that they cannot fulfill their ethical obligations in evaluating a particular manuscript, they should recuse themselves. Editors should ensure that their journals have a process in place that allows editors and peer reviewers to recuse themselves without interrupting or delaying the review process. Again, COPE's decision algorithms can be helpful to editors in creating such policies, and additional guidance is available from the ICMJE[2] and WAME.[3]

2.2.7 Peer Reviewers' Responsibilities

Peer reviewers should be familiar with the current research and other developments in their fields, and they should be committed to accepting manuscripts that would advance those fields, including studies with negative results. Reviewers also should be well versed in the editorial policies of the publications for which they review submissions.

Peer reviewers are ethically obligated to review only manuscripts that they believe they are qualified to evaluate, that they believe they can review objectively, and that they can review within the requested timeline. Reviewers are obligated to conduct reviews with diligence, without prejudice, and without compromising the confidential nature of the process. As part of their confidentiality obligations, reviewers should not communicate to anyone but the commissioning editor any information contained in the manuscripts they have been asked to review unless the editor agrees that doing so is necessary. Reviewers also should not use any knowledge gained from reviews for their own research until after the manuscripts are published.

If reviewers have knowledge of or even suspect author-related misconduct or other unethical circumstances, they should report their concerns to the journal's commissioning editor but to no one else. In such cases, it is insufficient for reviewers to simply decline to review the manuscripts.

2.3 ETHICAL BREACHES WITH LEGAL IMPLICATIONS

The Committee on Publication Ethics[4] (COPE) addresses the "core practices and policies that journals and publishers need to obtain the highest standards in publication ethics." To assist journals in adopting such practices and policies, COPE provides decision algorithms for assessing how specific ethical concerns should be approached.

Transparency and full disclosure are among the core values of science. In many of the ethical issues described below, the lack of transparency or full disclosure is the ethical violation. The same information properly disclosed, described, or referenced may be appropriate and valuable.

Authors should be willing and able to disclose every aspect of their research in the proper forum. Articles should accurately describe studies as they were conducted; the words, ideas, and other intellectual property of others should be identified as such; and research data sets should be made available to the scientific community with the proper safeguards.

Common ethical breaches in research and publishing are summarized in Tables 2.2 and 2.3.

2.3.1 Data Fabrication and Falsification

"Data fabrication" is the intentional creation or synthesis of data, whereas "data falsification" is the fraudulent manipulation of data. Unlike inadvertent errors in collecting or processing data, data fabrication and falsification are intentional, serious ethical breaches with legal implications.[37] Furthermore, falsifying the results of medical studies can have catastrophic health consequences for patients, including death.[38]

Allegations of data fabrication and falsification are often difficult to substantiate because the intent of the researchers may be difficult to determine and corroborating evidence may be lacking. Such claims must be pursued with diplomacy, discretion, and scrupulous attention to due process because such charges can seriously affect researchers' professional credibility and reputations.

An emerging source of falsified information is digital image manipulation, in which photo or graphic software programs are used to inappropriately modify images such as radiographs, normal-lens photographs, micrographs, blots, gels, and scans. Editors should specifically ask authors whether and how they have modified the images submitted with their manuscripts and insist on receiving the original files to verify that the images have been appropriately sized, appropriately cropped, and otherwise altered without becoming misleading.[39]

Table 2.2. Common Ethical Breaches in Research and Publishing

Ethical breach	Definition
Data fabrication	Intentionally creating or synthesizing data
Data falsification	Fraudulently manipulating data
Selective reporting	Deliberately reporting only select data to disguise, distort, or misinterpret results
Plagiarism	Taking intellectual ideas or words from others and mispresenting them as the authors' own without acknowledging the source
Copyright infringement	Reusing copyrighted material (eg, images, tables, and figures) without permission from the source
Undisclosed duplicate publication	Publishing essentially the same data in 2 or more publications without the consent of the publications involved
Undisclosed divided publication	Dividing the findings from a single study into 2 or more manuscripts and publishing them separately without making a useful contribution beyond what was provided in the first manuscript and without informing all the publications involved
Competing interests	Holding interests that could plausibly bias or that could be seen as biasing editorial decisions or research findings
Guest or gift authorship	Listing as authors those who have not met the criteria for authorship
Ghost authorship	Failing to list as authors those who have met the criteria for authorship
Animal welfare	Not confirming that the care and use of research animals met accepted standards
Institutional board approval	Confirming that studies of humans and animals were approved in advance by an appropriate institutional review board
Informed consent	Failing to confirm that informed consent was obtained from all participants in a clinical study before treatment began
Publication of protected information	Publishing protected health information of study participants, or publishing images or other identifying information of any individuals without obtaining specific consent from the individuals to publish those images

2.3.2 Selective Reporting of Research Results

"Undisclosed selective reporting" is the unethical practice of reporting only desirable outcomes (also referred to as "cherry picking") or of failing to report undesirable outcomes (also referred to as "lemon dropping").[40]

Another questionable way of analyzing data involves looking for statistically significant P values to add to a manuscript, a practice called "data dredging," "P-hacking," "looking for the pony," and "HARK," which stands for "hypothesizing after the results are known." When authors do not disclose that they engaged in data dredging, it is considered to be unethical selective reporting. However, when the practice is disclosed, it can be legitimately used to explore data sets.

Another practice that is unethical when not disclosed is "data trimming," in which the highest and lowest values of a distribution (ie, the outlying values) are discarded before data are analyzed. Removing the outliers has some statistical advantages, but the technique must be disclosed because the data have been manipulated.[41]

Table 2.3. Research Ethics Checklist

No.	Page[a]	NA	Topic
1.			The trial met the criteria of equipoise for the treatment.
2.			The trial adhered to the Nuremberg Code,[b] the Declaration of Helsinki,[c] and the Belmont Report.[d]
3.			Specific efforts were made to recruit underrepresented groups. Ensure that results are reported separately by sex and, if relevant, for other racial, ethnic, or socioeconomic groups.[e]
4.			The trial was registered before patients were enrolled.
5.			An appropriate institutional review board approved the trial before patients were enrolled.
6.			Animal welfare approval was obtained before the study began.
7.			Voluntary, informed consent (adults) or assent (minors below the age of consent) was obtained from all participants in a clinical study before treatment began.
8.			Patient records were kept confidential and, if appropriate, de-identified.
9.			Any harm that occurred during the research has been addressed.
10.			Strengths and limitations of the study are discussed.
11.			Sources of funding are identified.
12.			Plans are in place to notify participants of the results.

[a] Indicate in the manuscript where each item is addressed.

[b] See https://history.nih.gov/display/history/Nuremberg%2BCode.

[c] See https://www.wma.net/policies-post/wma-declaration-of-helsinki-ethical-principles-for-medical-research-involving -human-subjects/.

[d] See https://www.hhs.gov/ohrp/regulations-and-policy/belmont-report/read-the-belmont-report/index.html.

[e] See https://doi.org/10.1186/s41073-016-0007-6.

2.3.3 Plagiarism

Researchers commit plagiarism when they take ideas or words from the intellectual efforts of others and either misrepresent them as their own or simply use them without crediting the sources. This serious ethical deception places personal interests ahead of scientific integrity. Besides being an ethical violation, plagiarism can be a legal infraction if copyright infringement is involved (see Sections 2.3.4, "Copyright Infringement," and 3.3.5, "Infringement and Plagiarism"). The *AMA Manual of Style*[6] lists the following 4 types of plagiarism:

(1) Direct plagiarism: Verbatim lifting (ie, copying) of passages without enclosing the borrowed material in quotation marks and without crediting the original author.
(2) Mosaic: Borrowing the ideas and opinions from an original source and a few verbatim words or phrases without crediting the original author.
(3) Paraphrase: Restating a phrase or passage, providing the same meaning but in a different form without attribution to the original author.
(4) Insufficient acknowledgment: Noting the original source of only part of what is borrowed or failing to cite the source material in a way that allows the reader to know what is original and what is borrowed.

"Patch writing" is a milder form of plagiarism in which authors copy descriptions of standard research methods from a published article into their own. No credit is given or inferred because the descriptions are standard. Patch writing often occurs in the basic sciences and by authors for whom English is a secondary language. Although this practice is usually benign, the copied material may be detected by plagiarism software and included in plagiarism scores.

In addition, researchers "self-plagiarize" when they include material from their own published works without explicitly acknowledging those works. Although often done inadvertently, self-plagiarism can be misleading when it implies that the information is original to the new manuscript. Authors can avoid self-plagiarism simply by appropriately citing any of their previous works used in their new manuscript (ie, "self-citation").

Less clear-cut are rare situations in which similar ideas or words are arrived at independently at approximately the same time. In such cases, research institutions, universities, and professional societies must carefully determine whether the ideas or words were developed independently and simultaneously.

2.3.4 Copyright Infringement

When properly cited, brief excerpts from copyrighted materials can be used without permission under fair-use guidelines[42] (see Section 3.3.1, "Fair Use and Fair Dealing"). Besides being unethical, it is illegal in most countries to reprint larger portions of copyrighted material without permission. The consequences can range from having to issue a correction notice (ie, an erratum or corrigendum) and removing the infringed material to paying huge fines and damaging the reputations of the authors, journals, and publishers. Copyright holders are most likely to bring suit for copyright violations when they can show that the violations deprived them of income sufficient to make legal representation and court costs worth litigation.

In most countries, the moment text is recorded in a tangible medium, it is protected by copyright, regardless of whether the copyright is registered[43] (see Section 3.1.3, "Ownership"). Many publications stipulate that authors transfer their copyright to the publications so that anything published under their masthead becomes their property. Publications sponsored by professional societies often have authors transfer copyright ownership to the societies. In contrast, an increasing number of publications have adopted open-access models in which authors retain the copyright to their works and use Creative Commons licenses[44] to determine what permissions to automatically grant to interested parties (see also Section 3.3.4, "Creative Commons Licenses"). In addition, documents in the public domain, such as those developed by the US government and works for which copyright expired, can be used without permission as long as they are cited appropriately (see Section 3.1.4, "Public Domain").

Publishers that own the copyright to their published works grant licenses through several avenues and under many different rules. Some publishers grant their authors permission to reuse any content from their works in any other publication, provided that credit is given to the original publications. Some publishers participate in the publishers' agreement of the International Association of Scientific, Technical, and Medical Publishers (STM). In its simplified form, the STM agreement[45] allows all signatories the right to republish up to 3 figures from other signatories' material without fee, provided a license is obtained and credit is given as stipulated in the license. Other publishers charge to reuse their copyrighted material in other publications. Many of these licenses can be obtained through simple online interfaces, but others require specific communication with representatives of the copyright holders (eg, authorized staff in editorial offices, publishing houses, and professional societies).

2.3.5 Authors' Responsibilities

When authors intentionally misrepresent findings or other data, they violate their obligation to advance scientific knowledge (see Table 2.4). When scientists place job pressure, financial gain, or enhancement of their reputations above that obligation, they destroy their reputations and the credibility of their past and future research, as well as that of their coauthors. If their fabricated and falsified data are assumed to be valid, these researchers may even undermine the value of subsequent related research performed by other scientists. Fabricated and falsified data can waste time, misdirect future research, squander research funding, cause damage if practical applications are based on the data, lead to public distrust and disillusionment, and result in questionable treatments and the death of patients.

Although data fabrication and falsification are never appropriate, selective reporting can be, provided it is used only with great care, it is fully justified, and it is disclosed. Researchers should not use selective reporting to omit data that could be even remotely relevant to their findings simply to avoid complicating their explanations or weakening their conclusions. To ensure full disclosure of research and to thwart perceptions that might cast doubt on valid findings, researchers should at least summarize the process used for selective reporting and the reasons for omitting data. If the data are available in full elsewhere, that source should be referenced. Publishing the full data set as supplementary content or submitting research data to data-sharing repositories provides readers access to the data to use for such purposes as writing commentaries on the original research and conducting related research (see Section 2.2.3, "Data Sharing"). To advance scientific knowledge, publishers should support authors in their data-sharing efforts.

Authors should understand that "good" manuscripts are built not only on original ideas but also from original writing. Consequently, plagiarizing is unethical. Many journals check manuscripts for plagiarism and will act against authors who plagiarize. To avoid plagiarism, authors should keep detailed notes throughout their research, and they should carefully examine their manuscripts before submitting them to be sure they properly credit other investigators' findings, ideas, and other intellectual contributions. Authors should also carefully examine their own ideas and words to be certain that they were not subconsciously derived from other scientists' intellectual activities.

To avoid copyright infringement, authors should never submit content that is under copyright without clearly noting that it is reused with permission from the copyright holders. To obtain permission to use content under copyright, authors should request official licenses from the copyright holders. These licenses should stipulate the correct

Table 2.4. Ethical Responsibilities of Those Submitting and Publishing Research

Authors	• Strive to advance scientific knowledge. • Do not put personal interests before scientific or publication quality. • Do not fabricate, falsify, or otherwise misrepresent data or results. • Do not report only select data. Report, or be able to report, all details of the research with full transparency and disclosure. • Summarize any process used for selective reporting and the reasons for omitting data. • Submit manuscripts to only one publication at a time. • Do not engage in duplicate or divided publication unless you have disclosed that the same data were used in previous publications.	• Do not plagiarize others. Cite references for the ideas and words of others. • Cite your own work when appropriate (ie, not for name recognition) so that readers will know the ideas in the current work are not new. • Obtain written permission to use copyrighted information. • Document in manuscripts that written consent was obtained to publish images or other identifying information of study participants and other individuals. • Make data available to qualified researchers and the public and, when possible, place the data in an appropriate repository.
Journal editors	• Strive to prevent misconduct throughout the publication process. • Be alert for unsourced copyrighted material. At minimum, have authors confirm that they have permission to republish such material. • Ensure that authors correctly cite original publications and copyright holders. • Have authors confirm that they have received permission to republish copyrighted material. • Remove from manuscripts copyrighted material for which the author cannot document permission to reprint. • Have authors confirm that they received consent to publish in their manuscripts images or other identifying information of participants and other individuals. • Inform authors that although you are not responsible for adjudicating allegations of misconduct, you may cooperate with other groups authorized to investigate misconduct, such as authors' institutions.	• Evaluate allegations of fabrication, falsification, and plagiarism, and determine whether they are credible. If allegations are deemed credible, notify the researchers' employers and any appropriate oversight committees or review boards. • Provide all data necessary to conduct factfinding investigations should allegations of misconduct arise. • If misconduct is discovered after publication, correct the written record. Publish an expression of concern or a full retraction in a prominent place, and notify indexing and abstracting services. Do not remove a retracted article from the literature, but post a retraction statement prominently in the online version of the article. • Understand that peer review and related information are protected speech; they are proprietary and not subject to legal discovery.
Peer reviewers	• Be alert for instances of research or publication misconduct, both before and after publication. • Watch for citation omissions and inaccuracies in manuscripts under review.	• Report any concerns to the commissioning editor only. • When ethical questions arise, consider assembling other evidence that may aid in resolving the issues.

final disposition of the licensed material, which often includes the name of the publisher of the new manuscript, the title of the publication in which the manuscript will be published, the date of publication, and the circulation of the publication. If any of this information is unknown, authors should obtain it from the new publication's editorial office or publisher. Authors should meticulously abide by all the terms of the licenses, including required credit language and allowed uses. Authors should be prepared to submit copies of these licenses should the publisher or journal office request them.

2.3.6 Editors' Responsibilities

Journal editors' responsibility to prevent and detect misconduct begins long before manuscripts are submitted for consideration and lasts throughout the publication process
(see Table 2.4). For example, journals' instructions to authors should state that authors
may be asked to provide all data necessary to conduct fact-finding investigations should
allegations of misconduct arise. However, because few journals have the resources
to conduct extensive investigations, these instructions may note that editors are not
responsible for proving or disproving such allegations and that editors may cooperate
with other entities authorized to investigate misconduct, such as authors' institutions.

Once manuscripts are submitted or accepted, many journals make use of software
programs that detect plagiarism.[46,47] Journal staff, editors, peer reviewers, and sometimes the authors themselves can use such software to greatly reduce the chances of
publishing plagiarized text. These programs flag text strings that have appeared in other
publications, but not all flagged text is plagiarized copy. For example, many research
methods in the basic sciences employ standard language, or they can only be described
correctly in a few ways. Consequently, flagged text should be read by editors or their
staffs to determine whether it might have been plagiarized.

In addition, editors should be alert for material that is clearly from other copyrighted
sources. Editors should ensure that authors provide full citations to original publications
and copyright holders. At minimum, editors should have authors confirm that they have
received permission to republish the material. Alternatively, editors may request copies
of the licenses to confirm that the authors obtained permission to reuse the material
and to ensure that the publication adheres to the terms of the licenses.

Another line of defense that editors can rely on to alert them to misconduct is the
peer-review process (see Section 2.3.7, "Peer Reviewers' Responsibilities").

When receiving an allegation of fabricated, falsified, or plagiarized research, editors
should evaluate the allegation to determine whether the charge seems credible. This
step may include querying all authors of the work and the contributors acknowledged
in the text. Should editors determine that the matter is not simply an oversight, misunderstanding, or innocent error, they should notify the researchers' employers and any
appropriate oversight committees or review boards and provide them with evidentiary
information. Should the employers also be suspect, editors may have to rely on independent oversight committees or review boards. For example, in the United States, the
Office of Research Integrity assists in instances in which funding from the US Department
of Health and Human Services supported the research under scrutiny. In addition, the
United Kingdom Research Integrity Office provides editors with advice and guidance on
misconduct issues.

Should fabrication, falsification, or plagiarism go undetected until after publication,
a journal editor's primary responsibility is to correct the written record.[1] A retraction
should be obtained from the authors or their employers or, if that is not forthcoming,
from the journal's editor. The retraction should be published prominently within the
print and online editions of the publication, and it should be submitted to indexing
and abstracting services (eg, Web of Science, PubMed,[28] Google Scholar,[48] EBSCO,[49] and

JSTOR[50]). Not only will this public notification make readers aware of the misconduct, but it will also serve a disciplinary role in dealing with the authors' offense.

Although a journal editor may decide to retract a research article, the article should not be deleted from the publication's online version. The retracted work is still part of the historical record, and it may be of value to subsequent researchers (eg, those studying misconduct). Rather, the journal should label the online versions of the article to clearly indicate that it has been retracted. In addition, indexing services may have specific requirements about how the retraction should be indicated in the indexes so that it is properly linked to the original article.

As a precursor to retraction, some online journals post an "expression of concern," which alerts readers that the editors are concerned about some aspect of the data or the analytical integrity of the research. (See Section 3.5, "Correcting the Literature," in the Council of Science Editors' recommendations paper,[1] and see the Committee on Publication Ethics' retraction guidelines.[51])

In addition to addressing allegations of fabricated, falsified, and plagiarized research, journal editors should investigate concerns that authors have used copyrighted material without permission or without due credit. If editors determine that permission was not obtained to reuse the material, the material should be removed from the manuscript. If permission was given but credit language is missing, credit language should be added to the manuscript before it is published. If the work has already been published, its online version should be updated by adding the credit language. In the latter case, a correction notice should be issued in print and online indicating that the published article was corrected. For the online version of the publication, this notice should be included on the same web page as the article.

2.3.7 Peer Reviewers' Responsibilities

Peer reviewers should be on the alert for misconduct both before and after manuscripts are published (see Table 2.4). Peer reviewers should report any suspicions to journal editors along with any corroborating evidence. Reviewers should not directly contact the authors, the authors' professional peers, or the authors' employers about these suspicions. Instead, reviewers should rely on the editors to do so. However, reviewers may independently investigate the published record and assemble other evidence that could help editors resolve the matter.

Because of their familiarity with the subjects of the manuscripts they review, reviewers should watch for citation omissions and inaccuracies during the peer-review process and note them in their reviews. If omissions appear to be intentional, reviewers should note these and any plausible reasons for the omissions in confidential comments to the journal editors.

2.4 DUPLICATE, DIVIDED, AND JOINT PUBLICATION

In its simplest form, duplicate publication is the description of essentially the same study in 2 or more publications without the knowledge or permission of either journal.

A duplicate manuscript may even include new data or new findings, but if the practical relevance and import of the new material are modest or the findings remain basically unaltered, the publication may be considered redundant. However, if both journals are informed of the duplication, if the second journal obtains permission to publish the article, and if the duplication is clearly noted in the second publication, the duplicate publication may be appropriate. For example, 2 or more journals may agree to jointly publish the same article at approximately the same time. Joint publications are usually on topics with discipline-wide implications, such as consensus statements, reporting guidelines, and position statements.

A variant of duplicate publication is "divided publication" (also known as "fragmented publication," "piecemeal publication," and "salami science"). Divided publication occurs when authors separate what should be reported as a single study into 2 or more publications.[52]

In their most egregious forms, duplicate and divided publications are attempts by authors to increase their publication lists to enhance their reputations, prominence, and professional advancement. Authors who submit duplicate or divided publications unfairly disadvantage colleagues who more closely adhere to professional ethics. More often, duplicate manuscripts are submitted by inexperienced authors who may not fully understand the implications of their actions or by authors who erroneously conclude that their manuscripts fall into one of the special situations in which duplicate publication may be acceptable. (See Table 2.5 for a summary of responsibilities related to duplicate, divided, and joint publications.)

2.4.1 Drawbacks to Duplicate Publication

Duplicate publication is widely proscribed for reasons such as the following:

(1) It can mislead other researchers into counting the same findings twice, which can give those findings more weight than is due.
(2) It squanders limited scientific research funds.
(3) It needlessly expands the burgeoning scientific literature.
(4) It needlessly consumes time and resources that could be devoted to reviewing and publishing other articles.

2.4.2 Justified Duplicate Publication

Circumstances in which duplicate publication may be justifiable vary among journals, but they generally include the following. In all of these situations, both journals must agree to and announce the duplication.

(1) The second publication contains substantially new data and new findings. In the second publication, the new data and findings must be clearly distinguished from the original ones.
(2) Important findings published in one country and language warrant republication in another country and language to reach an audience that otherwise would not readily have access to the findings. For example, if a country requires that research funded by the government be published in the language of that country, the research may be translated

Table 2.5. Managing Duplicate, Divided, and Joint Publication

Authors	• Understand the problems created by duplicate and divided submissions. • Become familiar with editorial policies governing duplicate, divided, and joint publication. • Submit an article to only one journal at a time. • Confirm in the submission letter that the manuscript has not been published and that it is currently not being considered at another journal.	• Determine whether the journal automatically refers the manuscripts it rejects to lower-ranking journals (ie, "downstreaming"). • Disclose information on other publications based on the same data to both editors and readers. • Inform all involved journals of previous publications based on the same data. • Withdraw manuscripts if problems cannot be resolved in revision before or after peer review.
Journal editors	• Establish and enforce editorial policies governing duplicate, divided, and joint publication. Ensure that these policies are stated in the journal's instructions for authors. • State in instructions for authors that manuscripts can be submitted to only one journal at a time. • Specify that cover letters include a guarantee that manuscripts are not under consideration by any other journals. • When duplicate or divided publication is deemed appropriate, require authors to note this fact in the article. • Investigate credible claims of duplicate submission, duplicate publication, or divided publication by gathering information and requesting explanations from authors.	• If duplicate submissions are discovered during the review process, inform authors of the possible negative effects on journals and authors, and insist that the manuscript be withdrawn from all but one journal. • When duplicate or divided publication was not disclosed before publication, publish an expression of concern or a notice of retraction explaining the violation, which may include an explanation by the authors or their institutions. • If authors withdraw a manuscript because it has been accepted elsewhere, notify the editor of the other journal, and perhaps discuss appropriate corrective measures with that editor.
Peer reviewers	• Be familiar with journal policies on duplicate and divided publication. • Notify the commissioning editor of any publications similar to the manuscript under review.	• If possible duplication is discovered when reviewing for another journal, notify both editors.

for journals in other languages so that it is available to the rest of the scientific community. In such cases, besides agreeing to the duplication, all of the journals should address any related copyright issues before the manuscripts are republished. In addition, the titles on the duplicate publications should indicate that the articles are secondary to their primary publication, and the original publications should be cited.

(3) Important findings published in a journal of one specialty or profession sometimes warrant republication in a journal of another specialty or profession to reach an audience that otherwise might not have ready access to the findings. For example, a study of surgery for patients with pediatric heart conditions might be appropriate for journals in surgery, pediatrics, and cardiology.

(4) The duplicate publications are intended to supplant other published documents, such as preliminary or in-house reports, articles and abstracts reporting interim analyses, articles posted on preprint servers, and unreviewed symposium proceedings. In all such cases, journal editors should be informed of the previously published works in part to ensure that such updated manuscripts do not violate the journals' policies related to duplicate publication.

2.4.3 Authors' Responsibilities

Most journals have editorial policies that prohibit duplicate and divided publications and describe special situations and exceptions. Authors should become familiar with those policies, especially as they relate to disclosure. These policies typically require authors to attest in their submission letters that their manuscripts are not currently under review elsewhere and that the material has not been published elsewhere. These policies do not apply to a rejected manuscript, which authors can submit to another journal without noting that it was rejected elsewhere.

When submitting a manuscript that duplicates published results or reports new data and new findings from previously published research, whether in the primary or gray literature, authors should explain this circumstance to the editor in their submission letter. Authors should also explain why they believe the new manuscript is worthy of publication, and they should include a copy of the previous publication or the source information, preferably with a hyperlink to the material. In addition, authors should appropriately cite their previously published research in their new manuscript.

2.4.4 Editors' Responsibilities

Editors should ensure that their scientific journals have editorial policies on duplicate and divided publications. Such policies simplify editorial decisions, saving time and ensuring consistency. The policies should be published regularly and prominently in the journals' print editions, on their websites, and in their print and online instructions to authors.

When duplicate publication is deemed appropriate, editors should require authors to briefly note the initial publication prominently in the first or second paragraph of the text or in a footnote to the title of the new publication. That note should include the reason for duplication and cite the original publication.

Policies on duplicate and divided publications should require authors to disclose in their submission letter any related or previous publications of results from the research reported in the newly submitted manuscript.

Should an editor learn that authors have not disclosed that their manuscript is a duplication, the journal's policies should obligate the editor to gather information about the initial publication and to obtain an explanation from the authors. Situations judged to be oversights may warrant informing authors that their future submissions will be scrutinized with greater care, which will include alerting the reviewers to previous infractions. With regard to flagrant or repeat offenders, the policies should dictate whether the editor should report the misconduct to the authors' employers, the employers' relevant oversight bodies, or both.

If an investigation determines that a published article reiterates an earlier publication without the knowledge of the editors of both journals, the editor of the second journal should publish an expression of concern or a notice of retraction that explains the violation. This notification may include an explanation by the authors or the authors' institutions. Online publications can place such notices prominently within the articles. Such notices inform other researchers of the misconduct, protect the credibility of the journals and their editorial policies, and thwart professional benefits that might otherwise accrue to the authors.

2.4.5 Peer Reviewers' Responsibilities

Journal reviewers should be familiar with journal policies on duplicate publication, and they should notify journal editors of any publications similar to the manuscripts under review.

Another line of defense against duplicate publication consists of the manuscript review committees that many universities and some government agencies have to assess scholarly manuscripts before they are submitted to journals. The members of these review committees should be alert to duplicate manuscripts, and they should prevent their submission.

2.5 COMPETING INTERESTS

According to Section 2.1.3, "Conflicts of Interest," in the Council of Science Editors' recommendations paper on integrity,[1] conflicts of interest in scientific publishing are "conditions in which an individual holds conflicting or competing interests that could bias editorial decisions."

Preferably termed "competing interests," conflicts occur when an author's, reviewer's, or editor's judgment is potentially compromised by financial, professional, academic, ethical, political, or personal interests (see Table 2.6). Competing interests may be real or perceived, potential or actual, or inconsequential or harmful. Having a competing interest does not always imply or result in biased judgment.

Profit motive is one form of competing interest. This conflict occurs when researchers publish findings that were influenced by the financial interests of the researchers, the funders of the research, or the institutions contracting with the funders on behalf of the authors. For example, a conflict exists when a commercial funder influences the form or content of the reported research or determines which research material is published and which is withheld. Sometimes, direct funding and material support may qualify as competing interests, even if the funding only covers the costs of the research.[53,54] Similarly, questions can arise when authors own patent rights to products that receive favorable findings in the reported research and when authors own stock in companies that would benefit from favorable findings. Akin to profit motive is the competing interest that occurs when research is performed under contract with advocacy groups that hope the research will advance their interests.

Nonfinancial conflicts may stem from personal, political, academic, and religious interests.[1]

Most journals require that authors disclose all interests in relevant financial activities outside the submitted research, including the following:

- Employment
- Formal affiliations (eg, serving on companies' boards of directors)
- Research funding

Table 2.6. Managing Competing Interests

Authors	• Disclose any real, potential, or perceived competing interests related to any manuscript submitted for publication, even if you are certain that no conflict directly affects the work.	• Request that editors bypass reviewers who might be biased against you or the research, and explain those concerns.
Journal editors	• Publish competing interest policies, and identify the circumstances that require disclosure by authors and peer reviewers. • Confirm that the authors have reported competing interests in the article. • Consider whether the competing interests may have biased the research. • Consider authors' requests to bypass reviewers who might be biased for any reason. • Have peer reviewers submit a list of possible competing interests. • Send reviewers only abstracts at first so that they can decline to review manuscripts for which they believe they have competing interests. • Have reviewers report any competing interests when submitting reviews of manuscripts. • Consider whether your or peer reviewers' interests might compromise or appear to compromise your or their judgment.	• Recuse yourself from evaluating any manuscript if competing interests are likely to compromise your judgment or could appear to. • Handle cases of suspected serious competing interests with good judgment, and, if necessary, investigate the suspicion discreetly. • Determine whether a failure to disclose potential competing interests before publication was likely intentional. • When evaluating potentially biased research, determine what actions are appropriate, such as publishing the manuscript if the evidence for bias is weak, publishing with a description of concern, or rejecting the manuscript. • If competing interests are disclosed only after acceptance, determine why they were not disclosed earlier, and note this fact in the published article. • When competing interests are discovered after publication, ask the authors to justify why this information was not disclosed earlier.
Peer reviewers	• Inform editors of any competing interests when being considered as a reviewer, as well as before reviewing specific manuscripts.	• Recuse yourself if interests could affect your judgment of a manuscript or appear to do so.

- Material support
- Consultancies, royalties, and honoraria
- Payments to attend meetings or to testify at governmental or legal hearings
- Stock ownership
- Intellectual property rights and patents

The period during which competing interests are considered to be relevant varies by journal. Some journals are concerned only about the period during which the research was conducted. Other journals focus on the period between the conception and publication of the research. Still others are interested in the previous 3 to 5 years. However, the amount of financial compensation is irrelevant.

Competing interests can also arise in the review process. For example, reviewers or editors might have a positive or negative bias toward some authors, they may be competing with authors in some way, or they may have prejudices concerning the authors' institutions or countries of origin.

Substantiated misconduct involving conflicts of interest is rarely discovered and validated because journals don't research possible conflicts of interest and are not expected to "police" authors. Consequently, competing interests may occur often, and they may remain undetected even when they are apparent and readily confirmed.

2.5.1 Authors' Responsibilities

Authors should fully disclose any real, potential, or perceived competing interests related to any manuscript submitted for publication, even if the authors themselves are certain that no conflict directly affects the work. Depending on the journal's preference, competing interests should be disclosed in a submission letter, on the manuscript's title page, in a field provided on the journal's online submission site, or on a form provided by the journal.

Failure to inform editors of any potential competing interests at the time of submission often violates journal policies and could be incriminating should a question of motive arise during the review process or after publication.

Authors, in turn, can help prevent conflicts among peer reviewers by requesting that editors bypass certain reviewers if the authors believe those reviewers would be biased against the research or the authors. Authors should give precise reasons for such requests, and journal editors should carefully consider the requests and grant them if appropriate.

2.5.2 Editors' Responsibilities

Editors should ensure that their journals routinely publish their policies regarding competing interests and the circumstances that require disclosure. In addition, if editors are aware of competing interests that authors might have, editors should question the authors should they fail to disclose those interests.

When a potential competing interest is disclosed or discovered, the journal's editor must weigh the likelihood that the authors allowed the conflict to bias their research results. Even if the editor has no reason to suspect that the results have been tainted by the potential competing interest, it is frequently advisable to disclose the potential competing interest to readers. Editors may require that authors agree to incorporate such statements in their manuscripts before they are reviewed. Many journals follow this procedure for all published articles, either by publishing a disclosure section within each article or by providing hyperlinks to online documents that provide information on the authors' possible competing interests. Such disclosure permits readers to decide for themselves how much confidence to place in the research results.

When a serious competing interest is suspected to have tainted reported findings, the editor will have to rely on good judgment and may need to investigate discreetly. A number of outcomes are possible. The editor's investigation or the peer-review process may rule out the likelihood of bias, or the editor may find the evidence to be inconclusive and decide to publish the manuscript with a brief description of the concern. Other options include returning the manuscript to the authors without initiating a peer review.

When an undisclosed competing interest is discovered, the editor must determine whether the failure to disclose was intended or unintended and then take appropriate action. If the manuscript has not been published, the editor may choose one of the courses described above. If a competing interest is not discovered until after an article is published, the editor may elect to publish in a prominent place a notice explaining the circumstances and the potential competing interest. For online publications, that notice can be placed prominently within the article itself.

If authors wait to disclose potential competing interests until an article has been accepted and is being prepared for publication, the journal's editor should ask the authors to justify why this information was not disclosed earlier. All information related to the competing interests should be reviewed by the editor and, in some circumstances, the manuscript's peer reviewers before the information is added to the article prior to publication.

Editors must also be sensitive to their own competing interests and recuse themselves from editorial decisions when their objectivity could be compromised or questioned. For example, friendship, collaboration, or competition with authors could create competing interests for editors, as could research findings that could positively or negatively affect the editors' own professional or financial interests. Although editors need not step aside if they believe they can remain fully objective in such situations, they should consider whether the perception of impropriety could cast doubt on the efficacy and robustness of their journals' reputations and the authors' findings. When faced with such competing interests, editors can have associate or guest editors assume the editorial duties for the manuscripts. Many publishers find it helpful to establish policies on competing interests for their editors. Some publishers even disclose instances of editorial recusal annually.[55]

Besides being mindful of their own competing interests, editors should be aware that peer reviewers may have interests that could prejudice their reviews. To minimize bias in reviews, editors may provide reviewers with guidelines for evaluating competing interests and deciding under what circumstances to disclose these interests to editors.

Establishing and visibly posting clear policies that direct editors and reviewers to disclose competing interest and that define when recusal may be required can build trust with authors, readers, and the public.[56]

2.5.3 Peer Reviewers' Responsibilities

Peer reviewers should inform journal editors of any competing interests, biases, or personal conflicts that might prejudice their review of a manuscript. Reviewers should be willing to recuse themselves if they have competing interests.

Before assigning reviewers to manuscripts, some journals ask reviewers to submit a list of financial dealings with manufacturers or other entities related to their field. Journal editors, in turn, take this information into account when assigning reviewers and when making decisions based on reviewers' comments. In addition, most journals send only manuscript abstracts to reviewers at first so reviewers can decline to review the entire work if they believe they have a competing interest or another reason to recuse themselves. As a last line of defense, many journals ask reviewers to indicate any potential competing interests when they submit their reviews.

With regard to the manuscripts they evaluate, peer reviewers should inform editors of any knowledge they have of the authors' competing interests, especially ones that might have influenced the research process or findings in unethical ways, such as data falsification (see Section 2.3.1, "Data Fabrication and Falsification").

2.6 REVIEW PROCESS AND PRIVILEGED INFORMATION

Matters of confidentiality in the manuscript review process vary widely (see Table 2.7). Some journals allow or require reviewers to identify themselves to the authors in peer reviews, which is known as "open peer review." Other journals withhold the identities of the reviewers from the authors while allowing the reviewers to know who the authors

Table 2.7. Responsibilities of Dealing with Privileged Information

Authors	• Become familiar with the journal's policies regarding confidentiality during the review process. • Do not solicit secondary reviews or give copies of the manuscript to anyone else while the manuscript is being reviewed by the journal. • Do not discuss the manuscript with anyone known or suspected to be a reviewer. • Become familiar with the journal's policy about posting research on preprint servers and institutional websites, as well as self-archiving on personal websites.	• Disclose to the journal any previous posting of part or all of the research to institutional or personal websites or to prepublication archives. • Determine whether the journal embargos articles (ie, informs the media about manuscripts in advance of publication under the stipulation that the media not report that information until the manuscripts are published). If the journal embargos content, do not distribute copies of or information about the article until after it is published. • Follow the journal's policies on giving interviews to the media during the embargo period.
Journal editors	• Protect authors' interests throughout the review process. • In the case of open review, inform authors whether they may talk about their manuscripts directly with assigned peer reviewers. • Clearly state confidentiality requirements to authors and reviewers. • Ensure that manuscript assessment is unbiased, independent, and critical. • Publicize whether the peer review process is open or anonymous and whether the reviews are signed or unsigned.	• Keep the contents of both manuscripts and peer reviews confidential from anyone except the authors, reviewers, and appropriate members of the journal's editing team. • Honor authors' request to make or not make the reviews of rejected manuscripts available to other journals. • Maintain the confidentiality of the peer-review process if asked to provide recommendations for the professional advancement of authors, even if the journal has an open review process.
Peer reviewers	• Fully understand and accept each journal's confidentiality requirements. • If a journal's policies are unacceptable, decline to serve as a reviewer. • Understand that in addition to accepting a professional responsibility, you are being entrusted to protect proprietary and valued material. • Keep confidential all information in manuscripts and reviews, unless a journal uses open peer review.	• If approached by an author of a manuscript under review, neither confirm nor deny being a reviewer. • Inform editors if you know the authors of a manuscript, and outline the good-faith steps taken to prevent being influenced by personal knowledge of or any external information about the authors. • Destroy electronic and hard copies of the manuscripts after reviewing them.

are, or they may conceal the identities of both authors and reviewers; in both cases, this is known as "anonymous" review.

Regardless of which variation is used, it is customary to treat the materials under review with confidentiality. The work represents a substantial investment of time, effort, creativity, and funding, and it is usually an instrument for professional advancement and perhaps monetary return. Therefore, those participating in the review process should understand that the work is proprietary to the authors and that confidentiality needs to be respected until after publication.[1]

2.6.1 Authors' Responsibilities

When submitting a manuscript to a peer-reviewed journal, authors should become familiar with the journal's policies regarding confidentiality during the review process. If the journal requires confidentiality while the manuscript is under review, authors should not solicit secondary reviews from colleagues or provide copies of the manuscript to anyone else. Making changes on the basis of such secondary reviews and submitting revisions while the manuscript is under formal review could slow and complicate the journal's review process, as well as increase the burden on the journal's reviewers. In addition, authors should resist the temptation to discuss their manuscript with anyone whom they suspect or know to be a reviewer of the manuscript.

Authors should check whether the journal has an embargo on prepublication release of information. Under an embargo agreement, the journal makes forthcoming publications available to news media in advance of the formal publication date in exchange for the news media agreeing to wait to release the information until a specified date and time, which is usually the day of or the day before the journal's formal publication date.[6] If an embargo is in place, authors are expected to honor it too by refraining from distributing materials related to their forthcoming publications and refraining from discussing the details of their findings before they are published. A common exception to this restriction is presenting the research at scientific meetings. In addition, journals may ask authors to give interviews to reporters during the embargo period.

Before posting their draft manuscripts on their personal websites or on online prepublication archives (eg, *Nature Precedings*[57]), authors should be aware of the standards and expectations of the journals to which they plan to submit their manuscripts. In some disciplines, posting to preprint servers is encouraged (eg, arXiv[58] in physics, mathematics, computer science, and other fields). Such services allow the public to view and download manuscripts, comment on them, and cite them before they are formally submitted for publication. In other disciplines, using preprint servers is discouraged or even prohibited. Regardless of which practice is advocated by their discipline, authors should disclose to journals any previous posting of their research or portions of it to personal websites or prepublication archives.

Journals may require that the formally published versions of articles contain notices of previous publication and that citations to formally published articles be added to the previous postings. Conversely, journals may reject such manuscripts as being published already.

2.6.2 Editors' Responsibilities

Editors must protect authors' interests throughout the review process. As part of this obligation, editors should clearly delineate their journals' confidentiality requirements to authors and reviewers. For example, if a journal does not conceal reviewers' identities from authors, it should explain whether all contact regarding reviews should be made through the editor or be conducted independently between the authors and reviewers. Many journals do not condone the latter practice because it increases the risk that personalities and professional eminence will affect the outcome of reviews.

When developing confidentiality policies, editors must remember that their highest priority in publishing a journal is to advance science.[1] Hence, policies should ensure that the assessment of manuscripts is unbiased, independent, and critical.[59] In deciding what type of peer review to institute, editors should carefully consider whether open or anonymous review could result in inappropriate decisions because authors' good or bad reputations may be taken into account by the reviewers, even subconsciously. On the other hand, open peer review may discourage a reviewer from giving a biased or unnecessarily harsh criticism of a manuscript. Some editors, reviewers, and authors argue that if neither the authors' nor reviewers' identities are known, the review and the author reactions to the review will be more objective, leading to better science. However, in small scientific disciplines, the identities of authors and reviewers may be easily guessed despite being undisclosed. In addition, some leading editors consider the anonymous review processes to be unethical because the reviews are unsigned. Some propose that open review be made even more transparent by allowing readers to see the exchanges between peer reviewers and authors.

Regardless of which peer-review process they follow, journals are obligated to keep the contents of both manuscripts and peer reviews confidential from anyone other than the authors, reviewers, and editors. However, authors of rejected manuscripts may request that they be allowed to share their peer reviews with any additional journals the authors submit the manuscripts to.

Maintaining the confidentiality of the peer-review process extends even to situations in which editors are called on to provide recommendations for the professional advancement of authors. If editors know the authors solely on the basis of works submitted for review and publication, editors should avoid making recommendations based on anything other than the published record. Even if their review processes are open, editors should refrain from providing recommendations based on information gathered during peer review because this process may provide only a limited and, therefore, distorted picture of researchers' capabilities.

2.6.3 Peer Reviewers' Responsibilities

Peer reviewers should fully understand and accept each journal's confidentiality requirements. If they find the stipulations of a journal to be unacceptable, they must decline serving as reviewers for that journal. Reviewers should understand that they are not only accepting a professional service responsibility but they are also being entrusted by the editors and authors to protect proprietary and valued material.

Reviewers are expected to maintain confidentiality about the contents of the manuscripts they review and about the fact that they are reviewing a specific manuscript.[60] They should not disclose their identities to authors or colleagues before or after publication without permission from the assigning editor.[61] To improve the quality of a review, the editor may permit reviewers to consult with third parties if they have the expertise and agree to maintain confidentiality.

For journals that use an anonymous review process, the reviewers' obligation to keep their reviews confidential extends even to authors. If approached by an author of a manuscript under review, a reviewer should neither confirm nor deny being a participant in the review process.

If reviewers in an anonymous review process can guess who the authors of a manuscript are, reviewers should make a good-faith effort to review the manuscript on its own merits, not taking into account any external information about those they assume to be the authors. Reviewers should inform editors that they believe they know who the authors are, and reviewers should outline for the editors the steps they took to ensure they conducted the review in good faith.

Another step reviewers should take to ensure confidentiality is to not keep electronic or hard copies of manuscripts after reviewing them.

2.7 WITHDRAWAL OF MANUSCRIPTS AND DUPLICATE SUBMISSIONS

The term "withdrawal" refers to removing a manuscript from the editorial process after submission but before publication. If an article needs to be removed after publication, a "retraction" must be issued and published. (More information on retracted articles is available at Retraction Watch at https://retractionwatch.com.)

The act of withdrawing a manuscript should not be taken lightly by any party because the authors, reviewers, editors, and editorial staff would have already invested significant time and resources in the manuscript. However, withdrawing manuscripts is unavoidable in some circumstances, and it is preferable to retracting articles after publication.

Authors may choose to withdraw their manuscripts for several reasons, including their belief that reviewers criticized their work too harshly or their dissatisfaction with a journal's processes. Editors, in turn, may decide to withdraw manuscripts for such reasons as suspicion of ethical violations and inability to contact authors.

One of the most egregious reasons for withdrawing a manuscript is duplicate submission. Authors may be tempted to submit their manuscript to more than one journal at the same time in the hope of shortening the publication process, gambling that only one of the journals will accept the manuscript or that it will be accepted by more than one journal, allowing the authors to choose the highest-ranking journal. However, duplicate submission is considered improper within the conventions of scientific publication, and most journals have strict policies prohibiting duplicate publication (see Section 2.4, "Duplicate, Divided, and Joint Publication"). In fact, if a journal learns that a manuscript it has accepted has been simultaneously submitted elsewhere, the journal may reverse its decision to accept the manuscript on principle or as a precaution.

Another common reason for withdrawal is an authorship dispute that arises after submission. Such disputes are relatively common, and they can be contentious. Journals may be contacted by researchers claiming that they should have been listed as authors on submitted manuscripts but were not. Or researchers may complain that a named author does not meet the criteria for authorship and should be removed from the list of authors. Journals may be poorly positioned to resolve such conflicts, so they may request that the authors' institutions do so.

2.7.1 Authors' Responsibilities

Authors should do due diligence in researching the policies of any journals to which they submit their manuscripts, including the journals' position on duplicate publication and their general review practices. For example, when submitting manuscripts to a journal whose publisher has other, less-prestigious journals in the same or related discipline, authors should find out whether their journal of choice automatically refers (ie, "down-streams") the manuscripts it rejects to those lower-ranking titles. Such due diligence prevents authors from being surprised by aspects of the editorial or production process that might disappoint them and lead them to withdraw their manuscripts.

Authors should be aware of the problems created by duplicate submissions and the negative effects that such disregard for established conventions of scientific publication can have on limited scientific resources and on the authors' reputations. Authors must, therefore, submit each manuscript to only one journal at a time and wait to submit that manuscript elsewhere until it is rejected by the first journal or formally withdrawn by the authors from the journal.

If authors submit their manuscript in error to 2 or more journals at the same time, the authors should immediately inform the journals' editors of the situation to prevent problems that could lead to sanctions against the authors, to duplicate publication, and to copyright infringement (see Section 2.3.4, "Copyright Infringement").

Withdrawing manuscripts can be warranted if authors discover errors or deficiencies of such magnitude that they cannot be addressed in revising the manuscripts during the editorial process. Such withdrawals, however, should not be contrived to facilitate submitting the manuscripts to other journals.

2.7.2 Editors' Responsibilities

To prevent misunderstandings that may lead authors to withdraw their manuscripts, editors should ensure that their journals' ethical, peer-review, and downstreaming policies are stated clearly in their instructions for authors on the journals' websites and their online manuscript-submission systems.

Editors should state in their instructions for authors that manuscripts can be submitted to only one journal at a time. The instructions also may specify that cover letters include a guarantee that the manuscripts are not under consideration by any other journals.

If editors become aware of duplicate submissions made before or during their review of a manuscript, they should explain to the authors the mistake they made and the consequences it has for the journals and authors involved. Editors should insist that the authors withdraw the manuscript from all but one journal.

If authors withdraw a manuscript because it has been accepted elsewhere, the journal's editor should notify the other journal of the infraction and perhaps discuss appropriate corrective measures with that journal's editor.

2.7.3 Peer Reviewers' Responsibilities

If, in the course of a review, a reviewer becomes aware that authors have simultaneously submitted a manuscript to another journal, the reviewer should immediately notify the editor of the journal for which the reviewer is providing a review. Should the reviewer discover the duplicate submission because the reviewer was asked to review the manuscript by 2 or more journals, the reviewer should notify the editors of all the journals involved. The editors, in turn, may elect to suspend their review processes until the situation has been clarified.

2.8 AUTHORSHIP DISPUTES

Because many individuals contribute to modern scientific research, disputes over authorship can arise for numerous reasons (see Table 2.8). For example, authors may disagree about who should be listed as the primary or senior author, or they may argue about the authorship order. In addition, contributors judged not to be authors may argue that they should be listed as authors instead of being listed in the acknowledgments section. On the other end of the spectrum, colleagues are sometimes listed as contributors without their knowledge.

Disputes frequently occur because authors and other contributors are misinformed about authorship guidelines or because they are following outdated authorship practices. In some situations, varying interpretations of authorship guidelines produce acrimony. In addition, authorship may be defined differently depending on scientific discipline.

Sometimes, flagrant authorship omissions or inclusions are addressed by bodies such as journal editorial boards and panels assembled by the authors' employers. Some editors may be willing to serve as arbiters. However, if the situations are not readily resolved, editors may elect to delay review and publication of the manuscripts until the authorship disputes are resolved. The Committee on Publication Ethics' guidelines[4] for dealing with authorship issues can be useful in resolving such disputes.

2.8.1 Authors' Responsibilities

Authors should be aware of generally accepted authorship criteria and publication ethics, such as those developed by the International Committee of Medical Journal Editors (ICMJE).[62] They also should understand the authorship policies and procedures of the

Table 2.8. Responsibilities for Authorship and Disputed Authorship

Authors	• Be aware of the authorship criteria and publication ethics that are generally accepted in the field and that are specific to the journals to which manuscripts are submitted. • Identify each manuscript's authors and the order of authorship ideally before the research begins but no later than before the manuscript is written. • Be clear about which tasks each author is responsible for. • Keep updated written records of authorship, author order, and author tasks throughout the research. • Ensure that all those who are listed as authors meet the criteria for authorship and that all who meet the criteria are listed as authors.	• Resolve all author disagreements ideally before submitting the manuscript and certainly before publication. • Be prepared to sign the journal's form indicating that the authors meet the criteria for authorship. • Be prepared to justify any changes in authorship after the manuscript has been submitted. • If an author chooses to withdraw from authorship before publication, notify the journal editor of the author's intent and reason. • Give all those who contributed to the research but do not meet the authorship criteria the opportunity to be acknowledged as contributors. Do not do so without obtaining the contributors' consent.
Journal editors	• Publish instructions for authors and insist that authors follow them. • Confirm that all those listed as authors meet the criteria for authorship and that all those who meet the criteria for authorship are listed as authors. • Consider requesting that all authors itemize their contributions to the research and to the manuscript. • Require that any change in authorship after submission be approved by all authors before publication.	• Notify the authors and their employers that the journal is not required or equipped to resolve authorship issues, especially right before publication.
Peer reviewers	• Be familiar with each journal's authorship policies.	• Inform journal editors of authorship irregularities discovered during peer review.

journals to which they intend to submit their manuscripts. Authors should comply fully with those requirements and seek clarification from those journals if any of their requirements seem ambiguous.

Ideally, authorship issues should be determined when the original research plan is developed, even if the research is to be conducted by a large multiauthor group. All authors should be aware that they will be designated as authors, and they should agree to their specific tasks in the project and agree to approve all manuscript submissions. If the research plan and authorship tasks shift as the project evolves, all such changes should be recorded and agreed to by all authors. The authors of the work and their listed order must be confirmed before the manuscript is submitted for publication. Before submission, individual authors can withhold their names for such reasons as disagreeing with the manuscript's conclusions. After the manuscript is submitted, any changes in the number, names, or order of authors should be justified to the editor before publication proceeds.

In addition to confirming that all those who qualify for authorship are listed as authors, authors should contact all the people who contributed to the manuscript in some way but who do not qualify for authorship to request permission to include their names in the acknowledgments section. Like authors, contributors can withhold their names.

2.8.2 Editors' Responsibilities

Editors should develop instructions for authors for their journals. These instructions may be based on the generally accepted recommendations of the ICMJE or similar bodies. The ICMJE[62] has 4 requirements for authorship. Variations of those 4 criteria are embodied in the authorship guidelines of the American Physical Society[63]; the American Statistical Association[64]; the National Academies of Sciences, Engineering, and Medicine[65] in the United States; and the Society for Neuroscience.[66] Other disciplines may accept one or more of the ICMJE's criteria as defining authorship. In journals following the ICMJE definition, all 4 of the following authorship criteria must be met:

- Substantial contributions to the conception or design of the work or to the acquisition, analysis, or interpretation of data for the work
- Drafting the work or revising it critically for important intellectual content
- Final approval of the version of the manuscript to be published
- Agreement to be accountable for all aspects of the work by ensuring that questions related to the accuracy or integrity of any part of the work are appropriately investigated and resolved

The ICMJE has 3 optional criteria that may be helpful in resolving authorship disputes:

- Understanding the full scope of the work
- Being able to say which authors are responsible for each part of the research
- Having confidence in the integrity of the contributions of all authors

Authorship requirements prohibit gratuitous authorship (eg, gift and guest authorships) for those who provide funds or supervise research groups, such as laboratory directors. Authorship requirements also should be designed to reduce the relatively common practice of including as authors employees who worked solely on data collection. Professional scientific writers and editors, who draft manuscripts under the supervision of the researchers who conducted the research, also do not qualify for authorship because they are not responsible for the research that they report. Instead, such contributors should be included in the manuscripts' acknowledgments sections.

Many journal editors require that all listed authors sign a statement indicating that they have met the journals' authorship criteria. Many journals also require that authors describe their contributions to the research and the manuscript (see Table 2.9, and see the CRediT guidelines[67]). This information is often published in journal articles. In addition, journals often demand that changes in authorship or author order be approved by all authors before publication.

2.8.3 Peer Reviewers' Responsibilities

Peer reviewers should be familiar with journals' authorship policies. Should they become aware of possible improprieties in authorship, they should inform the journal editors of those concerns.

Table 2.9. Example of a Publication Ethics Checklist

Each author should initial each statement below to confirm that it is correct.										
1. Research involvement	Each author was substantively involved in the planning or conduct of the research.									
	Initials									
2. Assistance in investigation	Each author assisted in the investigation.									
	Initials									
3. Writing involvement	Each author was substantively involved in writing or critically revising the manuscript.									
	Initials									
4. Approval of manuscript	Each author has approved the final version of the manuscript to be published.									
	Initials									
5. Named authors meet criteria	Only those meeting criteria in items 1 through 4 are listed as named authors.									
	Initials									
6. Understanding of research scope	Each author understands the full scope of the research.									
	Initials									
7. Knowledge of all contributions	Each author can identify the contribution of each of the other authors.									
	Initials									
8. Coauthor integrity	Each author has confidence in the integrity of the contribution of each of the other authors.									
	Initials									
9. Accountability for work	Each author agrees to be accountable for all aspects of the work and to investigate and resolve any questions related to the work's accuracy or integrity.									
	Initials									
10. Competing interests	Each author has faithfully reported any competing interests.									
	Initials									
11. Exclusive submission	The manuscript is not currently being considered for publication elsewhere.									
	Initials									
12. Nothing plagiarized	All ideas and data in the manuscript that were created by others are appropriately acknowledged.									
	Initials									
13. No selective reporting	The manuscript does not withhold undesirable outcomes.									
	Initials									
14. No divided publication	The manuscript reports all substantive original ideas and data associated with the research.									
	Initials									
15. Privacy protection	No identifying information of study participants or other individuals is included in the submission without the consent of those individuals.									
	Initials									

Table 2.10. Responsibilities of Sponsoring Societies and Other Journal Owners and Publishers

- Support journal editors' "editorial independence."
- Understand that the decision to publish specific articles will be based on their scientific merit, not on the journal's financial success or reputation.
- Accept that the contribution of sponsoring societies and other journal owners is limited to financial and management issues, including business models and business policies.
- Govern relationship with journal editors by contracts specifying the responsibilities of each party and the circumstances under which editors may be terminated.
- Recognize that journals are responsible for the conduct of their editors, whereas institutions are responsible for the conduct of their authors.
- Understand that journals are responsible for safeguarding the publication process and ensuring the quality and integrity of scientific literature.
- Ensure that collaborations between owners and journals are open and transparent.

2.9 RESPONSIBILITIES OF SPONSORING SOCIETIES, JOURNAL OWNERS, AND RESEARCH INSTITUTIONS

Sponsoring societies and other journal owners and publishers are responsible for granting journal editors "editorial independence" and honoring that independence (see Table 2.10). Editors must have the freedom to decide which manuscripts to publish, as well as which other materials to include, such as editorials, news items, and even appropriate advertising.[1] These decisions should be based on the scientific importance of the journals' content and not on the journals' financial success or reputation.[59]

The responsibilities of sponsoring societies, journal owners, and other publishers are usually limited to financial and management issues, including business models and business policies.

Editors and publishers should agree to the responsibilities of each party in a clearly stated contract or terms-of-reference document. The document may outline the circumstances under which editors may be terminated by their publishers (eg, the editors are not carrying out their responsibilities).

All parties should work together with trust and respect.[1] For example, institutions and journals should collaborate in an effective and transparent manner to protect the quality and integrity of scientific research. Institutions should be responsible for the conduct of their researchers and should create and encourage a healthy research environment. Journals, on the other hand, should be responsible for the conduct of their editors, for safeguarding the publication process, and for ensuring the reliability of what they publish.

Excellent information on the relationships among these parties can be found in the Council of Science Editors' recommendations paper on integrity[1] and on the websites of the International Committee of Medical Journal Editors[2] and the Committee on Publication Ethics.[4,12]

CITED REFERENCES

1. Council of Science Editors, Editorial Policy Committee. CSE's recommendations for promoting integrity in scientific journal publications. Council of Science Editors; [accessed 2023 Jan 20]. https://www.council scienceeditors.org/recommendations-for-promoting-integrity-in-scientific-journal-publications

2. International Committee of Medical Journal Editors. https://www.icmje.org

3. World Association of Medical Editors. https://www.wame.org

4. Committee on Publication Ethics. https://publicationethics.org

5. EQUATOR Network. https://www.equator-network.org/

6. JAMA Network Editors. AMA manual of style: a guide for authors and editors. 11th ed. Oxford University Press; 2020. Also available at https://www.amamanualofstyle.com

7. Office of Research Integrity, Public Health Service (US). https://ori.hhs.gov

8. National Science Foundation (US), Office of Inspector General. https://oig.nsf.gov/

9. UK Research Integrity Office. https://ukrio.org/

10. Patwardhan B, Desai A, Chourasia A, Nag S, Bhatnagar R. Guidance document: good academic research practices. University Grants Commission (India); 2020. https://www.ugc.ac.in/e-book/UGC_GARP _2020_Good%20Academic%20Research%20Practices.pdf

11. Council of Scientific & Industrial Research (India). CSIR guidelines for ethics in research and in governance. Council of Scientific & Industrial Research (India); [accessed 2023 Jan 22]. https://www.csir .res.in/notification/csir-guidelines-ethics-research-and-governance

12. Wager E, Kleinert S on behalf of COPE Council. Cooperation between research institutions and journals on research integrity cases: guidance from the Committee on Publication Ethics (COPE). Committee on Publication Ethics; 2012 Mar. https://publicationethics.org/files/Research_institutions_guidelines_final.pdf

13. Atlas R et al. Statement on scientific publication and security [editorial]. Science. 2003;299(5610): 1149. https://doi.org/10.1126/science.299.5610.1149

14. Lishner DA. Sorting the file drawer: a typology for describing unpublished studies. Perspect Psychol Sci. 2021 Mar 1. https://doi.org/10.1177/1745691620979831

15. Paez A. Gray literature: an important resource in systematic reviews. J Evidence-Based Med. 2017;10(3):233–240. https://doi.org/10.1111/jebm.12266

16. Beall J. Predatory publishers are corrupting open access. Nature. 2012;489(7415):179. https://doi .org/10.1038/489179a

17. Kurt S. Why do authors publish in predatory journals? Learned Publishing. 2018;31:141–147. https:// doi.org/10.1002/leap.1150

18. Elmore SA, Weston EH. Predatory journals: what they are and how to avoid them. Toxicol Pathol. 2020;48(4):607–610. https://doi.org/10.1177/0192623320920209

19. Christopher J. The raw truth about paper mills. FEBS Lett. 2021;595(13):1751–1757. https://doi.org /10.1002/1873-3468.14143

20. Seifert R. How Naunyn-Schmiedeberg's Archives of Pharmacology deals with fraudulent papers from paper mills. Naunyn-Schmiedeberg's Arch Pharmacol. 2021;394(3):431–436. https://doi.org/10.1007 /s00210-021-02056-8

21. Systematic manipulation of the publishing process via "paper mills." Committee on Publication Ethics; 2020 Sep. https://publicationethics.org/resources/forum-discussions/publishing-manipulation-paper-mills

22. ClinicalTrials.gov. National Library of Medicine (US). https://clinicaltrials.gov

23. Clinical trials. International Committee of Medical Journal Editors. https://www.icmje.org /recommendations/browse/publishing-and-editorial-issues/clinical-trial-registration.html#two

24. National Research Council (US), Division on Earth and Life Studies, Board on Life Sciences, Committee on Responsibilities of Authorship in the Biological Sciences. Sharing publication-related data and materials: responsibilities of authorship in the life sciences. The National Academies Press; 2003. Also available at https://www.nap.edu/catalog/10613/sharing-publication-related-data-and-materials-responsibilities-of -authorship-in#

25. European Union Drug Regulating Authorities Clinical Trials database. European Medicines Agency. [updated 2022 Mar 3]. https://eudract.ema.europa.eu

26. RCSB PDB. Research Collaboratory for Structural Bioinformatics. https://www.rcsb.org/

27. International Nucleotide Database Collaboration. https://www.insdc.org

28. PubMed Central. National Library of Medicine (US). https://www.ncbi.nlm.nih.gov/pmc/

29. Europe PMC. https://europepmc.org/

30. Journal of Negative Results in Biomedicine. https://jnrbm.biomedcentral.com/

31. The missing pieces: a collection of negative, null and inconclusive results. PLOS. 2020 Sep 23. https://collections.plos.org/collection/missing-pieces/

32. The All Results Journals. http://arjournals.com

33. Journal of Articles in Support of the Null Hypothesis. https://www.jasnh.com

34. Journal of Pharmaceutical Negative Results. https://www.pnrjournal.com/

35. World Association of Medical Editors. Recommendations on publication ethics policies for medical journals; [accessed 2023 Aug 8]. https://wame.org/recommendations-on-publication-ethics-policies-for-medical-journals#Responding%20to%20Allegations

36. Committee on Publication Ethics. Allegations of misconduct; [accessed 2023 Aug 9]. https://publicationethics.org/misconduct

37. Fong EA, Wilhite AW, Hickman C, Lee Y. The legal consequences of research misconduct: false investigators and grant proposals. J Law Med & Ethics. 2020;48(2):331–339. https://doi.org/10.1177/1073110520935347

38. Rahman MS et al. The health consequences of falsified medicines—a study of the published literature. Trop Med Int Health. 2018;23(12):1294–1303. https://doi.org/10.1111/tmi.13161

39. Rosner M, Yamada KM. What's in a picture? The temptation of image manipulation. J Cell Biol. 2004;166(1):11–15. https://doi.org/10.1083/jcb.200406019

40. Fraser H, Parker T, Nakagawa S, Barnett A, Fidler F. Questionable research practices in ecology and evolution. PLOS ONE. 2018;13(7):e0200303. https://doi.org/10.1371/journal.pone.0200303

41. Trimmed mean definition. WallStreetMojo. https://www.wallstreetmojo.com/trimmed-mean/

42. Library of Congress, US Copyright Office. Copyright law of the United States (Title 17) and related laws contained in Title 17 of the United States Code. https://www.copyright.gov/title17

43. What does fixed in a tangible medium mean in United States copyright? USPatentLaw.cn. 2017 June 14. http://uspatentlaw.cn/en/what-does-fixed-in-a-tangible-medium-mean-in-united-states-copyright/

44. About the licenses. Creative Commons. https://creativecommons.org/licenses/

45. Permission guidelines. STM. https://www.stm-assoc.org/intellectual-property/permissions/permissions-guidelines/

46. Similarity Check. Crossref. https://www.crossref.org/services/similarity-check/

47. iThenticate. Turnitin, LLC. www.ithenticate.com

48. Google Scholar. https://scholar.google.com/

49. EBSCO. https://www.ebsco.com/

50. JSTOR. https://www.jstor.org/

51. COPE Council. Retraction guidelines—English. Version 2. Committee on Publication Ethics; 2019 Nov. https://publicationethics.org/files/retraction-guidelines-cope.pdf

52. Bevan, DR. Duplicate and divided publication. Can J Anaesth. 1991;38(3):267–269. https://doi.org/10.1007/BF03007612

53. Friedman LS, Richter ED. Relationship between conflicts of interest and research results. J Gen Intern Med. 2004;19(1):51–56. https://doi.org/10.1111/j.1525-1497.2004.30617.x

54. Romain PL. Conflicts of interest in research: looking out for number one means keeping the primary interest front and center. Curr Rev Musculoskelet Med. 2015;8(2):122–127. https://doi.org/10.1007/s12178-015-9270-2

55. Angell M. Acceptance address at Council of Science Editors meeting. Sci Ed. 2003;26(5):148–149.

56. Gottlieb JD, Bressler NM. How should journals handle the conflict of interest of their editors? Who watches the "watchers"? JAMA. 2017;317(17):1757–1758. https://doi.org/10.1001/jama.2017.2207

57. Nature Precedings. Springer Nature. https://www.nature.com/npre/

58. arXiv. Cornell University. https://arxiv.org

59. International Committee of Medical Journal Editors. Recommendations for the conduct, reporting, editing, and publication of scholarly work in medical journals. International Committee of Medical Journal Editors; [updated May 2022]. https://www.icmje.org/recommendations/

60. Min SK. Ethics and responsibilities of peer reviewers to the authors, readers, and editors. Vasc Specialist Int. 2021;37(1):1–3. https://doi.org/10.5758/vsi.213711

61. Transparent peer review for all. Nat Commun. 2022;13(1):6173. https://doi.org/10.1038/s41467-022-33056-8

62. International Committee of Medical Journal Editors. Defining the role of authors and contributors. International Committee of Medical Journal Editors. https://www.icmje.org/recommendations/browse/roles-and-responsibilities/defining-the-role-of-authors-and-contributors.html

63. American Physical Society. APS guidelines for professional conduct. American Physical Society. https://www.aps.org/policy/statements/02_2.cfm

64. American Statistical Association, Committee on Professional Ethics. Ethical guidelines for statistical practice. American Statistical Association. https://www.amstat.org/about/ethicalguidelines.cfm

65. National Academies of Sciences, Engineering, and Medicine (US), Committee on Science, Engineering, and Public Policy. On being a scientist: responsible conduct in research. 2nd ed. The National Academies Press; 1995. https://nap.nationalacademies.org/read/4917/chapter/1

66. Society for Neuroscience. Guidelines for responsible conduct regarding scientific communication. Society for Neuroscience. https://www.sfn.org/about/professional-conduct/guidelines-for-responsible-conduct-regarding-scientific-communication

67. CRediT (Contributor Roles Taxonomy). National Information Standards Organization (US). https://credit.niso.org/

ADDITIONAL REFERENCES

ACS Publications. Ethical guidelines to publication of chemical research. American Chemical Society; [updated 2021]. https://pubs.acs.org/pb-assets/documents/policy/EthicalGuidelines-1635271193737.pdf

American Mathematical Society. Policy statement on ethical guidelines. American Mathematical Society; [revised 2022]. https://www.ams.org/about-us/governance/policy-statements/sec-ethics

American Society for Biochemistry and Molecular Biology. Code of ethics. https://www.asbmb.org/about/code-of-ethics

Bulger RE, Heitman E, Reiser SJ, editors. The ethical dimensions of the biological and health sciences. 2nd ed. Cambridge University Press; 2002. Part 4: The ethics of authorship and publication.

DuMez E. The role and activities of scientific societies in promoting research integrity: report of a conference, April 10, 2000, Washington, DC. American Association for the Advancement of Science and the US Office of Research Integrity; 2000 Sep. Also available at https://ori.hhs.gov/documents/role_scientific_societies.pdf

Gabrielsson S, Eriksson S, Godskesen T. Predatory nursing journals: a case study of author prevalence and characteristics. Nurs Ethics. 2021;28(5):823–833. https://doi.org/10.1177/0969733020968215

Jones AH, McLellan F, editors. Ethical issues in biomedical publication. Johns Hopkins University Press; 2000.

Lafollette MC. Stealing into print: fraud, plagiarism, and misconduct in scientific publishing. University of California Press; 1992.

National Academies of Sciences, Engineering, and Medicine (US), Committee on Science, Engineering, and Public Policy, Panel on Scientific Responsibility and the Conduct of Research. Responsible science: ensuring the integrity of the research process. Vol 1. The National Academies Press; 1992. Also available at https://www.nap.edu/catalog/1864/responsible-science-ensuring-the-integrity-of-the-research-process-volume

Online Ethics Center for Engineering and Science. University of Virginia. https://onlineethics.org

Publication manual of the American Psychological Association. 7th ed. American Psychological Association; 2020.

The responsible researcher: paths and pitfalls. Sigma Xi, The Scientific Research Honor Society; 1999. Also available at https://w.astro.berkeley.edu/~kalas/ethics/documents/Sig_Chi_Researcher.pdf

Teich AH, Frankel MS. Good science and responsible scientists: meeting the challenge of fraud and misconduct in science. American Association for the Advancement of Science; 1992. (AAAS publication; No 92-13S).

3 The Basics of Copyright

Editor: Kelly Newton

3.1 COPYRIGHT

Copyright laws define 2 types of intellectual property rights: economic rights, which constitute copyright, and moral rights, which are associated with but separate from copyright. Economic rights enable creators to control the reproduction, performance, display, and distribution of their intellectual property as well as the creation of derivative works. Although historically most important for artistic and literary creations with commercial value, these economic rights have become increasingly important for other intellectual property, including compilations of research results, databases, computer software, architectural works, and hull designs for vessels.

Copyrights are comparable to patents and trademarks, which protect the creativity invested in inventions, slogans, and other distinctive goods and services. To encourage future creativity and to protect investment in past efforts, copyright provides authors, artists, and other creators with legal recourse to prevent their creations from being exploited and to recover damages or profits from unauthorized use.

Copyrights do not always belong to the creators of works, however. In many countries, for example, the rights to works made for hire automatically belong to the creators' employers or clients. Works created by government employees may be in the public domain. In addition, creators can reassign their rights.

The moral rights outlined in copyright laws help to ensure that proper credit is given to creators when their materials are copied or republished by others, and they offer creators some control over the integrity of their works. In general, moral rights can be waived but not transferred. In addition, moral rights vary depending on the country. For example, Canadian copyright law conveys more extensive moral rights than does US copyright law.

In the United States, the US Constitution[1] empowered Congress to create copyright law. The current US law is embodied in the US Code, Title 17, which includes the Copyright Act of 1976,[2] amendments to that act, the Berne Convention Implementation Act of 1988,[3] the Uruguay Round Agreements Act of 1994,[4] and the Digital Millennium Copyright Act of 1998.[5] The US Copyright Office of the Library of Congress administers the Copyright Act.

This chapter summarizes US law, with differences noted for the laws of Canada,[6] the United Kingdom,[7,8] Australia,[9] and New Zealand.[10] Although laws in other countries differ in some respects, some form of protection exists in the approximately 180 countries that have signed the Berne Convention for the Protection of Literary and Artistic Works of 1971,[11] the Universal Copyright Convention of 1971,[12] or revisions of either. All the nations in the European Union, for example, are parties to the Berne Convention, and most have signed the Universal Copyright Convention. In addition, the European Union has provisions to govern copyright among its member nations, which were updated and extended in the 2019 Directive on Copyright in the Digital Single Market.[13]

In general, the copyright law in the country in which a work is written, created, or published applies to that work, regardless of the language in which the work is written. For detailed instructions on publishing notices of copyright in works and registering works, consult the agency in charge of intellectual property in the country of interest. Seek legal counsel from someone familiar with the copyright law in the country of interest when uncertain about planned uses of copyrighted materials and when copyright may have been violated. This chapter in this style manual does not constitute legal advice on copyright issues.

3.1.1 Eligibility

Copyright laws assign to creators of original works exclusive rights to determine how and when such works are copied, performed, reproduced, or republished. To meet the minimum requirements for copyright, ideas, news, facts, data, and other material must result from independent intellectual effort. Table 3.1 lists the types of work that are and are not eligible for copyright protection.

Although ideas may be the result of independent intellectual effort, copyright is granted for the expression of those ideas, not for the ideas themselves. Likewise, copyright is granted for the description of the news, not for the news itself, and for the compilation of facts and data, not for the facts and data themselves. Impromptu speeches, performances, and other works that are not fixed in a tangible form are not copyrightable, although they would be as soon as they are written down or recorded. Other works may be protected by patents (eg, original inventions) or by trademarks (eg, words, phrases, symbols, and designs), and some may be protected by either patent or copyright (eg, computer software).

Table 3.1. Eligibility for Copyright in the United States[a]

Eligible: original works of authorship, including the following:
Literary works, including books, periodicals, manuscripts, poetry, computer software, and databases
Musical works, including any accompanying words
Dramatic works, including any accompanying music
Choreographic works and pantomimes
Visual arts, including photographic, pictorial, graphic, and sculptural works
Audiovisual works, including motion pictures, television programming, and radio broadcasts
Sound recordings

Not eligible: works that usually do not meet the minimum requirements of authorship,[b] including the following:
Ideas, procedures, processes, systems, methods of operation, concepts, principles, and discoveries
Data, facts, and news
Standard lists, calendars, and tables from public documents
Blank forms intended to be filled out
Computing and measuring devices
Names, titles, and short phrases, no matter how distinctive (seek trademark protection instead)
Inventions (seek patent protection instead)
Certain familiar symbols and designs (seek trademark protection instead)
Choreographic works, impromptu speeches, and other works that are not fixed in a tangible form (such works become eligible once they have been fixed in tangible forms such as notation and recording)

[a] Copyright basics.[16]
[b] Compilations and descriptions of some of these works may be eligible if they meet the test for originality.

3.1.2 Authorship and Duration of Rights

Authorship determines the duration of the copyright (see Tables 3.2 and 3.3), so the identity of the author or authors is crucial. If only one self-employed author is involved, there is no question who the author is. If there are 2 or more creators, authorship is shared unless there is agreement to the contrary. If the work is a film, a musical drama, or another work involving different kinds of authorship, determining authorship becomes more difficult.

In the United States, the creators of an original work are the authors no matter the kind of work, unless the work was made for hire. In Canada, the United Kingdom, Australia, and New Zealand, authorship depends on the type of material (eg, literary work, photograph, film, or sound recording).

In the United States, if the creators are employees and the work is the result of their employment, then the work was made for hire, and the employer is considered the author and the copyright owner (see Section 3.1.5, "Works Made for Employers"). In Canada, the United Kingdom, Australia, and New Zealand, those who create works for hire are the authors, but the employers mayown the copyright.

As signatories to the Berne Convention of 1971[11] and its revisions, the United States, Canada, the United Kingdom, Australia, and New Zealand are among approximately 180 countries that recognize that copyright takes effect as soon as a work is in fixed, tangible form, which can include a saved electronic form.

Unlike other personal property, copyright ends at a legally defined time. In the United

Table 3.2. History of Copyright Duration in the United States[a]

Date of creation	Description of copyright notice and registration	Duration of protection from date of original copyright[b]
Before 1 January 1950[c]	Work published without a copyright notice and not registered	None
	Published work with a copyright notice or unpublished, registered work that was not renewed in the 28th year	28 years
	Published work with a copyright notice or unpublished, registered work that was renewed before 31 December 1963	56 years
	Published work with a copyright notice or unpublished, registered work with a renewal term still in effect on 1 January 1978	95 years[d]
Between 1 January 1950 and 31 December 1963	Work published without a copyright notice and not registered	None
	Published work with a copyright notice or unpublished, registered work that was not renewed during the 28th year	28 years
	Published work with a copyright notice or unpublished, registered work that was renewed during the 28th year	95 years[d]
Between 1 January 1964 and 31 December 1977	Work published without a copyright notice or unpublished work that was not registered	None
	Published work with a copyright notice or unpublished, registered work, regardless of whether a renewal was registered during the 28th year	95 years[d]
Before 1 January 1978[e]	Unpublished, unregistered work created by individuals but not for hire	Life of the last-surviving author plus 70 years
	Unpublished, unregistered work created for hire, as well as any anonymous or pseudonymous works[f]	120 years from creation or 95 years from first publication, whichever is shorter
On or after 1 January 1978	Regardless of whether the work is published, includes a copyright notice, or is registered:	
	work created by individuals but not for hire	Life of the last-surviving author plus 70 years
	work created for hire, as well as anonymous or pseudonymous works[f]	120 years from creation or 95 years from first publication, whichever is shorter

[a] Based on *Duration of Copyright.*[22]

[b] Prior to 1978, terms of duration were measured from the date of publication or, in the case of unpublished works, the date of registration. Since 1978, the terms are calculated from the end of the year in which the last-surviving creator died or, in the case of work for hire, from the date the work was created or published.

[c] For works initially copyrighted between 19 September 1906 and 31 December 1918, the US Congress passed 9 acts that ultimately extended the second term from 28 to 41 years.

[d] The Copyright Act of 1976 increased the renewal term to 47 years, and the Copyright Term Extension Act of 1998 extended the renewal term another 20 years, establishing 95 years of protection from the date of the original copyright.

[e] These provisions for unpublished, unregistered works were retroactively created in the Copyright Act of 1976, which went into effect 1 January 1978. Previously, an unpublished work was protected only if its copyright was registered. In addition, the 1976 act stipulated that regardless of how long ago an unpublished, unregistered work was created, it would be protected by copyright until at least 31 December 2002. Furthermore, if such a work were published before the end of 2002, its copyright term would be extended until the end of 2047.

[f] For anonymous or pseudonymous works, should the authors' identities be revealed in Copyright Office records, the term of duration would change to 70 years after the last-surviving author's death.

States, Canada, the United Kingdom, Australia, and New Zealand, all terms of duration are measured from either the end of the year in which the last-surviving creator died or the end of the year in which the work was created or made public. With a few exceptions, new published works are protected for the life of the last-surviving author plus 70 years in the United States, Canada, the United Kingdom, and Australia and for the life of the last-surviving author plus 50 years in New Zealand. Works made for hire in the United States are protected for 95 years from the date of publication or for 120 years from the date of creation, whichever expires first.

However, works copyrighted before the US Copyright Act of 1976[2] went into effect on 1 January 1978 are covered by different rules. Before 1964, copyright lasted for 28 years from the date it was secured, and copyright could be renewed for another 28 years, for a total of 56 years. Through a series of extensions, the United States eventually increased copyright duration for works created before 1978 to 95 years, provided the works' copyright did not expire between extensions. To be protected by copyright in the United States before 1978, published works needed to include a copyright notice. In addition, only certain types of unpublished works could be protected under copyright, and only if their copyright was registered. In addition, before 1978, copyright in the United States was secured on the date a work was published or, in the case of an unpublished work, the date its copyright was registered.

3.1.3 Ownership

Under the terms of the Berne Convention of 1971[11] and its revisions, as soon as an original work of authorship is created in a fixed, tangible form, its creator automatically owns the copyright to the work, except for works made for hire in countries in which copyright falls to employers (see Section 3.1.5, "Works Made for Employers"). Coauthors of a work are co-owners of its copyright unless there is an agreement to the contrary.

With a collective work, such as an anthology or an issue of a periodical, each contribution is considered a distinct and independent work. In addition to the copyright for each contribution, the collective work has a separate copyright for the compilation as a whole. In the United States, each contribution and the compilation can be registered separately to the individual authors and the compilers, respectively (see Section 3.2.2, "Registration"). Similarly, the United Kingdom, Australia, and New Zealand grant separate copyrights for each contribution and for the layout of published editions. In these 3 countries, the copyright for the layout is owned by the publisher and lasts 25 years from date of publication.

3.1.3.1 Rights and Responsibilities of Copyright Owners

Copyright laws convey exclusive economic and moral rights to copyright owners. As economic rights, copyright allows owners to reproduce their works, to distribute copies, to prepare derivative products from their works, and to perform and display their works. As property, the economic rights can be sold, transferred, assigned, or licensed in whole or in part. For example, a copyright owner may sell or assign print distribution rights but retain all electronic distribution rights. The rights may be conveyed to the owner's heirs by will or probate, and they may be confiscated through legal action.

Table 3.3. Duration of Copyright Protection in Canada, Australia, the United Kingdom, and New Zealand

Type of work or right	Canada[a,b]	United Kingdom[b,c]	Australia[b,d]	New Zealand[b,e]
Literary, dramatic, musical, and artistic works unpublished at the time of the last-surviving author's death	Life of the last-surviving author plus 70 years or 50 years from first publication, whichever is later; transitional provisions apply to works unpublished prior to 31 December 1998	Until 31 December 2039 or the life of the last-surviving author plus 70 years, whichever is later	If published posthumously, 70 years after first publication	Life of the last-surviving author plus 50 years
Published literary, dramatic, musical, and artistic works, excluding photographs	Life of the last-surviving author plus 70 years Canada increased the duration from 50 to 70 years as of 30 December 2022. However, the change was not retroactive, so works of authors who died before 1972 are in the public domain. Prior to 1 January 1994, the exact date of death was used, and since then, Canada has measured from the end of the year of death.	Life of the last-surviving author plus 70 years	Life of the last-surviving author plus 70 years	Life of the last-surviving author plus 50 years
Anonymous and pseudonymous works	For an unpublished work, 75 years after creation For a published work, 75 years after publication or 100 years after creation, whichever is sooner	70 years from date of creation or publication, whichever is later	70 years after publication; if the work is made public within 50 years of its making, 70 years after first being made public	50 years after publication
Cinematographic films	If scripted and created, life of the last-surviving author plus 70 years If not scripted but created before 1 January 1994, 50 years after creation, regardless of whether the film was released If not scripted but created on or after 1 January 1994, 50 years after release	Life plus 70 years of the last-surviving principal (eg, the principal director, the last author of the screenplay and dialogue, or the composer of music specially created for the film)	70 years after the making of the film; if the film is made public within 50 years of its making, 70 years after first being made public If not released to the public, the life of the last-surviving author plus 70 years	50 years from creation or re-lease to the public, whichever is later

	Canada[a]	United Kingdom[c]	Australia[d]	New Zealand[e]
Photographs	If the photographer is an individual, the life of the photographer plus 70 years. If the creator is a corporation and a natural person does not hold the majority of shares in the corporation, the end of the calendar year in which the photograph was initially taken plus 50 years	Life of the photographer plus 70 years. For unpublished works, see the Copyright, Designs and Patents Act[7]	Life of the photographer plus 70 years	Life of the photographer plus 50 years
Sound recordings	70 years after the end of the calendar year in which the work was first recorded	50 years from creation or release	70 years after the making of the sound recording; if the sound recording is made public within 50 years of its making, 70 years after first being made public	50 years from creation or release to the public, whichever is later
Broadcasts and cable programs	[not specified]	50 years after creation	50 years from date of release to the public	50 years from date of release to the public
Published editions	[not specified]	25 years from publication	25 years from publication	25 years from publication
Computer-generated works	[not specified]	50 years after creation	Life of the last-surviving creator plus 70 years	50 years after creation
Crown copyright material	[not applicable]	[not applicable]	50 years after the making of the material	[not applicable]

---

[a] Based on Canada's Copyright Act,[6] R.S.C. 1985, c C-42, as amended, ss 6–12.

[b] Unless otherwise noted, terms of duration are measured from the end of the year in which the last-surviving creator died or the work was created.

[c] Based on the United Kingdom's Copyright, Designs and Patents Act 1988[7] (c 48), s 86, and Duration of Copyright and Rights in Performances Regulations 1995,[8] s 5, 6, 13B, and 14 [Statutory Instrument 1995 No. 3297].

[d] Based on Australia's Copyright Act 1968, Act No. 63 of 1968[9] as amended, s 34, s 95, and s 96.

[e] Based on New Zealand's Copyright Act 1994[10] as amended.

The moral rights conveyed by copyright laws are those of attribution and integrity. The right of attribution is the right to be acknowledged as the creator of a particular work. The right of integrity forbids anyone to distort, mutilate, or modify a work in a way that is prejudicial to the creator's honor or reputation without permission from the creator. Unlike copyrights, moral rights cannot be sold or transferred, but they may be waived in some countries (eg, the United Kingdom). In the United States, moral rights are restricted to certain types of visual art, while in Canada, moral rights apply to all works eligible for copyright. In the United Kingdom and Australia, these rights are conveyed to authors of literary, dramatic, musical, and artistic works, as well as to screenwriters, directors, and producers of films. In New Zealand, moral rights mimic those granted in the United Kingdom and Australia, except they do not extend to film producers.

Reproduction of a work by anyone other than the copyright owner requires written permission from the copyright owner, except as allowed under fair use and fair dealing (see Section 3.3.1, "Fair Use and Fair Dealing"). Owners of copyrighted material can control reproduction of that material in 4 ways:

(1) License the use for a royalty or fee, and impose appropriate conditions
(2) License the use and impose appropriate conditions without seeking a royalty or fee
(3) Deny the request
(4) Fail to reply, which must be interpreted as equivalent to denying the request

Copyright owners can use copyright collectives to facilitate permissions and royalties (see Section 3.3.3, "Permission to Republish Scientific Works"). However, copyright owners, not copyright collectives, are responsible for any monitoring they want conducted to identify unauthorized use of their materials by others.

3.1.3.2 Limitations on Rights of Owners

The rights of copyright owners are limited somewhat by exceptions to infringement. For example, the US Copyright Act[2] allows others to use copyrighted materials in the following ways without requiring permission from the copyright owners:

(1) Use parts of works if the principle of fair use is observed (see Section 3.3.1, "Fair Use and Fair Dealing")
(2) Sell or give away legally obtained copies, except for fair-use copies made for personal use
(3) Perform or display works in classrooms and certain other nonprofit settings
(4) Broadcast over licensed stations if no charge is made to the public to see or hear the transmission
(5) Make a limited number of ephemeral recordings of audiovisual works
(6) Reproduce and distribute copies of published, nondramatic literary works in specialized formats exclusively for use by people with visual or other disabilities
(7) Obtain compulsory or negotiated licenses by paying royalties to make and distribute recordings of nondramatic musical compositions or to perform or display works on television, radio, or digital audio transmissions

Copyright does not prevent the use of copyrighted materials in ways that would alter their original form and content such that the restructured materials no longer mirror the copyrighted materials in full or in part. For example, a table may be based on select data from a copyrighted source without violating copyright if the data are displayed in a substantially different manner and the copyrighted source is listed in the table's credit line (see Section 30.1.1.9, "Credit Lines"). Copyright, therefore, protects only against copying or other unauthorized uses, and that protection is of limited scope and duration.

In Canada, the rights of owners are limited in situations involving fair dealing (see Section 3.3.1, "Fair Use and Fair Dealing"), parody, public recitation of extracts, and reproduction of works permanently situated in public locations. Exceptions also apply for educational institutions, libraries, archives, and museums. In addition, Canadian copyright law allows users to make a single copy of a computer program either for backup or for adaptation to make the program compatible with a particular computer.

In the United Kingdom, Australia, and New Zealand, fair dealing is permitted for the purposes of research and study, criticism and review, and reporting of news. Free-use provisions also cover photographing, filming, or otherwise including publicly displayed sculptures and other works of artistic craftsmanship in broadcasts, as well as copying computer programs for specified purposes.

3.1.3.3 Transfer of Ownership

All or part of the copyright to a work can be transferred to someone else by written agreement, by will or intestate succession as personal property, or through legal action. The United States and Canada are among the countries in which transfers of exclusive copyrights may be recorded in national copyright offices.

Many publishers require authors to transfer at least some of the rights to their manuscripts upon acceptance for publication.

3.1.4 Public Domain

In the United States, the following are in the public domain and are not protected by copyright:

(1) Works that were published before 1978 without a copyright notice and that were not registered
(2) Works for which the copyright has expired (eg, for works published in 1927, copyright would have expired in 2022, provided it was renewed in 1955)
(3) Works that are considered to be common property (eg, facts, principles, standard calendars, and measurement tools)
(4) Works created by US federal employees in the course of their duties, although the US government may seek protection in other countries (see Section 3.1.5.3, "Government Employees and Contractors")

A useful guide to the kinds of works that are under copyright in the United States and the kinds that are in the public domain is maintained by the Cornell University Copyright Information Center.[14]

Similar to the United States, the countries of Canada, the United Kingdom, and Australia generally limit public domain to works for which copyright has expired and works that never met the test for copyright protection. In contrast, New Zealand does not have a formal concept of public domain. Works created and published by the Canadian, British,

Australian, and New Zealand governments are subject to Crown copyright and are not automatically in the public domain. Also not in the public domain are works created by independent contractors for the Canadian and Australian governments. Instead, they are owned by the creators unless there is agreement to the contrary.

Works in the public domain should be used carefully, cited appropriately, and not plagiarized. Public domain is the end point in the continuum of ownership of the copyright in a work. For example, an author who writes a manuscript today in the United States or the United Kingdom owns the copyright to that work from the time the words flow onto paper or into a word-processing file. When the manuscript is accepted for publication, the author may be required to transfer the copyright to the publisher. Regardless of whether the copyright is owned by the publisher or the author, the published manuscript remains covered by copyright for the life of the author plus 70 years. After the copyright expires, the manuscript becomes part of the public domain and can be reprinted freely without infringing on the copyright. Thus, modern publishers have long been able to reprint without permission the works of naturalist Charles Darwin, FRS, and microbiologist Louis Pasteur, PhD, both of whom died toward the end of the 19th century, when copyright duration was shorter in the United Kingdom and France than it is now. However, publishers will have to wait until the beginning of 2067 to reprint without permission the works of astronomer Carl Sagan, PhD, who died in 1996, and until early 2082 to reprint the works of Dr Sagan's first wife, evolutionary biologist Lynn Margulis, PhD, who died in 2011.

3.1.5 Works Made for Employers

3.1.5.1 Employees

In the United States, copyrightable materials created by an employee within the scope of the employment become the intellectual property of the employer. The employer—not the employee—becomes both the owner of record (see Section 3.1.3, "Ownership") and the author of record (see Section 3.1.2, "Authorship and Duration of Rights"). In the United States, these materials are called "works made for hire."

In contrast, in the United Kingdom, Australia, and New Zealand, employees remain the authors of works created during employment and retain the moral rights, while their employers own the copyrights. Similarly, in Canada, employers generally own the copyrights to works created by their employees, absent formal agreements to the contrary.

3.1.5.2 Independent Contractors

In the United States, material created by an independent contractor hired to create a product that is copyrightable becomes a work made for hire only if the parties expressly agree in writing that the product is a work made for hire. Section 101 of the US Copyright Act[2] lists 9 categories of works made for hire that can become an employer's intellectual property:

(1) Contribution to a collective work
(2) Part of a motion picture or other audiovisual work
(3) Translation

(4) Supplementary work (as in work done as an adjunct to the work done by the primary author)
(5) Compilation
(6) Instructional text
(7) Test
(8) Answer material for a test
(9) Atlas

If a work belongs to one of those categories, the contract should specify that the product is to be prepared as a "work made for hire" to ensure that no question arises as to the owner of the copyright. Even for contractual work that falls outside these categories, the copyright can be transferred to the entity that commissioned the work, provided the contract expressly states that copyright ownership will be transferred. A contract may also describe any limitations on the use of the copyrighted work.

In the United Kingdom and Canada, works prepared by independent contractors usually belong to the contractors in the absence of agreements to the contrary. However, in Canada, those who commission engravings, photographs, and portraits and who pay valuable consideration for those works generally own the copyrights to those works unless otherwise stated in commission agreements. In addition to allowing copyright to be assigned in agreements, Canada is among the countries that allows moral rights to be waived.

Similarly, most works prepared in Australia by independent contractors belong to the contractors in the absence of agreements to the contrary. However, the copyright that Australia grants commissioners of artwork extends beyond what Canada allows to include paintings, drawings, cinematographic films, and sound recordings. In addition, Australian newspaper and periodical publishers own the rights to reproduce commissioned works in books and facsimile editions of their newspapers and periodicals.

New Zealand goes even further than Australia by adding diagrams, maps, charts, plans, models, sculptures, and computer programs to the list of commissioned works whose copyrights are owned by those who commission the works.

3.1.5.3 Government Employees and Contractors

Because works produced exclusively by US federal employees within their official duties are not protected under US copyright law, they are in the public domain, unless access to those works is restricted on a statutory basis (see Section 3.1.4, "Public Domain"). However, works produced by independent contractors for the US government are not automatically in the public domain because contractors and grantees are not considered government employees for purposes of copyright. Thus, works produced under government contracts are protected under the US Copyright Act,[2] and copyright ownership depends on the terms of the contracts.

Likewise, works of joint authorship created by US employees and nonfederal colleagues are protected at least in part by US copyright. However, the law is not specific about how far copyright applies in this situation, so the law is open to interpretation. When the contributions of nongovernment authors can be isolated in an entire work, those contributions are likely copyrightable, and the contributions made by US government employees are likely to be in the public domain. When it is not possible to separate the contributions in a jointly written work, each coauthor shares equal ownership of the entire work, and it is likely copyrightable.

Works created by Canadian, British, Australian, and New Zealand government employees are subject to Crown copyright and are not in the public domain. Works created by independent contractors for the Canadian, British, and Australian governments belong to the contractors in the absence of agreements to the contrary. In contrast, works created by contractors for the New Zealand government are subject to Crown copyright.

3.1.6 Public Disclosure

Unlike the US government, state and local governmental bodies in the United States may copyright their materials, in which case, they cannot be reproduced without those bodies' consent. However, federal, state, and local laws ensure that public records are available to the public for inspection. In addition, members of the public can obtain a reasonable number of copies of these documents on request. However, that is quite different from an individual or company making unauthorized and unlimited copies of copyrighted materials owned by state and local bodies.

3.2 COPYRIGHT PROCEDURES

Under the terms of the Berne Convention of 1971[11] and its revisions, copyright protection begins automatically from the moment a work is created in a fixed form. Copyright begins without any notice, formality, process, or application.

3.2.1 Notice

The standard copyright notice (eg, "© 2023 Council of Science Editors") is not required to establish copyright in the United States, Canada, the United Kingdom, Australia, New Zealand, or any other signatory to the Berne Convention of 1971.[11] Nevertheless, a copyright notice should be included in a work if the copyright owner wants to ensure that users are aware that the material is copyright protected. As described in Section 3.2.2.1, "Benefits," a copyright notice provides additional benefits in the United States and affords protection in some countries that are not party to the Berne Convention.

Copyright notices for works produced within the United States must consist of all the following elements:

(1) On visually perceptible copies, the copyright symbol "©," the word "copyright," or the abbreviation "copr" (on sound recordings, the symbol "℗"—"P" in a circle—is used instead)
(2) The year of first publication of the work
(3) The name of the owner of the copyright

The letter "c" in parentheses—(c)—and other variations are not acceptable substitutes for the copyright symbol. Set the copyright notice in roman type with a space between the copyright symbol, word, or abbreviation and the year of first publication. The notice may include the phrase "all rights reserved," which is necessary to satisfy requirements in some countries.

A copyright notice should be "permanently legible to the ordinary user of the work and . . . not concealed from view upon reasonable examination."[15] On printed publications, customary positions include on the title page or the page opposite the title page; at the end of the text, especially for short documents; and as part of the masthead of serials. See the circular *Copyright Notice*[15] from the US Copyright Office for more details and examples of acceptable locations for copyright notices.

3.2.2 Registration

In the United States and Canada, registering the owner's copyright is beneficial and encouraged, although not required. There is no registration system in the United Kingdom, Australia, or New Zealand.

3.2.2.1 Benefits

Although not mandatory in the United States, registering copyright with the US Copyright Office creates a public record of a copyright claim and establishes prima facie evidence of the validity of the copyright and the facts stated on the copyright certificate. Infringement lawsuits cannot be filed in US courts until the copyright is registered for works of US origin. Depending on the timing of when they registered their works, copyright owners may receive statutory damages and lawyers' fees from court actions in addition to the actual damages and profits to which they may be entitled. Another benefit is that copyright registration allows for works to be registered with the US Customs Service to provide protection from the importation of infringing copies. See the circular *Copyright Basics*[16] from the US Copyright Office for more details.

In Canada, one of the benefits of registration is that it creates a presumption of validity—that is, a presumption that copyright subsists in the work and that the person or entity registering the work is the copyright owner.

3.2.2.2 Submission

There are 2 ways to register a work in the United States: online and by mail. Online registration is faster and less expensive. To register online, the following are needed:

(1) A completed application form, which can be filled out online
(2) Uploadable files of the materials being registered
(3) A nonrefundable filing fee

To register by mail, 3 elements are required:

(1) A completed copy of the appropriate registration form (eg, Form TX for most printed materials and Form SE for serials)
(2) The required number of copies of the materials being registered
(3) A nonrefundable filing fee

For instructions on online and hard-copy applications, see the US Copyright Office's circular *Copyright Registration*.[17]

To register a work in Canada, the application and filing fee must be submitted to the Canadian Intellectual Property Office (CIPO). Application forms and instructions for online and by-mail registration are available from the CIPO website.[18]

3.2.3 Deposit

If materials published in the United States include a copyright notice, the copyright owners have 3 months from the date of publication to send 2 complete copies of the material to the Register of Copyrights. For details, see the US Copyright Office's circular *Best Edition of Published Copyrighted Works for the Collections of the Library of Congress*.[19]

The Canadian Intellectual Property Office does not accept copies of works being registered. However, the Library and Archives of Canada Act[20] requires copies of every work published in Canada to be sent to the Library and Archives of Canada. The Australian Copyright Act[9] has a similar provision requiring printed materials to be deposited in specified libraries.

There is no formal deposit procedure in the United Kingdom or New Zealand. However, to establish that works existed as of specific dates, copyright owners may deposit copies with a bank or solicitor or send copies to themselves by special delivery and leave the envelopes sealed. These copies can be used later as evidence should copyrights be infringed.

3.3 USE OF COPYRIGHTED MATERIAL

3.3.1 Fair Use and Fair Dealing

Although copyright laws forbid unauthorized reproduction of protected works, US law does allow for fair use of copyrighted work for such purposes as criticism, comment, news reporting, teaching, scholarship, and research. Unfortunately, the distinction between fair use and infringement is not easily defined or described. Although there are guidelines for deciding between fair and unfair uses, no hard rules specify how many words or lines or what percentage of content is fair. Consider the following factors when determining whether a particular reuse is fair or unfair[2]:

(1) The purpose and character of the intended use, including whether such use is for personal research, nonprofit educational purposes, or commercial enterprise
(2) The nature of the copyrighted work
(3) The amount and substantiality of the portion used in relation to the copyrighted work as a whole
(4) The effect of the use on the potential market for the copyrighted work or on its value

In light of these factors, legal copies of materials copyrighted in the United States can ordinarily be made without the copyright owners' permission under the following situations:

(1) A single copy for personal use in education or noncommercial research
(2) A single copy of a small portion of a work (eg, a chapter, article, poem, or chart) for each member of a group assembled for educational purposes under certain conditions

Copies made without the owners' permission or without royalty payments would not be fair in the United States under the following situations:

(1) Copies of several articles from the same volume of a scholarly journal, even for personal or educational use
(2) Extensive passages from one work quoted in another, even with attribution
(3) Any substantial copying from a work in lieu of purchasing the work
(4) Posting portions of someone else's work on a website from which it can be printed at no charge, thereby reducing the market for the original work (some publishers and other copyright holders readily allow others to repost abstracts, but permission should be obtained from the copyright holders beforehand)

The concept of "fair dealing," which is used in Canada, the United Kingdom, Australia, and New Zealand, is somewhat narrower than "fair use." For example, although some copying for educational purposes is, the restrictions on educational use are more extensive than in the United States, and royalties apply.

Like fair use, fair dealing can be used as a defense against allegations of copyright violation. In determining whether a particular reproduction is fair, consideration is given to both the quality of the reproduction (ie, the value of the part of the work that is reproduced in relation to the value of the rest of the work) and the quantity that is copied. In addition, the reproduction must fit within one of the following permitted purposes: research, private study, criticism, review, or news reporting.

Although fair dealing and fair use are similar concepts, what might be considered fair use in the United States might not meet the criteria for fair dealing in the United Kingdom. By the same token, what might qualify for fair dealing in Australia might not be considered fair use in the United States. It is, therefore, never inappropriate to check with copyright holders when unsure of whether copying or reusing copyrighted material would be permitted.

3.3.2 Internet and Copyrights

Original materials posted on the internet are usually copyrighted automatically upon being posted, and they are subject to all protections afforded any other copyrighted material. Likewise, fair-use or fair-dealing provisions may apply, allowing limited copying. Posting copyrighted online materials owned by others could infringe on copyright unless the materials have been licensed from the copyright owners for that use. This is different from downloading materials for fair-use purposes, such as making a paper copy for subsequent reading and study.

The absence of a copyright statement or symbol should not be taken to mean that material on a website is not protected by copyright. Unless a statement is posted allowing free reuse of online content, permission should be obtained before copying or reusing material that falls outside a strict interpretation of fair use.

3.3.3 Permission to Republish Scientific Works

Republishing scientific manuscripts or parts of them (eg, figures and tables) requires permission from the copyright owners, provided that fair use or a Creative Commons license does not apply (see Sections 3.3.1, "Fair Use and Fair Dealing," and 3.3.4, "Creative Commons Licenses"). For the content in many scientific journals, copyright is held by the journals' sponsoring societies or publishers if the journals require authors to transfer their copyright or exclusive rights as a condition of publication. However, scientific journals, especially open-access journals, are increasingly offering options that allow authors to retain copyright. In addition, some content is in the public domain. For example, no copyright exists within the United States on manuscripts for which the authors were all federal employees, the work was developed within the scope of official US governmental duties, and the publishers did not add copyrightable contributions. Within the United States, such material can be used without permission, but the source should be cited. Outside the United States, the US government can claim copyright on such works. Additionally, works may be protected by copyright both within and outside the United States if the publishers added copyrightable contributions before publishing the manuscripts.[21]

When requesting permission from copyright owners or their designated agents, first look at the journals' or publishers' websites for their standard procedures for copying or republishing material. Many publishers' websites provide permission request forms that can be completed online. Some publishers deal with permissions in house, while others outsource this responsibility to collectives such as the Copyright Clearance Center. Regardless of the procedure, the following information will be needed to request permission:

(1) The intended use (eg, educational use or reuse for commercial enterprise)
(2) The duration of use
(3) Pertinent source information that clearly identifies the material to be reused (eg, year, volume, number, edition, pages or portions of a page, and table or figure number)
(4) Information about the publication in which the material is to be reused (eg, name of the publication, the publisher, and the expected number of print copies, as well as whether the content will be translated, whether it will be posted online, and whether the online version will be on a site that is password protected)

Copyright outsourcing centers can provide prompt authorization to copy or republish many copyrighted materials. These centers can bill for and receive royalty payments. In addition, they may assist in resolving questions regarding whether fair use or fair dealing would allow legal copying of materials without obtaining permission and paying royalties. Some prominent collectives are the Copyright Clearance Center in the United States (https://www.copyright.com), Access Copyright in Canada (https://www.access copyright.ca), the Copyright Licensing Agency in the United Kingdom (https://www.cla .co.uk), Copyright Agency Limited in Australia (https://www.copyright.com.au), and Copyright Licensing Limited in New Zealand (https://www.copyright.co.nz).

3.3.4 Creative Commons Licenses

Creative Commons licenses were established in 2002 to allow individual and institutional copyright holders to license their work to others in a manner simpler than traditional copyright licensing. Under Creative Commons, copyright owners decide how their material should be licensed throughout the duration of the copyright. Copyright holders then insert a statement, mark, or both into their works to indicate that they are covered by Creative Commons licenses. Those who wish to reuse material covered by Creative Commons licenses do not have to request permission from the copyright owners. Instead, the material can be reused automatically in the manner specified in each work's Creative Commons license.

Every Creative Commons license ensures that the creators of original works receive credit for their work. Based on copyright laws, these licenses are honored worldwide and last as long as the copyright.

Copyright holders can select from 6 different Creative Commons copyright licenses.*

Attribution (CC BY)

The most accommodating of the 6 licenses, the CC BY license allows others the right to distribute, adapt, and build on the original work as long as the original work and the copyright holder are acknowledged. Commercial reuse is permitted under this license.

Attribution-ShareAlike (CC BY-SA)

Like the basic CC BY license, this license allows others the right to distribute, adapt, and build on the original work as long as the original work and copyright holder are acknowledged. However, derivative works based on the original must carry CC BY-SA licenses also. Commercial reuse is permitted.

Attribution-NoDerivs (CC BY-ND)

This license allows others the right to redistribute the original work commercially or noncommercially as long as the work remains unchanged and whole and as long as the original work and copyright holder are acknowledged. No derivative works are allowed.

Attribution-NonCommercial (CC BY-NC)

This license allows others the right to distribute, adapt, and build on the original work for noncommercial purposes only as long as the original work and copyright holder are acknowledged.

Attribution-NonCommercial-ShareAlike (CC BY-NC-SA)

This license allows others the right to redistribute, adapt, and build on the original work for noncommercial purposes only as long as the original work and copyright holder are acknowledged. Derivative works based on the original must be licensed under the same CC BY-NC-SA terms.

Attribution-NonCommercial-NoDerivs (CC BY-NC-ND)

The most restrictive of the 6 licenses, the CC BY-NC-ND allows others the right to use and distribute the original work without alteration or modification for noncommercial

---

* BY = attribution; CC = Creative Commons; NC = noncommercial; ND = no derivatives; SA = share alike.

purposes as long as the work and its copyright holder are acknowledged. No derivative works are allowed.

Creative Commons also provides a tool by which creators of a work can waive all rights and place their work in the public domain. This tool is called CC0.

As a condition of funding or employment, many funding bodies and academic institutions have open-access policies that require work to be distributed under a specific Creative Commons license, most commonly CC BY.

More detailed information on Creative Commons licenses can be found on the Creative Commons website at https://creativecommons.org.

3.3.5 Infringement and Plagiarism

Infringement is any unauthorized use of copyrighted material. It is illegal, and depending on the severity of the offense, it is subject to civil remedies or to criminal prosecution and court action, including fines and imprisonment. Examples of infringement include the following:

(1) Photocopying too much of a book or journal
(2) Selling what was once a fair-use copy of a manuscript
(3) Making copies for sale or rent
(4) Distributing copies for the purpose of trade
(5) Exhibiting a work in public without permission
(6) Importing illegal copies of a book or sound recording for sale or rent
(7) Authorizing or facilitating infringement by another party

In contrast, plagiarism is the intentional act of copying someone else's work and claiming it to be one's own intellectual effort. If a plagiarized work is under current copyright protection, the act is also copyright infringement (see Section 2.3, "Ethical Breaches with Legal Implications"). Whether an original work is by polymath Nicolaus Copernicus, DCanL, and out of copyright; by paleontologist Stephen Jay Gould, PhD, and still under copyright; or by the National Aeronautics and Space Administration's computer scientist Annie Easley, BS, and in the public domain, falsely claiming authorship is plagiarism and, therefore, a violation of academic integrity. Plagiarism is also considered a breach of moral rights in Canada, the United Kingdom, Australia, and New Zealand.

Fair use and fair dealing are not forms of infringement, but the line between them and infringement is fine and sometimes difficult to draw. There are no simple rules to follow other than moderation and good judgment. When in doubt, ask for permission or pay royalties.

Copyright owners determine the degree to which they monitor the use and abuse of their works. Although copyright infringement may be a criminal offense, law enforcement officers will not act unless the affected owner files a complaint. Negotiation between parties may be sufficient to settle disputes over rights. If not, lawsuits may be necessary. To file lawsuits in the United States for works of US origin, the works must be registered with the US Copyright Office (see Section 3.2.2, "Registration").

3.4 INTERNATIONAL LAW

There is no such thing as an "international copyright" that protects works throughout the world. National laws differ, but most countries offer protection to foreign works under certain circumstances. The major international treaties covering copyright are the Berne Convention for the Protection of Literary and Artistic Works of 1971[11] and the Universal Copyright Convention of 1971.[12] In addition, bilateral agreements exist between some countries. To protect a work in a particular country, determine the extent of protection available in that country and consult someone familiar with the laws there.

The Universal Copyright Convention calls for using a copyright notice consisting of the copyright symbol, the date of first publication, and the name of the copyright owner (eg, "© 2023 Council of Science Editors"). Neither the word "copyright" nor its abbreviation "copr" without the symbol is valid outside the United States.[22] Some Latin American countries do not recognize the symbol or the word "copyright" and insist on the phrase "All rights reserved" instead. Both terms are recommended when full protection is needed.[23]

CITED REFERENCES

1. US Constitution, art I, sect 8, cl 8.

2. Copyright Act of 1976, Pub L No. 94-553, 90 Stat 2541 (1976 Oct 19).

3. Berne Convention Implementation Act of 1988, Pub L No. 100-568, 102 Stat 2853 (1988 Oct 31).

4. Uruguay Round Agreements Act of 1994, Pub L No. 103-465, 108 Stat 4809, 4973 (1994 Dec 8).

5. Digital Millennium Copyright Act of 1998, Pub L No. 105-304, 112 Stat 2860, 2887 (1998 Oct 28). https://www.copyright.gov/legislation/dmca.pdf

6. Copyright Act. RSC 1985, Chapter C-42. Department of Justice Canada; [updated 2019 Jun 17]. https://wipolex.wipo.int/en/legislation/details/19112

7. Copyright, Designs and Patents Act 1988 (c 48). Queen's Printer of Acts of Parliament (UK); 1988. https://www.legislation.gov.uk/ukpga/1988/48/contents

8. The Duration of Copyright and Rights in Performances Regulations 1995. Queen's Printer of Acts of Parliament (UK); 1995 [updated 2017 Jun 4]. https://www.opsi.gov.uk/si/si1995/Uksi_19953297_en_1.htm

9. Copyright Act 1968, Act No. 63 of 1968 as amended. Australian Government, Attorney-General's Department; [updated 2022 Aug 5]. (Amendments up 2021). https://www.legislation.gov.au/Details/C2022C00192

10. Copyright Act 1994, No. 143. Parliamentary Counsel Office (NZ); 1994, reprinted 2020 Aug 7. https://www.legislation.govt.nz/act/public/1994/0143/latest/DLM345634.html

11. Berne Convention for the Protection of Literary and Artistic Works, Paris Act of July 24, 1971, as Amended on September 28, 1979. https://wipolex.wipo.int/en/text/283698. (A list of the signatories can be found at https://www.copyrightaid.co.uk/copyright_information/berne_convention_signatories.)

12. Universal Copyright Convention as revised at Paris on 24 July 1971, with appendix declaration relating to Article XVII and resolution concerning Article XI 1971. United Nations Educational, Scientific and Cultural Organization. http://portal.unesco.org/en/ev.php-URL_ID=15241&URL_DO=DO_TOPIC&URL_SECTION=201.html

13. Directive (EU) 2019/790 of the European Parliament and of the Council of 17 April 2019 on copyright and related rights in the digital single market and amending Directives 96/9/EC and 2001/29/EC. Official

Journal of the European Union (Engl Ed). 2019 Apr 17; L 130:92–125. https://eur-lex.europa.eu/legal-content/EN/TXT/HTML/?uri=CELEX:32019L0790&from=EN

14. Hirtle P. Copyright term and the public domain in the United States. Cornell University Library; [updated 2021 Mar 17]. https://copyright.cornell.edu/publicdomain

15. Copyright notice. US Copyright Office; [revised 2021]. (Circular 3). https://www.copyright.gov/circs/circ03.pdf

16. Copyright basics. US Copyright Office; [revised 2021]. (Circular 1). https://www.copyright.gov/circs/circ01.pdf

17. Copyright registration. US Copyright Office; [revised 2022]. (Circular 2). https://www.copyright.gov/circs/circ02.pdf

18. Canadian Intellectual Property Office. Copyright. [modified 2022 Aug 18]. https://www.ic.gc.ca/eic/site/cipointernet-internetopic.nsf/eng/h_wr00003.html

19. Best edition of published copyrighted works for the collections of the Library of Congress. US Copyright Office; [revised 2021]. (Circular 7B). https://www.copyright.gov/circs/circ07b.pdf

20. Library and Archives of Canada Act. 2004, Chapter 11. Department of Justice Canada; [updated 2015 Feb 26]. https://canlii.ca/en/ca/laws/stat/sc-2004-c-11/latest/sc-2004-c-11.html

21. CENDI Copyright Working Group. Frequently asked questions about copyright and computer software: issues affecting the US Government with special emphasis on open source software. CENDI Secretariat; 2009 [updated 2022 Aug]. https://www.cendi.gov/pdf/CENDI_OpenSourceFAQ_2Dec2020.pdf

22. Duration of copyright. US Copyright Office; [revised 2021]. (Circular 15a). http://www.copyright.gov/circs/circ15a.pdf

23. Strong WS. The copyright book: a practical guide. 6th ed. MIT Press; 2014

2

General Style Conventions

4 Alphabets, Symbols, and Signs

Editors: Heather DiAngelis, MA, and Dana M. Compton

4.1 ALPHABETS

Most European languages are written and printed using the letters of the Roman alphabet. In general, English uses unmodified letters from the Roman alphabet. For special needs, scientific English uses some special characters formed from Roman and Greek letters, as well as some special typefaces.

4.1.1 Roman Alphabet

The classical Roman alphabet had 23 letters. Between the 15th century and the 18th century, 3 additional letters were developed: the consonant J (to distinguish the consonant I from the vowel I), the vowel U (to distinguish the vowel V from the consonant V), and the consonant W (to replace using 2 U's side by side). All 26 letters in the present-day Roman alphabet can be written in both capital (or uppercase) letters or as small (or lowercase) letters (see Table 4.1).

In some non-English European languages, Roman letters are modified by diacritics or combined in ligatures to form additional characters (see Table 4.2 and Section 5.6, "Punctuation Marks in Non-English Languages"). These additional characters are not generally needed in English scientific texts, except for proper names of people and places and for direct quotations of non-English words. For more detailed information about special characters in non-English languages, see *The Chicago Manual of Style*.[1]

4.1.2 Greek Alphabet

Most of the capital letters of the classical Greek alphabet can be related to Roman capital letters (see Table 4.1). Both uppercase and lowercase Greek letters are used in many nomenclatures and notations for scientific disciplines, notably astronomy, biochemistry, chemistry, mathematics, pharmacology, physics, and statistics. These uses are discussed in Part 3, "Special Scientific Conventions," of this manual.

Table 4.1. Roman and Greek Alphabets

Roman letters		Greek letters[a]		
Capital (uppercase)	Small (lowercase)	Name of letter	Capital (uppercase)	Small (lowercase)
A	a	alpha	A	α
B	b	beta	B	β
C	c			
D	d	delta	Δ	δ
E	e	epsilon	E	ε
F	f			
G	g	gamma	Γ	γ
H	h	eta	H	η
		theta	θ	θ
I	i	iota	I	ι
J	j			
K	k	kappa	K	κ
L	l	lambda	Λ	λ
M	m	mu	M	μ
N	n	nu	N	ν
O	o	omicron	O	o
P	p	pi	Π	π
Q	q			
R	r	rho	P	ρ
S	s	sigma	Σ	σ
T	t	tau	T	τ
U	u	upsilon	Y	υ
V	v			
W	w			
X	x	xi	Ξ	ξ
Y	y			
Z	z	zeta	Z	ζ
		phi	Φ	φ
		chi	X	χ
		psi	Ψ	ψ
		omega	Ω	ω

[a] In the normal order of the Greek alphabet, the letter gamma follows beta, zeta follows epsilon, and xi follows nu. The order in this table shows the relation of the Greek alphabet to the Roman alphabet.

Lowercase Greek letters are typically set in italic (or "slanted") type in mathematics, physics, and statistics. In other disciplines, lowercase Greek letters are typically set in roman type.

4.1.3 Hebraic Alphabet

Hebraic letters are rarely used in English-language scientific literature except for the aleph (א), which is occasionally needed in mathematics as a full-size symbol or a superscript character.

Table 4.2. Common Diacritics and Special Formations for the Roman Alphabet Used Primarily in Languages Other than English

Name	Mark	Example
Acute accent	´	é
Double acute accent	˝	ő
Grave accent	`	è
Breve	˘	Ğ
Caron, or wedge (háček, Czech for "little hook")	ˇ	ž
Cedilla	¸	Ç
Circumflex	ˆ	Ô
Dot	·	İ
Ligature	[no mark]	æ
Macron	ˉ	Ē
Ogonek (Polish for "little tail")	˛	ę
Over-ring (kroužek, Czech for "little circle")	˚	å
Bar, or stroke	/	Ø
Tilde	˜	ñ
Umlaut (diaeresis mark)	¨	ü

4.1.4 References for Non-English Language Sources

The titles of articles, serials, and books originally published in non-English languages are treated differently in end references depending on which alphabet those sources were written in.

(1) In the case of a non-English title in the Roman alphabet, use the original title, followed by the English translation placed within square brackets.
(2) For a title in Greek or Cyrillic characters, romanize the letters, and place the English translation within square brackets after the original title.
(3) For a title in a character-based alphabet (eg, Chinese or Japanese), use only an English translation, but enclose the translated title within square brackets.

For more on using non-English titles in end references, see Section 29.3.3.2.1, "Format of Titles."

4.2 SYMBOLS AND SIGNS

Notations representing quantities, objects, and actions have been used through all of recorded history (eg, numerals for representing counted objects). As the pace of scientific discovery and description accelerated in the Renaissance and in succeeding centuries, new notations were developed to efficiently represent the much more complex concepts that were being formulated. This led to the current extensive systems of symbolic notation for chemicals, genes, animal functions, mathematical operations, and other subjects of scientific inquiry (see Part 3, "Special Scientific Conventions").

Many symbolic notations in science were developed in a logical and coherent scheme for specific functional needs. They may represent what cannot be as economically expressed by terms, or they may represent functional relations among various items, such as mathematical equations. In contrast, commonly used abbreviations (eg, PhD and DNA) are simply shortened forms of terms that take the place of the entire terms and save space (see Chapter 11, "Abbreviations").

Graphic notations that stand for relations and operations are often called symbols or signs. Examples are the plus symbol (+) and the minus symbol (−). In its standards handbook on quantities and units, the International Organization for Standardization (ISO)[2] distinguishes between the terms "sign" and "symbol" but does not define the distinction.

In this style manual, the term "symbol" refers to a character or other graphic unit representing a quantity, unit, element, unit structure, relation, or function. Mathematical notations, for example, are generally referred to as symbols. This usage is consistent with the style manuals of the American Institute of Physics[3] and the American Mathematical Society,[4] as well as with the standard for electronic manuscript markup of the National Information Standards Organization[5] in the United States.

In contrast, the term "sign" is used for a notation that conveys a direction for readers (eg, a footnote sign is a lowercase letter or other character that directs readers to footnotes).

4.2.1 Types of Symbols

Symbols fall into 3 general categories:

(1) Single alphabetical or numerical characters or an unspaced combination of 2 or more alphabetical, numerical, or alphanumeric characters. Such symbols may be abbreviated forms of the terms they represent (eg, "V" for "volume" and "cos" for "cosine"), but need not be (eg, "Φ" for "electric flux"). A few symbols are graphic units derived from an alphabetical origin (eg, "-C" for the "Cambrian geologic period").
(2) Punctuation marks, particularly in mathematics (eg, "!" for a factorial and ":" for a ratio).
(3) Graphic units designated as specific representations (eg, "=" for "equals," ">" for "greater than," and "<" for "less than").

A symbol may stand alone, or it may be incorporated into a larger symbol. For examples of various scientific symbols, see Table 4.3.

An abbreviation can function as a symbol when it is part of a group of symbols that together are used for the same conceptual purpose. For example, abbreviations of names for geologic time (eg, "Pz" for "Paleozoic" and "J" for "Jurassic") serve as symbols when they are used with other alphabet-derived symbols (eg, "from -C to J"). Some chemical symbols are abbreviated versions of element names (eg, "C" for "carbon" and "O" for "oxygen"), while others indicate relations among chemical elements in a unit structure (eg, "CO_3^{-2}" for "carbonate ion"). When an abbreviation does not serve any of these purposes, it remains designated as an abbreviation (see Chapter 11, "Abbreviations").

In this style manual, symbols for several scientific fields are explained in the chapters dedicated to those fields in Part 3, "Special Scientific Conventions" (see Table 4.4).

Table 4.3. Examples of Scientific and Common Symbols

Symbol	Name of symbol	Type of subject
1	one	unit
x	unknown quantity	quantity
n	number	quantity
m	mass	quantity
P	probability	quantity
V	volume	quantity
CO	cardiac output	rate
3	$1 + 1 + 1$	unit structure
m	male (eg, in a family tree)	unit structure
Al	aluminum, or aluminium	unit, chemical
$AlBr_3$	aluminum bromide	unit structure, chemical
sis	oncogene related to simian sarcoma	unit structure, genetic
Pz	Paleozoic Era	unit, geologic time
m	meter, or metre	unit, measurement
$	dollar	unit, monetary
€	euro	unit, monetary
cos	cosine	mathematical function
+	plus (for addition)	mathematical function

4.2.2 Type Styles for Symbols

The appropriate uses of roman and italic type and other aspects of styling symbols are specified in the ISO standards handbook *Quantities and Units* (see Table 4.5).[2] For example, for symbols for vectors and tensors, ISO's *Quantities and Units* specifies using a serif font with either boldface italic type (eg, *v*) or nonbold italic below a right-facing arrow (eg, $\vec{v}$) (see Section 12.3.1.3, "Vectors, Scalars, and Tensors"). In addition, Fraktur, or blackletter script, is used for some conventions in mathematics and physics.[3,4]

In general, multiletter symbols should be set in roman type rather than italic, even when they represent quantities. An example is "CO" for "cardiac output," a variable quantity used with the unit liters per minute. Use of roman type for multiletter variables avoids potential misinterpretation of the symbol as a combination of 2 single-letter variables, in this case "*C*" and "*O*." Nonetheless, the American Chemical Society[6] recommends italicizing 2-letter variables defining chemical transport properties (eg, "*Bi*" for "Biot number"). With this convention, special care is needed in indicating multiplication involving such variables to avoid misinterpretation.

For more on styling conventions, see Table 4.5 and Chapter 10, "Type Styles, Excerpts, Quotations, and Ellipses."

4.3 FONTS FOR MATHEMATICAL AND OTHER SCIENTIFIC USE

In electronic workflow, problems can arise if special characters and symbols in more than one font set are imported into a document. Characters taken from different font

Table 4.4. Scientific and Common Symbols Detailed in This Style Manual

Subject	Section	Table
Algebra	12.3.1, "Font Style Conventions"	12.10
Allergens	23.5.7, "Renin–Angiotensin System"	
Amino acids	17.13.2, "Amino Acids, Peptides, and Polypeptides"	17.13
Astronomy	26.3, "Celestial Coordinates"	26.3
Atomic and molecular states	16.6, "Symbols for Electronic Configurations of Atoms," and 16.7, "Symbols for Molecular States"	
Biothermodynamics	18.2, "Units in Biothermodynamics"	18.1
Blood groups	23.4.3, "Blood Groups," and 23.5.2.1, "Blood Groups"	
Bone histomorphometry	23.5.3, "Bone Histomorphometry"	
Calculus	12.3.1.2, "Calculus"	12.11
Carbohydrates	17.13.4, "Carbohydrates"	
Chemical elements	16.4, "Chemical Elements"	16.3
Chemical formulas	16.5, "Chemical Formulas," and 17.2, "Formulas"	
Chemical kinetics	18.3.1, "Variables," and 18.3.3, "Processes"	
Chromosomes	21.8 "Viruses," through 21.15, "Animals"	21.2, 21.32, and 21.34
Chromosomes: components and units	21.2, "Genetic Units, Measures, and Tools"	
Complement	23.5.6.2, "Complement"	
Crystallography	25.6, "Crystals"	25.1 and 25.2
Density of water	25.11.2.4, "Density"	
Drug receptors	20.3, "Symbols for Receptors of Drugs and Other Humoral Mediators"	
Electrocardiographic recordings	23.5.4.2, "Clinical Notations"	
Elementary particles	16.2, "Elementary and Composite Particles"	16.1
Enzyme kinetics	18.7, "Enzyme Kinetics"	18.4
Enzymes	17.13.7, "Enzymes"	
Genes and phenotypes	21.8.1, "Bacteriophages (Phages)," through 21.15, "Animals"	21.4–21.10, 21.13–21.21, 21.24–21.30, and 21.32
Hemodynamic functions	23.5.4.1, "Hemodynamics"	23.3
Hemoglobins	23.5.2.2, "Hemoglobins"	
Immunoglobulins	23.5.6.1, "Immunoglobulins"	
Immunologic systems	23.4.4, "Immunologic Systems," and 23.5.6, "Immunologic Systems"	
Isotopic modifications	17.10, "Isotopically Modified Compounds"	
Lipids	17.13.9, "Lipids"	
Mathematical functions	12.3.1, "Font Style Conventions," and 12.4, "Mathematics in Text and Display"	12.9
Mathematical operators	12.3.1.1, "Mathematical Operators"	12.10

Table 4.4. Scientific and Common Symbols Detailed in This Style Manual (*continued*)

Subject	Section	Table
Mathematical symbols	12.3.1, "Font Style Conventions"	12.10
Matrices	12.3.1.4, "Matrices and Determinants"	12.13
Monetary units	12.1.6, "Monetary Units"	12.2
Non-SI units	12.2.2, "Outdated Units"; 12.2.3, "Other Measurement Systems"; and 25.13, "Non-SI Units"	12.5, 12.7, and 12.8
Nuclear reactions	16.3, "Nuclear Reactions"	16.2
Nucleic acids and related compounds	17.13.10, "Nucleic Acids and Related Compounds"	
Nuclides	16.4, "Chemical Elements"	
Pedigrees	23.5.11, "Reproduction: Inheritance and Pedigrees"	23.6
Pharmacokinetics	20.5, "Pharmacokinetics"	20.2
Plant physiology	23.2, "Plant Anatomy and Physiology"	23.1 and 23.2
Prenols	17.13.13, "Prenols"	
Prostaglandins	17.13.14, "Prostaglandins"	
Proteins	17.13.15, "Proteins"	
Purines and pyrimidines	17.13.10.1, "Nucleic Acids"	17.16
Renal function	23.5.8, "Kidney and Renal Function"	
Respiratory functions	23.5.9, "Respiration"	23.4
Retinoids	17.13.16, "Retinoids"	17.17
Rocks and minerals	25.5, "Rocks and Minerals"	
Roman numerals	12.1.1, "Arabic and Roman Numerals"	12.1
Set theory notations	12.3.1.2, "Calculus"	12.12
SI base units, derived units, and prefixes	12.2.1, "Système International d'Unités (SI)"	12.3, 12.4, and 12.6
Soil horizons	25.7.1, "Pedon Descriptions"	25.3
Stereochemistry	17.11, "Stereochemical Nomenclature"	17.11
Steroids	17.13.17, "Steroids"	17.18
Statistics	12.5, "Statistics"	12.14
Thermal regulation	23.5.10, "Thermal Physiology"	23.5
Vitamins	17.13.19, "Vitamins"	

SI, Système International d'Unités (International System of Units)

sets may vary in style, positioning, and size. Beyond the problems related to how such variable characters appear in print, other difficulties may arise when documents are posted online. For example, some characters may not be rendered correctly on screen if the same fonts are not installed on the user's computer.

To address these problems, the Scientific and Technical Information Exchange (STIX), a group of 6 major scientific publishers, developed a comprehensive set of fonts for special characters used in mathematical, medical, and other scientific publishing. The STIX fonts are designed so that they can be used in all stages of the publishing process, from manuscript creation through final print and electronic publication.[7] More information

Table 4.5. Style Conventions for Alphanumeric Symbols

Symbol category	Typeface	Examples and notes
Quantity	Italic	Examples: y and V
		Notes: Usually a single Roman or Greek letter with no following punctuation mark except as needed for normal punctuation within a sentence. May be modified by a subscript. For a frequent exception to using italic, see Section 4.2.2, "Type Styles for Symbols." If the text must be in italic type, quantity symbols that would normally be set in italics should be set in roman type.
Subscript modifiers		
For a quantity symbol	Italic	Example: P_x for the probability (P) of an unknown quantity (x)
For other symbols	Roman	Example: V_L for volume (V) of the lung (L)
Unit of measurement	Roman	Examples: m (meter) and kg (kilogram)
		Notes: Used in roman type regardless of the type used for the rest of the text. When used in conjunction with a numeric value, the unit follows the numeric value and is separated from it either by one space when used as a noun or by a hyphen when part of a compound modifier. Use the singular form of units for plurals. Do not follow with a punctuation mark except as needed for normal punctuation within a sentence. Usually lowercase letters except if the unit name is derived from a proper name (see Section 12.1.4.1, "Numbers").
Number	Roman	Examples: 11 and 64 (arabic numerals); XVIII and xviii (roman numerals)
Mathematical operator	Roman	Examples: + and ×
		Note: There are rare exceptions to the roman convention, such as $O(\ )$ for "of the order of."
Subatomic particle	Italic or roman	Examples: π^- and μ^+
		Note: Although italic type is commonly used, there is no accepted standard regarding whether to use italic or roman type for these symbols (see Section 16.2, "Elementary and Composite Particles"). Regardless of whether particle symbols are in italics or roman, their nonalphabetic superscripts and subscripts should be in roman.
Chemical element	Roman	Examples: K, Rb, Sn, and $CaCl_2$
		Note: Use roman for subscripts and superscripts in chemical compounds (see Sections 16.4, "Chemical Elements," and 16.5, "Chemical Formulas").
Chromosome	Roman	Example: 47, XXY Note: See Chapter 21, "Genes, Chromosomes, and Related Molecules."
Gene	Italic	Examples: *sis* and *HRCT* Note: See Chapter 21, "Genes, Chromosomes, and Related Molecules."
Phenotype	Roman	Example: HRCT Note: See Chapter 21, "Genes, Chromosomes, and Related Molecules."

about the STIX font-creation project, including information on downloading the STIX fonts, is available on the project's website at https://www.stixfonts.org.

For the online environment, additional resources are available to properly display mathematical and other scientific content. MathML allows publishers and others to use Extensible Markup Language tags to display equations across the web and other applications.[8] MathJax uses JavaScript to display mathematics, and it can be used to produce MathML content.[9]

CITED REFERENCES

1. The Chicago manual of style: the essential guide for writers, editors, and publishers. 17th ed. The University of Chicago Press; 2017. Also available at https://www.chicagomanualofstyle.org

2. International Organization for Standardization. Quantities and units—part 2: mathematics. International Organization for Standardization; 2019. (ISO 80000-2:2019).

3. AIP style manual. 4th ed. American Institute of Physics; 1990. https://publishing.aip.org/wp-content/uploads/2021/03/AIP_Style_4thed.pdf

4. Letourneau M, Wright Sharp J. AMS style guide: journals. American Mathematical Society; 2017.

5. National Information Standards Organization. Electronic manuscript preparation and markup. NISO Press; 1995. (ANSI/NISO/ISO 12083-1995 [R2002]; [updated 2013 Jan 16]).

6. Coghill AM, Garson LR, editors. The ACS style guide: effective communication of scientific information. 3rd ed. American Chemical Society; 2006.

7. STIX fonts project. Ver. 2.13 b171. STI Pub Companies. http://www.stixfonts.org/

8. MathML 3. W3C; 2021. https://www.w3.org/Math/

9. MathJax. Ver. 3.1.4. https://www.mathjax.org/

ADDITIONAL REFERENCES

Bringhurst R. The elements of typographic style. 4th ed. Hartley & Marks; 2013.

Diringer D, Minns E. The alphabet: a key to the history of mankind. Kessinger Publishing, LLC; 2010.

McArthur T, Lam-McArthur J, Fontaine L, editors. The Oxford companion to the English language. 2nd ed. Oxford University Press; 2018.

5 Punctuation and Related Marks

Editors: Trevor Lane, DPhil, PGCELT; Joanna Odrowaz; and Dara R. Rochlin

5.1 OVERVIEW

For many centuries, punctuation was applied in English mainly as a guide for readers to stop or pause, which is punctuation's rhetorical use. Beginning in the late 17th century, the grammatical use of punctuation began to replace the rhetorical use. Today, punctuation serves 2 main purposes: clarifying the relation of prose elements to one another, such as periods demarcating the end of sentences, and supporting the meaning of statements, such as question marks signaling that questions are being asked and exclamation points indicating that statements are being loudly or emphatically expressed. Scientific literature also has contributed to the evolution of punctuation, resulting from the need for conventions for numerical and mathematical expressions and for specialized symbolization and nomenclature.

Clear discussions of the logic and development of punctuation can be found in *Making a Point: The Pernickety Story of English Punctuation*.[1] In addition, *The Elements of Typographic Style*[2] offers a complete catalog of analphabetic characters (ie, punctuation marks and related characters).

This chapter covers 3 main groups of punctuation marks:

- terminal marks, which are used mainly to indicate the close of sentences
- marks used within sentences or words
- marks indicating relations between 2 or more lines of text (see Table 5.1)

The sections of this chapter outline general uses of punctuation and, if appropriate, specialized uses in science. Quotation marks, ellipses, and the relations between quotation marks and other punctuation marks are discussed in Section 10.2, "Excerpts and Quotations." The use of conventional punctuation marks in markup languages such as Standard Generalized Markup Language and Extensible Markup Language is discussed in detail in Chapter 31, "Typography and Manuscript Preparation."

In deciding whether punctuation is needed, 3 general principles can be applied:

- Use specific punctuation marks when called for by well-established and internationally accepted conventions of style, bearing in mind that established uses can change over time and vary among countries.
- When the need for a mark is unclear, use the mark if it will reduce or eliminate ambiguity.
- If a format or text structure logically arranges text elements in a way that makes the meaning clear, omit punctuation that might otherwise be used (eg, after short introductory phrases and at the end of each line in vertical lists).

5.2 TERMINAL MARKS

The primary role of terminal marks is to indicate the end of sentences. Regardless of which terminal mark is used, only one space should be used to separate sentences. The practice of using 2 spaces between sentences has long been outdated.

Table 5.1. List of Punctuation Marks

Name	Mark	Sections in this chapter
Ampersand	&	5.4.8, "Ampersand"
Angle brackets	‹ ›	5.3.9, "Angle Brackets"
Apostrophe	'	5.4.4, "Apostrophe"
Asterisk	*	5.4.7, "Asterisk"
"At" symbol	@	5.4.9, "'At' Symbol"
Braces—single and paired	{ { }	5.3.8, "Braces," and 5.5.1, "Brace"
Colon	:	5.3.1, "Colon"
Comma	,	5.3.3, "Comma"
Diacritics		5.4.6, "Diacritics"
Ditto marks	"	5.5.2, "Ditto Marks"
Dots as leaders	.....	5.5.3, "Dots as Leaders"
Em dash	—	5.3.5.2, "Em Dash"
Two-em dash	——	5.3.5.3, "Two-Em Dash"
Three-em dash	———	5.3.5.4, "Three-Em Dash"
En dash	–	5.3.5.1, "En Dash"
Exclamation point	!	5.2.3, "Exclamation Point"
Guillemets and chevrons (quotation marks in French and Spanish)	« »	5.6, "Punctuation Marks in Non-English Languages"
Hash mark	#	5.4.10, "Hash Mark"
Hyphen	-	5.4.1, "Hyphen"
Introductory interrogation mark in Spanish	¿	5.6, "Punctuation Marks in Non-English Languages"
Paragraph mark	¶	5.5.4, "Paragraph Mark"
Parentheses	( )	5.3.6, "Parentheses"
Period	.	5.2.1, "Period"
On-the-line period	.	5.2.1.2, "Specialized Uses: On-the-Line Period"
Raised period	·	5.2.1.3, "Specialized Uses: Raised Period"
Prime symbol—single and double	′ ″	5.4.5, "Prime Symbol—Single and Double"
Question mark	?	5.2.2, "Question Mark"
Quotation marks—double	" "	5.3.4.1, "Double Quotation Marks"
Quotation marks—single	' '	5.3.4.2, "Single Quotation Marks"
Semicolon	;	5.3.2, "Semicolon"
Slash	/	5.4.2, "Slash"
Square brackets	[ ]	5.3.7, "Square Brackets"
Vertical bar	|	5.4.3, "Vertical Bar"
Vinculum	——	5.3.10, "Vinculum"

5.2.1 Period (.)

The period, also known as a "full stop," is fundamental to punctuation. It has several general and scientific uses.

5.2.1.1 General Uses

(1) To close a declarative or imperative sentence. (This use includes closing a stand-alone declarative or imperative sentence within parentheses.)

Four of the soluble vitamins of the B complex have precise roles in the citric acid cycle.
The first rat died. (It refused to eat.)
Filter the solution while it is hot.

Similarly, use a period to end an elliptical sentence, which is a declarative sentence that has been shortened but the missing text is implied.

What is the worst ethical sin in science? Fraud. What next? Plagiarism.

(2) To separate introductory elements such as numbers and letters from text in an ordered list. (Parentheses may also be used for this purpose, as noted in Section 5.3.6, "Parentheses.")

1. Biochemistry	a. São Paulo
2. Chemistry	b. Buenos Aires
3. Physics	c. Rio de Janeiro

(3) To separate host names from other elements in uniform resource locators (URLs) for websites. (For this purpose, periods are usually called "dots.")

http://www.councilscienceeditors.org

Periods are also used in the subset of URLs called digital object identifiers. Better known as "DOIs," these identifiers are often included in end references.

https://doi.org/10.1140/epja/s10050-021-00622-5
https://doi.org/10.1515/jom-2020-0238
https://doi.org/10.5281/zenodo.5681252

For other general functions for the period, see Sections 10.3, "Ellipses," 5.5.3, "Dots as Leaders," and 5.2.1.3, "Specialized Uses: Raised Period."

5.2.1.2 Specialized Uses: On-the-Line Period

(1) For certain abbreviations for which periods are widely recommended in the scientific nomenclature, such as the scientific names of organisms (see Chapter 22, "Taxonomy and Nomenclature"). A period should also be used with the abbreviation for "number" to distinguish the word "no" from the abbreviation "No."

species *becomes* sp. (singular) and spp. (plural) *not* sp and spp
Staphylococcus aureus becomes *S. aureus* not *S aureus*
Number 1 *becomes* No. 1 *not* No 1

This style manual recommends omitting periods from other abbreviations, including acronyms and other initialisms, as well as from general Latin phrases (see Section 11.2, "Punctuation and Typography").

eg   ie   NASA   HIV   Dr   PhD

(2) As the decimal point, especially in US, British, and Canadian usage (see also Sections 5.3.3.5, "Specialized Uses," and 12.1.3, "Format of Numbers").

The infant weighed 10.7 kg at birth.

(3) To indicate hierarchical divisions represented numerically (as in the numbering of section headings in this style manual).

5.3.1 Colon   5.3.2 Semicolon   5.3.3 Comma

(4) In end references to indicate the end of components, such as author names and article titles (see Section 29.3.3, "Components of End References").

Aghajamali A, Karton A. Comparative study of carbon force fields for the simulation of carbon onions. Aust J Chem. 2021;74(10):709–714.

(5) To indicate ring size in chemical nomenclature.

spiro[2.3]hexane   bicyclo[2.2.2]octane

5.2.1.3 Specialized Uses: Raised Period (·)

The raised period is also known as the "dot operator," "raised dot," "middle dot," "centered dot," and "half-high dot." Its specialized uses are as follows:

(1) As a multiplication symbol in equations and other mathematical expressions and in compound units (see Sections 12.3.1.1, "Mathematical Operators," and 12.2.1.1, "SI Rules," and Table 12.10).

$k·g(a+2)$ $1\,C = 1\,A·s$

(2) As an ellipsis symbol in mathematical expressions (see Section 12.4.3, "Punctuation and Spacing").

$x_1 + x_2 + \cdots + x_n$

(3) To indicate associated base pairs of nucleotides (see Section 17.13.10.5, "Conventions for 1-Letter Symbols").

G·C *for* guanine and cytosine

(4) To indicate connection with adducts (eg, water of hydration) in a chemical formula.

$Na_2B_4O_7·10H_2O$

(5) As a group of 3 raised periods to indicate chemical associations of unspecified type.[3]

$F···H-NH_3$ $C···Pt$

5.2.2 Question Mark (?)

In general, the question mark indicates interrogation or uncertainty, regardless of context. In addition, it is used as a character in URLs for internet addresses.

(1) To close a freestanding question whether or not it is a complete sentence.

What is the government's policy on pollution of river headwaters?
The policy was overturned. Why?

(2) To close a declarative sentence ending with a complete question set off by a comma, even if the question is not a direct quotation set off with quotation marks. If the question is not a direct quotation, do not capitalize the first word of the question. If a comma is not used before the question, end the sentence with a period, not a question mark.

The committee asked, why were so many species killed?
The committee asked, "Why were so many species killed?"
The committee asked why so many species were killed.

(3) To indicate uncertainty about an element within a sentence or within a notation, such as a chromosome or chromosomal structure (see Table 21.33).

Girolamo Fracastoro (1483?–1553) was the father of the concept of infectious disease.
45,XXY,–?8 (45 chromosomes, XXY sex chromosomes, missing chromosome that is probably chromosome 8)

Using a question mark with a date indicates that the date is unknown or in doubt among experts. Position the question mark to clearly indicate the queried element.

1172?–1221 *not* ?1172–1221 (when the start date is in doubt)
1546–?1627 *not* 1546–1627? (when the end date is in doubt)
1822?–?1873 *not* ?1822–1873? (when both start and end dates are in doubt)

Authors should not use a question mark to communicate that they are uncertain of the information. Instead, authors should briefly explain the uncertainty.

5.2.3 Exclamation Point (!)

The exclamation point, also known as an "exclamation mark," is usually used to convey a loud or sudden utterance or emphatically expressed emotion. In addition to its general uses, the exclamation point has a number of scientific uses.

5.2.3.1 General Uses

(1)   To close a declarative or imperative sentence that is intended to convey rhetorical emphasis, such as expressing strong emotions, indicating surprise, or issuing urgent orders. Such use is usually not justified in reports of scientific findings, but exclamation points may be appropriate in commentaries on findings or in documents about science. When using an exclamation point is justified, use only one exclamation point.

> The "science" of Sigmund Freud, MD, was pure metaphysics!
> *not* The "science" of Sigmund Freud, MD, was pure metaphysics!!!

(2)   To close an interrogative sentence in place of a question mark in the rare situation in which the rhetorical emphasis of a sentence is more relevant than the inquiry.

> How do they expect researchers to finish that on time!

5.2.3.2 Specialized Uses

(1)   As a factorial symbol in mathematics (see Table 12.10).

> $4!$ *to represent* $4 \times 3 \times 2 \times 1$

(2)   In botanical writing to indicate examined specimens.

> Lectotype here designated: P!; isolectotypes: K!, NY!.

(3)   In some search engines, to truncate a term to retrieve variant word endings (other search engines may use an asterisk or question mark instead).

> cathod! *to retrieve* cathode, cathodes, cathodic, cathodically, cathodal, and cathodally

5.3 INTRASENTENCE MARKS

5.3.1 Colon (:)

Although not a frequently used intrasentence mark, the colon indicates a pause that is stronger than that of the semicolon or comma but weaker than that of the period. The colon's function is to clearly direct readers to the material that follows.

5.3.1.1 General Uses

(1)   To separate 2 independent clauses, specifically if the second clause amplifies or clarifies the first. The first word after the colon should be capitalized if the second clause is a formal statement or long quotation.

> We had second thoughts about the first run: The data lacked the needed precision.
> This is the rule: On closing a file, back it up on an external hard drive.

If the second independent clause is related to the first but is not directly explanatory, a period or semicolon is more appropriate (see Sections 5.2.1, "Period," and 5.3.2, "Semicolon").

(2) To introduce a quotation, especially when the quotation clarifies or amplifies the preceding independent clause. In general, the quotation should be no more than 2 sentences. Longer quotations should be treated as block quotations (see Section 10.2.2, "Block Quotations"). For short verbatim quotations, a comma can be used as an alternative to the colon (see item No. 4 in Section 5.3.3.2, "Other General Uses").

> In his book *The Selfish Gene*, Richard Dawkins, DPhil, DSc, coined the word "meme" and likened memes to genes: "Just as genes propagate themselves in the gene pool by leaping from body to body via sperms or eggs, so memes propagate themselves in the meme pool by leaping from brain to brain via a process which, in the broad sense, can be called imitation."

(3) To introduce a list in a series after an independent clause. Lowercase the first word after the colon unless it is a proper noun or unless the series consists of independent clauses. Do not insert a colon between a sentence's verb and object or between a preposition and its object, even if the list is in vertical format.

> The lectures covered 3 topics: fiber optics, microelectronics, and robotics.
> *or* The lecture series addressed the following topics: fiber optics, microelectronics, and robotics.
> *not* The lectures covered: fiber optics, microelectronics, and robotics.
>
> Ethanoic acid is also known as "acetic acid," "ethylic acid," and "methane carboxylic acid."
> *not* Ethanoic acid is also known as: "acetic acid," "ethylic acid," and "methane carboxylic acid."

When a colon introduces a vertical list, omit closing punctuation for items, except if each item forms a complete sentence.[4]

> The survey was carried out with 3 purposes in mind:
> - to estimate the size of rodent populations
> - to estimate environmental influences on population sizes
> - to identify previously unrecognized species

5.3.1.2 Specialized Uses

(1) To separate elements in titles, such as the main title and subtitle of a book, journal, chapter, article, table, or figure.

> *The CSE Manual: Scientific Style and Format for Authors, Editors, and Publishers*
> "On-site Chiropractic Care as an Employee Benefit: A Single-Location Case Study"
> Table: Calculus Symbols

Note that the words "Chapter," "Table," and "Figure" and any accompanying numbers can also be separated from titles by other punctuation, such as a comma, period, or an em dash, or just by extra spaces.

> Chapter 3: "Epiphytes and Phorophytes"
> Chapter 3, "Epiphytes and Phorophytes"
> Chapter 3. "Epiphytes and Phorophytes"
> Chapter 3—"Epiphytes and Phorophytes"
> Chapter 3   "Epiphytes and Phorophytes"

(2) In end references, to separate the main title and a subtitle of an article or book, even if the original title did not include punctuation, and to separate issue number and page range (see Sections 29.3.3.2.2, "Punctuation," and 29.3.3.8.1, "Location within a Work"). Although the first words of subtitles are capitalized in the titles of articles and books, they are lowercased in end references.

> Cunsolo A, Harper SL. Climate change and health: a grand challenge and grand opportunity for public health in Canada [editorial]. Health Promot Chronic Dis Prev Can. 2019;39(4):119–121. https://doi.org/10.24095/hpcdp.39.4.01
> Balsam LB, Hoffman WD. Commentary: the molecular pandemonium of coronavirus disease 2019. J Thorac Cardiovasc Surg. 2021 Feb;161(2):e227–e228. https://doi.org/10.1016/j.jtcvs.2020.06.003

(3) To separate parts of a ratio (see Section 12.3.4, "Ratios, Proportions, and Percentages").

> The ratio of women to men was 3:1.
> We recommended a 5:10:5 fertilizer.

(4) To separate hours, minutes, and seconds in a time unit (see Section 13.2.2, "Clock Time"). However, when a time unit such as a second or hour is measured in a decimal fraction, use a decimal point rather than a colon.

> **24-hour system**
> 2345:15 *or* 23:45:15 *for* 23 hundred hours, 45 minutes, and 15 seconds
> 0615:32.7 *or* 06:15:32.7 *as decimal notation for* 6 hundred hours, 15 minutes, 32 seconds, and 7/10 of a second

> **12-hour system**
> 12:38 AM *for* 38 min after midnight
> 7.25 PM *as decimal notation for* 15 min after 7 in the evening (60 min × 0.25 = 15 min)

(5) To symbolize a chromosomal break in detailed descriptions. In addition, a double colon symbolizes a break and reunion, the disruption of one gene by another, or the integration of one gene beside another without disruption (see various tables in Chapter 21, "Genes, Chromosomes, and Related Molecules").

> 46,XX,del(1)(pter→q21::q31→qter), which symbolizes the breakage and reunion of bands 1q21 and 1q31 in the long arm of chromosome 1
> *ade6::ura4*, which symbolizes the disruption of *ade6* by *ura4*
> *BCR::ABL1*, which symbolizes fusion of the *BCR* and *ABL1* genes

(6) To group the locants in chemical names (ie, letters and numbers identifying the location of an atom or group in a molecule).

> 2,3:4,5-bis-*O*-(phenylmethylene)-*altro*-hexodialdose
> (3,4:3,9:5,6:6,7:7,8-penta-μH)-(3-*endo*-H)-*nido*-nonaborane
> pyrido[1′,2′:1,2]imidazo[4,5-*b*]quinoxaline

(7) To separate indications of originating and sanctioning authors to protect fungal names designated as "sanctioned." Note that in this case, the colon is preceded and followed by a space (see Section 22.4.2.2, "Sanctioned Names").

> *Boletus piperatus* Bull : Fr

5.3.2 Semicolon (;)

As a punctuation mark used for linking clauses and phrases, the semicolon is stronger than the comma. Its role is to indicate closely related clauses.

5.3.2.1 General Uses

(1) To separate with rhetorical force 2 or more independent clauses. However, in most cases, the period is preferred to the semicolon for this purpose.

When the semicolon is used, the clauses are often joined by a conjunctive adverb such as "however," "besides," or "therefore." The semicolon is generally not used between clauses joined by coordinating conjunctions, such as "and." In such cases, the clauses are usually separated by a comma. Similarly, semicolons should not be used instead of commas to separate short clauses in a series.

> The patient was first treated in the emergency department; however, she soon had to be transferred to the intensive care unit.
> Slow-stop mutants complete the round of replication; they cannot start another.
> *not* Bohrium was first reported by a Soviet research team in 1976; but it was officially discovered 5 years later by a German team. (Use comma instead.)

not The failure rate was high because many students did not prepare for the written examination; the questions were too difficult; and time allotted was too short. (Use commas instead.)

(2) To separate 2 or more independent clauses connected by a coordinating conjunction when at least one clause has internal commas.

> The uppermost formation, first identified by Smith, is sandstone; and the next lower, identified by Jones, is shale.
> All of the selected organic solvents, consisting of ketones, alkanes, and alcohols, were used without further purification; but the deionized water was distilled and then microfiltered.

(3) To separate 2 clauses when the second is structurally parallel.

> The first peak appeared at 3.5 min; the last, at 10.5 min.

(4) To separate the elements of a complex series in which at least one of the elements has internal commas.

> The survey was conducted in multiple settings: restaurants and bars; schools, technical institutes, and universities; and health centers, medical clinics, and dental clinics.
> His ethnographic studies concentrated on 3 groups: Chinese, Japanese, and Taiwanese; French, Germans, and Austrians; and Inuit, Mexicans, and Peruvians.

When semicolons separate a complex series in the middle of a sentence, the last item in the series does not end with a semicolon.

> Two inner planets, Venus and Mars; 2 outer planets, Saturn and Neptune; and the dwarf planet Pluto are being studied at the observatory.
> The US conference brought together physicians, both MDs and DOs; dentists, both DMDs and DDSs; and nurses, both RNs and LPNs, to discuss ways to improve patient care.

5.3.2.2 Specialized Uses

(1) In end references, to separate the year from the volume for a journal article (see Section 29.3.3.6, "Date"). In this usage, no spacing is used between the semicolon and the volume number.

> Asplund-Samuelsson J, Hudson EP. Wide range of metabolic adaptations to the acquisition of the Calvin cycle revealed by comparison of microbial genomes. *PLOS Comput Biol.* 2021;17(2):e1008742. https://doi.org/10.1371/journal.pcbi.1008742
> Soulier N, Bryant DA. The structural basis of far-red light absorbance by allophycocyanins. Photosynth Res. 2021;147(1):11–26.

(2) To symbolize chromosomal rearrangements (see various tables in Chapter 21, "Genes, Chromosomes, and Related Molecules").

> 46,XX,t(12;?)(q15;?), which symbolizes the rearranged chromosome 12 in which the segment of the long arm (q) distal to band 12q15 could not be identified.

(3) To separate symbols for genes on different chromosomes or to separate linked genes (see various tables in Chapter 21, "Genes, Chromosomes, and Related Molecules"). In this usage, the semicolon is followed by a space.

> *bw; e; ey*   *Sh2/sh2; Bt2/bt2*

(4) To separate subscripts in selectively labeled chemical compounds.

> $[1-^2H_{1,2}]SiH_3OSiH_2OSiH_3$

(5) In chemical names, to separate sets of locants in which colons have already been used.

> benzo[1″,2″:3,4;5″,4″:3′,4′]dicyclobuta[1,2-*a*:1′,2′-*a*′]diindene

5.3.3 Comma (,)

The most commonly used intrasentence mark, the comma serves mainly to clarify grammatical structures in the following ways:

- by separating items in a series
- by separating independent clauses in the same sentence
- by setting off introductory material
- by separating other elements of a sentence or an expression
- by bracketing material

5.3.3.1 Serial Comma

In a simple series of 3 or more elements, the serial comma is used before the coordinating conjunction to separate the last word, phrase, or clause from the preceding element. Also known as the "Oxford comma" and the "Harvard comma," the serial comma is preferred by most English-language scientific publications. Nonscientific publications are less likely to prefer the serial comma because many of those publications consider a coordinating conjunction like "and" or "or" to be a sufficient pause between the last 2 elements in a series. Even publications that do not prefer the serial comma will use it when necessary to prevent ambiguity.

> Seeds of tomato, bean, and pepper were planted in April.
> The patient is not allergic to penicillin, ampicillin, or erythromycin.
> The student was inspired by her parents, Albert Einstein, and Rosalind Franklin. (Without the serial comma, the sentence could be misinterpreted as indicating that the student's parents were Albert Einstein and Rosalind Franklin.)
> Our framework considers network structure, detects errors, and provides timely feedback.

5.3.3.2 Other General Uses

(1) To separate 2 or more independent clauses in a compound sentence. The comma before the last independent clause is followed by a coordinating conjunction (eg, "and," "but," "or," "nor," "for," or "yet").

> Lithium-ion batteries dominate the energy market for electric vehicles, but these batteries have serious limitations when operated in cold regions.
> The linear sizes of the quasars have a direct power–law relationship with their redshifts, and the converse is the case for their galaxy counterparts.

The comma can be omitted between 2 independent clauses if the clauses are short and the absence of a comma would not cause ambiguity (see Section 5.3.3.3, "Optional Uses").

> The experiment was simple, but the results were significant. *or* The experiment was simple but the results were significant.

A major pitfall associated with using a comma before a coordinating conjunction to join 2 clauses is that a sentence fragment can be created if the second clause is subordinate to the first rather than the clauses being independent of each other. This can happen when a clause lacks a subject or a subject and a verb or when an entire clause is subordinate to a main clause (see also Section 5.3.3.4, "Incorrect Uses").

> We finished surveying in June and published the map in December.
> *not* We finished surveying in June, and published the map in December.

(2) To set off introductory phrases or clauses from independent clauses, such as the following:

- clauses that begin with subordinating conjunctions (eg, "if," "although," "because," "since," "when," "where," and "while")

> When 2 equally efficacious and safe treatments are available, the patient's preference should be taken into account.

However, do not use a comma when clauses with subordinating conjunctions are followed by subject–auxiliary verb inversions.

Not until a systems approach is taken can research culture be improved.

- adjunctive adverbial phrases, such as those that indicate time or location

Between 2000 and 2023, the journal published 1,627 original research articles.
By the end of 2022, all 900 of the experiments had been completed.
In the last section, we will discuss some future research directions.

However, do not use a comma when adjunctive adverbial phrases are followed by subject–auxiliary verb inversions.

Only recently have flight restrictions been lifted.

- conjunctive adverbial phrases, such as those that indicate logical relationships

However, the researchers determined that the data were inconclusive.
Consequently, the university submitted additional grant applications.

- disjunctive adverbial phrases and clauses, such as those that indicate stance

In theory, the process should be simple.
Unsurprisingly, the results of the new study mirrored those of studies conducted during the previous 20 years.
To be honest, we did not originally consider that possibility, but we have now included it in the model.

- premodifying clauses

To determine whether a conflict of interest exists, the editors questioned the authors and examined the evidence.
Prompted by questionable data from the pilot study, the researchers decided to revise the protocol before launching a full-scale clinical trial.
Given that the International Space Station is in low orbit, it is regularly visible to the naked eye from Earth.

A major concern with premodifying clauses is that if they are constructed incorrectly, the clauses could end up becoming dangling modifiers.

not After collecting the data, a normality test was performed. (Sentence implies that the data performed the test.)
instead After collecting the data, we performed a normality test.

(3) To set off nonrestrictive (ie, nonessential) postmodifying clauses from independent clauses.

They used several qualitative methods, including focus group discussions and in-depth interviews.
The new journal agreed to publish the controversial study, ending the authors' 3-year effort to get their work accepted for publication.

(4) To separate quotations from attribution, regardless of whether the quotation is verbatim and placed between quotation marks or paraphrased and used without quotation marks. The comma is omitted if the quotation is introduced by the conjunction "that." For short verbatim quotations, a colon can be used as an alternative to the comma (see item No. 2 in Section 5.3.1.1, "General Uses").

"The meeting of two personalities is like the contact of two chemical substances: If there is any reaction, both are transformed," Carl Jung, MD, proposed.
According to Edwin Hubble, PhD, humans explore the universe with their 5 senses and call that adventure science.
Marie Curie, PhD, said, "Nothing in life is to be feared; it is only to be understood." *but* Marie Curie, PhD, said that nothing in life should be feared but instead it should be understood.

(5) To set off transitional or parenthetical words or phrases.

The opposite, however, is not necessarily true.
The next geological survey, of course, has to be conducted with more precise instruments.

In the end, 8 samples may not be enough.
Keys provide, except in the most specialized works, a useful means of identification.

(6) To set off a short word of address or emphasis interjected into a sentence.

Well, correlation does not imply causation.
Remember, students, you have come here to learn.
Penicillin, alas, was in short supply in the region.

(7) To separate elements to clarify meaning.

In all, 8 experiments were performed. *not* In all 8 experiments were performed.

(8) To separate a nonrestrictive descriptive phrase or clause from the rest of the sentence. Place commas both before and after the phrase or clause if it is in the middle of a sentence. However, use a comma only before the phrase or clause if it ends a sentence.

The cells, which came from the institute's depository, were infected.
The report, currently under review, will be completed soon.
The structure consists of 5 layers, which are indicated in Figure 4 by arrows.

(9) To separate a nonrestrictive appositive from the rest of the sentence. Commas are placed before and after the nonrestrictive appositive if it is in the middle of the sentence.

Mazin Butros Qumsiyeh, a zoologist, founded the Palestine Museum of Natural History.
Physicists Hendrik "Hans" Kramers and Werner Heisenberg shared a mentor, Niels Bohr.

(10) To separate 2 or more adjectives before a noun or noun phrase. Do not place a comma between an adjective and a noun or between an adjective and a noun phrase. Whether a comma is needed between adjectives can be determined by substituting the word "and" for the comma and checking the sense.

The star is a main-sequence, low-mass, K-class red dwarf.
not The star is a main-sequence, low-mass, K-class, red, dwarf.

(11) To separate contrasting expressions and interdependent clauses.

The more packaging that is recycled, the less that will be discarded in landfills.

(12) To indicate an elliptical construction.

The first round of review usually takes 3 months; the second, 3 weeks.

(13) To separate different elements of geographic designations, including the elements in addresses. In general, when the names of municipalities are followed by the names of regions, such as states, providences, and nations, commas are placed before and after the regional names. In addresses, omit the comma between regional names and postal codes.

The Aswan Museum in Elephantine, Egypt, is located near the ruins of Abu.
The main address for the National Institutes of Health is 9000 Rockville Pike, Bethesda, MD 20892.

(14) To set off the year in a date expressed as month, day, and year. Commas are used both before and after the year when the date includes the month and day, but the commas are omitted when only the month and year are stated.

She remembered that November 18, 2022, was a special day.
She remembered that November 2022 was a special month.

The month-day-year convention is not recommended for scientific writing. Instead, dates should be written in the sequence of day, month, and year or year, month, and day (see Section 13.3.1, "Days and Months"). Commas are not used in either of those preferred conventions.

15 April 2022 *or* 2022 April 15 *not* 15 April, 2022 *and not* 2022, April 15

When a day of the week precedes a date, place a comma before and after the date.

Monday, 5 June 2023, *or* Monday, 2023 June 5, *not* Monday, 5 June 2023 *and not* Monday 2023 June 5

(15) To separate every 3 digits of numbers (see Section 12.1.3.1, "Breaks within a Number").

12,578,896   782,378   6,193

Although this style manual prefers this convention, it is not a universal style. The Système International d'Unités[5] (SI), for example, does not separate the digits in numbers with 4 or fewer digits (ie, ≤9999). For larger numbers, the SI separates every 3 digits with a thin space instead of a comma (ie, 64 721 493).

(16) To separate abbreviations of academic degrees, certifications, fellowships, and other credentials from names.

> C. Everett Koop, MD, DSc, FACS, was among the most recognized surgeons general of the US Public Health Service.
> Apollo astronaut Harrison Schmitt, PhD, was the first scientist to land on the moon.
> In a June 1970 article, Jeanne Pougiales, CRNA, chronicled the early reliance on nurse anesthetists at the Mayo Clinic.

5.3.3.3 Optional Uses

A comma is optional in the following instances.

(1) To separate 2 relatively simple and short independent clauses connected by a coordinating conjunction if the lack of a comma would not produce ambiguity.

> The survey was completed, and we went home. *or* The survey was completed and we went home.

(2) To set off a short introductory phrase or clause if the comma would not contribute to clarity or ease of reading.

> Despite the pause, the run was successful. *or* Despite the pause the run was successful.

(3) To separate 2 subordinating conjunctions (eg, between "that" and "for" and between "that" and "because") when the second one introduces a dependent clause that, in turn, modifies another dependent clause. In this situation, the comma between the 2 conjunctions is commonly omitted.

> The physicists stressed that for the results to be reproducible, maximum force needs to be calculated. *or* The physicists stressed that, for the results to be reproducible, maximum force needs to be calculated.

5.3.3.4 Incorrect Uses

A comma should not be used for the following purposes.

(1) To separate a restrictive subordinate clause from a preceding independent clause, except if needed for clarity.

> The authors celebrated because their paper was finally accepted.
> *not* The authors celebrated, because their paper was finally accepted.

(2) To separate 2 dependent clauses joined by a coordinating conjunction.

> The authors realized that their abstract was too long and they would need to rewrite it.
> *not* The authors realized that their abstract was too long, and they would need to rewrite it.

> The study was delayed while the researchers waited for funding and the institutional review board assessed the protocol.
> *not* The study was delayed while the researchers waited for funding, and the institutional review board assessed the protocol.

Inserting a comma between 2 dependent clauses is among the most frequent mistakes made with commas. This error, which creates sentence fragments, is especially common in sentences with long dependent clauses. To avoid this mistake, determine whether the subject and verb of the main clause apply to the 2 dependent clauses or whether the second clause is a complete statement by itself. In the first example above, both dependent clauses stem from the sentence's main clause, "The authors realized that," while in the second example, both dependent clauses flow from "The study was delayed while."

> The archaeologists explained that excavated bone is traditionally cleaned by using a neutral detergent or applying alcohol or other chemical solvents.
> *not* The archaeologists explained that excavated bone is traditionally cleaned by using a neutral detergent, or applying alcohol or other chemical solvents.

> We excluded patients who did not attend the follow-up visit or who discontinued treatment.
> *not* We excluded patients who did not attend the follow-up visit, or who discontinued treatment.

However, when 3 or more dependent clauses are used in a series, they should be separated by commas.

> Microbial growth increased when the temperature was increased, pH was decreased, and nutrient levels were increased.

(3) To separate a subordinating conjunction of an embedded dependent clause from a preceding coordinating conjunction that joins 2 independent clauses.

> The researchers initiated the second reaction, but because the stopwatch was broken, they had to abandon the test.
> *not* The researchers initiated the second reaction, but, because the stopwatch was broken, they had to abandon the test.

(4) To signal the start of an embedded adverbial phrase or nonfinite clause after a coordinating conjunction that joins 2 independent clauses.

> We will recruit participants by email, and after checking their eligibility, we will randomly assign them to the 2 study groups.
> *not* We will recruit participants by email, and, after checking their eligibility, we will randomly assign them to the 2 study groups.

(5) To separate a subject clause from the predicate of the sentence or to introduce an object clause starting with "that."

> Where the plant grew was fertile ground. *not* Where the plant grew, was fertile ground.
> He said that the bird was a flicker. *not* He said, that the bird was a flicker.

(6) To set off a restrictive appositive (ie, a word or phrase added to a generic term to convey a specific meaning).

> The species *Bombyx mori* is distinguished from others in the genus by . . .
> *not* The species, *Bombyx mori*, is distinguished from others in the genus by . . .

Without commas, the first sentence indicates that *Bombyx mori* is one of many species. With commas, the second sentence indicates that *Bombyx mori* is the only species.

(7) To separate digits in page numbers, address numbers, ZIP Codes or other postal codes, years, and accession numbers used for continuous DNA segments of genomes.

> p 6984 *not* p 6,984
> 10779 Glenwood Avenue *not* 10,779 Glenwood Avenue
> ZIP Code 64114 *not* 64,114
> CE 1066 *not* CE 1,066
> 4000 BCE *not* 4,000 BCE
> CAAA01119629 *not* CAAA,0111,9629

(8) To separate family name suffixes from surnames.

> Earl Wilbur Sutherland Jr *not* Earl Wilbur Sutherland, Jr
> Edmund Converse Peirce II *not* Edmund Converse Peirce, II

5.3.3.5 Specialized Uses

(1) To separate author names in end references and the components of an organization author in references (see Sections 29.3.3.1.1, "Personal Authors" and 29.3.3.1.2, "Organizations as Authors").

> Husain N, Hara T, Sullivan P. Wind turbulence over misaligned surface waves . . .
> American Academy of Pediatrics, Council on Sports Medicine. Care of the young athlete . . .

(2) Used without spaces to separate elements in symbolic representation of human chromosome aberrations (see Table 21.34).

> 46,XX,t(4;13)(p21;q32)

(3) Followed by a space, to separate linked genes in some organisms.

> *CHX1, EST1/CHX1, EST1*

(4) Used without spaces to separate locants in chemical formulas (see Section 17.3.4, "Locants").

> 4,5-Difluoro-2-nitroaniline   *N,N,N′,N′*-Tetramethyldiaminomethane

(5) In patent numbers, to separate every 3 digits of numbers.

> US patent 10,660,875

(6) In some European conventions for numbers, as the decimal point. The International Organization for Standardization recommends this convention for quantities and units.[6] In contrast, SI and the National Institute of Standards and Technology in the United States recommend using an on-the-line period or a comma for the decimal marker based on the customary use in the language and context of the manuscript.[5] This style manual recommends using an on-the-line period instead of a comma as the decimal marker (see Section 12.1.3.1, "Breaks within a Number").

5.3.4 Quotation Marks

As outlined in Section 10.2.1, "Run-in Quotations," American and British conventions vary in how they use quotation marks. The American convention uses double quotation marks for a primary quotation and single quotation marks for a quotation within a quotation. In contrast, British style uses single quotation marks for a primary quotation and double quotation marks for a quotation within a quotation.

In addition, in American usage, periods and commas at the end of quoted material are always placed before the closing quotation mark, regardless of whether the quoted material is a full or partial sentence. In British style, periods and commas are placed before the closing quotation mark when the quoted material is a full sentence and after the quotation marks when the quoted material is a sentence fragment, phrase, or single word. In both conventions, colons, semicolons, and dashes are placed after quotation marks, unless those punctuation marks are part of the quoted material. The American and British styles agree that question marks and exclamation points are placed before closing quotation marks if the quoted material constitutes an entire interrogative, exclamatory, or imperative statement and outside quotation marks if the quote is a sentence fragment, phrase, or single word.

This style manual recommends that each publication select either the American or British convention and consistently adhere to it. This manual applies the American convention throughout to demonstrate how to consistently adhere to one style.

5.3.4.1 Double Quotation Marks (" ")

Also known as "double speech marks" and "double inverted commas," double quotation marks are used mainly to delineate quoted words, terms, or longer elements of text. These uses are discussed in detail in Section 10.2, "Excerpts and Quotations." A few specialized uses are noted here.

(1) To indicate the title of a journal article, a book chapter, or a series title in running text. However, quotation marks are omitted from these titles in end references.

> The chapter "Geographic Designations" is at the end of Part 2 of this manual.

(2) To suggest a nonstandard or ironic use of a word or term.

> The politicians told us that they "cared" about big science.

(3) To indicate a word or phrase used as such (eg, as an example or explanation). This style manual recommends using quotation marks instead of italics for words as words and phrases as phrases. This convention eliminates the potential for confusion with the italicization required for scientific conventions. Moreover, the content is easier to read when it is in quotation marks rather than italics.

> Avoid the sloppy use of "impact," "case," and "individual."

5.3.4.2 Single Quotation Marks (' ')

Also known as "single speech marks" and "single inverted commas," single quotation marks are used primarily to enclose a quotation within another quotation (see Section 10.2, "Excerpts and Quotations"). When a quotation within a quotation begins or ends at the same point as the encompassing quotation, separate the single quotation mark from the double quotation mark with a thin space, which can be created with Unicode 2009.

> To support his case, "Fletcher Watson, PhD, attributed to Harvey H. Nininger 'half of the meteorite discoveries in the world.' "
>
> In their landmark paper proposing the structure of DNA, James Watson, PhD, and Francis Crick, PhD, FRS, commented, "The configuration of the sugar and the atoms near it is close to Furberg's 'standard configuration.' "

A scientific use of single quotation marks is to enclose the names of plant cultivars, also known as "fancy names" (see Section 22.4.1.5, "Names of Cultivated Plants").

> 'American Dewberry'   'Era'   'Proctor'

5.3.5 Dashes

Four lengths of dashes are used in printed text: the en dash, the em dash, the 2-em dash, and the 3-em dash. This style guide recommends the American convention of using dashes without spaces, as opposed to the British convention of placing a space before and after a dash.

5.3.5.1 En Dash (–)

The en dash, also known as the "en rule," is approximately the width of the capital letter "N." It can be created with Unicode 2013 under the symbols tab in most word-processing programs.

Similar to the hyphen, the en dash can serve as a connector (see Section 5.4.1, "Hyphen"). Unlike the hyphen, the en dash also indicates interruptions or omissions, as do the other 3 kinds of dashes. Whether to use an en dash or a hyphen is more of a concern for manuscript editors than authors. Authors can rely on manuscript editors to spot any hyphens that should be converted to en dashes.

The en dash has the following main uses.

(1) To link 2 words or terms representing items of equal rank, including compound modifiers. In this construction, the en dash is interpreted to mean "and" or "to."[7]

    author–editor relationship   hexane–benzene solvent
    cost–benefit analysis   north–south avenues
    gas–liquid chromatography   the Río San Juan–Lake Nicaragua route

For this purpose, en dashes also are used with the adjectival forms of country names to indicate a connection between 2 or more countries. However, neither en dashes nor hyphens should be used for terms such as "African American," "Mexican American," and "French Canadian" when referring to people living within a country (see Section 8.3, "Human Groups").

    Italian–Canadian relations *but* the Italian Canadian immigrant population
    Chinese–Australian agreements *but* Chinese Australians

(2) To connect names in eponymous terms attributed to 2 people. However, do not add an en dash to a nonhyphenated proper name.

    Michaelis–Menten kinetics *named after* Leonor Michaelis, MD, and Maud Leonora Menten, MD, PhD
    Mann–Whitney *U* test *named after* Henry B. Mann, PhD, and Donald R. Whitney, PhD
    but Bence Jones protein *named after* Henry Bence Jones, MD, MAFRCP, FRS

(3) As a connector to join a separable affix or other word to a modifier that consists of more than one word, regardless of whether the modifier itself is hyphenated.

    the Winston-Salem–Raleigh group of scientists (from 2 cities in North Carolina in the United States: Winston-Salem and Raleigh)
    ex–editor in chief
    non–double-barreled names
    a sugar maple–dominated forest *but* a maple-dominated forest
    pre–Song Dynasty ceramics

In contrast, when modifiers of equal rank are linked with an en dash, the separable affix should use a hyphen, not an en dash.

    non-Michaelis–Menten kinetics
    non-win–win situation

(4) Where space is limited (eg, tables), to link numbers representing a range of values. However, use the word "to" instead of an en dash if either number in the range is negative or if the range of numbers follows the word "from." Similarly, if the range follows the word "between," use "and" instead of an en dash.

    10–14 g *but* between 10 and 14 g
    20–45 mm *but* from 20 to 45 mm
    −5 to 25 °C *not* −5–25 °C
    −3.5 to +4.7 kg *not* −3.5–+4.7 kg
    pages 6–10 *or* from page 6 to page 10 *not* from page 6–10

(5) For year ranges. For this purpose, use the entire year for both the first and second date in a range. Use "and" or "to" instead of an en dash if the range of years follows the word "between" or "from," respectively. Similarly, use "and" or "to" in place of an en dash if the years span 2 eras (eg, from a year before the common era to one in the common era).

    She served as the academy's 2021–2023 president.
    not She served as the academy's 2021–23 president.

> Between 1997 and 2017, the *Cassini* spacecraft traveled approximately 7.8 Tm during its mission to Saturn.
> *not* Between 1997–2017, the *Cassini* spacecraft traveled approximately 7.8 Tm during its mission to Saturn.

> Classical antiquity was roughly 776 BCE to CE 476.
> *not* Classical antiquity was roughly 776 BCE–CE 476.

(6) For page ranges in end references (see Section 29.3.3.8.1, "Location within a Work").

> J Athl Train. 2013;48(5):716–720.

(7) As a minus symbol if a separate minus symbol (eg, Unicode 02212) is not available in a word-processing system.

(8) To indicate an empty cell in a table.

(9) To represent chemical bonds.

> $C_6H_5CO-O-COCH_3$

5.3.5.2 Em Dash (—)

As implied by its name, the em dash, or "em rule," is approximately the width of the capital letter "M." The correct character is created by Unicode 2014, which is usually available in word-processing software under the symbols tab. Alternatively, the em dash may be represented by 2 adjacent hyphens with no spacing.

The em dash has several main uses.

(1) To set off elements within a sentence that express a parenthetical break in meaning. The setoff statement usually defines, elaborates, emphasizes, explains, or summarizes the preceding information in the sentence. In this function, em dashes are akin to commas and parentheses, but the em dash is the most emphatic of those 3 kinds of punctuation.

> Deep learning—a subtype of advanced machine learning typically involving artificial neural networking algorithms—has the potential to profoundly affect all sectors of society.
> Osteoporosis—perhaps the most common postmenopausal disorder—cannot be diagnosed without specialized imaging equipment.

(2) To make a sharp break from the central message of the sentence. The content between em dashes can be an afterthought, an abrupt shift in thought, an emphasized concept, or other extra information.

> Traveling to neighboring solar systems—contrary to what science fiction movies suggest—would take decades, provided we had the technology.

(3) To set off content that contains parentheses, colons, or commas.

> In scientific writing, it is essential to distinguish sets from subsets—for example, patients with acquired immunodeficiency syndrome are a subset of patients infected with the human immunodeficiency virus—so that subsets are not treated as if they are independent from the sets they belong to.

(4) To indicate interruptions in speech.

> The dean said, "I am greatly impressed—maybe 'astonished' would be more accurate—with our newest research."

(5) To set off introductory elements from a sentence that explains the significance of the introductory elements.

> The emerald ash borer beetle, the starling, and the spongy moth—these pests all came from abroad.

(6) To separate a stand-alone quotation (eg, an epigraph on a book's title page) or editorial statement from its source.

> Nothing in science has any value to society if it is not communicated, and scientists are beginning to learn their social obligations.—Anne Roe, PhD, *The Making of a Scientist*
> The opinions expressed in this letter are those of the author and do not represent journal policy.—The Editor

5.3.5.3 Two-Em Dash (——)

The 2-em dash, created with 2 em dashes or 4 hyphens in a row, is used to represent unknown or missing letters in words.

(1) To indicate illegible letters in the original text.

> My dear Mr Darwin, Your st—— about our descent is . . .

(2) To preserve a person's anonymity, such as the name of a study participant in a clinical research trial. However, it may be preferable to use wording that avoids the need to imply a name.

> Patient M—— was the first in this series to develop the complication of . . . *or* The first patient in this series to develop the complication of . . .

5.3.5.4 Three-Em Dash (———)

Created with 3 em dashes or 6 hyphens in a row, the 3-em dash has limited uses.

(1) To complete incomplete spoken statements.

> His chronic harping on ecology was a steady pain in my ——— .

(2) To represent the names of authors in end references when the authors are the same as those in the preceding reference. In this use, the 3-em dash is followed by a period. This style manual does not recommend this use for the 3-em dash.

> Carson R. The edge of the sea . . .
> ———. Under the sea-wind, a naturalist's picture of ocean life . . .

5.3.6 Parentheses ( )

Parentheses are also called "parenthesis marks" and "round brackets." In the United Kingdom, parentheses are commonly called "brackets."

5.3.6.1 General Uses

(1) To enclose a parenthetical word, phrase, or sentence.

> The most common use of parenthesis marks (parentheses) is . . .
> Urinary incontinence (and disorders discussed in other chapters) is a frequent problem . . .

The text within parentheses is grammatically independent of the rest of the sentence. As a consequence, in the second example above, the singular verb "is" is correct because the subject is the singular "urinary incontinence," not "urinary incontinence and disorders."

When the interruption in the text needs a greater or lesser emphasis than that provided by parentheses, use em dashes or commas, respectively (see Sections 5.3.5.2, "Em Dash" and 5.3.3, "Comma").

(2) To introduce an initialism, another abbreviation, or a symbol after an entire term so that the abbreviation or symbol can be used later in place of the entire term. This convention

should be used only to introduce abbreviations and symbols that are used later in the manuscript.

> Arteriovenous malformation (AVM) occurs most frequently in the brain and spinal cord. AVMs are congenital defects between arteries and veins.
> The National Library of Medicine (NLM) in the United States is part of the National Institutes of Health (NIH). The NLM and NIH's other institutes fall under the umbrella of the US Department of Health and Human Services.

(3) Used with numerals or letters to introduce elements of a list or series, usually in a vertical presentation. Although the single right parenthesis, as in "1)" and "A)," also can be used for this purpose, this style manual recommends using full parentheses.

> Three projects were funded:
> (1) Philadelphia: Archaeology of the Late 17th Century
> (2) New York: Sites of Black Cemeteries
> (3) San Francisco: Pre–Gold-Rush Buildings

> The inclusion criteria were as follows: (1) minimum patient age of 18 years, (2) histologically proven glioblastoma, and (3) complete data sets.

(4) To indicate an in-text reference using the name–year citation system (see Section 29.2.1.2, "Name–Year")

> As noted recently (Evans and Armus, 2021), the . . .

(5) To enclose directive text, which instructs readers to take specific action (eg, a cross-reference directing readers to text elsewhere in the document).

> As discussed elsewhere (see Section 17.3, "General Conventions of Nomenclature"), do not capitalize a chemical name unless it begins a sentence, title, or section heading.

(6) To indicate an optional plural for a noun that could be either singular or plural.

> Ask your coauthor(s) to sign the declaration.
> The relevant guideline(s) can be found in each scientific field's recognized style manual(s).

Although this convention is often used when writers are unaware of how many persons or things are involved, it results in awkward sentences and may suggest that fact-checking was insufficient. It is preferable to find out how many persons or things are involved to determine whether singular or plural nouns are appropriate. When the number involved cannot be determined, using plural nouns is generally preferred.

(7) To enclose the 3 digits that make up the area code in a telephone number in the United States and Canada.

> +1 (555) 555-5555

5.3.6.2 Specialized Uses

(1) To enclose such elements as issue numbers, series names, and country abbreviations in end references (see Sections 29.4.1.6, "Volumes and Issues of Journal Articles," 29.3.3.9, "Series," and 29.3.3.1.2, "Organizations as Authors").

> Yazdani S, Helwany S, Olgun CG. The mechanisms underlying long-term shaft resistance enhancement of energy pile in clays. Can Geotech J. 2021;58(11):1640–1653. https://doi.org/10.1139/cgj-2019-0236
> Asimellis G. Wave optics. SPIE; 2020. 396 p. (Lectures in optics; vol 3). https://doi.org/10.1117/3.2506314
> O'Mara SM, editor. Collective memory. Elsevier; 2022. (Progress in brain research; vol 274.) https://www.sciencedirect.com/bookseries/progress-in-brain-research/vol/274/issue/1
> National Academies of Sciences, Engineering, and Medicine (US). Radioactive sources: applications and alternative technologies. The National Academies Press; 2021. https://doi.org/10.17226/26121

(2) To enclose mathematical elements to indicate their order of operation (see Section 12.3.2.1, "Fences").

$$z = k(a + b + c) \qquad y = 3.47(x + z)^3$$

(3) To show that a compound's chemical formula contains 2 or more of a specific polyatomic subunit. The parentheses enclosing the chemical symbol of the subunit are followed by a subscript number indicating the number of such subunits in the compound. Similarly, for homopolymers, parentheses surround the condensed structural formula of the monomer, followed by the subscript "n."

$$K_4Fe(CN)_6 \qquad Ca(NO_3)_2 \qquad (CH_2CHCH_3)_n$$

(4) To indicate branching in condensed structural formulas in organic chemistry.

$$CH_3CH(CH_3)CH_2CH_3 \qquad CH_3CH(OH)CH_3$$

(5) When needed for clarity, to indicate the chemical formula of a polyatomic anion, with the anion's overall charge indicated as a superscript number after the parentheses. Note that the superscript number is not used if it is "1."

$$(SO_4)^{2-} \qquad (MnO_4)^-$$

(6) To enclose roman numerals serving as oxidation numbers for positive elements of a compound. For this use, no space is used before the opening parenthesis, but a space is used after the closing parenthesis.

Pb(IV)   Pb(IV) oxide   lead(IV) oxide

(7) To enclose side chains substituted in amino acids (see Section 17.13.2.5, "Side Chains").

Cys(Et)

(8) To enclose the stereochemical descriptors "R," "S," "E," and "Z" (see Section 17.11.1, "Organic Compounds").

(2S)-alanine   (E)-2-butene

(9) For immunoglobulin notations (see Section 23.5.6.1, "Immunoglobulins").

$F(ab')_2$ IgG(Pr)

(10) For notation of specificities of histocompatibility antigens.

HLA-Bw56(w22)

(11) To enclose symbols indicating structural alteration of a chromosome (see Tables 21.31 and 21.33).

46,XX,t(4;13)(p21;q32)

(12) To enclose the name of the author of an original taxonomic description when a species is transferred to another genus (see Sections 22.4.1.1.1, "Transfer of a Species to Another Genus," and 22.5.3, "Transfer of a Species to Another Genus").

Tetraneuris herbacea Greene, renamed as *Hymenoxys herbacea* (Greene)
Taenia dunubyta Rudolphi, renamed as *Hymenolepis diminuta* (Rudolphi)

5.3.7 Square Brackets [ ]

Although the term "brackets" is often used for this pair of marks, the term "square brackets" clearly describes these punctuation marks and distinguishes them from "round brackets," more commonly known as "parentheses"; the "bracket pair," often called "curly brackets" or "braces"; and "angle brackets."

5.3.7.1 General Uses

(1) To demarcate text or letters added to quoted text to amplify or clarify the original text.

> Harvey Cushing, MD, commented, "When [William] Osler moved [to Baltimore], he was not risking his future."
> The last entry in his journal was: "I took lunch with D[arwin] and spotted P[otter] with his mistress E[laine Smythe]."

(2) To demarcate editorial comments and other nonoriginal material inserted into quotes, reprints, and anthologies.

> "When I was in London, I briefly met Darwynne [sic], that horrible chap who claims we are descended from monkeys."

"Sic," which means "thus" in Latin, is often included in square brackets as an editorial comment to point out an error in the original writing, such as a misspelled word or a grammar mistake. One drawback to this convention is that it assumes readers know the correct form. If the correction for a mistake is not readily apparent, consider following an error with the correction enclosed in square brackets.

> "I briefly met Darwynne [Charles Darwin], that horrible chap who claims we are descended from monkeys."

(3) To enclose a parenthetical statement within a parenthetical statement that is enclosed by parentheses.

> Members of the public should check that online information on COVID-19 comes from reliable sources (eg, the World Health Organization [WHO]).

(4) To show a translation of original text.

> The article was titled "Repenser la Compréhension de l'Ordonnance: l'Exemple des Soins aux Sourds" ["Rethinking the Prescription's Comprehension: An Example of Care Centers for Deaf People"].

(5) To enclose a phonetic transcription of a difficult-to-pronounce word.

> *Carcharodon* [kär′karə͵dän] is a genus of shark.
> Othniel C. Marsh, MA, the first professor of paleontology in the United States, named the dinosaur *Theiophytalia* [THEE-oh-fie-TAL-ya].

5.3.7.2 Specialized Uses

(1) For multiple bracketing in mathematical expressions and chemical names (see also Sections 12.3.2.1, "Fences," and 17.3.3, "Enclosing Marks," respectively).

> $z = k[(a + b) - y(c + d)]$
> bis(bicyclo[2.2.2]octadiene)platinum

The general sequence for multiple bracketing in mathematics is {[( )]}, in which formulas in parentheses are enclosed by square brackets and formulas in square brackets are enclosed by braces. This is the reverse of the usual sequence in text.

In some chemical usages, internal brackets must be used, and parentheses may surround these square brackets, as in ([ ]). For more on the sequence of enclosures in chemistry, see Section 17.2.3, "Structural Formulas."

(2) Used with parentheses to indicate interval notation as follows:

> [a, b] means an interval of real numbers between and including a and b
> (a, b) means an interval of real numbers between but excluding a and b
> (a, b] means an interval excluding a but including b
> [a, b) means an interval including a but excluding b

An alternative system uses only square brackets: [a, b];]a, b[;]a, b]; and [a, b[, respectively.

(3) As part of chemical concentrations.

   [Na$^+$]   [HCO$_3^-$]

(4) To enclose isotopic prefixes (see Section 17.10.2, "Isotopically Labeled Compounds").

   [^{14}C$_2$]glycolic acid   [^{32}P]AMP

(5) To enclose complex substituent prefixes containing internal parentheses.

   N,*N*-bis[(phenylmethoxy)carbonyl]alanine

(6) To enclose connecting points in fusion nomenclature.

   benz[*a*]anthracene

(7) To enclose numbers used as ring-size indicators in spiro and bicyclo names (see Section 17.6.3.2, "Bicyclic and Polycyclic Ring Systems").

   spiro[2.3]hexane   bicyclo[2.2.2]octane

(8) To denote DNA and RNA sequence variations.[7]

   NC_000003.12:g.63912687AGC[(50_60)] (square brackets indicate that 50 to 60 AGC copies are contained on one allele with DNA variants)
   LRG_199t1:c.[296T>G;476T>C;1083A>C];[296T>G;1083A>C] (square brackets enclose multiple substitutions on both alleles with DNA variants)
   LRG_199t1:r.[76a>u;103del] (square brackets enclose a substitution and a deletion on one allele with RNA variants)
   NM_004006.2:r.[76a>u];[76a>u] (square brackets indicate the same substitution on both alleles with RNA variants)

(9) To denote proteins encoded by DNA sequence variants.[7]

   NP_003997.1:p.[Ser68Arg;Asn594del] (square brackets enclose both a substitution and a deletion on one allele with variants)
   NP_003997.1:p.[Ser68Arg];[Asn594del] (square brackets enclose a substitution on one allele with variants and a deletion on another allele with variants)

5.3.8 Braces { }

Braces are also known as "curly brackets" and "curly braces." In mathematics and chemistry, paired left and right braces are used to signify aggregation, enclosing elements already enclosed by both parentheses and square brackets (see Sections 12.3.2.1, "Fences," and 17.3.3, "Enclosing Marks"). In addition, a single multiline brace can be used to aggregate elements in adjacent lines, but this use is uncommon outside mathematics (see Section 5.5.1, "Brace").

5.3.9 Angle Brackets ‹ ›

Angle brackets are used in specific contexts to signify instructions. For example, they may be used to enclose the name of a computer key to be pressed in program instructions.

   Press ‹Enter› to start searching.

In addition, angle brackets are used in coding and metadata tagging.

Note that the angle brackets used as enclosures in mathematics and physics (ie, "⟨ ⟩") are narrower and taller than standard angle brackets. The opening mathematical angle bracket ⟨ is created with Unicode 27E8, and the closing bracket ⟩ is created with Unicode 27E9.

5.3.10 Vinculum (‾‾‾‾)

The vinculum (meaning "a bond," from "vincire," the Latin word for "to bind") can be used in mathematics to indicate a fraction.

$$\frac{a-b}{c+d}$$

The vinculum also can perform the same function as parentheses in mathematics, but parentheses are preferred.

$$a - \overline{b - c} \text{ or } a - (b - c)$$

A vinculum over a single letter can represent the arithmetic mean of a variable, the complex conjugate, or the complement of a set. In statistics, the mean value of a characteristic in a sample of the population is known as "x-bar" or "$\bar{x}$."

Another use for the vinculum is to indicate a line segment in geometry.

$\overline{AB}$ (denotes a straight line extending from point A to point B)

5.4 TERM AND WORD MARKS

5.4.1 Hyphen (-)

The hyphen is akin to the slightly longer en dash in that both serve as connectors. However, the hyphen cannot be used to indicate an interruption or an omission, while the en dash can (see Section 5.3.5.1, "En Dash"). In general, authors need not belabor whether to use a hyphen or an en dash. Hyphens will suffice in submitted manuscripts for the purposes of both punctuation marks. Manuscript editors will convert any hyphens that should be en dashes.

The hyphen is also covered in Sections 6.2, "Formation of Scientific Terms," and 6.3, "Word and Term Division."

5.4.1.1 General Uses

The 2 main uses for the hyphen are dividing and joining.[1] In their dividing function, hyphens mark word divisions for line breaks in text. Authors should not insert hyphens to divide words in manuscripts because line endings will differ when the manuscripts are typeset. Hyphens inserted at the end of lines during the manuscript stage may end up misplaced, in the middle of lines in the typeset stage. Once the manuscript is typeset, manuscript editors and proofreaders will correct hyphenation in line endings when software fails to adhere to the following guidelines:

(1) For words and terms that are spelled with hyphens, break lines after the hyphens (eg, "proto-oncogene" and "cost-effective treatment").
(2) For lines that end with words containing prefixes or suffixes, break the words after their prefixes or before their suffixes, if possible (eg, "bio-marker" and "electro-lyte").
(3) Break other words before or after the syllables with either the primary or secondary stress mark as designated in pronunciation guidelines of the specialized or standard dictionary adopted by the intended publisher of the manuscript (eg, "cardio-vascular" and "hy-drogen").

In their joining function, hyphens form terms by connecting some prefixes and suffixes to stem words, by uniting 2 or more modifiers that precede a noun, and by converting verbs into nouns. There is also a tendency in English for 2 closely related stem words to evolve into a single term. Such terms often undergo a transitional period during which the 2 stem words are separated by a hyphen (eg, "health care" to "health-care" to "healthcare").

(1) To connect prefixes and suffixes to stem words in such situations as the following: The stem word is capitalized (eg, "pre-Colombian"); the prefix is "ex," "quasi," or "self" (eg, "ex-husband" and "self-inflicted"); and the suffix is "-elect" (eg, "president-elect").

(2) To prevent visual confusion when letters are repeated between a stem word and its prefix or suffix (eg, meta-analysis and shell-like). However, in American English, hyphens are often omitted in such terms (eg, "preeclampsia" and "coordinates"). For additional information, see Sections 6.2.1.1, "Nonhyphenated Prefixes," and 6.2.2, "Formations with Suffixes in Scientific Terms." To resolve uncertainties about whether to hyphenate a specific word with a prefix or suffix, consult a general or specialized dictionary, preferably the one used by the publication to which the manuscript will be submitted.

(3) For some compound terms, especially if using a space instead of a hyphen between the terms would give one of the stem words a meaning inappropriate to the context.

> light-year (the hyphenated compound is the distance light travels in 1 year; without the hyphen, "light year" could appear to mean a year that is not heavy with work or other responsibilities)
> cure-all (by adding a hyphen, the verbal phrase "cure all" is converted to a noun)
> has-been (the hyphen converts the present perfect progressive auxiliary verb "has been" to a noun)

(4) For verbs needing hyphens to convey their correct meanings.

> She re-covered the petri dish. *but* He recovered quickly from the operation.
> Such patients are usually re-treated. *but* The water retreated from the structures.

(5) For a compound modifier in which the second element is a past or present participle.

> all-encompassing strategy   well-established rules
> ill-advised procedure   well-known physicist
> seizure-inducing drugs

Omit the hyphen if the compound modifier follows the verb (ie, a predicate modifier) and the first component is an adverb (eg, "well"). In such constructions, it is clear that the adverb modifies the participle after it, not the subject of the sentence.

> The journal's instructions are well established.   He was well known in this field.

However, hyphens should be used in compound modifiers that are hyphenated in standard dictionaries, regardless of where the terms appear in sentences.

> The theory was well-founded.   Their study design is ill-defined.

Do not hyphenate adverbial or adjectival elements in compound modifiers if the adverb ends in "ly."

> the widely applauded conservation plan   a clearly described new species

(6) For compound modifiers when the nonhyphenated form could be ambiguous.

> a large-bowel obstruction (to distinguish an obstruction in the large bowel from a bowel obstruction that is large)
> low-frequency amplitudes (to distinguish amplitudes at low frequencies from frequency amplitudes that are low)

(7) When 2 or more hyphenated compound modifiers have the same root word, omit all but the last instance of the root word but retain the hyphens.

low-, medium-, and high-frequency amplitudes   sodium- and potassium-conserving drugs

This rule does not apply to unhyphenated compound words with the same root when the words are in a series. Such words should not be abbreviated by replacing the root with a hyphen in all but the last compound word.

not hypo- and hyperthermia *but* hypothermia and hyperthermia
not under- and overcharged *but* undercharged and overcharged

In addition, avoid splitting words that have a common hyphenated prefix.

not self-assessment and -evaluation *but* self-assessment and self-evaluation

(8) For modifiers consisting of both numeric values and units. Age terms require 2 hyphens.

a 5-g dose   a 10-woman team   a 3-planet solar system   a 50-year-old patient

(9) Between the numerator and denominator in a spelled-out fraction less than 1, unless either integer is already hyphenated.

one-third of the population   thirty-seven hundredths

For fractions larger than 1, see Section 12.1.2.3, "Fractions."

(10) For compound cardinal and ordinal numbers from 21 through 99 when spelled out at the beginning of a sentence.

Eighty-five samples were collected.
Twenty-first–century scientists have access to more data than did their predecessors.

For en dash usage with hyphenated compounds, see Section 5.3.5.1, "En Dash."

(11) For most terms in which the letter "e" is used at the beginning to stand for "electronic." The term "email" is an exception because it is widely used as a single word without a hyphen.

e-book   e-commerce   e-business   e-alert   email

(12) To separate the 3-digit exchange prefix from the 4-digit line number in telephone numbers in the United States and Canada.

+1 (555) 555-0000

Using hyphens to join words and terms is also discussed in Sections 6.2.1.2, "Hyphenated Prefixes," and 6.2.2.2, "Words as Suffixes." Additional guidance is available in such style guides as *The Chicago Manual of Style*,[4] *New Oxford Style Manual*,[8] *Editing Canadian English*,[9] the *United States Government Publishing Office Style Manual*,[10] and the *AMA Manual of Style*.[11]

5.4.1.2 Unnecessary Uses

Hyphens are not required in the following instances:

(1) For well-established compound terms. To determine which terms are well established, consult standard dictionaries, preferably the scientific and general dictionaries recommended by the publication to which the manuscript will be submitted.

freezing point determination   amino acid residues
potassium chloride absorption   Sertoli cell analysis

(2) In compounds with comparative or superlative adjectives.

better adjusted children   most evolved system

(3)   For Latin phrases used adjectivally.

> a post hoc hypothesis   in vitro testing   in situ cancer

(4)   For letters used as modifiers in scientific terms. But hyphens are used when these modifiers are part of adjectival phrases.

> LE cells *but* LE-cell rosettes
> T lymphocytes *but* T-cell lymphocyte antigen
> S curve *but* S-curve characteristics

(5)   To join a number, letter, spelled-out Greek letter, or combination thereof to the preceding word.

> type 2 diabetes mellitus
> level B2 proficiency
> amphotericin B
> interleukin 1 beta *but* IL-1β
> coronavirus disease 2019 *but* COVID-19

(6)   To separate year, month, and day in a date. Although the International Organization for Standardization[12] recommends formatting dates as YYYY-MM-DD with hyphens between the arabic numerals for year, month, and day, this style manual recommends writing months out fully or as 3-letter abbreviations and using either the sequence of day, month, and year or the sequence of year, month, and day (see Section 13.3, "Dates").

> 6 August 2023 *or* 2023 August 6 *not* 2023-08-06
> 18 Apr 1955 *or* 1955 Apr 18 *not* 1955-04-18

5.4.1.3 Specialized Uses

(1)   To represent single bonds in chemical or molecular formulas or names. (For double bonds, which use hyphens and double lines, see Section 17.11.1.1, "Double Bonds.")

> $(CH_3)_2$-CH-CH$_2$-CH(NH$_2$)-COOH

(2)   Between an element symbol (eg, "C" for "carbon") and the numeral designating a particular atom, as well as between an element's spelled-out name and the mass number of an isotope.

> C-3   iodine-131

(3)   Between 3-letter symbols representing amino acid residues in a peptide with a known sequence. Each hyphen represents a peptide bond (see Section 17.13.2, "Amino Acids, Peptides, and Polypeptides").

> Gly-Lys-Ala-His

(4)   To separate a chemical compound from a prefix that specifies the compound's molecular configuration, denotes an isomer, or serves as a locant (see Section 17.3.4, "Locants").

> *S*-benzyl-*N*-phthaloylcysteine   o-cresol   9-(1,3-dihydroxy-2-propoxymethyl)guanine

(5)   As one option for showing linkages of nucleotides in polynucleotides (see Section 17.13.10.2, "General Principles for Describing Chains").

> pG-A-C-C-T-T-A-G-C-A-A-T-Gp

5.4.2 Slash (/)

The slash has several synonyms, including "forward slash," "slant," "slant line," "virgule," "oblique, "oblique bar," "oblique mark," and "oblique stroke."

The solidus, or "shilling mark," is similar to the slash, except that the solidus has a greater slant than the slash (ie, " ⁄ " as opposed to "/"). The solidus was traditionally used for separating the numerator and denominator in a fraction (eg, 13 / 16) and for separating units of traditional British currency (ie, pounds, shillings, and pence). Both uses are uncommon today. Except for fractions inserted as special characters by word-processing programs (eg, ½ and ¾), fractions are typically created in text by using the slash instead of the solidus (eg, 3/5 and 7/19). In addition, in 1971, the United Kingdom changed its pound–shilling–pence system of currency to a decimal system for pounds and pence (eg, UK£ 1.25).

5.4.2.1 General Uses

(1) To indicate 2 terms that represent the elements of a ratio.

> The scientist/administrator ratio on the project was 3:1.

(2) To connect 2 hyphenated terms of equal significance.

> left-brain/right-brain task differentiation

(3) As a character in URLs for specific web pages.

> https://www.councilscienceeditors.org/publications
> https://press.uchicago.edu/ucp/books/book/chicago/S/bo13231737.html

5.4.2.2 Unnecessary Uses

Although the slash is used by many writers to indicate alternatives, better options are usually available. This style manual does not recommend using slashes for the following:

(1) For such terms as "and/or" and "he/she" when actual circumstances are in question, even though such terms appear in standard dictionaries.

    The imprecise term "and/or" should be avoided in scientific writing in favor of wording that clearly conveys the intended meaning. For example, "and/or" can be replaced by either "and" or "or," depending on which conjunction is more accurate for the situation. Another option is to list both alternatives followed by a comma and "or both."

> The patients were admitted for chronic obstructive pulmonary disease or congestive heart failure.
> *or* The patients were admitted for chronic obstructive pulmonary disease, congestive heart failure, or both.
>
> *not* The patients were admitted for chronic obstructive pulmonary disease and/or congestive heart failure.

    Rather than use the constructions "he/she" or "his/her" for a person of unspecified gender, it is preferable to rephrase a sentence as a plural construction if more than one person is involved. If only one person is involved, the "singular they" may be used if the individual's gender is unknown, undisclosed for confidentiality or study-masking purposes, or otherwise inappropriately described by the pronoun "he" or "she" (see Section 7.4.2, "Singular They"). The "singular they" may be used if a person self-identifies as "they," provided using that plural pronoun will not distort the scientific results.

> The study participants were asked whether they were allergic to penicillin.
> *not* Each study participant was asked whether he/she was allergic to penicillin.
>
> Any astronomer can locate Jupiter's great red spot if they have the right telescope.
> *not* Any astronomer can locate Jupiter's great red spot if he/she has the right telescope.

(2) For punctuating a series of 3 or more items. For this purpose, commas are preferred to slashes (see Section 5.3.3, "Comma").

> The geology tour of Pennsylvania included Pittsburgh, Philadelphia, and York.
> *not* The geology tour of Pennsylvania included Pittsburgh/Philadelphia/York.

(3) For connecting coordinate modifiers. For this purpose, en dashes or hyphens are preferred to slashes (see Sections 5.3.5.1, "En Dash," and 5.4.1, "Hyphen").

> The physician–patient relationship is exemplified by . . .
> *not* The physician/patient relationship is exemplified by . . .
>
> The university has launched a geology–geophysics program.
> *not* The university has launched a geology/geophysics program.

(4) To separate day, month, and year in a date consisting strictly of arabic numerals. This convention is not recommended for scientific writing. Instead, dates should be written in the sequence of day, month, and year or year, month, and day (see Section 13.3.1, "Days and Months"). Slashes are not used in either of those preferred conventions. In both conventions, the names of months are spelled out or abbreviated to the first 3 letters.

> 29 January 2023 *or* 2023 January 29 *not* 29/01/2023 *not* 2023/29/01 *and not* 01/29/2023
> 20 Jul 1969 *or* 1969 Jul 20 *not* 20/07/1969 *not* 1969/20/07 *and not* 07/20/1969

(5) To separate years in a range. Instead, this style manual recommends separating years with an en dash and using the entire year for both the first and second date in a range (see Section 5.3.5.1, "En Dash").

> 2020–2021 *not* 2020/2021 *and not* 2020/21
> 1698–1704 *not* 1698/1704

5.4.2.3 Specialized Uses

(1) As a symbol of division, either to convey an operation in a formula or to imply division by a unit of measure (see Sections 12.3.1.1, "Mathematical Operators," and 12.2.1, "Système International d'Unités (SI)," as well as Table 12.10). In the context of a mathematical operation, the slash means "divided by." When used with a unit of measure, the slash means "per."

> $1/4\,y = 3.5x/(a+b)$ kg·m/s^2 (for kilogram-meter per second squared)

(2) In expressions of rate or concentration (see Section 12.1.4.6, "Rates").

> 5 m/s   20 mol/L

To avoid mathematical ambiguity, use no more than a single slash in an expression of rate or concentration.

> 1.5 pCi/(km^2·yr) *or* 1.5 pCi · km^{-2}·y^{-1} *not* 1.5 pCi/km^2/yr

(3) To separate symbols for mutated genes on homologous chromosomes (see various tables in Chapter 21, "Genes, Chromosomes, and Related Molecules").

> *y w f*/*B*

(4) To separate clones of unique karyotypes in human chromosome nomenclature, with bracketed numbers indicating the number of cells in each clone.[13]

> 47,XX +21 [15]/46,XX [6]   50,X,+X,−Y,+10,+14,+17,+21[5]/46,XY[15]
> 45,X[5]/47,XYY[5]/46,XY[10]   46,XY,del(5)(q13q33),−7,+8[2]/46,XY[18]

(5) To separate symbols for homologous genes or alleles (see various tables in Chapter 21, "Genes, Chromosomes, and Related Molecules").

> *ac17*/*AC17*   *dhfrts-3*/*DHFRTS*   *BTU2*/*BTU2*

5.4.3 Vertical Bar (|)

The vertical bar, or "vertical line," has specialized uses in mathematics and in chemical and electrochemical notation. The vertical bar is usually used in pairs. For example, double vertical bars are used to enclose a matrix in running text (see Section 12.3.1.4,

"Matrices and Determinants"). Single vertical bars enclosing a term indicate its modulus, or absolute value or magnitude.

$$\| a_i b_i c_i \|$$
$$| -x | \text{ is equivalent to } | x |$$

A pair of single solid vertical bars indicates a phase boundary at each of the 2 electrodes in an electrochemical cell. In addition, a central broken bar indicates a porous partition, and a central double-broken bar indicates a salt bridge.[14]

$$Zn|Zn^{2+} \vdots Cu^{2+}|Cu \qquad Cu|Cu^{2+} \vdots\vdots Ag^{+}|Ag$$

In geometry, double vertical bars denote parallelism between 2 lines or line segments.

$$AB \parallel CD$$

5.4.4 Apostrophe (')

Apostrophes are used to form possessives (eg, "meteorologist's forecast" and "anthropologists' field trip"). Apostrophes are also used in forming contractions (eg, "I'd prefer" for "I would prefer"), but contractions may be inappropriate in some formal scientific writing. For more details on using the apostrophe, see Section 6.5, "Possessives."

5.4.5 Prime Symbol—Single and Double (' and ″)

The single prime symbol is not interchangeable with the single quotation mark or the apostrophe, nor is the double prime symbol interchangeable with a double quotation mark.

The double prime symbol should not be created by using 2 single prime characters in a row. The correct characters are available in most word-processing programs as special characters (ie, Unicode 2032 for prime and Unicode 2033 for double prime). If substitutes must be used for these characters, the substitutions should be indicated in the manuscript so that the publisher can change them during typesetting.

The most common general use for single and double prime symbols is to indicate feet and inches, respectively (eg, 5′4″). This style manual recommends using metric units instead (see Section 12.2.1, "Système International d'Unités (SI)," and Table 12.3).

The prime sign has 3 main scientific uses.

(1) With locants in chemical names.

   N,N'-dimethylurea

(2) In geographic coordinates to indicate divisions of a degree, with the single prime symbol used for minutes and the double prime used for seconds (see Section 14.2.1, "Latitude and Longitude").

   The meteorite was found at lat 52°33′05″N, long 33°21′10″E.

   Note that there is no space between the numerical value and the unit symbol for degree, minute, and second (ie, °, ′, and ″, respectively).

(3) In genetics to indicate directionality of nucleic acid sequences (eg, messenger RNA [mRNA] and the "sense," or nontemplate, strand of DNA), based on the numbering of carbon atoms in the pentose sugar molecule within nucleotide monomers. The letters representing nucleotide bases are read from the 5′ (pronounced "five-prime") end to the 3′ ("three-prime") end.

   mRNA: **DNA:**
   5′-CCACAACACGUA-3′   5′-CCACAACACGTA-3′

5.4.6 Diacritics

For some words of non-English origin, diacritics can be either retained or omitted in English usage.

> résumé *or* resume   façade *or* facade   naïve *or* naive

Some other diacritics are dropped, and the letters are replaced with their closest anglicized equivalents. However, the original pronunciation is retained without otherwise altering the spelling.

> roentgen *from* Röntgen   hotel *from* hôtel   smorgasbord *from* smörgåsbord

In other cases in which diacritics are dropped, the spelling is altered to reflect the original pronunciation.

> canyon *from* cañón   dim sum *from* dímsām

For current usage of words with diacritics, consult a standard English-language dictionary, preferably the one adopted by the intended publisher of the manuscript. For the most widely used diacritics in European languages, see Table 4.2.

In personal names and place names, retain diacritics if the names have not been anglicized.

> paleontologist Dorothée Le Maître, PhD
> crater Šafařík
> physicist Wojciech Świętosławski

Although word-processing programs offer a wide variety of special characters combining letters with diacritics, such characters must be checked after typesetting to ensure that the desired characters were generated.

5.4.7 Asterisk (*)

The asterisk is used in gene symbolization (see Tables 21.13, 21.26, 21.28, 21.29, and 21.32).

> *MDH-B*1*   *Ho*hl/Ho*hl*

In addition, the asterisk can be used to indicate a footnote in text, tables, or figures. However, this style manual recommends using superscript lowercase letters for this purpose (see Section 30.1.1.8, "Footnotes").

Do not use asterisks to indicate omitted text or missing letters in a word. Instead, use ellipsis points or 3-em dashes for omitted text (see Sections 10.3, "Ellipses," and 5.3.5.4, "Three-Em Dash"). Use 2-em dashes for missing letters in a word (see Section 5.3.5.3, "Two-Em Dash").

5.4.8 Ampersand (&)

Retain the ampersand when it is part of a proper name, such as the name of a company or corporation (eg, "Johnson & Johnson" and "John Wiley & Sons"). Otherwise, do not use an ampersand as a substitute for the coordinating conjunction "and" in titles, text, or other components of scientific publications.

> history and physical *not* history & physical
> latitude and longitude *not* latitude & longitude
> hematoxylin and eosin stain *not* hematoxylin & eosin stain

5.4.9 "At" Symbol (@)

5.4.9.1 General Uses

(1) In email addresses, to separate usernames from domains (see also Section 5.7, "Punctuation Marks in Email and URL Addresses").

    CSE@councilscienceeditors.org   info@nigms.nih.gov   info@rsc-src.ca

(2) At the beginning of usernames for social media such as Facebook, X (formerly Twitter), and Instagram.

    @CouncilofScienceEditors   @CScienceEditors

Do not use the symbol "@" to represent the word "at" in titles, text, or other components of scientific publications.

5.4.9.2 Specialized Uses

(1) Placed after a gene symbol to indicate a gene cluster (see Table 21.32).

    IGH@   *IGL@*

(2) In nanotechnology, to show that a chemical composite consists of an atom, ion, or molecule encased by other atoms in the format of either "guest@host" or "core@shell."

Endohedral fullerenes:	Coated nanoparticles:
$La@C_{60}$	$Au@Pt$
$Xe@C_{60}$	$Ag@SiO_2$
$Sc_3N@C_{78}$	$Au@CoFe$

5.4.10 Hash Mark (#)

The hash mark is also known as the "octothorpe," "numeral symbol," "pound symbol," and "space symbol." It is used in some URLs, and it is used to create hashtags to categorize content in social media.

Other uses for the hash mark are not applicable to scientific publishing. For example, the hash mark should not be used to indicate a footnote in text, tables, or figures. Instead, superscript lowercase letters should be used for this purpose (see Section 30.1.1.8, "Footnotes"). Similarly, the hash mark should not be used as the symbol for "village" in cartography, the "pound" symbol in the avoirdupois system of weights, or the "number" symbol in general contexts.

5.5 MARKS FOR LINE RELATIONS

5.5.1 Brace

A multiline brace can be used as a fence for aggregation in mathematics.

$$\left\{ \begin{array}{l} x = a + b \\ y = c + d \\ a = e + f \end{array} \right\}$$

To simplify manuscript preparation, avoid using braces to cluster related text elements. Instead, use a tabular list or a simple sentence.

Typical Differences in Spelling
US: color, leukemia, and theater
British: colour, leukaemia, and theatre

or

Differences in spelling are typified by the US spellings "color," "leukemia," and "theater" and the British spellings "colour," "leukaemia," and "theatre."

not

Typical Differences in Spelling

$$US \begin{Bmatrix} color \\ leukemia \\ theater \end{Bmatrix} British \begin{Bmatrix} colour \\ leukaemia \\ theatre \end{Bmatrix}$$

5.5.2 Ditto Marks (〃)

Ditto marks are used in lists and tables to indicate repetition of the word or term appearing directly above the ditto marks. However, ditto marks should be avoided in scientific publications to prevent ambiguity. If ditto marks are used, create the marks with Unicode 3003 instead of using double quotation marks or the double prime symbol.

5.5.3 Dots as Leaders (. . . . .)

Graphic designers sometimes use a line of closely spaced periods to lead readers from a text element at or near the left-hand margin to a text element at or near the right-hand margin, as is often done in a table of contents.

Period 5.2.1
Colon 5.3.1
Comma 5.3.3

See Section 10.3, "Ellipses," for how to use spaced dots to indicate an omission in a quotation.

5.5.4 Paragraph Mark (¶)

The paragraph mark, also known as the "pilcrow," is a footnote sign used in text, tables, and figures. However, this style manual discourages using such symbols for designating footnotes. Instead, superscript lowercase letters should be used for this purpose (see Section 30.1.1.8, "Footnotes").

5.6 PUNCTUATION MARKS IN NON-ENGLISH LANGUAGES

Some punctuation marks not used in English may have to be preserved within English scientific texts, especially if the punctuation marks are used in quotes taken from non-English documents. Examples are the guillemets, or "chevrons," used in French and Spanish to mark quotations and the inverted question mark that opens an interrogative statement in Spanish.

N'oubliez pas ces mots: «Guérir quelquefois, soulager souvent, consoler toujours.»
¿Quién o qué estaba «interpretando»?

Consult *The Chicago Manual of Style*[4] for more on punctuation in non-English languages.

5.7 PUNCTUATION MARKS IN EMAIL AND URL ADDRESSES

Email addresses use the @ symbol to separate usernames from domain names and a period as the final punctuation mark before the code designating institution type or country of origin (eg, ".com," ".edu," and ".uk"). In addition, periods, hyphens, and underscores may appear within both usernames and domains.

> a-smith@companyA.com
> c.a.jones@organization.org
> b_jones@universityB.edu
> mainoffice@society-in-Canada.ca

URLs use a variety of punctuation marks and other symbols. To avoid ambiguity regarding correct URL addresses and to ensure that hyperlinked URLs work, authors should not insert any line breaks in URLs in their manuscripts no matter how irregular line endings become. However, where line breaks are needed for publication in print or PDF, typesetters should make these breaks as follows: after the "http://" or "https://" at the beginning of a URL; before a single slash, period, comma, colon, hyphen, tilde, underscore, question mark, number sign, or percent symbol; or before or after an equal sign or ampersand. If possible, avoid breaking at the period in "www." or "org." In addition, a hyphen should never be added to create a line break, nor should a hyphen that is part of a URL appear at the end of a line. If necessary, a long word within a URL may be broken between words or syllables without hyphenation.

> https://www.councilscienceeditors.org/resource-library
>
> *may break in the following ways*
>
> https://
> www.councilscienceeditors.org/resource-library
>
> https://www.council
> scienceeditors.org/resource-library/
>
> https://www.councilscience
> editors.org/resource-library
>
> https://www.councilscienceeditors.org
> /resource-library
>
> https://www.councilscienceeditors.org/resource
> -library
>
> *but not*
>
> https://www.councilscience-
> editors.org/resource-library
>
> *and not*
>
> https://www.councilscienceeditors.org/resource-
> library
>
> *avoid if possible*
>
> https://www
> .councilscienceeditors.org/resource-library
>
> *and*
>
> https://www.councilscienceeditors
> .org/resource-library

CITED REFERENCES

1. Crystal D. Making a point: the pernickety story of English punctuation. Profile Books; 2016.

2. Bringhurst R. The elements of typographic style. 4th ed. Hartley & Marks; 2013.

3. Banik G, Baysinger G, Kamat P, Pienta N. ACS guide to scholarly communication. American Chemical Society; 2020. https://pubs.acs.org/doi/book/10.1021/acsguide

4. The Chicago manual of style: the essential guide for writers, editors, and publishers. 17th ed. The University of Chicago Press; 2017. Also available at https://www.chicagomanualofstyle.org

5. Newell DB, Tiesinga E, editors. The international system of units (SI): United States version of the English text of the ninth edition (2019) of the International Bureau of Weights and Measures publication Le Système International d'Unités (SI). National Institute of Standards and Technology; 2019. https://www.doi.org/10.6028/NIST.SP.330-2019

6. International Organization for Standardization. Quantities and units. Part 1: General. International Organization for Standardization; 2009. (ISO 80000-1:2009).

7. Sequence variant nomenclature. Human Genome Variation Society; 2020. https://varnomen.hgvs.org

8. New Oxford Style Manual. 3rd ed. Oxford University Press; 2016

9. Virag K, editor. Editing Canadian English: a guide for editors, writers, and everyone who works with words. 3rd ed. Editors' Association of Canada; 2015.

10. US Government Publishing Office. Style manual: an official guide to the form and style of federal government publishing. US Government Publishing Office; 2016. https://www.govinfo.gov/collection/gpo-style-manual

11. JAMA Network Editors. AMA manual of style: a guide for authors and editors. 11th ed. Oxford University Press; 2020. Also available at https://www.amamanualofstyle.com

12. International Organization for Standardization. Date and time—representations for information interchange—part 2: extensions. International Organization for Standardization; 2019. (ISO 8601-2:2019).

13. McGowan-Jordan J, Hastings RJ, Moore S, editors. ISCN 2020: an international system for human cytogenomic nomenclature (2020). Karger; 2020.

14. Inczédy J, Lengyel T, Ure AM. International Union of Pure and Applied Chemistry. Compendium of analytical nomenclature: definitive rules 1997. Blackwell Scientific Publications; 1998. Also available at https://media.iupac.org/publications/analytical_compendium/

6 Spelling, Word Formation and Division, Plurals, and Possessives

Editors: Rachel J. Lowery and Elizabeth A. Arnold, ELS

6.1 SPELLING

6.1.1 American and British Differences

Although scientific English is almost uniform throughout the world, there are differences in some of its details, notably spelling. Unlike the Académie française (the French Academy), which specifies the spelling of French words, no national or international authority governs the spelling of English words. Hence, many English words have variant forms, many of which reflect American and British preferences (eg, American preference for "toward" and "focused" versus British preference for "towards" and "focussed").

British forms tend to be preferred in the Commonwealth countries, such as the United Kingdom, Australia, Canada, and New Zealand. However, some American conventions have been adopted in Canada (eg, Canadian and American preferences agree on "aluminum" instead of "aluminium," but they disagree on "enrolment" and "enrollment").

Some of the British and American preferences can be rationalized, while others relate to historical influences and cultural trends. In general, the British forms reflect the continuing influence of French, stemming from the Norman Conquest in AD 1066, and an education system focused on Latin and Greek, whereas many American forms reflect trends toward clearer representation of the pronunciation of words and simplified spelling.

Both groups of variant forms are usually listed in the major English dictionaries published in the United States and the Commonwealth countries. An extensive summary of differences can be found under the heading "American English and British English" in *The Oxford Companion to the English Language*.[1]

In choosing whether to use American or British spellings, publications should be governed by the forms that most of their readers prefer, and authors should follow the preferences selected by the publication to which they submit their manuscripts. Journal offices can select a specific American, British, Canadian, or Australian dictionary as the standard for spelling in submitted manuscripts, and authors can easily adhere to those standards by using word-processing programs that include spellcheckers that adhere to specific national preferences.

6.1.1.1 Nouns with Variant Endings

American and British forms of English include many nouns with variant endings (see Table 6.1).

6.1.1.2 Verbs Ending in "ize," "ise," "yze," and "yse"

The "ize" ending is widely used in both American and British English, but "ise" is more common in British usage.[2] Whereas "yze" is more common in American English, "yse" is more common in British English.

Preferences depend in part on etymology and pronunciation. For example, British usage prefers the "yse" ending for verbs derived from nouns that end in "lysis," while American usage prefers "yze" (eg, "analyse" and "analyze," respectively, as the noun derivatives of "analysis"). In contrast, the "ise" ending is used in both British and American usage for many words pronounced with a "z" rather than an "s" sound (eg, "advertise," "advise," and "compromise").

American	British
catalyze *from* catalysis	catalyse *from* catalysis
civilize *from* civilization	civilise *from* civilisation
organize *from* organization	organise *from* organisation
rationalize *from* rationalization	rationalise *from* rationalisation

6.1.1.3 Verbs Ending in "l" or "ll"

When used as the terminal letter in verbs ending with a stressed syllable, "l" may be single or double in both American and British usage, although the British tendency is toward the single "l." In both usages, a single "l" is doubled when a suffix (eg, "able," "ant,"

Table 6.1. Nouns with Variant Spellings in American and British English

Category of noun	American	British	Notes
	Examples		
	American	British	
Ending in "ction" or "xion"	connection	connexion	
	deflection	deflexion	
Ending in "er" or "re"[a]	center	centre	Some words have the same form in both American and British English (eg, manager, interpreter, and mediocre)
	fiber	fibre	
	liter	litre	
	maneuver	manoeuvre	
	meter	metre (for SI unit) and meter (for instrument)	
Ending in "or" or "our"[a,b]	behavior	behaviour	For some words, "or" is preferred in both American and British English (eg, stupor, terror, director, monitor, pallor, and tremor)
	color	colour	
	tumor	tumour	
Ending in "ense" or "ence"	defense	defence	"defensible" in both
	license	licence	"license" as a verb in British English
	offense	offence	"offensive" in both
	practice	practice	"practise" as a verb in British English
			Adjectives "immense" and "intense" in both
Ending in "log" or "logue"[a]	analog	analogue	British forms also accepted in American English
	catalog	catalogue	
			"dialogue," "ideo-logue," and "secret-agogue" in both
With digraphs "ae" and "oe"	anesthesia	anaesthesia	British preferences reflect Latin and Greek origins
	cesium	caesium	
	celiac	coeliac	
	diarrhea	diarrhoea	
	edema	oedema	
	esophagus	oesophagus	
	estrogen	oestrogen	
	etiology	aetiology	
	fetus	foetus	
	gynecology	gynaecology	
	hematology	haematology	
	leukemia	leukaemia	
	orthopedics	orthopaedics	
	pediatrics	paediatrics	
Derived from verbs ending in "e"	acknowledgment	acknowledgement	
	aging	ageing	
	judgment	judgement	
	likable	likeable	

Table 6.1. Nouns with Variant Spellings in American and British English (*continued*)

	Examples		
Category of noun	American	British	Notes
Other variants	aluminum	aluminium	
	artifact	artefact	
	check	cheque	
	draft	draught	
	mold	mould	
	program	programme	
	sulfur	sulphur	

SI, Système International d'Unités (International System of Units)

[a] American endings generally reflect phonetically based forms, while British endings generally reflect French derivation (eg, "honour" from "honneur").

[b] For words derived from nouns ending in "or" or "our," the simpler spelling is preferred for some words in both American and British English (eg, coloration, honorary, and laborious).

"ation," "ed," and "ing") is added to verbs ending in a stressed syllable. Derivative nouns formed by adding "ment" use the double "l" in American usage but not in British usage.

American	British
compel, compelling	compel, compelling
distill, distilled	distil, distilled
enroll, enrolling, enrollment	enrol, enroling, enrolment
forestall, forestallment	forestall, forestalment
fulfill, fulfillable	fulfil, fulfillable
propel, propellant, propelled	propel, propellant, propelled
install, installment	install, instalment

To determine the stressed syllable in such words or just to check their spellings when suffixes are added, consult a standard dictionary, preferably the one used by the publication of choice.

For guidelines on when to double other single terminal consonants, see Section 6.1.4, "Doubled Consonants before Suffixes."

6.1.2 Words Derived from Languages Other than English

Words transliterated or derived from languages other than English can have variant spellings. In general, the variant that is phonetically unambiguous is preferred. For example, "leukemia" or "leukaemia," derived from the Greek *leukos*, is preferred to "leucemia" or "leucaemia."

When the spelling of a word in English is identical to that in the language of origin except for one or more diacritics and the word appears in standard English dictionaries, the diacritics are generally omitted. When these words are anglicized, their pronunciations usually change (eg, the French words "brassière" and "émotion" have become "brassiere" and "emotion" in English and are pronounced differently in French and English). For non-English words and phrases less widely used in English, their original spellings and diacritics are retained.

aide-mémoire   chargé d'affaires   hors-d'œuvres   pièce de résistance

6.1.3 Verbs Beginning with "en" or "in"

Some verbs beginning with "en" or "in" and their noun derivatives are spelled with either opening syllable, but the resulting words may have different meanings.

Same meaning:	**Different meanings:**	**No variant:**
enclose, inclose	ensure, insure	enact *not* inact
endorse, indorse		insurance *not* ensurance
endue, indue		

6.1.4 Doubled Consonants before Suffixes

The following practices on whether to double terminal consonants when forming derivative words with such suffixes as "ed," "er," "ence," "est," and "ing" are based on *Hart's Rules for Compositors and Readers at the University Press, Oxford*.[3] These guidelines have exceptions, so if the spelling of a word is questionable, consult a standard dictionary, preferably the one used by the intended publication. For guidance on forming scientific words with suffixes, see Section 6.2.2, "Formations with Suffixes in Scientific Terms."

(1) Double a single terminal consonant when it follows a single vowel in a monosyllabic word and the suffix begins with a vowel.

    big, bigger   hot, hottest   rot, rotted   sag, sagging

    Exceptions include "bus, busing."
    When the suffix begins with a consonant instead of a vowel, do not double the consonant.

    commit, commitment   content, contentment   state, statement

(2) Double a single terminal consonant for a polysyllabic word with a final syllable with a short vowel sound and a suffix beginning with a vowel. This rule applies to polysyllabic words regardless of whether the final syllable is accented.

    control, controlled   format, formatting   input, inputting   occur, occurred

(3) Do not double a single terminal consonant if the terminal accent moves from the last syllable in the stem to the first syllable in the derived form.

    infer, inference   refer, reference

(4) A single terminal consonant may or may not be doubled if the final syllable is unaccented in the stem.

    travel, traveler (American preference), traveller (British preference)
    cancel, canceled (American preference), cancelled (British preference)

(5) Do not double the terminal consonants "h," "k," "w," and "x."

    watch, watching   ask, asking   flaw, flawed   sex, sexed

6.2 FORMATION OF SCIENTIFIC TERMS

The rapid growth of knowledge in the sciences has generated many new terms and words. Some may be single words coined from Latin or Greek roots (eg, "idiopathic" from the Greek "idio" and "pathos"). More often, they are compound terms created from existing stem words or stem words combined into single words.

Over time, the spelling of these terms may morph. For example, some terms that were originally compound terms, or open compounds, have shifted to single words, or closed compounds. As a consequence, some scientific terms may become closed compounds while others with the same stems may remain open compounds.

Closed compound	Open compound
bloodstream	blood vessel
database	data set
headphone	head shield

New scientific terms also may be formed by adding prefixes or suffixes to stem words. The rules governing prefixes and suffixes are not firm and unequivocal, and usage tends to shift over time (see Section 6.2.1, "Formations with Prefixes in Scientific Terms," and Section 6.2.2, "Formations with Suffixes in Scientific Terms," respectively).

Some scientific fields have specific rules for vernacular terms, such as the plant sciences (see Section 22.4.1.6, "Vernacular Names for Plants") and the animal sciences (see Section 22.5.7, "Vernacular Names for Animals").

Concise summaries of general practices for forming new words can be found in *The Chicago Manual of Style*,[4] the US Government Publishing Office's style manual,[5] *New Hart's Rules*,[2] *Editing Canadian English*,[6] "Writing Tips Plus,"[7] and the *AMA Manual of Style*.[8]

6.2.1 Formations with Prefixes in Scientific Terms
6.2.1.1 Nonhyphenated Prefixes
Many scientific terms are formed by adding nonhyphenated prefixes to stem words (see Table 6.2 for examples). In general, these prefixes indicate action, character, location, number, state, or time. Many of these prefixes do not stand alone as words (eg, "iso" and "para"), and some of those prefixes are derivatives of adjectives (eg, "chemo" from "chemical" and "geo" from "geographic").

When uncertainty exists about whether to hyphenate new terms formed with such prefixes, the preference is to not hyphenate the terms. However, if the last letter of a prefix and the first letter of a stem are the same vowel, a hyphen is generally used between the prefix and the stem (eg, "anti-inflammatory," "semi-independent," and "extra-axial"). Over time, it may become acceptable to remove the hyphen between double vowels, as happened with "coordination" and "microorganism."

When forming words with prefixes, a common error is confusing homonymic or near-homonymic prefixes. Most such errors can be identified by spellcheckers in word-processing programs.

ante and anti:	for and fore:
antediluvian *not* antidiluvian	forward *not* foreward
antifreeze *not* antefreeze	foreword *not* forword

6.2.1.2 Hyphenated Prefixes
Under the following circumstances, hyphens should be used to join prefixes to stem words:

(1) The stem is capitalized.

pre-Columbian civilization   post-Copernican astronomy   sub-Saharan Africa

(2) Omitting the hyphen changes the meaning.

Re-cover the flask after adding the reagent. *but* You may not recover enough precipitate.

Table 6.2. Examples of Scientific Terms Formed with Nonhyphenated Prefixes

Prefix	Example	Prefix	Example
ab	abduction	meta	metagenesis
ad	adduction	micro	microfossil
aero	aerostatics	mid	midbrain
after	aftershock	milli	millisecond
ante	antepartum	mini	minicomputer
anti	anticodon	multi	multiprocessor
astro	astrophysics	non	nonconductor
auto	autoimmunity	over	overdose
bi	bivalve	para	paramyxovirus
bio	biomechanics	photo	photochemistry
brady	bradycardia	physio	physiotherapy
chemo	chemotherapy	phyto	phytopathology
co	coenzyme	poly	polyarthritis
counter	counterimmunoelectrophoresis	post	postprecipitation
de	denitrification	pre	prediabetes
di	diketone	pro	proinsulin
dys	dysphasia	pseudo	pseudopod
eco	ecosystem	re	recombination
electro	electrosurgery	semi	semiconductor
exo	exopathogen	socio	socioeconomic
extra	extrasystole	stereo	stereochemistry
geo	geochemistry	sub	subspecies
hemi	hemianesthesia	super	supersaturation
hemo	hemodialysis	supra	suprascapular
hyper	hyperventilation	tachy	tachycardia
hypo	hypomenorrhea	trans	transpolarizer
in	incoordination	tri	tribromoethanol
infra	infrared	ultra	ultracentrifuge
inter	intercostal	un	unconformity
iso	isohexane	under	undernutrition
macro	macroflora		

For recommendations on hyphenation of compound nouns and adjectives, see Section 5.4.1, "Hyphen."

6.2.2 Formations with Suffixes in Scientific Terms

6.2.2.1 Suffixes That Are Not Words

Many nouns and adjectives are formed by adding suffixes that are not complete words. These suffixes are not hyphenated. Because some of these suffixes are similar, the words they form may be easily misspelled. Most such misspellings can be caught by spellcheckers in word-processing programs (see also Section 7.5.1, "Homophones and Near Homophones").

able	ible	eous	ous
ance	ence	erous	orous
ant	ent	ful	full
ative	ive	ified	yfied
cede	ceed	efy	ify

6.2.2.2 Words as Suffixes

Many nouns have been formed by adding a suffix that is itself a complete word to a stem verb or noun. In such formations, the hyphen is often but not always omitted.

Suffix	Example
away	runaway reactor
down	breakdown of the varieties of species
in	cave-in at the excavation site (hyphen generally desirable)
like	wormlike lower vertebrate
off	heavy runoff causing a flood
out	turnout greater than expected
over	turnover in the faculty
up	breakup of the iceberg

Whether a suffix is unhyphenated and whether it is used as a separate word may depend on whether the term that the suffix is part of is used as a noun, verb, or adjective.

The announcement of the shutdown was delayed several days. ("Shutdown" is a noun.)
They shut down the reactor. ("Shut down" is a phrasal verb.)
The shut-down reactor needed extensive repairs. ("Shut-down" is an adjective.)
The researchers will follow up with all the patients in the study. ("Follow up" is a phrasal verb.)
The patient scheduled a follow-up appointment. ("Follow-up" is an adjective.)

In addition, a hyphen is needed between stem words that end in 2 identical letters and suffixes that start with the same letter (eg, "shell-like"). A hyphen is also needed before suffixes used with some stem words ending in consonants that are ascenders, such as "d" and "l" (eg, "mold-like").

6.2.2.3 Variant Suffixes ("ic" and "ical")

Some word pairs ending in the suffixes "ic" and "ical" convey the same meaning. Consult the specialized or general dictionary specified by the publication of choice to determine which word in a synonymous pair is preferred. If the dictionary does not indicate a preference, select one spelling to use consistently throughout the manuscript.

etiologic, etiological   histologic, histological   microscopic, microscopical

Other pairs ending in "ic" and "ical" carry different meanings. Consult the intended publication's preferred specialized or general dictionary to determine whether the words in a pair have the same or different meanings.

economic theory *but* an economical method of collecting samples
a demanding work ethic *but* a violation of ethical principles
national historic sites *but* an outmoded historical theory
the statistic thus calculated *but* a statistical analysis

In addition, some word variants are not idiomatic (eg, "chemic" is not a shorter form of "chemical," and "publical" is not a longer form of "public").

Of special note for scientific manuscripts, stem words ending with "ology" that describe fields of study should be converted to adjectives ending in "ic" or "ical" to describe particular entities within those disciplines rather than using the stem words themselves.[8]

> histologic diagnosis *not* histology (to refer to a lesion diagnosed by histologic examination)
> pathologic lesion *not* pathology (to refer to an abnormality)

6.3 WORD AND TERM DIVISION

Word-processing programs and computer-driven composition have eliminated the need for authors and manuscript editors to be concerned about word divisions at the end of lines when preparing manuscripts for submission. Instead, authors and editors only need to insert hyphens in compound words needing hyphenation.

However, authors and editors still need to be concerned with end-of-line hyphenation when reviewing typeset galleys and page proofs. For consistency in dividing words at the end of lines in these documents, apply the divisions indicated in the publication's specialized or general dictionary of choice. American divisions are generally based on syllable structure, while British divisions are usually based on etymologic elements. When dictionaries indicate that a word can be broken in more than one place, use the guidelines below to decide which break is preferable.

(1) Break hyphenated words and terms after the hyphen if possible to avoid using another hyphen.

> one-   *not*   one-di-
> dimensional scan   mensional scan

(2) Divide long chemical names by syllables, by etymologic units, or according to other hyphen rules, but at least 4 characters should appear on each line. Do not break at a hyphen connecting a locant or descriptor prefix.

> 2-acetyl-   *not*   2-
> aminofluorene   acetylaminofluorene

(3) The part of the word carried over to the next line should not look like a separate word.

> path-   *not*   patho-
> ologic   logic

(4) Do not divide single-syllable and very short polysyllabic words.

> health   *not*   he-
>   alth
> also   *not*   al-
>   so

(5) Do not separate numbers from associated unit designations.

> a concentration of   *not*   a concentration of 250
> 250 g/L   g/L

See *The Chicago Manual of Style*[4] and *New Hart's Rules*[2] for more detailed guidance on word division. For recommendations about line breaks within internet addresses, see Section 5.7, "Punctuation Marks in Email and URL Addresses."

6.4 PLURALS

6.4.1 General Principles

For most common nouns, form the plural by adding "s."

catalysts	electrons	parasites
chemicals	organisms	stars

Use "es" for nouns that end in sibilants, such as soft "ch," "s," "sh," "x," and "z."

batches	ostriches	solonetzes
gases	rashes	thoraxes
microswitches		

For nouns whose singular form ends in "i," the plural is usually formed by adding "s." Some exceptions add "es" after "i," and other exceptions use the same spelling for their singular and plural forms.

fermis	tsunamis	alkalies	bonsai

For nouns ending in "o" preceded by another vowel, form the plural by adding "s" to the singular. For nouns ending in "o" preceded by a consonant, there is no strict rule. Some of these nouns take only "s," others take "es," and yet others use both forms or neither.

ratios	potatoes
infernos	tomatoes
quangos	buffalo (secondary spellings buffaloes and buffalos)
embargoes	tornadoes (secondary spelling tornados)

Form the plural of nouns ending in "y" preceded by a consonant or a consonant sound by changing the "y" to "i" and adding "es."

bibliographies	flies	histories	pregnancies	technologies
discrepancies	frequencies	libraries	specialties	universities

For most nouns ending in "y" preceded by a vowel, add "s" only to form the plural.

airways	alloys	buoys	monkeys	wiseguys

6.4.2 Proper Nouns

Form the plural of most proper nouns in the same manner as for common nouns. Do not alter the spelling for proper nouns ending in "i," "o," or "y," except for the shortened forms of geographic terms like mountains.

All India Institutes of Medical Sciences	the Fermis	2 hot Julys
Africa Centres of Excellence	the Hardys	the Alleghenies
National Institutes of Health	the Waksmans	Canadian Rockies

If adding "s" or "es" to a proper noun would result in a false pronunciation, the plural is the same as the singular. This situation arises mainly with French names that end in an unpronounced "s," "z," or "x." In their plural forms, such names are pronounced as if "s" or "es" were added to their singular forms.

the Agassiz *pronounced but not spelled as* the Agassizes
the Dubos *pronounced but not spelled as* the Duboses
the Roux *pronounced but not spelled as* the Rouxes

6.4.3 Irregular Forms

Many nouns have irregular plural forms, notably English nouns originating before the Norman Conquest. There are no formal rules for these plurals.

child, children	louse, lice
ox, oxen	goose, geese

For some single-syllable nouns, the plural is formed by changing vowels within the word. The plurals of compound words that end in these words are formed in the same way.

foot, feet	clubfoot, clubfeet
man, men	woman, women
mouse, mice	dormouse, dormice

The plurals of words ending in "man" that are not compounds are formed simply by adding "s."

human, humans	German, Germans

Although the plurals of most nouns that end in "f," "ff," and "fe" are formed in the regular way, a few change their endings to "ves" in their plural form.

calf, calves	half, halves	self, selves
life, lives	midwife, midwives	quarterstaff, quarterstaves

See Section 6.4.6, "Nouns with 2 Plural Forms."

6.4.4 Nouns with One Form

Some nouns have the same spelling for both their singular and plural forms. For these nouns, whether they are singular or plural depends on the context in which they are used.

aircraft	matter (meaning "substance")	series
corps	pains (meaning "effort")	spacecraft
forceps	progeny	species
goods (meaning "material things")	remains	sperm

Similarly, the names of some nationalities have one form that can be either singular or plural.

Chinese	English	French	Japanese	Portuguese

Some nouns cannot represent a plural concept and, therefore, have no plural form.

horsepower	information	sunlight	water

For some nouns, the sole form appears to be plural because it ends in "s" or "es," but the form is actually singular.

herpes	news
measles	rickets
mumps	shingles (the skin eruption herpes zoster)

For the names of certain animals, the singular is used to denote both an individual animal or 2 or more.

deer	moose
fish (and individual species, such as trout, cod,	sheep
haddock, and herring)	swine

For some of these nouns, as well as some that generally have only singular forms (eg, "oat" and "wheat"), the regularly formed plural is used to indicate more than one species, strain, or variety.

3 fishes of interest   6 experimental wheats   10 mutant oats

For the names of some large mammals and other organisms, either the singular or the regularly formed plural may be used to indicate the plural. When in doubt as to whether the singular form is acceptable as the plural, consult the specialized or general dictionary of choice of the intended publication, and select the preferred spelling.

antelope, antelopes	elk, elks	lobster, lobsters
caribou, caribous	giraffe, giraffes	walrus, walruses

6.4.5 Plural Endings from Languages Other than English

English scientific vocabularies have many terms that are taken directly from other languages. The plurals of such words are formed by the rules of the original language (see Table 6.3). In some cases, anglicized forms that follow the general rules for plural formation are also acceptable (see Section 6.4.6, "Nouns with 2 Plural Forms"). The acceptability of the anglicized plural forms differs with the type of publication and its audience. Each publication should create its own list of preferred plural forms to include in its instructions to authors. For plural forms not listed in the instructions, consult the publication's specialized or general dictionary of choice to determine the preferred forms.

6.4.6 Nouns with 2 Plural Forms

For some nouns with 2 or more plural forms, the different forms have different meanings.

brothers (male siblings), brethren (members of a society)
dies (devices for stamping), dice (gambling pieces)
indexes (alphabetical lists of topics or names), indices (numerical expressions)
ossa (bones), ora (mouths)
staffs (groups of employees or assistants), staves (poles)

For other nouns, the different forms of the plural are identical in meaning. Specialized and general dictionaries may approach these forms in 3 ways: They may suggest which form is preferred, they may indicate that 2 or more forms are acceptable, or they may list only a single option. In addition, publications may have their own lists of preferred forms (see Section 6.4.4, "Nouns with One Form"). In the examples below, the first plural form is the variant recommended by this style manual.

Singular	Plural
behavior	behavior, behaviors
biceps	biceps, bicepses
biomass	biomass, biomasses
femur	femora, femurs
gladiolus	gladioli, gladioluses
hoof	hooves, hoofs
thorax	thoraxes, thoraces

6.4.7 Letters and Numbers

Form the plural of a single letter by adding an apostrophe and "s," regardless of whether the letter is capitalized or lowercased. This formation avoids confusion with true words (eg, "as" and "is") and abbreviations (eg, "Ps" for "photosynthesis").

P's and Q's   *x*'s and *y*'s

Table 6.3. Plural Endings from Non-English Languages

Endings		Examples	
Singular	Plural[a]	Singular	Plural[a]
a (Latin)	ae	abscissa	abscissae *or* abscissas
		alga	algae *or* algas
		formula	formulae *or* formulas
eu, eau (French)	eux, eaux	milieu	milieux *or* milieus
		rouleau	rouleaux *or* rouleaus
en (Latin)	ina	foramen	foramina *or* foramens
		rumen	rumina *or* rumens
ex, ix (Latin)	ices	appendix	appendices (for anatomical structures) *or* appendixes (for closing sections of books)
		fornix	fornices *or* fornixes
		index	indices (in economics and mathematics) *or* indexes (for alphabetic lists of topics and names in the closing sections of books)
is, es (Greek or Latin)	es	analysis	analyses
		hypothesis	hypotheses
		ascites	ascites
itis (Greek or Latin)	itides	arthritis	arthritides
		meningitis	meningitides
o (Italian)	i	virtuoso	virtuosi *or* virtuosos
on, oan (Greek)	a	criterion	criteria *or* criterions
		mitochondrion	mitochondria
		protozoan and protozoon	protozoa *or* protozoans
um (Latin)	a	addendum	addenda *or* addendums
		bacterium	bacteria
		corrigendum	corrigenda
		datum[b]	data
		erratum	errata
		medium	media *or* mediums
		phylum	phyla
		symposium	symposia *or* symposiums
us (Latin)[c]	i	bacillus	bacilli
		focus	foci *or* focuses
		fungus	fungi *or* funguses
	uses	apparatus	apparatuses
		prospectus	prospectuses

[a] For many terms, the plural may also be formed by adding "s" or "es."
[b] For tidal benchmarks, see Section 25.11.4.1, "Tidal Datum."
[c] The plural of "genus" is "genera," not "geni" or "genuses."

In contrast with forming the plural of a letter, only add "s" to form the plural of a number or date written in numerals. Placing an apostrophe before "s" would make the number or date singular possessive instead of plural.

> expressed in 100s *not* expressed in 100's
> the 2020s *not* the 2020's
> patients in their 50s *not* patients in their 50's

Adhere to the rules in Section 6.4.1, "General Principles," when forming the plurals of numbers expressed as words or when writing expressions that indicate numbers.

> ones   in her teens   hundreds of specimens
> at sixes and sevens   counting by tens   thousands of species

For general guidance on plurals of initialisms and other abbreviations, see Section 11.1.3, "Plurals of Abbreviations."

6.4.8 Units of Measurement

The symbols for most units of measurement are the same in their singular and plural forms (see Section 12.2.1.1, "SI Rules").

> 1 kg, 5 kg   1 mL/min, 18 mL/min   1 m, 3 m   1 °C, 45 °C

6.4.9 Plurals of Compound Words and Terms

For compounds that are spelled as a single word, the plural is formed regularly by adding the appropriate ending to the word (see also Section 6.4.3, "Irregular Forms").

> checkup, checkups   landform, landforms
> dormouse, dormice   spacesuit, spacesuits

In contrast, hyphenated and open compounds take the plural form of the noun that is the basis of the term.

> chief of staff, chiefs of staff   Portuguese man-of-war, Portuguese men-of-war
> editor in chief, editors in chief   president-elect, presidents-elect
> passerby, passersby   runner-up, runners-up
> physician author, physician authors   surgeon general, surgeons general

If the compound contains no nouns or if none of the nouns are significant in the context, "s" is added to the last component.

> forget-me-not, forget-me-nots   go-between, go-betweens

If the components of a compound term are more or less equivalent, the plurals of both are used.

> woman scientist, women scientists

6.4.10 Common Names Taken from Scientific Names

For some plants, the vernacular English names are the genus names set in roman type without an initial capital letter (see Section 22.4.1.6, "Vernacular Names for Plants"). The plural forms of these vernacular names are usually formed according to the general rules for plurals.

> camellia, camellias
> crocus, crocuses (secondary spelling croci)
> iris, irises
> gladiolus, gladioli (secondary spelling gladioluses)

Some microorganisms have vernacular plural designations based on the name of each microorganism's genus.

chlamydiae (for organisms in the genus *Chlamydia*)
mycobacteria (for organisms in the genus *Mycobacterium*)
pseudomonads (for organisms in the genus *Pseudomonas*)
salmonellae (for organisms in the genus *Salmonella*)
staphylococci (for organisms in the genus *Staphylococcus*)
streptococci (for organisms in the genus *Streptococcus*)
treponemes (for organisms in the genus *Treponema*)

For microorganisms without a generally used vernacular English name, add a descriptive term to the italicized genus name (eg, "*Escherichia* bacteria," "*Candida* yeasts," and "*Bacillus* species").

6.5 POSSESSIVES

6.5.1 General Principles

To form possessives, add an apostrophe and "s" to most singular common and proper nouns, as well as some indefinite pronouns.

the patient's condition   everyone's responsibility
China's satellite   one's own diagnosis
the wolf's territory   someone's stethoscope
Pettigrew's study   nobody's error

Some indefinite pronouns, such as "any," "few," "many," "none," and "such," do not have a possessive form.

6.5.2 Singular Nouns That End in "s"

The general principle of adding an apostrophe and "s" holds for most singular nouns that end in "s," including proper nouns. Pronunciation can serve as a guide: If the possessive "s" is pronounced, it should be part of the written form.

Anders Celsius's temperature scale   the grass's texture
Enid Charles's commentary   the moss's reproductive capacity
Carl Friedrich Gauss's formulas   the lens's properties

If the double sibilant sounds awkward, recast the expression to avoid the possessive form.

the texture of the grass   the properties of the lens

By tradition, the possessive forms of the names of "Jesus" and "Moses" are formed by adding an apostrophe only, as are Greek and hellenized proper names ending in "s." Another exception based on tradition is the possessive for the "United States," which is formed by adding only an apostrophe, even though the "United States" is a singular proper noun.

Archimedes' screw   Hippocrates' teachings   Ramses' tomb   Zeus' temple

For proper names that end in a silent "s," "x," or "z," an apostrophe and "s" must be added to produce the correct pronunciation.

Louis Agassiz's theories of glaciation   Arkansas's geography
René Descartes's philosophical essays

6.5.3 Plural Nouns

Form the possessive of common and proper plural nouns that end in "s" by adding only an apostrophe.

> the animals' behavior
> the lenses' characteristics
> the patients' histories
> the physicians' privileges
> the Cousteaus' family tradition in oceanography
> the Curies' 5 Nobel Prizes

Form the possessive of plural nouns that do not end in "s" by adding an apostrophe and "s."

> the bacteria's growth patterns
> the men's pulmonary capacity tests
> the children's eating habits
> the strata's varying rocks

When plural nouns end in a sibilant sound, the resulting phrases may sound awkward. In such cases, recast the sentences to avoid the possessive form.

> the mice's nesting material *becomes* the nesting material used by mice
> the geese's migratory formation *becomes* the migratory formation of geese
> this series' end *becomes* the end of this series

6.5.4 Pronouns

Apostrophes are not used in forming the possessives of pronouns, but "s" often is.

> his *for* he and him   our and ours *for* we
> hers and her *for* she and her   your and yours *for* you
> its *for* it   their and theirs *for* they
> mine and my *for* I and me

Do not confuse the possessive pronoun "its" with the contraction "it's" for "it is."

6.5.5 Compound Expressions

Form the possessive of singular compound expressions by adding an apostrophe and "s" to the final elements. For plural compound expressions that end in the letter "s," add an apostrophe only to the final elements.

> someone else's proposal   the surgeons general's warnings
> the editor in chief's guidelines   the chiefs of staff's responses
> everybody else's preferences   institutional review committees' decisions

For nouns in a series, the form of the possessive is determined by whether joint or individual ownership is intended. For joint ownership, add an apostrophe and "s" only to the final element. For individual ownership, add an apostrophe and "s" to each element.

> James Watson and Francis Crick's DNA model (James Watson, PhD, and Francis Crick, PhD, created the model together)
> James Watson's and Francis Crick's memoirs (Watson and Crick wrote separate memoirs)
> the student and her tutor's appointment (the student and tutor have the same appointment)
> the student's and her tutor's telephone numbers (the student and tutor have separate telephone numbers)

In a possessive expression that incorporates one or more possessive pronouns, all nouns and pronouns take the possessive form regardless of whether the expression is conveying joint or individual ownership.

Linda B. Buck's and your research
my student's and my assessment of the data

6.5.6 Eponymic Terms

Medical and other scientific vocabularies contain many eponymous terms, compound terms that incorporate the proper names of physicians, researchers, theoreticians, patients, or places. These eponymous terms cover a wide variety of laws, theories, methods, anatomical parts, conditions, diseases, reagents, syndromes, tests, and other entities. Traditionally, many terms named for researchers incorporated the names in their possessive form, whereas those referring to patients and places used the nonpossessive form. This style manual recommends that the possessive form be eliminated from all eponymic terms to clearly differentiate them from true possessives.

Crohn disease *not* Crohn's disease
Carvallo sign *not* Carvallo's sign
Down syndrome *not* Down's syndrome
McCune–Albright syndrome *not* McCune–Albright's syndrome

As an exception to this rule, continue to use the possessive form for the official names of organizations that incorporate eponymic terms as possessives. Consult the organizations' websites or other reliable sources to confirm the correct usage.

Alzheimer's Association
Crohn's & Colitis Foundation
Cushing's Support & Research Foundation
The Michael J. Fox Foundation for Parkinson's Research

For capitalization guidelines for eponymic terms, see Section 9.4.3, "Eponymic Terms."

6.5.7 Organization Names

Form the possessive of an organization's full name by adding an apostrophe and "s" to the last element of the name or an apostrophe only if the last element is a plural that ends in "s." For acronyms and other initialisms that end in capital "S," add an apostrophe and "s," not just an apostrophe (see also Section 11.1.4, "Possessives of Initialisms").

the American Psychological Association's journal *American Psychologist*
the Council of Science Editors' publication *Science Editor*
the IAS's more than 1,000 fellows ("IAS" *for* "Indian Academy of Sciences")

The names of many organizations and institutions incorporate possessives or plurals. Determine correct usage by verifying the official names of the organizations against their websites and other reliable sources. For names that include the possessive of plural nouns that do not end in "s" (eg, "children"), an apostrophe and the letter "s" are likely to be incorporated into those names.

American Medical Writers Association ("Writers" is plural)
Editors' Association of Canada ("Editors'" is possessive)
Children's Health Queensland ("Children's" is possessive)

6.5.8 Proper Names Set in Italic Type

To form the possessive of the names of books and journals, set the name in italic type and the apostrophe and "s" in roman type.

Thinking, Fast and Slow's premise   *The Lancet*'s reputation   *Surgery*'s 170th volume

6.5.9 Expressions of Duration

Expressions of duration based on the genitive case are analogous to possessives and are formed with an apostrophe. In many instances, these expressions can be rewritten without possessives if desired.

> a week's vacation, a week of vacation, a one-week vacation
> in 3 days' time, in 3 days
> after many years' experience, after many years of experience

If the word "of" cannot be inserted into the expression, the possessive is incorrect.

> 6 months pregnant *not* 6 months' pregnant   8 days late *not* 8 days' late

CITED REFERENCES

1. McArthur T, Lam-McArthur J, Fontaine L, editors. The Oxford companion to the English language. 2nd ed. Oxford University Press; 2018.

2. New Hart's rules: the Oxford style guide. 2nd ed. Oxford University Press; 2014.

3. Hart's rules for compositors and readers at the University Press, Oxford. 39th ed. Oxford University Press; 1983.

4. The Chicago manual of style: the essential guide for writers, editors, and publishers. 17th ed. The University of Chicago Press; 2017. Also available at https://www.chicagomanualofstyle.org

5. US Government Publishing Office. Style manual: an official guide to the form and style of federal government publishing. US Government Publishing Office; 2016. https://www.govinfo.gov/content/pkg/GPO-STYLEMANUAL-2016/pdf/GPO-STYLEMANUAL-2016.pdf

6. Virag K, editor. Editing Canadian English: a guide for editors, writers, and everyone who works with words. 3rd ed. Editors' Association of Canada; 2015.

7. Public Services and Procurement Canada, Translation Bureau. Writing tips plus; [modified 5 May 2022]. https://www.noslangues-ourlanguages.gc.ca/en/writing-tips-plus/index-eng

8. JAMA Network Editors. AMA manual of style: a guide for authors and editors. 11th ed. Oxford University Press; 2020. Also available at https://www.amamanualofstyle.com

7 Prose Style and Word Choice

Editors: Beva Nall-Langdon and Caitlyn Trautwein

7.1 SCIENTIFIC PROSE

Effective scientific prose is precise, succinct, and fluent. These qualities depend on myriad details, including the choice of words, the length and flow of individual sentences, and the way in which sentences and paragraphs relate to each other. To increase the likelihood that manuscripts will be accepted for publication and to decrease revisions after acceptance, authors need to adhere to the standards of scientific prose adopted by their discipline of study.

This chapter addresses a number of common style issues in scientific prose, but it is not all-inclusive. Authors seeking additional guidance in writing scientific prose can find help not only in other chapters in this style manual but also in the books listed in this manual's bibliography in the sections "Style Manuals" and "Usage and Prose Style."

7.2 GRAMMATICAL ERRORS

7.2.1 Subject and Predicate Agreement in Number

The number of the noun that serves as a sentence's subject must agree with the number of the sentence's predicate (ie, verb and its modifiers). In other words, singular subjects should have singular predicates, and plural subjects should have plural predicates.

In English, most plural subjects are formed by adding "s" or "es" to the nouns' singular form. In contrast, the present tense of many plural predicates is formed by removing the letter "s" from the end of the verbs' singular form. Forming the past tense of predicates is less of a concern because their singular and plural forms are usually identical.

> NaCl consists of a sodium ion and a chloride ion.
> Astrophysicists agree that Earth is a planet.
> China invented gunpowder.
> Geneticists around the world worked on the Human Genome Project.

7.2.1.1 Collective Nouns

In general, collective nouns take singular verbs as predicates when these nouns refer to everyone or everything in them as one whole.

> The institutional review board announces decisions every Friday.

In the example above, all of the announcements are made by the entire board; the board's members do not individually make each announcement.

Although collective nouns may take plural verbs when referring to the individuals in the groups, the preferred approach is to replace the collective nouns with clearly plural terms.

> All members of the research committee review grant requests. *instead of* The research committee review all grant requests.

When abbreviated, collective nouns should be treated as singular; when spelled out, they should be treated as plural.

> The CIHR invests in research on chronic diseases. ("CIHR" stands for the "Canadian Institutes of Health Research," which is treated as singular.)
> The National Academies in the United States consist of 3 nonprofit organizations. ("National

Academies" stands for the "National Academies of Sciences, Engineering, and Medicine," which is a plural term.)

7.2.1.2 Nouns Derived from Latin and Other Languages

A common mistake that results in subject-predicate disagreement in scientific writing is failing to distinguish between singular and plural forms of English nouns derived from Latin and other languages. In general, singular Latinate nouns that end in "um," "us," and "a" are pluralized by changing the endings to "a," "i," and "ae," respectively. Hence, the plural of "datum" is "data," the plural of "nucleus" is "nuclei," and the plural of "alga" is "algae." For examples of the singular and plural forms of scientific terms derived from Latin and other languages, see Table 6.3.

> The most widely used culture media are . . . ("media" is the Latin plural for "medium")
> The criteria include . . . ("criteria" is plural for the English derivative for the Greek word for "criterion")
> The milieux of 3 multicellular organisms were compared . . . ("milieux" is the French plural for "milieu")

Although substituting the plural form of a Latinate or other non-English term for its singular form may not hinder understanding in popular, nonscientific writing, scientific prose requires precision, and thus such distinctions should be maintained. "Data," for example, is commonly treated as singular in nonscientific writing, but in most scientific disciplines, "datum" and "data" should be used as the term's singular and plural forms, respectively. Some scientific disciplines make an exception to this rule by forming plurals by adding the letter "s" to the singular form of Latinate terms. For example, the US Geological Survey[1] specifies that "datums" rather than "data" should be used to refer to base elevations of bodies of water (see Section 25.11.4.1, "Tidal Datum").

7.2.2 Dangling Participles

Present and past participles, which are formed by ending certain verbs with "ing" and "ed," begin adjectival phrases that modify the subjects of sentences or clauses. Unless a participle unambiguously modifies the subject or clause that performs the action described by the participle, the participle is said to be "dangling." This happens because the participle modifies the wrong part in the sentence or because no subject is apparent.

> Reviewing the available data, the cause of the accident was mechanical, not chemical.

As written, the sentence above implies that the "cause" was "reviewing the available data," which is clearly not the intended meaning. The subject modified by the participle is clearly identified in the following version of the sentence:

> Reviewing the available data, the committee concluded that the cause of the accident was mechanical, not chemical.

In scientific writing, the word "based" frequently creates a dangling participle.

> Based on the evidence, the accident was caused by a mechanical, not a chemical, failure.

The construction of this sentence indicates that the "accident" is based on the evidence, but the intention is more likely that a conclusion, judgment, or decision is based on the evidence.

Potential revisions:
Basing its decision on the evidence, the committee concluded that the accident was caused by a mechanical, not a chemical, failure.
Based on the evidence, the committee's conclusion was that the accident was a mechanical failure rather than a chemical one.

One way to determine whether a participle is dangling is to move it to the interior of the sentence so that it follows the subject noun:

The accident, based on the evidence, was caused by . . .

According to this test, the participle is dangling because the accident was not in fact based on the evidence.

Not all participles are used at the beginning of sentences, which can make spotting them even more difficult. In the following sentence, which ends in a participle, it is unclear who is using the altimeter.

The county was surveyed using digital altimeters.

Possible revisions:
Using digital altimeters, the workers surveyed the county.
The workers used digital altimeters to survey the county.

7.2.2.1 Dangling Participles Used with the Passive Voice

Dangling modifiers are especially common when present participles are used in sentences written in the passive voice (see Section 7.3.2, "Active versus Passive Voice"). In such a sentence, the subject is absent, so the participle mistakenly modifies the closest noun instead.

After examining the patient, the data were entered into a computer.

The passive sentence above indicates that the data examined the patient. Once the sentence is converted from the passive to the active voice, its participle logically modifies its subject.

After examining the patient, the physician entered the data into a computer.

The present participle "using" is among the most frequently misused modifiers in sentences that are written in the passive voice.

Using a programmed sequence of maneuvers, samples were collected from the Martian soil.

Because the participial phrase "using a programmed sequence of maneuvers" is adjacent to the noun "samples," the sentence above suggests that the samples used the programmed sequence. Again, revising the sentence from the passive to the active voice can correct the mistake.

Using a programmed sequence of maneuvers, the Mars rover collected soil samples.

7.2.2.2 Correcting Dangling Participles

Here are 4 options for correcting dangling participles. Each option is used to correct the dangling modifier in the sentence below.

Using a 4-mm punch, biopsies were taken from 3 skin lesions.

(1) Add a noun or pronoun for the participle to modify. If the verb in the main clause is passive, change it to an active verb.

Using a 4-mm punch, we took biopsies from 3 skin lesions.

In this revised version, the phrase "using a 4-mm punch" clearly modifies the adjacent word, the pronoun "we."

(2) Convert the dangling modifier into another construction, such as a clause with a subject and verb. If the verb in the main clause is passive, change it to an active verb.

We used a 4-mm punch to take biopsies from 3 skin lesions.

In this revised version, the participial phrase "using a 4-mm punch" is converted into the clause "we used a 4-mm punch."

(3) Modify the participle into a gerund (ie, a verb form used as a noun).

Using a 4-mm punch allowed us to take biopsies from 3 skin lesions.

In this revised version, the participial phrase "using a 4-mm punch" becomes a gerund and the subject of the sentence.

(4) Convert the participial phrase into another construction, such as a clause, but retain the passive voice. This option is especially useful in the Methods of research reports, which frequently use passive voice (see Section 1.2.2.2, "Methods").

A 4-mm punch was used to take biopsies from 3 skin lesions.

Although this version is in the passive voice, the "dangling" modifier has been eliminated by making "a 4-mm punch" the subject of the sentence.

See also Section 7.3.2, "Active versus Passive Voice."

7.3 APPROPRIATE VERB TENSES AND VERB FORMS

Writing direct and natural scientific prose is highly dependent on choice of verb tense and verb form. Choosing verb tense is relatively straightforward because it is dictated by scientific conventions. More challenging is using active verbs, gerunds, participles, and infinitives to create lively and less formulaic prose.

7.3.1 Verb Tenses in Scientific Reports

Verb tenses typically vary among the 4 standard sections of research reports, which are typically titled the Introduction, Methods, Results, and Discussion (commonly known as the IMRAD format). In general, the past tense is used to describe the actions that are outlined in the Methods and Results because these actions have already taken place. In the Introduction and Discussion, the choice of verb tense varies, depending on whether results, conclusions, or suggestions for further study are being described.

Although the next 4 sections outline the standard conventions for verb tenses in scientific reports, some research fields and some types of studies may follow different conventions. To determine which conventions to follow for a specific report, authors should review similar reports published in their targeted journal of publication.

7.3.1.1 Verb Tenses in the Introduction Section

Verb tenses vary in the Introduction based on the different elements in the section. The present tense is used to describe the current state of knowledge of the problem addressed in the study, which can include background information such as previous publications relevant to the problem. In contrast, the past tense is used to summarize the study's methods in the Introduction. In addition, if the Introduction summarizes the results and the conclusions, the past tense and present tense are used, respectively.

In the example below of an Introduction, relevant background information is in the present tense in the first paragraph. In the second paragraph, the study's methods and results are summarized in the simple past tense, which is used to describe completed action.

> Introduction
>
> Raziline is an antihypertensive drug that inhibits angiotensin-converting enzyme (ACE), which converts angiotensin I to the active vasoconstrictor angiotensin II. Treatment with ACE inhibitors increases concentrations of the receptor for angiotensin II on surfaces of epithelial cells in pulmonary alveoli. Angiotensin II receptors also serve as receptors for the novel coronavirus SARS-CoV-2, which enters cells upon binding to these receptors, thus precipitating viral infection. Increased concentration of angiotensin II receptors in patients treated with raziline may predispose these patients to infection with the virus SARS-CoV-2.
>
> To test this hypothesis, we compared the incidence of SARS-CoV-2 viral infection in patients treated with raziline with that of untreated control patients. Our analysis showed a slight increase in the incidence of viral infection in patients treated with raziline compared with controls. This difference was not statistically significant.

7.3.1.2 Verb Tense in the Methods Section

In the Methods, the verbs used to describe what was done in the study or experiment are in the simple past tense.

> To prevent contamination from RNA, we used RNAse to digest the samples.

7.3.1.3 Verb Tenses in the Results Section

Results are reported in the simple past tense because they are events that occurred in the past.

> Calcium added to resting cultures stimulated phospholipase.

An exception to this rule is that present tense verbs such as "show," "represent," "depict," "illustrate," "display," and "present" are used to refer to data in tables and figures in the Results. The present tense is used in this context because the authors are referring to the content of the manuscript rather than to past events.

> Figure 1 shows the results of the hematological tests.

7.3.1.4 Verb Tenses in the Discussion Section

In the Discussion, conclusions and generalizations about the study are stated in the present tense.

> Genlibimab shows promise as an antineoplastic agent because it can inhibit angiogenesis in acidic environments, as found in developing tumors.
>
> Our study shows that cholesterol levels rise when patients consume a diet high in saturated fat.

In contrast, the simple past tense is used for the results in the Discussion.

> Cholesterol levels dropped.

When referring to results published in other publications, authors can use the simple past or present perfect in their Discussion. As a rule of thumb, the simple past tense can be used to describe research results obtained in the intermediate to distant past (eg, 2 or more years ago). The present perfect tense can be used to describe studies completed more recently (eg, within the past 2 years).

Simple past tense:
Liu et al demonstrated that rosuvastatin significantly reduced blood lipid levels.

Present perfect tense:
David et al have demonstrated that pollinating kiwifruit with honeybees improves yield without affecting maturity.

The Discussion frequently ends with suggestions for future research to address questions raised by the current study. Authors should use the future tense to describe such proposed activities. In the example below, verb tense changes from the present tense for stating the authors' conclusion to the future tense for suggesting future studies.

> Our findings indicate that silicon used as the anode material in lithium-ion batteries increases power capacity significantly compared with similar batteries that use graphite as the anode material. The main limitation of silicon as anode material is its inherent instability. Future research will likely focus on developing materials science technologies to increase the stability of silicon anodes in lithium-ion batteries.

7.3.2 Active versus Passive Voice

Active voice is preferred to passive voice because it produces sentences that are more direct and easier to understand. Active voice focuses sentences on the actors and the actions they are taking. In scientific prose, active voice is often used in the Introduction to describe the current state of the problem or other situation being studied, while passive voice is typically used in the Methods to describe the processes that were followed.

With active voice, the subject of a sentence performs the action of the verb, while with passive voice the subject receives the action of the verb.

Active voice:
The chemist prepared the samples.

Passive voice:
The samples were prepared by the chemist.

In the first example above, "chemist" is the subject, and "samples" is the object. In the second example, "samples" is the grammatical subject of the passive verb "were prepared."

Sentences in the passive voice always include a form of the verb "to be" (eg, "is," "are," "was," and "were") and a past participle (eg, "prepared," "centrifuged," and "analyzed"). The actor in a passive sentence may be identified in a prepositional phrase beginning with "by" (as in "by the chemist" in the example of passive voice above), or the actor may not even be mentioned (as in, "The samples were prepared").

For more information on verbs and their conjugations, refer to *Barron's 501 English Verbs*.[2]

7.3.3 Gerunds ("ing" Verbs)

Gerunds are verb forms ending in "ing" that function as nouns in a sentence. Gerunds strengthen scientific prose by conveying action, unlike nouns.

Gerunds can serve as the subject of a sentence, as "running" does in the example below.

> Running the ultracentrifuge overnight enabled us to complete the experiment in 24 hours.

Gerunds can be the object of a sentence, as "sequencing" is below.

> The study included sequencing the entire genome.

Gerunds also can be the object of a preposition, as "dieting" is the object of the preposition "by" in this example.

> The subjects were able to lower their cholesterol by dieting.

7.3.4 Participles as Noun Modifiers

Both present participles and gerunds are verb forms ending in "ing," but they have different functions. Instead of serving as a noun like a gerund, a participle functions as a modifier of a noun like an adjective (see Section 7.2.2, "Dangling Participles"). In the example below, the present participle "confounding" modifies the noun "data."

> The confounding data were omitted.

In addition to present participles, which end in "ing," this verb form includes past participles, which end in "ed" or "en." Here, the past participle "centrifuged" modifies the noun "cells."

> The centrifuged cells were transferred to another tube.

7.3.5 Infinitives (to + verb)

The infinitive form of verbs can be used as nouns. Infinitives consist of the word "to" followed by a verb. Like gerunds, infinitives convey action.

In the example below, the infinitive "to complete" is the subject of the sentence.

> To complete the survey on time required new personnel.

Below, the infinitive "to conduct" is the object of the sentence.

> They wanted to conduct the experiment during the growing season.

7.3.6 Verbs that Require Gerunds or Infinitives

In English, correct idiomatic usage requires that certain verbs be immediately followed by a gerund. Similarly, other verbs must be immediately followed by an infinitive. Still other verbs should be followed by either a gerund or an infinitive.

Examples of verbs followed by a gerund include "suggest" and "consider."

> Dr Jackson suggests using different primers.
> Our committee considered purchasing new software.

Verbs that are followed by an infinitive include "agree" and "seem."

> She agreed to resubmit the paper.
> The mice seemed to respond to the drug.

"Remember" and "prefer" are examples of verbs that should be followed immediately by either a gerund or an infinitive.

> The oceanographer remembered deploying the buoy.
> The oceanographer remembered to deploy the buoy.
> The 3 astrophysicists prefer using Macs to PCs.
> The 3 astrophysicists prefer to use Macs instead of PCs.

Authors and editors can refer to the *Longman Dictionary of Contemporary English*[3] and *Barron's 501 English Verbs*[2] to determine whether a specific verb should be followed by a gerund, an infinitive, or either.

7.3.7 Replacing Latinate Nouns with Active Verbs and Gerunds

A common source of unnecessarily long sentences and dull, static prose is the use of abstract nouns, typically Latinate nouns formed from verbs and ending in such suffixes as "ion," "tion," "ment," and "ance." Examples of such abstract nouns are "production" from "produce," a derivative of the Latin word "producere," and "interpretation" from "interpret," a derivative of the Latin term "interpretari." Besides being lengthy, these abstract nouns contribute to dullness because they require linking verbs (eg, "am," "are," "is," "was," and "were") or passive verbs, as well as prepositions.

Authors and editors can often remedy this by converting abstract nouns to their verb roots.

> **Original:**
> A direct correlation between serum vitamin B_{12} concentration and mean nerve conduction velocity was seen.

> **Revision:**
> The mean velocity of nerve conduction correlated directly with the vitamin B_{12} concentration in serum.

Replacing abstract nouns with their equivalent verbs not only changes the voice from passive to active, it also clarifies the subjects of sentences and makes sentences more specific and vivid.

> **Original:**
> Following termination of exposure to pigeons and resolution of the pulmonary infiltrates, there was a substantial increase in lung volume, some improvement in diffusing capacity, and partial resolution of hypoxemia.

> **Revision:**
> After the patient stopped keeping pigeons, her pulmonary infiltrates partly resolved, her lung volume greatly increased, her diffusing capacity improved, and her hypoxemia partially resolved.

Similar results can be obtained by replacing Latinate nouns with gerunds. In the example below, gerunds are substituted for 4 Latinate "tion" nouns, and the main verb is changed from the passive to active voice.

> **Original:**
> In the conduction of clinical trials of experimental medications, randomization of the study population, confounding variable identification and documentation, and rigorous statistical analysis are strongly recommended.

> **Revision:**
> In conducting clinical trials of experimental medications, we strongly recommend randomizing the study population, identifying and documenting all confounding variables, and rigorously analyzing the statistics.

7.4 INCLUSIVE LANGUAGE

Stereotypic language and other biased vocabulary should be avoided both because they reduce scientific accuracy and because they ignore and obscure the complexity of scientific questions. In addition, such vocabulary might insult or otherwise offend readers and the subjects of the discourse.

This section of this style manual provides a few examples to sensitize authors and editors to inclusive language, also known as "conscious language." Additional guidance on diversity, equity, and inclusion (DEI) in scientific publishing is available from the Council of Science Editors' Diversity, Equity, and Inclusion Scholarly Resources.[4] Entire books also have been written on identifying and avoiding biased language,[5] and relevant advice is available in some other style manuals.[6–9]

7.4.1 Grammatical Gender

English nouns and pronouns have traditionally been categorized into 4 genders: masculine, feminine, common, and neuter. Masculine and feminine gender is determined by the sex of the individual described by a noun or pronoun (eg, "man" and "he" are masculine, while "woman" and "she" are feminine). Common gender applies to nouns and pronouns that do not refer to a specific sex (eg, "scientists" "postdoctoral students," and "they"), while neuter gender applies to nouns and pronouns that do not have a sex (eg, "experiment," "planets," and "it"). Masculine, feminine, and common terms are used to describe human beings, as well as other animals with personal names. Neuter terms are used for objects, as well as animals without personal names.

Grammatical difficulties arise with gender-specific nouns and pronouns when the sex of an individual is unknown or if the individual self-identifies as other than male or female. Both circumstances can result in awkward terms, cumbersome constructions, subject-object disagreements, and other concerns. Such problems can often be avoided by substituting common-gender terms for masculine and feminine ones.

> A business executive approved 3 research grants. *not* A businessman or businesswoman approved 3 research grants.
> A camera operator should be assigned to the research department. *not* A cameraman or camerawoman should be assigned to the research department.
> A firefighter extinguished the fire in the lab. *not* A fireman or firewoman extinguished the fire in the lab.
> We have a new mail carrier. *not* We have a new mailman or mailwoman.
> A meteorologist can focus on research or forecasting. *not* A weatherman or weatherwoman can focus on research or forecasting.
> Today, a police officer is more likely to have a degree in criminal justice than 50 years ago. *not* Today, a policeman or policewoman is more likely to have a degree in criminal justice than 50 years ago.
> A repair technician fixed our electron microscope. *not* A repairman or repairwoman fixed our electron microscope.
> Journals without sales representatives are likely to have limited ad revenue. *not* Journals without salesmen or saleswomen are likely to have limited ad revenue.

Also use common-gender terms in place of male referents like "mankind" that refer to groups that include more than men.

> The analysis ignored the environmental concerns of ordinary people. *not* The analysis ignored the environmental concerns of the man in the street.
> Scientific discoveries in the past 100 years have advanced the knowledge of all human beings. *not* Scientific discoveries in the past 100 years have advanced the knowledge of mankind.

However, whenever possible, avoid coining awkward terms by substituting "person" for "man" or "woman" at the end of terms. Sometimes, rewriting the text to describe the subject's role with a verb instead of a noun may offer a solution.

> Gordon R. England chaired the National Academy of Engineering from 2016 to 2020. *instead of* Gordon R. England was the 2016–2020 chairperson of the National Academy of Engineering. Judy Woodruff began co-anchoring "PBS NewsHour" in 2011. *instead of* Judy Woodruff became a co-anchorperson of "PBS NewsHour" in 2011.

Another way to avoid awkward wording and grammatical errors in such situations is to use plural nouns and pronouns when an antecedent can be either plural or singular. This solution eliminates resorting to constructing cumbersome objects such as "he or she" or using the singular they (see Section 7.4.2, "Singular They").

> All of the patients signed their consent forms. *not* Each patient signed his or her [or their] consent form.
> Corresponding authors are responsible for keeping their fellow authors informed of the status of their submissions. *not* The corresponding author is responsible for keeping his or her [or their] fellow authors informed of the status of their submission.

7.4.2 Singular They

"Singular they" refers to using the plural third-person pronoun or one of its derivatives (eg, "them" and "their") with a singular antecedent, often because the sex of the antecedent is unclear. The "singular they" and its derivatives should be used sparingly and judiciously. The "singular they" often can be avoided by rewriting sentences to use the plural they or to avoid pronouns altogether.

> All of the physicians found that they had difficulty executing the protocol. *instead of* Each physician found that they had difficulty executing the protocol.
> Only one participant was dropped from the study because of noncompliance with the protocol. *instead of* Only one participant was dropped from the study because they were out of compliance with the protocol.

In cases in which the antecedent is only one person, the "singular they" may be used if that person self-identifies with the pronoun "they." It may also be used if the person's gender is unknown, undisclosed for confidentiality or study-masking purposes, or otherwise inappropriately described by the pronouns "he" or "she."

> When the patient self-identified as nonbinary during their recent history and physical, they also noted that they had a nonbinary sibling.
> Of the 4 men and 1 woman who participated in the pilot vaccine study, 1 person did not obtain immunity according to their antibody level.

Special considerations may apply to scientific studies in which the participants' sex is an essential variable. Under such circumstances, study authors may need to report whether they used sex as assigned at birth or as self-reported.

7.4.3 Race, Ethnicity, Nationality, and Citizenship

Because the term "race" does not have a precise definition in biological terms, its use is subject to limitations in scientific context. When the term is applied to human beings, its use depends mainly on judgments about physical characteristics that can vary widely. When applied to other organisms, "race" is an infraspecific category that is defined on the basis of geographic range, physiologic traits, and other factors.

"Ethnicity" is based on such factors as culture, history, nationality, and religion. Both "nationality" and "citizenship" describe a person's relationship to a specific country.

Nationality may be defined by a variety of factors, including citizenship and place of birth. Citizenship has a narrower definition based on legal status in a sovereign state.

Whenever possible, descriptions of human populations or large social groups should draw on sharply definable criteria (eg, country of birth or residence and self-described identity). If race, ethnicity, nationality, citizenship, or religion is pertinent to a scientific study (eg, the condition under investigation has a higher prevalence in a genetically related population than in the population at large), this relevance should be justified explicitly in the study report, and the method of measuring the variable should be stated (eg, self-identification).[10]

When race or ethnicity is pertinent to a study, terms based on color and customary usage should be capitalized, including "Black," White," "Asian," "Pacific Islander," "Indigenous," "Native Hawaiian," "Appalachian," and "Highlander."[11] Also capitalize "Latino," which can refer to an individual man, a group of men, or a group of men and women; "Latina," which can refer to an individual woman or group of women; and "Latinx," a proposed gender-neutral term. Terms based on color and customary usage are ever changing, so authors and editors should be aware that the terms used in this style manual may evolve. For example, the term "Afro-American" has changed to "African American," "Eskimo" has generally been replaced with "Inuit," and "Oriental" has been replaced with "Asian."

In addition, pseudoprecise terms should not be used. For example, "Caucasian" is no more scientifically precise than "White," and many consider "Caucasian" to be archaic.

The American Medical Association's Manual of Style Committee offers more detailed guidelines on using race and ethnicity in biomedical publications.[12] Also see Section 8.3, "Human Groups."

7.4.4 Inappropriate Categorization of Individuals with Health Problems

In referring to individuals with health problems, avoid such depersonalizing terms as "an alcoholic," "a diabetic," "an epileptic," "a schizophrenic," and "a disabled person." Instead, use "person-first descriptions," such as "a patient with diabetes mellitus," "a child with epilepsy," and "a person with a disability."

It is especially important to avoid using the term "handicap" when referring to a person with a disability. "Handicap" is a judgmental term, whereas the term "disability" refers to a condition that limits the ability of a person to carry out some common activity or function. However, "handicap" can be used to refer to environmental and attitudinal barriers to usual functioning. ("The outmoded design of the stairway in the building's entrance is a handicap to wheelchair users.")

Also avoid such terms as "mentally ill," "insane," and "mental defect." Instead, specify the diagnosed condition as part of a person-first description (eg, "a person with bipolar disorder.")

7.5 CONFUSED AND MISUSED TERMS

7.5.1 Homophones and Near Homophones

"Homophones" are words with the same sound but different meanings and spellings. One of the most commonly misused set of homophones is "your" and "you're." The former, which is the possessive of "you," is often mistakenly used in place of the latter, which is the contraction for "you are" (eg, "your welcome" is a frequent mistake). "Near homophones" are words that are pronounced nearly the same. "Immigrate," which refers to moving to another place, is often confused with its near homophone, "emigrate," which refers to moving from a place.

Authors may end up using the wrong homophone or near homophone if they are uncertain of the correct spelling of the intended word. Spellcheckers in word-processing programs may not detect such misuse. In fact, word-processing programs sometimes recommend the wrong homophone.

The following are other common homophones and near homophones:

albumen, albumin	hear, here	read, red
affect, effect	hour, our	shear, sheer
are, our	its, it's	tenant, tenet
cell, sell	one, won	their, there
complement, compliment	poor, pour	to, too
discreet, discrete	principal, principle	weather, whether

7.5.2 Similar-Sounding Last Syllables

A spelling error akin to selecting wrong homophones is spelling words with similar-sounding but incorrect ending syllables. "Able" and "ible" are often incorrectly interchanged, as are "ance" and "ence" and "ant" and "ent." Here are the correct spellings of some words whose ending syllables are frequently misspelled in scientific writing:

abundance	dispensable	intransigence	prominent
acceptable	divisible	occurrence	relevant
accessible	existence	permanence	resemblance
compatible	feasible	preferable	resilient
consistent	independent	preference	resistant
credible	inevitable	prevalence	solvent

7.5.3 Imprecisely Applied Words

When preparing scientific publications, authors and editors should strive to select words that accurately and precisely convey the intended meaning. The examples below are commonly used words in scientific writing that have similar but distinct definitions. The definitions below focus on usage in scientific contexts and on distinctions that can clarify meaning. Some of these words may have more specific meanings in some scientific disciplines. These meanings can be found in scientific dictionaries, such as the *McGraw-Hill Dictionary of Scientific and Technical Terms*.[13] For common words, *The American Heritage Dictionary of the English Language*[14] discusses nuances and preferences in its synonym and usage notes. If authors know which publisher they will submit their work to, they should defer to the scientific and general dictionaries preferred by that publisher.

a, an, the: The articles "a," "an," and "the" precede and function to identify nouns either as one member of a group or as a specific item. The articles "a" and "an" are used correctly when they precede a noun that is one member of a group. For example, in the sentence "A new drug was approved for treating sickle cell disease," the new drug is one member of a group of drugs for sickle cell disease that exist or may be identified in the future. If the noun is not part of a group and is likely to remain so, the article "the" is appropriate. For example, in the sentence "The Central Air Data Computer may have been the world's first microprocessor," the first "the" precedes the name of a specific device while the second "the" refers to a one-of-a-kind ranking. See also Section 7.10, "Use of Articles."

abnormality: See **disease, abnormality, condition**.

absorbance, absorptance, absorptivity: "Absorbance" is the logarithm of the ratio of the intensity of light entering a solution to the intensity of light transmitted through the solution. "Absorptance" and "absorptivity" refer to the ratio of energy absorbed by a body to the energy striking it.

absorption, adsorption: "Absorption" is the ongoing or completed process of taking up by capillary, osmotic, chemical, or solvent action. "Adsorption" is the holding of something by the surface of a solid or liquid through physical or chemical forces.

accuracy, precision: "Accuracy" is the degree of correctness of a measurement or of a statement. When applied to a measurement, "precision" is the degree of refinement with which the measurement is made or stated. For example, the number 3.43 is more precise than 3.4, but it is not necessarily more accurate. When applied to a statement, "precision" implies the qualities of definiteness, specificity, and succinctness.

adduce, deduce, induce: To "adduce" is to bring forward as an example or as evidence for proof in an argument. To "deduce" is to reason to a conclusion or infer it from a principle. To "induce" is to reach a conclusion through inductive reasoning, such as using particular facts to establish a principle. To "induce" also means to bring about an effect, as in "to induce labor."

affect, effect, impact: As a verb, "affect" means to influence. ("The budget cut affected all of the university's research programs.") As a noun, "affect" means the impression of feeling or emotion conveyed by a person's demeanor, action, or speech. ("The diagnostic feature in this case was the patient's flat affect.") "Effect" as a verb means to bring about a change. ("By adopting the scientific method, researchers effected a scientific revolution.") As a noun, "effect" refers to the result of some action. ("The effects of climate change are worldwide.") Although often used as a synonym for "affect," "impact" as a verb should be reserved for the striking of one body against another. ("Humans have long contemplated what would happen should an unstoppable force impact an unmovable object.) "Impact" as a noun can refer to a collision of 2 objects or a strong effect of an action. ("The impact of the meteorite resulted in a crater." "The impact of the coronavirus pandemic was drastic.")

after, following: "After" can mean simply later than a particular time or event. ("He died of anaphylactic shock soon after you saw him.") "After" can also imply cause and effect. ("He died of anaphylactic shock after swallowing a capsule of the wrong antibiotic.") Although "following" is a synonym of "after," it is best reserved as an indicator of position not related to time. ("The following authors are frequently cited in genetics textbooks: Charlotte Auerbach, PhD; Gregor Mendel; and Sewall Wright, ScD.")

aggravate, exacerbate, irritate: To "aggravate" and "exacerbate" mean to worsen an existing condition. To "irritate" means to evoke a reaction (eg, cause inflammation). As a consequence, "irritate" refers to a new condition.

aliquot, aliquant, sample: An "aliquot" is a part of a total amount of a gas, liquid, or solid that divides evenly into the whole. For example, 10 mL is an aliquot of 100 mL. An "aliquant" is a part that does not divide evenly into the total. For example, 8 mL is an aliquant of 20 mL. A "sample" is a part taken as representative of its source for analysis or study, such as a blood sample obtained for analysis.

alternate, alternative: As both a verb and an adjective, "alternate" describes the successive passage back and forth from one state, action, or place to another. ("Water alternates between liquid and solid states in this temperature range." "The alternate states were liquid and solid.") "Alternative" is used as both a noun and adjective to represent a choice between 2 or more mutually exclusive states or places. ("An alternative method is to use a catalyst.")

although, whereas, while: "Although" is best used in scientific writing to mean "in spite of." ("Although he was treated with antibiotics, he died of the infection.") Like "although," "whereas" can be used to indicate contrast. ("An electric vehicle can be carbon-neutral if it obtains electricity from a power grid that is 100% renewable, whereas a vehicle with an internal combustion engine cannot be carbon-neutral.") However, "whereas" is best used to introduce a conditional statement, and in that sense, it is similar to "because." ("Whereas Earth's magnetic north pole shifts with time, its location differs from that of the planet's geographic north pole.") "While" is best used to indicate a period of time. ("While the patient was waiting for surgery, his angina pectoris became steadily more frequent.")

among, between: The preposition "among" indicates a relation involving 3 or more units of the same kind, with an emphasis on the units' relationships to an entire group. ("Among the antibiotics in this class, azithromycin is the best choice.") The preposition "between" emphasizes one-on-one relations involving units of the same kind. ("The study examined the interaction between predator and prey.") Although typically used with only 2 units, "between" can be used to describe 3 or more units, provided it is describing the one-to-one relations between the units instead of each unit's relationship to the entire group. (In the sentence "The choice was between penicillin, ampicillin, and erythromycin," the one-on-one relations are between penicillin and ampicillin, between ampicillin and erythromycin, and between erythromycin and penicillin, as opposed to comparing each antibiotic to the rest of the group.)

an: See a, an, the.

analog, analogous, homolog, homologous, homoeolog, homoeologous: The noun "analog" and the adjective "analogous" refer to organs or other structures that are similar in function but dissimilar in origin. In evolutionary biology, for example, sharks and dolphins both have dorsal fins to provide them with stability while swimming, but these fins are analogous structures because sharks are fish and dolphins are mammals. "Homolog" and "homologous" refer to organs and other structures that correspond in structure, position, origin, or other characteristics but not necessarily in function, such as homologous chromosomes and homologous proteins. "Homoeolog" and "homoeologous" refer to partially homologous chromosomes or genes. ("In *Drosophila*, genes X and Y are homoeologs.")

anatomy, morphology, structure: "Anatomy" is the study of structure, especially of living things. In addition, "anatomy" refers to structural makeup. "Morphology" is the study of shape, structure, or formation of living things, especially external shape and form. "Morphology" also refers to structural characteristics. ("The morphology of the human knee results from the internal anatomy of this joint.") "Structure" refers to the parts of living or nonliving things as they relate to each other. ("Suspension bridges and cable-stayed bridges are similar in structure because the decks of both types of bridge are supported by cables connected to towers.")

ante, anti: The prefix "ante" indicates before in a sequence or location (eg, "antedate" and "antebrachial artery"). The prefix "anti" indicates against, opposed to, or in contrast to (eg, "antibody," "antiparticle," and "anti-inflammatory").

as, because, since: To avoid ambiguity, use the conjunction "as" only to convey a sense of time. ("As we were completing the paper, new evidence came to light.") Avoid using "as" to convey cause. (*Not* "As we were away in Africa, he thought he could pilfer our data.") Instead, use the conjunction "because" to show a causal relationship. ("Because our results were inconclusive, we were unable to publish our study.") Like "as," the conjunction "since" should only be used to show a temporal relation. ("Since then, there have been 3 additional outbreaks.") To avoid ambiguity, "since" should not be used in a causal sense. (In the sentence "Since the initial finding was validated, our study has been discontinued," it is unclear whether the study was discontinued because the initial finding was validated or whether the study was discontinued after the finding was validated.)

assess, determine, evaluate, examine, measure: These terms are frequently used synonymously, but they have distinct meanings. "Assess" should be reserved to mean estimating a value, as in taxation, or setting the amount of a payment, as in a fine. Although commonly used as a synonym for "measure," the verb "determine" means to set a limit on or to establish conclusively. "Evaluate" means to ascertain or fix a value on the object of the action, whereas the meaning of "examine" is to look at, inquire into, or test closely. "Measure" means to examine an object quantitatively, as in measuring the concentration of blood glucose.

assure, ensure, insure: "Assure" connotes providing surety, as in making a promise, pledge, or guarantee. ("The dean assured the faculty members that research would remain their top priority.") "Ensure" means making sure or removing doubt. ("To ensure consistent results, the researchers created an extensive checklist.") "Insure" should be used when the intended meaning is related to a financial guarantee or pledge related to property or life. ("In the United States, Medicare insures more than 60 million elderly.")

average, characteristic, typical: Although the term "average" can refer to "statistical mean," average should generally be avoided in scientific usage because the term can be ambiguous. "Mean" should be used instead (see the entry for "mean, median, mode"). "Characteristic" and "typical," which convey a sense of representativeness, are often suitable in place of "average" for nonstatistical purposes.

axenic, gnotobiotic: The adjective "axenic" refers to organisms kept in isolation from other living things or an environment for such isolation. The adjective "gnotobiotic" refers to germ-free and other laboratory animals that live in the presence of only known species of microorganisms.

based on, on the basis of: "Based on" should be used as a verb, not as an adverb. ("They based their conclusion on the evidence from multiple trials.") "On the basis of" is an adverbial form. ("The treatment rate was decreased on the basis of effective results in field trials.")

because: See as, because, since.

because of, due to: "Because of," which is a conjunction with a preposition, means as a result of or owing to. ("The problem occurred because of a mechanical failure.") "Due to," which is an adjective with a preposition, means attributable to. ("The problem was due to a mechanical failure.") The correct choice for a sentence can be revealed by substituting "as a result of" for "because of" and "attributable to" for "due to."

believe, feel, think: These subjective verbs connote different degrees of conviction or persuasion. "Believe" implies a definite conviction regardless of the strength of evidence. "Feel" implies an intuitive or not fully reasoned conviction. "Think" implies a view based on evidence or logic. Because no one can be sure of what someone else thinks, authors should reserve these terms to convey their own thoughts. In quoting others, authors should use a neutral term such as "said."

benchmark, criterion, standard: Although these 3 nouns can be used synonymously, they do have some differences in meaning. "Criterion" is a benchmark without an assigned value, whereas "standard" is a benchmark that has a specific value. When used in the context of judgment, "criterion" refers to a specification for favorable or unfavorable judgment, whereas "standard" is a specific value against which a judgment is made.

between: See among, between.

bi, semi: The prefix "bi" can indicate 2, as in "binary" and "bipolar," or intervals of 2 units, as in "binary" and "bicentennial." The prefix "semi" indicates half, as in "semicircle" and "semidiurnal," or partial, as in "semiretirement" and "semiconscious." Notable exceptions to this distinction between the 2 prefixes are that "bimonthly" can mean either every 2 months or semimonthly and "biannual" and "semiannual" both mean twice a year. The term for every 2 years is "biennial," not "biannual." To avoid ambiguity, "semimonthly" and "semiannual" are preferred to "bimonthly" and "biannually," respectively, and the phrase "every 2 years" is preferred to "biennial." Also see quasi, semi.

can, may: "Can" designates ability. ("The hospital can accommodate 200 patients.") "May" designates permission. ("You may reference previously published materials if you cite your sources.")

carry out: See perform, carry out, execute.

case, patient: A "case" is an instance, example, or episode. ("She had a case of measles" and "The worst case would be 2 earthquakes within 2 days.") A "patient" is a person. ("We saw 12 patients in the clinic.") In medical literature, a case is "reported" in terms of the description of presentation, diagnosis, treatment, and outcome for a patient with a particular condition, but the patient with the condition is "described."

cause: See etiology, cause.

characteristic: See average, characteristic, typical.

circadian, diurnal: The adjective "circadian" refers to an occurrence approximately every 24 h. The meaning of the adjective "diurnal" varies by scientific field. In meteorology, "diurnal" refers to processes that are completed within 24 h and recur every 24 h. In biology, the term refers to processes that occur daily or every day in the daytime. In botany, "diurnal" refers to the characteristic of being open in the daytime and closed at night, as occurs with many flowers.

common, frequent, regular: "Common" means appearing often or repeatedly (eg, "a common finding in this type of study"). "Common" also refers to vernacular names of organisms, as opposed to their scientific names. "Frequent" means occurring often or at relatively short intervals. ("Rainy days are frequent during the spring.") "Regular" indicates ordered or consistent. It also can mean at fixed times or points. ("Halley's comet makes a regular appearance every 75 to 76 years.")

comparable, similar: Reserve "comparable" to indicate comparisons between items. ("After we studied demographic data, we concluded that the mortality statistics of Sweden and Germany are comparable.") The adjective "similar" should be used to indicate likeness. ("The colors fuchsia and magenta are similar hues.")

compared to, compared with, than: "Compared to" focuses on similarities between dissimilar objects. Often, these comparisons are metaphoric or fanciful. ("He compared the bird's flight to running through the air.") More frequently used in scientific writing is "compared with," which implies looking for similarities or differences in objects that fall within the same classification. ("In the clinical trial, a low-fat diet was compared with a high-fat diet.") However, "compared with" is often mistakenly used to mean "than," so before using "compared with," authors should determine whether "than" is sufficient ("Penicillin has fewer serious side effects than amoxicillin." *not* "Penicillin has fewer serious side effects compared with amoxicillin.") Also see versus, vs, v.

complement, compliment: As both a verb and noun, "complement" conveys the sense of completing something. ("We proposed a follow-up survey to complement the original study." "The follow-up survey was a complement to the original study.") When used as a scientific term, the noun form of "complement" may refer to a group of proteins in serum that have activity in the immune system. In contrast, as both a verb and noun, "compliment" refers to a favorable comment. The adjective "complimentary" frequently has the same meaning, although it can also describe a good or service that is provided free of charge.

compose, comprise, consist of, constitute: When used as active verbs after plural subjects, "compose" and "constitute" mean to form, to make up a single object, or to go together. The subjects of these verbs are individual parts that make up whole entities. ("Seven elements compose [or constitute] group 1 of the periodic table.") Another meaning for "constitute" is to amount to, to equal, to set up, or to establish. ("Twenty-seven moons constitute the known natural satellites of the planet Saturn.") In contrast, when used as passive verbs, "compose" and "consist" are synonymous with the active verb "comprise" in that they mean to include, to contain, or to be made up of. As a consequence, the subjects of passive "compose" and "consist" and active "comprise" are whole entities, not individual parts, while the objects are the subjects' individual parts, not whole entities. ("The periodic table consists of [or is composed of or comprises] 118 elements.") In sentences that use the active voice, "constitute" is generally less cumbersome than "compose," whereas "consists" is preferable in sentences that use the passive voice.

condition: See disease.

congenital, genetic: The adjective "congenital" means born with or present at birth. The adjective "genetic" refers to genes, chromosomes, or their effects in producing phenotypes or determining heritable characteristics. A disease or abnormality caused by genetic effects is not necessarily apparent at birth, nor is a congenital defect necessarily caused by genes, chromosomes, or their effects.

conjecture: See law, theory, hypothesis, conjecture.

connote, denote: "Connote" implies a meaning beyond the usual, exact definition of a thing. ("The new budget connotes the university's growing commitment to research.") "Denote" indicates the presence or existence of a thing. ("Rapidly rising CO_2 levels denote increased industrial activity.")

consist: See compose, comprise, consist of, constitute.

constitute: See compose, comprise, consist of, constitute.

contagious, infectious, infective: "Contagious" means capable of being transmitted from one organism to another. "Infectious" means harboring an infecting agent or caused by an infecting agent ("an infectious disease"). "Infective" describes an agent that can cause infection. ("Not all the bacteria found in this environment are infective.")

continual, continuous: "Continual" means a prolonged succession or recurrence over time with or without interruption ("continual state of arousal" and "continual encounters within the enclosure"). "Continuous" implies never ceasing even briefly. ("From the surface of the moon, the view of Earth is continuous.")

conventional: See customary, conventional, traditional, normal, norm.

criterion: See benchmark, criterion, standard.

currently, at present, presently: Although both "currently" and "at present" mean now, "currently" is the preferred term. In contrast, "presently" means in the near future.

customary, conventional, traditional, normal, norm: "Customary" means long used, commonly practiced for a long period, or used habitually. "Conventional" means an established and generally agreed-on practice or characteristic. "Traditional" means long used or applied, and it tends to connote long-standing and general acceptance in a professional or social group or community. "Normal" is an adjective indicating that the noun it modifies has the characteristics of a satisfactory or desirable majority or lacks any abnormality. However, "normal" and its antonym "abnormal" should not be used to describe the health status of people. "Normal" also has the following statistical meaning: A "normal" distribution of values is centered on the mean and the median, and it has a symmetrical shape, which is often called a bell-shaped curve. "Norm" implies a desired characteristic, not just a normal characteristic.

data, database, data set: The plural noun "data" refers to individual items of information. Its singular form is "datum." A "database" is a formal structure, such as a computer file or printed document, that contains data on a conceptually coherent subject that is organized for retrieval and analysis. A "data set" is a particular coherent body of data maintained in a database.

deduce: See adduce, deduce, induce.

definite, definitive: "Definite" means clearly limited or firmly established, whereas "definitive" means conclusive or defining.

demonstrate, exhibit, reveal, show: Although "demonstrate" and "exhibit" are often used as inflated versions of "show," "demonstrate" should be reserved for a deliberate action intended to illustrate a procedure, while "exhibit" should be used for a deliberate action to make something visible. ("The technician demonstrated how to operate the pH meter." "The mineralogists exhibited their new specimens at the last congress.") Neither term should be used to mean passively having something apparent. (Use "The patient had a rash" rather than "The patient demonstrated [or exhibited] a rash.") Note that inanimate agents cannot demonstrate anything, so avoid such constructions as the following: "The data demonstrated an increase in blood pressure when the dose was lowered." "Reveal" represents an action to make visible what had been hidden or unapparent. "Reveal" should not be used as a synonym for "report." The verb "show" is commonly used in scientific publications to describe data in tables and figures. ("Figure 5 shows the patient's status after the surgery.")

denote: See connote, denote.

determine: See assess, determine, evaluate, examine, measure.

different, diverse, disparate: "Different" means having at least some dissimilar characteristics. ("The fossil evidence establishes that this outcropping is a different formation.") "Diverse" means having a notable range of differences. ("University professors have diverse interests and skills.") "Disparate" means distinctly different. ("The researchers reached disparate conclusions after analyzing the same evidence.") Also see varying, differing, different, diverse, various.

different from, different than: The phrase "different from" is preferred if it precedes a noun form (ie, a noun, pronoun, noun phrase, or noun clause). ("The results are different from those that we expected.") The phrase "different than" is preferred if it precedes an adverbial clause. ("The results are different than we expected.")

differing: See varying, differing, different, diverse, various.

digit: See number, numeral, digit.

disease, abnormality, condition: "Disease" describes a specific set of characteristics that define a disorder in humans, other animals, or plants. In humans, for example, these characteristics can include symptoms, signs, and anatomical changes. "Abnormality" is a deviation from normal characteristics that may or may not be related to a disease. Although a human can have an "abnormality," a human should not be called an "abnormality." "Condition" is a broader term in that it takes into account the overall state of a human, other animal, or plant, including diseases and abnormalities.

disinterested, uninterested: "Disinterested" means to have no stake in the outcome, whereas "uninterested" means unconcerned or indifferent.

disparate: See different, diverse, disparate.

distinctive: See unique, unusual, distinctive, rare.

diurnal: See circadian, diurnal.

diverse: See different, diverse, disparate.

dosage, dose: In medicine, "dosage" implies not only the amount of a medication or treatment but also the frequency of its administration. ("The dosage of analgesic was changed to 50 mg every 4 h.") "Dosage" is not synonymous with "dose," which means the amount of a drug administered at one time. ("Each dose of the analgesic is 50 mg.") "Dose" is also used to refer to the amount of radiation administered in radiotherapy and to the amount of chemicals applied in agriculture.

due to: See because of, due to.

effect: See affect, effect, impact.

employ: See use, utilize, employ.

ensure: See assure, ensure, insure.

epidemic, epiphytic, epizootic: All 3 terms refer to outbreaks of disease or associated conditions, usually widespread in a population of hosts or occurring at a frequency much higher than normal. "Epidemic" refers to a disease in a human population, "epiphytic" to an outbreak of infectious disease in plants, and "epizootic" to a disease in an animal population other than humans.

etiology, cause: "Etiology" should be reserved to mean the study and description of a cause. It should not be used as a synonym for "cause," which is the reason why something occurred.

evaluate: See assess, determine, evaluate, examine, measure.

exacerbate: See aggravate, exacerbate, irritate.

examine: See assess, determine, evaluate, examine, measure.

execute: See perform, carry out, execute.

exhibit: See demonstrate, exhibit, reveal, show.

farther, further: As either an adverb or adjective, "farther" describes a more distant physical point. "Further" can be used as a transitive verb, an adverb, or an adjective to mean moving along or developing conceptually. ("The theory did little to further our knowledge of galaxies farther away than Andromeda.")

feel: See believe, feel, think.

few, fewer, less: "Few" and "fewer" are adjectives indicating small and smaller in number, respectively. These 2 terms are used with counted or countable items. ("We have fewer astrologers per capita now than we did in 1900.") "Less," a cognate of "little," is used with uncounted or uncountable quantities, and it means a smaller total amount. ("We have less information on the genome of the horse than on that of the human.")

flammable, inflammable, nonflammable: Both "flammable" and "inflammable" are adjectives meaning capable of combustion. "Nonflammable" means noncombustible. "Flammable" is preferred to "inflammable" because the prefix "in" usually means "the opposite of," which can result in "inflammable" being confused for "nonflammable."

fungus, fungal, fungous, fungoid: The noun "fungus," whose plural is "fungi," refers to an organism with nucleated cells having rigid walls but no chlorophyll. The adjectives "fungal" and "fungous" mean relating to or caused by a fungus ("a fungal [or fungous] infection"). However, "fungal" is preferred because "fungous" can be confused with the noun "fungus." The adjective "fungoid" means resembling a fungus.

further: See farther, further.

general, generally, generic, generically, usual, usually: The adjective "general" and the adverb "generally" describe a broad or group-typical application or characteristic. ("Scientists generally devote their careers to specific fields of study.") Similar to their strictly scientific meaning of relating to genus, "generic" and "generically" can refer to items that are of the same category. ("Biology, chemistry, physics, and mathematics fall under the generic category of basic science.") "Generic" also refers to the nonbranded names of brand products. ("The generic name for the Alzheimer disease drug Aduhelm is aducanumab-avwa, while Velcro is generically described as a hook-and-loop fastener.") Also see Section 20.1.1, "Nonproprietary (Generic) Drug Names." "Usual" and "usually" connote situations that are likely, expected, or relatively frequent. ("Influenza cases usually increase in the fall and winter months.")

genetic: See congenital, genetic.

gnotobiotic: See axenic, gnotobiotic.

haphazard: See random, haphazard.

healthy, healthful: "Healthy" denotes good health in a living organism. "Healthful" means conducive or supportive of good health. ("A healthful lifestyle can help to keep a person healthy.")

heterogenous, heterogeneous: See homogenous, homogeneous, heterogenous, heterogeneous.

homoeolog, homoeologous: See analog, analogous, homolog, homologous, homoeolog, homoeologous.

homogenous, homogeneous, heterogenous, heterogeneous: "Homogeneous" and "heterogeneous" are the preferred terms in these 2 sets of synonyms. The first set refers to having closely similar or identical characteristics, such as components, structure, or origin, while the second set refers to having dissimilar characteristics. The prefix "homo" means "the same," while "hetero" means "different" or "other." "Homogeneous" has the additional meaning of having a uniform quality throughout, while heterogeneous has the additional meaning of having a mixture of components that are not uniform.

homolog, homologous: See analog, analogous, homolog, homologous, homoeolog, homoeologous.

hypothesis: See law, theory, hypothesis, conjecture.

hypothesize, hypothecate: To "hypothesize" means to form a hypothesis. To "hypothecate" means to pledge property as security without transferring the rights to that property.

identical to, identical with: Although the 2 forms are regarded as equally acceptable in some usage guides, this style manual prefers "identical to." ("Ice is identical to frozen water.")

immunize: See vaccinate, immunize.

impact: See affect, effect, impact.

imply, infer: "Imply" indicates or suggests that evidence itself leads to a conclusion, regardless of whether the conclusion is correct. ("These data imply that the sensors are defective.") "Infer" indicates that a person has drawn a conclusion from evidence, regardless of whether the conclusion is correct. ("I infer from the data that the sensors are defective.")

incidence, prevalence, point prevalence, period prevalence: In botany and agriculture, "incidence" refers to the frequency of a disease or condition, as opposed to its severity. In epidemiology, "incidence" is the number of new cases occurring in a population of stated size during a stated period, while "prevalence" is the total number of cases existing in a population of stated size at a particular time. "Point prevalence" refers to the number of cases registered on a particular date, while "period prevalence" refers to the number of cases registered during a stated period. For more detailed definitions, see relevant entries in *A Dictionary of Epidemiology*.[15]

individual, person: As a noun or adjective, "individual" indicates a specific and distinguished unit in a group, such as one woman or man among humans. A "person" connotes a particular human being with a unique personality. "Person" is preferred to the simpler and dehumanized term "individual." Also see people, persons.

induce: See adduce, deduce, induce.

infectious, infective: See contagious, infectious, infective.

infer: See imply, infer.

infested, infected with: In medicine, "infested" means harboring or carrying lower organisms, such as worms and insects, which often produce adverse consequences in the host. In plant pathology, "infested" refers to microorganisms present in or on nonliving substrates, such as dead plant material, soil, tools, and containers. In both medicine and plant pathology, "infected with" means harboring or carrying microorganisms that cause disease.

inflammable: See flammable, inflammable, nonflammable.

inherent, intrinsic: In general, these are synonymous adjectives that mean a characteristic necessarily in and of the noun they modify. However, in anatomy, "intrinsic" means entirely within a structure or organism.

insure: See assure, ensure, insure.

irritate: See aggravate, exacerbate, irritate.

law, theory, hypothesis, conjecture: These are terms for concepts with decreasing degrees of certitude. In science, a "law" is a concept with a high degree of certitude for predicting phenomena. A "theory" is a broad concept based on extensive observation, experimentation, or reasoning that is expected to account for a wide range of phenomena. A "hypothesis" is a narrow concept, generally postulated as a potential explanation for phenomena that needs to be tested by experiment or observation. Akin to a hypothesis, a "conjecture" is speculation proposed for testing.

less: See few, fewer, less.

localize, locate: "Localize" means to confine, restrict, or attribute to a particular place or to have the characteristic resulting from such action. ("The infection localized in the antecubital space.") "Locate" means to specify, place, or find in a particular location. ("We finally located the infection in the right pleural cavity.") "Localize" should not be used to mean "find." (*Not* "We localized the primary site of the disseminated cancer in the pancreas.")

majority, most: "Majority" is frequently used as a synonym for "most," but the simpler term "most" is preferred when an accurate quantity is not needed but a number greater than half needs to be implied. ("Most physicians trained in the United States today take the United States Medical Licensing Examination.")

mean, median, mode: The "mean" is the arithmetic average of a set of values, whereas the "median" is the midpoint of the distribution of values. Mode is the most frequent value that appears in a set of data. In some cases, these values are identical, as can happen in a normal distribution like a symmetrical bell-shaped curve. ("In a set of measurements consisting of 1 cm, 6 cm, 6 cm, 7 cm, and 10 cm, the mean, median, and mode are all 6 cm.")

measure: See assess, determine, evaluate, examine, measure.

median: See mean, median, mode.

meiosis, mitosis, miosis: Both "meiosis" and "mitosis" are types of cellular division. In "meiosis," cellular division produces daughter cells with a haploid number of chromosomes, whereas "mitosis" results in daughter cells with a diploid number of chromosomes. Unrelated to cell division, "miosis" refers to an abnormally constricted (small diameter) ocular pupil.

method, methodology, technique: "Method" and "technique" are both widely used to mean an analytic, quantitative, observational, or other kind of procedure. A valuable distinction is to reserve "method" for procedure and "technique" for the skill applied in carrying out a procedure. ("This bioassay is a reliable method when the analyst carefully applies the technique.") A "methodology" is a system or combination of methods or techniques.

mode: See mean, median, mode.

morphology: See anatomy, morphology, structure.

most: See majority, most.

mucus, mucous, mucoid: "Mucus" is a thick, slimy secretion produced by body membranes and glands. "Mucous" is an adjective indicating that something has the characteristics of mucus or produces mucus, as in "mucous membranes." "Mucoid" means resembling mucus but not necessarily being identical to it.

mutant, mutation: As a noun, "mutant" refers to an organism carrying or expressing one or more genetic mutations. It may also be used as an adjective, as in a "mutant gene." "Mutation" is a stable and heritable change in a nucleotide sequence of DNA or RNA.

need: See require, need.

nonflammable: See flammable, inflammable, nonflammable.

norm, normal: See customary, conventional, traditional, normal, norm.

number, numeral, digit: "Number" is the count of some class of objects. A "numeral" is a single character in the group of numbers from 0 to 9. ("The number 345 is represented by the arabic numerals 3, 4, and 5 in that sequence.") "Digit" can be used as a synonym for "numeral," but it also can be used to refer to the number of numerals and their representation of magnitude in the decimal system. ("He reported his data with 3 digits.") In addition, "digits" refers to fingers and toes.

nutrition, nutritional, nutritious: "Nutrition" is the discipline concerned with foodstuffs and feeding, including the study of deficiencies and toxic effects associated with food. "Nutrition" may also refer to a desirable diet. "Nutritional" means of or relating to nutrition. "Nutritious" describes substances that yield desirable nutrition.

outbreak: This imprecise term is often applied to mean a sudden appearance of a disease, especially an infectious disease, or the sudden appearance of a social phenomenon. More precise alternatives include "episode," "sudden occurrence," "epidemic," and "epizootic."

parameter: A "parameter" is a variable to which a particular value can be assigned to determine the value of other variables. "Parameter" should not be used as a synonym for any of the following terms: "variable," which is a quantity that can assume any of a set of values; "index," which is something that leads to a particular conclusion; "indicator," which is something that indicates a set of conditions; and "guideline," which is a recommendation.

pathology: Reserve "pathology" to mean the discipline that studies diseases, disorders, and other abnormalities in humans, other animals, and plants. It is not a synonym for "abnormality," "disease," "disorder," or "lesion."

patient: See case, patient.

people, persons: "People" means a group of human beings with some characteristic in common, such as nationality (eg, "the French people") or location (eg, "the people living east of the Volga River"). Styles vary on whether to use "persons" or "people" to emphasize individuality. This style manual recommends "persons." ("Persons with visual impairment can use a variety of software tools to access the internet.") This style manual also prefers "persons" to "people" when referring to a numbered group. ("An ambulance took 3 injured persons to the same hospital.")

percent, percentage: "Percent" represents units per 100 units (eg, 45% means 45 units per 100 units). "Percentage" refers to a quantity or rate expressed as the unit percent (eg, 45% is a percentage). The difference between 2 percentages should be stated as a difference in percentage points, not as a percent (eg, the difference between 25% and 50% is 25 percentage points, not a 25% difference). Also see Section 12.3.4, "Ratios, Proportions, and Percentages."

perform, carry out, execute: These verbs can often be replaced by more specific verbs, such as "analyze" and "operate." ("He analyzed [*not* performed an analysis of] the possible orbits in a gravitational field.")

person: See individual, person and see people, persons.

precision: See accuracy, precision.

presently, at present: See currently, at present, presently.

prevalence: See incidence, prevalence, point prevalence, period prevalence.

proven, proved: The irregular past participle "proven" should generally be replaced by "proved," a past participle with regular form. ("It has been proved [*not* proven] that a retrovirus is responsible.") However, "proven" can be used as an adjective. (The oceanographers relied on a proven strategy.)

quasi, semi: The prefix "quasi" modifies a stem term to indicate a less-than-full degree or extent (eg, "quasiscientific"). The prefix "semi" has the same meanings as "quasi," but it also carries the more specific meaning of half (eg, "semicircle"). Also see bi, semi.

random, haphazard: "Random" refers to a method of sampling in which members of a population have a known chance of being selected, as in numbers randomly generated by computers. "Haphazard" means without plan or direction.

rare: See unique, unusual, distinctive, rare.

regime, regimen: "Regime" means a regular pattern, as in seasonal weather patterns. It also refers to a government in power. "Regimen" refers to a stipulated program for treatment or activity, as in a dietary or therapeutic regimen.

regular: See common, frequent, regular.

relationship, relation: In general, reserve "relationship" to mean dealings between 2 or more persons. "Relation" is usually adequate to describe a connection between inanimate objects. ("There is a cause-and-effect relation between the human immunodeficiency virus and acquired immunodeficiency syndrome.")

require, need: Use "require" as a transitive verb that means actively setting obligatory or compelling expectations. ("The journal requires authors to submit their manuscripts online.") As a verb, "need" is appropriate for describing what is essential for passive agents, such as plants and animals. ("Green-leaved plants need sunlight.")

reveal: See demonstrate, exhibit, reveal, show.

sample: See aliquot, aliquant, sample.

semi: See bi, semi.

sensitivity, specificity: When applied to diagnostic methods, "sensitivity" is the ability of a test to correctly identify persons who have a specific disease or other condition. A highly sensitive diagnostic method or test will generate very few false-negative results but may generate some false-positive results. "Specificity" is the capability of a method to correctly identify patients who do not have the disease or condition. A highly specific diagnostic method or test will generate very few false-positive results but may generate some false-negative results. For more detailed explanations, see *A Dictionary of Epidemiology*.[15]

show: See demonstrate, exhibit, reveal, show.

sign, symptom: "Signs" are objective manifestations of disease (eg, swelling, redness, and fever). "Symptoms" are subjective manifestations of disease (eg, pain, itching, and nausea).

significant: "Significant" should be used to mean indicating or pointing to. In addition, "significant" has a precise meaning in statistics in which reaching a predefined numeric threshold points to a specific statistical conclusion. ("The mean blood pressure was significantly lowered, with a *P* value of less than 0.05.") In medicine, the term "clinically significant" is used to describe when the effect of a treatment is greater than what is likely to occur by chance. A number of calculations exist for determining clinical significance. When "indicating," "pointing to," or a statistical meaning are not intended, replace "significant" with an adjective such as "great," "important," "influential," "major," "valuable," "useful," or "desirable."

similar: See comparable, similar.

since: See as, because, since.

specificity: See sensitivity, specificity.

standard: See benchmark, criterion, standard.

structure: See anatomy, morphology, structure.

symptom: See sign, symptom.

technique: See method, methodology, technique.

than: See compared to, compared with, than.

that, which, who: "That" and "which" should be used with clauses that describe inanimate objects and animals without names, whereas "who" should be reserved for clauses that describe human beings and animals with names. In American usage, "that" introduces restrictive clauses, also known as essential or required clauses. ("This is the research laboratory that the university built in 2021.") In American usage, "which" introduces nonrestrictive clauses, also known as nonessential or nonrequired clauses. ("This research laboratory, which the university built in 2021, is used by the biology and chemistry departments.") In British usage, the roles of "that" and "which" are reversed. In both American and British usage, "who" can be used in both restrictive and nonrestrictive clauses. ("An engineering professor who retired last year oversaw the construction of the new research lab." "The retired engineering professor, who holds a doctorate in civil engineering, oversaw the construction of the new research lab.") Regardless of which pronoun is used to introduce them, restrictive clauses are used without commas, whereas nonrestrictive clauses are enclosed with commas.

the: See a, an, the.

theory: See law, theory, hypothesis, conjecture.

think: See believe, feel, think.

traditional: See customary, conventional, traditional, normal, norm.

trophic, tropic: When used as part of an adjective, the suffix "trophic" indicates a stimulating, nourishing, or supporting function in growth or development. ("Anabolic steroids increase muscle mass and are therefore myotrophic.") In addition, when combined with prefixes such as "hyper" and "hypo," the suffix "trophic" indicates rate or extent of growth or development. ("A hypertrophic scar is thicker than the surrounding normal skin.") In ecology, "trophic" refers to obtaining nutrition, as in "phagotrophic." In contrast, the suffix "tropic" indicates capability to respond to a modifying or changing agent. ("Chemotropic algae respond to chemical stimuli.")

type: "Type" is often used synonymously with "kind," but in science, "type" should be reserved to mean an inanimate object or a specific animal, plant, or microorganism representative of a larger group of closely related objects, animals, plants, or microorganisms. In this sense, "type" is similar to "category." ("The blue whale is a type of cetacean.") A "type specimen" is a specimen used to establish a species name. ("The type specimen of the raccoon-like species olinguito is housed in the American Museum of Natural History.")

typical: See average, characteristic, typical.

uninterested: See disinterested, uninterested.

unique, unusual, distinctive, rare: "Unique" means one of a kind. It is not a synonym for "unusual" or "distinctive," both of which mean uncommon. Nor is "unique" a synonym for "rare," which means occurring seldom.

unusual: See unique, unusual, distinctive, rare.

use, utilize, employ: These verbs are often used synonymously, but in most cases, "use" is adequate to mean applying or drawing on for a purpose. When consumption is implied, the more specific terms "use up" and "consume" are preferred. The verbs "use" and "utilize" have similar meanings. However, "use" is preferred to "utilize" because "utilize" may appear overly formal. "Utilize" should be reserved for describing using an item in a novel or unusual application. ("Off-the-shelf software was utilized to solve this complex problem.") "Employ," in turn, should be reserved to mean putting a person to work or an object to use.

usual, usually: See general, generally, generic, generically, usual, usually.

vaccinate, immunize: Although these 2 terms are often used as synonyms, "vaccinate" means to purposely expose a person or another animal to an antigen in hopes of eliciting protective antibodies, whereas "immunize" implies that exposure to an antigen through infection or vaccination successfully elicits protective antibodies.

varying, differing, different, diverse, various: Use "varying" to mean changing, and use "differing," "different," and "diverse" to mean having unlike characteristics. In the sentence "Philadelphia, Pennsylvania, and New York, New York, in the United States have varying mean annual temperatures in a cycle," "varying" means that each city has a mean temperature that changes in a cycle. In the sentence "Sydney, Australia, and London, United Kingdom, have differing mean annual temperatures," "differing" means that the cities' mean annual temperatures are not the same. Unlike "varying," "various" is a near-synonym for "differing." ("Various racial and ethnic groups make up the population of Los Angeles, California.")

versus, vs, v: "Versus" and its abbreviations, "vs" and "v," mean against. Whether the entire word or one of its abbreviations is used depends on the accepted idiom in a given field. In medicine, for example, "versus" is the idiomatic form (eg, "graft-versus-host disease"), while the abbreviation "v" is idiomatic for the titles of legal cases (eg, *The State of Tennessee v John Thomas Scopes*"). When a field does not have an established idiomatic form for the term, spell out "versus." However, the uses for "versus," "vs," and "v" are fairly limited in scientific literature. For example, they should not be used for comparing and contrasting. Use "compared with" instead, as when reporting the results of different drugs in clinical trials. ("In this study, we compared penicillin with ampicillin for treating patients for pneumococcal pneumonia." *not* "We studied penicillin versus ampicillin for treating patients for pneumococcal pneumonia.")

whereas: See although, whereas, while.

which: See that, which, who.

while: See although, whereas, while.

7.6 EXCESSIVELY LONG COMPOUND TERMS

Excessively long compound terms, also known as "stacked nouns," can make scientific writing difficult to read and comprehend. These long terms are strings of adjectives and nouns that end with the subject or object of a sentence. Some of the terms may be single modifiers that modify adjacent nouns while others may be compound modifiers that modify the sentence's subject or object. As the 3 examples below show, it can be unclear as to which terms modify which nouns in such strings.

> a percentage transmission recording ultraviolet light absorption meter
> a new type motor skills college performance test
> the National Institutes of Health National Institute of Neurological Disorders and Stroke Division
> of Extramural Research in the United States

The meaning of long terms often can be clarified by rephrasing some of the relationships and hyphenating directly related modifiers. A good rule of thumb is to include no more than 3 adjectives or adjectival nouns in a row. In cases in which a compound modifier modifies an adjectival noun, use a hyphen between the terms in the compound modifier and use the longer en dash (created with Unicode 2013) to link the compound modifier to the adjectival noun. Note that a hyphen should not follow an adverb ending in "ly" (*not* "a swiftly-acting drug" or "a newly-discovered–gene mutation").

> an ultraviolet-light–absorption meter for recording percentage transmission

Another method for clarifying relationships between nouns is to reverse their order and separate them with prepositions, such as "of," "for," "at," and "in."

> a new college performance test for motor skills
> the Division of Extramural Research in the National Institute of Neurological Disorders and Stroke at the National Institutes of Health in the United States

Long strings of adjectival nouns and nouns are especially burdensome for readers when the terms are repeated in their entirety elsewhere in the text. For subsequent references for such terms, the main noun, by itself or accompanied by a few modifiers, is usually sufficient to remind readers of the entire term.

> the ultraviolet-light–absorption meter
> the new motor-skills test
> the Division of Extramural Research

7.7 EXCESSIVE ABBREVIATION

Heavy use of abbreviations can make reading and understanding scientific literature difficult. This is especially true of abbreviations that are used only a few times in a manuscript and those that are created just for the manuscript. Numerous abbreviations slow reading, and unfamiliar abbreviations may force readers to return repeatedly to the first references to those abbreviations in the text to remind themselves of the definitions. Judicious use of abbreviations is justified if the abbreviations are well recognized in a publication's scientific field or, in the case of new abbreviations, are central to the topic under discussion.

Rather than abbreviating terms that are used infrequently in a manuscript or that consist of just a few words, the terms can be written out in full or shortened to their key elements. The following paragraph is an example of how excessive abbreviation can be problematic.

> Elevated serum cholesterol (SC) levels have been shown to be an independent risk factor (RF) for coronary artery disease (CAD). The risk of CAD is most closely associated with the level of low-density lipoprotein cholesterol (LDL-C), although total SC level itself is a definite RF. LDL is a so-called "carrier" protein (CP) with which cholesterol, itself a steroid-like lipid, is associated. How LDL-C in the serum results in CAD is a very important question. One possible explanation is that cholesterol somehow is transported by the LDL (the CP) specifically to the coronary arteries, where it is deposited in the artery walls.

> **Revised version:**
> Elevated serum cholesterol levels have been shown to be an independent risk factor for coronary artery disease. The risk of coronary artery disease is most closely associated with the level of low-density lipoprotein cholesterol (LDL-C), although total serum cholesterol level itself is a definite risk factor. LDL is a so-called "carrier" protein with which cholesterol, itself a steroid-like lipid, is associated. How LDL-C in the serum results in coronary artery disease is a very important question. One possible explanation is that cholesterol somehow is transported by the carrier protein specifically to the coronary arteries, where cholesterol is deposited in the artery walls.

In the second version, the number of abbreviations is reduced from 4 to 1. The terms "serum cholesterol," "risk factor," and "coronary artery disease" are not used enough times to justify abbreviating them, plus they are relatively short terms to begin with. The abbreviation "LDL-C" is well recognized within the medical profession, and its abbreviation is less cumbersome to use than repeating the entire term throughout the paragraph.

If abbreviations are necessary, spell them out in each new section of text if they are newly coined or are likely to be unfamiliar to the audience.

> Introduction: Dural arteriovenous fistulas (dAVFs) are abnormal passageways that occur between arteries and veins located in the dura mater . . .
> Methods: This retrospective study of dural arteriovenous fistulas (dAVFs) . . .

As an alternative to spelling out terms in each section, a list of abbreviations may be given at the beginning of the work.

See Chapter 11, "Abbreviations," for a detailed discussion of this topic.

7.8 UNNECESSARY WORDS AND PHRASES

Unnecessary words and phrases, also known as verbiage, slow down reading. Such verbiage as "it is interesting to note that" adds no information and can be omitted. Other passages containing verbiage can be reconstructed to remove unnecessary words. For example, sentences in the passive voice often can be shortened by converting them to the active voice.

> It is reported by Michael M. Patterson, PhD . . . *becomes* Michael M. Patterson, PhD, reported . . .

Unneeded modifiers are another source of unnecessary words. Although used frequently in popular speech for emphasis, many modifiers are not needed for clear meaning, especially modifiers that convey concepts inherent in the unmodified terms.

> Careful hemodynamic monitoring is needed to prevent . . .

"Careful" can be omitted from the sentence above because the author is unlikely to mean "careless monitoring."

Listed below are examples of wordy phrases that can be shortened.

Wordy	Concise
a majority of	most
a number of	many, numerous, several, some
absolutely essential	essential
accounted for the fact that	explained
actual fact	fact
along the lines of	like
an innumerable number of	innumerable, countless, many
an order of magnitude of 10	10 times
and also	and
are of the same opinion	agree
as a consequence of	because of
ascertain the location of	find
at a rapid rate	rapidly
at no time	never
at the conclusion of	after
at the present moment, at this point in time	now, currently
based on the fact that	because

Wordy	Concise
based upon	based on
bright green in color	bright green
but while, yet while	while
by means of	by, with, using
caused injuries to	injured
completely filled	filled
conducted inoculation experiments on	inoculated
definitely proved	proved
despite the fact that	although
due to the fact that	because
during the course of	during, while
during the time that	when, while
end result	result
fewer in number	fewer
final destination	destination
for the purpose of examining	to examine
for the reason that	because
foreseeable future	future
future plans	plans
general public	public
give rise to	cause
has the capability of	can, is able to
help increase, help reduce, help alleviate, help lessen	increase, reduce, alleviate, lessen
his own personal, her own personal, their own personal, your own personal	his, her, their, your
I believe, I think, we think, we believe	[Often can be omitted without changing intended meaning]
if conditions are such that	if, when
in a satisfactory manner	satisfactorily
in an adequate manner	adequately
in all cases	always, invariably
in case	if
in close proximity to	near
in connection with	about, concerning
in order to	to
in the course of	during, while
in the event that	if
in the near future	soon
in the vicinity of	near
in view of the fact that	because
is in a position to	can, may
it has been reported by the research director	the research director reported
it is believed that	we believe [or omit altogether]
it is often the case that	often
it is possible that the cause is	the cause may be
it is this that	this
it is worth noting that, it is worth pointing out that	note that
it would thus appear that	apparently
lacked the ability to	could not
large amounts of	much, many
large in size	large
large numbers of	many
lenticular in character	lenticular

Wordy	Concise
located in, located near	in, near
necessitates the inclusion of	necessitates, needs, requires
of such hardness that	so hard that
on account of	because
on behalf of	for
on the basis of	from, by, because
on the grounds that	because
original source	source
oval in shape, oval-shaped	oval
owing to the fact that	because
past history	history
period of time, time period	period
pre-existing	existing
preplanning	planning, advance planning, early planning
prior to [in time]	before
referred to as	called, titled, named
results so far achieved	results so far, results to date
round in shape	round
serves the function of being	is
smaller in size	smaller
subsequent to	after
take into consideration	consider
the object in question	this object
the question as to whether	whether
the treatment having been performed	after treatment
there can be little doubt that this is	this probably is
through the use of	by, with
throughout the entire area	throughout the area
throughout the whole of the experiment	throughout the experiment
2 equal halves	halves
was of the opinion that	thought
with a view to getting	to get
with reference to	about
with regard to	regarding, concerning, about
with the result that	so that
within the realm of possibility	possible

7.9 JARGON

All scientific and other specialized fields have jargon, which includes both technical vocabulary and informal idioms used in those fields. Jargon may be acceptable in formal reports as long as its meaning is clearly understood by the intended audiences for the reports. However, when a word or phrase obscures meaning for some of the target readers, that word or phrase should be replaced by a term that clearly conveys the intended meaning.

Another concern about jargon is that many terms may not translate well from one language to another. For example, jargon used in a German-language article might not make sense to many American scientists. In addition, the meaning of jargon may vary even by regions in the same country.

The following are among the more common types of jargon found in English:

(1) shortened forms of words that arise in conversation and may enter formal communication

> The patient's crit was determined through lab testing. *for* The patient's hematocrit was determined through laboratory testing.

(2) verb-object relations that are ignored

> We stocked trout in the stream. *for* We stocked the stream with trout.

(3) nouns that are not used with their proper meaning

> No pathology was found in the lung. *for* No abnormalities were found in the lung.

(4) euphemisms that are used to soften harsh realities

> The patient expired. *for* The patient died.
> The rats were sacrificed. *for* The rats were killed.

7.10 USE OF ARTICLES

Articles modify nouns. In English, "the" is the only definite article, and "a" and "an" are the only indefinite articles.

Whether to use an article and which one to use depend in part on whether the noun the article modifies is countable or noncountable.

Countable nouns are concrete items that can be counted as discrete units. In their singular form, countable nouns (eg, "pipette," "specimen," and "sample") should be preceded by the indefinite article "a" or "an" on their first use in the document. In their plural form, countable nouns are used without articles on their first use, although they can be preceded by modifiers (eg, "some"). (See Section 6.4, "Plurals," for a more detailed discussion of plural forms of nouns.)

Noncountable nouns (eg, "ethanol," "water," and "oxygen") cannot be counted as discrete units. When first used in a document, noncountable nouns are used without articles. Like plural countable nouns, noncountable nouns can be used with modifiers. They also can be used with expressions of quantity (eg, "a piece of" and "a lot of.")

In general, the definite article "the" is used before a noun only after that noun has been introduced earlier in the document. This rule applies to both countable and noncountable nouns.

> **Countable noun:**
> Chemicals were mixed to form a precipitate. The precipitate was removed by filtration.

> **Noncountable noun:**
> Ground-level ozone is a pollutant created by a photochemical reaction. The ozone can cause respiratory symptoms.

Good sources for guidance on using articles include the *Longman Dictionary of Contemporary English*[3] and *Understanding and Using English Grammar Student Book with Pearson Practice English App.*[16]

7.11 SCIENTIFIC ENGLISH FOR MULTILINGUAL AUTHORS

Different cultures have different rhetorical conventions, and academic styles may differ by culture. Studies in the field of contrastive rhetoric indicate that such style differences relate more to academic traditions of writing within cultures than to any inherent differences between languages.

Whereas the norm in English-language manuscripts in Western science journals is to be relatively direct and concise, the norm in some other cultures is to be circumlocutory and to rely on inference.

This entire chapter offers guidance to authors who are unfamiliar with Western traditions in scientific English. Especially helpful to such authors should be Section 7.3, "Appropriate Verb Tenses and Verb Forms." In addition, the following sections address some of the basic concepts that such authors need to take into consideration when drafting manuscripts for publishers that follow Western traditions in scientific English.

7.11.1 Digressive Writing

To demonstrate their command of English, multilingual authors may use long sentences with overly elaborate sentence structures and long and unusual words. In adopting such digressive writing, authors also may build up to their main points slowly, or they may skirt around the main points and expect readers to infer them. In addition to affecting word choice and sentence structure, digressive writing influences content, paragraph logic, and even transitions between paragraphs and sentences.

In contrast, in Western academic tradition, the preference is for explicit, unambiguous discussion of arguments. Thus, paragraphs begin with clear topic sentences, followed by supporting examples. In addition, transitions between sentences and paragraphs are commonly used to clearly link the elements of arguments.

The following 2 examples of a conclusion in a Discussion demonstrate some of the differences between digressive writing and more direct scientific English in the Western academic tradition.

Example 1: Digressive writing

Researchers have suspected for years that glucagon-like peptide-1 (GLP-1) agonists might function much like glucagon and, therefore, could suppress appetite enough to result in long-term weight loss. Our study involved human participants who were obese but otherwise were healthy and who ate a standardized diet. The only intervention was that the experimental group was given weekly doses of the GLP-1 agonist semaglutide by subcutaneous injection. After 12 weeks of the study, the mean loss of body weight among experimental subjects was statistically significant and exceeded the accepted standard of weight loss intervention, which is widely held to be weight loss of 5% of overall body weight. It is concluded that, under these experimental conditions, semaglutide was effective as an aid for losing weight.

In the first example, the author starts with background information and then discusses methods and results, all of which would have been stated already in the Introduction, Methods, and Results. In addition, the conclusion is not stated until the end of the paragraph.

Example 2: Direct scientific English

Our results clearly support the hypothesis that semaglutide is a safe and effective weight loss drug. Like endogenous glucagon-like peptide-1 agonists, semaglutide suppresses appetite, which is its presumed mechanism of action in weight loss. The mean 15.1% weight loss (compared with placebo) in the study participants was significant because it far exceeds the widely accepted standard for weight loss intervention (5% or more loss of body weight).

The second example is concise, and it exemplifies "patterns of linearity," a tight flow of logic that is easy for readers to follow. The paragraph begins with a topic sentence, which provides the conclusion of the study. The topic sentence is immediately followed by an explanation of the significance of the results

For more on writing direct scientific English, see Section 7.3.2, "Active versus Passive Voice." For more on basic word choices, see Sections 7.5.3, "Imprecisely Applied Words," and 7.8, "Unnecessary Words and Phrases."

7.11.2 Grammatical and Syntactical Problems

7.11.2.1 Prepositions

The meanings of generally corresponding prepositions in English and other languages do not always coincide. The appropriate use of a preposition in a non-English language may not be idiomatic in English. A good example is that the English preposition "of" is not always the best translation for "de" in Spanish.

> soy de Valencia *means* "I am from Valencia" *not* "I am of Valencia"

Two valuable resources for idiomatic use of prepositions in English are *Longman Dictionary of Contemporary English*[3] and *Understanding and Using English Grammar Student Book with Pearson Practice English App.*[16]

7.11.2.2 Position of Verbs

In some non-English languages, predicates (ie, verbs indicating action or state) are placed at the beginning or end of sentences, which is not idiomatic in English. In English, predicates usually are placed between the subjects and objects of sentences, except in interrogative sentences, which ask questions. Because this is not the case in many other languages, some authors fluent in those languages may inadvertently misplace predicates when describing their scientific work in English.

> **Nonidiomatic:**
> Found by Smith important activity against some dimorphic fungi.

> **Idiomatic:**
> Smith found important activity against some dimorphic fungi.

7.11.2.3 Progressive Tense of Verbs

Although progressive tense is rarely used in the American and British academic traditions for scientific English, some variants of English, notably those spoken in India and its neighboring countries, use progressive tense extensively.

> **Nonidiomatic:**
> We are finding fossil evidence that this formation belongs in the Cretaceous Period.

> **Idiomatic:**
> We have found fossil evidence that this formation belongs in the Cretaceous Period.

7.11.2.4 Phrasal (Two-Word) Verbs

English phrasal verbs are sometimes used in scientific writing for figurative or metaphoric meanings. These phrases consist of a verb (usually of action or movement) followed by a second word (an adverbial or prepositional particle). A common idiomatic mistake is omitting the particle in a phrasal verb, which usually changes the meaning of the verb.

Nonidiomatic:
Tensions related to funding made it difficult to patch the research community's spirit.

Idiomatic:
Tensions related to funding made it difficult to patch up the research community's spirit.

In the example above, "patch up" is used as a metaphoric equivalent of "repair." Omitting "up" distorts the metaphoric meaning.

Good sources on phrasal verbs include *Oxford Phrasal Verbs*,[17] *The Oxford Dictionary of Idioms*,[18] and *Oxford Idioms: Dictionary for Learners of English*.[19] Also see *Barron's 501 English Verbs*,[2] which lists phrasal verbs as well as regular and irregular verbs.

7.11.2.5 Inappropriate Carryover of Gender from Non-English Languages

In English, a noun is generally assigned feminine or masculine gender only when it represents a person or other animal of female or male sex. This linguistic use of gender is known as natural gender. Many other languages assign grammatical gender to nouns seemingly in a random way and without reference to natural gender. Authors who are accustomed to using grammatical gender should be mindful that idiomatic English assigns only natural gender to nouns.

Nonidiomatic:
This new product is naftifine, and his potency is similar to clotrimazole against strains of *Candida albicans*.

Idiomatic:
This new product is naftifine, and its potency is similar to that of clotrimazole against strains of *Candida albicans*.

Although gender is assigned in popular speech and print to some English nouns lacking natural gender (eg, ships and countries), this idiom is rarely used in scientific writing.

The deepwater research vessel RV *Tangaroa* uses Wellington, New Zealand, as its home port.
not The deepwater research vessel RV *Tangaroa* uses Wellington, New Zealand, as her home port.

For more on the gender of English nouns, see Section 7.4.1, "Grammatical Gender."

7.11.3 Vocabulary Problems

Many terms from other languages cannot be directly translated into English without losing their meanings. These include "false cousins," which are cognates that have totally different meanings in their respective languages, and "transfer coinages," which are non-English terms mistakenly thought to exist in English.

7.11.3.1 False Cousins

Many English nouns have cognates in other languages, which originated from the same ancestral language as the English nouns but have different meanings. Known as "false cousins," these cognates are common in Romance languages because like English, these languages have Latin heritages. For example, in English, the noun "raisin" refers to a dried grape. In French, "un raisin" is a freshly harvested grape, whereas "raisins secs" refers to dried grapes. In English, "exit" means the way out of a room or situation, whereas the Spanish word "exito" means "success." Some false cousins do originate from non-Romance languages. For example, "gift" refers to a present or something given in English, whereas it means "poison" in German.

7.11.3.2 Transfer Coinages

Authors for whom English is not a primary language, especially those whose native tongue is a Romance language, may translate nouns into apparently English equivalents that are, in fact, not English words at all.

> His work for many years was on the causalism of infectious diseases.

"Causalisme" is a French term meaning "theory of causation." Its equivalent medical term in English is "etiology."

7.11.4 Inappropriate and Missing Capitalization

Some non-English languages routinely capitalize common nouns, while some other languages do not capitalize proper nouns. Authors whose primary language is other than English may apply that language's capitalization rules to their works in English. It is also common even for authors whose primary language is English to mistakenly capitalize common nouns, especially job titles.

> The university named a new research director. *not* The University named a new Research Director.
> The journal's editor holds a doctorate in planetology. *not* The Journal's Editor holds a Doctorate in Planetology.

For the basic rules on which words to capitalize and which to lowercase in English, see Chapter 9, "Capitalization."

7.11.5 Spacing with Punctuation Marks

Some non-English languages use different spacing with punctuation marks than does English. Such conventions may mistakenly find their way into English writing. For the basic rules of punctuation in English, see Chapter 5, "Punctuation and Related Marks."

7.11.6 Editorial Assistance for Idiomatic Problems

Multilingual authors may wish to have their manuscripts reviewed by one or more readers with a strong knowledge of English idiom. Alternatively, authors may have their manuscripts revised by professional manuscript-editing services that specialize in ensuring that scientific papers are written in idiomatic English. Authors should wait until late in the writing process to have their manuscripts reviewed for language, preferably after their scientific content is deemed satisfactory and after the content has been organized in the correct sequence.

CITED REFERENCES

1. Hansen WR, editor. Suggestions to authors of the reports of the United States Geological Survey. 7th ed. United States Geological Survey; 1991. 289 p. https://doi.org/10.3133/7000088

2. Beyer TR. Barron's 501 English verbs. 3rd ed. Barron's Educational Series; 2013.

3. Longman dictionary of contemporary English. 6th ed. Pearson; 2015.

4. Council of Science Editors, Committee of Diversity, Equity, and Inclusion. Diversity, equity, and inclusion scholarly resources. Council of Science Editors; [accessed 2023 Jan 12]. https://www.councilscienceeditors .org/resource-library/diversity-equity-and-inclusion-in-scholarly-publishing/

5. Maggio R. The dictionary of bias-free usage: a guide to nondiscriminatory language. Oryx Press; 1991.

6. The Chicago manual of style: the essential guide for writers, editors, and publishers. 17th ed. The University of Chicago Press; 2017. Also available at https://www.chicagomanualofstyle.org

7. Virag K, editor. Editing Canadian English: a guide for editors, writers, and everyone who works with words. 3rd ed. Editors' Association of Canada; 2015.

8. JAMA Network Editors. AMA manual of style: a guide for authors and editors. 11th ed. Oxford University Press; 2020. Also available at https://www.amamanualofstyle.com

9. Publication manual of the American Psychological Association. 7th ed. American Psychological Association; 2020.

10. International Committee of Medical Journal Editors. Recommendations for the conduct, reporting, editing, and publication of scholarly work in medical journals. International Committee of Medical Journal Editors; [updated May 2022]. https://www.icmje.org/recommendations/

11. Council of Science Editors, Board of Directors. Scientific Style and Format update: capitalize racial and ethnic group designations. Council of Science Editors; 2020. https://doi.org/10.36591/SE-D-4303-97

12. Flanagin A, Frey T, Christiansen SL, AMA Manual of Style Committee. Updated guidance on the reporting of race and ethnicity in medical and science journals [editorial]. JAMA. 2021;326(7):621–627. https://doi.org/10.1001/jama.2021.13304

13. McGraw-Hill dictionary of scientific and technical terms. 6th ed. McGraw-Hill; 2003.

14. The American Heritage dictionary of the English language. 5th ed, updated. Houghton Mifflin Harcourt; 2015.

15. Porta M, editor. A dictionary of epidemiology. 6th ed. Oxford University Press; 2014.

16. Azar B, Haden S. Understanding and using English grammar student book with Pearson practice English app. 5th ed. Pearson Education ESL; 2016.

17. McIntosh C, editor. Oxford phrasal verbs: dictionary for learners of English. 2nd ed. Oxford University Press; 2006.

18. Ayto J. The Oxford dictionary of idioms. 4th ed. Oxford University Press; 2020.

19. Parkinson D, Francis B, editors. Oxford idioms: dictionary for learners of English. 2nd ed. Oxford University Press; 2006.

ADDITIONAL REFERENCES

Academic phrasebank. The University of Manchester; [accessed 2023 Jan 12]. https://www.phrasebank .manchester.ac.uk/

Conscious style guide. Quiet Press; [accessed 2023 Jan 12]. https://consciousstyleguide.com/

Garner BA. Garner's modern English usage: the authority on grammar, usage, and style. 5th ed. Oxford University Press; 2022.

Skillin ME, Gay RM. Words into type. Prentice Hall; 1974.

8 Names and Personal Designations

Editors: Trevor Lane, DPhil, PGCELT; Dara R. Rochlin; and Brit Stamey

8.1 PERSONAL NAMES

All publications have a responsibility to ensure that personal names are accurate in bylines, text, bibliographic citations, and references. Consistently using personal names accurately—and where appropriate, academic titles and other credentials—contributes to a publication's reputation for being stringent in checking facts.

In scientific publications, accurate names in citations and references ensure that the appropriate authors are acknowledged for their past contributions and that other researchers can retrieve those contributions. Likewise, accurate author names in bylines enable correct tagging for electronic indexing, which further aids in authorship attribution and article retrieval.

8.1.1 Differences in Naming Formats Based on Context

To the extent possible, publications should use the spelling, punctuation, diacritics, and other formatting preferences that authors have for their personal names. However, names may be rendered in different formats depending on the context. Sections 8.1.1.1, "Names in Bylines," through 8.1.1.4, "Names in End References and Indexes," cover the typical formats used.

8.1.1.1 Names in Bylines

In article bylines, both forename (ie, given name) and surname (ie, family name) are provided for each author. The forename is usually spelled out in full. It may be followed by up to 2 initials for middle names, also known as "additional forenames." Middle initials are used with periods but without a space between initials. Some authors have compound forenames or surnames, and some prefer to use a middle name as their main forename.

Surnames are often followed by abbreviations of the authors' highest academic degrees and other credentials. Lower academic degrees in the same discipline are omitted (eg, list only the PhD of an author who earned a master of science degree and a doctorate of philosophy in the same discipline, but list both MS and PhD degrees earned in different disciplines). Additional credentials may include national honors, such as "OBE" for officer of the Order of the British Empire; recognitions from peers, such as "FAAFP" for fellow of the American Academy of Family Physicians; and licensure, such as "MRCP(UK)" for member of the Royal Colleges of Physicians of the United Kingdom. The abbreviations for degrees and other credentials are used without periods.

> Timothy H. Rainer, MD
> Irene O.L. Wong, PhD, MMedSc
> Eliseo J. Pérez-Stable, MD
> R.V. Paul Chan, MD, MSc, MBA, FACS

For a byline with 2 author names, the names are separated by the word "and," as well as a comma if the first author's name ends with an abbreviation for a degree or honor. Additionally, 3 or more author names in the same byline are separated by semicolons instead of commas if at least one of the names ends with an abbreviated degree or honor. However, if only the final name ends in an abbreviated degree, no comma is needed to separate 2 names in a byline, and commas are sufficient to separate 3 or more names.

> Monica Webb Hooper, PhD, and Anna María Nápoles, PhD, MPH
> Nita G. Valikodath, MD; Michael F. Chiang, MD; and Gabriel M. Leung, MD, MPH

8.1.1.2 Names in Text of Research Articles

When citing published research articles, authors' names are usually provided in text only when the name–year citation system is used for references (see Section 29.2.1.2, "Name–Year"). In contrast, when the citation–sequence system or the citation–name system is used, authors' names are usually provided only in the end references (see Sections 29.2.1.1, "Citation–Sequence," and 29.2.1.3, "Citation–Name," respectively).

When the name–year system is used, each in-text citation consists of the author's surname followed by the publication year without punctuation. In addition, the surname and year are placed between parentheses within the sentence. When citing a work with 2 authors, both surnames are separated by the conjunction "and." If the cited work has 3 or more authors, only the first author's surname is provided followed by "et al" to indicate that the work has additional authors.

> The deepest place in Earth's oceans is the Challenger Deep in the Mariana Trench (Stewart and Jamieson 2019).
> Gait asymmetry was reduced in male study participants following osteopathic manipulative treatment (Hill et al 2022).

An alternative method of the name–year system incorporates authors' surnames into sentences as grammatical subjects or other nouns. With this alternative, only publication years are enclosed in parentheses. However, this alternative is not recommended by this style manual.

> Woodruff et al (2022) found evidence of an avian-style respiratory disorder in a nonavian dinosaur.
> In the study by Harris et al (1969), the use of cell fusion demonstrated the existence of tumor suppressor genes.

When personal communication is cited in text, place the first and middle initials and surnames of the person who wrote the communication within parentheses (see Section 29.4.16.2, "Personal Communications"). Because personal communications are not included in end references, their in-text citations are more extensive than those for published research. These citations should include the nature of the cited information, and they should clearly indicate that the citations are not included in the reference list.

> . . . local and systemic reactions were expected to be stronger after the second dose (2022 letter from TD Perez to author; unreferenced, see "Notes")

However, if an article's text refers to personal communication from one or more of the article's own authors, the names should be abbreviated to their initials (eg, "DPB" would be used for Dimitri P. Bertsekas, PhD).

8.1.1.3 Names in Text of Other Articles

In biographical essays, news articles, historical articles, obituaries, and other articles that do not formally report the results of research studies, full names should be used on first mention, followed by abbreviations for academic degrees and other credentials. On subsequent references in the same article, only surnames are used, preceded by such titles as "Dr" for "doctor" or "Prof" for "professor," if appropriate.

8.1.1.4 Names in End References and Indexes

In bibliographic references and indexes, surnames are listed first followed by initials for given names and middle names. Surnames are separated from initials by a space, periods are omitted from initials, and spaces are not placed between initials. Authors' names are separated from each other with commas. Unlike in-text references in the name–year citation system, which limit authors' names to 2 before using "et al," end references list up to 5 authors before using "et al" (see Section 29.3.3.1.1, "Personal Authors"). This rule applies to end references in the citation–sequence, citation–name, and name–year systems. For both in-text and end references, "et al" is used with only the first author's name. (For differences between the 3 systems for end references, see Section 29.2.2, "Advantages and Disadvantages of the Systems.")

> **Citation–sequence and citation–name systems:**
> Samaranayake LP, Fakhruddin KS, Ngo HC, Chang JWW, Panduwawala C. The effectiveness and efficacy of respiratory protective equipment (RPE) in dentistry and other health care settings: a systematic review. Acta Odont Scand. 2020;78(8):626–639. https://doi.org/10.1080/00016357.2020.1810769
> Lee SMC et al. Efficacy of gradient compression garments in the hours after long-duration spaceflight. Front Physiol. 2020;11:784. https://doi.org/10.3389/fphys.2020.00784
>
> **Name–year citation:**
> Samaranayake LP, Fakhruddin KS, Ngo HC, Chang JWW, Panduwawala C. 2020. The effectiveness and efficacy of respiratory protective equipment (RPE) in dentistry and other health care

settings: a systematic review. Acta Odont Scand. 78(8):626–639. https://doi.org/10.1080
/00016357.2020.1810769
Lee SMC et al. 2020. Efficacy of gradient compression garments in the hours after long-duration
spaceflight. Front Physiol. 11:784. https://doi.org/10.3389/fphys.2020.00784

8.1.2 Conventions for Authors' Names

Publications should honor authors' preferences for their names. Therefore, author bylines and similar content should clearly indicate the formal name of each author, which
may include middle names or initials. Of particular importance are surnames, also
known as "family names." Authors are best known in the research community by their
surnames, which are not always authors' last names. Consequently, some publishers
use typography to highlight surnames (eg, by placing surnames in all capital letters or
by using an initial capital followed by small capitals).

In many countries, including the United States, Canada, the United Kingdom, and
most members of the European Union, the most common naming convention is that
surnames follow given and middle names.

It is relatively simple to identify surnames when they consist of one element or a
hyphenated compound (eg, the surname of cytogeneticist Barbara McClintock, PhD,
was "McClintock," and the surname that developmental biologist Christiane Nüsslein-
Volhard, PhD, uses is "Nüsslein-Volhard").

However, in cases in which compound surnames are unhyphenated, it may be difficult
to correctly identify the surnames (eg, biochemist Sir John Cowdery Kendrew, PhD, used
only "Kendrew" as his surname, but physicist Maria Goeppert Mayer, PhD, used "Goeppert
Mayer" as hers). A similar difficulty is identifying surnames in names that incorporate
prefixes, also known as "particles." The prefixes "von," "de," and "los," for example, are
treated differently depending on an author's country or region of origin (eg, the surname
of French mathematician and lawyer Pierre de Fermat, LLB, was "Fermat," whereas the
surname of English physicist Richard de Villamil was "de Villamil"). For a selection of
such conventions, see Table 8.1.[1-4]

When starting a sentence with a surname that contains a prefix, capitalize the prefix.
In bibliographic references, prefixes are used if they are part of surnames but omitted
when they are not. If prefixes are lowercase in surnames, they remain lowercase in
references. In addition, such prefixes determine alphabetic order.

> English biomedical gerontologist Aubrey de Grey, PhD, *becomes* de Grey A *not* Grey AD
> German chemist Justus von Liebig, PhD, *becomes* Liebig J *not* Liebig JV *or* von Liebig J

In some Asian and other cultures, surnames are determined differently. However,
for professional purposes, some authors from such cultures follow the convention of
placing their surnames last,[1,5] and they may even anglicize the spelling of their names
(eg, Japanese cell biologist Ōsumi Yoshinori, DSc, uses "Yoshinori Ohsumi" in his published works, and Chinese paleontologist Zhāng Mímàn, PhD, uses "Meemann Chang").

Authors whose surnames are not readily apparent should clearly indicate their
preference in the manuscripts they submit for publication, and they should adhere to
that preference consistently in their submissions. Editors, in turn, should request this
information if authors do not provide it.

To increase the likelihood that their names are consistently used and recognized in

Table 8.1. Guidelines for Determining Formal Surnames According to Their Countries of Origin[a]

Origin of name	Convention for surname used in text	Convention for bibliographic reference
Europe, except Hungary, Portugal, and Spain	For a single surname (family name), use the last element of the name.	
	Röntgen *for* physicist Wilhelm Röntgen, PhD	Röntgen W
	Zernike *for* physicist Frits Zernike, PhD	Zernike F
	For compound surnames, use both parts of the compound, regardless of whether they are hyphenated.	
	Milne-Edwards *for* zoologist and physician Alphonse Milne-Edwards	Milne-Edwards A
	Kamerlingh Onnes *for* physicist Heike Kamerlingh Onnes, PhD	Kamerlingh Onnes H
	For surnames that contain a prefix and that originated in English-speaking countries and Italy, include the prefix in the surname, and stylize the prefix as it is used in the full name.	
	Van Allen *for* nuclear physicist James Van Allen, PhD	Van Allen J
	D'Alonzo *for* osteopathic pulmonologist Gilbert E. D'Alonzo Jr, MS, DO	D'Alonzo GE Jr
	de Bruyne *for* aircraft engineer Norman de Bruyne, PhD, FRS	de Bruyne N
	de Finetti *for* probabilist statistician Bruno de Finetti, MS	de Finetti B
	du Sautoy *for* mathematician Marcus du Sautoy, DPhil, FRS, OBE	du Sautoy M
	Di Pippo *for* astrophysicist Simonetta Di Pippo, MSc	Di Pippo S
	For surnames that originated in France, exclude the prefix "de," but include all other prefixes, which are usually capitalized.	
	Broglie *for* physicist Louis de Broglie, PhD	Broglie L
	La Salle *for* physician and chemist François Poulletier de La Salle	La Salle FP
	Le Chatelier *for* chemist Henry Louis Le Chatelier, PhD	Le Chatelier HL
	d'Alembert *for* mathematician and physicist Jean le Rond d'Alembert, FRS	d'Alembert JLR
	Du Châtelet *for* natural philosopher and mathematician Émilie du Châtelet	Du Châtelet É
	For surnames that originated in Germany, Austria, and the Netherlands, include the following prefixes: am, de, del, della, delle, des, di, du, l', la, las, le, les, li, los, ver, vom, zum, and zur.	
	zur Hausen *for* virologist Harald zur Hausen, MD, DSc	zur Hausen H
	de Bary *for* surgeon and botanist H. Anton de Bary, MD	de Bary HA
	de Vries *for* botanist and geneticist Hugo de Vries, PhD (when standing alone in text, use "De Vries")	de Vries H
	However, omit the following prefixes from surnames that originated in Germany, Austria, and the Netherlands: den, op de, ten, ter, van, van den, van der, von, and von der.	

Table 8.1. Guidelines for Determining Formal Surnames According to Their Countries of Origin[a] (*continued*)

Origin of name	Convention for surname used in text	Convention for bibliographic reference
	Helmholtz *for* physicist and physician Hermann von Helmholtz, MD	Helmholtz H
	Leeuwenhoek *for* microscopist Anton van Leeuwenhoek, FRS (when standing alone in text, use "Van Leeuwenhoek")	Leeuwenhoek A
	Waals *for* theoretical physicist and thermodynamicist Johannes van der Waals, PhD (when standing alone in text, use "Van der Waals")	Waals J
	For surnames that originated in Denmark, Norway, and Sweden, include prefixes, except for the prefixes that are omitted in Germanic names, such as "den," "van," and "von."	
	de Laval *for* engineer Gustaf de Laval	de Laval G
	la Cour *for* meteorologist Poul la Cour, MS	la Cour P
	Euler *for* botanist, geologist, and chemist Astrid Cleve von Euler, PhD	Euler AC
Hungary	Surnames precede given names, unless authors transposed their names for use in English-language manuscripts.	
	Irinyi *for* chemist Irinyi János (in English contexts, the full name is often inverted to "János Irinyi")	Irinyi J
Portugal	The last element of the family name is used as the surname.	
	Nascimento *for* linguist Maria Bacelar do Nascimento, PhD	Nascimento MB
	Graça *for* ethnologist António dos Santos Graça	Graça AS
	However, if the last element is a qualifier indicating family relationship (eg, Filho, Neto, and Sobrinho), include the next-to-last element with the last element.	
	Coimbra Filho *for* biologist and primatologist Adelmar Faria Coimbra Filho	Coimbra Filho AF
	Castro Neto *for* physicist Antonio H. Castro Neto, PhD	Castro Neto AH
Spain	For a compound surname consisting of the father's surname followed by the mother's birth surname, use all elements as the family name.	
	Salas Falgueras *for* biochemist and molecular geneticist Margarita Salas Falgueras, PhD	Salas Falgueras M
	Ramón y Cajal *for* neuroscientist and pathologist Santiago Ramón y Cajal, MD	Ramón y Cajal S
	However, for a married woman who adds her husband's surname to hers or who replaces her mother's surname with her husband's, use only her husband's surname, which may be preceded by the preposition "de."	
	de Escobar *for* chemist Gabriella Morreale de Escobar, PhD	de Escobar GM
	de San Blas *for* molecular biologist and biochemist Gioconda Cunto de San Blas	de San Blas GC

Table 8.1. Guidelines for Determining Formal Surnames According to Their Countries of Origin[a] (*continued*)

Origin of name	Convention for surname used in text	Convention for bibliographic reference
	If a surname consists of one name with a prefix that is an article (eg, el, la, las, and los), include the prefix and capitalize it.	
	El Ferreiro *for* archaeologist Pepe El Ferreiro	El Ferreiro P
	If the surname has a prefix that is either a preposition or a preposition and an article, omit the prefix, except for married women who use "de" before their husbands' surnames.	
	Ulloa *for* astronomer and sociologist Antonio de Ulloa, FRS	Ulloa A
	Fuente *for* biochemist Gertrudis de la Fuente, PhD	Fuente G
India	In modern usage, the surname is the last element, but if "Das" or "Sen" precedes an Indian name, include it as part of the surname.	
	Raman *for* physicist Chandrasekhara Venkata Raman, MA, FRS	Raman CV
	Sen Gupta *for* psychologist Narendra Nath Sen Gupta, PhD	Sen Gupta NN
	Das Gupta *for* astronomer Mrinal Kumar Das Gupta, PhD, FNI	Das Gupta MK
China, Japan, North Korea, South Korea, and Vietnam	The surname is usually the first element of a person's name. However, if an author has adopted the Western convention of putting the surname last, treat the name as an English name.	
	Yang *for* theoretical physicist Yang Chen-Ning, PhD, FRS	Yang CN
	Kao *for* electrical engineer and physicist Charles K. Kao, PhD, FRS (whose Chinese-style name was Kao Kuen)	Kao CK
Arabic states	For a name without a prefix, use the last name as the surname.	
	Zewail *for* chemist Ahmed Hassan Zewail, PhD	Zewail AH
	If a person's name includes the prefix "al-" or "el-," use what follows the prefix as the surname.	
	Biruni *for* polymath Abu Rayhan al-Biruni (but when standing alone in text, use "al-Biruni")	Biruni AR
	Include other prefixes (eg, abd, abu, and ibn) in the surname.	
	Ibn Firnas *for* polymath Abbas Ibn Firnas	Ibn Firnas A

[a] Based on *The Chicago Manual of Style*,[1] *American Journal of Neuroradiology*,[2] *The MIBIS Manual*,[3] and *The New York Public Library Writer's Guide to Style and Usage*.[4]

citations to their works, authors can obtain a persistent digital identifier from Open Researcher and Contributor ID[6] (ORCID) and list all their works in their ORCID record. Authors should use their preferred name in their ORCID record. They can also list name variants and aliases if they desire. Authors can include their ORCID identifier with article submissions, grant applications, and other scholarly documents. Conversely, publishers and funding agencies may request authors' ORCID identifiers to verify authors' information.

In situations in which it may not be possible to consult authors directly, editors should explore the following methods to try to determine the correct format for authors' names.[7]

(1) Review the submitted version of the article to check whether the authors have used a distinguishing text format to indicate their surnames (eg, boldface, underlining, or all uppercase).

(2) Check whether the authors have cited any of their previous works in their manuscript's reference list, and follow the name format used there. In addition, if the name–year citation system was used in the manuscript, any in-text citations provided to the authors' previous works can be used to confirm surnames.

(3) Check the bibliographic citations on the title pages of other works the authors have written. Bibliographic citations should list authors' surnames first, followed by the authors' initials (see Section 27.7.1.5, "Bibliographic Reference").

(4) Determine whether the authors have ORCID records, and if they do, follow the preferred name format listed in those records.

(5) Follow the name format used in the authors' curricula vitae, their institutional websites, or their professional network profile pages (eg, in ResearchGate and LinkedIn).

(6) Follow the name format most commonly used in the bibliographic records in indexes and databases that list the authors' works.

If an author's preference cannot be determined from the author or other sources, treat the last name in an unhyphenated compound name as the surname and the first in the compound name as the author's middle name, treat a prefix as part of the surname, and treat as the surname the last name given by an Asian author. For authors of other national origins, *The Chicago Manual of Style*[1] and other sources provide comprehensive guidelines for determining the names to be used in bibliographic systems and indexes, but these sources may disagree on some recommendations.

8.1.3 Suffixes in Names

If generational suffixes such as "Junior" (Jr), "Senior" (Sr), the "Second" (II), and the "Third" (III) are part of authors' names, place those suffixes in their abbreviated forms immediately after the surnames. Do not use a comma to separate a name from a generational suffix, and do not use periods in the abbreviation. In bylines and text, use such designations only with complete personal names (ie, given names combined with surnames) and before abbreviations for academic degrees and other credentials. In bibliographic references, generational suffixes follow the initials for given and middle names, and academic and other credentials are omitted. Whereas roman numerals are used with some generational suffixes in text, arabic ordinal numerals are used in bibliographic references (see Section 29.3.3.1.1, "Personal Authors").

On first reference in text:	For subsequent references in text:	In a bibliographic reference:
Charles E. Still Sr, DO	Dr Still *or* Still	Still CE Sr
William P. Murphy Jr, MD	Dr Murphy *or* Murphy	Murphy WP Jr
James F. Bell III, PhD	Dr Bell *or* Bell	Bell JF 3rd

If a work mentions 2 members of a family whose names differ only in generational suffixes, terms such as "the senior," "the junior," "the older," and "the younger" can be used to distinguish the 2 after the first full mentions of their names (eg, "the senior Dr Still" and "the younger Dr Murphy").

8.1.4 Academic Degrees, Honors, and Other Credentials

In text, a wide range of degrees, honors, and other designations may be appended to personal names. These modifiers, which are usually abbreviated, indicate achieved or honorary academic degrees, recognition of peers, licensure, and other credentials. See Table 8.2 for examples of degrees and honors. More extensive lists can be found in the *Acronyms, Initialisms and Abbreviations Dictionary*[8] and the *AMA Manual of Style*.[9] In addition, many standard dictionaries include academic degrees and their abbreviations.

When degrees and other designations follow a personal name, use their abbreviations without periods, and capitalize them according to Table 8.2 or another appropriate source. Use commas before and after such designations. If in doubt about a qualification's abbreviation, check with the entity that bestows the degree or honor, or check with the author.

Limit designations to those that are relevant to the topic of the manuscript (eg, a source's doctorate in chemistry would be relevant to a manuscript on atomic structure, but a master's degree in economics probably would not). Relevant designations can include fellowships in scientific disciplines. Fellowships are typically designated with the letter "F" followed by the initialisms for the organizations bestowing the fellowships.

> FAAAS *for* fellow of the American Association for the Advancement of Science
> FAS *for* fellow of the Nigerian Academy of Science
> FCSSA *for* fellow of the College of Surgeons of South Africa
> FNAS *for* fellow of the National Academy of Sciences of the Republic of Korea
> FRCPI *for* fellow of the Royal College of Physicians of Ireland
> FRS *for* fellow of the Royal Society (which is located in the United Kingdom)

Various methods exist for sequencing degrees and other designations. For scientific manuscripts, this style manual recommends sequencing degrees from lowest to highest academic level, regardless of the order in which they were earned. Do not list a lower degree if it is a prerequisite for a higher degree unless the degrees are in different subject areas or specialties. If designations other than academic degrees are used, place abbreviations for academic degrees first and list abbreviations for other designations in increasing order of distinction.

> Niels Bohr, PhD *not* Niels Bohr, BS, MS, PhD
> C. Everett Koop, MD, DSc *not* C. Everett Koop, BA, MD, DSc

In many anglophone countries, the following sequence is usually used for academic degrees:

- bachelor's degrees, such as BS and BA
- master's degrees, such as MS and MPH
- professional doctoral degrees, such as DDS and PharmD
- academic doctoral degrees, such as PhD and DSc
- honorary degrees, such as DMSc (Hon) and LLD (Hon)

The academic status of some degrees differs depending on the awarding country. For example, in the United States and Canada, "MD" is a professional doctoral degree, while in many other English-speaking countries, "MD" is an academic doctoral degree earned after obtaining an "MBBS," an "MB ChB," or another predoctoral medical degree that is equivalent to the US and Canadian MD. Doctor of osteopathic medicine (DO) degrees earned in the United States are professional doctoral degrees that are educationally and

Table 8.2. Abbreviations of Select Academic Degrees and Other Professional Credentials

Abbreviation	Term	Abbreviation	Term
AuD	doctor of audiology	MA	master of arts
BA or AB	bachelor of arts	MB or BM	medicinae baccalaureus (bachelor of medicine)
BN	bachelor of nursing		
BPharm	bachelor of pharmacy	MBA	master of business administration
BS or BSc	bachelor of science		
BVSc	bachelor of veterinary science	MBBS, BMBS, MB ChB, MB BCh, MB BChir, or BM BCh	medicinae baccalaureus, baccalaureus chirurgiae (bachelor of medicine, bachelor of surgery)
ChD or DCh	chirurgia doctor (doctor of surgery)		
		MBiochem	master of biochemistry
CM, ChM, MC, MCh, or MChir	chirurgia magister (master of surgery)	MBiol	master of biology
		MChem	master of chemistry
DC	doctor of chiropractic	MCompSci	master of computer science
DDS	doctor of dental surgery		
DHyg	doctor of hygiene	MD or DM	medicinae doctor (doctor of medicine)
DMD	doctor of medicine in dentistry		
		MEd	master of education
DMSc	doctor of medical science	MEng	master of engineering
DNP	doctor of nursing practice	MGeog	master of geography
DO	doctor of osteopathic medicine	MLS	master of library science
		MMedSci or MMedSc	master of medical science
DPM	doctor of podiatric medicine	MPH	master of public health
DrPH	doctor of public health	MPhil	master of philosophy
DSc or DS	doctor of science	MPhys	master of physics
DSW	doctor of social work	MS, MSc, or SM	magister scientia (master of science)
DVM	doctor of veterinary medicine		
		ND	naturopathic doctor
DVSc	doctor of veterinary science	NP	nurse practitioner
EdD or DEd	doctor of education	OD	doctor of optometry
ELS	editor in the life sciences	PE	professional engineer
ELS(D)	diplomate editor in the life sciences	PharmD, DP, or DPharm	doctor of pharmacy
ELS(H)	honored editor in the life sciences	PhD or DPhil	philosophiae doctor (doctor of philosophy)
JD or JuD	juris doctor (doctor of jurisprudence or doctor of law)	RN	registered nurse
		RPh	registered pharmacist
LLB	legum baccalaureus (bachelor of laws)	ScD or SD	scientia doctor (doctor of science)
LLD	legum doctor (doctor of laws)	VMD	veterinary medical doctor

legally on par with the US and Canadian MD degrees, while osteopathy degrees earned outside the United States are bachelor's or master's degrees.

Many publications limit academic degrees to the 2 highest. If authors hold more than one degree at the same level, list them in the order in which they were earned. If that cannot be determined, list the degrees in alphabetic order.

Do not use abbreviated honorific titles such as "Dr" and "Prof" before full names that include academic degrees. In general, use honorific titles before last names only and only after complete names with degrees and other designations have been provided. Alternatively, publishers may decide to omit honorific titles altogether. Publishers should be consistent within publications on whether to use honorifics.

On first reference in text:	For subsequent references in text:
Antonia C. Novello, MPH, MD *not* Dr Antonia C. Novello, MPH, MD	Dr Novello *or* Novello
Ahmed H. Zewail, PhD *not* Prof Ahmed H. Zewail, PhD	Dr Zewail *or* Zewail

When using degrees and other designations by themselves without personal names, generally spell out the designations rather than abbreviating them.

> A master of science degree is not necessarily a prerequisite for entering a doctoral program.
> *not*
> An MS degree is not necessarily a prerequisite for entering a PhD program.

For bibliographic and indexing purposes, omit degrees and other designations after names, as well as honorific titles like "Dr" and "Prof" before names.

On first reference in text:	For subsequent references in text:	In a bibliographic reference:
Elizabeth H. Blackburn, AC, PhD, FRS, FAA	Dr Blackburn *or* Blackburn	Blackburn EH

8.1.5 Military Titles

Because the official abbreviations of military ranks may be difficult for civilian readers to understand, military titles denoting rank or position should be spelled out instead of abbreviated in the bylines and text of documents intended for primarily civilian audiences. As with civilian titles, military ranks should be capitalized when they immediately precede personal names. Military ranks should be lowercased when used elsewhere, including when used in apposition to names and when used on their own.

> Lieutenant General Frank F. Ledford Jr, MD, served as the 37th surgeon general of the US Army.
> The US Navy's first female rear admiral, Alene Duerk, RN, commanded the Navy Nurse Corps.

In the United States, members of the commissioned corps of the US Public Health Service (USPHS) and the National Oceanic and Atmospheric Administration (NOAA) also hold military rank.

Conventions for rank may differ from service to service within the same country, and among those services, similar titles can designate different ranks. For example, the rank of captain in the US Army, US Air Force, and US Marines is equivalent to the junior-officer rank of lieutenant in the US Navy, US Coast Guard, USPHS, and NOAA. In contrast, the rank of captain in the latter 4 services is equivalent to the senior-officer rank of colonel in the first 3.

Conventions for rank also differ among countries. The military services in Canada and the United Kingdom hyphenate most compound ranks (eg, "lieutenant-commander"), whereas those in the United States and Australia do not (eg, "lieutenant commander"). See *Army Dictionary and Desk Reference*[10] for US military ranks and *The Canadian Press Stylebook*[11] for ranks in the Canadian services.

> Brigadier-General Frank Melville Lott, DDS, PhD, served as the colonel commandant of the Royal Canadian Dental Corps for 9 years.

Major General Samuel Roy Burston, MBBS, led the Royal Australian Army Medical Corps from 1942 to 1948.

In tables, addresses, and other places in which space is limited and many military titles are used, military ranks and titles may be abbreviated. For civilian readers, the unofficial abbreviations suggested in *The Chicago Manual of Style*[1] are more readily understandable than official military abbreviations.

Brig Gen *in place of* the military abbreviation BG
Lt Col *in place of* the military abbreviation LTC
Capt *in place of* both the junior-officer rank of CPT and the senior-officer rank of CAPT

In the bylines, text, tables, and other elements of documents intended for military audiences, use military abbreviations of ranks and titles. The abbreviations precede personal names, and academic degrees are omitted after the names and replaced with abbreviations for branch of service. Commas are placed before and after abbreviations for branch of service.

Rear Admiral Murray Goldstein, MPH, DO, *becomes* RADM Murray Goldstein, USPHS
Maj Gen Malcolm C. Grow, MD, *becomes* MG Malcolm C. Grow, MC, USAF (with "MC" for "Medical Corps" and "USAF" for "United States Air Force")

When referencing personnel who have retired from active duty, add "(Ret)" after their names if the document relates to the uniformed services. If the document is unrelated to such matters, do not include rank, branch of service, or "(Ret)" in manuscripts published after the personnel retired from active duty. These designations should also be omitted for authors and others who separated from the uniformed services before retirement.

8.1.6 Retroactive Name Changes

To maintain correct attribution and promote equity and inclusion, publishers are increasingly granting authors' requests to change their names in digital articles and records after publication without publishing public correction notices. Among the reasons why authors request that their bylines be updated is that they legally altered their names because of changes in marital status or gender identity. Although international guidelines related to retroactive name updates due to life events are still evolving, the Committee on Publication Ethics[12,13] has drafted guiding principles that journals may use in developing their own policies.

8.2 NAMES OF ORGANIZATIONS

The first time an organization's name is mentioned in a document, the complete, accurate name should be used. Subsequently, the name may be abbreviated to an initialism (eg, "BAS" for the "Brazilian Academy of Sciences" and "SCJ" for the "Science Council of Japan") or shortened by inversion, colloquialization, or truncation (eg, "Haffkine Institute" for the "Haffkine Institute for Training, Research and Testing" in India and "National Academies" for the "National Academies of Sciences, Engineering, and Medicine" in the United States).

When initialisms are used, they should be introduced as part of the first mention by placing them in parentheses after the full names of the organizations, except in cases in which the initialisms are well known to the intended readers (see Section 11.3, "Usage").

> The National Institute of Arthritis and Musculoskeletal and Skin Diseases (NIAMS) in the United States is part of the National Institutes of Health (NIH). Besides the NIAMS, NIH consists of 20 other institutes and 6 centers. (If the intended audience consists of health care professionals, the parenthetical use of "NIH" after the agency's full name can be omitted. However, NIAMS is not well known enough to omit using its initialism in conjunction with its full name on first mention.)

Another option for referring to an organization after its full name has been introduced is to use a common noun that is part of the full name (eg, "academy" for the American Academy of Audiology, "council" for the International Council of Chemical Associations, and "university" for the National University of Trujillo in Peru). These common nouns need not be capitalized in subsequent references. However, for some audiences and some publications—notably, those affiliated with the named organization—capitalizing the shortened form may be justified.

> On 1 July, the Survey will release the *Geological Atlas of Mars and Venus*. (in a publication of the US Geological Survey)
> After mapping Mars, the survey issued its first planetary atlas. (in an astronomy journal)

8.3 HUMAN GROUPS

Although race does not have a precise definition in biological terms, when race is pertinent to a study, terms based on race should be capitalized, including "Black" and "White."[14] In addition, pseudoprecise terms should not be used. For example, "Caucasian" is no more scientifically precise than "White," and many consider "Caucasian" to be archaic. For more on referencing race in scientific literature, see Section 7.4.3, "Race, Ethnicity, Nationality, and Citizenship," and the Council of Science Editors' Diversity, Equity, and Inclusion Scholarly Resources.[15]

Do not hyphenate "American" when it is part of a compound proper noun or when it is an integral part of other 2-word terms.

> an African American   North American species   10 Irish Americans
> Native Americans   a Mexican American   Latin American studies

Capitalize terms referring to Indigenous people. Because terms like "American Indians" and "Native Americans" encompass many Indigenous communities, it is preferable to refer to individuals by their specific communities or nations.

> Sioux, Choctaw, Shoshone, and Iroquois (United States)
> First Nations, Inuit, and Métis (Canada)
> Aboriginal Australians and Torres Strait Islander peoples (Australia)
> Sami (Norway, Sweden and Finland, as well as adjacent land in Russia)
> Ainu (Japan)

When a name refers to an Indigenous society or a tribe as a whole, use the singular form. In general, add "s" to the name when referring to 2 or more individuals of the same society or tribe. Exceptions include "Inuk," whose plural and adjectival form is "Inuit." The adjectival form of most societal and tribal names is the same as the names themselves. Adding "n" or "an" at the end of these names usually denotes language groups (eg, the Mayan language group and the Piman language).

Indigenous society as a whole:
The Maya inhabited Tikal in AD 800.
The Chinook and the Quinault are among the tribes that make up the Quinault Indian Nation.

Specific number of people:
Awaiting Hernán Cortés were 25 Mayas.
Smallpox killed hundreds of Chinooks in 1770.

Adjectival use:
The Maya calendar begins at the year equivalent to 3114 BC.
Chinook canoes were seaworthy.

Language:
The Mayan languages are numerous.
Speakers of Chinookan lived along rivers.

Capitalize the names of groups of humans if they are derived from proper names of geographic entities, if they are names for ethnically or culturally homogeneous groups, and if they are names for adherents to organized bodies.

Derived from geographic entities:

Asian	Nordic
Hispanic	New Yorker
Latino, Latina, and Latinx	Highlander

Ethnically or culturally homogeneous groups:

Kurds	Yorubas
Mongols	Welsh

Member or adherent of organized bodies:

Free Gardener	Jaycee
Freemason	Optimist
Odd Fellow	

8.4 SCIENTIFIC NAMES

Many authoritative scientific bodies have developed style conventions that govern the formation of scientific names in their disciplines and that define the proper usage of those names. Table 8.3 lists chapters in this manual that discuss nomenclature in various scientific disciplines.

8.5 GEOGRAPHIC NAMES

In general, a geographic entity should be referred to by the name accepted in the country where it is located. Online databases of official names are maintained by some government agencies (eg, the Geographic Names Information System[16] for the United States and the Canadian Geographical Names Data Base[17]). An alternative is to consult an authority such as the US Central Intelligence Agency's World Factbook.[18]

When a populated place or major geographic feature has an anglicized name that has become accepted through long use, the anglicized version is often preferable in English text and reference lists, unless there is a risk of the name being misinterpreted.

Cantabrian Range *for* Cordillera Cantábrica
Cologne *for* Köln
Florence *for* Firenze
Munich *for* München

Table 8.3. Scientific Names: Relevant Chapters in This Style Manual

Subject	Chapter	Chapter title
Animals	22	"Taxonomy and Nomenclature"
Astronomical objects	26	"Astronomical Objects and Time Systems"
Bacteria	22	"Taxonomy and Nomenclature"
Chemical compounds	17	"Chemical Formulas and Names"
Chemical elements and their components	16	"Subatomic Particles, Chemical Elements, and Related Notations"
Chromosomes	21	"Genes, Chromosomes, and Related Molecules"
Diseases	24	"Disease Names"
Drugs	20	"Drugs and Pharmacokinetics"
Fossils	22	"Taxonomy and Nomenclature"
Genes	21	"Genes, Chromosomes, and Related Molecules"
Geographic entities	14	"Geographic Designations"
Particle physics	16	"Subatomic Particles, Chemical Elements, and Related Notations"
Plants	22	"Taxonomy and Nomenclature"
Rock formations	25	"Earth"
Soils	25	"Earth"
Units of measure	12	"Numbers, Units, Mathematical Expressions, and Statistics"
Viruses	22	"Taxonomy and Nomenclature"

> The Rhine *for* Der Rhein
> Pearl River Estuary *for* Zhujiang Kou
> Takla Makan Desert *for* Taklimakan Shamo

However, for some place names, spellings that more closely approximate local-language names have replaced traditional anglicized versions.

> Beijing *for* Peking   Kolkata *for* Calcutta   Mumbai *for* Bombay
> Nanjing *for* Nanking   Chennai *for* Madras   Qingdao *for* Tsingtao
> Bengaluru *for* Bangalore   Kyiv *for* Kiev   Yangon *for* Rangoon

In most scientific contexts, short forms of country names and of the names of constituent parts are sufficient to identify research sites.

> China *for* the People's Republic of China
> Mexico *for* the United Mexican States
> Russia *for* the Russian Federation
> South Africa *for* the Republic of South Africa
> Switzerland *for* the Swiss Confederation
> United States *for* the United States of America

However, in references to a government or its agencies, the full formal name of a country or region may be needed.

> Democratic People's Republic of Korea *for* North Korea
> Republic of Korea *for* South Korea
> Gabonese Republic *for* Gabon
> Hashemite Kingdom of Jordan *for* Jordan
> Hong Kong Special Administrative Region of the People's Republic of China *for* Hong Kong

A country's complete current name can usually be found on the country's official government website. Additionally, the formal names of many nations can be found in the US Central Intelligence Agency's World Factbook.[18]

For more information on using geographic names, see Chapter 14, "Geographic Designations," and Section 9.4.4, "Geographic Designations."

8.6 TRADE NAMES, TRADEMARKS, AND OTHER COMMERCIAL NAMES

See Section 9.4.8, "Trademarks," for a discussion of trade names, trademarks, and other commercial names.

CITED REFERENCES

1. The Chicago manual of style: the essential guide for writers, editors, and publishers. 17th ed. The University of Chicago Press; 2017. Also available at https://www.chicagomanualofstyle.org

2. Castillo M. Authors' names in a globalized American Journal of Neuroradiology. Am J Neuroradiol. 2009;30(3):441–442. https://doi.org/10.3174/ajnr.A1367

3. Di Lauro A, Brandon E. The MIBIS manual: preparing records in microcomputer-based bibliographic information systems. 2nd ed. International Development Research Centre; 1995. https://idl-bnc-idrc.dspace direct.org/bitstream/handle/10625/16224/IDL-16224.pdf?sequence=1

4. The New York Public Library writer's guide to style and usage. HarperCollins Publishers; 1994.

5. Sun XL, Zhou J. English versions of Chinese authors' names in biomedical journals: observations and recommendations. Sci Ed. 2002;25(1):3–4. https://www.councilscienceeditors.org/wp-content/uploads /v25n1p003-004.pdf

6. Open Researcher and Contributor ID (ORCID). ORCID; [accessed 2023 Jan 13]. https://orcid.org

7. Lee C. What's in a name? Two-part surnames in APA style. APA Style Blog. American Psychological Association. 2017 May 4. https://blog.apastyle.org/apastyle/2017/05/whats-in-a-name-two-part-surnames -in-apa-style.html

8. Miskelly M, editor. Acronyms, initialisms & abbreviations dictionary: a guide to acronyms, abbreviations, contractions, alphabetic symbols, and similar condensed appellations. 55th ed. 11 vols. Gale Research, Inc; 2021.

9. JAMA Network Editors. AMA manual of style: a guide for authors and editors. 11th ed. Oxford University Press; 2020. Also available at https://www.amamanualofstyle.com

10. Zurick T. Army dictionary and desk reference. 4th ed. Stackpole Books; 2010.

11. McCarten J, editor. The Canadian Press stylebook: a guide for writers and editors. 19th ed. Canadian Press; 2021.

12. Tanenbaum TJ et al. A vision for a more trans-inclusive publishing world:high level principles for name changes in publishing. Committee on Publication Ethics; 2021 Jan 13. https://publicationethics.org /news/vision-more-trans-inclusive-publishing-world

13. Committee on Publication Ethics, Author Name Changes Working Group. Update on COPE guidance regarding author name changes. Committee on Publication Ethics; 2021 Jan 13. https://publicationethics .org/news/update-cope-guidance-regarding-author-name-changes

14. Council of Science Editors, Board of Directors. Scientific Style and Format update: capitalize racial and ethnic group designations. Council of Science Editors; 2020 Oct 20. https://www.councilscienceeditors .org/wp-content/uploads/Changes-to-SSF8-Style-Recommendations-10.20.20.pdf

15. Council of Science Editors, Committee of Diversity, Equity, and Inclusion. Diversity, equity,

and inclusion scholarly resources. Council of Science Editors; [accessed 2023 Jan 12]. https://www
.councilscienceeditors.org/resource-library/diversity-equity-and-inclusion-in-scholarly-publishing/

16. Geographic Names Information System. US Geological Survey. http://geonames.usgs.gov/

17. Canadian Geographical Names Data Base. Natural Resources Canada; [accessed 2023 Jan 13]. https://
open.canada.ca/data/en/dataset/e27c6eba-3c5d-4051-9db2-082dc6411c2c

18. The World Factbook. US Central Intelligence Agency. http://cia.gov/the-world-factbook/

ADDITIONAL REFERENCES

Conscious style guide. Quiet Press; [accessed 2023 Jan 12]. https://consciousstyleguide.com/

A guide to names and naming practices. Financial and Banking Information Infrastructure Committee (UK);
2006. https://www.fbiic.gov/public/2008/nov/Naming_practice_guide_UK_2006.pdf

9 Capitalization

Editors: Andrea Rahkola, ELS, and Heather Poirier, MA

9.1 GENERAL GUIDANCE

Some capitalization conventions, such as those for capitalizing proper nouns, are widely known and accepted. Others, such as those for position titles, are still debated and are dependent on the preferences of individual authors and publishers. Like *The Chicago Manual of Style*,[1] this style manual recommends a spare style of capitalization, restricting initial capital letters to instances in which they are clearly warranted and refraining from using small capitals.

This chapter includes some general recommendations for capitalization in English writing and a summary of recommendations in some scientific fields. More recommendations on capitalization for specific scientific disciplines are provided in relevant chapters in Part 3, "Special Scientific Conventions," of this manual, while additional general guidance on capitalization conventions is available in *The Chicago Manual of Style*.[1]

9.2 SYNTACTIC CAPITALIZATION
9.2.1 First Word of a Sentence

Capitalize the first word of every sentence, except when a sentence follows a semicolon or is within parentheses or brackets in another complete sentence.

The patient was offered a choice between physical therapy or surgery.
We delineated 6 study plots measuring 3 m × 3 m; no plots were adjoining.
Subjects were assigned to groups based on age and symptoms (sex was recorded for each person but was not a factor in determining group placement).
Are there enough data to support the authors' conclusions?

If a sentence begins with a scientific term that has an initial lowercase letter, a symbol, or a number, restructure the sentence, if possible, to move the term to later in the sentence.

The *t* test was applied to each set of data. *preferred to* *t* tests were used in the analyses. *or* *t* Tests were used in the analyses.
The pH must be carefully controlled. *preferred to* pH must be carefully controlled.

If rewording a sentence would create wordiness or change the meaning of the sentence, capitalize the root term of such a compound word, but do not change its initial lowercase letter, symbol, or number.

α-Toluene can be used for this purpose.
p53 control of CDK2 kinase activation is clear during apoptosis.
3T3 cells were originally obtained from Swiss mouse tissue.

Prefixes (eg, "anti-," "bi-," "non-," "pre-," and "uni-") are integral parts of the words they modify, so capitalize a prefix if it appears at the beginning of a sentence. Even if a compound term is created with a hyphenate or en dash, do not capitalize the second part of the term unless it is a proper noun (see Section 9.2.4, "Prefixes Used with Proper Nouns").

Bicarbonate concentrations were abnormally low in the blood.
Anti-amyloid therapies have been shown to slow cognitive decline.
Pre–Industrial Revolution society was predominantly rural.

If the first word of a sentence is a proper noun that usually begins with a lowercase letter, capitalize the first letter.

Du Pont argued this point in an earlier paper.
but
We agree with the conclusion reached by du Pont.

The recommendations outlined in this section also apply to footnotes in tables, even if the footnotes are not grammatically complete sentences.

9.2.2 First Word after a Colon

Capitalize the first word after a colon if it introduces one or more sentences that modify but are independent of the sentence before the colon. Colons are frequently used in this way to introduce direct quotations, dialogue, and questions and to introduce statements that further explain the previous statement.

The lead investigator addressed the nomenclature committee: "On the basis of new evidence, we propose that these 2 taxa be merged."
The evidence was enough to answer our question: Did these 2 fossil remnants represent the same organism?

If the words that follow the colon do not form a complete sentence, do not capitalize the first word unless it is a proper noun.

Two alternatives for treatment were proposed: surgical resection and relief of pain through medication.

9.2.3 Quotations and Excerpts

Capitalize the first word of a direct quotation when the quotation is a complete sentence within another sentence (ie, in adjacent syntactic relation to the rest of the sentence) and when the quotation is formally introduced or followed by an attribution term, such as "said," "explained," or "declared," or preceded by a colon (see Section 9.2.2, "First Word after a Colon").

> Theoretical physicist Stephen Hawking, PhD, said, "Scientists have become the bearers of the torch of discovery in our quest for knowledge."
> "When you've seen one Earth, you've seen them all," Apollo 17 astronaut Harrison Schmitt, PhD, quipped while he was too busy collecting Moon rocks to look up at his home planet.

However, do not capitalize the first word if the quotation is a partial quote within a sentence (ie, a syntactic part of the sentence).

> Claude Lévi-Strauss, DrE, argued that rather than giving the right answers, a scientist "asks the right questions."

Follow the same guidelines when taking excerpts from printed material. If an excerpt is not grammatically joined to what comes before or after it, capitalize the first word of the excerpt, regardless of whether the word was originally uppercased or lower-cased. However, if the excerpt is a syntactic part of the sentence, do not capitalize the first word, even if the word was originally uppercased. If the case of the first letter must be modified, it is not necessary to enclose the modified first letter in square brackets.[1]

> **Complete sentence excerpted, with formal introduction:**
> The review summarized the requirements: "To remain certified, the farm must undergo 2 inspections and laboratory testing of fish each year."
>
> **Complete sentence excerpted, forming integral part of sentence:**
> The review noted that "to remain certified, the farm must undergo 2 inspections and laboratory testing of fish each year." *not* The review noted that "[t]o remain certified . . ."

Different guidelines regarding change of case are usually applied for legal and other works that frequently reference quoted material. Those guidelines are described in detail in *The Chicago Manual of Style*.[1] For additional information about using excerpts in scientific publishing, see Section 10.2, "Excerpts and Quotations."

9.2.4 Prefixes Used with Proper Nouns

In general, lowercase such prefixes as "anti-," "ex-," "pre-," and "un-" when hyphens or en dashes are used to combine them with proper nouns as long as the prefixes are not used at the beginning of sentences.

> mid-July   anti-American   post–World War II

However, some variants of this principle have evolved due to common use. In some cases, a prefix may be capitalized before a proper noun (eg, Pan-American), incorporated into a single unhyphenated capitalized word (eg, Precambrian), or incorporated into an unhyphenated word in which the prefix and proper noun are lowercase (eg, transatlantic). When the proper form is in doubt, consult the intended publication's specialized or general dictionary of choice.

9.3 TITLES AND HEADINGS WITHIN TEXT

The following capitalization rules for titles of works (eg, books, journals, articles, poems, and lectures) and headings within parts of works (eg, chapters and sections) are referred to as "headline style," "title case," and "upstyle." These rules apply primarily to titles and headings used within running text to refer to other documents or other sections within the same document. Headline-style capitalization is preferred for running text because it helps readers distinguish titles and headings from adjacent text. More detail on headline style is provided in *The Chicago Manual of Style*.[1]

In contrast, for the titles and headings on covers, cover pages, and the top of chapters in the works themselves, it is acceptable to use either headline-style capitalization or "sentence-style" capitalization, also known as "downstyle." For titles of works in reference lists, sentence-style capitalization is recommended by this manual in accordance with the principle of spare style of capitalization (see Sections 9.1, "General Guidance," 9.7, "Reference Lists," and 29.3.3.2.1, "Format of Titles").

9.3.1 Titles
9.3.1.1 Multiterm Titles

In titles, capitalization is generally dependent on word class. However, always capitalize the first and last words of a title, regardless of the words' classifications. Capitalize all words in the following classes: nouns, pronouns, verbs, adjectives, and adverbs, as well as subordinate and adverbial conjunctions. Do not capitalize articles (ie, "a," "an," and "the"), coordinate conjunctions (eg, "and," "or," "nor," and "but"), or prepositions (eg, "with," "in," and "through") unless they appear at the beginning or end of a title or are part of a phrasal verb. In headline-style capitalization, word length is irrelevant. For example, "is" and "are" are capitalized because they are verbs, whereas "without" and "through" are lowercased because they are prepositions.

> *Sexual Behavior in the Human Male*   *Reality Is Not What It Seems*
> *Public Health under Siege*   *Algorithms to Live By*

In a title that is broken into 2 or more parts by a colon, question mark, or other punctuation, capitalize the words immediately before and after the punctuation, regardless of the words' classifications.

> *The Cosmic Connection: An Extraterrestrial Perspective*
> *What Is Life? The Physical Aspect of the Living Cell*

Do not capitalize the word "to" in infinitives within a title.

> *Learning to Read*   *How to Tread Lightly on the Earth*

Do not alter the form of locants and similar prefixes in chemical names in titles, but capitalize the root terms.

> *β-Agonists in Respiratory Medicine*   *L-Erythrose and Related Sugars*

Capitalize scientific names as they would appear in running text (see Section 22.2.2.1, "Binomial System for Species Names"). For example, do not capitalize specific epithets in botany and microbiology or specific names in zoology.

> *The Metabolism of Escherichia coli*   *Homo sapiens and Predecessors*

Capitalize the components of temporary compounds and coordinate terms created with hyphens or en dashes. (These terms are modifiers that would not otherwise use hyphens or en dashes.)

> *Behavior of Well-Adjusted Children in the Classroom*
> *Determining Acid–Base Status* (in which "Acid–Base" is a coordinate term)
> *Nitrogen-Fixing Bacteria and Legumes*

Do not capitalize the second and subsequent components of a term that would normally be hyphenated, except for proper nouns or proper adjectives.

> *A Helter-skelter Approach*
> *Prevalence and Risk Factors for Pre-eclampsia in Indian Women: A National Cross Sectional Study*
> *Attitudes toward Abortion among Non-Christian Women in New York*

9.3.1.2 Single-Word Titles

In general, capitalize a single word serving as a title. For chemical names and hyphenated terms, follow the conventions in Section 9.3.1.1, "Multiterm Titles."

> *Opticks*   *Physica*

9.3.2 Headings

In running text, capitalize the titles of book chapters and journal articles according to the principles in Sections 9.3.1.1, "Multiterm Titles," and 9.3.1.2, "Single-Word Titles." Enclose the titles in quotation marks.

> The last chapter, "Future Prospects," was originally published in *New Scientist*.
> The *Proceedings of the National Academy of Sciences* published the article "DNA Sequencing with Chain-Terminating Inhibitors" in December 1977.

For cross-references to headings for the basic components of articles, books, and other documents (eg, Preface, Results, and Discussion), follow headline-style capitalization but do not enclose the heading in quotation marks.

> . . . according to criteria outlined in Materials and Methods . . .
> . . . who wrote the Preface for this volume . . .

However, if part or chapter headings are modified from their original form or used as an adjective, the headings should be lowercased and used without quotation marks.

> A complete description can be found in the methods section.
> This discrepancy is analyzed further in the discussion section.
> . . . in the chapter on living with chimpanzees . . .

9.3.3 Parts in a Series

Capitalize nouns and abbreviations such as "volume" and "fig" that are used with numerals or letters to designate their positions in a series of similar parts of a book, chapter, journal, article, or supplement. Use without quotation marks.

> Volumes 2–4   Figure 6
> Vol 5   Figs 9–12
> Part A   Table 3
> Parts 3 and 4   Tables 6 and 7
> Plate III   Supplement 3
> Plates I and II   Suppl 16

Do not capitalize nonheading components of works like "page" and "line," and do not abbreviate these words in running text.

> The photograph on page 678 is a prime example.

9.4 NAMES

9.4.1 Proper Nouns

Capitalize all proper nouns, including the names of people, places, organizations, institutions, departments, political entities, and events. In addition, some adjectives based on proper nouns should be capitalized. In running text, do not capitalize the article "the" preceding an organizational name, unless it is part of the official name.

> As a Jewish scientist fleeing Nazi Germany, Albert Einstein, PhD, accepted an offer to join the Institute for Advanced Study in Princeton, New Jersey, in 1933.
>
> Vancouver, British Columbia, Canada, is located on the Burrard Peninsula, along the Strait of Georgia.
>
> Founded in 1957, the Council of Science Editors was originally named the Conference of Biology Editors.
>
> Only submissions to *The Lancet* in 2017 were included in the final count.
>
> In Australia, The University of Melbourne offers doctorates in medicine, dental surgery, optometry, psychological sciences, and physiotherapy.
>
> Sydney, Australia, hosted the 20th International Conference on Soil Mechanics and Geotechnical Engineering.

Use lowercase for truncated terms containing only generic terms without indicating specific names (see also Section 8.2, "Names of Organizations").

> the hospital   the state   the institute   the conference

In a plural construction for organizational entities, use lowercase for the generic part of the names.

> Harvard and Princeton universities
> Toronto Western and Mount Sinai hospitals
> the departments of botany, forestry, and zoology

Do not capitalize the name of a discipline or specialty except when it is part of a proper noun, such as the official name of a department.

> Five of the 17 students are pursuing a career in astrophysics.
> The Department of Astrophysics will host a seminar.

For capitalization of trade names, see Section 9.4.8, "Trademarks."

9.4.2 Sociocultural Designations

See Sections 8.3, "Human Groups," and 7.4.3, "Race, Ethnicity, Nationality, and Citizenship," for recommendations on capitalizing the names of human groups (eg, racial, ethnic, cultural, and geographic groups). Included in both sections are recommendations for capitalizing Black and White when they refer to race.

9.4.3 Eponymic Terms

Capitalize the first letter of the proper noun in an eponymic term, but do not capitalize derivative or adjectival forms of these terms.

Finkelstein reaction	duct of Müller *but* müllerian duct
Hodgkin disease	Gasser ganglion *but* gasserian ganglion
Noble agar	Parkinson disease *but* parkinsonian tremor
Down syndrome	Gram stain *but* gram-negative bacteria
Wilms tumor	Hunter sore *but* hunterian chancre
cesarean section	linnaean system of classification
fallopian tube	graafian follicle

For recommendations on avoiding the possessive forms of eponymic terms, see Section 6.5.6, "Eponymic Terms."

Several other kinds of eponymic terms should not be capitalized, including the names of units established by authoritative bodies; terms derived from proper nouns that relate to objects, especially apparatuses; and terms in well-established common usage. If in doubt, consult the intended publication's specialized or general dictionary of choice or a resource appropriate to the subject matter.

ampere	burley tobacco	joule	plaster of paris
angstrom	congo red	lambert	roman type
benday process	coulomb	merino sheep	rutherford
bessemer steel	curie	paris green	timothy grass
bunsen burner	gauss	petri dish	venturi tube

9.4.4 Geographic Designations

Capitalize the names of the great divisions (eg, zones) of Earth's surface and the names of distinct regions or districts of Earth. Lowercase adjectives derived from these terms, as well as truncated versions of the nouns containing only generic terms without indicating specific names.

the Antipodes *but* antipodean
the Great Divide
the Tropic of Cancer *but* the tropics
North Temperate Zone *but* the temperate zone
the Equator *but* equatorial

In most cases, do not capitalize the names of the points of the compass and derived adjectives. However, capitalize those terms if they are part of generally accepted names designating specific areas.

northern Saskatchewan
an eastern exposure
Eastern Hemisphere
western songbirds
located in the South *but* traveled south in the United States
just north of the South Pole
East Africa *and* North Africa *but* central Africa
the American Midwest *but* midwestern
Southeast Asia *but* southeastern Asia *and* central Asia

Capitalize generic geographic terms, such as "lake," "river," "ocean," and "mountain," when they form part of proper nouns. However, lowercase those terms if they precede proper nouns along with the article "the."

Atlantic Ocean	English Channel	Monte Pissis *but* the mountain Pissis
Canadian Shield	Grouse Mountain	Nile River *but* the river Nile
Canoe Lake	Mount Everest	Terrigal Lagoon *but* the lagoon Terrigal

Do not capitalize a generic term that follows a capitalized generic term.

> the Fraser River valley *but* the Fraser Valley
> the Nile River basin *but* the Nile Basin
> the Rio Grande valley (as a Spanish word for "river," "Rio" is the generic capitalized term in this
>   example)

In a plural construction, lowercase generic nouns that would be capitalized in singular constructions.

> Adriatic and Aegean seas   Jostedal and Hardangerjøkulen glaciers
> Great Slave and Lesser Great Slave lakes   Vancouver and Saltspring islands

Capitalize the article "the" only when it is a formal part of a geographic name, as it is for The Bahamas, officially the Commonwealth of The Bahamas, and The Gambia, officially the Republic of The Gambia.[2,3] For names of geographic entities, consult online geographic databases and countries' official websites. Other resources include comprehensive atlases and geographic dictionaries, such as *The Columbia Gazetteer of the World*.[4]

For further recommendations on geographic names, see Chapter 14, "Geographic Designations."

9.4.5 Taxonomic Names

Capitalize the Latin, or scientific, names of phyla, classes, orders, families, genera, and their subdivisions. Although the genus of a species is capitalized, use lowercase for specific epithets in botany and microbiology and for specific names in zoology, as well as for designations of infraspecific taxa (see Section 22.2.2.1, "Binomial System for Species Names"). In general, use lowercase for vernacular, or common, names of organisms (see Section 22.2.3, "Vernacular Names").

> Lepidoptera   *Homo sapiens*   *Alnus viridis* ssp. *Crispa*   common juniper

Neither capitalize nor italicize adjectives and English nouns derived from scientific names.

> salmonid   orthopteran   pneumocystis pneumonia

A few terms for organisms have more than one meaning, which may create difficulties in determining whether they should be capitalized. For example, capitalize "Metazoa" and "Protozoa" when they are used as the names of divisions, but use lowercase for the common-noun designation "metazoan" and its plural, "metazoans," and for "protozoan" and its plural, "protozoa." Use lowercase when the plural forms of these words are used with the article "the" to represent individual members of the division, not the division itself.

> The protozoa moved in an almost synchronized fashion.
> The metazoans attract the attention of most zoo visitors.

A similar situation exists with the word "primates," which serves as the name of an order and the name of individual members of the order.

> Gorillas, chimpanzees, and baboons belong to the order Primates.
> It is common for primates to adapt to communal living.

For more details on capitalizing the names of organisms and viruses, see Chapter 22, "Taxonomy and Nomenclature." For capitalization rules for chromosomes and genes, see Chapter 21, "Genes, Chromosomes, and Related Molecules."

9.4.6 Geologic Names

In general, capitalize the accepted names for geologic and stratigraphic time periods in formal usage. However, because the second element of these terms is inconsequential, it may be lowercased in running text, or it may be omitted altogether when the first term is preceded by the article "the."

> The Archean Eon followed the Hadean Eon.
> Mammals dominate the Cenozoic era.
> *Homo sapiens* is the only human species to survive the Pleistocene.

Capitalize the names of geologic formations only if they have been formally published according to the rules set out in the *North American Stratigraphic Code.*[5] Several online sources provide searchable lists of names that adhere to that code.[6–8] See Chapter 25, "Earth," for further information on capitalizing geologic names and related terms.

9.4.7 Astronomical Names

Capitalize the names of galaxies, constellations, stars, planets and their satellites, asteroids, and other celestial objects.

> Milky Way   North Star *or* Polaris   Halley's Comet
> Solar System   Mars   asteroid 4 Vesta

Capitalize the words "sun," "earth," and "moon" when used in connection with other astronomical terms. In this context, do not use the article "the" with "Earth." Lowercase these 3 terms when used in general.

> The Moon was the first destination for space travelers, and the feasibility of humans traveling to Mars from Earth is now under investigation.
> Some of the Sun's solar winds hit Earth's atmosphere.
> The sun provides energy for photosynthesis.
> Saturn has more than 80 moons.

A rule of thumb for the word "earth" is to lowercase it when preceded by the definite article, "the."

> the 4 corners of the earth
> the salt of the earth
> The planets closest to the Sun are Mercury, Venus, Earth, and Mars.

Use lowercase for the adjectival form of astronomical bodies' proper names.

> jovian *for* Jupiter   lunar *for* Moon   solar *for* Sun

See Section 26.2, "Capitalization and Lowercase," for more on capitalizing astronomical terms.

9.4.8 Trademarks

Trademarks are proprietary product names that are legally registered to companies or persons. Unless a particular brand is being discussed in a document, use the product's generic term. If the brand name must be used, capitalize it.

> bandage *for* Band-Aid   clear acrylic plastic *for* Plexiglas   petroleum jelly *for* Vaseline

A product's trademark protection may vary from country to country. For example, the trademark Aspirin is still protected in such countries as Germany, Canada, and Mexico, but it is no longer protected in such countries as Australia, New Zealand, the United

Kingdom, and the United States. For publications produced in countries where the name Aspirin is protected, the drug should be treated as a brand name for acetylsalicylic acid. See Section 20.2, "Use of Nonproprietary (Generic) or Proprietary (Trade) Names."

A number of databases can be accessed to determine whether a name is trademarked.[9-11] In the United States, for example, the legitimacy of a trademark can be ascertained by using the Trademark Electronic Search System of the US Patent and Trademark Office.[10] In addition, the World Intellectual Property Organization maintains a brand database with information from more than 70 national and international collections.[11]

When used in marketing materials, trademark names typically include trademark designations (ie, ® for trademarks that have been officially registered and ™ for marks that have not been registered but that manufacturers wish to identify as their own). These designations are unnecessary in formal scholarly writing,[12] provided the trademark names are spelled correctly and their uses are identified.

In contrast, in experimental drug studies, the drugs should be identified by their proprietary, capitalized names and their manufacturers, as well as by their generic, lowercased names. To determine whether drug names are proprietary and to find the drugs' nonproprietary names, consult such resources as the *United States Pharmacopeia*[13] and the Canadian Pharmacists Association's *Compendium of Pharmaceuticals and Specialties*.[14] See Chapter 20, "Drugs and Pharmacokinetics," for additional information about drug names.

Retain internal capitalization for trademark names that use bicapitalization, or intercaps, in which one or more letters within names are capitalized in addition to the names' first letter (eg, ScienceDirect and AstraZeneca). However, convert trademark names to conventional capitalization when they use all capital or all lowercase letters.

> House of Science *not* HOUSE OF SCIENCE
> Alzheimer's Association *not* alzheimer's association

9.4.9 Websites and Electronic Addresses

Use headline-style capitalization for the titles of websites but lowercase for the websites' uniform resource locators, better known at URLs.

> Council of Science Editors' website *but* councilscienceeditors.org
> Committee on Publication Ethics' website *but* publicationethics.org
> International Committee of Medical Journal Editors' website *but* icmje.org

In addition, use lowercase for email addresses.

> cse@councilscienceeditors.org
> scienceeditor@councilscienceeditors.org
> custserv@press.uchicago.edu

9.4.10 Names for Abstractions

The names of abstract ideas and objects, including the seasons, may be capitalized when personified. Thus, "Mother Nature" is always capitalized, but "nature" is capitalized only when it is clearly personified.

> Three varieties are found in nature, but more have been developed in the laboratory.
> While living near the Arctic Circle, he learned that Nature can be a harsh taskmistress.

9.5 ABBREVIATIONS

For recommendations on the general and nuanced capitalization of abbreviations, see Section 11.1.1, "Contraction Abbreviations," and Section 11.2, "Punctuation and Typography." For recommendations on capitalizing abbreviated academic degrees and other credentials, see Section 8.1.4, "Academic Degrees, Honors, and Other Credentials," and Table 8.2.

9.6 TABLES

See Section 30.1, "Tables," for recommendations on using capitalization in tables.

9.7 REFERENCE LISTS

In some respects, capitalization guidelines for reference lists and bibliographies differ from those used for other parts of scientific publications and books. In particular, titles in end references and bibliographies are set in sentence-style capitalization instead of headline style. See Chapter 29, "References," for recommendations on capitalization in end references and bibliographies.

9.8 INDEXES

Reserve capitalization in indexes for terms that have initial capitals in running text, such as proper nouns. This practice should help index users to distinguish common nouns, including nonproprietary names, from proper nouns, including trade names.

hexylresorcinol	(nonproprietary name for an antiseptic drug)
Hibiclens	(Mölnlycke Health Care's trademark name for its version of the drug chlorhexidine gluconate)
Zoloft	(Pfizer Inc's trademark for its brand of sertraline hydrochloride)
histidine	(nonproprietary name for an amino acid)

Even if a publication adopts an index style that calls for every term to have initial capitalization, the following terms should have initial lowercase letters.

(1) compound surnames in which the prefixes begin with a lowercase letter (eg, von Willebrand disease and du Pont)
(2) italicized prefixes of chemical compounds (eg, p-Aminobenzoic acid and o-Toluic acid)
(3) standard symbols and abbreviations that begin with a lowercase letter (eg, pH, pK, and mRNA)

See also Section 30.3, "Indexes."

CITED REFERENCES

1. The Chicago manual of style: the essential guide for writers, editors, and publishers. 17th ed. The University of Chicago Press; 2017. Also available at https://www.chicagomanualofstyle.org

2. Ministry of Defense (UK). Guidance: Country names: The Permanent Committee on Geographical Names for British official use; [updated 2021 May 14]. https://www.gov.uk/government/publications/country-names

3. Department of State (US), Office of the Historian. All countries. https://history.state.gov/countries/all

4. Cohen SB, editor. The Columbia gazetteer of the world. 2nd ed. Columbia University Press; 2008. Also available at http://www.columbiagazetteer.org/main/Home.html

5. North American Commission on Stratigraphic Nomenclature. North American stratigraphic code. Stratigraphy. 2021;18(3):153–204. https://ngmdb.usgs.gov/Geolex/resources/docs/NACSN_Code_2021.pdf

6. US Geological Survey. National Geologic Map Database: Geolex Search. US Geological Survey and Association of American State Geologists; [modified 2021 Aug 24]. https://ngmdb.usgs.gov/Geolex/search

7. Government of Canada. WEBLEX Canada: Lexicon of Canadian geological names on-line. Government of Canada; [modified 2021 Feb 1]. https://weblex.nrcan.gc.ca

8. Gobierno de México. Léxico Estratigráfico de México https://www.sgm.gob.mx/Lexico_Es/

9. Canadian Intellectual Property Office. Trademarks; [modified 2019 Aug 16]. http://www.ic.gc.ca/eic/site/cipointernet-internetopic.nsf/eng/h_wr00002.html

10. United States Patent and Trademark Office. Trademarks. United States Patent and Trademark Office; [modified 2019 Oct 21]. http://www.uspto.gov/trademark

11. Word Intellectual Property Organization. Trademark Database Portal. Word Intellectual Property Organization; [accessed 2021 Aug 23]. https://www.wipo.int/amc/en/trademark/

12. JAMA Network Editors. AMA manual of style: a guide for authors and editors. 11th ed. Oxford University Press; 2020. Also available at https://www.amamanualofstyle.com

13. United States Pharmacopeial Convention. USP–NF. United States Pharmacopeial Convention. https://www.uspnf.com/purchase-usp-nf

14. Canadian Pharmacists Association. Compendium of pharmaceuticals and specialties: the Canadian drug reference for health professionals. Canadian Pharmacists Association; 2020. Annual.

10 Type Styles, Excerpts, Quotations, and Ellipses

Editors: Heather DiAngelis, MA, and Dana M. Compton

10.1 TYPE STYLES FOR STYLE CONVENTIONS

Virtually all scientific literature in English is printed in a roman typeface, either a serif face, such as Times New Roman, or a sans serif face, such as Helvetica. Serif fonts contain serifs, which are short lines stemming from the upper and lower ends of letters that are intended to help pull the reader through the text. Sans serif fonts do not have these decorative features.

Variants of the basic faces are used for specific conventions in prose style or for specific symbolic meaning. The 2 main variants are italics and boldface. Other conventions call for capital (uppercase) letters, small capitals, superscript characters (raised above the line of type), and subscript characters (positioned below the line). Superscript and subscript characters are generally set smaller than the main text. For more detail on the characteristics of type, see Chapter 31, "Typography and Manuscript Preparation."

10.1.1 Italic Type
10.1.1.1 General Uses
In general publications, italic type has 8 widely applied uses. This style manual recommends only the first 5 for scientific publications. Using the other 3 in scientific publications could create confusion with italics used for special scientific conventions (see Section 10.1.1.2, "Scientific Uses").

(1) In running text for the titles of books, journals, and other complete documents. In running text, quotation marks should be used to distinguish titles of chapters and other

book sections from the titles of whole books and to distinguish the titles of journal articles from journal titles (see Section 5.3.4, "Quotation Marks").

> His article, "The Gene Responsible for Multiple Sclerosis," was published in *The New England Journal of Medicine.*
> The chapter "Quotations and Dialogue" in *The Chicago Manual of Style* covers the topic in detail.

(2) For many non-English words and phrases and their abbreviations.

> The researchers categorized *los competidores* by age and years of athletic training.

This rule does not apply to non-English proper names.

> While studying classic architecture, she visited Sacré-Cœur, the Cathédrale Notre-Dame de Paris, and Les Invalides in Paris.

Similarly, italics should not be used with widely used non-English phrases (such as "a priori" and "in vitro"), amputated phrases (such as "post hoc" for *post hoc ergo propter hoc*, as in "a post hoc argument"), and abbreviations (such as "ie" for *id est* [that is] and "et al" for *et alii* [and others]). Although taken from Latin, these phrases and abbreviations are now considered standard English in science.

(3) For a letter or number in either running text or a legend that refers to a corresponding letter or number in an illustration, whether the character in the illustration is in italic or roman font. The in-text letter should match the letter in the illustration with regard to capitalization.

> We could not account for the outlier *h* in Figure 3.

(4) For a short preface or an explanatory note by a publisher or editor that must be clearly distinguished from the author's text.

> (*The following short paper was submitted anonymously, but the editors felt it merited publication nonetheless.—Editor in chief*)

(5) For "see" and "see also" references in indexes (see Section 30.3.6, "Cross-referencing").

> **In an index reference:**
> coronary heart disease
> *see* ischemic heart disease

(6) For a word or phrase when introduced in text for definition, explanation, or discussion.

> The concept of *stratum* took a long time to develop in the evolution of geology.

For scientific publications, this style guide instead recommends placing such words and phrases within quotation marks rather than in italics.

(7) For a word or phrase that must be represented as such, which is often referred to as "words as words" and "phrases as phrases."

> In desktop publishing, the term *typeface* has given way to *font.*

As with the previous rule, this style guide recommends that scientific publications place such words and phrases within quotation marks rather than in italics (see also Section 5.3.4.1, "Double Quotation Marks").

(8) For explanatory words such as "for" and "but" used within running text.

> **In running text in general publications:**
> NIH *for* National Institutes of Health   PVC *for* polyvinyl chloride

Although this convention is followed in examples throughout this style manual, it is not recommended for scientific publications. Instead, the explanatory words should be in roman type, and the words being explained should be enclosed in quotation marks as if they were words as words.

In running text in scientific publications:
"NIH" for "National Institutes of Health"   "PVC" for "polyvinyl chloride"

10.1.1.2 Scientific Uses

All scientific publications should adhere to the scientific uses for italics itemized below. This list is not exhaustive.

(1) For single-letter symbols standing for known or unknown quantities and variables (see Table 4.5 and Section 4.2.2, "Type Styles for Symbols"). In contrast, multiletter symbols for quantities are typically in roman typeface (see Section 4.2.2, "Type Styles for Symbols"). For applications in mathematics and statistics, see Tables 12.9, 12.11, and 12.14 and Section 12.3.1, "Font Style Conventions"; in chemical kinetics, Tables 18.1 to 18.4; in pharmacokinetics, Table 20.2 and Section 20.5, "Pharmacokinetics"; in cardiovascular and respiratory physiology, Tables 23.3 and 23.4 and Sections 23.5.4, "Circulation," and 23.5.9, "Respiration."
(2) For single-letter modifiers (usually in subscript) of symbols when the modifiers represent quantities (see Table 4.5).
(3) For letters symbolizing atomic particles (see Tables 16.1 and 16.2 and Section 16.2, "Elementary and Composite Particles").
(4) For the scientific name of a genus, species, subspecies, or variety. The name of a higher taxon is usually not italicized. See Section 22.2, "Taxonomy and Nomenclature: Commonalities across Hierarchical Systems."
(5) For letters and numerals used as symbols for genes and alleles in most systems of gene symbolization (see Chapter 21, "Genes, Chromosomes, and Related Molecules").
(6) For prefixes of chemical names designating structural and configurational relations (see Section 17.3, "General Conventions of Nomenclature").
(7) For some crystallographic symbols (see Section 25.6, "Crystals").
(8) For the names of manufactured satellites after they have been christened (see Section 25.12.4, "Names of Environmental Satellites").

Scientific text that should be in italic type may be changed to roman type when the surrounding text is italicized for design reasons, such as accommodating explanatory notes and headings.

Editor's note: *Sablefish*, Anoplopoma fimbria, *is sensitive to handling* . . .

10.1.2 Boldface Type

Boldface type has no standard use in general literature. However, some general and scientific publications use boldface for specific elements in indexes (such as the page numbers of the main text on subjects) and in bibliographic references (such as volume and year numbers).

Another use for boldface is to highlight the letters in a term that are used to form an initialism, especially when some of the letters in the initialism are not initial letters of words in the term.

The acronym "AIDS" stands for "**a**cquired **i**mmuno**d**eficiency **s**yndrome."

In a scientific context, boldface is also sometimes used for symbols for vectors and matrices in mathematical expressions (see Section 12.3.1, "Font Style Conventions").

10.1.3 Capital Letters

10.1.3.1 General Uses

The general uses of initial capital letters are described in detail in Chapter 9, "Capitalization."

10.1.3.2 Scientific Uses

In a scientific context, initial capital letters are used for the following purposes.

(1) For the taxonomic name of a phylum, class, order, family, genus, or subdivision but not for a specific or subspecific taxon, except where permitted by international codes (see Chapter 22, "Taxonomy and Nomenclature").

> *Pleurotus ostreatus*

(2) For symbols such as those used for genes, chromosomes, blood groups, cardiovascular physiology, and respiratory physiology (see Chapters 21, "Genes, Chromosomes, and Related Molecules," and 23, "Anatomical and Physiological Descriptions for Macroorganisms").

(3) For the formal names of historical epochs, geologic ages and strata, and zoogeographic zones, as well as for other terms used in classifications (see Chapter 25, "Earth," and Section 9.4.6, "Geologic Names").

(4) For the proper names of astronomical bodies (see Sections 9.4.7, "Astronomical Names," and 26.2, "Capitalization and Lowercase").

(5) For proper-name components in the vernacular names of organisms (see Section 22.2.3, "Vernacular Names").

> American black currant   Arizona white oak   Hawaiian bud moth   Virginia pine sawfly

10.1.4 Small Capital Letters

Small capital letters, often called "small caps," are generally designed to be the height of a typeface's lowercase "x." For typefaces that do not have small capital fonts, small caps are often created by reducing the size of capital letters.

10.1.4.1 General Uses

Small capital letters are often used to aid readers in distinguishing some parts of text from other parts. Some publications use small capitals for abbreviations for academic degrees (eg, MD), eras (eg, BC and CE), and time of day (eg, AM and PM). However, this style manual does not recommend using small capital letters for general purposes because small caps do not convey any meaning beyond that of regular capital letters.

10.1.4.2 Scientific Uses

Small capital letters convey a few specific scientific meanings.

(1) Configurational prefixes D and L for amino acids and carbohydrates (see Sections 17.13.2.1, "Amino Acids," and 17.13.4, "Carbohydrates").

(2) Modifiers indicating the anatomic site of a gas phase in symbols of respiratory function (see Table 23.4).

10.1.5 Superscript and Subscript Characters

Superscripts are widely used in scientific notation as qualitative and quantitative modifiers to words, phrases, and symbols in running text. Although most commonly used

Table 10.1. Examples of Superscript Symbols

Item represented	Symbol	Section or table in this style manual
Celestial coordinate for right ascension	$14^h17^m13^s$	Section 26.3, "Celestial Coordinates"
Citation by reference number (citation–sequence and citation–name systems)	. . . as previously reported[12] . . .	Sections 29.2.1.1, "Citation–Sequence," 29.2.1.3, "Citation–Name," and 29.2.4, "Placement of In-Text References"
Degrees, geographic	$75°00'15''W$	Section 14.2.1, "Latitude and Longitude"
Degrees, temperature	$50°C$	Table 12.4
Electric charge of a particle	p^+	Section 16.3, "Nuclear Reactions"
Exponent	e^x	Table 12.9
Ionic charge	Na^+, Al^{3+}	Section 16.4, "Chemical Elements"
Mass number	^{14}C	Section 16.4, "Chemical Elements"
Oxidation number	Mn^{VII}	Section 16.4, "Chemical Elements"
Thermodynamic state	$G°$	Section 18.3.2, "States"

Table 10.2. Examples of Subscript Symbols

Item represented	Symbol	Section or table in this style manual
Atomic number	$_7N$	Section 16.4, "Chemical Elements"
Constant designation	K_i	Table 18.4
Derivative specified	$D_x y$	Table 12.11
Member of chemical group	vitamin D_3	Section 17.13.19, "Vitamins"
Number of atoms in a chemical compound	H_2O	Sections 16.5, "Chemical Formulas," and 17.2.1, "Empirical Formulas"
Number of molecular units	$\alpha_2\lambda_2$	Section 23.5.6.1, "Immunoglobulins"
Ring systems, chemical	$1_2,2_2{:}2_5,3_2$-terpyridine	Section 17.6.3.2, "Bicyclic and Polycyclic Ring Systems"
Scalar designation	F_x, F_y	Section 12.3.1, "Font Style Conventions"
Thermodynamic quantities	C_p	Section 18.3.3, "Processes," and Table 18.1

for references in the citation–sequence and citation–name systems, superscripts have several other purposes (see Table 10.1 for examples). These numeric, alphabetic, and graphic characters are positioned above the main line of type, and they are in a smaller font than that used for the main type. In some symbols, superscripts may themselves have superscripts.

Like superscripts, subscripts are widely used as qualitative and quantitative modifiers to symbols, and subscripts are in a smaller font than the font of the main type. However, whereas superscripts are above the main line of text, subscripts are below the main line. Table 10.2 illustrates some examples of subscripts.

Use roman type for superscript and subscript characters that serve as qualitative modifiers (ie, descriptive modifiers). Use italic type for superscript and subscript characters that serve as symbols for quantities.

$$e^+ \qquad S^{2-} \qquad n\theta^\xi \qquad C_1 \qquad K_T \qquad V_{CO_2}$$

10.2 EXCERPTS AND QUOTATIONS

Short passages taken from another text can usually be placed between quotation marks within the body of the paragraph in which they are quoted. Such quotations are sometimes called "run-in quotations." In contrast, long passages should be distinguished from the text into which they are inserted by placing the passages in separate paragraphs and varying the typography. These quotations are often called "block quotations," "extracts," or "setoff quotations."

10.2.1 Run-in Quotations

Quotation marks are used at the beginning and end of run-in quotations, and quoted material is separated from other text by appropriate punctuation. The styles for run-in quotations differ between American and British usage, mainly in the following 2 ways:

(1) American style uses double quotation marks for a primary quotation and single quotation marks for a quotation within a quotation. British style uses single quotation marks for a primary quotation and double quotation marks for a quotation within a quotation.
(2) In American usage, periods and commas at the end of quoted material are always placed before the closing quotation mark, regardless of whether the quoted material is a full or partial sentence. In British style, periods and commas are placed before the closing quotation mark when the quoted material is a full sentence and after the quotation mark when the quoted material is a sentence fragment, phrase, or single word.

In general, publications[1-5] in the United States and Canada follow the American style, whereas publications[6-7] in the United Kingdom, Australia, and other Commonwealth nations follow the British style. Both styles are described in this section and in Sections 10.2.3, "Quotation Marks in Relation to Other Punctuation Marks," and 5.3.4, "Quotation Marks." This style manual recommends that each publication select either the American or British convention and consistently adhere to it. To demonstrate how to consistently adhere to one style, the American style is applied throughout the rest of this manual.

The examples below demonstrate the differences between American and British styles in using primary quotation marks to designate a portion of a quotation incorporated into a sentence.

> **American style:**
> The self-taught meteoriticist Harvey Harlow Nininger described "tiny droplets of melted country rock," and he felt he had tangible proof of an explosion.

> **British style:**
> The self-taught meteoriticist Harvey Harlow Nininger described 'tiny droplets of melted country rock', and he felt he had tangible proof of an explosion.

In the next set of examples, secondary quotation marks enclose quotations within quotations, also known as "internal quotations."

> **American style:**
> According to the manuscript, "Nininger noticed tiny 'bombs' of what he described as 'yellow-green-brown slag.'"

> **British style:**
> According to the manuscript, 'Nininger noticed tiny "bombs" of what he described as "yellow-green-brown slag".'

Tertiary quotations (ie, quotations within secondary quotations) are rarely needed in scientific writing. When tertiary quotations are needed, use the primary quotation marks again.

American style:
Our professor of meteoritics said, "Fletcher G. Watson, PhD, wrote, 'Nininger is responsible for "half" of the meteorite discoveries in the world.' "

British style:
Our professor of meteoritics said, 'Fletcher G. Watson, PhD, wrote, "Nininger is responsible for 'half' of the meteorite discoveries in the world." '

If a quotation is incorporated into the natural syntax of the quoting sentence, lower-case the first letter even if the quoted text begins with a capital in the original document.[1] It is not necessary to enclose the lowercase letter in square brackets.

American style:
Francis Weld Peabody, MD, pointed out that "one of the essential qualities of the clinician is interest in humanity."

British style:
Francis Weld Peabody, MD, pointed out that 'one of the essential qualities of the clinician is interest in humanity'.

If the quoted text is not seamlessly incorporated into the quoting sentence's syntax, capitalize the first letter even if the quoted text begins with a lowercase letter (see also Section 9.2.3, "Quotations and Excerpts"). It is not necessary to enclose the capital letter in square brackets.

American style:
Peabody began by saying, "One of the essential qualities of the clinician is interest in humanity."

British style:
Peabody began by saying, 'One of the essential qualities of the clinician is interest in humanity.'

10.2.2 Block Quotations

"Block quotations" are paragraphs that consist only of quoted material. Various typographic devices can be applied to identify a block of text as quoted text. The most commonly used method is to indent the entire quoted text from both the left and right margins and to use a smaller font than that of the main text. Block quotations are generally not enclosed in quotation marks. The sentence introducing the block quotation should end with a colon.

In *The Body Fluids*,[12] J. Russell Elkinton, MD, and T.S. Danowski, MD, insist that all physicians will have to deal with such problems as fluid management and electrolyte disturbances:

Many disease states, which are otherwise unrelated, have certain features in common such as starvation, dehydration, vomiting, diarrhea, sweating, and renal dysfunction. These symptoms can exert profound influences upon the volume and composition of body fluids and solutes.

If the text preceding the excerpt does not clearly indicate its source, the source should be placed within parentheses after the excerpt's closing punctuation mark.

. . . volume and composition of body fluids and solutes. (Elkinton and Danowski, *The Body Fluids*[12])

Determining which device to use for block quotations is usually best done by layout specialists. The standing style for excerpts can vary among journals, books, and reports.

However, this style manual recommends block quotations consistently adhere to the following rules.

(1) Do not indent the first line of an excerpt, even if the first sentence in the original was indented.

> The second paragraph of Benjamin Lewin's text[17] goes directly to the heart of the matter:
>
>> The crucial feature of Mendel's work, a century ago, was the realization that the gene is a distinct entity. The era of molecular biology began in 1945 when Schrödinger developed the view that the laws of physics might be inadequate to account for . . . stability during innumerable generations of inheritance.

(2) The initial letter of an excerpt should be capitalized, even if the quoted material begins in the middle of a sentence. It is not necessary to enclose the capital letter in square brackets.

(3) Although ellipses should be omitted at the beginning and end of excerpts, ellipses should be used to indicate where original content has been omitted within quoted material. Use only 3 dots for an ellipsis, regardless of whether the omitted material is within a sentence or between sentences (see Section 10.3, "Ellipses"). If omitted content has been replaced with equivalent text, enclose the replacement within square brackets.

> These symptoms can exert profound influences upon . . . body fluids and solutes.
> These [disorders] can [disturb] body fluids and solutes.

(4) Reference citations in the original material may be omitted, but these omissions should be indicated by ellipses.

(5) An obvious typographic error in the original may be corrected, but idiosyncratic spellings and phrasings in historical sources should be preserved. If such an idiosyncrasy could be interpreted by readers as an error rather than an accurate reflection of the original text, add the term "sic" within square brackets or, optionally, replace the error with the correct form, also within square brackets.

> The physiological analyses of Claude Barnard [sic] are often pointed to as . . .
> The physiological analyses of Claude [Bernard] are often pointed to as . . .

(6) Preserve typographic devices used for specific scientific meaning, such as boldface for vectors in mathematical and physics texts and italics for species names. However, words or terms that are bolded in the original for emphasis may be reproduced in roman type.

(7) If in quoting a source's text an author adds italics for emphasis, that change should be explained by a phrase such as "italics added" within square brackets.

> Schmidt *proved* [italics added] that Cohen's conjecture was . . .

(8) Non-English words, phrases, and sentences in excerpts should preserve the conventions of the original language, including diacritics and accent marks (eg, "les médicaments"). However, guillemets (ie, "chevrons" used in French as quotation marks) should be omitted at the beginning and end of block quotations and replaced with quotation marks in the middle of the quoted material.

(9) A translation of a word, phrase, or longer element that is not part of the excerpt should appear within square brackets after the original content.

> The housing projects were managed under the Üretici Belediyecilik [Producer Municipality] model, which advocates that municipalities and cooperatives take an active role in housing production.

10.2.3 Quotation Marks in Relation to Other Punctuation Marks

As outlined in Section 10.2.1, "Run-in Quotations," American and British styles differ on where to place periods and commas in relation to end quotation marks. American style places periods and commas before quotation marks regardless of whether the quotes are single words, phrases, sentence fragments, or one or more full sentences. In contrast, British style places the periods and commas after end quotation marks for quotes shorter than a full sentence. For quotations of one or more sentences, British style places periods and commas before the quotation marks, just like American style.

American style:
He saw the ecological effects as "an utter disaster."
At her retirement dinner, she said, "My work is finished," and then she said, "But your work has just begun."

British style:
He saw the ecological effects as 'an utter disaster'.
At her retirement dinner, she said, 'My work is finished,' and then she said, 'But your work has just begun.'

The American and British styles agree that the placement of question marks and exclamation points depends on whether the quoted material constitutes an entire interrogative, exclamatory, or imperative statement. If the quoted material does, question marks and exclamation points are placed before the end quotation marks. If the quoted material does not, question marks and exclamation points are placed after the quotation marks.

American style:
The physicist asked, "Did we just discover a new element?"
I am outraged that the physicist said, "We discovered a new element"!

British style:
The physicist asked, 'Did we just discover a new element?'
I am outraged that the physicist said, 'We discovered a new element'!

Both the American and British styles agree that colons and semicolons should always be placed after the end quotation marks.

American style:
He listed the elements of "an adequate study design"; then he dismissed his students.

British style:
He listed the elements of 'an adequate study design': statement of question, definition of . . .

More on American style can be found in *The Chicago Manual of Style*.[1] Additional details on British style can be found in *New Hart's Rules: The Oxford Style Guide*.[6]

10.3 ELLIPSES

In both run-in and block quotations, indicate any material omitted within quoted text with ellipsis points, not with dashes or asterisks. However, ellipsis points are not needed to indicate material omitted immediately before or after the quoted text.

Ellipsis points are on-the-line dots (ie, periods) usually separated from each other and from adjacent characters by single spaces. While some style manuals[1] use a "3- or 4-dot method," this style manual recommends using only 3 ellipsis dots, regardless of whether material is omitted in the middle of a sentence or between sentences and regardless of the number of sentences omitted.

With the 3- or 4-dot method, 3 dots are used for an omission within a sentence, and

4 dots are used for an omission of one or more sentences from a quote. In the 4-dot method, the first dot serves as the closing period for the prior sentence, so there is no space between the first dot and the last word in the prior sentence.

3-dot method:

"There are mutations that affect the ability of *E. coli* cells to engage in DNA repair . . . The major known pathways . . . are the *uvr* excision-repair system and the *dam* replication mismatch-repair system."

3- or 4-dot method:

"There are mutations that affect the ability of *E. coli* cells to engage in DNA repair. . . . The major known pathways . . . are the *uvr* excision-repair system and the *dam* replication mismatch-repair system."

Retain other punctuation preceding or following an omission if it will make the sense of the passage clearer.

"This is what he concluded . . . : The crater was probably not meteoric."
"I can't believe he reached that conclusion . . . ! That crater is clearly meteoric."

CITED REFERENCES

1. The Chicago manual of style: the essential guide for writers, editors, and publishers. 17th ed. The University of Chicago Press; 2017. Also available at https://www.chicagomanualofstyle.org

2. Virag K, editor. Editing Canadian English: a guide for editors, writers, and everyone who works with words. 3rd ed. Editors' Association of Canada; 2015.

3. JAMA Network Editors. AMA manual of style: a guide for authors and editors. 11th ed. Oxford University Press; 2020. Also available at https://www.amamanualofstyle.com

4. Public Services and Procurement Canada, Translation Bureau. Writing tips plus; [modified 2022 May 5]. https://www.noslangues-ourlanguages.gc.ca/en/writing-tips-plus/index-eng

5. US Government Publishing Office. Style manual: an official guide to the form and style of federal government publishing. Government Publishing Office; 2016. https://www.govinfo.gov/collection/gpo -style-manual

6. New Hart's rules: the Oxford style guide. 2nd ed. Oxford University Press; 2014.

7. Australian Public Service Commission. Style manual. Commonwealth of Australia; c2020. https:// www.stylemanual.gov.au/

ADDITIONAL REFERENCE

Bringhurst R. The elements of typographic style. 4th ed. Hartley & Marks; 2013.

11 Abbreviations

Editors: Audrey Daniel Lusher and Elizabeth A. Arnold, ELS

11.1 FORMATION OF ABBREVIATIONS

Abbreviations are shortened forms of common words, phrases, organization names, and even complex ideas and events (eg, the abbreviation "CNS" refers to the central nervous system). Abbreviations can also be symbols, which are notations used to represent something else, such as variables, concepts, and mathematical relations. Examples include chemical symbols (eg, "Na," the chemical symbol for sodium, from the Latin "*natrium*") and units of measure (eg, "kg" for kilogram). However, other symbols take graphic or numerical forms (see Section 4.2, "Symbols and Signs").

Abbreviations have been discovered in inscriptions that go as far back as ancient Greece,[1] and they are widely used today. In one study's analysis of 24 million article titles and 18 million article abstracts, more than 1 million unique abbreviations were identified.[2] The history and current practice of abbreviating are summarized in such works as *The Oxford Companion to the English Language*[3] and *The Barnhart Abbreviations Dictionary*.[4]

Although the English language does not have formal and consistent rules for constructing abbreviations, English-language abbreviations are generally of 2 basic types: contraction abbreviations and suspension abbreviations. However, many variations of abbreviations for the same terms can be found throughout English prose, and the punctuation of abbreviations is subject to wide variation (see Section 11.2, "Punctuation and Typography").

11.1.1 Contraction Abbreviations

Contraction abbreviations fall into 3 general categories.

(1) In its simplest form, a contraction abbreviation consists of the initial letter of a word with or without one or more letters from within the word.

Alberta *becomes* AB	isoleucine *becomes* Ile
Celsius *becomes* C	milligram *becomes* mg
friable *becomes* fr	number *becomes* No. (from the Latin *numero*)
inch *becomes* in.	tuberculosis *becomes* TB

(Note that in the abbreviations "in." and "No.," a closing period is required to avoid ambiguity with the preposition "in" and the negative statement "no.")

(2) Another form of contraction abbreviation is an initialism, which uses the initial letters of 2 or more words, occasionally combined with internal letters from the words. In such abbreviations, all of the letters are usually capitalized.

National Institutes of Health *becomes* NIH
European Academy of Sciences *becomes* EURASC
ethylenediaminetetraacetic acid *becomes* EDTA

Occasionally, contraction abbreviations use a mix of uppercase and lowercase letters. To avoid confusion from one scientific document to another, editors and authors should follow the original authors' stylization of such abbreviations.

MeSH *not* MESH *for* medical subject headings
MoMA *not* MMA *for* Molecular Motion Algorithms

When an initialism is a pronounceable word, it is referred to as an acronym (eg, "COVID-19" for coronavirus disease 2019). Of note, some abbreviations that appear to be pronounceable words are instead pronounced letter by letter (eg, "WHO" for "World Health Organization" is pronounced as "W-H-O" rather than "who").

In viral nomenclature, the term "sigla" is used for initialisms that use more than one letter from some or all of the words in compound terms to create a pronounceable word. For example, "*Comovirus*" comes from "cowpea mosaic virus" (see Section 22.7.4, "Abbreviated Viral Designations").

Some acronyms eventually cease to be capitalized (eg, "laser" for "**l**ight **a**mplification by **s**timulated **e**mission of **r**adiation" and "scuba" for "**s**elf-**c**ontained **u**nderwater **b**reathing **a**pparatus").

(3) Hybrid abbreviations are a third type of contraction abbreviation. Although formed the same way as other contraction abbreviations, hybrid abbreviations are pronounced as a combination of letters and a word.

ANOVA *pronounced* ay-nova *stands for* analysis of variance
TLIF *pronounced* tee-lif *stands for* transforaminal lumbar interbody fusion
OPEC *pronounced* oh-peck *stands for* Organization of the Petroleum Exporting Countries

Generally, when pronounceable hybrid abbreviations and acronyms are used as nouns, they are not preceded by the article "the" (eg, "NASA" for "National Aeronautics and Space Administration"). "The" is typically used with other initialisms when they are used as nouns and pronounced by their individual letters (eg, "the CAE" for "the Canadian Academy of Engineering").

11.1.2 Suspension Abbreviations

A suspension abbreviation omits the terminal letters of the word (eg, "approx" for "approximately"). Although less common than contraction abbreviations in scientific publishing, suspension abbreviations achieve the same goal of shortening common words or phrases to achieve brevity in publication.

11.1.3 Plurals of Abbreviations

In general, to form the plural of abbreviations that do not contain periods, add only "s" with no apostrophe (eg, DVMs and DDSs). However, an apostrophe may be needed to form the plural of an abbreviation if adding "s" alone creates confusion (eg, "n's" to distinguish "numbers of units" in subgroups of study samples from "NS" for "not significant"). If an abbreviated term is itself a plural, do not add the "s" (eg, "CDC" *not* "CDCs" *for* "Centers for Disease Control and Prevention" in the United States). In such instances, the abbreviation may be treated grammatically as a collective noun.

In scientific writing, do not add "s" to a symbol for a unit of measure to form its plural. (See also Section 12.1.4, "Placement of Units and Symbols," and Section 12.2, "Units and Systems of Measure.")

> df *not* dfs *for* degrees of freedom   kg *not* kgs   lb *not* lbs   mm *not* mms

11.1.4 Possessives of Initialisms

To form the possessive of acronyms and other initialisms, follow the guidelines in Section 6.5, "Possessives." In general, add an apostrophe and the letter "s" to initialisms that are singular or collective nouns (eg, "NZIC's" for the possessive of "New Zealand Institute of Chemistry"). This rule applies even when an initialism ends in a capital "S" (eg, "The GPS's signal was weak"). For plural initialisms that end in a lowercase "s," add an apostrophe after the "s" (eg, "The DOIs' prefixes are assigned by Crossref and other DOI registration agencies").

11.2 PUNCTUATION AND TYPOGRAPHY

American practice has generally been to place a period (ie, full stop) after a contraction abbreviation of a single word, whereas the period is omitted in British practice (eg, "Dr." vs "Dr" for "Doctor"). In contrast, American practice has been to omit periods in initialisms created from the first letters of the words in full terms, and British practice tends to use a period after each letter (eg, "IUPAP" and "I.U.P.A.P.," respectively, for "International Union of Pure and Applied Physics"). This style manual recommends eliminating periods in all forms of abbreviations whenever reasonable. The following guidelines cover applying punctuation and typography conventions to abbreviations:

(1) In general, avoid periods, or full stops, within or at the end of abbreviations.

> Mr *not* Mr.   MSc *not* M.Sc.   PhD *not* Ph.D.   WHO *not* W.H.O.

(2) Retain periods if they are widely recommended in the nomenclature of the scientific field that is the focus of the document, as in the scientific names of organisms (see Chapter 22, "Taxonomy and Nomenclature").

> species *becomes* sp. (singular) and spp. (plural) *not* sp and spp
> *Staphylococcus aureus* becomes *S. aureus* not *S aureus*

(3) Omit periods after initials standing for personal names in end references (see Section 29.3.3.1.1, "Personal Authors").

> Hoffmann M, Kleine-Weber H, Schroeder S, et al. SARS-CoV-2 cell entry depends on ACE2 and TMPRSS2 and is blocked by a clinically proven protease inhibitor. *Cell.* 2020;181(2):271-280.e8. https://doi.org/10.1016/j.cell.2020.02.052

However, when identifying authors in the methods section of a journal article and other running text, include periods with the authors' initials.

Two authors (J.G.W. and H.D.) measured the samples at each time point.

(4) Omit commas before and after abbreviations representing parts of personal names (eg, "Jr" and "III"). However, use commas before and after abbreviations for academic degrees and other credentials that are readily recognized as such by readers (see Section 8.1.4, "Academic Degrees, Honors, and Other Credentials").

While Thomas Midgley Sr is known mainly for his inventions in the field of automobile tires, Thomas Midgley Jr was instrumental in developing leaded gasoline and early chlorofluorocarbons, which were later identified as contributing greatly to pollution.

Lonnie Bristow, MD, and William G. Anderson I, DO, were the first Black presidents of the American Medical Association and the American Osteopathic Association, respectively.

Henry F. Schaefer III, PhD, pioneered the field of computational quantum chemistry.

(5) In general, capitalize all letters in abbreviations of multiword terms (ie, acronyms and other initialisms), even if the unabbreviated terms are not usually capitalized in running text. Do not use small capitals.

AM *not* am *and not* AM *for* ante meridiem (before noon)   DNA *for* deoxyribonucleic acid
AI *for* artificial intelligence

(6) Use roman, or upright, type for abbreviations, unless the abbreviations must follow a widely recognized formal nomenclatural convention that italicizes certain abbreviations. In general, reserve italics for abbreviations of terms that are expected to be italicized in their unabbreviated forms.

The Chinese Academy of Sciences (CAS) is the world's largest research organization.
not
The Chinese Academy of Sciences (*CAS*) is the world's largest research organization.

The *Oxford English Dictionary* (*OED*) is a widely used resource.
not
The *Oxford English Dictionary* (OED) is a widely used resource.

The dominant bacteria, *S. aureus* and *E. coli*, were identified by microscopy.
not
The dominant bacteria, S. *aureus* and E. *coli*, were identified by microscopy.

11.3 USAGE

11.3.1 General Rules

Limit abbreviations in running text to those that are widely used in the scientific discipline that is the focus of the document (see exceptions to this rule in Sections 11.3.2, "Scientific Usage," and 11.4, "Abbreviations of Study Names"). Use standardized forms whenever possible rather than nonstandard variants.

OD *not* DO *for* doctor of optometry   Texas A&M University *not* Texas AM University

Most abbreviations have different meanings even within the same field.[5]

LSD *can stand for* lysergic acid diethylamide *or* least significant difference
SARS *stands for* severe acute respiratory syndrome *but* SARs *stands for* structure activity relationships *and* SAR *stands for* systemic acquired resistance in plants

Because few abbreviations have unambiguous meanings, abbreviations are usually introduced parenthetically immediately following the first time the full terms are used in a work's main body and any other part of the work that may be expected to stand on its own, such as the work's abstract and figures (see Section 30.1.1.7, "Expansion of Abbreviations," for the rules for defining abbreviations in tables). Alternatively, a list of abbreviations used in a scholarly work may be given at the beginning of the work. Once an abbreviation is defined in the main body, the abstract, a figure, or a table, it can be used by itself throughout the rest of that part of the manuscript.

> The International Livestock Research Institute (ILRI) is located in Nairobi, Kenya. The ILRI conducts research to increase the production of animal-sourced foods.
>
> The National Institute of Mental Health (NIMH) in the United States established the Biobehavioral Research Awards for Innovative New Scientists, better known as BRAINS. The NIMH is using BRAINS to support the research of exceptional scientists during the formative stages of their careers.

If a term is used only once in a manuscript, do not provide its abbreviation either parenthetically in the text or in a list of abbreviations. However, if an abbreviation is well known but the full term is not, the abbreviation can be used with or in place of the full term, regardless of whether the term is used once or multiple times.

Even widely used abbreviations are not always appropriate to use in running text. For example, whenever possible, avoid using lowercase abbreviations at the start of a sentence, and do not use symbols for units of measure unless numbers precede them.

> We isolated mtDNA. *not* mtDNA was isolated.
>
> We measured the distance in megaparsecs, not megalight-years. *not* We measured the distance in Mpc, not Mly.
>
> Morning changes in eating habits were observed each day. *not* AM changes in eating habits were observed each day.

To determine accepted forms of abbreviations, consult the discipline-specific and general dictionaries preferred by the publication to which the document will be submitted.

11.3.2 Scientific Usage

Abbreviations are usually justified on the grounds that they make the text easier to read, yet some long-used unabbreviated terms may be more readily recognized than new abbreviations. Furthermore, new and nonstandard abbreviations may present problems for users searching for particular abbreviations in electronic documents and databases. For guidance, the following recommendations are offered (see also Section 7.6, "Excessively Long Compound Terms"):

(1) Widely known scientific abbreviations (eg, "DNA" for deoxyribonucleic acid) may be used in titles, abstracts, text, headings, figures, and tables without definition. However, this rule applies to few abbreviations because most have more than one definition.

(2) Abbreviations not widely known throughout science should not appear undefined. If such abbreviations are well established in the discipline served by a specialized publication, they may be used in the publication's titles and abstracts without explanation, provided they are defined in the text, figures, and tables.

(3) Abbreviations not acceptable under criterion 1 or 2 should not be used in titles and headings, but they may be used in abstracts, text, and figures if parenthetically defined on first use. (See Section 30.1.1.7, "Expansion of Abbreviations," for the rules for tables.)

(4) Abbreviations, especially those allowed under criterion 3, should be avoided if they are needed no more than 3 times in the text. Instead, the full terms could be used each time, or pronouns or general terms could be used after the full terms are introduced (eg, Aus-

tralia's National Clinical Evidence Taskforce could be referred to as "the taskforce" after first mention of the full name).

11.4 ABBREVIATIONS OF STUDY NAMES

Using acronyms and other initialisms created from the official names of studies can be a convenient way to refer to those studies. However, to create desired acronyms, some researchers capitalize letters in the middle of words and lowercase major words in their studies' titles. Known by such terms as "bicapitalization" and "intercaps," this violation of generally accepted rules of capitalization is done to produce pronounceable and easy-to-remember acronyms, even though the studies' full titles can be visually awkward.

> Multidisciplinary drifting Observatory for the Study of Arctic Climate (MOSAiC)
> OSTEOPAThic Health outcomes In Chronic low back pain (OSTEOPATHIC)
> Preferred Reporting Items for Systematic reviews and Meta-Analyses (PRISMA)

When referring to studies that were published with awkwardly capitalized names, modify the names to adhere to the capitalization guidelines in this style manual (see Section 9.3.1, "Titles"). However, retain the original investigators' stylization of the names' abbreviations to avoid confusion as to which studies are being discussed.

> INTrEPID *not* INTEPWAID *for* Investigation of Non–Transplant-Eligible Patients Who Are Inotrope Dependent
> RELAx-AMI *not* RELAAMITPCI *for* Randomized Early versus Late Abciximab in Acute Myocardial Infarction Treated with Primary Coronary Intervention

11.5 ABBREVIATIONS OF COMMON LATIN TERMS

Many abbreviations of common Latin terms remain in the vocabulary of scholarly publishing, reflecting the many centuries in which all scholarly literature was written in Latin. Many of these abbreviations are typically used in parenthetical statements in text, as well as in footnotes and endnotes.

Among the most well established of these abbreviations in scientific literature are "ie" for *id est* (that is), "eg" for *exempli gratia* (for example), "et al" for *et alii* (and others), and "vs" for *versus* (against). In adherence to the guideline in Section 11.2, "Punctuation and Typography," to eliminate periods in abbreviations whenever reasonable, this style manual recommends using common Latin abbreviations without periods.

Alternatively, these Latin abbreviations may be replaced in text with their English equivalents. This is highly recommended when describing dosages to ensure that directions are clearly communicated in text (eg, use "nothing by mouth" instead of "npo" from the Latin *nil per os*, and use "twice per day" instead of "bid" from the Latin *bis in die*).

Note that if common Latin abbreviations are used, care should be taken to convey the intended meanings. In particular, some authors confuse "ie" and "eg." The content that follows "ie" should be all-inclusive, whereas the examples that follow "eg" should be a sampling of a larger whole.

> The 2 oldest known planets in the Solar System (ie, Jupiter and Saturn) . . .
> The physical sciences (eg, physics and chemistry) . . .

CITED REFERENCES

1. Abbreviation. In: Encyclopaedia Britannica. Encyclopaedia Britannica, Inc; [accessed 2023 Jan 16]. https://www.britannica.com/topic/abbreviation

2. Barnett A, Doubleday Z. Meta-research: the growth of acronyms in the scientific literature. eLife. 2020;9:e60080. https://doi.org/10.7554/eLife.60080

3. McArthur T, Lam-McArthur J, Fontaine L, editors. The Oxford companion to the English language. 2nd ed. Oxford University Press; 2018.

4. Barnhart RK, editor. The Barnhart abbreviations dictionary. John Wiley & Sons; 1995.

5. Davis NM. Medical abbreviations: 32,000 conveniences at the expense of communications and safety. 15th ed. Neil M Davis Associates; 2011.

ADDITIONAL REFERENCES

Acronyms, initialisms & abbreviations dictionary: a guide to acronyms, abbreviations, contractions, alphabetic symbols, and similar condensed appellations. 55th ed. 11 vols. Gale Research, Inc; 2021.The American heritage abbreviations dictionary. 3rd ed, updated. Houghton Mifflin; 2007.

The Chicago manual of style: the essential guide for writers, editors, and publishers. 17th ed. The University of Chicago Press; 2017. Also available at https://www.chicagomanualofstyle.org

Davis NM. Medical abbreviations: 55,000 conveniences at the expense of communication and safety. 16th ed. Neill M. Davis Associates; 2019.

Kotyk A. Quantities, symbols, units, and abbreviations in the life sciences: a guide for authors and editors. Humana Press; 1999.

Medical subject headings. National Library of Medicine (US); 1999 [reviewed 2021 Dec 6]. https://www.nlm.nih.gov/mesh/meshhome.html

Modern Language Association. MLA handbook. 9th ed. Modern Language Association of America; 2021.

The Oxford essential dictionary of abbreviations. Berkeley Books; 2004.

Stedman's medical abbreviations, acronyms & symbols. 5th ed. Lippincott Williams & Wilkins; 2012.

12 Numbers, Units, Mathematical Expressions, and Statistics

Editors: Thomas A. Lang, MA; Jessica S. Ancker, MPH, PhD; and Raymond Lambert, MLS, ELS

12.1 NUMBERS

A "number" is "an arithmetical value, expressed by a word, symbol, or figure, representing a particular quantity and used in counting, calculating, and showing order in a series or for identification."[1] Formatting guidelines are needed because numbers can be presented in a variety of ways (eg, 547.2, 6 million, 1.54×10^6, ¼, and $7 \ g \cdot m^{-1} \cdot d^{-1}$). The guidelines in this style manual are widely applicable. In instances in which they are not, common sense and editorial judgment should be exercised to ensure clarity, precision, logic, and consistency within documents.

For additional guidance, *Medical Uses of Statistics*[2] and the *Handbook of Writing for the Mathematical Sciences*[3] provide broad discussions on using numbers in text. Another valuable resource is the International Organization for Standardization's Standard 80000,[4] which is recognized as the international standard for expressing quantities and units.

12.1.1 Arabic and Roman Numerals

Numerals are the symbols used alone or in combination to represent numbers. Arabic numerals are more commonly used than roman numerals in both scientific and general writing in English. As names of numbering systems, "arabic" and "roman" are used without capitalization unless beginning a sentence.

There are 10 arabic numerals: 0, 1, 2, 3, 4, 5, 6, 7, 8, and 9. In most scientific contexts, these numerals are combined according to the decimal system, which expresses numbers in base 10. One of these 10 digits is used in each place, and each place value is a power of 10 (eg, 0.45, 13, and 352). In some contexts, another base may be used, such as the hexadecimal system (base 16) used in some computer applications.

Roman numerals are not used for data, but they are used in such situations as formally expressing dates (eg, copyright dates), numbering sporting events (eg, the "XXXII Olympiad" and the "XXIV Olympic Winter Games"), numbering elements in certain classification systems (eg, "CN IV" for the fourth cranial nerve and "clotting factor II"), and designating royalty succession (eg, "King Charles III" of the United Kingdom and "Queen Margrethe II" of Denmark). In addition, pages in the front matter of books and other publications are often numbered with lowercase roman numerals (see Sections 28.4.2, "Front Matter," and 28.5.1, "Pagination"). Chapters and sections are sometimes numbered with uppercase roman numerals, but this practice has been declining in favor of using arabic numerals because they are more easily understood.

Roman numerals consist of combinations of up to 7 letters, each of which has a specific value (see Table 12.1). A given letter may appear up to 3 times within a roman numeral,

Table 12.1. Values of Roman Numerals

Roman numeral[a–c]	Arabic equivalent
I, i	1
V, v	5
X, x	10
L, l	50
C, c	100
D, d	500
M, m	1,000

[a] Uppercase and lowercase letters should not be combined in a single expression.
[b] Lowercase letters are often used in the front matter of books before the text begins on page 1.
[c] A bar, or "overline," above a letter multiplies the value by 1,000 (eg, V = 5 and $\bar{V}$ = 5,000).

and the corresponding values are summed. When 2 or more of the same numeral sum to a value of a higher numeral, the higher numeral is used (eg, use "X," not "VV"). For values that require 4 or more of the same letter, the numeral is formed by combining multiple letters in the following manner: When a letter of smaller value follows one of equal or higher value, the values are summed. When a letter of smaller value precedes one of higher value, the smaller value is subtracted from the larger. In the latter case, the value of the smaller numeral cannot be less than one digit place below the value of the larger number (eg, "I" can be used only before "V" or "X," "X" can be used only before "L" or "C," and "C" can be used only before "D" or "M").

XXI = 21	iv *not* iiii
xxiv = 24	XV *not* VVV
XXVI = 26	xc *not* lxxxx
xliv = 44	m *not* dd
MCM = 1,900	CMXCIX *not* IM

12.1.2 Numerals or Words—Modern Scientific Number Style

Modern scientific number style uses numerals to represent whole numbers, with a few exceptions that apply mainly to the numbers zero and one. This style differs from the most common practice in literary writing, in which the single-digit whole numbers 0 to 9 are generally spelled out in words and larger numbers are written in numerals. One advantage of the modern scientific style is that most quantities are expressed in a similar manner. In addition, quantities stand out more in running text when expressed in numerals instead of words.

Modern scientific number style makes exceptions for zero and one because they can indicate more than just quantities. For example, "one" can be used as a personal pronoun or a synonym for "you" (eg, "one must never forget that") and as an indefinite pronoun ("this one is preferred"). "Zero" and "one" are also used in ways that are more like figures of speech than precise quantifications (eg, "in one or both cases," "in any one year," and "a zero-tolerance policy"). In addition, the numeral "1" can be easily confused with the letters "l" and "I," particularly in running text, and the value "0" can be confused with the letter "O" or the variable "o." For more on "zero" and "one," see items 3 and 4 in Section 12.1.2.1, "Cardinal Numbers."

12.1.2.1 Cardinal Numbers

In scientific text, titles, headings, tables, and figure captions, use cardinal numerals rather than words to express whole numbers greater than one and all decimal numbers.

3 hypotheses	52 trees	4 times
7 samples	328 amino acids	0.5 mm

Also use cardinal numerals to designate mathematical relationships, such as ratios and multiplication factors.

5:1	3-dimensional *not* three-dimensional
at 100× magnification	2-fold *not* twofold

Exceptions to using numerals fall into 5 categories.

(1)   If a number begins a sentence, title, or heading, spell out the number. However, in many cases, this convention can be avoided either by rewording the sentence, title, or heading so that the number appears elsewhere or by creating a compound sentence by joining a sentence that starts with a number to the previous sentence.

> Twenty milligrams is the desired amount, but 15 mg is enough.
> The desired amount is 20 mg, but 15 mg is enough.
> The drug is administered in a single dose; 20 mg is the desired amount, but 15 mg is enough.

(2)   When 2 numbers are adjacent, spell out one of the numbers, and leave the other as a numeral. In general, retain as a numeral any number that precedes a unit of measurement. Do not separate 2 such numbers with a comma (eg, "sixty 75-mg doses," not "sixty, 75-mg doses"). Often, a better option is to reword the sentence to separate the numbers.

> The sample was divided into eight 50-g aliquots. *not* The sample was divided into 8 fifty-gram aliquots.
> The sample was divided into 8 aliquots of 50 g each.

(3)   For most general uses, spell out "zero" and "one." For example, "one" should be spelled out when used as an indefinite pronoun, as a personal pronoun, or as a synonym for "you."

> **Indefinite pronoun:** This one is preferred.
> **Personal pronoun:** In supporting scientific ethics, one is obliged to . . .
> **Personal pronoun substituted for "you":** One must never forget that . . .

In addition, both "zero" and "one" should be spelled out when used more like figures of speech than precise quantifications.

> zero-based budgeting
> zero-inflated binomial distribution
> values approaching zero
> the zeros were . . .
> in one or both of the . . .
> in any one year
> one of the subspecies
> at one time
> on the one hand
> one physician
> in one such instance
> Of the possible avenues of research, this one is the most promising.
> Here was one alternative we should have examined.

However, express the whole numbers zero and one as numerals when they are directly connected to a unit of measure and when they specify assigned or calculated values.

> 1 year   when β is less than 1   1 mm   a mean of 0
> 1 J   when $z = 0$ with q fixed at 1

Similarly, express zero and one as numerals when they are part of a series and when they are closely or intermittently linked with other numbers.

> **Series:** 0, 1, 5, and 9 were . . .
> **Closely linked:** 1 of 4 subspecies
> **Closely linked:** 2 applications instead of 1 were . . . between 0 and 2
> **Intermittently linked:** Of these, 3 samples were . . . , and 1 sample was . . . The last 5 were . . .

(4)   When a whole number greater than one is used idiomatically or within a figure of speech, spell it out, just like zero and one.

However, when a figure of speech is not likely to be readily understood by readers unaccustomed to idiomatic English, recasting the phrase is generally a better option.

Idiomatic prose	Rewording
This fact tells us a thing or two about . . .	This fact tells us several things about . . .
Of two minds	Undecided
A thousand and one possibilities	Innumerable possibilities
The one and only reason	The only reason *or* The specific reason

A number may be used in such a way that the exact numeric quantity is secondary to the overall meaning. In scientific writing, reword to avoid the number altogether or to use it appropriately.

Idiomatic prose	Rewording
Among the four of us	Among our group *or* Among our group of 4
We three	We
The two of them	Both of them
One or two of these	Some of these
An additional week or two of growth	An additional 1 to 2 weeks of growth

Whether a number has an idiomatic or enumerative purpose may not always be clear. For example, in the phrase "We deleted those five data points," "five" could be considered descriptive (indicating *which* specific data points were deleted). As such, "five" may be better expressed as a word. In contrast, in the phrase "We deleted 5 data points," "5" is clearly enumerative: It indicates *how many* data points were deleted but not which ones in particular. The words "the" and "those" immediately preceding numbers generally indicate that the numbers may be serving an idiomatic rather than an enumerative purpose.

(5)   When a cardinal number is used as a plural, it should be spelled out.

> tens of thousands of people *not* 10s of 1,000s of people
> trillions of stars *not* 1,000,000,000,000s of stars *or* 10^{12}s of stars
> counting tens and twenties *not* counting \$10s and \$20s

12.1.2.2 Ordinal Numbers

Ordinal numbers generally convey rank order, not quantity. Rather than being expressly enumerative (ie, defining "how many"), ordinal numbers often describe "which," "what," or "in what sequence." Because ordinal numbers serve as adjectives and adverbs, their function is more prose-oriented than quantitative, so the modern scientific number system treats ordinals slightly differently than cardinals. Ordinal numbers should not be confused with the ordinal "level of measurement" (see Section 12.5.1, "Descriptive Statistics").

(1)   In general, spell out all single-digit ordinals, which correspond to the numbers 1 to 9.

> the ninth time   were first discovered
> a third wave of immigrants   the fifth duckling emerged

(2)   Express as numerals ordinal numbers corresponding to 10 and higher. The "st," "nd," "rd," and "th" suffixes for these ordinals should be written "on the line" rather than as superscripts.

> the 21st century *not* the 21st century   for the 63rd time
> the 42nd test run   the 355th sample

(3)   As a major exception to expressing single-digit ordinals in words, use the numeric form of these ordinals when they are intermittently linked with larger ordinals, such as in a series within a sentence.

The 5th, 8th, and 10th hypotheses . . . *not* The fifth, eighth, and tenth . . . *and not* The fifth, eighth, and 10th . . .

We used the 2nd through 14th samples to test different protocols and reserved the 1st and 15th as controls. *not* We used the second through 14th samples to test different protocols and reserved the first and 15th as controls.

(4) As an option to distinguish individual elements, single-digit ordinals can be used in numerical form when ordinals are used repeatedly.

We examined the 4th study participant. Next, we looked at the 5th. Then we returned to the 1st, 2nd, and 3rd.

Although the general policy for ordinals dictates that words be used in the above example, the numeric form does a better job of distinguishing the references to the individual study participants. Whichever style is chosen for ordinals in such situations, use it consistently throughout the document.

(5) When an ordinal number is used as a plural, spell it out.

We measured the insects' lengths to tenths of a centimeter.

12.1.2.3 Fractions

In general, spell out fractions in running text. Hyphenate all 2-word fractions, whether used as adjectives or nouns.

one-half of the samples *or* half of the samples   a third of the study plots
nearly three-quarters of the population   a two-thirds majority

For fractional quantities greater than one, use mixed fractions if the precise value is not intended. Set a "built-up fraction" (eg, a fraction set as a single typographic character, such as "⅓") next to a whole number (eg, "5⅓"), but insert a space between the whole number and a fraction that is "knocked down" (ie, a fraction consisting of 2 numbers separated by a virgule, as in "7 13/16").

was followed for 3⅓ years   1¼ km away   2 3/32 yards

When a precise value needs to be conveyed, the decimal form or percent form is preferred.

3.5 L   27% of stars   a study area measuring 1.25×3.0 km

For complex fractions in which the numerator or denominator is a mathematical expression, the knocked-down form is preferred to avoid miscommunication.[2]

$(a+b)/c$ *for* $\dfrac{a+b}{c}$ *not* $a+b/c$ *and not* $a+\dfrac{b}{c}$

12.1.3 Format of Numbers
12.1.3.1 Breaks within a Number

Separate every 3 digits in whole numbers, as well as in the numbers to the left of the decimal point in decimal numbers. American convention uses commas to separate whole numbers and a period as the decimal point. British convention uses commas or spaces to separate whole numbers and an elevated dot as the decimal point. Many European countries use spaces to separate whole numbers and a decimal comma instead of a decimal point.

American	British	European
3,968	3,968 *or* 3 968	3 968
145,000	145,000 *or* 145 000	145 000
243,568,981	243,568,981 *or* 243 568 981	243 568 981
3.14	3·14	3,14
0.32147	0·32147	0,32147
47,938.275	47,938·275	47 938,275

For certain numbers, a comma between digits is inappropriate (eg, street address numbers, US ZIP Codes and Canadian postal codes, and telephone numbers) (see item 7 in Section 5.3.3.4, "Incorrect Uses").

12.1.3.2 Zeros before Decimals and Decimals after Whole Numbers

For positive numbers less than 1 and negative numbers greater than −1, always use an initial zero to the left of the decimal point, rasied dot, or decimal comma. An initial zero should also be used before correlation coefficients and *P* values. An initial zero removes any ambiguity about whether a digit was erroneously omitted before the decimal point. It also improves the readability of numbers in tables. (See also Section 12.5, "Statistics.") Never follow a whole number (ie, an integer) with a decimal point unless the whole number is at the end of a sentence.

0.497 *not* .497   −0.72 *or* −0·72 *or* −0,72 $P = 0.04$ *not* $P = .04$ 74 dogs *not* 74. dogs

12.1.3.3 Significant Digits

Numerals in text, tables, and figures should usually reflect the precision with which the numbers were derived. In estimating numbers of ducks within flocks, for example, biologists might count in units of 10 individuals, so estimates of the average flock size should reflect that level of precision, not more or less precision. Likewise, instruments measure values accurately to some level of precision, and that level should be reflected in the numbers provided. For example, if a duck's wingspread is measured in tenths of a centimeter, the wingspread should be expressed to tenths of a centimeter (eg, 75.8 cm), not to centimeters (eg, 76 cm) or hundredths of a centimeter (eg, 75.79 cm). Likewise, body mass determined with a scale that is accurate to 0.1 g should be expressed in tenths of a gram (eg, 5.0 g), not to grams (eg, 5 g) or hundredths of a gram (eg, 5.04 g).

When expressing means, standard deviations, and other calculated values, however, the value can be calculated up to one significant digit beyond the accuracy of the original measurement. In the case of duck wingspreads measured to the nearest tenth of a centimeter, the mean might be mathematically calculated as 75.3333 cm, but it should be rounded to 75.33 cm or even 75.3 cm for reporting purposes. For the average size of duck flocks measured in tens of individuals, the standard deviation of mean flock size would be reported in whole numbers (eg, 34, not 33.5).

In reporting a quantity, the number of significant digits must be commensurate with scientific importance, not strictly the precision of the measurement. It is unnecessary and distracting to report significant digits that have little scientific meaning or import. For example, percentages expressed to one decimal place may be acceptable if sample sizes are large (at least 100), whereas with smaller sample sizes, percentages should be rounded to whole numbers. Calculating to too many decimal points can lead to "spurious precision." For example, if the average age of a sample of adult patients is 24.3 years, rounding to 24 is almost always sufficient. However, if the patients are newborns, a mean age of 24.3 days may be more meaningful than a mean age of 24 days.

12.1.3.4 Rounding Numbers

12.1.3.4.1 RULES FOR ROUNDING

The following rules apply to rounding a calculated quantity to no more than one significant digit beyond the original measurement (see Section 12.1.3.3, "Significant Digits"). Although the original numbers in all the examples below have 3 significant digits, the same rules apply to rounding numbers with more or fewer significant digits.

(1) If the second digit to the right of the decimal point is less than 5, delete that digit, and leave the first digit to the right unchanged.

   4.24 *becomes* 4.2   3.21 *becomes* 3.2

(2) If the second digit to the right of the decimal point is 5 or greater, delete that digit, but round up the first digit to the right by 1.

   4.28 *becomes* 4.3   3.76 *becomes* 3.8

(3) Rounding is not the same as truncating, in which unwanted digits are simply deleted without changing the remaining number. Authors should specify whether values have been rounded or truncated.

   4.28 *becomes* 4.3 when rounded but 4.2 when truncated

When reporting numbers, round them to 2 significant digits unless the science requires more precision. However, do not round numbers until after analyzing them. For an alternative system, consult "Guide to the Rounding of Numbers," Annex B in the International Organization for Standardization's *Quantities and Units. Part 0: General Principles (31-0[E])*.[5]

12.1.3.4.2 REASONS FOR ROUNDING

A number of principles govern the extent to which numbers are rounded.

(1) Statistical reasons to round—In testing the statistical reproducibility of results to the first, second, and third decimal places of correlation coefficients, Arthur G. Bedeian, DBA, and colleagues[6] determined that 2 digits are statistically meaningful only for samples of more than 500 and that 3 digits are meaningful only for samples of more than 100,000.
(2) Cognitive reasons to round—The human mind processes numbers most effectively when they have at no more than 2 significant digits.[7] Thus, if a procedure costs $32,833, rounding to $33,000 should improve comprehension and recall.
(3) Pragmatic reasons to round—Rounding is often common sense. For example, if the mean age of a sample is 32.81 years, it may make sense to round down to 32.8 years, which is not quite 4 days fewer than 32.81 years, or even to round up to 33 years, which is barely 69 days more than 32.81 years. This reasoning applies to many other demographic and clinical characteristics.

(4) Clinical reasons to round—Reporting precision beyond the smallest meaningful differ-
     ence on a measurement scale is uninformative. For example, on the 63-point Beck De-
     pression Inventory, a 5-point difference is considered to be the smallest clinically impor-
     tant difference.[8] Reporting a mean change of 4.2 points may be mathematically correct,
     but it isn't helpful because the change is not large enough to be clinically important.

(5) Methodological reasons for deciding against rounding—Terminal-digit bias is the ten-
     dency to round to the nearest 0, even number, or 5.[9-11] This bias often occurs in med-
     icine. For example, terminal-digit bias is often reported in manual measurements of
     blood pressure,[12,13] discharge times in emergency departments,[14] birth weights,[15] self-
     reported age among cancer patients from developing countries,[16] and self-reported year
     of the onset of menopause.[17]

     Terminal-digit bias can have marked adverse effects on study results. In one study on
     blood pressure, changing the definition of hypertension from "equal to or greater than
     140 mm Hg" to "greater than 140 mm Hg" reduced the prevalence of hypertension from
     26% to 13% in the overall group.[9]

12.1.4 Placement of Units and Symbols

12.1.4.1 Numbers

The symbols used with numbers may be alphabetical abbreviations of units (eg, mm, kV,
g, qt, and ft) or nonalphabetic marks (eg, %, $, ≥, and ±). When used without numbers,
these symbols are generally spelled out (eg, "millimeter" and "percent," not "mm" and
"%," respectively).

When numbers and symbols are combined, various rules determine whether they
should be separated.

(1) Separate a number that precedes an alphabetical unit symbol with a single space.

     an overall length of 130 mm
     closure occurred 106 s after

(2) Generally, do not separate a number from a nonalphabetic unit symbol with a space,
     whether the symbol precedes or follows the number. This rule for unit symbols differs
     from the rule for separating mathematical operators from numbers (see Section 12.3.1.1,
     "Mathematical Operators").

     44% $98
     12° (for angles, not temperature)   £12,000
     +2.5   50× magnification
     <1,500 kg

     Alternatively, replace the symbols with words, and use spaces between the words and
     numbers.

     greater than 2 A   negative 7.3   at 100 times lens objective

     However, a few nonalphabetic symbols, such as the symbol for degrees Celsius, are
     separated from numbers by spaces.

     12 °C

(3) Do not use spaces between the elements of a geographic coordinate. Designating latitude
     and longitude is optional because the compass direction provides that information (see
     also Section 14.2, "Geographic Coordinates").

     38°45′N, 77°12′26″E   lat 38°45′N, long 77°12′26″E

(4) If a number and symbol start a sentence, write out the complete term in words.

Five milliliters of supernatant were extracted. *not* Five mL of supernatant were extracted.
Fifty-seven percent of the samples were contaminated. *not* Fifty-seven % of the samples were contaminated.

Often, moving the number and symbol to later in the sentence is a better option.

We extracted 5 mL of supernatant.
We determined that 57% of the samples were contaminated.

12.1.4.2 Numeric Ranges

(1) When expressing a range of numbers in text, always use the word "to" or "through" to connect the numbers, not a hyphen or the hyphen's longer cousins, the en dash (ie, "–") and the em dash (ie, "—"). Exceptions include using en dashes when comparing 2 or more sets of ranges in running text and when listing page ranges and year ranges in bibliographic end references.

In text:
yielded −0.3 to +1.2 differences *not* −0.3–+1.2 differences
yielded 0.8 to −2.7 differences *not* 0.8−−2.7 differences
7 June to 15 June *not* 7 June–15 June *and not* 7–15 June
7 to 12 aliquots *not* 7–12 aliquots
from 240 to 350 participants *not* from 240–350 participants

In a reference:
Bailor JC, Mosteller F. Medical uses of statistics. 2nd ed. NEJM Books; 1992. p 385–389.

When space is limited in tables, 2 numbers in a range can be separated with an en dash if the numbers are placed next to each other. However, if the numbers in tables are separated by words, mathematical operators, or symbols, use "to" or "through" instead of an en dash.

Note about en and em dashes: In general, authors need not trouble with coding en dashes or em dashes, which use Unicodes 2013 and 2014, respectively. Authors can use a single hyphen for an en dash and 2 consecutive hyphens (ie, "--") for an em dash. Editors, designers, or other publications staff will code these punctuation marks before documents are typeset.

(2) When a range includes numbers of several digits, do not truncate the second number by omitting the leading digits that the numbers have in common. Although some publications allow the second number to be truncated in inclusive page numbers, this style guide discourages that practice (see Section 29.3.3.8.1, "Location within a Work").

1938 to 1954 *not* 1938 to '54   1466 to 1472 km *not* 1466–72 km

(3) When a range of numbers is accompanied by an alphabetical symbol, the symbol may be placed either after just the second number or after both numbers. The same options apply to the few nonalphabetic symbols that are separated from numbers by a space (eg, degrees Celsius). Whichever option is selected should be used consistently throughout the same journal, book, or other publication.

23 to 47 kV *or* 23 kV to 47 kV   2 to 7 °C *or* 2 °C to 7 °C
50 to 250 W/m² *or* 50 W/m² to 250 W/m²

However, when a range is accompanied by a nonalphabetic symbol that needs to be adjacent to numbers (eg, the percent symbol), the symbol should be used with both numbers.

10% to 15% *not* 10 to 15% $60 to $80 *not* $60 to 80

(4) When the numbers in a range use 2 different units of measurement, convert both numbers to the same unit.

between 0.002 g and 1 g *not* between 2 mg and 1 g
16 days to 180 days *not* 16 days to 6 months

(5) If the first number in a range begins a sentence, spell out both numbers in the range, but use numerals for all other numbers within the same sentence. Generally, the accompanying unit of measurement should appear only after the second number, and the unit should be spelled out. If the unit must appear with both numbers, the unit should be spelled out both times. Frequently, a better option is to rewrite the sentence to position the range later in the sentence so that both numerals and symbols can be used.

> **Beginning with range:** Twenty-three to twenty-five kilometers . . . *not* Twenty-three to 25 km . . .
> **Alternative:** The test range was 23 to 25 km . . .
> **Beginning with range:** Twelve to fifteen grams were added. *not* Twelve to 15 g were added.
> **Alternative:** We added 12 to 15 g.
> **Beginning with range:** Seven percent to eleven percent of the samples . . . *not* Seven percent to 11% of the samples . . .
> **Alternative:** Of the samples, 7% to 11% . . .

(6) When a range is preceded by the word "between," separate the numbers by the word "and," not by "to," "through," or an en dash.

> between 1 and 12 June *not* between 1 to 12 June *and not* between 1–12 June

(7) To prevent misunderstanding, avoid using the word "by" before a range. "By" can convey the amount of change from an original value rather than a range of values.

> Growth increased 0.1 to 0.3 g/d (a range in which growth could be as low as 0.1 g/d or as high as 0.3 g/d). *not* Growth increased by 0.1 to 0.3 g/d (a change in which growth increased from 0.2 g/d to a new level of 0.3 g/d).

Similarly, be careful when expressing 2 numbers preceded by words such as "increase," "decrease," and "change." In those situations, a range may be intended, but readers might misinterpret the first value as an initial value and the second as the new value.

> increased from 10 g/d to 18 g/d (Was an unspecified amount increased 10 to 18 g, or was the original amount of 10 g/d increased 8 g?)

Unless context makes the meaning unambiguous, qualification may be needed.

> **To express range:**
> increased by a range of 10 to 18 g/d
> decrease ranged from 15 m to 9 m
> changes ranged from 7 to 20 °C

> **To express initial and final values:**
> increased from an initial value of 10 g/d to a final value of 18 g/d
> decreased by 6 m to 9 m
> temperature changed from 7 °C to 20 °C

(8) When changes are from one range to a new range, using an en dash or a hyphen within each range may make sentences easier and quicker to interpret than using words both to express each range and to compare one range to the other.

> increased from 10–23 mm to 12–27 mm *not* increased from 10 to 23 mm to 12 to 27 mm

12.1.4.3 Dimensions

For dimensions, use either a multiplication symbol (ie, "×," not a lowercase letter "x") or the word "by" to separate the measurements. The accompanying alphabetical abbreviation for the unit of measurement may be placed either after just the last dimension or after each dimension.

> 10 × 55 × 5 mm   10 mm × 55 mm × 5 mm   10 by 55 by 5 mm

12.1.4.4 Series

For a series of numbers, place the symbol after the last numeral only, except when using a nonalphabetic symbol that cannot be separated from numbers by a space.

> 12, 17, 43, and 66 kV   £15, £22, or £31
> diameters of 5 and 9 mm   38%, 55%, and 29%
> categories of <3, 3–7, and >7 g

12.1.4.5 Descriptive Statistics

For descriptive statistics, place the symbol either after the last numeral or after each one.

> The mean (SD) was 10 (3.2) mg/dL. *or* The mean (SD) was 10 mg/dL (3.2 mg/dL).
> The mean (range) was 10 (2 to 16) mg/dL. *or* The mean (range) was 10 mg/dL (2 to 16 mg/dL).
> The median (interquartile range) was 23 (15 to 30) mg/dL. *or* The median (25th to 75th percen-
> tiles) was 23 mg/dL (15 to 30 mg/dL).

12.1.4.6 Rates

Express simple rates (1 or 2 units) either with a slash or with negative exponents. Use only negative exponents to express complex rates (more than 2 units). However, use a slash, not negative exponents, for units that are themselves rates (eg, "A/m," a measure of magnetic field strength).

> 256/h   0.15 $g \cdot m^{-1} \cdot d^{-1}$ 5 A/m *not* 5 $A \cdot m^{-1}$

12.1.5 Scientific Notation

In scientific writing, express very large and very small numbers in powers of 10 (scientific notation).

> 2.6×10^8 *not* 260,000,000 7.41×10^{-6} *not* 0.00000741
> 4.23×10^{11} *not* 423,000,000,000 9.13×10^{-9} *not* 0.00000000913

When expressing a range of values that have the same power of 10, write out both limits of the range in full, or enclose the range within parentheses followed by the power of 10.

> 2.6×10^7 to 9.7×10^7 *or* $(2.6$ to $9.7) \times 10^7$ *not* 2.6 to 9.7×10^7

If the limits of the range have different exponents, both limits must be shown in full.

> 3.7×10^{15} to 5.9×10^{16}

When expressing a mean with a standard deviation (SD), place the SD in parentheses. The mean and SD can be written out in full, or the root numbers of the mean and SD can be written before their common power of 10.

> 7.4×10^3 (SD, 0.4×10^3) *or* 7.4 (SD, 0.4) $\times 10^3$ *not* 7.4 (SD, 0.4×10^3)

For approximations rather than precise experimental quantities, use numerals for those below 1 million and a combination of numerals and words for those equal to or greater than 1 million. However, do not go beyond 2 decimal points when combining numbers and words, and do not use such expressions with scientific units of measure.

> 125,000 study participants *not* 125 thousand study participants
> 3 million people
> $13.9 million
> 1.5×10^6 km *not* 1.5 million km
> 4.3×10^4 g *not* 43 thousand g

When used in documents that will have international readership, terms greater than "million" can cause confusion because they can have differing meanings. For example, the term "billion" means 1,000 million (1,000,000,000) in the United States, but historically, it meant 1 million million (1,000,000,000,000) in the United Kingdom.[18] Although a million million has not been the standard meaning of 1 billion in the United Kingdom since the mid-1970s, it is still found in rare situations. When confusion is a possibility, use scientific notation, or write such numbers out completely in digits.

12.1.6 Monetary Units

For monetary sums, use each nation's appropriate currency symbol in combination with numerals. For sums in the millions, the word "million" may be substituted for the last 6 numbers (see Section 12.1.5, "Scientific Notation").

$73	€125,000
$245 million *or* $245,000,000	¥4.95 trillion *or* ¥4,950,000,000,000
£85,000	

To designate thousands, millions, or larger amounts, do not mix units of currency with scientific notation or with the prefix symbols of the Système International d'Unités.

$58,000 *not* $5.8 × 10⁴ *and not* $58K

Maintain this distinction in tables, illustrations, and other places where abbreviated forms are otherwise acceptable (see Chapter 30, "Tables, Figures, and Indexes").

Table column heading: $, millions *or* $, 000,000 *or* millions of $ *not* $M, *not* $, M *and not* M$

Do not add a space between nonalphabetic currency symbols and numerals, but do use a space to separate alphabetic-only currency abbreviations from numerals. Table 12.2 provides the symbols for many national currency units.[19]

$749	£749	CHF 749	749 RUB	Col$749	€749

The dollar symbol ($) is used as the unit of currency in many nations, as is the pound sterling symbol (£). Generally, include the country qualifier or prefix each time dollars and pound sterling currency are mentioned in documents (see Table 12.2).

A$749 (Australia)	GB£749 (United Kingdom)
Can$749 (Canada)	SY£749 (Syria)
NZ$749 (New Zealand)	GI£749 (Gibraltar)
US$749 (United States)	FK£749 (Falkland Islands)

To represent the dollar and peso, this style manual recommends using the "$" symbol with one vertical line instead of "$" with 2 vertical lines or "$" with an interrupted vertical line.

12.2 UNITS AND SYSTEMS OF MEASURE

12.2.1 Système International d'Unités (SI)

In scientific writing, metric measure is the accepted form for expressing physical and chemical quantities. The currently recommended system is the Système International d'Unités, or SI (the International System of Units), which was developed by the Comité International des Poids et Mesures, or CIPM (the International Committee for Weights and Measures).

Table 12.2. Symbols for Select National Currencies[a]

Country	Monetary unit[b]	Common symbol[c,d]	Previous currency[d]
Argentina	Argentine peso	$	
Australia	Australian dollar	A$ and $A[e]	
The Bahamas	Bahamian dollar	B$	
Belgium	euro	€	Belgian franc (BF)
Bermuda	Bermuda dollar	Bd$	
Brazil	Brazilian real	R$	
Canada	Canadian dollar	Can$	
Chile	Chilean peso	Ch$	
China	yuan renminbi	¥	
Colombia	Colombian peso	Col$	
Costa Rica	Costa Rican colón	₡	
Denmark	Danish krone	DKr	
Dominican Republic	Dominican peso	RD$	
Ecuador[f]	United States dollar	US$ and $[e]	sucre (S/)
Egypt	Egyptian pound	E£ and £E	
Finland	euro	€	markka (mk)
France	euro	€	French franc (F)
Germany	euro	€	deutsche mark (DM)
Greece	euro	€	drachma (D_p)
Guyana	Guyana dollar	G$	
Hungary	forint	Ft	
India	rupee	Rs	
Ireland	euro	€	Irish pound (IR£)
Italy	euro	€	Italian lira (£)
Japan	yen	¥	
Mexico	Mexican peso	Mex$	
New Zealand	New Zealand dollar	NZ$ and $NZ	
Norway	Norwegian krone	NKr	
Peru	sol	S/	
Poland	zloty	Zl	
Russia	ruble	RUB and ₽	
Spain	euro	€	euro pesata (Pts)
Sweden	Swedish krona	Sk	
Switzerland	Swiss franc	CHF	
Taiwan	Taiwanese dollar	NT$	
Turkey	Turkish lira	TL and T£	
United Kingdom	pound sterling	£, GB£, and stg[g]	
United States of America	United States dollar	$ and US$[h]	

[a] More extensive tables are available in the International Organization for Standardization's currency standard.[19]

[b] Capitalization in this column reflects correct use of uppercase and lowercase letters.

[c] In total, 19 countries have converted to the euro as of 2015. For more information on the euro, see http://www.euro.ecb.int.

[d] Only symbols with Unicode numbers are shown. The symbols in the native script are not shown. Unicodes for the symbols of current currencies in this table include Unicode 20AC for the euro, Unicode 00A3 for the pound sterling, Unicode 00A5 for the Chinese yuan and the Japanese yen, and Unicode 20A1 for the colón.

[e] This style guide recommends the first form for consistency in style for the many symbols for national currencies that are based on a dollar unit. However, the form $A is recommended by the Australian Government Publishing Service.[20]

[f] In 2000, Ecuador discontinued using the sucre and adopted the US dollar as its national currency.

[g] The third symbol is uncommon.

[h] The first form is recommended for US publications having mainly US readers. The second form is recommended for US documents that refer to several dollar currencies, whether they have a US or worldwide readership. The second form also is recommended for publications with high readership outside the United States.

Table 12.3. Système International d'Unités Base Units and Their Symbols

Quantity[a]	Unit[b]	Symbol[b]	Definition
Length	meter	m	Length of the path traveled by light in a vacuum during a time interval of 1/299,792,458 of a second
Mass[c]	kilogram	kg	Unit of mass equal to the mass of the international prototype of the kilogram
Time	second	s	Duration of 9,192,631,770 periods of radiation corresponding to the transition between the 2 hyperfine levels of the ground state of the cesium 133 atom
Electric current	ampere	A	Constant current that would produce a force equal to 2×10^{-7} N/m if maintained between 2 straight parallel conductors of infinite length and of negligible circular cross-section that are placed 1 m apart in a vacuum
Thermodynamic temperature	kelvin	K	1/273.16 of the thermodynamic temperature of the triple point of water
Amount of substance	mole	mol	Amount of a substance that contains as many elementary entities as there are atoms in 0.012 kg of carbon 12 (when the term "mole" is used, the elementary entities must be specified: atoms, molecules, ions, electrons, other particles, or specified groups of particles)
Luminous intensity	candela	cd	Intensity of a source that emits monochromatic radiation of frequency 540×10^{12} Hz in a given direction and that has a radiant intensity in that direction of 1/683 W/sr[d]

[a] The basic units in the Système International d'Unités (SI) are often used to derive other units. Some examples of derived units are square meter (m^2) for area, cubic meter (m^3) for volume, kilograms per cubic meter (kg/m^3) for density or mass density, and amperes per square meter (A/m^2) for density of electrical current. Some other units derived from basic SI units have been given special names (see Table 12.4).
[b] Capitalization in these columns reflects correct use of uppercase and lowercase letters.
[c] The word "weight" is often used for "mass" in everyday language, but in science, "weight" is a force, for which the SI unit is the newton (see Table 12.4).
[d] W/sr = watts per steradian

The SI provides a coherent system of measure constructed from 7 base units, each of which is precisely defined (see Table 12.3). From these base units, several additional units have been mathematically derived. Many of these additional units are algebraic combinations of the base units (eg, "square meter," or m^2, for area; "cubic meter per second," or m^3/s, for flow; and "amperes per meter," or A/m, for magnetic field strength). Others have been given special names and symbols (eg, "pascal," or Pa; "joule," or J; and "hertz," or Hz) (see Table 12.4). In turn, more units have been derived from algebraic combinations of the unique units (eg, "pascal second," or Pa·s; "joule per mole Kelvin," or J/(mol·K); and "newton meter," or N·m).

Some units of measure that are not part of the SI are nevertheless recommended by the CIPM for use with SI units, and still other non-SI units are currently accepted because they are widely used and important in the sciences (see Table 12.5).

Standard prefixes such as "giga" and "centi" are used with the SI units to designate quantities much larger or smaller than a given unit (see Table 12.6). For such quantities, prefixes are combined with the symbols for the base and derived SI units. One exception is the base unit kilogram, which is already prefixed. In this case, prefixes are attached to the unit stem "gram" rather than to "kilogram." Except for liter, curie, and roentgen, prefixes are not used with non-SI units.

Table 12.4. Derived Units in the Système International d'Unités That Have Special Names and Symbols[a]

Derived quantity	Name[b]	Symbol	Derivation via SI base units	Derivation via SI-derived units
Radioactivity	becquerel	Bq	s^{-1}	
Electric charge, quantity of electricity	coulomb	C	$s \cdot A$	
Celsius temperature	degree Celsius[c]	°C	K	
Capacitance	farad	F	$m^{-2} \cdot kg^{-1} \cdot s^{4} \cdot A^{2}$	C/V
Absorbed dose of radiation, specific energy (imparted), kerma (an acronym for "kinetic energy released to matter")	gray	Gy	$m^{2} \cdot s^{-2}$	J/kg
Inductance	henry	H	$m^{-2} \cdot kg \cdot s^{-2} \cdot A^{-2}$	Wb/A
Frequency	hertz	Hz	s^{-1}	
Energy, work, quantity of heat	joule	J	$m^{-2} \cdot kg \cdot s^{-2}$	N·m
Catalytic activity	katal	kat	$s^{-1} \cdot mol$	
Luminous flux	lumen	lm	$m^{2} \cdot m^{-2} \cdot cd = cd$	cd·sr
Illuminance	lux	lx	$m^{2} \cdot m^{-4} \cdot cd = m^{-2} \cdot cd$	lm/m^{2}
Force	newton	N	$m \cdot kg \cdot s^{-2}$	
Electric resistance	ohm	Ω	$m^{2} \cdot kg \cdot s^{-3} \cdot A^{-2}$	V/A
Pressure, stress	pascal	Pa	$m^{-1} \cdot kg \cdot s^{-2}$	N/m^{2}
Plane angle	radian[d]	rad	$m \cdot m^{-1} = 1$	
Electric conductance	siemens	S	$m^{-2} \cdot kg^{-1} \cdot s^{3} \cdot A^{2}$	A/V
Dose equivalent[e]	sievert	Sv	$m^{2} \cdot s^{-2}$	J/kg
Solid angle	steradian[d,f]	sr	$1\ m^{2}/m^{2}$	
Electric potential difference, electromotive force	volt	V	$m^{2} \cdot kg \cdot s^{-3} \cdot A^{-1}$	W/A
Power, radiant flux	watt	W	$m^{2} \cdot kg \cdot s^{-3}$	J/s
Magnetic flux	weber	Wb	$m^{2} \cdot kg \cdot s^{-2} \cdot A^{-1}$	V·s

SI, Système International d'Unités (International System of Units)

[a] The derived units are listed in alphabetical order based on their names in the table's second column.

[b] Capitalization in this column reflects correct use of uppercase and lowercase letters.

[c] For this unit, SI prefixes are used with the term "degree" and its symbol, "°," not with the term "Celsius" or its symbol, "C" (eg, millidegree Celsius, or m°C).

[d] The radian and steradian may be used advantageously in expressions for derived units to distinguish between quantities of a different nature but the same dimension. In practice, the symbols "rad" and "sr" are used where appropriate, but the derived unit "1" is generally omitted. In photometry, the unit name "steradian" and the unit symbol "sr" are usually retained in expressions for derived units.

[e] Other quantities expressed in sieverts are ambient dose equivalent, directional dose equivalent, personal dose equivalent, and organ equivalent dose.

[f] "Steradian" is also known as "square radian."

In medicine, several traditional units of measurement are still preferred to their newer SI-equivalent units. Concern that health care professionals would misunderstand the SI units and make errors in treating patients has led to the continued use of traditional units. For example, the preferred measurement for blood pressure continues to be "millimeters of mercury" (mmHg), not the SI's "kilopascals" (kPa). Normal blood pressure is usually expressed as 120/80 mmHg, not as 16/10.6 kPa. Concentrations traditionally expressed as ng/mL or mg/mL are seldom expressed in the SI's mmol/L. Likewise, calories are preferred to joules and pH to nmol/L of hydrogen.

Table 12.5. Other Units of Measurement Accepted for Use with Units of the Système International d'Unités[a]

Name[b]	Symbol[b]	Value in SI units
Units accepted for use with SI units		
minute (time)	min	1 min = 60 s
hour	h	1 h = 60 min = 3,600 s
day	d	1 d = 24 h = 86,400 s
degree (angle)	°	$1° = (\pi/180)$ rad
minute (angle)	′	$1′ = (1/60)° = (\pi/10,800)$ rad
second (angle)	″	$1″ = (1/60)′ = (\pi/648,000)$ rad
liter	L	$1\ L = 1\ dm^3 = 10^{-3}\ m^3$
metric ton *or* tonne	t	$1\ t = 10^3\ kg$
neper	Np	1 Np = 1
bel[c]	B	$1\ B = (1/2)\ \ln 10\ Np$
electron volt	eV	$1\ eV = 1.60218 \times 10^{-19}\ J$, approximately[d]
unified atomic mass unit	u	$1\ u = 1.66054 \times 10^{-27}\ kg$, approximately[d]
astronomical unit	ua	$1\ ua = 1.49598 \times 10^{11}\ m$, approximately[d]
Units accepted for use with SI units but not encouraged[e]		
nautical mile		1 nautical mile = 1,852 m
knot		1 nautical mile per hour = (1,852/3,600) m/s
are	a	$1\ a = 1\ dam^2 = 10^2\ m^2$
hectare	ha	$1\ ha = 1\ hm^2 = 10^4\ m^2$
bar	bar	$1\ bar = 0.1\ MPa = 100\ kPa = 1,000\ hPa = 10^5\ Pa$
ångström	Å	$1\ Å = 0.1\ nm = 10^{-10}\ m$
barn	b	$1\ b = 100\ fm^2 = 10^{-28}\ m^2$
curie[f]	Ci	$1\ Ci = 3.7 \times 10^{10}\ Bq$
roentgen[f]	R	$1\ R = 2.58 \times 10^{-4}\ C/kg$
rad[f]	rad	$1\ rad = 1\ cGy = 10^{-2}\ Gy$
rem[f]	rem	$1\ rem = 1\ cSv = 10^{-2}\ Sv$

SI, Système International d'Unités (International System of Units)

[a] Except for curie, roentgen, rad, and rem, the units in this table are accepted by the SI to accommodate the needs of specialized scientific, legal, and commercial interests.

[b] Capitalization in these columns reflects correct use of uppercase and lowercase letters.

[c] More commonly used with the SI prefix "deci-" (ie, "decibel").

[d] These values can only be obtained by experiment and are, therefore, not known exactly.

[e] The SI recommends that the relationship of these units to SI units be defined in every document in which they appear. Although nautical mile and knot are accepted by the Comité International des Poids et Mesures, use of these 2 units is discouraged by the International Association for the Physical Sciences of the Ocean (see Section 25.13, "Non-SI Units").

[f] The curie, roentgen, rad, and rem are widely used in the United States and are accepted by the National Institute of Standards and Technology in the United States for use with the SI (see also Table 12.7).

Additional information on employing SI units and converting non-SI units to SI units is available in the Bureau International des Poids et Mesures' *The International System of Units (SI)*[21] and other publications.[22–24]

12.2.1.1 SI Rules

The CIPM adopted the following rules for writing unit symbols with numerals and for combining SI units with SI prefixes.

Table 12.6. Prefixes Used with Units of the Système International d'Unités[a]

Factor[b]	Name[c]	Symbol[c]	Factor[b]	Name[c]	Symbol[c]
10^{-1}	deci	d	10^1	deka	da
10^{-2}	centi	c	10^2	hecto	h
10^{-3}	milli	m	10^3	kilo	k
10^{-6}	micro[d]	μ	10^6	mega	M
10^{-9}	nano	n	10^9	giga	G
10^{-12}	pico	p	10^{12}	tera	T
10^{-15}	femto	f	10^{15}	peta	P
10^{-18}	atto	a	10^{18}	exa	E
10^{-21}	zepto	z	10^{21}	zetta	Z
10^{-24}	yocto	y	10^{24}	yotta	Y

[a] Scientific notation may be used instead of prefixes (eg, 4 MW can be written 4×10^6 W).

[b] In scientific notation, 10^{-1} is 1/10, 10^{-2} is 1/100, etc, and 10^1 is ×10, 10^2 is ×100, etc.

[c] Capitalization in these columns reflects correct use of uppercase and lowercase letters.

[d] Although "μ" is the SI prefix for micro, the abbreviation for micrograms, "μg," is frequently replaced with "mcg" in handwritten pharmaceutical prescriptions because "μg" has been misinterpreted as "mg" for "milligram," which can have catastrophic consequences. In print, however, the symbol "μg" is rarely mistaken for "mg."

(1) The symbols for SI units are printed in lowercase roman type except for the units with names derived from proper nouns, in which case the first letter of the symbol is capitalized (eg, "W" for watt, "T" for tesla, and "Hz" for hertz).

14 s	4 kg	17.6 W
37 lx	6 m	0.076 Pa
0.59 cd	45 °C	386 K

SI symbols are separated from the numbers on their left by one space. In this regard, the degree symbol in Celsius (ie, °C) differs from the non-SI degree symbols for plane angles (eg, a 45° angle) and for longitude and latitude (eg, 45°30′N) (see Section 12.1.4.1, "Numbers").

(2) SI unit symbols are identical whether they are used as singular or plural, and they are not followed by periods, except at the end of sentences. In addition, singular verbs are used with SI measurements of 1 or less, and plural verbs are used with measurements greater than 1.

10 g *not* 10 gs   6 mm *not* 6 mms   4 mL were added *not* 4 mL was added

(3) Indicate a compound unit that is the product of 2 or more SI units with a centered dot (raised period or half-high dot) without spaces around it.[25] A single space between SI symbols is no longer considered acceptable because the centered dot reduces confusion between compound units and SI units with prefixes (eg, while "m·N" clearly refers to meter newton, "m N" could be confused with "mN," which means millinewton). In addition, do not substitute the times symbol (×) for the centered dot.

V·s *not* V s *and not* V×s

(4) Use a slash, a horizontal line, or a negative exponent to express a derived unit formed by dividing 2 other units.

m/s *or* $\frac{m}{s}$ *or* m·s^{-1}

(5) Set prefixes in roman type with no space between the prefixes and the unit symbols.

ng   kV   mJ   MHz

(6) Use only one prefix per unit symbol.

> 8 ng *not* 8 mµg 1×10^6/s *not* 1 M/s

12.2.1.2 Other Recommendations for Writing SI Units

In addition to adhering to the CIPM rules in Section 12.2.1.1, "SI Rules," scientific documents should conform with the following guidelines, many of which are recommended by the National Institute of Standards and Technology[26] of the US Department of Commerce.

(1) Use SI prefixes when the numerical values are less than 0.1 and greater than 1,000 times the base and derived SI units. Alternatively, express very large and very small numbers with scientific notation, in which case 10^x or 10^{-x} substitutes for the prefix. This alternative can also be used with non-SI units, to which prefixes cannot be applied.

> 70.5 km *or* 7.05×10^4 m *not* 70,500 m
> 12 pg *or* 1.2×10^{-13} g *not* 0.00000000000012 g
> 1.3×10^6 ua *not* 1.3 Mua (non-SI astronomical units)

(2) Spell out the names of units when they are used in text without accompanying numerical values.

> measured to the closest millimeter *not* measured to the closest mm

However, to save space in tables, illustrations, and parenthetical text notations, unit symbols may be used with or without numerical values. The chosen style should be used consistently throughout each table, illustration, or parenthetical text notation in the document.

> **Table column heading:** Length in millimeters *or* Length (mm) *not* Length (in mm)

(3) In general, when an SI unit is used with a numerical value, the value should be in arabic numerals, not spelled out, and the SI unit should be expressed by its symbol, not its spelled-out name.

> 7 mm   9 V   48 s   400 K

However, when a sentence starts with a numerical value with an SI unit, spell out both the number and the SI unit, or reword the sentence to move the number and SI unit to later in the sentence (see item 1 in Section 12.1.2.1, "Cardinal Numbers," and item 4 in Section 12.1.4.1, "Numbers").

> Six kilograms were used. *or* We used 6 kg. *not* Six kg were used.

(4) When a prefix is added to an SI unit, adhere to the 2 previous rules by using symbols for both the prefix and SI unit or by spelling both out.

> The device required 6 kV. *not* The device required 6 kvolt.
> Six kilovolts were needed. *not* Six kiloV were needed.
> The device requires kilovolts, not megavolts. *not* The device requires kV, not Mv.

(5) Do not mix unit symbols and unit names. Do not apply mathematical operations to unit names.

> 10 m³/s *not* 10 cubic meters/s *and not* 10 cubic meters/second

(6) Avoid placing other words between 2 or more SI symbols or names, unless the information is an essential part of at least one of the units. If the latter is the case, spell out all SI units rather than using their symbols, and place the other words carefully to ensure they are associated with the intended SI units.

> larval density averaged 30.2 g/m² *not* larval density averaged 30.2 g of larvae/m² *and not* larval density averaged 30.2 grams of larvae per square meter

larval density of 26.4 grams per lineal transect kilometer *not* larval density of 26.4 g/lineal
transect km

(7) Units for 2 or more related quantities must be unambiguous. If necessary, include the
unit with every numeral. Even when unnecessary, it is acceptable to use the unit with
every numeral, provided that this style is applied consistently. Avoid mixing units of
measurement, unless one or more of the numbers is less than 0.1 or greater than 1,000
times the most common SI unit used.

from 10 s to 75 s *or* from 10 to 75 s *not* from 10 s to 1.25 min
10 m × 30 m *or* 10 × 30 m *not* 1 dm × 30 m
5 cm by 133 m *not* 0.05 by 133 m

(8) When a prefix is required for a derived unit used in a fraction, attach the appropriate
prefix to the term in the numerator, not the denominator.

4,000,000 N/m^2 *becomes* 4 MN/m^2, *not* 4 N/Mm^2

(9) Use SI's base and derived units for weight and volume instead of parts per thousand
(ppt), parts per million (ppm), and parts per billion (ppb) (eg, 14 mL/L rather than
14 ppt). See the conversion table below.

Concentration	Weight to weight	Weight to volume	Volume to volume
ppt	g/kg *or* mg/g	g/L *or* mg/mL	mL/L *or* μL/mL
ppm	mg/kg *or* μg/g	mg/L *or* μg/mL	μL/L *or* nL/mL
ppb	μg/kg *or* ng/g	μg/L *or* ng/mL	nL/L *or* pL/mL

(10) When spelling out the names of compound units, separate the words by a space or a
hyphen.[25] Do not use a centered dot, even though it is used with SI symbols to indicate
the product of 2 or more SI units (see also item 3 in Section 12.2.1.1, "SI Rules").

volt second *or* volt-second *not* volt·second *but* V·s

(11) When expressing a quotient of units without a number, spell out the unit names and the
word "per." However, unit symbols without a number may be allowed within parentheses, in which case use the slash rather than the word "per."

was measured in kilojoules per hour *not* was measured in kilojoules/hour
was measured (kJ/h) just before *not* was measured in kJ/h just before

(12) When expressing a quotient of units with a number, use a slash or a negative exponent
rather than the word "per" or the abbreviation "p." An exception to using "p" is "kph"
for "kilometers per hour."

6 kJ/h *or* 6 kJ·h^{-1} *not* 6 kJ per h *and not* 6 kJph
70 km/h *or* 70 km·h^{-1} *or* 70 kph *not* 70 km per h

(13) Multiple slashes in a mathematical expression are ambiguous. For example, "8/2/4"
might mean "(8/2)/4," which equals 1, or "8/(2/4)," which equals 16. It could also be
misunderstood as a date, as in "2 August 2004." Therefore, when an expression has
more than one unit in the denominator, only one slash should be used. To accomplish
that, use parentheses and negative exponents in combination with the slash to render
the expression correctly.

0.3 kg/(mg·h) *or* 0.3 kg·mg^{-1}·h^{-1} *not* 0.3 kg/mg/h

12.2.2 Outdated Units

Several non-SI units (Table 12.7) are now outdated. Some of these outdated units are
based on the centimeter-gram-second, or CGS, system. This style manual recommends
not using them. If they are used, their relation to their corresponding SI units should
be defined.

Table 12.7. Outdated Units of Measurement Not Endorsed by the Système International d'Unités[a]

Name	Symbol	Value in SI units
Based on the centimeter-gram-second system		
erg	erg	$1 \text{ erg} = 10^{-7} \text{ J}$
dyne	dyn	$1 \text{ dyn} = 10^{-5} \text{ N}$
poise	P	$1 \text{ P} = 1 \text{ dyn·s/cm}^2 = 0.1 \text{ Pa·s}$
stokes	St	$1 \text{ St} = 1 \text{ cm}^2/\text{s} = 10^{-4} \text{ m}^2/\text{s}$
gauss[b]	G	$1 \text{ G} \triangleq 10^{-4} \text{ T}$
oersted[b]	Oe	$1 \text{ Oe} \triangleq (1{,}000/4\pi) \text{ A/m}$
maxwell[b]	Mx	$1 \text{ Mx} \triangleq 10^{-8} \text{ Wb}$
stilb	sb	$1 \text{ sb} = 1 \text{ cd/cm}^2 = 10^4 \text{ cd/m}^2$
phot	ph	$1 \text{ ph} = 10^4 \text{ lx}$
gal[c]	Gal	$1 \text{ Gal} = 1 \text{ cm/s}^2 = 10^{-2} \text{ m/s}^2$
Other non-SI units		
curie[d]	Ci	$1 \text{ Ci} = 3.7 \times 10^{10} \text{ Bq}$
roentgen[d]	R	$1 \text{ R} = 2.58 \times 10^{-4} \text{ C/kg}$
rad[d,e]	rad *or* rd[f]	$1 \text{ rad} = 1 \text{ cGy} = 10^{-2} \text{ Gy}$
rem[d,e]	rem	$1 \text{ rem} = 1 \text{ cSv} = 10^{-2} \text{ Sv}$
X unit[g]		$1 \text{ X unit} \approx 1.002 \times 10^{-4} \text{ nm}$
gamma[e]	γ	$1 \text{ γ} = 1 \text{ nT} = 10^{-9} \text{ T}$
jansky	Jy	$1 \text{ Jy} = 10^{-26} \text{ W·m}^{-2}\text{·Hz}^{-1}$
fermi		$1 \text{ fermi} = 1 \text{ fm} = 10^{-15} \text{ m}$
metric carat[h]		$1 \text{ metric carat} = 200 \text{ mg} = 2 \times 10^{-4} \text{ kg}$
torr	Torr	$1 \text{ Torr} = (101{,}325/760) \text{ Pa}$
standard atmosphere	atm	$1 \text{ atm} = 101{,}325 \text{ Pa}$
Calorie (at 15 °C)[i]	cal	$\text{cal}_{15} = 4.1855 \text{ J}$
Calorie (International Table)[i]	cal_{IT}	$1 \text{ cal}_{IT} = 4.1868 \text{ J}$
Calorie (thermochemical)[i]	cal_{th}	$1 \text{ cal}_{th} = 4.184 \text{ J}$
micron[j]	μ	$1 \text{ μ} = 1 \text{ μm} = 10^{-6} \text{ m}$

SI, Système International d'Unités (International System of Units); T, tesla

[a] These outdated units are not appropriate for current scientific writing. They are listed here because they are often found in older publications. The first 10 are centimeter-gram-second (CGS) units.

[b] Because the gauss, oersted, and maxwell are 3-dimensional units, they cannot strictly be compared with their corresponding SI units. Consequently, the mathematical symbol "$\triangleq$" for "corresponds to" is used in converting these CGS units to SI units.

[c] Used in geodesy and geophysics, "Gal" expresses acceleration caused by gravity.

[d] The curie is a special unit used in nuclear physics to express the activity of radionuclides. The roentgen expresses exposure to x-radiation and γ radiation. The rad expresses an absorbed dose of ionizing radiation. The rem is used in radioprotection to express dose equivalent. Because these 4 units are widely used in the United States, they are accepted by the National Institute of Standards and Technology in the United States for use with SI. However, these units are not accepted by the Comité International des Poids et Mesures.

[e] These non-SI units are exactly equivalent to SI units with appropriate submultiple prefixes.

[f] When "rad" could be confused with the symbol for radian, use the alternative symbol "rd."

[g] The X unit was used to express the wavelengths of x-rays.

[h] The "metric carat" was adopted for commercial dealings in diamonds, pearls, and precious stones.

[i] "Calorie" was used in multiple ways.

[j] Originally adopted in 1879, the "micron" was discontinued in 1968. In its place, use the SI unit "micrometer" and its symbol, "μm."

12.2.3 Other Measurement Systems

Other systems of measurement are still in use, even though almost all countries support SI. For example, parts of the avoirdupois, troy weight, and apothecaries' systems are still in common use, particularly in the United States (see Table 12.8). Although these other units of measure may be appropriate for certain audiences, this style manual recommends using only SI units and other CIPM-approved units in scientific writing.

For further discussion of the metric system and other common systems of measure, see *Dictionary of Scientific Units: Including Dimensionless Numbers and Scales*.[23] For an extensive set of unit conversion factors, see *Conversion Tables of Units in Science and Engineering*.[27] Additionally, most dictionaries have tables of weights and measures that provide conversions, and conversion programs are available on the internet.[21]

12.3 MATHEMATICAL SYMBOLS AND EXPRESSIONS

The International Organization for Standardization (ISO) developed the current standard for mathematical symbols in the natural sciences and technology.[28] The third edition of the ISO's standards handbook *Quantities and Units*[29] and publications from the American Mathematical Society[30,31] also can be helpful in determining how to use mathematical symbols.

New and idiosyncratic notations, which are often needed to advance the field of mathematics, should be carefully defined in manuscripts, even if they are defined in the ISO standards.

12.3.1 Font Style Conventions

Use arabic numerals in mathematical expressions. In both displayed equations and text, use italics for scalar variables and constants represented by a single letter (eg, A, M, x, and y). Such variables and constants may incorporate subscripts or superscripts, which should be placed to the right of the variable without a separating space. Use roman type for abbreviations and symbols consisting of several letters. Similarly, abbreviations for mathematical functions (eg, "d" for "derivative") should be set in roman type, not italic (see examples in Table 12.9).

The symbols for many mathematical quantities and operations have embellishments (ie, distinguishing accessory marks above or below the characters). Many of these embellishments are difficult, expensive, and sometimes impossible to set in type, so alternatives have been developed for some of the most common marks. Vectors, which are traditionally represented by a right arrow above the variable ($\vec{v}$), may be alternatively represented in print with boldface italic letters (**v**). Similarly, the boundaries of sigma-class symbols are traditionally represented in displayed mathematics by indices placed directly above and below the symbols. However, when these symbols are used in running text, the indices can be placed adjacent to the symbol as subscripts and superscripts so that the indices fit better in lines of text.

$$\sum_{i=1}^{n} becomes\ \Sigma_{i=1}^{n} \qquad \int_{\pi}^{\infty} becomes\ \smallint_{\pi}^{\infty}$$

Table 12.8. Select Weights and Measures Used in the United States and the United Kingdom and Their Metric Equivalents

Unit[a]	Metric equivalent	Other equivalents
acre	0.4047 ha	4,840 square yards
board foot (bd ft)	2.36 dm^3	1 ft × 1 ft × 1 in
bushel		
US	35.2 dm^3	32 qt or 4 pecks
Imperial	36.4 dm^3	4 imperial quarts
calorie (cal)[b]	4.18 J	
carat, metric	0.200 g	3.086 gr
cord	3.625 m^3	128.01 ft^3
drachm (apothecaries' system; UK)		
dry[c]	3.888 g	60 gr
fluid	3.552 mL	
dram (apothecaries' system; US)		
dry[c]	3.888 g	60 gr
liquid	3.697 mL	0.23 in^3
fathom	1.829 m	6 ft or 2 yards
foot (ft)	30.48 cm	12 in
furlong	201.2 m	40 rods or 660 ft
gallon (gal)		
US, liquid	3.785 L	4 qt
UK	4.546 L	4 imperial quarts
grain (gr)[d]	0.065 g	
inch (in)[e]	2.54 cm	1,000 mils
mil	0.0524 mm	0.001 in
mile (mi)		
statute	1.609 km	5,280 ft or 1,760 yards
nautical[f]	1.852 km	1.151 statute miles or 6,076 ft
ounce (oz)		
apothecaries' system	31.10 g	480 gr
avoirdupois system	28.35 g	437.5 gr
troy	31.10 g	480 gr
US fluid	29.57 mL	1.805 in^3
UK fluid	28.41 mL	
peck, dry		
US	8.810 dm^3	8 qt
UK	9.092 dm^3	
pint (pt)		
US dry	550.6 cm^3	33.6 in^3
US liquid	0.4732 L	16 fluid ounces
UK	568.3 cm^3	
pound		
apothecaries' system (lb)	373.2 g	12 oz or 5,760 gr
avoirdupois system (lb)	453.6 g	16 oz or 7,000 gr
troy (lb)	373.2 g	12 oz or 5,760 gr

Table 12.8. Select Weights and Measures Used in the United States and the United Kingdom and Their Metric Equivalents (*continued*)

Unit[a]	Metric equivalent	Other equivalents
quart (qt)		
US liquid	0.9464 L	2 pt or 57.75 in³
US dry	1,101 cm³	2 pt or 67.20 in³
UK dry	1,137 cm³	69.35 in³
rod	5.029 m	16.5 ft
ton (avoirdupois system)		
long	1,016 kg	2,042.5 lb
short	907.2 kg	2,000 lb
yard	0.9144 m	3 ft or 36 in

[a] All of the unit names in this column are lowercased.

[b] The value of a calorie in metric units beyond the second decimal place depends on the definition of "calorie" being used (see Table 12.7).

[c] The avoirdupois dram in both the United States and the United Kingdom is equal to 1.772 g.

[d] The grain is the smallest, or base, unit in the troy weight, avoirdupois, and apothecaries' systems of measure. In all 3 systems, it is equal to 0.0648 g.

[e] The unit "inch" is preferably spelled out. The abbreviation "in" without a period is acceptable but only if it is accompanied by an arabic numeral.

[f] Although nautical mile is approved for use with units in the Système International d'Unités (see Table 12.5), it is included in this table for comparison to the statute mile and for conversions to kilometers, statute miles, and feet.

Table 12.9. Mathematical Functions

Symbol[a,b]	Meaning	Remarks
$\exp x$ *or* e^x	exponential of x	
$\log_a x$	logarithm to base a of x	When the subscript is omitted, base 10 is assumed.
$\ln x$ *or* $\log_e x$	natural (Napierian) logarithm of x	
$\sin x$, $\cos x$, $\tan x$, $\cot x$, $\sec x$, and $\csc x$	trigonometric functions of x	
$\sinh x$, $\tanh x$, etc	hyperbolic functions of x	
$\arcsin x$ *or* $\sin^{-1} x$, $\arctan x$, $\operatorname{arcsec} x$, etc	inverse trigonometric functions of x	
$\operatorname{arsinh} x$ *or* $\sinh^{-1} x$, $\operatorname{artanh} x$, $\operatorname{arsech} x$, etc[c]	inverse hyperbolic functions of x	
$\lim_{x \to a} y$	the limit of y as x approaches a	$\overline{\lim}$, least upper limit $\underline{\lim}$, greatest lower limit

[a] All of the unit names in this column are lowercased.

[b] For any of these functions, parentheses may be used for greater clarity—for example, "exp(x)," "ln($x+y$)," and "cos(x_1)."

[c] For inverse hyperbolic functions, the International Union of Pure and Applied Physics[35] recommends dropping the letter "c" in "arc." The symbols thus become "arsinh," "arsech," "arcosh," etc.

A few mathematical symbols have 2 or more levels of embellishments (eg, $\widetilde{\overline{P}}$). Try to replace such notations with alternatives, such as those found in *Mathematics into Type: Updated Edition*[31] and other publications from the American Mathematical Society.[30]

Some mathematical presentations use roman letters in a script typeface (eg, $\mathscr{R}$), and some others use non-English characters. Greek letters are common, but occasionally, Fraktur (German blackletter) or Hebraic letters are used. For further discussions of such uses, see *Dealing with Non-English Alphabets in Mathematics*[32] in addition to *Mathematics into Type*.[31]

12.3.1.1 Mathematical Operators

12.3.1.1.1 UNICODES

Whenever possible, use the special symbols available in most word-processing software for mathematical operators instead of keystrokes. For example, use Unicode 2212 for the minus symbol (−) rather than a hyphen or en dash, and use Unicode 00D7 for the multiplication symbol (×) instead of lowercase x. However, if using Unicodes is problematic for authors, publishers can replace keystroke-created operators with Unicodes during the editing or typesetting stage.

Avoid using the asterisk (*) to denote multiplication except when used appropriately as a computer-programming symbol. Also do not use the centered dot (·) for the multiplication symbol because this dot is used as the decimal point in British convention and can otherwise be confused with the decimal point.

Some mathematical operators closely resemble each other but have different meanings. For example, the equality, identity, and congruency symbols take on negative meanings when a slash is placed through them. For example, "=," which uses Unicode 003D, means "equals," but "≠," which uses Unicode 2260, means "does not equal." Similarly, while the meaning of the wavy line symbol "≈" (Unicode 2248) is "approximately equal to," the symbol "~" (Unicode 007E) should be reserved for indicating "is similar to" in plane geometry and "is equivalent to" in matrix calculus. In addition, in physics, "≃" (Unicode 2243) means "asymptotically equal to."

All of the wavy symbols should be used only in mathematical expressions. In text, the word "approximate" is preferred to the symbols. However, the term "circa" and its abbreviation "ca" should be avoided.

> The temperature of the system was approximately 45 °C. *not* The temperature of the system was ~45 °C.

12.3.1.1.2 SPACING

Use single spaces to set off numbers and variable symbols from common mathematical operators (eg, +, −, ×, and ÷) and from equality and inequality symbols (eg, =, <, >, and ≥) (see Table 12.10). Publishers may replace these spaces with thin spaces (ie, Unicode 2009) during the editing or typesetting stage.

> 4.6×10^9 kg $x > y > z$

However, omit spacing between numerals and the following operators:

- the centered dot, or product dot, indicating multiplication, as in "3·5," *not* "3 · 5"
- parentheses to denote multiplication, as in "(7 + 2)5," *not* "(7 + 2) 5"
- exponents designating power, as in "10^{14}," *not* "$10^{\ 14}$"
- the slash to denote division, as in "48/4," *not* "48 / 4"
- the factorial symbol, as in "6!," *not* "6 !"

Two or more mathematical operators should not appear side by side. (However, see Section 12.3.1.3, "Vectors, Scalars, and Tensors.")

> at greater than −10 °C *not* at >−10 °C

In addition, do not use these operators between words in running text. Spell them out instead.

> the target zone equals the optimum plus the . . . *not* the target zone = the optimum + the . . .

Table 12.10. Common Operators in Arithmetic, Algebra, and Number Theory[a]

Symbol	Meaning	Unicode	Remarks
+	plus	002B	
−	minus	2212	
× or ·	times	00D7 for × and 00B7 for the centered dot ·	$x \times y$ or $x \cdot y$ also shown as xy and $(x)y$
/ or ÷	divided by	2215 for / and 00F7 for ÷	x/y or $x \div y$ also shown as $\frac{x}{y}$ and xy^{-1}
=	equals *or* is equal to	003D	
≠	does not equal *or* is not equal to	2260	
≡	is identical with *or* identically equal to	2261	
≅	is approximately equal to *or* is congruent with	2245	physics and plane geometry use ≈ (Unicode 2248).
≃	asymptotically equal to	2243	physics
~	is similar to	223C	plane geometry
∽	is equivalent to	223D	matrix calculus
>	is greater than	003E	double symbols ≫ mean "much greater than" in physics (Unicode 226B)
<	is less than	003C	double symbols ≪ mean "much less than" in physics (Unicode 226A)
≥	is greater than or equal to	2265	
≤	is less than or equal to	2264	
∝	is proportional to	221D	
!	factorial	0021	eg, 4! represents $4 \times 3 \times 2 \times 1$

[a] Some mathematical symbols are frequently called "signs," but this style manual uses the word "symbol," not "sign," for operators (see Section 4.2, "Symbols and Signs").

The rules governing spacing for mathematical operators differ somewhat from the rules for nonalphabetic unit symbols (see Section 12.1.4.1, "Numbers"). Also see Section 12.3.2.2, "Spacing of Mathematical Symbols."

12.3.1.2 Calculus

The symbols used in calculus are shown in Table 12.11, and some of the common notations for set theory are presented in Table 12.12. More complete lists and definitions of mathematical symbols can be found in *CRC Standard Mathematical Tables and Formulae*[33] and *Mathematics Dictionary*.[34] For the specialized mathematical symbols used in physics, see the recommendations of the International Union of Pure and Applied Physics.[35]

The Association of American Publishers[36] has detailed directions for employing Standard Generalized Markup Language (SGML) to incorporate these specialized symbols into manuscripts. The association has also developed extensive tables of mathematical symbols, entity references for markup, and verbal descriptions of the symbols.

Table 12.11. Calculus Symbols

Symbol	Meaning	Remarks
Sigma-class (often used with the upper and lower limits displayed)		
Σ	summation of terms	Use Unicode 2211.
Π	product of terms	Use Unicode 220F.
$\int$	integration of terms	Use Unicode 222B.
$\oint$	curvilinear integration	Use Unicode 222E.
Other classes		
Δ	delta; a finite increment of a function	As in Δx. Use Unicode 2206 for Δ.
D *or* d	derivative symbol; an infinite increment of a function[a]	As in dx.
∂	partial derivative symbol; a variation in a function	Use Unicode 2202.
lim	limit	
dy/dx *or* $D_x y$	derivative of y with respect to x	Where $y = f(x)$.[b]
$\partial u/\partial x$ *or* $D_x u$	partial derivative of u with respect to x	Where $u = f(x,y)$. Use Unicode 2202 for ∂.
$\partial^2 u/\partial xy$ *or* $D_y(D_x u)$	the second partial derivative of u, the first with respect to x and the second with respect to y	Use Unicode 2202 for ∂.
∇	del *or* nabla	Del is used as an operator on vector functions or, with superscript 2, as the Laplacian operator (ie, ∇^2). Use Unicode 2207.
Grad f *or* ∇f	gradient of f	Use Unicode 2207 for ∇.
div A *or* $\nabla \cdot A$	divergence of A	Use Unicode 2207 for ∇.
$\nabla \times v$	curl of v	Use Unicode 2207 for ∇.
∇^2 *or* Δ	the Laplacian operator	Use Unicode 2207 for ∇ and Unicode 2206 for Δ.

[a] The derivative symbol should be in roman type in accordance with the recommendation of the International Union of Pure and Applied Physics.[35] The italic form, occasionally used in the United States, is discouraged because the derivative symbol represents a function, not a variable or quantity.
[b] f is a function of one or more variables.

Table 12.12. Notations Used in Set Theory

Symbol	Meaning	Unicode	Remarks
$\in$	is an element of	2208	$x \in M$; x is an element of set M
$\notin$	is not an element of	2209	$y \notin M$; y is not an element of set M
$\ni$	contains as an element	220B	$M \ni z$; set M contains z as an element
$\supset$	contains as a proper subclass	2283	$M \supset N$; set M contains set N as a proper subclass
$\subset$	is contained as a proper subclass within	2282	$N \subset M$; set N is contained as a proper subclass within set M
$\supseteq$	contains as a subclass	2287	$C \supseteq E$; set C contains set E as a subclass
$\subseteq$	is contained as a subclass within	2286	$E \subseteq C$; set E is contained within set C as a subclass
$\cup$	union or sum of	222A	$A \cup B$; the union of set A and set B
$\cap$	intersection of	2229	$A \cap B$; the intersection of set A and set B
$\emptyset$ *or* $\wedge$	empty (or null) set	2205 *or* 2227	a set containing no members

12.3.1.3 Vectors, Scalars, and Tensors

The symbols for vectors are usually lowercase letters set in bold roman type. The components of vectors are scalars, which are set in lowercase letters in italics (not bold) in the same typeface as vectors. Although tensors are sometimes represented by bold roman characters, they are less likely to be misread as vectors if set in bold italics of a sans serif typeface.[35] However, this representation of tensors is not universal in the mathematics and physics literature, and exceptions and other symbols are common.

> Vector: $\mathbf{f}$
> Scalar components of vector: f_x, f_y
> Tensor: $\boldsymbol{S}$

In the multiplication of vectors and tensors, some of the traditional mathematical symbols, such as centered dots and multiplication symbols, have special meanings.

$\mathbf{a \cdot b}$	scalar product of 2 vectors	(multiplication dot is used)
$\mathbf{a \times b}$	vector product of 2 vectors	(multiplication cross is used)
$\mathbf{ab}$	dyadic product of 2 vectors	(closed up, no multiplication symbol)
$\boldsymbol{S} \cdot \mathbf{b}$	product of a tensor and a vector	(multiplication dot is used)
$\boldsymbol{S} : \boldsymbol{U}$	scalar product of 2 tensors	(colon is used)
$\boldsymbol{P} \otimes \boldsymbol{U}$	tensor product of 2 tensors	(also sometimes $P \otimes U$ with tensors in just italics, not bold italics)

12.3.1.4 Matrices and Determinants

Matrices and determinants are arrays of elements in columns and rows.

In running text, an overall symbol for a matrix may be used. Traditionally, the symbol is a capital, boldface letter (eg, matrix $\mathbf{A}$). Alternatively, general symbols representing a matrix may be shown between double vertical bars or within parentheses or braces (see Table 12.13). In contrast, when matrices are expressed in displayed mathematics, they are enclosed in double vertical bars, brackets, or sometimes large parentheses.

In text   In display

$$\mathbf{A} = \left\| a_i b_i c_i \right\| = \begin{bmatrix} a_1 & b_1 & c_1 \\ a_2 & b_2 & c_2 \\ a_3 & b_3 & c_3 \end{bmatrix} = \left\| \begin{matrix} a_1 & b_1 & c_1 \\ a_2 & b_2 & c_2 \\ a_3 & b_3 & c_3 \end{matrix} \right\|$$

A determinant is a similar array of elements set between single vertical bars. Because vertical bars are also used to indicate the absolute value of a real number and the modulus of a complex number, the text notation for a determinant should be in the form "det $\mathbf{A}$" rather than $|\mathbf{A}|$.

In text   In display

$$\det \mathbf{B} = \det(x_{ij}) = \begin{vmatrix} x_{1,1} & x_{1,2} & x_{1,3} & x_{1,4} \\ x_{2,1} & x_{2,2} & \cdots & \\ x_{3,1} & x_{3,2} & \cdots & \\ x_{4,1} & x_{4,2} & x_{4,3} & x_{4,4} \end{vmatrix}$$

12.3.2 Aggregating and Spacing of Symbols

12.3.2.1 Fences

Enclosures such as parentheses and brackets, called "fences" in mathematics, are used to aggregate groups of symbols. In general, the order in which common fences are used in mathematics is as follows: braces on the outside followed by brackets within braces

Table 12.13. Matrix Notations

Symbol	Meaning
$\mathbf{A}$ or (a_{ij}) or $\|a_{ij}\|$[a] (in physics, A)	symbol for a matrix
$\mathbf{A}^{-1}$	inverse of matrix $\mathbf{A}$
$\mathbf{A}'$ or $\mathbf{A}^{\mathrm{T}}$ (in physics, A$^{\sim}$)[b]	transpose of matrix $\mathbf{A}$
$\bar{\mathbf{A}}$[c] (in physics, A*)	complex conjugate of matrix $\mathbf{A}$
$\mathbf{A}^{\mathrm{H}}$ (in physics, A'; do not use A†)[d]	the Hermitian conjugate of matrix $\mathbf{A}$
det $\mathbf{A}$ or $|\mathbf{A}|$[e]	determinant of matrix $\mathbf{A}$
$tr\,\mathbf{A}$ (in physics, Tr A)	trace of matrix $\mathbf{A}$

[a] The double vertical lines ‖ are created with Unicode 2016.
[b] The operator $\sim$ is created with Unicode 223C, which is then made into a superscript.
[c] Ā is created by inserting Unicode 0305 right after the letter "A."
[d] The dagger † is created with Unicode 2020, which is then made into a superscript.
[e] The single vertical line | is created with Unicode 007C.

and then parentheses within brackets. This order is different than the order of enclosures in nonmathematical prose, which consists of parentheses on the outside followed by brackets within parentheses and then parentheses again within brackets.

mathematics: $\{[(\,)]\}$ prose and other nonmathematical uses: $([(\,)])$

If more levels of fences are needed for mathematical expressions, use additional parentheses, brackets, and braces in the same order. Each new set of 3 enclosers should be set in larger type as they are added to the left and right sides of expressions.

$$\{[(\{[(\,)]\})]\}$$

Other mathematical fences include angled brackets and single and double vertical bars, which have special meanings and should not be used to extend the basic set of fences. Also avoid using angled brackets for aggregation, as in "$\langle a + b \rangle$." Such usage could be ambiguous because angled brackets are commonly used as "less than" and "greater than" operators; as a way to indicate an ordered set of objects, as in "$\langle x, y, z \rangle$"; and as an alternative to the horizontal bar above a symbol to mean an average value.

Fences are usually used in matching pairs, but in physics and higher mathematics, unlike members may be used as pairs (eg, "$|z\rangle$" for the state z of a system). Occasionally, only a left-hand fence is used.

12.3.2.2 Spacing of Mathematical Symbols

A space should be placed before and after the operator symbols listed in Table 12.10 when they are used to represent mathematical operations.

$x = -4y - 1$ *not* $x=-4y-1$ $0 < y < zw$ *not* $0<y<zw$ $(x + p)a \ge y^3 z(1 - 3r)$

Also use a space on either side of symbols for trigonometric functions, logarithms, and exponential and limit functions.

$b \sin x$ $\log x$

However, no space should be used if the quantities preceding or following these symbols are enclosed by fences; the function carries a superscript or subscript; or the function itself is part of a superscript, subscript, or limit of a sigma-class symbol.

$(ac)\sin^3 2y$ $\exp(a+2b)$ $y^{\sin x}$

Spaces are also not appropriate in a variety of other mathematical expressions. In the following situations, terms are closed up (ie, no spaces):

(1) between quantities multiplied together when the multiplication operator is not shown

   $2b$ ac $6yz\beta$

(2) between fences and the variables on either side of them

   $(a-1)y$ $(4p-4bc)(1-a)$ $a|x|$

(3) between terms and their subscripts or superscripts and between subscripts or super-scripts and the following terms

   $\cos^3 y$ $(a-1)y^3 z$ $c^{x-2}d$

(4) between the symbols for plus, minus, or plus/minus and the numbers or variables to which they apply, when used to designate positive or negative values of those numbers or variables

   $-2x$ the values +13, −7, or ±2

(5) between combinations of superscripts and subscripts and the variables to which they apply

   m_n^{2r}

In such combinations, set the superscript directly above the subscript, unless this is typographically impossible. When typography will not accommodate the format above, the subscript or the superscript may be set next to the variable, followed by the other member of the pair.

   $m^{2r}{}_n$

In higher mathematics (eg, tensors), the alternatives above are not allowed, and no spaces may appear between a variable and either index. Such notations should only be altered after consulting the authors.

12.3.3 -Fold, Factor, and Times

The terms "-fold," "factor," and "times" are easily misinterpreted when used to express increases and decreases in base values. Sentence structure and word choice can produce doubt as to whether those terms refer to the amount of the increase or decrease (ie, the difference between the new and the old) or to the new value in terms of the multiplier.

In the set of examples below, the intention is to describe the change from a base volume of 10 mL to a final volume of 30 mL. The example labeled "Ambiguous" can be misinterpreted as saying that the amount of the increase was 10 mL × 3, or 30 mL. However, in this case, the new volume would be 40 mL, not 30 mL. Such ambiguity can generally be overcome by careful word choice. One of the easiest approaches is to avoid using the word "increase" altogether and instead let the mathematics do the explaining. Another option is to include the actual values.

> **Ambiguous:**
> The increase in the initial volume of 10 mL was 3-fold. (The wording could be interpreted to mean either that the volume increased by 30 mL for a final volume of 40 mL or that the final volume is 30 mL.)

> Unambiguous:
> The initial volume was 10 mL, and the final volume was 30 mL.
> The final volume was 30 mL, 20 mL greater than the initial volume of 10 mL.
> The final volume was 3 times the initial volume of 10 mL.
> The final volume was 300% of the initial volume of 10 mL.
> The final volume represented a 3-fold increase, from 10 to 30 mL.

Because decreases cannot exceed 100% of the original value (except in discussions involving positive and negative numbers), percentages should be used in place of the terms "factor" and "times." As with the word "increase" for positive changes, "decrease" should be avoided if possible when describing negative changes. When "decease" cannot be avoided, include the actual values.

> Unambiguous:
> The final volume was 33% of the initial volume of 30 mL.
> The volume decreased by a factor of 3, from 30 to 10 mL.

However, "-fold" may be appropriate in some situations because any given "-fold" value is the same for both positive and negative changes. For example, if a base value is decreased by 50%, a 100% increase is required to get back to the original value, but when this sequence is expressed in terms of "-fold," the same value applies in either direction (ie, 2-fold). In addition, "-fold" has a logarithmic logic, whereby 2-fold equals 50%, 4-fold equals 25%, etc.

12.3.4 Ratios, Proportions, and Percentages

Ratios, proportions, and percentages convert data to a common scale that simplifies data comparisons.[37] When presenting ratios, proportions, or percentages, be sure both the numerator and denominator are readily apparent. If they are not, data comparisons cannot be fully interpreted (eg, 50% can mean 2 of 4 or 5,000 of 10,000). Knowing the actual values used to calculate the comparisons is necessary for full and accurate interpretation.

A "ratio" consists of 2 numbers separated by a colon. The 2 numbers represent mutually exclusive groups; the quantity in the numerator is not part of the quantity of the denominator (eg, a ratio of girls to boys, not of girls to all children). Do not use spaces between the numbers and the colon. Do not express a ratio as a single number in which "1" is understood to be the denominator. In addition, if the ratio is reduced to its simplest form, relevant data should be included so that the ratio can be interpreted accurately.

> The ratio of positive to negative results was 105:35.
> The ratio of positive to negative results was 3:1 (105:35). (In this case, the ratio is reduced to its simplest form, but the relevant data are provided.)
> *not* The ratio of positive to negative results was 3. (In this case, readers would be expected to understand that the denominator is "1.")

In the conventions used in epidemiology and medicine, risk, odds, and hazard ratios less than 1 indicate a beneficial effect, and those greater than 1 indicate an adverse effect. The numerators and denominators should be placed according to whether the researchers' intent is to focus on the beneficial or adverse effect. For example, if the risk of death with treatment is 2 and the risk without treatment is 4, the ratio could be expressed either as "2/4 = 0.5" or as "4/2 = 2," depending on whether researchers want to focus on the treatment reducing the risk of death by 50% or on lack of treatment increasing the risk by 200%.

A proportion is calculated by division. Proportions differ from ratios in that the quantity in the numerator is a subset of the quantity in the denominator (eg, a numerator with the number of men with prostate cancer and a denominator with the number of men at risk for prostate cancer).

> The proportion of study participants with negative results was 0.75 (105/140). The remaining 0.25 (35/140) of the participants had positive results.

A percentage is calculated by multiplying a proportion by 100. Converting to percentages is especially useful when proportions are small because doing so often eliminates the leading zero before the decimal point (eg, a proportion of 0.014 becomes 1.4%). As with a ratio or proportion, the numerator and denominator of a percentage must be readily apparent to avoid misleading readers.[37] However, sometimes providing only the percentage and denominator is sufficient if readers can easily calculate the numerator.

> In the first cohort, 75% of the 140 study participants had negative results.
> In the first cohort, 75% (105/140) of the study participants had negative results.
> Of the 140 study participants in the first cohort, 25% had positive results.

Report proportions and percentages to the degree of precision appropriate for the measurement (see Section 12.1.3.3, "Significant Digits"). For denominators less than 20, reporting the actual numerical data may be more useful or less misleading than using either proportions or percentages (eg, "2 out of 4 specimens" is preferable to "50% of specimens," which can imply a much larger number than 2). Generally, when proportions or percentages derived from a mix of large denominators (ie, >100) and small denominators (ie, <100) are in close proximity in a manuscript, report all to the decimal level appropriate for the smallest denominator.

Indicate clearly whether a change in percentages is the difference between 2 percentages or a percentage of the initial value. The difference between 2 percentages is expressed as "percentage points" using the following formula: final percentage – initial percentage. In contrast, when reporting the *percentage change* from an initial value, use the following formula: [(final value – initial value)/initial value] × 100. Thus, the difference between a final value of 60% and an initial value of 80% is a decrease of 20 percentage points, but the percentage change is a 25% decrease.

> **Values for calculations**
> final value = 60%   final value = 75%
> initial value = 80%   initial value = 25%
>
> **Calculation for percentage difference**
> 60% – 80% = –20%   75% – 25% = 50%
>
> **Calculation for percentage change**
> (60% – 80%)/80% = –0.25 × 100 = –25%   (75% – 25%)/25% = 2 × 100 = 200%
>
> **In-text descriptions**
> • a decrease of 20 percentage points from   • an increase of 50 percentage points from
>   80% to 60% (preferred)   25% to 75% (preferred)
> • the final value decreased to 60% from   • the final value increased to 75% from an
>   the initial value of 80%   initial value of 25%
> • a 25% decrease from 80% to 60%   • a 200% increase from 25% to 75%

12.4 MATHEMATICS IN TEXT AND DISPLAY

12.4.1 Guidelines for Text

When mathematical expressions are written within lines of text (ie, running text), limit the dimensions (ie, height and length) of those expressions as much as possible to maintain normal spacing between words and between text lines. Although how mathematical expressions are displayed in text is primarily the concern of manuscript editors and page designers, authors can reduce problems by limiting fractions, complex exponents, and large symbols used in expressions. For example, authors can use slashes or exponents for fractions rather than vertical stacking, and they can convert graphic symbols such as the square root symbol to an exponential notation. However, should authors prefer to use graphic symbols, they may need to reach agreement with manuscript editors on how to typeset the expressions.

$$a/b \text{ or } ab^{-1} \text{ not } \frac{a}{b} \qquad (b-d)^{1/2} \text{ not } \sqrt{b-d}$$

In running text, use only one slash in any given component of an expression.

$$(a/b)/c \text{ not } a/b/c$$

Exponents with more than one level become unwieldy within text lines. The exponential form can be converted to the "exp" form for use in text.

In display   In text

$$e^{x^2-1} \qquad\qquad \exp(x^2-1)$$

However, large exponents may be better shown in display format than in text format.

In display   In text

$$e^{\frac{a-b}{c+d}} \text{ or } \exp\left(\frac{a-b}{c+d}\right) \qquad \exp[(a-b)/(c+d)]$$

When an equation carries over from one line of text to the next (ie, a "continued expression"), breaking after a mathematical operator is preferred as a hint to readers that the expression continues on the next line. Avoid breaking expressions within a pair of fences whenever possible. As with maintaining normal spacing, breaking equations at line endings is more of a concern for manuscript editors and page designers than of authors. Authors, however, should keep the rules for breaking equations in mind when reviewing galleys and page proofs.

The most basic guidelines for breaking equations at mathematical operators are given below in order of preference. For a more complete treatment of this subject, see *Mathematics into Type: Updated Edition*.[31]

(1) after or, if necessary, before an equals symbol

> All of the variables in the following equation should be set in italics: $x - 2 = 3by^3z - 4m$.

(2) after or, if necessary, before a plus or a minus symbol as long as the symbol is not within a set of fences

> For an approximation to the solution, we used $b_s t_r = -K[(u/y) + (v/r)]^n + ba[(u_m/y) + (v_m/r)]$.

(3) after or, if necessary, before a multiplication symbol or between sets of fences in which a multiplication symbol is used between the sets

The authors represented this complex equation as $R_e = (d^n u^{2-n} \rho / 8^{n-1} K) \times (4n/3n + 1)^n$.

(4) before a sigma-class symbol

We used the Fourier series $f(x) = a_0 +$

$$\sum_{n=1}^{\infty} \left(a_n \cos \frac{n\pi x}{L} + b_n \sin \frac{n\pi x}{L} \right).$$

Such line breaks are more important when the text lines are left-justified than when they are fully justified. Left-justified lines can end short of the right margin, so equations are more likely to look as if they have ended at the end of such lines.

12.4.2 Guidelines for Display

In general, break displayed equations according to the same principles used for text equations (see Section 12.4.1, "Guidelines for Text"). Unlike text equations, however, displayed equations can include mathematical operations in which a single term may take up several lines. In addition, breaks occur before, not after, operators in displayed equations. The first operator on the next line is termed the "continuing operator." Another difference between breaking text equations and displayed equations is that displayed equations need to align vertically from one line to the next, as described below.

(1) For a sequence of equations in which the left side is unchanged, vertically align the equals symbols with each other on each line.

$$2u_0 v_0 = u_0^2 + v_0^2 - (u_0 - v_0)^2$$
$$= k - (u_0 - v_0)^2$$

(2) For continued expressions in which the left side is long, vertically align the equals symbol on the second line with the first operator on the first line.

$$[(a_1 + ia_2) + (a_{11}s_1 + a_{21}s_2)] / [(b_1 + ib_2) + (b_{11}s_1 + b_{21}s_2)]$$
$$= f(x)g(y) + \dots$$

(3) For continued expressions in which the right side is long, align the continuing operator with the first term to the right of the equals symbol on the line above.

$$f(x) = 2k(a_2 + 5b_1)(3c - b_2 c)$$
$$+ 4ac\{a_1 b_1 + [(4 - b_2)^2 (ab + 4ck - b_2 c)]\}$$

(4) If splitting an equation within fences is unavoidable, vertically align the continuing operator with the first symbol within the fences.

$$f(x)g(y) = \sin ab[R(2k \cos b) - 2R_0(2k \cos b)$$
$$+ R_1(b \sin ab) + \cos b]$$

12.4.3 Punctuation and Spacing

Regardless of whether mathematical expressions are written in sentences or placed in display format, follow the same rules for punctuation such as periods and commas as used for nonmathematical text. Syntax, not presentation, should determine punctuation after mathematical expressions.

One difference between punctuation in mathematics and nonmathematical text is that 3 dots centered above the line indicate operations and relations.

$$x_1 + x_2 + \cdots + x_n$$

Otherwise, the standard ellipsis on the baseline is used, such as for omitted elements in lists.

$$x_1, x_2, \ldots, x_n$$

12.5 STATISTICS

The international standard for statistical definitions and symbols is the International Organization for Standardization's standard 3534-4.[38] Common symbols are listed in Table 12.14. Additional tables of statistical symbols can be found in *Medical Uses of Statistics*.[2] Common errors and pitfalls in statistical reporting are summarized in *How to Report Statistics in Medicine: Annotated Guidelines for Authors, Editors, and Reviewers*.[37] In addition, *Basic Statistical Reporting for Articles Published in Clinical Medical Journals: The SAMPL Guidelines*[39] provides basic statistical reporting guidelines that are suitable for including in a journal's instructions for authors.

12.5.1 Descriptive Statistics

Descriptive statistics summarize data sets to simplify their reporting and sometimes their analysis. Categorical data—nominal (unranked) categories or ordinal (ranked) categories—are summarized by the number, proportion, or percentage of observations in each category. Continuous data (ie, measurements on a scale that has an infinite number of possible values) form distributions, which are described most often with 2 or 3 numbers. A "measure of central tendency" indicates the value on the distribution around which the bulk of the data is found, most commonly the mean (ie, the arithmetic average), the median (ie, the value at the 50th percentile), or the mode (ie, the most common value). Additionally, a "measure of dispersion" indicates the variability of the distribution.

The most common measure of dispersion for normally distributed data is the standard deviation (SD). The SD is a calculated value that is subtracted from and added to the mean of a distribution. The SD identifies the range of values that encompasses approximately 68% of the data, half of which (ie, 34% of the data) is below the mean (ie, the negative SD) and half of which is above the mean (ie, the positive SD). Two SDs above and below the mean encompass approximately 95% of the values, and 3 SDs above and below the mean encompass approximately 99% of the values.

The SD and these percentages are often found in statistical manuscripts because they are associated with the standard normal distribution, an important distribution in the field of statistics. When reporting the SD, do not use the "±" symbol, which is redundant because the SD extends equally on both sides of the mean. For many publications, "SD" is among the few initialisms that can be used without expansion.

Table 12.14. Symbols Used in Statistics[a]

Population symbol	Sample symbol	Explanation
	F	statistic derived in F test, which is a ratio of 2 variances
H_0		null hypothesis
H_1		alternative hypothesis
N		number of individuals or subjects in a population or lot size
	n	number of individuals or subjects in a sample of a population
	P	probability of obtaining a difference as great as or greater than the one observed under the null hypothesis (ie, statistical probability)
	R	coefficient of multiple correlation, range of a sample
ρ	r	Spearman rank-correlation coefficient[b]
	r^2	coefficient of determination[c]
	R^2	coefficient of multiple determination[c]
σ	s	standard deviation[d]
σ^2	s^2	variance
	SE	standard error of the mean[e]
	CV	coefficient of variation
	t	statistic derived in Student's t test
$\bar{X}$	$\bar{x}$	arithmetic mean
	α	probability of a type I error; significance level
	β	probability of a type II error
	$1 - \beta$	statistical power
	χ^2	statistic derived from a chi-squared test
	ν	number of degrees of freedom[f]
θ	$\hat{\theta}$	parameter (unknown constant) and estimator (function of a data set)

[a] For additional symbols, see the International Organization for Standardization's standard 3534-1.[38]

[b] Besides the Spearman rank-correlation coefficient, rho (ρ), for 2 continuous variables of any distribution, there are several other measures of correlation: Pearson product–moment correlation coefficient, r, for 2 normally distributed, continuous variables; Kendall rank correlation coefficient, tau (τ), for 2 ordinal variables or 1 ordinal and 1 continuous variable; point biserial correlation coefficient, r (sometimes given as r_{pbi}), for a continuous variable and a categorical variable with 2 levels ("recovery status": recovered or not); and intraclass and interclass correlation coefficients (ICC) to assess agreement within and between observers.

[c] The coefficient of determination, r^2, indicates the amount of variation in the response variable that is explained by the explanatory variable in a simple linear regression model. The coefficient of multiple determination, R^2, indicates the amount of variation in the response variable that is explained by the explanatory variables in a multiple linear regression model.

[d] The abbreviation "SD" for "standard deviation" is often used without expansion on first mention.

[e] The abbreviation "SE" for "standard error of the mean" is often used without expansion on first mention, but the abbreviation "SEM" is not used for this term.

[f] The abbreviation "df" for "number of degrees of freedom" is often used without expansion on first mention.

Other common measures are the range, the interquartile range (IQR), and the inter-percentile range, which are suitable for describing both skewed and normal (bell-shaped) distributions.

- The range is the difference between the minimum and maximum values of the distribution.
- The IQR is the difference between the values at the 25th and 75th percentiles (ie, the values at the first and third quartiles). Like SD, "IQR" is among the initialisms that many publications use without expansion.

- The interpercentile range compares differences in values between other percentiles (eg, the 10th and 90th percentiles).

For all 3 of these measures, only the values themselves are usually reported, not the difference between them.

All measurements have variability. When reporting variability, the units and measure of dispersion should be unambiguous (see Section 12.1.4.2, "Numeric Ranges").

> mean (SD) = 44% (3%) *or* mean of 44% (SD, 3%) *not* SD = 44 3% *and not* 44% (±3%)
> mean of 104 mm (SD, 11 mm) *not* mean of 104 11 mm *and not* mean of 104 mm (SD 11)
> median (25th to 75th percentile) survival of 7 years (4.5 to 9.5 years)

Unlike the SD, the standard error (SE) is not a descriptive statistic. Instead, it is a measure of the precision of the population estimate—here, the population mean. It is an approximately 68% confidence interval. Nevertheless, some fields of science traditionally use the SE as a descriptive statistic, reporting it along with the mean. When reporting the SE, label it appropriately so that it is not confused with the SD or the confidence interval. The distinction also needs to be made for error bars in graphs.

12.5.2 Estimates and Confidence Intervals

An estimate is a probable value for a population that is projected from a measured value of a sample from that population. In other words, studies of samples estimate population values. In contrast, a study that includes all the members of a population is called a "census." A census identifies the actual values instead of estimating them. Hence, in a census, the mean value of data can be known, but in a sample, the mean value can only be estimated.

The results of most studies are based on samples and, therefore, are estimates. Consequently, study results should usually be accompanied by a measure of precision, usually a confidence interval (CI).

The *P* values that are often reported with the results of studies have a yes-or-no interpretation: The results are either statistically significant or not.

However, because the results of most studies are estimates, not definitive yes-or-no answers, CIs are often preferred. CIs can indicate the precision of any estimate, including differences; risk, hazard, and odds ratios; regression coefficients; correlation coefficients; percentages; and differences. Narrower CIs indicate more precise estimates than do wider CIs. The most common CI is 95%, but other coefficients can be used. For small samples in clinical research, for example, CIs are sometimes set at 90%.

Because CIs are expressed in the same units as the estimates they accompany, they focus attention on the estimates (ie, the outcome) and away from *P* values. Thus, CIs are increasingly preferred to *P* values, even though *P* values may still be reported.

> The increase in mean survival was 7 months (95% CI, –2.3 to +5 months).
> Median concentration increased from 43 to 155 mg/dL, a difference of 112 mg/dL (95% CI, 64 to 160 mg/dL; *P* = 0.09).
> The odds ratio for structural failure with untreated material was 16.5 (95% CI, 12.5 to 20.5).

12.5.3 Hypothesis Testing

In hypothesis testing, the ultimate product is the probability (*P*) value. The *P* value should not be confused with the alpha (α) level, which defines the threshold of statistical significance (the most common value being $\alpha = 0.05$). P values greater than α levels are not statistically significant by definition. Only those less than α are statistically significant. P values should be reported for each hypothesis tested, and if possible, these values should be expressed as equalities, not inequalities.

> $P = 0.002$ *not* $P > 0.05$ *and not* $P = $ NS for "not significant"
> widths did not differ greatly $(P = 0.13)$ *not* $(P \approx 0.13)$

Place a zero to the left of the decimal when reporting P values (see also Section 12.1.3.2, "Zeros before Decimals and Decimals after Whole Numbers").

> $P = 0.005$ *not* $P = .005$

For most applications, the smallest P value that need be reported is $P < 0.001$. For genetic associations, however, P values may be smaller orders of magnitude[1] (see Section 12.1.3.3, "Significant Digits").

In scientific writing, reserve "significance" for its statistical meaning. Use "marked," "important," "substantial," and similar words for nonstatistical descriptions. In addition, in sentences containing significant P values, describing the results as "significant" is redundant.

> The association was important . . . *not* The association was significant . . .
> The difference was 72 mL $(P = 0.03)$. *not* The 72-mL difference was significant $(P = 0.03)$.

For many of the hypothesis tests used in frequentist statistics (which use P values), the following statistical and clinical information should generally be reported to document the tests in the context of their use in the text:

(1) The hypothesis tested and the purpose of the analysis
(2) The appropriate descriptive statistics summarizing each group in the analysis
(3) Whether the data were paired (ie, related) or independent
(4) The minimum difference or change in the outcome variable deemed to be important
(5) Whether a power analysis was conducted, and if so, the type I error (α), the type II error (β), the difference of interest (δ), and the estimated percentage of attrition, all of which are used to calculate the minimum sample size
(6) The name of the statistical test or procedure used
(7) Whether the test was one- or two-tailed
(8) The threshold of statistical significance (ie, the α level)
(9) Whether the analysis was adjusted for multiple comparisons (ie, whether multiple hypothesis tests were conducted on the same data)
(10) Whether the data conformed to the assumptions of the tests used to analyze them. The most common assumptions that should be verified are as follows:

  - whether the data were normally distributed or were successfully transformed to a more normal distribution and a description of how normality was assessed
  - whether the variances of the distributions were similar
  - whether the data were analyzed with parametric or nonparametric tests
  - in linear regression analyses, whether the assumption of linearity was verified
  - in linear regression analyses, whether the variables were assessed for colinearity and interaction

(11) The statistical software used in the analysis and its version number
(12) The P value associated with the test
(13) The test statistic and, if applicable, its degrees of freedom (df)
(14) The result of the analysis (eg, the outcome, estimate, or "effect size," such as differences between groups, the absolute risk reduction, or the slope of a regression line)

(15) A measure of precision, such as a 95% CI, for at least primary outcomes, such as an estimate

(16) The relevance and implications of the result

Statistical results are often reported parenthetically in manuscripts. The individual elements of a test are normally separated by commas, and tests are normally separated by semicolons.

> This 9.5-degree difference was clinically important (95% CI, 7.23 to 11.73 degrees; 2-tailed Student's t test, $t = 8.43$, df = 49, $P = 0.13$).

The test statistic and, for some tests, its df can be reported in any of the following 3 ways:

$$t = 8.43, \text{df} = 49 \qquad t_{49} = 8.43 \qquad t(49) = 8.43$$

12.5.4 Bayesian Statistical Analyses

For reporting the results of Bayesian statistical analyses, see John K. Kruschke, PhD,[40] and J.B. van Doorn, MSc, and colleagues.[41]

CITED REFERENCES

1. Altman GA, Bland JM. Presentation of statistical data. BMJ. 1996;312(7030):572. https://doi.org/10.1136/bmj.312.7030.572

2. Bailor JC, Mosteller F. Medical uses of statistics. 2nd ed. NEJM Books; 1992. p 385–389.

3. Higham NJ. Handbook of writing for the mathematical sciences. 2nd ed. Society for Industrial and Applied Mathematics; 1998.

4. International Organization for Standardization; International Electrotechnical Commission. ISO/IEC 80000 2009 ISO/IEC international standards 80000, parts 1 to 14. International Organization for Standardization; 2009.

5. International Organization for Standardization. Quantities and units. Part 0: general principles. 3rd ed. International Organization for Standardization; 1992. (ISO 31-0[E]). Amendment 1, 1998; Amendment 2, 2005.

6. Bedeian AG, Struman MC, Streiner DL. Decimal dust, significant digits, and the search for stars. Organiz Res Meth. 2009;12(4):687–694. https://doi.org/10.1177/1094428108321153

7. Ehrenberg ASC. The problem of numeracy. Am Statistician. 1981;35(2):67–71. https://www.jstor.org/stable/2683143

8. Hiroe T et al. Gradations of clinical severity and sensitivity to change assessed with the Beck Depression Inventory-II in Japanese patients with depression. Psychiatry Res. 2005;135(3):229–235. https://doi.org/10.1016/j.psychres.2004.03.014

9. Wen SW, Kramer MS, Hoey J, Hanley JA, Usher RH. Terminal digit preference, random error, and bias in routine clinical measurement of blood pressure. J Clin Epidemiol. 1993;46(10):1187–1193. https://doi.org/10.1016/0895-4356(93)90118-k

10. Butler KR, Minor DS, Benghuzzi HA, Tucci M. Terminal digit bias is not an issue for properly trained healthcare personnel using manual or semi-automated devices—biomed 2010. Biomed Sci Instrum. 2010;46:75–80.

11. Harrison WN, Lancashire RJ, Marshall TP. Variation in recorded blood pressure terminal digit bias in general practice. J Hum Hypertens. 2008;22(3):163–167. https://doi.org/10.1038/sj.jhh.1002312

12. Mengden T et al. Use of automated blood pressure measurements in clinical trials and registration studies: data from the VALTOP Study. Blood Press Monit. 2010;15(4):188–194. https://doi.org/10.1097/MBP.0b013e328339d516

13. Villar J, Repke J, Markush L, Calvert W, Rhoads G. The measuring of blood pressure during pregnancy. Am J Obstet Gynecol. 1989;161(4):1019–1024. https://doi.org/10.1016/0002-9378(89)90777-1

14. Locker TE, Mason SM. Digit preference bias in the recording of emergency department times. Eur J Emerg Med. 2006;13(2):99–101. https://doi.org/10.1097/01.mej.0000195677.23780.fa

15. Edouard L, Senthilselvan A. Observer error and birthweight: digit preference in recording. Public Health. 1997;111(2):77–79. https://doi.org/10.1016/S0033-3506(97)90004-4

16. Denic S, Saadi H, Khatib F. Quality of age data in patients from developing countries. J Public Health (Oxf). 2004;26(2):168–171. https://doi.org/10.1093/pubmed/fdh131

17. Crawford S et al. Digit preference in year at menopause. Data from the study of women's health across the nation [poster abstract]. Ann Epidemiol. 2000;10(7):457. https://doi.org/10.1016/S1047-2797 (00)00140-X

18. Billion bites the dust [editorial]. Nature. 1992;358(6381):2. https://doi.org/10.1038/358002b0

19. International Organization for Standardization. Codes for the representation of currencies and funds. International Organization for Standardization; 2001. (ISO 4217). https://www.iso.org/iso-4217 -currency-codes.html

20. Australian Public Service Commission. Style manual. Commonwealth of Australia; c2020. https:// www.stylemanual.gov.au/

21. Bureau International des Poids et Mesures. The international system of units (SI). 9th ed. Organisa-tion Intergouvernementale de la Convention du Mètre; 2019. English translation available in PDF format at https://www.bipm.org/en/publications/si-brochure/

22. ASTM Committee E-43 on Metric Practice. Standard practice for use of the International System of Units (SI): the modernised metric system. American Society for Testing and Materials; 1993. (ASTM E380–89a).

23. Jerrard HG, McNeill DB. A dictionary of scientific units: including dimensionless numbers and scales. 6th ed. Springer; 1992.

24. Taylor BN, editor. The international system of units (SI). 2001 ed. National Institute of Standards and Technology; 2001 Jul. 74 p. (NIST special publication; SP 330 2008 ED).

25. American National Standards Institute. Metric practice. American National Standards Institute; 1992. (ANSI/IEEE standard 268-1992).

26. National Institute of Standards and Technology (US), Physics Laboratory. The NIST reference on constants, units, and uncertainty. National Institute of Standards and Technology; 1998 Feb [updated 2003 Dec]. Definitions of the SI base units; [about 2 p]. http://physics.nist.gov/cuu/Units/current.html

27. Horvath AL. Conversion tables of units in science and engineering. Elsevier; 1986.

28. International Organization for Standardization. Quantities and units. Part 2: mathematical signs and symbols to be used in the natural sciences and technology. International Organization for Standardization; 1992. (ISO 80000-2:2009[E]).

29. International Organization for Standardization. Quantities and units. 3rd ed. International Organi-zation for Standardization; 1993. ISO standards handbook

30. American Mathematical Society. A manual for authors of mathematical papers. American Mathe-matical Society; 1990.

31. Swanson E, O'Sean A, Schleyer A. Mathematics into type. Updated ed. American Mathematical Society; 1999.

32. Burton BW. Dealing with non-English alphabets in mathematics. Tech Commun. 1992;39(2):219– 225. https://www.jstor.org/stable/43090277

33. Zwillinger D. CRC standard mathematical tables and formulae. 32nd ed. CRC Press; 2011.

34. James R. Mathematics dictionary. 5th ed. Van Nostrand Reinhold; 1992.

35. International Union of Pure and Applied Physics, Commission for Symbols, Units, and Nomenclature, Atomic Masses and Constants. Symbols, units, nomenclature and fundamental constants in physics: 1987 revision. Physica. 1987;146A(1–2):1–68. Prepared by Cohen ER, Giacomo P.

36. Association of American Publishers. Markup of mathematical formulas. Ver 2.0. Revised ed. Asso-ciation of American Publishers; 1989.

37. Lang TA, Secic M. How to report statistics in medicine: annotated guidelines for authors, editors, and reviewers. 2nd ed. American College of Physicians; 2006.

38. International Organization for Standardization. Vocabulary and symbols. Part 1: general statistical terms and terms used in probability. International Organization for Standardization; 2006. (ISO 3534-1).

39. Lang T, Altman D. Basic statistical reporting for articles published in clinical medical journals: the SAMPL guidelines. In: Smart P, Maisonneuve H, Polderman A, editors. Science editors' handbook. European Association of Science Editors; 2013.

40. Kruschke JK. Bayesian analysis reporting guidelines. Nat Hum Behav. 2021 Oct;5(10):1282–1291. https://doi.org/10.1038/s41562-021-01177-7

41. van Doorn J et al. The JASP guidelines for conducting and reporting a Bayesian analysis. Psychon Bull Rev. 2021 Jun;28(3):813–826. https://doi.org/10.3758/s13423-020-01798-5

ADDITIONAL REFERENCES

AIP style manual. 4th ed. American Institute of Physics; 1990. https://publishing.aip.org/wp-content/uploads/2021/03/AIP_Style_4thed.pdf

American National Standards Institute. IEEE/ASTM standard for use of the International System of Units (SI): the modern metric system. ANSI; 2002. (SI10-2002).

Baron DN. Units, symbols, and abbreviations: a guide for biological and medical editors. 4th ed. Royal Society of Medicine (GB); 1988.

Cook JL. Conversion factors. Oxford University Press; 1991.

Darton M, Clark J. The Macmillan dictionary of measurement. Macmillan; 1994.

Hansen WR, editor. Suggestions to authors of the reports of the United States Geological Survey. 7th ed. Geological Survey (US); 1991. 289 p. https://doi.org/10.3133/7000088

McGraw-Hill dictionary of mathematics. 2nd ed. McGraw-Hill; 2003. 307 p.

Monteith JL. Consistency and convenience in the choice of units for agricultural science. Exp Agric. 1984;2:105–107. https://doi.org/10.1017/S0014479700003227

Oxford Dictionaries. http://oxforddictionaries.com/definition/english/number

The Karhu D, Vanzieleghem M. Significance of digits in scientific research. AMWA J. 2013;28(2):58–60.

Wingfield D, Freeman GK, Bulpitt CJ, on behalf of the General Practice Hypertension Study Group. Selective recording in blood pressure readings may increase subsequent mortality. QJM. 2002;95(9):571–577. https://doi.org/10.1093/qjmed/95.9.571

Young DS. Implementation of SI units for clinical laboratory data: style specifications and conversion tables. Ann Intern Med. 1987;106(1):114–129. https://doi.org/10.7326/0003-4819-106-1-114

Young DS, Huth EJ, editors. SI units for clinical measurement. American College of Physicians; 1998.

13 Time, Dates, and Age Measurements

**Editors: Jacob Kendall-Taylor and
Stanley N.C. Anyanwu, MBBS, FMCS, FWACS, FACS, FICS**

13.1 STANDARDS

This chapter provides general guidance on formatting dates and time, as well as specialized conventions used in some disciplines. For additional specialized conventions for dates and time, see Section 25.2, "Geologic Time Units," and Section 26.6, "Astronomical Dates and Time."

Another resource is the International Organization for Standardization, which has a comprehensive standard for conveying time and dates,[1,2] as well as a standard for quantities and units of space and time.[3] *The Chicago Manual of Style*[4] also provides detailed guidance on using time and dates in published documents.

13.2 TIME
13.2.1 Units and Symbols

In general, spell out units of time in running text when they appear without a numerical value.

> The year can be measured accurately to millionths of a second.
> Solar systems exist for billions of years.

For units of time accompanied by numerical values, use standard symbols, preferably those recommended by the Système International d'Unités, or SI (the International System of Units) (see Section 12.2.1, "Système International d'Unités (SI)," and Tables 12.3 and 12.5). The SI's base unit for time is the second, which uses the standard symbol "s." In addition, the Bureau International des Poids et Mesures,[5] which developed the

SI, has designated the following 3 units as acceptable for designating time, even though these units are not officially part of the SI: day (d), hour (h), and minute (min). Although SI prefixes can be used with "second" (eg, "millisecond," or "ms"), these prefixes are never used with "day," "hour," or "minute."

> Data travel from the retina to the brain in 20 ms.
> The experiment was conducted over the course of 240 h.
> Traveling from Earth to the Moon by spacecraft can take as few as 3 d.

SI offers no suggestions for time units larger than a day (eg, "week," "month," and "year"). However, where abbreviated forms are needed, it is appropriate to use "wk," "mo," and "y." In such cases, check publishers' guidelines because they vary regarding when abbreviations can be used (eg, some publishers never permit abbreviations in text but do allow abbreviations in tables, figures, and figure legends).

> The samples were refrigerated for 1 wk.
> Funding was extended for 2 y.

If the numerical value involves only one unit of time, insert a space between the number and symbol. If the value calls for more than one unit, place the numbers and symbols together without spacing. Again, check with the intended publisher to ascertain the appropriate convention.

> 27 h   22h3min   2h15min4s

For fractions of hours, minutes, and seconds, use decimals.

> 3.21 s   15.8 min = 15min48s *not* 15min8s   12.3 h = 12h18min *not* 12h30min

For values of time, do not use the prime symbol (') for minute or the double prime symbol (") for second. Those symbols should be reserved for the geographic coordinates latitude and longitude, respectively.

13.2.2 Clock Time

Two systems are used to designate the time of day. In the 24-h system, the day begins at midnight, which is designated as 0000 (pronounced "zero hundred"), and it ends the next midnight, which is designated as 2400 (pronounced "24 hundred"). In the 12-h system, the day is divided into 12-h portions: midnight through 11:59 AM and noon to 11:59 PM.

The 24-h system, which is used by most countries outside the United States and is commonly known as "military time" in the United States, obviates ambiguity as to which part of the day is meant. Time is expressed as 4-digit numbers. The first 2 digits in the 24-h system represent hours, and the last 2 represent minutes. No colon or period is used between the hours and minutes.

From midnight to noon, the 24-h and 12-h systems parallel each other (eg, "0836" in the 24-h system is "8:36 AM" in the 12-h system). After noon, time is converted from the 24-h system to the 12-h system by subtracting 12 h (eg, "1530" in the 24-h system is "3:30 PM" in the 12-h system).

> 0602 = 6:02 AM   1802 = 6:02 PM
> 0028 = 28 min past midnight   1228 = 28 min past noon

Because the last 2 digits represent minutes, not decimal fractions of an hour, do not use the unit of measurement "h" after the 4-digit number. However, if the 4-digit numbers of the 24-h format would produce ambiguity, add the word "hours," but spell it out rather than abbreviate it.

> We stopped counting at 1530 hours. (Without "hours," it is unclear whether 1530 refers to a counted number or the time.)

The International Organization for Standardization (ISO)[1] recommends an alternative format for the 24-h system in which hours, minutes, and seconds are separated by colons and a decimal point is used for fractions of a second. Unlike the other components of this time format, the fractions of seconds are based on the decimal system, not on 60 subunits (eg, "0.3" represents 3/10 of a second, not 30/60 of a second). In addition, the ISO format can be truncated to just hours and minutes for less precise representations.

> 23:59:59.5 = 0.5 s before midnight   23:59 = 1 min before midnight

In contrast, in the 12-h system, the day is split in half. The abbreviations "AM," which stands for "ante meridiem" (ie, "before noon"), and "PM," which stands for "post meridiem" (ie, "after noon"), are used to distinguish between the halves of the day.

> 12:01 AM = 1 min after midnight *or* 12:01 in the morning
> 12:01 PM = 1 min after noon *or* 12:01 in the afternoon

In the 12-h system, minutes are separated from hours by a colon.

> 2:01 AM   11:54 PM

When the time is the start of an hour, do not use a colon and 2 zeros to designate minutes unless the time is part of a series that includes times that fall between the start of 2 hours. Similarly, use the terms "midnight" and "noon" by themselves unless they are part of a series of times, in which case they should be designated as "12 AM" and "12 PM" or "12:00 AM" and "12:00 PM." Never use the redundancies "12 midnight" and "12 noon."

> 3 PM, 5 PM, and 7 PM *not* 3:00 PM, 5:00 PM, and 7:00 PM
> 4:17 AM, 7:27 AM, 12:00 PM, and 5:00 PM *not* 4:17 AM, 7:27 AM, midnight, and 5 PM
> between midnight and noon *not* between 12 AM and 12 PM

This style manual recommends that the abbreviations "AM" and "PM" be capitalized. Some publishers' guidelines, however, call for using small capitals, and others favor using lowercase for these abbreviations. Regardless of which convention is used, these abbreviations should be separated from the time by a space.

With the 12-h system, time can be expressed using either the abbreviations "AM" and "PM" or the informal and ambiguous term "o'clock," but not both. This style manual recommends "AM" and "PM" over "o'clock." If "o'clock" is chosen, it is useful to add such phrases as "in the morning," "in the afternoon," and "in the evening" to eliminate ambiguity.

> 10 PM *or* 10 o'clock at night *not* 10 o'clock PM
> 6 AM *or* 6 o'clock in the morning *not* 6 o'clock AM

13.2.3 Time Zones

International time zones were legally established in the 19th century. The standard time system partitions Earth into 24 major international time zones. Most time zones are in increments of 15° of longitude, making the length of each approximately 1,665 km (ie, 1,035 miles) at the equator.

Greenwich, London, United Kingdom, was chosen for 0° of longitude, or the prime meridian. The prime meridian determines the eastern boundary of the time zone called "mean solar time," formerly known as "Greenwich mean time." Mean solar time is equivalent to coordinated universal time, which is also known as "universal time coordinated" (UTC) (see also Section 26.6.2.2, "Coordinated Universal Time").

Most international time zones differ from UTC by a whole number of hours, with hours increasing to the east of UTC and decreasing to the west until they meet at the International Date Line on the opposite side of the world from UTC. Some local time zones, however, differ by half hours, as does the Newfoundland zone, which encompasses Newfoundland and Labrador and is 0.5 h later than Atlantic time.

In addition, time zones deviate from their meridians at some points, mainly because of political boundaries. China, for example, geographically spans 5 international time zones, but the entire country stays on the same time zone, namely Beijing time.

Africa has 6 time zones; Asia, 11; Australia, 3; Europe, 7; North America, 6; and South America, 5. The 24 major time zones converge in Antarctica and the Arctic.

In North America, international time zones span from mean solar time on the eastern shores of Greenland to Hawaii-Aleutian time in the Pacific Ocean. The more commonly referenced time zones in North America are the Atlantic time zone, which begins approximately at the 60° meridian west of Greenwich; the eastern zone, at 75°; the central zone, at 90°; the mountain zone, at 105°; and the Pacific zone, at 120°.

Do not capitalize the names of time zones when writing them out in full, except when the name includes a proper noun (eg, Atlantic standard time) or when the name begins a sentence. Capitalize initialisms for time zones. Do not use periods in the initialisms and separate the time from the initialisms with a space.

> eastern standard time   Pacific standard time
> 11 AM CST   0500 AST

Some countries or parts of countries divide the year into standard time and daylight saving time. During daylight saving time, time is advanced by 1 h to lengthen the period during the evening that is in daylight. Observed roughly between early spring and mid-fall, daylight saving time differs from country to country, and generally, it falls at the opposite time of the year in the Northern Hemisphere than in the Southern Hemisphere.

> When it is 1 PM eastern daylight time (EDT) in New York, New York, United States, it is 9 AM Alaska daylight time (AKDT) in Anchorage, Alaska, United States; 6 PM West Africa time (WAT) in Lagos, Nigeria; and 3 AM Australian eastern standard time (AEST) the next day in Sydney, New South Wales, Australia.

In time zone initialisms, the initials "ST" do not always stand for "standard time." In some countries, daylight saving time is called "summer time," which is abbreviated as "ST." Standard time, in turn, is referred to as "normal time" or "winter time."

> When it is 1530 British summer time (BST) in London, United Kingdom, it is 1630 Central European summer time (CEST) in Berlin, Germany; 2100 Indian standard time (IST) in Mumbai, India; and 1030 Venezuelan standard time (VET) in Caracas, Venezuela.

For the names and initialisms of time zones around the world, Wikipedia is a reliable resource (search for "lists of time zones"). For current times around the world, consult the World Clock feature at https://www.timeanddate.com.

13.3 DATES

13.3.1 Days and Months

Capitalize and spell out the names of days of the week and months. These may be abbreviated to their first 3 letters if they appear in tables, graphs, end references, and other locations where short forms are needed.

For a complete date, use the full name or the 3-letter abbreviation for the month, rather than the numerical equivalent, to avoid any ambiguity related to differing conventions. (See Section 13.3.3, "Sequence of Date Elements," for recommendations for ordering the elements in a date.)

> 5 January 2022 *or* 5 Jan 2022 *or* 2022 January 5 *or* 2022 Jan 5
> *not* 5/1/2022 (European style) *and not* 1/5/2022 (US style)

If the day of the week is included in a date, place the day, month, and year in apposition to the day of the week by placing a comma after the day of the week and another comma after the full date.

> On Tuesday, 2 November 2000, the first long-term crew docked at the International Space Station.
> Wednesday, 16 May 2018, was the 300th anniversary of the birth of Italian mathematician Maria Gaetana Agnesi.

In astronomy, the Julian day is used instead of the calendar date. The Julian day represents the number of solar days since 1 January 4713 BC at noon mean solar time (for more detail, see Section 26.6.1.1, "Julian Day").

> The calendar date of midnight 24 June 2012 is the Julian day 2456224.5000.
> The calendar date of noon 2022 Oct 19 is the Julian day 2459882.0000.

13.3.2 Years

Express years in numerals. In scientific publications, avoid abbreviating years to 2 digits preceded by an apostrophe (eg, class of '19), even though this form is sometimes used in informal writing.

> The first report on this species was published in 2023.
> the 2000s and 2020s *not* the '00s and '20s *and not* the 00's and 20's

13.3.3 Sequence of Date Elements

In scientific publications, write dates in the sequence of day, month, and year (a form common in Europe and Canada, as well as with US military organizations) or year, month, and day (a form common in end references and astronomical and aeronautical literature). Although the month can be shortened to its first 3 letters for tables, graphs, end references, and other locations where short forms are needed, do not truncate the year to 2 digits.

> 23 April 2022 *or* 23 Apr 2022 *or* 2022 April 23 *or* 2022 Apr 23

This style manual prefers the 2 formats above to the numerical format of year-month-day (YYYY-MM-DD) recommended by the International Organization for Standardization. This manual also recommends avoiding the sequence of month, day, and year, which is commonly used in the United States. If this form must be used, spell out the month, and set the year off with commas. However, if the day is excluded, do not place punctuation between the month and year.

> The total solar eclipse of July 11, 1991, passed directly over the world's largest telescopes on the Mauna Kea volcano in Hawaii.
>
> An eclipse passed over the South Atlantic Ocean in June 1992.

For information about dates in end references, see Section 29.3.3.6, "Date."

13.4 SPANS OF TIME

13.4.1 Divisions of Historic Time

To designate eras, such as the Gregorian and Julian BC and AD, the Islamic BH and AH, the Jewish AM, and the relatively recent BCE and CE, use numerals for years, preceded or followed by era designations (see Table 13.1). The era designations of AD, AC, and AH precede the year, while all others follow the year. Note that there is no year zero.

The choice of era designation should agree with the convention used by the publication to which a manuscript is submitted or with the convention of the specific scientific discipline.

In general, the era designations of AD, CE, and AS can be omitted from dates if the context is clearly related to the current era. However, these designations should be used when the content also refers to BC or BCE dates or when the context mixes era designations of different kinds.

> In 1769, French explorer and botanist Jeanne Baret became the first woman to circumnavigate the world, and US astronaut and aeronautical engineer Neil Armstrong, MS, became the first human to step on the Moon in 1969.
>
> Although the first recorded archaeological dig dates back to the reign of Babylonian King Nabonidus between 555 and 539 BCE, the first modern archaeologist might have been John Aubrey, FRS, who lived between CE 1626 and 1697.
>
> The Greek physician Hippocrates died in the Roman year 384 AUC, which is 370 BC in the Gregorian calendar and 3714 AM in the Jewish calendar.

Also see Section 13.5.1, "Absolute Dating" and Section 25.2, "Geologic Time Units."

Table 13.1. Eras: Names, Abbreviations, and Origins

Name of era	Abbreviation	Source
After Christ	AC	Christian
Anno Domini (in the year of our Lord)	AD	Christian
Anno hegirae (in the year of the Hegira in 622 CE)	AH	Islamic
Anno mundi (in the year of the world) (also *anno Hebraico* [AH])	AM	Jewish
Anno salutis (in the year of salvation)	AS	Latin
Ab urbe condita (from the founding of the city of Rome in 753 BCE)	AUC	Roman
Before Christ	BC	Christian
Before the common era (equivalent to BC)	BCE	German
Before the Hegira (dates before 622 CE)	BH	Islamic
Before present (with "present" designated as the origin of practical radiocarbon dating in 1950)	BP	Science based
Common era (equivalent to AD)	CE	German

Table 13.2. The 3-Age Chronologic System for Describing Prehistoric Ages[a]

Age	Artifacts and cultural characteristics
Stone Age	

Paleolithic Age, or Old Stone Age
  Upper Paleolithic
  Middle Paleolithic (Mousterian)
  Lower Paleolithic
	Chipped stone tools only

Mesolithic Age, or Middle Stone Age	Chipped stone tools and microliths, as well as the beginnings of pottery and ground stone in some areas
Neolithic Age, or New Stone Age	Ground and polished stone tools, pottery, and agriculture
Chalcolithic, Eneolithic, or Copper Age	Period between the Neolithic Age and the Bronze Age when unalloyed copper was used rather than true tin bronze
Bronze Age	Tools and weapons of tin bronze and the emergence of an urban way of life
Iron Age	Tools and weapons of iron

[a] Some cultures remained "prehistoric" during historical periods of main civilizations (eg, the culture of Aboriginal Australians).

13.4.2 Names of Periods

The 3-age system, published in 1836 by antiquarian Christian Jürgensen Thomsen and used primarily in archaeology, divides prehistoric time into 3 successive periods. Each of these periods is defined by the main material for making tools in that period: stone, bronze, and iron (see Table 13.2). These periods also are defined by cultural developments. The actual dates of each period differ among regions of the world.

Progress from one age to another is demonstrated by artifacts uncovered at different levels in the ground. The term "upper," as in "Upper Paleolithic," refers to strata closer to the surface, which are newer. Hence, "upper" is equivalent to "late" and "lower" to "early" (see also Section 25.2, "Geologic Time Units"). The 3-age system is not useful in Africa, where bronze was not used south of the Sahara, or in the Americas, where bronze was never important and iron was not used until it was introduced by Europeans.

Capitalize the names given to cultural periods, eras, and epochs, as well as the words "age," "early," "middle," and "late." However, whether to capitalize the terms "period," "era," and "epoch" depends on whether they are used with formal or informal geologic times (see Section 25.2, "Geologic Time Units").

> Late Bronze Age
> Chalcolithic Age
> Early Woodland period
> Viking period
> The Cenozoic Era occurred during the Quaternary period of the Holocene epoch.

13.4.2.1 Ice Ages

The generic terms in names of ice ages (eg, "glaciation") should be lowercased, except for the "Glacial Epoch" (ie, the Pleistocene Epoch).

> Illinoian glaciation   Riss glaciation

The term "neoglacial" refers to the small-scale glacial advance that occurred during the "Holocene," the current epoch (see Section 13.4.2.2, "Recent and Present Time"). The most recent advance in the Holocene was the Little Ice Age, which took place from AD 1550 to 1850. Unlike "neoglacial," "neoglaciation" is an informal term used to designate glacial expansions later than the Holocene climatic optimum, which took place 7,000 to 3,000 years ago.

13.4.2.2 Recent and Present Time

The term "Holocene" refers to the current epoch, which began approximately 11,700 years ago. "Holocene" replaced the term "Recent," which now connotes any recent time of unspecified duration. In its new context, the term "recent" is lowercased.

For any geochronologic date consisting of a year and an era (eg, 41 BC and 1990 BP), do not add a term such as "age" or "before the present."

13.5 AGE MEASUREMENT

13.5.1 Absolute Dating

Absolute dating, also known as chronometric dating, refers to measuring age on a specific time scale that dates from or to a fixed reference point (see Table 13.1). In parts of the world that are predominantly Christian, that reference point is often the birth of Christ, which for chronometric purposes, has been observed as the year AD 1. Years are counted backward before Christ (BC) and forward after Christ, which can be designated "AD" from the Latin term "*Anno Domini*," "AC" for "after Christ," or "CE" for "common era." There is no year zero in the BC, AD, AC, and CE ages.

In contrast, "before the present" (BP) is an international system that does not reference a particular calendar system. In this system, AD 1950 usually represents the "present," and the higher the year number, the farther back in time the date (eg, the date 400 BP is the equivalent of AD 1550, and 1949 BP is the equivalent of AD 1).

13.5.2 Radiometric Dating

Radiometric dating consists of all methods of age measurement that rely on the nuclear decay of naturally occurring radioactive isotopes, whether they are short-life radioactive elements (eg, carbon-14) or long-life radioactive elements and their decay products (eg, the decay of 40-potassium to 40-argon, of uranium-238 to thorium-230, and of uranium-235 to protactinium-231).

Radiocarbon (ie, carbon-14) dating is still the most frequently used tool for dating organic samples as old as 50,000 years, whereas potassium-argon dating is commonly used for determining the age of rock samples at least 100,000 years old. For strata and artifact assemblages between those ages, dating is far less reliable.

Depending on the audience for a manuscript, it may be necessary to convert radiocarbon ages to calendar dates. Making this conversion requires taking into account temporal variations in the natural radiocarbon content of atmospheric carbon dioxide. For up-to-date calibrations, consult the website of the journal *Radiocarbon*.[6]

Unlike radiometric dating, calibrated radiometric dating checks dates against another dating method (eg, dendrochronology, in which dating is determined by annual growth rings in timber and tree trunks). With calibrated dates, eras are sometimes designed "Cal BC" or "Cal AD." In contrast, with uncalibrated radiometric dates, eras may be designed in lowercase (eg, ad, bc, and bp).

| **Calibrated:** | AD 1400 | 3700 BC | Cal AD 1550 |
| **Uncalibrated:** | ad 1500 | 3000 bc | 400 bp |

Both calibrated and uncalibrated radiometric dates are commonly reported with ranges of error using the plus or minus symbol (eg, "±"). These ranges are based on confidence levels.

Another method of dating is archaeomagnetic dating, which is the study and interpretation of the signatures of Earth's magnetic field recorded in archaeological materials. In early practice, archaeomagnetic dating was reported in the same form as carbon-14 dates, as a mean value plus or minus a number of years (eg, AD 1015 ± 35). It is now considered more accurate to report dates as ranges (eg, AD 980 to 1050).

However, if the plus or minus designation is used, include the appropriate following symbol to indicate what the numerical value of the deviation represents: "*s*" for "standard deviation," "*2s*" for "2 standard deviations," or "S_x" for "standard error of the mean" (see Section 12.5.1, "Descriptive Statistics").

CITED REFERENCES

1. International Organization for Standardization. Date and time—representations for information interchange—part 1: basic rules. International Organization for Standardization; 2019. (ISO 8601-1: 2019[en]). https://www.iso.org/obp/ui/#iso:std:iso:8601:-1:ed-1:v1:en

2. International Organization for Standardization. Date and time—representations for information interchange—part 2: extensions. International Organization for Standardization; 2019. (ISO 8601-2: 2019[en]). https://www.iso.org/obp/ui/#iso:std:iso:8601:-2:ed-1:v1:en

3. International Organization for Standardization. Quantities and units—part 3: space and time. International Organization for Standardization; 2019. (ISO 80000-3:2019[en]). https://www.iso.org/obp/ui /#iso:std:iso:80000:-3:ed-2:v1:en

4. The Chicago manual of style: the essential guide for writers, editors, and publishers. 17th ed. The University of Chicago Press; 2017. Also available at https://www.chicagomanualofstyle.org

5. Bureau International des Poids et Mesures. Le Système International d'Unités (SI). 9th ed. Bureau International des Poids et Mesures; 2019. English translation available in PDF format at https://www.bipm .org/en/publications/si-brochure

6. Radiocarbon. Arizona Board of Regents. Vol 1, 1959–. Available at https://www.cambridge.org/core /journals/radiocarbon/all-issues

14 Geographic Designations

Editor: Kenneth April

14.1 GEOGRAPHIC NAMES

The geographic locations in which research is conducted and descriptions of geographic features are important components in the reports of many scientific studies. The spelling of all geographic locations should be verified by the authors and the manuscript editors. Because location names change frequently in some countries and regions of the world, the spelling of all such names should be confirmed in online geographic name databases, atlases, gazetteers, and dictionaries.

14.1.1 Geographic Names in Text

For guidance on capitalizing geographic names, see Section 9.4.4, "Geographic Designations."

14.1.1.1 Municipalities

In general, when the name of a city or other municipality is used in running text, follow it with the unabbreviated name of the state, territory, possession, province, or other regional subunit and the full name of the country. Use commas before and after the names of states, other regional subunits, and countries when they follow city names.

> Medicine Hat, Alberta, Canada, is home to the Suffield Research Centre.
> Swan Hill, Victoria, Australia, is inhabited by the Wemba-Wemba people.

Meridian, Idaho, United States, is the birthplace of physical chemist and Gore-Tex creator Bill Gore, MS.

Publications should document in detail any exceptions they make to this rule so that those exceptions can be strictly followed. For example, publications may choose to omit the names of regional subunits, countries, or both for well-known cities (eg, "Los Angeles," "Tokyo," and "Rome"). Additionally, publications may leave off the name of the country in which they are published, provided that most readers are from that country (eg, a Canadian publication might use "Swift Current, Saskatchewan," while a US publication is likely to refer to the city as "Swift Current, Saskatchewan, Canada"). Likewise, publications may choose to omit regional subunits that readers are unlikely to be aware of and use only country names with city names (eg, an Australian publication might use "Tlaquepaque, Mexico," while a Mexican publication might use "Tlaquepaque, Jalisco"). However, country names and regional subunit names should not be omitted if doing so has a reasonable possibility of confusing readers (eg, do not use "London" for "London, Ontario, Canada").

> The Institut Curie in Paris conducts extensive research. (Paris is a well-known city, so the country name of France is unlikely to be needed unless a publication's readers could confuse the city with a Paris elsewhere in the world, such as Paris, Texas, United States.)
>
> Evidence of a new fault line was found 8 km south of Turkington, Missouri, United States. (Publications with readers primarily located in the United States may choose to omit "United States" but not "Missouri.")
>
> Although it is Russia's largest space launch facility, the Baikonur Cosmodrome is located in the city of Baikonur, Kazakhstan. (Publications may decide to omit the regional subunit "Kyzylorda Region" if their readers are primarily outside Central Asia and Eastern Europe. For readers in Central Asia and Eastern Europe, the city name alone may suffice.)

Where brevity is needed, as in tables and end references, the names of states, provinces, territories, and other regional subunits may be shortened by using widely recognized abbreviations, such as the postal abbreviations used in the United States, Canada, and Australia (see Table 14.1). When regional subunits are abbreviated for tables, end references, and other purposes, this style should be applied consistently throughout the publication for those purposes. However, do not abbreviate the names of regional subunits when they are used without the names of municipalities.

Canadian city	Year founded		Canadian city	Year founded
St John's, Newfoundland and Labrador	1497	*or*	St John's, NL	1497
Quebec City, Quebec	1608		Quebec City, QC	1608
Province	**No. of volcanoes**		**Province**	**No. of volcanoes**
Ontario	9	*not*	ON	9
Quebec	7		QC	7

Use English names for foreign cities, except when an institution's name given in a foreign language includes a city name (eg, "Milan," not "Milano," but "Università degli Studi di Milano"). Use the current official names of cities that previously were well known by other names (eg, "Mumbai" and "Kolkata" in India, not "Bombay" and "Calcutta"; "Beijing" and "Nanjing" in China, not "Peking" and "Nanking"; and "St Petersburg" and "Volgograd" in Russia, not "Leningrad" and "Stalingrad"). Publishers

Table 14.1. Postal Symbols of States, Provinces, and Territories of the United States, Canada, and Australia for Tables and End References

State, province, or territory	Symbol	State, province, or territory	Symbol
US states			
Alabama	AL	New Mexico	NM
Alaska	AK	New York	NY
Arizona	AZ	North Carolina	NC
Arkansas	AR	North Dakota	ND
California	CA	Ohio	OH
Colorado	CO	Oklahoma	OK
Connecticut	CT	Oregon	OR
Delaware	DE	Pennsylvania	PA
District of Columbia	DC	Rhode Island	RI
Florida	FL	South Carolina	SC
Georgia	GA	South Dakota	SD
Hawaii	HI	Tennessee	TN
Idaho	ID	Texas	TX
Illinois	IL	Utah	UT
Indiana	IN	Vermont	VT
Iowa	IA	Virginia	VA
Kansas	KS	Washington	WA
Kentucky	KY	West Virginia	WV
Louisiana	LA	Wisconsin	WI
Maine	ME	Wyoming	WY
Maryland	MD		
Massachusetts	MA	**US territories**	
Michigan	MI	American Samoa	AS
Minnesota	MN	Federated States of Micronesia	FM
Mississippi	MS	Guam	GU
Missouri	MO	Marshall Islands	MH
Montana	MT	Northern Mariana Islands	MP
Nebraska	NE	Palau	PW
Nevada	NV	Puerto Rico	PR
New Hampshire	NH	Virgin Islands	VI
New Jersey	NJ		
Canadian provinces			
Alberta	AB	Quebec	QC
British Columbia	BC	Saskatchewan	SK
Manitoba	MB		
New Brunswick	NB	**Canadian territories**	
Newfoundland and Labrador	NL	Northwest Territories	NT
Nova Scotia	NS	Nunavut	NU
Ontario	ON	Yukon	YT
Prince Edward Island	PE		

Table 14.1. Postal Symbols of States, Provinces, and Territories of the United States, Canada, and Australia for Tables and End References (*continued*)

State, province, or territory	Symbol	State, province, or territory	Symbol
Australian states		**Australian territories**[a]	
New South Wales	NSW	Ashmore and Cartier Islands	[none]
Queensland	QLD	Australian Antarctic Territory	TAS
South Australia	SA	Australian Capital Territory	ACT
Tasmania	TAS	Christmas Island	WA
Victoria	VIC	Cocos (Keeling) Islands	WA
Western Australia	WA	Coral Sea Islands	[none]
		Heard Island and McDonald Islands	[none]
		Jervis Bay Territory	ACT
		Norfolk Island	NSW
		Northern Territory	NT

[a] Only 2 Australian territories have unique postal symbols: the Australian Capital Territory and the Northern Territory. Of the remaining 8 territories, 4 share the postal symbols of the states they are nearest to, 1 shares another territory's postal symbol, and 3 uninhabited or nearly uninhabited territories do not have postal symbols.

should have policies regarding whether to use diacritics in foreign city names that are otherwise spelled the same in English (eg, "Montréal" or "Montreal" and "São Paulo" or "Sao Paulo").

14.1.1.2 Nations

Owing to internal and external strife and country reorganizations in many parts of the world, the names of some countries change over time. Check with online geographic databases or with countries' official websites to determine the current names of countries.

For nations that are better known by shortened versions of their official names, use the shortened names throughout manuscripts, as recommended in Section 8.5, "Geographic Names" (eg, "United States" instead of "United States of America" and "Spain" instead of "Kingdom of Spain"). However, if referencing the full name of a country is desirable, provide the full name with or near the first reference to the country and use the shortened name thereafter.

> Archaeological studies at the Castle of Gung Ye are hampered because the castle is in the demilitarized zone between South Korea, officially the Republic of Korea, and North Korea, officially the Democratic People's Republic of Korea.
>
> Microbiologists from the People's Republic of China and the Russian Federation met to discuss cooperation between China and Russia on identifying the new virus.
>
> The Republic of China (Taiwan), also known as Chinese Taipei, has a fellowship program for scholars from other countries to conduct research in Taiwan.
>
> Biotechnology is a research priority for both the Federal Republic of Germany and the United Mexican States, better known as Germany and Mexico, respectively.

When country names are used as nouns, always spell them out rather than using acronyms or other abbreviations (eg, "United States," not "US," and "New Zealand," not "NZ"). Use countries' acronyms or other abbreviations only as adjectives and only after they have been introduced on first use.

In the United Kingdom (UK), the National Health Service is overseen by the UK Department of Health and Social Care.

The Costa Rica (CR) Institute of Technology spearheaded building the first CR satellite, which monitored CR forests from space.

For most country names that are preceded by the article "the," do not capitalize the article.[1] Only capitalize "the" when it is a formal part of a geographic name, as it is for The Gambia, officially the Republic of The Gambia (see also Section 9.4.4, "Geographic Designations").

| the Cook Islands | the Czech Republic | the Netherlands | the Philippines |

14.1.1.3 Multinational Regions

Multinational regions typically share a common feature, such as geography, language, or culture. These regions are not always strictly defined, and some may overlap in composition. For example, North America consists of all the land from Panama north to the Arctic, including the West Indies, Greenland, and most of the Aleutian Islands. In contrast, Central America consists of the portion of North America from Panama north to Belize and Guatemala, excluding the West Indies.

The Middle East has no precise definition but is generally thought to extend from Turkey in the north to the Arabian Peninsula in the south and from Egypt in the west to Iran in the east. "Near East" is a dated term for the same region, and it should be used only in historical references.

Consult geographic dictionaries and databases to determine the composition of multinational regions and to determine whether their names should be capitalized.

| African savanna | Far East | Western Hemisphere |
| Amazon rainforest | eastern Europe | West Bank |

14.1.1.4 Geographic Features

Capitalize generic geographic terms, such as "river" and "mountain," when they are part of place names according to an appropriate resource, such as an atlas or gazetteer. Do not capitalize generic terms if they appear on their own or if they are plural and follow 2 or more proper names.

Nile River (but "the river" if used subsequently without the proper name)
Black and Caspian seas (but "Black Sea" and "Caspian Sea" if used individually)
Himalayan Mountains (but "the mountains" or "the Himalayas" in subsequent references)

Some geographic names contain foreign words that are the equivalents of English generic terms (eg, "rio" means "river," "mauna" and "yama" mean "mountain," "sierra" means "mountains" or "mountain range," and "sahara" is derived from "desert"). Do not add the generic English term to such names.

Rio Grande	*not*	Rio Grande River
Fujiyama *or* Mt Fuji	*not*	Fujiyama Mountain *or* Mt Fujiyama
Mauna Loa	*not*	Mauna Loa Mountain *or* Mt Mauna Loa
Sierra Madre	*not*	Sierra Madre Mountains
the Sahara	*not*	Sahara Desert

Generally, use Indigenous names rather than non-Indigenous names for geographic features (eg, "Denali," not "Mt McKinley"). On first mention of a geographic feature, its non-Indigenous name may be used in apposition to its Indigenous name, especially if readers are more likely to recognize the non-Indigenous name (eg, "Mosi-oa-Tunya, also known as Victoria Falls"). Government-established entities with names that include geographic features should be referred to by their official names, regardless of the language used for the official name (eg, "The Mosi-oa-Tunya National Park in Zambia neighbors the Victoria Falls National Park in Zimbabwe").

Many geographic regions have informal names that may or may not be recognized as proper nouns. Such terms should be used with care in scientific publications. If such terms are used, readers unfamiliar with the informal geographic regions could be deprived of important information. Furthermore, the boundaries of such regions may not be clearly defined. To eliminate confusion, specify locations by geographic coordinates (see Section 14.2, "Geographic Coordinates") or, if more appropriate for the context, by map-indicated political boundaries.

If informal regional names are used, capitalize them.

> the Great Plains (the plains region in Canada and the United States east of the Rocky Mountains)
> the National Capital Region (in Canada)
> the Mezzogiorno (for southern Italy)
> the Red Centre (for the central Australian desert region)

Many geographic regional names include descriptive terms such as "east," "west," "lower," "upper," and "central." These terms should be capitalized only if they are consistently used to denote defined regions that are recognized as proper names. *The Chicago Manual of Style*[2] lists a variety of such terms and their variations. Online and print atlases, gazetteers, and dictionaries may also be used to determine whether such terms are considered proper names.

> lower St Lawrence River   Southern California   Upper Peninsula (in Michigan)
> northern Italy   upper Euphrates   western California

Some geographic names are also used as generic terms. For example, as a proper noun, "Arctic" is the region of Earth north of the Arctic Circle, but as an adjective, it can refer either to that same geographic region or to very low temperatures. As an adjective, "Arctic" should be capitalized when referring to the region (eg, "Arctic communities"), but it should be lowercased when referring to temperature (eg, "arctic gale"). Established names of Arctic flora and fauna are also usually lowercased (eg, "arctic char"). Follow the same set of rules for "Antarctic" and "antarctic."

14.1.2 Authorities for Names

In addition to relying on geographic dictionaries, atlases, and gazetteers to standardize geographic names, publications can adhere to the recommendations of one or more national depositories for geographic names.

Publishers should have policies in place regarding the circumstances under which they will honor authors' preferences to use location names that differ from the recommendations of national depositories, dictionaries, atlases, and gazetteers.

14.1.2.1 United States

The Geographic Names Information System (GNIS),[3] developed by the US Geological Survey in cooperation with the US Board on Geographic Names, contains more than

2 million records for names of physical and cultural geographic features, both current and historical. The GNIS lists official names of places within the United States as specified by state boards of geographic names. The database also includes records of named features that do not have official status.

The GNIS is available as a searchable database.[3] When searching for geographic names in this database, narrow the search either by selecting the "Exact Match" checkbox from the drop-down menu in the "Names Search Mode" field or by selecting the type of geographic feature from the drop-down menu in the "Feature Classes" field. For example, for municipalities, select "Populated Places" from the drop-down menu; for rivers, select "Stream"; and for mountains, select "Range" or "Summit."

The database also maintains a separate search for geographic features in Antarctica.

14.1.2.2 Canada

In Canada, provincial and territorial governments accept or reject geographic names within their borders, except for the names of federally administered lands, such as reserves for Indigenous populations, national parks, and military reserves. The Geographical Names Board of Canada, in turn, acts as a central registry. This national board enters all official and some unofficial names into the Canadian Geographical Names Database,[4] which covers all types of geographic features, including undersea features. For place names in the province of Quebec, the database uses their French spellings.

14.1.2.3 Other Parts of the World

Many other countries have official bodies with responsibility for standardizing geographic names. The United Nations Group of Experts on Geographical Names[1] lists hyperlinks for searchable geographic name databases of approximately 30 countries.

A US source for place names outside the United States is the Geographic Names Server,[5] a repository of place names sanctioned by the US Board on Geographic Names.

14.1.3 Numbers

For most place names that include numbers, spell out the numbers, and do not hyphenate the names.

> Two Rivers, Wisconsin, United States
> Three Mile Island in Pennsylvania, United States
> Fourteen Mile Point, Michigan, United States
> *but* 100 Mile House, British Columbia, Canada

For cities in which dates are part of their names, do not spell out the numbers in the dates. When placing such names in alphabetical order, treat them as though the numbers were spelled out in English.

> 20 de Junio, Argentina   25 de Diciembre, Peru

14.1.4 Multiword Names

Capitalize all elements of multiword geographic names except prepositions, conjunctions, and articles. Consult geographic databases to determine whether hyphens should be used.

> Stratford-upon-Avon, United Kingdom
> Sainte-Anne-de-la-Pérade, Quebec, Canada
> Fond du Lac, Wisconsin, United States
> Truth or Consequences, New Mexico, United States

Some multiword geographic names include possessive words, and others have plural terms. Unfortunately, geographic databases may not always correctly indicate whether such place names use apostrophes or not. In such cases, rely on a recognized geographic dictionary, atlas, or gazetteer, or on the locations' official websites.

> Devil's Island in French Guiana *but* Devils Island in the Philippines
> Martha's Vineyard (island in Massachusetts, United States) *but* Marthas Bath (lake in New South Wales, Australia)
> Joe Batt's Arm, Newfoundland and Labrador, Canada *but* Batts Rock Bay in Barbados

14.1.5 Prefixes

Spell out and capitalize most generic geographic terms that are part of place names.

> Port Arthur, Tasmania, Australia
> Point Lobos State Natural Reserve in California, United States
> Le Havre, France
> Punta del Este, Uruguay
> San Luis Potosí, Mexico

In English-language location names, abbreviate "Saint" as "St" except for locations that officially use the spelled-out form (eg, "Saint John, New Brunswick, Canada"). Additionally, "Mount" should be abbreviated as "Mt" in the name of a mountain but not in the name of a populated place. As with most abbreviations, "St" and "Mt" are used without periods (see also Section 11.2, "Punctuation and Typography").

> St Louis, Missouri, United States
> St John's, Newfoundland and Labrador, Canada
> Mt Erebus (mountain in Antarctica)
> Mount Willoughby (town in South Australia, Australia)

For French-language location names in parts of the world where French is an official language, do not abbreviate "Saint," "Sainte," or "Mont."

> Mont-Tremblant, Quebec, Canada
> Saint-Étienne, France
> Mont-Sainte-Aldegonde, Belgium

14.2 GEOGRAPHIC COORDINATES

14.2.1 Latitude and Longitude

Latitude is the distance north or south of the equator. It is designated by parallels that have traditionally been measured in degrees, minutes, and seconds, beginning with 0° at the equator and progressing to 90° north or south of the equator. Longitude is the distance east or west of the reference meridian that is maintained by the International Earth Rotation and Reference Systems Service (IERS). Also known as the International Reference Meridian and as the prime meridian, the IERS Reference Meridian passes through Greenwich, London, United Kingdom. Longitude is designated by meridians, and it is traditionally measured in degrees, minutes, and seconds from 0° to 180°.

Latitude–longitude coordinates for many US geographic entities are available in the Geographic Names Information System.[3] Coordinates for non-US geographic entities can be found on the National Geospatial-Intelligence Agency's Geographic Names Server.[5]

Spell out "latitude" and "longitude" when used in text without numerical designations, but abbreviate them as "lat" and "long" when they are part of coordinates.

Several conventions exist for representing a geographic coordinate with latitude and longitude. Some of the variants were developed to facilitate storing coordinates in databases. For example, coordinates may be stripped of nonnumerical characters, but other aspects of coordinates remain standard, such as listing latitude first.

When space and special characters are not limiting factors, a geographic coordinate should begin with the latitude, followed by a comma and then by the longitude. The abbreviation "lat" or "long" precedes degree, minute, and second, which are followed by the directional designation of "N" or "S" for latitude and "E" or "W" for longitude. Do not insert spaces between the numbers, symbols, and directional indicators, and use a leading zero for degrees, minutes, and seconds fewer than 10. For minutes, use the special character for the prime symbol (ie, Unicode 2032), not the single quotation mark or the apostrophe. For seconds, use the double prime symbol (ie, Unicode 2033), not the double quotation mark.

  lat 43°15′09″N, long 116°40′18″E   lat 04°59′17″S, long 01°02′03″W

When space is limited, one option is to eliminate the abbreviations "lat" and "long" and rely on the directional indicators "N" and "S" to imply latitude and "E" and "W" to imply longitude.

  43°15′09″N, 116°40′18″E   04°59′17″S, 01°02′03″W

Another option is to treat latitudes north of the equator and longitudes east of the prime meridian as positive and latitudes south of the equator and longitudes west of the prime meridian as negative. With this convention, the abbreviations "lat" and "long" and the directional indicators are omitted. The "S" and "W" designations at the end of coordinate units are replaced with negative symbols at the beginning of the units. While the "N" and "E" designations are similarly removed, they are not replaced with plus symbols.

  43°15′09″, 116°40′18″   −04°59′17″, −01°02′03″

Regardless of which option is chosen, it should be used consistently throughout the manuscript.

Authors should not insert hyphens or hard returns in coordinates to adjust line breaks in word-processed manuscripts. However, in typeset text, breaking coordinates at line endings is not always unavoidable. The best place to break a coordinate is right after the comma. If that is not possible, break the coordinate after a symbol in one of the coordinate units, and use a hyphen to indicate the break.

  lat 43°15′09″N, long 116°-   lat 04°59′-
  40′18″E   17″S, long 01°02′03″W

Latitude and longitude may be reported in "decimal degrees" instead of degrees, minutes, and seconds. Decimal degrees are easier to use in databases and spreadsheets. In this system, the degree value appears to the left of the decimal point, and the minutes and seconds are converted to a decimal value that appears to the right of the decimal point. To compute the decimal value, divide the minutes by 60, divide the seconds by 3600, and add the 2 resulting values. The degree, minute, and second symbols are omitted, and the minus symbol is added to the beginning of southern latitudes and western longitudes. Latitude and longitude can be changed from one system to another by using web-based conversion programs such as the one offered by the Federal Communications Commission[6] in the United States.

> lat 43°15′09″N, long 116°40′18″E *becomes* 43.2525, 116.6717
> lat 04°59′17″S, long 01°02′03″W *becomes* −04.9881, −01.0342

14.2.2 Other Coordinate Systems

Some coordinate systems are not based on latitude and longitude. However, all coordinate systems are based on known reference points. These reference points should be stated at the outset of a manuscript if they are not commonly known. For an overview of coordinate systems, see *Coordinate Systems and Map Projections*.[7]

14.2.2.1 Plane Coordinate Systems

Plane coordinate systems (also called "plane rectangular coordinate systems" and "plane cartesian coordinate systems") are calculated by making linear measurements from a pair of fixed axes, each of which goes in 2 directions from where the axes intersect. When grid lines drawn from points along the axes intersect at right angles, the coordinates identifying the intersection are referred to as "rectangular coordinates." These coordinates are commonly written as a pair of numbers separated by a comma, within parentheses: (x, y).

Coordinates on the x axis of the grid are positive to the right of the y axis and negative to the left of the y axis. Similarly, coordinates on the y axis are positive above the x axis and negative below the x axis.

Usually, the first number in a coordinate pair represents a point along the x, or east–west, axis, commonly called the "easting." The second number is a point along the y, or north–south, axis, usually called the "northing." Consequently, the coordinate pairs in a plane coordinate system are in reverse order to geographic coordinate pairs, in which north–south latitude is listed before east–west longitude.

Although ground measurements in rectangular coordinates are usually in meters, grid units are arbitrary. For example, the Universal Transverse Mercator coordinate system, which is a commonly used plane coordinate system, divides the world into 60 north–south zones of 6° of longitude each. In addition, in the plane coordinate systems used by US states, the origin is shifted so that most of the state or locality covered by each system lies to the right of the y axis and above the x axis. Consequently, most coordinates in these systems are positive.

14.2.2.2 Polar Coordinate Systems

Like rectangular coordinates, polar coordinates define each position with a pair of coordinates, but the first coordinate in the pair is a linear measurement (r) and the second is an angular measurement (θ). In addition, instead of using 2 axes, polar coordinate systems use a single axis, which is known as the "polar axis" or "initial line." This axis passes through the origin, or pole. The position of a point is defined as the distance of the radius from the pole to that point and the angle that radius makes with the polar axis.

Angles in polar coordinates are usually expressed in degrees or centesimal units (grad or grade units), although radians are required in some theoretical map projections. Which angles are positive in polar coordinates and which are negative differ among disciplines. In mathematics, angles measured in a counterclockwise direction from the polar axis are positive, but in surveying, cartography, and navigation, positive angles are those measured clockwise from the axis.

14.2.2.3 National Grid Systems

Many areas and locales within the United States can be designated by their location in the US Public Land Survey System. Established in 1785 to survey much of the land west of the Ohio River, this grid system uses a number of principal north–south meridians and east–west baselines as its reference points. The intersection of a meridian and a baseline is referred to as an "initial point." Running north and south of the baseline are 6-mile-square townships, each of which has a township number. In addition, each township is numbered east and west of the meridian with a range number. Townships are divided into 1-mile-square sections numbered from 1 to 36, beginning with the most northeastern section of each township. Sections, in turn, are divided into quarter sections.

The description of a location on a grid begins with its location within a quarter section, followed by the section number of the township, and ending with the township's relation to the grid's meridian (see Section 25.9, "Tracts of Land").

> N1/2 SW1/4 sec 7, T 5 S, R 3 E, of the Boise Meridian
> (designates the north half of the southwest quarter of section 7 in a township that is 5 townships south and 3 ranges east of the Boise Meridian)

The US Geological Survey uses symbols on these grids to identify such features as boundaries, contours, bodies of water, vegetation, railroad lines, streets, and buildings.

Some maps of the British Ordnance Survey carry grid-square alphanumeric identifiers based on the British National Grid. With these identifiers, a location in Great Britain can be specified with increasing precision as the number of letters and digits in the identifier increases.

The first letter in an identifier in the British National Grid designates a grid square of 500 km^2. The second letter designates a 100-km^2 section within the 500-km^2 square. The letters are followed by 4, 6, 8, or 10 numbers that narrow down the location within the 100-km^2 grid square, using the southwest corner of each successively smaller square for easting and northing. If an identifier has 6 digits, the first and fourth digits indicate the location of a 10-km^2 section in the 100-km^2 section, the second and fifth digits indicate a 1-km^2 section, and the third and sixth digits designate a 100-m^2 section. For example, the grid reference SE 603 522 specifies the location of the Cathedral and Metropolitical Church of St Peter in York. Commonly known as York Minster, the cathedral is in the 100-km^2 section "SE," which covers much of Yorkshire, United Kingdom. Each set of digits in the grid reference can be used to drill down on the cathedral's location within the SE grid square. The first digit, 6, and the fourth digit, 5, indicate that York Minster is located in the 10-km^2 section identified as SE 65, which is 6 squares to the right and 5 squares above the 10-km^2 section in the southwest corner of the SE grid square. Within SE 65, the cathedral is located in the 1-km^2 section designated as SE 60 52, which is 2 grid squares above the 1-km^2 section in the southwest corner of SE 65. Finally, York Minster's 100-m^2 location is 3 grid squares to the right and 2 squares above the 100-m^2 square in the southwest corner of SE 60 52.

14.3 ADDRESSES

14.3.1 For Use in Manuscripts

When mailing addresses are provided in running text or figures, ensure that each address includes the following information:

(1) the contact's full name, including abbreviations of academic degrees and other credentials
(2) the contact's department and institution
(3) the institution's street address
(4) the municipality
(5) the state, province, or other regional unit, if applicable (for an address in the United States, Canada, or Australia, see the postal abbreviations for states and provinces in Table 14.1)
(6) the postal code, called "ZIP Code" in the United States (depending on the country, the postal code may follow the regional unit, precede the municipality, or fall between the municipality and regional unit)
(7) the country with the entire name spelled out (except for the United States of America, which can be abbreviated as "USA" and which can be omitted if the publication's primary audience is from the United States)

If in doubt about the proper form of a country name, use the spelling specified by the national postal agency of origin of the publication. Many national postal agencies—such as the US Postal Service (https://pe.usps.com/), Canada Post (https://www.canadapost -postescanada.ca), and the United Kingdom's Royal Mail (https://www.royalmail.com)— provide detailed guidance on addressing mail for domestic and international delivery.

14.3.2 For Corresponding Authors

Scientific journals usually ask authors to designate one author to receive correspondence on behalf of all the authors of an article. For this purpose, journals typically publish only the corresponding author's email address, which appears on the article's first or last page. For journals that also require the corresponding author's mailing address, follow the guidelines in Section 14.3.1, "For Use in Manuscripts."

CITED REFERENCES

1. United Nations Group of Experts on Geographical Names. United Nations Statistics Division; 2018. https://unstats.un.org/unsd/geoinfo/ungegn/geonames.html

2. The Chicago manual of style: the essential guide for writers, editors, and publishers. 17th ed. The University of Chicago Press; 2017. Also available at https://www.chicagomanualofstyle.org

3. Geographic Names Information System (GNIS). US Geological Survey; [accessed 2023 Jan 19]. https://edits.nationalmap.gov/apps/gaz-domestic/public/search/names

4. Search the Canadian Geographical Names Database (CGNDB). Natural Resources Canada; [accessed 2023 Jul 30]. https://geonames.nrcan.gc.ca/search-place-names/search

5. Geographic Names Server. National Geospatial-Intelligence Agency; [accessed 2023 Jan 19]. https://geonames.nga.mil/geonames/GeographicNamesSearch/

6. Degrees minutes seconds to/from decimal degrees. Federal Communications Commission; 2016. https://www.fcc.gov/media/radio/dms-decimal

7. Maling DH. Coordinate systems and map projections. 2nd ed. Pergamon Press Ltd; 1992.

ADDITIONAL REFERENCES

Cohen SB, editor. The Columbia gazetteer of the world. 2nd ed. Columbia University Press; 2008. Also available at http://www.columbiagazetteer.org/main/Home.html

Hansen WR, editor. Suggestions to authors of the reports of the United States Geological Survey. 7th ed. United States Geological Survey; 1991. https://doi.org/10.3133/7000088

National Geographic atlas of the world. 11th ed. National Geographic Society; 2019.

US Government Publishing Office. Style manual: an official guide to the form and style of federal government publishing. US Government Publishing Office; 2016. https://www.govinfo.gov/collection/gpo-style-manual

3

Special Scientific Conventions

15 The Electromagnetic Spectrum

Editor: Mary Warner, CAE

15.1 UNITS OF MEASURE

The electromagnetic spectrum is a continuum of radiated electromagnetic energy characterized by the basic relation $c = \lambda v$, in which "c" is the speed of light (approximately 3×10^8 m/s), "λ" is the wavelength of the radiation in meters (m), and "v" is the frequency of the radiation in hertz (Hz). Along this continuum, frequency increases from Hz to exahertz (EHz), while wavelength shortens from megameters (Mm) to femtometers (fm). The practical bounds of the electromagnetic continuum are from extremely low-frequency radio waves to high-frequency gamma rays produced by primary cosmic rays. Given that the spectrum is a continuum, the boundaries between segments of the spectrum are not exact. Instead, each segment overlaps neighboring segments (see Table 15.1).

"Electromagnetic waves" and "quantized photons" are equivalent concepts because of wave–particle duality. The energy of a photon is given by $E = hv$, in which "h" is Planck's constant. As radiation frequency increases in hertz, the energy of the radiation quanta (ie, discrete units) increases.

Dimension sets vary by scientific discipline, so publications should use the units that are standard for their disciplines.

Radio waves are usually characterized by frequency in hertz (Hz) or by wavelength in meters (m), centimeters (cm), or millimeters (mm). Visible, infrared, and ultraviolet (UV) light is often described by its wavelength in nanometers (nm) or micrometers (μm). Historically, wavelengths were measured in angstroms (Å), but that unit is no longer commonly used.

Infrared spectroscopists often use the inverse of the wavelength ($1/\lambda$), known as the "wave number" ($\bar{v}$). Wave numbers are expressed in units of cm^{-1}.

Any wavelength can be expressed as photon energy, for which the electronvolt (eV) is the base unit. Photon energy is most often used in measuring X-rays and gamma rays, which generate thousands of electronvolts (keV) to millions of electronvolts (MeV). However, at times, spectroscopists measure the photon energy of millimeter-wave, infrared,

Table 15.1. Electromagnetic Spectrum

Region	Frequency range		Wavelength		
	Conventional units	Hertz	Conventional units	Meters	Other common units
Radio waves					
Marine communications					
Extremely low frequency (ELF)	3 Hz to 30 Hz	3×10^0 to 3×10^1	100 Mm to 10 Mm	10^8 to 10^7	
Super low frequency (SLF)	30 Hz to 300 Hz	3×10^1 to 3×10^2	10 Mm to 1 Mm	10^7 to 10^6	
Ultra low frequency (ULF)	300 Hz to 3 kHz	3×10^2 to 3×10^3	1 Mm to 100 km	10^6 to 10^5	
Very low frequency (VLF)	3 kHz to 30 kHz	3×10^3 to 3×10^4	100 km to 10 km	10^5 to 10^4	
Broadcasting and other communications					
Low frequency (LF)	30 kHz to 300 kHz	3×10^4 to 3×10^5	10 km to 1 km	10^4 to 10^3	
Medium frequency (MF)—used for amplitude modulation (AM) radio broadcasts	300 kHz to 3 MHz	3×10^5 to 3×10^6	1 km to 100 m	10^3 to 10^2	
High frequency (HF)—used for shortwave radio	3 MHz to 30 MHz	3×10^6 to 3×10^7	100 m to 10 m	10^2 to 10^1	
Very high frequency (VHF)—used for frequency modulation (FM) radio and television broadcasts	30 MHz to 300 MHz	3×10^7 to 3×10^8	10 m to 1 m	10^1 to 10^0	
Ultra high frequency (UHF)—used for television broadcasts	300 MHz to 3 GHz	3×10^8 to 3×10^9	1 m to 100 mm	10^0 to 10^{-1}	

Microwaves

	Frequency	Frequency (Hz)	Wavelength	Wavelength (m)	
Communication and radar					
Super high frequency (SHF)	3 GHz to 30 GHz	3×10^9 to 3×10^{10}	100 mm to 10 mm	10^{-1} to 10^{-2}	
Extremely high frequency (EHF)	30 GHz to 300 GHz	3×10^{10} to 3×10^{11}	10 mm to 1 mm	10^{-2} to 10^{-3}	
Submillimeter wavelengths (micrometer waves or THz waves)					
Tremendously high frequency (THF)	300 GHz to 3 THz	3×10^{11} to 0.003×10^{11}	1 mm to 0.1 mm	10^{-3} to 0.010^{-3}	
Optical wavelengths					
Infrared	300 GHz to 400 THz	3×10^{11} to 4×10^{14}	1 mm to 760 nm	10^{-3} to 7.6×10^{-7}	10,000 cm^{-1} to 10 cm^{-1} (wave number)
Visible light	400 THz to 750 THz	4×10^{14} to 7.5×10^{14}	760 nm to 400 nm	7.6×10^{-7} to 4×10^{-7}	
Ultraviolet	750 THz to 3 PHz	7.5×10^{14} to 3×10^{15}	400 nm to 100 nm	4×10^{-7} to 10^{-7}	
Short wavelengths					
X-ray	30 PHz to 6 EHz	3×10^{16} to 6×10^{18}	10 nm to 50 pm	10^{-8} to 5×10^{-11}	1 keV to 100 keV 0.1 Å to 100 Å
Gamma ray	6 EHz to 600 EHz	6×10^{18} to 6×10^{20}	50 pm to 500 fm	5×10^{-11} to 5×10^{-13}	100 keV to 10 MeV

Hz, hertz; kHz, kilohertz; MHz, megahertz; GHz, gigahertz; THz, terahertz; PHz, petahertz; EHz, exahertz; Mm, megameter; km, kilometer; m, meter; mm, millimeter; nm, nanometer; pm, picometer; fm, femtometer; cm, centimeter; keV, kiloelectronvolt; MeV, megaelectronvolt; Å, angstrom visible, and UV radiation, which ranges from millielectronvolts (meV) for millimeter-wave and most infrared wavelengths to eV for the shortest infrared to UV wavelengths.

15.2 LONG WAVELENGTHS

At the low-energy end of the electromagnetic spectrum, where wavelengths are the longest, wavelengths can be as long as 10^8 m and frequencies as low as 3 Hz. Frequencies between 3 Hz and 3,000 GHz are defined as "radio bands" by the United Nations' International Telecommunications Union (ITU) (see Table 15.1). Among these bands are the ITU's "ultra high frequency" and "super high frequency" bands, which are often collectively referred to as "microwaves." Also included are the ITU's "extremely high frequency" and "tremendously high frequency" bands, which are often referred to as the "millimeter band" and the "submillimeter band," respectively. These radio bands are used for communications, electronics, and instrumentation.

The submillimeter band from 3×10^{12} Hz to 3×10^{13} Hz (3 THz to 30 THz) merges into the longest wavelengths of the infrared region of the electromagnetic spectrum (see Section 15.3, "Optical Wavelengths"). Researchers working in the submillimeter band often blend conventions and terminologies from microwave and optical spectroscopy.

15.3 OPTICAL WAVELENGTHS

Optical radiation consists of the infrared (IR), visible, and ultraviolet (UV) regions of the electromagnetic spectrum. Optical radiation is more often described by wavelength than by frequency.

15.3.1 Infrared Wavelengths

IR radiation has wavelengths as long as 1 mm and as short as 760 nm with frequencies between 3×10^{11} Hz and 4×10^{14} Hz. The International Commission on Illumination (CIE) and the International Organization for Standardization (ISO) have different schemes for dividing the IR spectrum.

The CIE divides the IR spectrum into 3 segments, labeled "C," "B," and "A." In standard CIE abbreviations for infrared, "IR" is followed by a hyphen and the letter designating the appropriate segment. The following are the CIE's approximate boundaries for IR segments:

IR-C	1 mm to 3 µm
IR-B	3 µm to 1.4 µm
IR-A	1.4 µm to 760 nm

In contrast, the ISO divides the IR spectrum into 3 bands labeled "far infrared" (FIR), "mid infrared" (MIR), and "near infrared" (NIR). Although not part of the terminology used in the ISO standard, these bands are often called "LWIR" for FIR, "MWIR" for MIR, and "SWIR" for NIR. In these alternate designations, the initial letters stand for "long-wave," "midwave," and "shortwave," respectively.

Designation	Abbreviation	Wavelengths (µm)
Far infrared	FIR	1,000 to 50
Mid infrared	MIR	50 to 3
Near infrared	NIR	3 to 0.78

Conventional IR spectroscopy of vibrational modes in molecules is performed with light in the wave number range of 4,000 cm^{-1} to 200 cm^{-1}, which corresponds to wavelengths from 50 µm to 2.5 µm. This range essentially overlaps the mid-infrared band.

A subset of MIR from 15 µm to 8 µm is known as the "thermal IR" band. This band roughly corresponds to the black-body emission curve, the characteristic spectrum of electromagnetic radiation emitted by all objects at a temperature higher than absolute zero (0 K). The black-body curve for each object has a peak, which is defined as the wavelength with the strongest emission. The higher the absolute temperature of the emitting object, the shorter the wavelength at which the black-body emission curve peaks. For temperatures on Earth, the black-body curve peaks at approximately 10 µm.

15.3.2 Visible Light

Visible light occupies the segment of the electromagnetic spectrum with wavelengths of approximately 760 nm to 400 nm. The human eye perceives wavelengths between approximately 700 nm and 400 nm, which constitute the familiar spectrum of colors from red to violet.

The colors of the visible spectrum overlap, so boundaries for color ranges often vary slightly among scientific disciplines. The approximate ranges of color wavelengths are as follows:

Color	Wavelengths (nm)
Red	700 to 630
Orange	630 to 590
Yellow	590 to 530
Green	530 to 480
Blue	480 to 440
Violet	440 to 400

15.3.3 Ultraviolet Wavelengths

UV radiation occupies the segment of the electromagnetic spectrum between 400 nm and 100 nm. UV radiation is traditionally divided into the following 4 ranges:

Range	Wavelengths (nm)
Near UV	400 to 300
Middle UV	300 to 200
Far UV	200 to 100
Extreme UV	<100

A commonly used alternative division uses 3 ranges:

Name	Abbreviation	Wavelengths (nm)	Photon energy (eV)
Ultraviolet A	UV-A	400 to 315	3.10 to 3.94
Ultraviolet B	UV-B	315 to 280	3.94 to 4.43
Ultraviolet C	UV-C	280 to 100	4.43 to 12.4

The short-wavelength limit of extreme UV is cited in various sources as between 40nm and 10 nm. Because wavelengths shorter than 200 nm are absorbed in air, studies of far and extreme UV radiation are conducted in another gas or in a vacuum, resulting in the alternative name "vacuum ultraviolet." The region from 185 nm to 120 nm is also called the "Schumann region," after the first known investigator of this region, the physicist and spectroscopist Victor Schumann.

Avoid using the initialisms "UVR" and "UVL." Instead, use "UV radiation" and "UV light," respectively.

15.4 SHORTER WAVELENGTHS

Beyond the ultraviolet portion of the electromagnetic spectrum are X-rays, gamma rays, and cosmic rays. Hyphenate the term "X-ray," and capitalize the "X." Do not hyphenate "gamma ray" or "γ ray" except when either version of the term is used as an adjective.

In these portions of the spectrum, it is common to use photon energy rather than wavelength or frequency to describe electromagnetic radiation.

X-rays occupy the spectrum segment from approximately 3×10^{16} Hz (30 petahertz [PHz]) with an approximate wavelength of 10 nm to approximately 6×10^{18} Hz (6 exahertz [EHz]) with an approximate wavelength of 50 picometers (pm). X-rays with frequencies lower than 3×10^{18} Hz (3 EHz) are known as "soft" X-rays. Those with higher frequencies are "hard" X-rays. Unlike electromagnetic radiation with slower frequencies and longer wavelengths, X-rays have enough energy to cause ionization when passing through matter (ie, "ionizing radiation").

Gamma rays occupy the next segment of the electromagnetic spectrum. "Soft" gamma rays share an ambiguous boundary with hard X-rays, starting near 6×10^{18} Hz with a wavelength of 50 pm and energy of 25 kiloelectronvolts (keV). "Soft" gamma rays extend to approximately 6×10^{19} Hz with a wavelength of 5 pm and energy of 250 keV. "Hard" gamma rays extend to frequencies of approximately 6×10^{20} Hz (600 EHz) with a wavelength of 500 femtometers (fm) and energy of 2.5 million keV, which is roughly the lower limit of energy for the gamma rays produced from atmospheric collisions of primary cosmic rays.

REFERENCES

National Institute of Standards and Technology (US). NIST reference on constants, units, and uncertainty. [accessed 2023 Feb 17]. http://physics.nist.gov/cuu/Constants/

Haynes WM, editor. CRC handbook of chemistry and physics. 93rd ed. CRC Press; 2012.

16 Subatomic Particles, Chemical Elements, and Related Notations

Editor: Mary Warner, CAE

16.1 SOURCES OF RECOMMENDATIONS

The authorities for nomenclature, symbolization, and other notations for subatomic particles and chemical elements are the International Union of Pure and Applied Physics and the International Union of Pure and Applied Chemistry,[1-4] whose websites can be accessed at https://iupap.org and https://iupac.org, respectively.

16.2 ELEMENTARY AND COMPOSITE PARTICLES

The smallest known units of matter are the elementary particles called leptons, quarks, hadrons, and gauge bosons (see Table 16.1). Leptons and quarks have corresponding antileptons and antiquarks, which have the same mass as the leptons and quarks but opposite quantum numbers. Quarks and antiquarks may combine to create protons and neutrons. Mesons and baryons, collectively known as hadrons, are also composed of quarks and antiquarks, but because they cannot be separated into their constituents, they are considered elementary particles. Gauge bosons (ie, photons, W and Z bosons, gluons, and gravitons) mediate interactions among elementary particles, including electrons. Consequently, gauge bosons act as force carriers.

Elementary and composite particles are represented by Greek and Roman letters, some with superscript or subscript modifiers. The typical symbols for elementary particles can be found in Table 16.1. Common convention italicizes the symbols for particles, except for the symbols' nonalphabetic superscripts and subscripts. This convention allows particle symbols to be distinguished readily in text.

Quarks are known by their names (eg, "up," "down," and "strange") or by their symbols (eg, "u," "d," and "s") (see Table 16.1).

Table 16.1. Elementary Particles: Examples of Symbols

Particles	Symbols[a]	Charges
Gauge bosons	γ, Z, and g	0
	W	+1 and −1
Leptons		
Electron neutrino	v_e	0
Electron	e	−1
Muon neutrino	v_μ	0
Muon	μ	−1
Tau neutrino	v_τ	0
Tau *also* tauon	T	−1
Quarks[b]		
Down	d	−⅓
Up	u	+⅔
Strange	s	−⅓
Charm	c	+⅔
Bottom *also* beauty	b	−⅓
Top *also* truth	t	+⅔
Hadrons (particles composed of quarks and antiquarks)		
Mesons (containing 1 quark and 1 antiquark)		
Containing only u and d particles	π^0, π^+, and π^-	0, +1, and −1
Containing 1 s particle[c]	K^+, K^0, K^-, and $\bar{K}^0$	+1, 0, −1, and 0
Containing s and $\bar{s}$	Φ	0
Containing 1 c particle[c]	D^+, D^0, and D_s^+	+1, 0, and +1
Containing c and $\bar{c}$	Ψ	0
Containing 1 b particle[c]	B^+, B^0, B^-, $\bar{B}^0$, B_s, and B_c	+1, 0, −1, 0, 0, and +1
Containing b and $\bar{b}$	Y	0
Baryons (containing 3 quarks)	p, n, and Λ_c^+	+1, 0, and +1
	Λ, Σ^+, Σ^0, and Σ^-	0, +1, 0, and −1
	Ξ^0, Ξ^-, Ξ_c^0, and Ξ_c^+	0, −1, 0, and +1
	Ω^-	−1

[a] The superscript plus and minus symbols and the superscript zero character indicate charge. The subscripts "s" and "c" indicate strange and charm quarks. Distinct antiparticles are indicated by an overbar, or overline.

[b] Antiquark symbols are the same as the quark symbols except that they have an overbar, or overline. The charge of an antiquark has the opposite sign to that of the corresponding quark.

[c] In combination with 1 other particle.

16.3 NUCLEAR REACTIONS

The symbols for the particles that are projectiles or products in nuclear reactions are set as Greek and Roman letters (see Table 16.2).

Indicate the electric charge of such a particle by adding a superscript with the plus or minus symbol or the zero character (eg, π^0, π^+, π^-, and e^-).

The symbol "p" is used without an indication of charge to refer to the positive proton. The symbol "e$^-$" in roman type is used to indicate the electron particle, while an

Table 16.2. Projectiles and Products in Natural and Artificial Nuclear Reactions

Name	Nuclear symbol	Chemical symbol
Gamma photon	γ	
Neutrino	ν, ν_e, ν_μ, and ν_τ	
Electron	e^- also β^-	
Positron	e^+ also β^+	
Tau *also* tauon	T	
Pion	Π	
Nucleon[a]	N	
Neutron	n	
Muon	μ^+	Mu^+
Muonium	μ^+e^-	$Mu^{\cdot}$
Muonide	$\mu^+(e^-)_2$	Mu^-
Proton	p^+	$^1H+$
Protium	p^+e^-	$^1H^{\cdot}$
Protide	$p^+(e^-)_2$	$^1H^-$
Deuteron	d^+	$^2H^+$ (D^+)
Deuterium	d^+e^-	$^2H^{\cdot}$ ($D^{\cdot}$)
Deuteride	$d^+(e^-)_2$	$^2H^-$ (D^-)
Triton	t^+	$^3H^+$ (T^+)
Tritium	t^+e^-	$^3H^{\cdot}$ ($T^{\cdot}$)
Tritide	$t^+(e^-)_2$	$^3H^-$ (T^-)
Hydron		H^+
Hydrogen		$H^{\cdot}$
Hydride		H^-
Helion	h	$^3He^{2+}$
Alpha particle	a	$^4He^{2+}$

[a] Often used either with mass within parentheses after the symbol for the resonances—for example, $N(1440)$—or with other particle symbols to indicate the products of a reaction—for example, $\pi\pi N$.

italicized *e* with a superscript is used to indicate the numerical value of the electron's charge.

Antiparticles are indicated by adding a bar above the symbol for the corresponding particle (eg, $\bar{\mu}$ and $\bar{c}$). The symbol "e⁺" is preferred for the positron, the equivalent of the electron's antiparticle. For the antiproton, the symbol is "$\bar{p}$."

An electron (e⁻) and a positron (e⁺) occurring in a bound state are collectively known as "positronium" (Ps). A hydrogen-like atom in which the proton is replaced by a positive muon (ie, a μ^+ with an e⁻ bound to it) is known as "muonium" and is given the notation "Mu·" ("Mu" with a superscript dot).

The notation for a nuclear reaction uses the following sequence of symbols with no spaces between the symbols: the initial nuclide (^{14}N in the example below), an open parenthesis, the incoming particle or photon (*a* in the example), a comma, the outgoing particle(s) or photon(s) (*p* in the example), a close parenthesis, and the final nuclide (^{17}O in the example).

$$^{14}N(a,p)^{17}O$$

Table 16.3. Atomic Numbers, Names, and Symbols of the Chemical Elements

Atomic number	Name[a]	Symbol	Atomic number	Name[a]	Symbol
1	hydrogen	H	44	ruthenium	Ru
2	helium	He	45	rhodium	Rh
3	lithium	Li	46	palladium	Pd
4	beryllium	Be	47	silver (argentum)	Ag
5	boron	B	48	cadmium	Cd
6	carbon	C	49	indium	In
7	nitrogen	N	50	tin (stannum)	Sn
8	oxygen	O	51	antimony (stibium)	Sb
9	fluorine	F	52	tellurium	Te
10	neon	Ne	53	iodine	I
11	sodium (natrium)	Na	54	xenon	Xe
12	magnesium	Mg	55	cesium[b] *also* caesium[c]	Cs
13	aluminum[b] *also* aluminium[c]	Al	56	barium	Ba
14	silicon	Si	57	lanthanum	La
15	phosphorus	P	58	cerium	Ce
16	sulfur[b] *also* sulphur[c]	S	59	praseodymium	Pr
17	chlorine	Cl	60	neodymium	Nd
18	argon	Ar	61	promethium	Pm
19	potassium (kalium)	K	62	samarium	Sm
20	calcium	Ca	63	europium	Eu
21	scandium	Sc	64	gadolinium	Gd
22	titanium	Ti	65	terbium	Tb
23	vanadium	V	66	dysprosium	Dy
24	chromium	Cr	67	holmium	Ho
25	manganese	Mn	68	erbium	Er
26	iron (ferrum)	Fe	69	thulium	Tm
27	cobalt	Co	70	ytterbium	Yb
28	nickel	Ni	71	lutetium	Lu
29	copper (cuprium)	Cu	72	hafnium	Hf
30	zinc	Zn	73	tantalum	Ta
31	gallium	Ga	74	tungsten (wolfram)	W
32	germanium	Ge	75	rhenium	Re
33	arsenic	As	76	osmium	Os
34	selenium	Se	77	iridium	Ir
35	bromine	Br	78	platinum	Pt
36	krypton	Kr	79	gold (aurum)	Au
37	rubidium	Rb	80	mercury (hydrargyrum)	Hg
38	strontium	Sr	81	thallium	Tl
39	yttrium	Y	82	lead (plumbum)	Pb
40	zirconium	Zr	83	bismuth	Bi
41	niobium	Nb	84	polonium	Po
42	molybdenum	Mo	85	astatine	At
43	technetium	Tc			

Table 16.3. Atomic Numbers, Names, and Symbols of the Chemical Elements (*continued*)

Atomic number	Name[a]	Symbol	Atomic number	Name[a]	Symbol
86	radon	Rn	103	lawrencium	Lr
87	francium	Fr	104	rutherfordium[d]	Rf
88	radium	Ra	105	dubnium[e]	Db
89	actinium	Ac	106	seaborgium	Sg
90	thorium	Th	107	bohrium	Bh
91	protactinium	Pa	108	hassium	Hs
92	uranium	U	109	meitnerium	Mt
93	neptunium	Np	110	darmstadtium	Ds
94	plutonium	Pu	111	roentgenium	Rg
95	americium	Am	112	copernicium	Cn
96	curium	Cm	113	nihonium	Nh
97	berkelium	Bk	114	flerovium	Fl
98	californium	Cf	115	moscovium	Mc
99	einsteinium	Es	116	livermorium	Lv
100	fermium	Fm	117	tennessine	Ts
101	mendelevium	Md	118	oganesson	Og
102	nobelium	No			

[a] Names in parentheses are Latin or other names on which the symbol is based.
[b] American spelling.
[c] British spelling.
[d] Also known as kurchatovium (Ku) before approval of the official name in 1997.
[e] Also known as hahnium (Ha) and nielsbohrium (Ns) before approval of the official name in 1997.

16.4 CHEMICAL ELEMENTS

The symbols for most chemical elements are derived from their English, Latin, Greek, or systematic names. These symbols consist of 1, 2, or 3 letters (eg, "H" for hydrogen; "K" for kalium, now called potassium; "Na" for natrium, now called sodium; "Ca" for calcium; and "Uue" for Ununennium, the numerical, or systematic, name for the yet-to-be-discovered element 119). See Table 16.3 for the approved symbols and names of the chemical elements through atomic number 118. Other useful resources are the Royal Society of Chemistry's interactive periodic table and the American Chemical Society's periodic table.

The symbols for elements should be printed in roman type, except when used in text or headings that are italicized. The initial letter of each symbol should be capitalized. Do not capitalize the spelled-out names of elements except at the beginning of a sentence, in titles and other headings that use headline-style capitalization, and for other appropriate reasons.

Interim systematic names are assigned to elements that have not yet been named or discovered. These elements also have 3-letter symbols based on the first 3 syllables of their interim names. Element 118, for example, used the systematic name "ununoctium" and the symbol "Uuo" until the name "oganesson" and the symbol "Og" were adopted in 2016.

Systematic names are currently used for undiscovered elements with atomic numbers 119 and above. After elements are discovered, they may be referred to in scientific literature by their interim systematic names for several yeas until they are assigned their official permanent names. For example, although tennessine was discovered in 2010, it was referred to as "ununseptium" until its official name was adopted in 2016. The following roots are used for forming the first 3 syllables of interim systematic names:

Numeral	Syllable
1	un
2	bi
3	tri
4	quad
5	pent
6	hex
7	sept
8	oct
9	enn
0	nil

These roots replace each of the corresponding 3 digits in theatomic numbers of undiscovered elements. The syllable "ium" is then added to the resulting combination of roots. The symbols of undiscovered elements, in turn, are formed by joining the first letters of the 3 roots and capitalizing the first letter. For example, the interim systematic name for element 119 is formed as follows: un [1] + un [1] + enn [9] + ium = ununennium. The symbol, in turn, is "Uue." Similarly, should element 274 ever be discovered, its interim systematic name will be "biseptquadium," formed by bi [2] + sept [7] + quad [4] + ium. Its symbol will be "Bsq."

In addition to consisting of 1 to 3 letters, the symbols for elements can be modified with other symbols that indicate atomic number, mass number, charge or oxidation number, numbers of atoms per molecule, and other information. Furthermore, an asterisk is sometimes used to denote an excited state, and a dot or bullet is used to denote a radical.

In the example below, "E" represents the element symbol, and the letters "a" through "d" indicate the positions of the modifying numbers defined in the key below the example. Few elements use all 4 of the modifying numbers.

$$^{b}_{a}E^{c}_{d}$$

a = atomic number (proton number)
b = mass number (nucleon number or baryon number)
c = charge number, oxidation number, or other information
d = the number of atoms of this element in a molecular formula

When the left superscript for mass number ("b" in the example) is omitted, the symbol is interpreted as including all isotopes in natural abundance.

The right superscript ("c" in the example) denotes the ionic charge with the net number of positive or negative charges followed by the appropriate charge sign. When the net charge is 1, only the charge sign is used. In addition, regardless of the net number of positive or negative charges, multiple plus or minus symbols are not used in the right superscript.

Na^{+} *not* Na^{1+} Al^{3+} *not* Al^{+++} S^{2-} *not* S^{--}

The right superscript also can be used for oxidation numbers, which are indicated by positive or negative roman numerals or the arabic zero.

Mn^{VII} O^{-II} Pt^{0}

A nuclide is an atom of specified atomic number (ie, proton number) and mass number (ie, nucleon number). Isotopic nuclides, or isotopes, are 2 or more nuclides with the same atomic number but different mass numbers. For example, the 3 nuclides of carbon all have atomic number 6 (ie, $_{6}C$), but they have different mass numbers: ^{12}C, ^{13}C, and ^{14}C. In contrast, isobaric nuclides, or isobars, are nuclides that have the same mass number but different atomic numbers. For example, ^{14}C and ^{14}N are isobars with mass number 14, although their atomic numbers are 6 and 7, respectively.

16.5 CHEMICAL FORMULAS

Chemical formulas represent entities composed of more than one atom (eg, molecules, complex ions, and groups of atoms).

N_2 *for the molecule dinitrogen*
$CaSO_4$ *for the compound calcium sulfate*
CH_3OH *for the compound methanol*
$PtCl_4^{2-}$ *for the dianion in potassium tetrachloroplatinate(II)*

In running text, consecutive sequencing of subscript and superscript designations, as in $PtCl_4^{2-}$ above, is preferable to stacking (eg, placing the superscript $^{2-}$ directly over the subscript $_4$). In consecutive sequencing, the subscript designation should precede the superscript designation. Reversing the order of the subscript and superscript would change the notation's meaning. For example, the notation "I_3^{-}" indicates that the entity I_3 bears a single negative charge, whereas I_3^{-} indicates an unlikely molecule consisting of 3 I^{-} entities.

To prevent ambiguity, parentheses, square brackets, and braces are needed to identify many complex groups. The specific enclosures are often mandated by the type of compound, complex, or group. Those recommendations are detailed by the International Union of Pure and Applied Chemistry (see also Section 17.2.3, "Structural Formulas").

For more detailed information about chemical formulas, see Chapter 17, "Chemical Formulas and Names."

16.6 SYMBOLS FOR ELECTRONIC CONFIGURATIONS OF ATOMS

The electronic configuration of an atom is indicated by the number of electrons in each of the atom's orbital subsets, or suborbitals, which are mathematical estimates of electron locations. The orbital located closest to an atom's nucleus has one suborbital. The farther orbitals are located from the nucleus, the more suborbitals they have, until they reach a maximum. Each suborbital, in turn, accommodates a maximum number of electrons. The orbital subsets are designated by the following letters, which follow the same order in each orbital.

s, which can have up to 2 electrons
p, which can have up to 6 electrons
d, which can have up to 10 electrons
f, which can have up to 14 electrons

Each orbital, in turn, is numbered with arabic numerals. The currently known elements have between 1 and 7 orbitals. Each suborbital is labeled with the number of its orbital followed by the letter of the suborbital, and the number of electrons in the suborbital is indicated by an arabic numeral in superscript (eg, "$5f^{14}$" is the last of 4 suborbitals in the 5th orbital, and it has 14 electrons).

To indicate the electronic configuration of an atom, each suborbital is listed generally from closest to farthest from the nucleus. Each subset is placed in parentheses, and the superscript with the number of electrons in the subset is placed after the closing parenthesis.

boron	$(1s)^2(2s)^2(2p)^1$	($1s^2 2s^2 2p^1$ is acceptable)
carbon	$(1s)^2(2s)^2(2p)^2$	($1s^2 2s^2 2p^2$ is acceptable)
nitrogen	$(1s)^2(2s)^2(2p)^3$	($1s^2 2s^2 2p^3$ is acceptable)

Once the 3rd orbital is reached, the 3d and 3f suborbitals need not be full before the 4s suborbital is populated. Similarly, once the 6th orbital is reached, the suborbitals in the 4th through 7th orbits may intermix.

manganese $(1s)^2(2s)^2(2p)^6(3s)^2(3p)^6(3d)^5(4s)^2$ (3d is not full, and 3f is empty)

lutetium $(1s)^2(2s)^2(2p)^6(3s)^2(3p)^6(3d)^{10}(4s)^2(4p)^6(4d)^{10}(5s)^2(5p)^6(4f)^{14}(5d)^1(6s)^2$
(4f follows 5p, 5d is not full but precedes 6s, and 5f is empty)

16.7 SYMBOLS FOR MOLECULAR STATES

The electronic states of molecules are labeled with the symmetry species of the wave function in the molecular point group. Greek or Roman nonitalic capital letters are used for these labels.

The electronic configurations of orbitals are labeled with lowercase letters in a manner analogous to that for atoms.

ground state of OH $(1\sigma)^2(2\sigma)^2(3\sigma)^2(1\pi)^3$

CITED REFERENCES

1. International Union of Pure and Applied Chemistry, Physical Chemistry Division. Quantities, units, and symbols in physical chemistry. 3rd ed. RSC Publishing; 2007. Prepared for publication by Cohen ER et al. Also known as IUPAC Green Book. https://doi.org/10.1039/9781847557889

2. International Union of Pure and Applied Chemistry, Clinical Chemistry Division. Compendium of terminology and nomenclature of properties in clinical laboratory sciences, recommendations 2016. RSC Publishing; 2017. Prepared for publication by Férard G, Dybkaer R, Fuentes-Arderiu X. Also known as IUPAC Silver Book. https://doi.org/10.1039/9781782622451

3. International Union of Pure and Applied Chemistry. Compendium of chemical terminology. 2nd ed. ICT Publishing House; 1997. Prepared for publication by McNaught AD, Wilkinson A. Updated 2019 July 1 by Nic M, Jirat J, Kosata B. Also known as IUPAC Gold Book. https://goldbook.iupac.org/

4. Hibbert DB, editor. Compendium of terminology in analytical chemistry. 4th ed.Royal Society of

Chemistry, 2023. Prepared for International Union of Pure and Applied Chemistry. Also known as IUPAC Orange Book. Also available at https://doi.org/10.1039/9781788012881

ADDITIONAL REFERENCES

Banik GM, Baysinger G, Kamat P, Pient N. The ACS guide to scholarly communication. American Chemical Society; 2020. https://doi.org/10.1021/acsguide
International Union of Pure and Applied Chemistry. IUPAC periodic table of elements. International Union of Pure and Applied Chemistry; 2022. Updates of the table are available at https://iupac.org/what-we-do/periodic-table-of-elements/
Rumble J, editor. CRC handbook of chemistry and physics. 102nd ed. CRC Press; 2021.

17 Chemical Formulas and Names

Editor: Mary Warner, CAE

17.1 RESOURCES

Chemicals can be represented in numerous ways in scientific literature. Although it is typical for a chemical to have multiple names and multiple formulas, not all of the nomenclature and formula systems described in Section 17.2, "Formulas," through Section 17.8, "Other Nomenclature Systems," apply to all chemicals. To determine which systems are appropriate for a particular chemical, consult the sections in this chapter that are relevant to the specific type of chemical. Those sections begin with Section 17.9, "Radicals and Ions," and end with Section 17.13, "Compounds of Biochemical Importance."

Nomenclature rules, recommendations, and publication styles for inorganic, organic, and biochemical compounds are established by the International Union of Pure and Applied Chemistry (IUPAC) and the International Union of Biochemistry and Molecular Biology (IUBMB).

These rules and recommendations can be found in IUPAC's "color books," namely, *Nomenclature of Inorganic Chemistry* (the Red Book),[1] *Nomenclature of Organic Chemistry* (the Blue Book),[2] *Compendium of Polymer Terminology and Nomenclature* (the Purple Book),[3] and *Biochemical Nomenclature* (the White Book).[4] Updates are published as PDFs on IUPAC's website at https://www.iupac.org. In addition, brief guides to organic, inorganic, and polymer nomenclature are available on that website as PDFs.

Guidelines for biochemical and enzyme nomenclature can be found on IUBMB's website at https://iubmb.qmul.ac.uk. In addition, the Royal Society of Chemistry has published *Principles of Chemical Nomenclature: A Guide to IUPAC Recommendations*.[5]

Extensive sections on chemical nomenclature can be found in the widely used reference handbooks *CRC Handbook of Chemistry and Physics*,[6] *Lange's Handbook of Chemistry*,[7] and *The ACS Guide to Scholarly Communication*.[8] The Chemical Abstracts Service

(CAS) maintains a registry database of information on naturally occurring and synthetic substances. Located at https://www.cas.org, the CAS registry assigns a unique identification number to each substance. Registry numbers can be linked to substance names (eg, trivial, systematic, and semisystematic names), formulas, diagrams, and other identifying data, regardless of the nomenclature system used. Authors are encouraged to include the CAS registry numbers in descriptions of the compounds being studied.

17.2 FORMULAS

17.2.1 Empirical Formulas

Empirical formulas are the simplest formulas for expressing chemical composition, and they are often used in indexes. Empirical formulas are also used when the exact molecular composition is not necessary or when the identity and ratio of atoms in a compound is known but the precise molecular composition is not. In the Hill system, which was devised by chemist Edwin A. Hill, PhD, empirical formulas are formed by citing in alphabetical order the element symbols together with appropriate subscripts unless organic groups are present or other ordering criteria apply. In the presence of organic groups, C and H are cited first.

$$OSi \qquad ClHg \qquad C_{10}H_{10}CaClFeO_4S$$

17.2.2 Molecular Formulas

Molecular formulas denote compounds' atomic composition by ordering constituents according to relative electronegativities, with the more electropositive constituents cited first. If a compound contains more than one electropositive or electronegative constituent, the sequence within each class of compounds is the alphabetical order of their symbols. In molecular formulas, the number of atoms or groups, the designation of oxidation state, and the indication of ionic charge follow the recommendations in Chapter 16, "Subatomic Particles, Chemical Elements, and Related Notations."

17.2.3 Structural Formulas

Structural formulas, which consist of line and stereo formulas, provide information about the ways the atoms and bonds of molecules connect in rings. Stereo formulas show the 3-dimensional distribution of atoms in space, whereas line formulas are 2-dimensional representations in which atoms are shown joined by lines representing single or multiple bonds, without indicating the spatial direction of the bonds. For line formulas with only single bonds, bonds need not be represented with symbols. For formulas with different types of bonds, indicate a single bond with an en dash (ie, "–," which is created with Unicode 2013), a double bond with the equals symbol (ie, "=," created with Unicode 003D), and a triple bond with the identical-to symbol (ie, "≡," created with Unicode 2261).

$$CH_3CH_2CH_2CH_3 \qquad CH_3-CH=CH-CH_2-C\equiv CH$$

Use parentheses to indicate groups that branch above or below the line.

$$CH_3CH_2-P(S)(OH)-OCH$$

In organic formulas, use square brackets to enclose multiples of chain segments and parentheses to enclose atoms and groups not in the chain.

$$CH_3[CH_2CH(CH_3)]_4CH_3$$

Because stereo formulas show the 3-dimensional distribution of atoms in space, they provide more detail on the connection of atoms than do 2-dimensional line formulas. In stereo formulas, lines of normal weight denote bonds approximately in the plane of the ring (ie, the circular structure of an atom). For a bond extending above the plane, IUPAC recommends using a thick line (eg, ▬▬) or a solid wedge (eg, ◀▬). For a bond extending below the plane of the ring, IUPAC recommends a thick hashed line (eg, ⊪⊪⊪⊪⊪⊪). Unknown stereochemistry can be indicated by a wavy line. To imply stereochemistry, indicate double bonds with accurate angles (ie, approximately 120°) if possible. Otherwise, use linear representation.

Another form of stereo formula uses fused rings. The rules for stereo formulas are complex, and they are only summarized here. Proper orientation of fused ring systems is essential for correct numbering. Orient fused ring systems with as many *ortho*-fused rings with vertical common bonds as possible in the same horizontal row. The *ortho*-fused rings are the rings that are adjacent to the primary carbon on the main ring. When additional rings are fused to the horizontal row, preference should be given to orientations that do not involve overlapping bonds, to those with the maximum number of rings in the upper right quadrant, to those with the minimum number of rings in the lower left quadrant, and to those that allow low locants to fusion positions. A 6-membered ring may be modified on one side to create a 5-membered ring or stretched in the middle to insert additional pairs of horizontal bonds, as needed.

9*H*-dibenzo[*de,rst*]pentaphene   2*H*-pentaleno[2,1-*b*]pyran

When a family of chemical structures is being discussed, the letter "R" may be used to designate part of the molecule. In some texts, "R" may represent the moiety that is the same in all members of the family, which would place the focus on the functional group that changes from one member of the family to another. For example, the structural formula below represents the common moiety of retinoids. Here, "R" represents the organic structure minus the characteristic group. Thus retinoic acid would be R–COOH.

Compounds related to retinoic acid can be represented by "R" plus the appropriate functional group.

R–CH$_2$OH (retinol)
R–CHO (retinal)
R–COOH (retinoic acid)
R–CH=NOH (retinal oxime)

where R =

17.3 GENERAL CONVENTIONS OF NOMENCLATURE

17.3.1 Case and Type Style

Do not capitalize a chemical name in the text unless it begins a sentence or section heading or is used in a title that uses headline-style capitalization. For such titles, capitalize the first letter of each major word, but do not capitalize any descriptor, positional prefix, or locant other than elemental symbols. Capitalize chemical trade names wherever they appear unless they were trademarked with initial lowercase letters.

Set trivial, systematic, and semisystematic names in roman type except for those affixes and locants that the nomenclature rules specify should be in italic or Greek letters.

17.3.2 Prefixes

Numerical prefixes (eg, "8") and multiplicative prefixes derived from Greek or Latin (eg, "di") are closed up to the rest of the name (ie, no hyphen or space) and are not italicized when used to describe a multiplicity of identical features in the name of a chemical compound (see Table 17.1). Structural prefixes that are integral parts of the name (eg,

Table 17.1. Simple Numerical Terms

Number	Prefix	Number	Prefix	Number	Prefix	Number	Prefix
1	mono[a]	10	deca *or* deci	100	hecta	1,000	kilia
		11	undeca[a]	101	henhecta[a]	1,001	henkilia[a]
		12	dodeca[b]	102	dohecta	1,002	dokilia[b]
2	di *or* bi[b]	20	icosa *or* eicosa	200	dicta[b]	2,000	dilia[b]
		21	henicosa[a]	201	hendicta[a]	2,001	hendilia[a]
		22	docosa[b]	202	dodicta[b]	2,002	dodilia[b]
3	tri *or* ter	30	triaconta	300	tricta	3,000	trilia
4	tetra *or* quarter	40	tetraconta	400	tetracta	4,000	tetralia
5	penta *or* quinque	50	pentaconta	500	pentacta	5,000	pentalia
6	hexa *or* sexi	60	hexaconta	600	hexacta	6,000	hexalia
7	hepta *or* septi	70	heptaconta	700	heptacta	7,000	heptalia
8	octa *or* octi	80	octaconta	800	octacta	8,000	octalia
9	nona *or* novi	90	nonaconta	900	nonacta	9,000	nonalia

[a] When alone, the numerical prefix for 1 is "mono." In association with other numerical terms, the prefix for 1 is "hen," except for 11, in which case the term is "undeca."

[b] When alone, the numerical prefix for 2 is "di." In association with other numerical terms, the term for 2 is "do," except for 200 and 2,000, in which cases the prefixes are "dicta" and "dilia," respectively.

"cyclo") are also closed up to the rest of the name and are not italicized. Nonintegral structural descriptors (eg, "-*iso*-") are italicized and set off by hyphens.

dibenzo[*c,e*]oxepine	*p*-aminobenzoic acid	isopentane
cyclohexane	polyester	8-*iso*-prostaglandin E$_1$

17.3.3 Enclosing Marks

Chemical nomenclature uses 3 kinds of enclosing marks: braces, { }; square brackets, [ ]; and parentheses, ( ). These enclosing marks are used slightly differently in formulas than they are in coordination and organic names.

17.3.4 Locants

Locants describe particular structural features or positions in chemical structures. Locants are designated with arabic numbers, Greek letters, and italicized Roman letters and words. Set locants off from the rest of chemical structures with hyphens or square brackets, and set them off from each other with commas. No space should come after those punctuation marks.

N-methylethanamine	*myo*-inositol	spiro[4.5]dec-6-ene
O-phosphono-*enol*-pyruvate	2*H*-pyran-3(4*H*)-thione	5α-androstane

Capitalize symbols for atoms being added or substituted and set them in italic type.

O-methylhydroxylamine indicates that a methyl group replaced one of the hydroxyl hydrogens on the oxygen atom of hydroxylamine.

6*H*-1,2,5-thiadiazine indicates that a hydrogen atom was added at the site specified by the locant.

Never capitalize structural descriptors such as "*as*" or "*o*" at the beginning of a sentence. Instead, capitalize the first nonitalicized letter in the name (eg, "*as*-Triazine" and "*o*-Methylphenol").

17.3.5 Levels of Complexity in Nomenclature

Historically, different chemical nomenclature conventions have been developed for different levels of molecular complexity. Those conventions include the following:

(1) Trivial names are the common names of chemicals, including traditional names (eg, "acetone," "benzene," and "urea") and trade names (eg, "Clorox" and "Lucite"). Trivial names also include names given for convenience to compounds of uncertain structure, especially those of biological origin (eg, "solanine" for members of the plant family *Solanaceae* and "penicillin" for members of the genus *Penicillium*). Trivial names carry little, if any, structural information, and they are usually derived from their source material. After the structure of a compound is determined, the compound's trivial name should be used only when that name adequately and unambiguously describes a well-established compound.

(2) Systematic names are based on various sets of rules. For organic compounds, systematic names are constructed from appropriate names of parent hydrides, stereodescriptors, prefixes, and suffixes (eg, "phosphane," "(2*E*)-but-2-ene," and "cyclobutane"). Depending on the context and the nomenclature convention used, a compound may have more than one appropriate systematic name.

(3) Semisystematic names are usually simplified alternatives to systematic names. Ex-

amples of semisystematic names are "methane," "cubane," "benzoic acid," and "phenol."

Any manuscript describing a complex molecule—especially a new one—should use the molecule's full systematic name at least once. Elsewhere, the manuscript can use the semisystematic name, the trivial name, or an abbreviation.

17.4 BINARY NOMENCLATURE

The binary approach to nomenclature bases names on the ionic nature of the constituents, with electropositive components listed first followed by electronegative components. Separate all terms for ions and neutral segments with spaces. Although the term "binary" literally refers to 2 components, this nomenclature system is readily extended to multiple components. Compounds should be referred to by either their chemical formulas or binary names but not by hybrid names.

Chemical formula	Binary name
MgKF	magnesium potassium fluoride *not* Mg potassium fluoride
NaCl	sodium chloride *not* Na chloride
N_2O	dinitrogen oxide *not* N_2 oxide
OF_2	oxygen difluoride *not* oxygen F_2
$Pt(NH_3)_2Cl_2$	platinum diammine dichloride *not* platinum diammine Cl_2
$Zn(OH)I$	zinc hydroxide iodide *not* Zn hydroxide iodide

17.5 COORDINATION NOMENCLATURE

In coordination nomenclature, compounds are regarded as having central atoms to which ligands (eg, ions, neutral atoms, and molecular groups) are attached. Coordination nomenclature is an additive system in which associated ligand names are added to the name of a central atom. However, the central atom is cited last in the name, preceded by the ligands in alphabetical sequence.

Chemical formula	Coordination nomenclature
$Pt(NH_3)_2Cl_2$	diamminedichloroplatinum(II)
$[Fe(CN)_5(NO)]^{2-}$	pentacyanonitrosylferrate(2−)
$K_4[Fe(CN)_6]$	tetrapotassium hexacyanoferrate(4−)
$[Co(NH_3)_5Cl]Cl_2$	pentamminechlorocobalt(2+) dichloride

To specify precisely the atoms attached to a metal, use italic letter locants.[1]

17.6 SUBSTITUTIVE NOMENCLATURE

The substitutive approach to naming chemical compounds is based on the concept of parent hydrides, which are compounds with skeletal structures consisting of carbon atoms, atoms of metallic or nonmetallic elements called "heteroatoms," and attached hydrogen atoms. The name of a compound is consistent with the skeletal structure and the bonding requirements of each skeletal atom. Heteroatoms in organic compounds and inorganic parent hydrides have standard bonding numbers. Other bonding numbers are indicated by the λ-convention, in which constituents are named according to their

Table 17.2. Suffixes for Derivation of Prefix Names from Parent Hydrides[a]

Monovalent	Divalent	Trivalent	Tetravalent
yl	diyl	triyl	tetrayl
	ylidene	ylidyne	ylylidyne
		ylylidene	diylidene
			diylylidene

[a] Suffixes beyond the tetravalent level are modeled on these forms.

Table 17.3. Characteristic Groups Cited Only as Prefixes

Characteristic group	Prefix	Characteristic group	Prefix
$-Br$	bromo	$-I$	iodo
$-Cl$	chloro	$-IO$	iodosyl
$-ClO$	chlorosyl	$-IO_2$	iodyl
$-ClO_2$	chloryl	$=N_2$	diazo
$-ClO_3$	perchloryl	$-N_3$	azido
$-F$	fluoro	$-NO$	nitroso
		$-NO_2$	nitro

distance from the primary (α) carbon using Greek letters. For example, the carbon next to the α carbon is termed the "β carbon," the next is the γ carbon, and so on. Hydrogen atoms of parent hydrides or other parent compounds are replaced by other atoms or groups called "substituents."

For organic compounds, the underlying principle is to dissect the structure into characteristic groups and fragments that can be represented by parent hydride structures. One of the parent hydride structures is chosen as the main part of a compound's name, and the other parts are cited as substituents to it. A substituent prefix is formed from the name of a parent hydride by adding the suffixes in Table 17.2.

Characteristic atoms and groups attached to a parent hydride that are not themselves derived from parent hydrides are cited as either prefixes or suffixes to the name of the parent hydride. Table 17.3 gives the names of some "compulsory prefixes," which are characteristic groups that are cited only as prefixes. See Table 17.4 for the names of prefixes and suffixes for some common characteristic groups.

When mononuclear parent hydrides (see Table 17.5) are modified, the names of the resulting compounds reflect the substitution. For example, 3 chlorine atoms may be substituted for the 3 hydrogen atoms of phosphane (PH_3) to create trichlorophosphane (PCl_3).

17.6.1 Mononuclear Parent Hydrides

The names of mononuclear parent hydrides are given in Table 17.5.

17.6.2 Acyclic Parent Hydrides

Names of unsubstituted acyclic (ie, straight-chain) parent hydrides have a stem and a characteristic element term, except for carbon chains. These hydrides also have char-

Table 17.4. Characteristic Group Suffixes and Prefixes

Formula	Suffix	Prefix
–COOH	carboxylic acid	carboxy
–(C)OOH	oic acid	. . .
–COO⁻	carboxylate	carboxylato
–(C)OO⁻	oate	. . .
–SO₂OH	sulfonic acid	sulfo
–SO₂O⁻	sulfonate	sulfonato
–COXᵃ	carbonyl halide	halocarbonyl *and* carbonohalidoyl
–(C)OXᵃ	oyl halide	. . .
–CONH₂	carboxamide	carbamoyl *and* aminocarbonyl
–(C)ONH₂	amide	. . .
–C(=NH)NH₂	carboximidamide *and* carboxamidine	carbamimidoyl *and* amidino
–(C)(=NH)NH₂	amidine	. . .
–CN	carbonitrile	cyano
–(C)N	nitrile	. . .
–CHO	carbaldehyde *and* carboxaldehyde	formyl
–(C)HO	al	. . .
>(C)=O	one	oxo
–O⁻	olate	oxido
–OH	ol	hydroxy
–SH	thiol	sulfanyl *and* mercapto
–S⁻	thiolate	sulfido *and* sulfanidyl
–NH₂	amine	amino
=NH	imine	imino

ᵃ X = F, Cl, Br, I, N₃, CN, NC, NCO, NCS, NCSe, and NCTe

Table 17.5. Mononuclear Parent Hydrides

Formula	Name	Formula	Name
BH₃	borane	AsH₃	arsane (arsineᵇ)
CH₄	methane	AsH₅	λ⁵-arsane (arsoraneᵇ)
SiH₄	silane	SbH₃	stibane (stibineᵇ)
GeH₄	germane	SbH₅	λ⁵-stibane (stiboraneᵇ)
SnH₄	stannane	BiH₃	bismuthane (bismuthineᵇ)
PbH₄	plumbane	BiH₅	λ⁵-bismuthane
NH₃	azaneᵃ	OH₂	oxidaneᵃ
PH₃	phosphane (phosphineᵇ)	SH₂	sulfaneᵃ
PH₅	λ⁵-phosphane (phosphoraneᵇ)		

ᵃ Not normally used as a parent hydride.
ᵇ Less preferred name still in use.

Table 17.6. Representative Acyclic Parent Hydrides

Name	Stem	Element term	Ending	Number of chain atoms; degree of saturation
ethane	eth[a]		ane	2 carbon atoms; fully saturated
propane	prop[a]		ane	3 carbon atoms; fully saturated
butane	but[a]		ane	4 carbon atoms; fully saturated
pentane	pent		ane	5 carbon atoms; fully saturated
disilane	di	sil	ane	2 silicon atoms; fully saturated
diazene	di	az	ene	2 nitrogen atoms; 1 double bond
tetrasulfane	tetra	sulf	ane	4 sulfur atoms; fully saturated
octa-1,3-diene	octa[b]		diene	8 carbon atoms; 2 double bonds at positions 1 and 3
nona-2,4,7-triyne	nona[b]		triyne	9 carbon atoms; 3 triple bonds at positions 2, 4, and 7
deca-1,3-dien-7-yne	deca[b,c]		dienyne	10 carbon atoms; 2 double bonds at positions 1 and 3; 1 triple bond at position 7
tetraaza-1,2-diene	tetra	aza	diene	4 nitrogen atoms; 2 double bonds at positions 1 and 2

[a] Trivial, or nonnumerical, stems are retained for compounds containing 1 to 4 carbon atoms.
[b] For euphony, an "a" is inserted between the stem and the characteristic ending when the stem ends in a consonant and the ending begins with a consonant.
[c] An "e" in endings denoting the state of saturation is elided, or omitted, when followed by a vowel, including the letter "y" when used as a vowel (eg, decadienyne = dien(e) + yne, with the "e" that is enclosed in parentheses elided).

acteristic endings, specifically "-ane" to designate fully saturated chains (ie, chains containing only single bonds) and "-ene" and "-yne" to designate unsaturated chains (ie, chains containing double or triple bonds). The "-ene" ending indicates the presence of a double bond, and the "-yne" ending indicates the presence of a triple bond. Numerical prefixes describe the number of unsaturations in a chain, and locants indicate their positions in the chain when known. Examples are shown in Table 17.6.

17.6.3 Cyclic Parent Compounds

Cyclic structures in which the ring contains only carbon and hydrogen atoms are called "hydrocarbons." Cyclic structures in which the ring contains atoms other than carbon are called "heterocycles." Atoms other than carbon in such a structure are called "heteroatoms." Structures with only one ring containing carbon and noncarbon atoms are called "heteromonocycles" or "heteromonocyclic rings."

17.6.3.1 Monocyclic Rings

All saturated monocyclic rings may be named by attaching the prefix "cyclo" to the name of the corresponding acyclic hydrocarbon. Unsaturation is described by the endings "-ene" and "-yne."[3]

cyclo + butane = cyclobutane
cyclo + octane + 1,3,5-trien-7-yne = cycloocta-1,3,5-trien-7-yne
cyclo + hexasilane = cyclohexasilane

Heteromonocyclic rings are named using 3 different methods:

Table 17.7. Prefixes for Hantzsch–Widman System of Skeletal Replacement ("a")

Element	Bonding number (valence)	Prefix	Element	Bonding number (valence)	Prefix
fluorine	1	fluora	phosphorus	3	phospha
chlorine	1	chlora	arsenic	3	arsa
bromine	1	broma	antimony	3	stiba
iodine	1	ioda	bismuth	3	bisma
oxygen	2	oxa	boron	3	bora
sulfur	2	thia	silicon	4	sila
selenium	2	selena	germanium	4	germa
tellurium	2	tellura	tin	4	stanna
mercury	2	mercura	lead	4	plumba
nitrogen	3	aza			

Table 17.8. Stems for Hantzsch–Widman System

Ring size	Unsaturated	Saturated	Ring size	Unsaturated	Saturated
3	irene (irine)	irane (iridine)	6C[c]	inine	inane
4	ete	etane (etidine)	7	epine	epane
5	ole	olane (olidine)	8	ocine	ocane
6A[a]	ine	ane	9	onine	onane
6B[b]	ine	inane	10	ecine	ecane

[a] 6A elements: O, S, Se, Te, Bi, and Hg
[b] 6B elements: N, Si, Ge, Sn, and Pb
[c] 6C elements: B, F, Cl, Br, I, P, As, and Sb

(1) The Hantzsch–Widman system combines terms for each heteroatom in the ring in the order given in Table 17.7, followed by an ending that indicates the ring size and that indicates whether the ring is saturated or unsaturated, as given in Table 17.8.

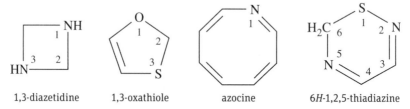

1,3-diazetidine   1,3-oxathiole   azocine   6H-1,2,5-thiadiazine

(2) Skeletal replacement nomenclature uses the "a" prefixes given in Table 17.7 to denote the replacement of the carbon atoms of a monocyclic hydrocarbon. This system may also be used for silicon heterocycles (eg, silabenzene), and it is mandatory for monocyclic rings with more than 10 ring members (eg, 1,4-dioxacyclododecane).

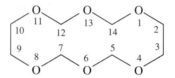

1,4,6,8,11,13-hexaoxacyclotetradecane

(3) A trivial system known as "cyclo(ab)$_x$" is a method in which heteroatoms are designated by "a" and "b" (see Table 17.7). A numerical term defines the number of the "a" heteroatoms.

$$\begin{array}{c}
\text{SiH}_2 - \text{O} \\
\text{O} \quad 8 \quad 1 \quad \text{SiH}_2 \\
7 \qquad 2 \\
\text{H}_2\text{Si} \quad 6 \qquad 3 \quad \text{O} \\
5 \quad 4 \\
\text{O} - \text{SiH}_2
\end{array}$$

cyclotetrasiloxane

17.6.3.2 Bicyclic and Polycyclic Ring Systems

There are several nomenclature systems for bicyclic and polycyclic ring systems.

(1) In fused ring systems, one component ring or ring system is selected as the parent system, and the bonds between carbon atoms 1 and 2, 2 and 3, 3 and 4, etc, are lettered in italics as "*a*," "*b*," "*c*," etc, and placed within brackets. For ring systems with more than 26 sides, the bonds between carbon atoms 26 and 27 through 51 and 52 are labeled with subscript numbers as "a_1" through "z_1," and the bonds between carbon atoms 52 and 53 through 77 and 78 are labeled "a_2" through "z_2," etc. Bonds common to a pair of components are indicated by the appropriate numerical locants for the attached components and by letter locants for the parent component. The locants and components are separated by a hyphen. Letter locants are run together when there is more than one, and they are listed in the direction in which they were assigned. Numerical locants are separated by commas and follow the same direction as the letter locants. Locants may be omitted in simple structures, provided doing so will not create ambiguity.

> benz[*a*]anthracene
> 1*H*-benzo[*a*]cyclopent[*f*]anthracene
> 9*H*-dibenzo[*de,rst*]pentaphene (see Section 17.2.3, "Structural Formulas")
> 2*H*-pentaleno[2,1-*b*]pyran (see Section 17.2.3, "Structural Formulas")

(2) Bridged ring systems fall into 2 types: bridged fused ring systems and so-called "von Baeyer ring systems," each of which has its own style and format considerations.

  Bridged fused ring systems are named by attaching names for bridges to the names for fused ring systems as described above. The locants indicate where the bridges are located in the ring systems.

> 1,4-methanonaphthalene   4,7-epoxyfuro[2,3-*d*]oxepin

  The von Baeyer ring systems are more correctly referred to as "polycyclic bridged ring systems" that cannot be treated as bridged fused ring systems. Polycyclic ring systems are named using the extended von Baeyer system, beginning with prefixes that indicate the number of rings (eg, "bicyclo" and "tricyclo"). Numbers indicating the bridge lengths are in decreasing order, they are separated by periods rather than commas and they are enclosed in square brackets. Locants of the attachment points for secondary bridges in polycyclic systems are indicated by pairs of superscript numbers. The lower of each pair is cited first, and the paired superscript numbers are separated by commas. Unsaturation is indicated by endings like "-ene" and "-diene," and heteroatoms are denoted by skeletal replacement prefixes (see Table 17.7).

> bicyclo[3.2.1]octane
> 1,3-diazatetracyclo[2.2.0.0^{2,6}.0^{3,5}]hexane
> tricyclo[3.3.2.0^{2,4}]dec-6-ene

(3) Spiro ring systems have 2 or more rings, at least 2 of which have only a single atom in common. This atom is called the "spiro atom." For nomenclature purposes, there are 2 distinct types of spiro compounds.

  A spiro compound composed only of saturated monocyclic rings uses the name of the corresponding acyclic hydrocarbon as its root. Depending on the number of spiro atoms present, the spiro compound has a prefix like "spiro" or "dispiro." The prefix is followed by a series of numbers that correspond to the number of atoms between the spiro atoms. The numbers are separated by periods and enclosed in brackets. These numbers are cited in ascending order within each ring, starting with one of the terminal rings and proceeding to the other terminal ring and then back to the first ring. The system should be numbered in the same order that the descriptors are cited, with the spiro atoms receiving the lowest locants. In systems with 3 or more rings, each time a spiro atom is reached for a second time, its locant should be cited as a superscript to the number describing the linking atoms. Such treatment is necessary to distinguish between linear systems and branched systems. Unsaturation is indicated by endings like "ene" and "diene," and heteroatoms are denoted by skeletal replacement prefixes (see Table 17.7).

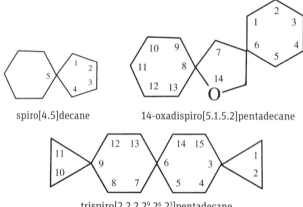

spiro[4.5]decane   14-oxadispiro[5.1.5.2]pentadecane

trispiro[2.2.2.2^9.2^6.2^3]pentadecane

Spiro compounds in which at least one component is a polycyclic ring system are named by enclosing the names of the components within brackets, cited in order beginning with the component that is earliest alphabetically, and adding prefixes like "spiro" and "dispiro" according to the number of spiro atoms. The locants of the second, third, and higher ring or ring system are serially primed. The positions of the spiro atoms are denoted by the locants of each ring, thelocants are separated by commas, and the locants are inserted between the names of the appropriate rings or ring systems.

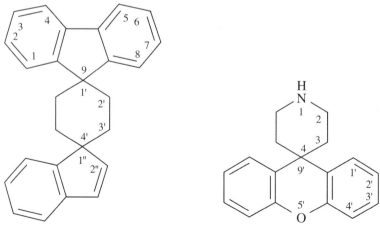

dispiro[fluorene-9,1'-cyclohexane-4',1"-indene]   spiro[piperidine-4,9'-xanthene]

(4) Ring assemblies are 2 or more rings or ring systems that meet the following conditions: The rings or ring systems are joined by only single bonds or only double bonds, and the

number of direct ring connections is one fewer than the number of rings or ring systems involved. Assemblies of identical rings or ring systems are named using substituent prefixes (see below) or numerical prefixes such as "bi," "ter," and "quarter" (see Table 17.1). Locants denoting the attachment points on each component are cited before the numerical prefixes, and sets of locants are separated by colons. Traditionally, locants for the first component are unprimed, and locants for succeeding components are primed serially. However, to avoid long strings of primes, the rings or ring systems may be numbered as if they were chains, and the positions on each component may be denoted by superscript numbers corresponding to the locant numbers of the ring systems.

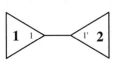

1,1'-bicyclopropane
1,1'-bicyclopropyl
$1^1,2^1$-bicyclopropane

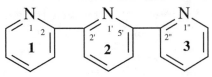

2,2':5',2"-terpyridine
$1^2,2^2:2^5,3^2$-terpyridine

Nonidentical ring assemblies are named by substitutive nomenclature principles (see Section 17.6, "Substitutive Nomenclature"). One ring or ring system is chosen as the parent, and the other rings or ring systems are cited as substituents.

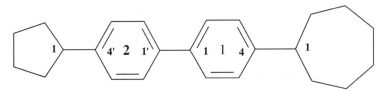

4-cycloheptyl-4'-cyclopentylbiphenyl

17.6.4 Fullerenes

Informally called "buckyballs," fullerenes are closed-cage, fully unsaturated carbon macrocycles having more than 20 carbon atoms with each carbon bound to 3 other carbon atoms. Because of this structure, fullerenes are devoid of hydrogen atoms. Nomenclature for fullerenes and their derivatives is based on organic nomenclature principles.

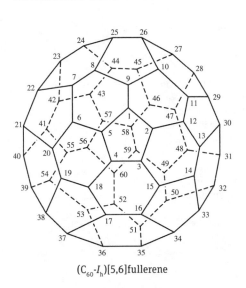

$(C_{60}\text{-}I_h)$[5,6]fullerene

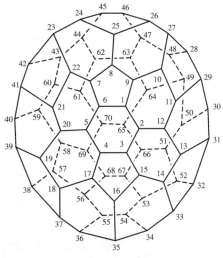

$(C_{70}\text{-}D_{5h(6)})$[5,6]fullerene

17.6.5 Phane Systems

Phane nomenclature may be used for the large number and variety of large rings and ring systems directly connected, or linked, by atoms or chains of atoms. In this system, rings and ring systems are "collapsed" to a single "atom," called a "superatom," which is represented by large solid dots in the phane skeletal structure. The resulting monocyclic or polycyclic ring or ring system is named and numbered as a saturated monocyclic hydrocarbon, a saturated bicyclic or polycyclic hydrocarbon, or a saturated spiro hydrocarbon (see Section 17.6.3.2, "Bicyclic and Polycyclic Ring Systems"), except that the ending of the name is "-phane" rather than "-ane." In phane names, prefixes derived from the names of the ring or ring system define the nature of the superatoms.

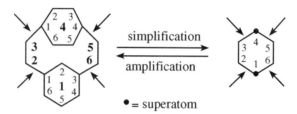

Phane parent hydride name: 1(1,3),4(1,4)-dibenzenacyclohephane
Simplified phane skeletal name: cycloheptaphane

17.6.6 Structural Affixes

The following structural affixes are used for organic compounds. The affixes are set in italics with a hyphen separating them from the rest of the name, except for "epi," "iso," "poly," and "pseudo," which are set in roman type and closed up to the rest of the name without a hyphen or space.

aci-	the acid form, as in "*aci*-acetoacetate," $CH_3C(OH)=CH-C(O)O-R$, and *aci*-nitro, $HON(O)=$
as-	asymmetric, as in "*as*-trichlorobenzene" and "*as*-triazine"
endo-	prefix indicating a bridging group
epi	prefix indicating a bridging group
gem-	geminal, 2 groups attached to the same atom, as in "*gem*-diol" or "*gem*-dimethyl-"
iso	denoting a single, simple branch at the end of a straight chain, as in "isopentane" and "isobutyl"
m-	*meta*, a 1,3-positional relationship in benzene
n-	abbreviation for normal, as in "*n*-butane"
o-	*ortho*, a 1,2-positional relationship in benzene
p-	*para*, a 1,4-positional relationship in benzene
peri-	a 1,8-positional relationship in naphthalene
poly	many, as in "polysulfide"
pseudo	indicating resemblance or relation, as in "pseudohalogen"
s- and *sym-*	symmetric, as in "*s*-trichlorobenzene" and "*s*-triazine"
sec-	secondary, indicating a single, simple branch next to the termination of a hydrocarbyl group, as in "*sec*-butyl"
t- and *tert-*	tertiary, indicating a double branch next to the termination of a hydrocarbyl group, as in "*t*-Bu" or "*tert*-butyl"
v- and *vic-*	vicinal, 3 adjacent groups, as in "*vic*-triazole"

Table 17.9. Functional Replacement Infixes and Prefixes

Infix	Prefix	Replaced atoms	Replaced atoms
amido	amido	−OH	−NH$_2$
azido	azido	−OH	−N$_3$
bromido	bromo	−OH	−Br
chlorido	chloro	−OH	−Cl
cyanatido	cyanato	−OH	−OCN
cyanido	cyano	−OH	−CN
dithioperoxo	dithioperoxy	−OH	−SS−
fluorido	fluoro	−OH	−F
iodido	iodo	−OH	−I
isocyanatido	isocyanato	−OH	−NCO
isocyanido	isocyano	−OH	−NC
nitrido	nitrido	=O *and* −OH	=N−$\begin{cases} or\ -N< \\ or\ =N- \end{cases}$
thiocyanatido	thiocyanato	−OH	−SCN
isothiocyanatido	isothiocyanato	−OH	−NCS
imido	imido	=O *or* −O−	=NH *or* −NH−
hydrazido	hydrazido	−OH	−NHNH$_2$
peroxo	peroxy	−OH	−OO−
seleno	seleno	=O *or* −O−	=Se *or* −Se−
telluro	telluro	=O *or* −O−	=Te *or* −Te−
thio	thio	=O *or* −O−	=S *or* −S−
thioperoxo	thioperoxy	−O−	−OS *or* −SO−

17.7 FUNCTIONAL REPLACEMENT NOMENCLATURE

Functional replacement nomenclature uses either prefixes or infixes to indicate the replacement of oxygen atoms or hydroxy groups by other atoms or groups in parent compounds (see Table 17.9). Italics are used for symbols for atoms being added or substituted.

—C(O)SH   thiobenzoic *S*-acid *or* benzenecarbothioic *S*-acid

CH$_3$P(O)(NH$_2$)(SeH)   *P*-methylphosphonamidoselenoic *Se*-acid

17.8 OTHER NOMENCLATURE SYSTEMS

Although substitutive nomenclature is the basis for forming systematic names for organic compounds, other systems are also used, sometimes in combination with substitutive nomenclature. One of the most common is functional class nomenclature, previously known as "radicofunctional nomenclature." With functional class nomenclature, a compound is named on the basis of a class of compounds, such as alcohols, amines, ethers, or ketones, and the appropriate form of the name of the parent hydride. In the examples below, each of the names contains a class name representing a functional group: $-OH$, $-O-$, or $-CO-$. No hydrogen atoms are replaced. However, the organic part of the structure is usually expressed by means of a substituent group name, which was called a "radical" when this naming system was called "radicofunctional nomenclature."

CH_3-OH	methyl alcohol
$CH_3CH_2-O-CH_2CH_3$	diethyl ether
$CH_3CH_2-CO-CH_3$	ethyl methyl ketone

17.9 RADICALS AND IONS

In substitutive nomenclature, a radical is indicated by a superscript dot and, if appropriate, a charge sign. In complex formulas, use square brackets or parentheses. The suffixes and substituent prefix endings for radicals and ions used in substitutive nomenclature are listed in Table 17.10. For more complex examples, see the International Union of Pure and Applied Chemistry's comprehensive publication *Revised Nomenclature for Radicals, Ions, Radical Ions and Related Species.*[9]

$$H^{\cdot} \qquad {}^{\cdot}CH_3 \qquad (SO_2)^{\cdot-} \qquad [CH_2{=}CH{-}CH_3]^{\cdot+}$$

Zwitterions (ie, structures with internally compensating charges) are named by combining appropriate operational suffixes at the end of the name of a neutral parent hydride in the order of "-ium," "-ylium," "-ide," and "-uide."

$$(CH_3)\overset{+}{\underset{2}{N}}{-}\overset{-}{\underset{1}{N}}H_3 \qquad \text{3-(2,2,2-trimethylhydrazin-2-ium-1-ide)}$$

17.10 ISOTOPICALLY MODIFIED COMPOUNDS

An isotopically modified compound has an isotopic ratio of nuclides for at least one element that deviates measurably from that occurring in nature. It is either an isotopically substituted or an isotopically labeled compound.

17.10.1 Isotopically Substituted Compounds

An isotopically substituted compound is one in which essentially all the molecules of the compound have only the indicated nuclide at each of the designated positions. The formulas for these compounds should be written with the appropriate nuclide symbol. The names, in turn, should consist of the nuclide symbol in parentheses with locants if necessary, followed by the name of the compound or the group being substituted.

${}^{14}CH_4$	(${}^{14}C$)methane
CH_3CH^2H-OH	(1-${}^2H_1$)ethanol

17.10.2 Isotopically Labeled Compounds

Isotopically labeled compounds may be either specifically labeled or selectively labeled. In formulas and names for both, use square brackets around the appropriate nuclide symbols and multiplying subscripts, if any.

A specifically labeled compound is one in which a unique isotopically substituted compound is formally added to an analogous isotopically unmodified compound. In

Table 17.10. Suffixes and Substituent Prefix Endings for Radicals and Ions Derived from Parent Hydrides

Class	Operation	Suffix	Example	Substituent prefix ending or prefix	Example	Replacement nomenclature suffix	Example
Radicals	Loss of H·	yl, etc (see Table 17.2)	methyl CH₃·	ylo	ylomethyl–CH₂·	...	...
Cations	Addition of H⁺	onium	ammonium NH_4^+ sulfonium SH_3^+	oniumyl	ammoniumyl $-NH_3^+$ sulfoniumyl $-SH_2^+$	onia	azonia >N⁺< thionia >S⁺–
		ium	methanium CH_5^+ azanium NH_4^+	iumyl	methaniumyl $-CH_4^+$ azaniumyl $-NH_3^+$		
	Loss of H⁻	ylium	methylium CH_3^+ λ⁵-phosphanylium PH_4^+	yliumyl	methyliumyl $-CH_2^+$ λ⁵-phosphanyliumyl $-PH_3^+$	ylia	λ⁵-phosphanylia >P⁺<
Anions	Loss of H⁺	ide	methanide CH_3^- λ⁵-phosphanide PH_4^-	idyl	methanidyl $-CH_2^-$ λ⁵-phosphanidyl $-PH_3^-$	ida	λ⁵-phosphanida >P⁻<
		ate	R-carboxylate R-COO⁻	ato	carboxylato -COO⁻	...	...
		olate	methanolate CH₃-O⁻	ido	oxido -O⁻	...	...
	Addition of H⁻	uide	boranuide BH_4^-	uidyl	boranuidyl $-BH_3^-$	uida	boranuida —B⁻—

contrast, a selectively labeled compound is one in which a mixture of isotopically substituted compounds is formally added to an isotopically unmodified compound in such a way that the position of each labeling nuclide is defined but the number of each nuclide may not be.

	Formula	Name
Specifically labeled:	$[^{14}C]H_4$	$[^{14}C]$methane
	$CH_2[^2H_2]$	$[^2H_2]$methane
	$CH_3CH[^2H]-OH$	$[1-^2H_1]$ethanol
	$CH_3CH_2-[^{18}O]H$	$[^{18}O]$ethanol
	$CH_3CH_2-O[^2H]$	ethan$[^2H]$ol
Selectively labeled:	$[^2H]CH_4$	$[^2H]$methane
	$[1-^{14}C,^{18}O]CH_3CH_2-OH$	$[1-^{14}C,^{18}O]$ethanol

A nonselectively labeled compound is one in which both the positions and the numbers of the labeling nuclides are undefined.

$[^{14}C,^2H]CH_3CH_2-OH$ $[^{14}C,^2H]$ethanol

In general labeling, all positions of the designated element are labeled but not necessarily in the same isotopic ratio. The symbol "G" in roman type may be used in place of locants to indicate general labeling.

$[G-^{14}C]$pentanoic acid   D-$[G-^{14}C]$glucose

In uniform labeling, all positions or all specified positions of the designated element are labeled in the same isotopic ratio. The symbol "U" in roman type may be used in place of locants to indicate uniform labeling.

$[U-^{14}C]$pentanoic acid   D-$[U-^{14}C]$glucose   D-$[U-1,3,5-^{14}C]$glucose

17.10.3 Isotopically Deficient Compounds

With an isotopically deficient compound, one or more of the nuclides are present in less than the natural ratio. The deficiency is indicated with the italic term "*def*" in square brackets, which immediately precedes the appropriate nuclide symbol.

$[def]^{13}CHCl_3$ $[def^{13}C]$chloroform

17.11 STEREOCHEMICAL NOMENCLATURE

Many special italicized symbols and terms are used in chemical nomenclature to describe the stereochemical configuration of a variety of compounds. Some specialized fields have developed their own systems.

17.11.1 Organic Compounds

Descriptors frequently used in organic stereochemical nomenclature are listed in Table 17.11.

17.11.1.1 Double Bonds

Configurations at double bonds are described by the italic letters "*Z*" and "*E*," enclosed in roman parentheses and placed immediately in front of the part of the name to which they refer. "(*Z*)" stands for "*zusammen*," which means "together." It describes a config-

Table 17.11. Descriptors for Organic Stereochemical Nomenclature

Descriptor	Definition and use	Examples
(+)-	rotation of the plane of polarization to the right	(+)-glucose
(–)-	rotation of the plane of polarization to the left	(–)-fructose
(±)-	racemic form	(±)-glucose *and* (±)-4-(2-aminopropyl)phenol
ambo-	molecule with 2 or more chiral elements that are present as a mixture of the 2 racemic diastereoisomers in unspecified proportions (set in italics)	L-alanyl-*ambo*-leucine (the dipeptide formed from L-alanine and DL-leucine)
anti-	similar to *trans* (ie, on the opposite side of a defined structural feature) (set in italics)	*anti*-benzaldoxime *and* *anti*-7-bromobicyclo[2.2.1] hept-2-ene
cis-	atoms or groups on the same side of a reference plane (set in italics)	*cis*-cyclohexane-1,4-diol *and* *cis*-but-2-ene
D-	absolute configuration to the right at the α-carbon in an amino acid compared with serine or at the highest-numbered center of chirality in a carbohydrate compared with glyceraldehyde when displayed in a Fischer projection (set as a small capital letter) (compare with L-)	D-serine *and* D-glucose
DL-	mixture of equimolar amounts of D- and L- enantiomers (set as small capital letters)	DL-leucine *and* DL-glucose
d, dextro-	**obsolete term** for the rotation of the plane of polarization to the right; the preferred prefix is "(+)-"	(no longer used)
endo-	an inner position as opposed to an outer position; used mainly in bicyclic and polycyclic ring systems in which an *endo* substituent points toward the interior part of the structure (set in italics) (compare with *exo-*)	*endo*-2-methylbicyclo[2.2.1]heptane
ent-	abbreviation for "enantio" or "enantiomer," indicating that the steric configurations are the opposite of that described or implied by the descriptors in the name (set in italics)	*ent*-symplocosigenol
epi-	abbreviation for "epimer," indicating diastereoisomers that have the opposite configuration at only 1 of 2 or more tetrahedral stereogenic centers present in the respective molecular entities (set in italics) (This prefix has other meanings in other contexts. See Appendix K of the *Naming and Indexing of Chemical Substances for Chemical Abstracts 2007*.[10])	16-*epi*-vobasan *and* 2-*epi*-α-tocopherol
exo-	an outer position as opposed to an inner position; used mainly in bicyclic and polycyclic ring systems in which an *exo* substituent points away from the interior part of the structure (set in italics) (compare with *endo-*)	*exo*-2-dimethylbicyclo[2.2.2]octane
L-	absolute configuration to the left at the α-carbon in an amino acid compared with serine or at the highest-numbered center of chirality in a carbohydrate compared with glyceraldehyde when displayed in a Fischer projection (set as a small capital letter) (compare with D-)	L-alanine *and* L-arabinose
l, levo-	**obsolete term** for the rotation of the plane of polarization to the left; the preferred prefix is "(–)-"	(no longer used)
meso-	a term describing achiral members of a set of diastereoisomers that also includes one or more chiral members	*meso*-tartaric acid
rac-	abbreviation for "racemic," a mixture of equal amounts of enantiomers (set in italics)	*rac*-leucine
syn-	similar to *cis* (ie, on the same side of a defined structural feature) (set in italics)	*syn*-benzaldoxime *and* *syn*-7-bromobicyclo[2.2.1]hept-2-ene
trans-	atoms or groups on opposite sides of a reference plane (set in italics)	*trans*-but-2-ene *and* *trans*-cinnamic acid

uration in which the preferred atoms or the groups attached to each atom with a double bond lie on the same side of a reference plane that is perpendicular to the plane defined by the atoms of the double bond and its adjacent atoms. "(*E*)" stands for "*entgegen*," meaning "opposite." It describes a configuration in which the preferred atoms or the groups attached to each atom with a double bond lie on opposite sides of the reference plane. Unless doing so would create ambiguity, the prefixes "*cis*" and "*trans*" (see Table 17.11) may be substituted for "*Z*" and "*E*," respectively.

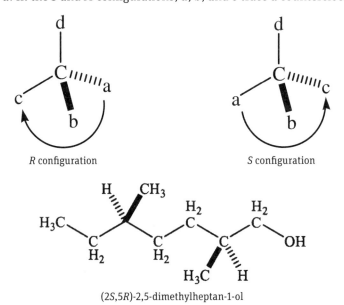

(*Z*)-2-chlorobut-2-ene   (*E*)-but-2-ene
cis-2-chlorobut-2-ene   *trans*-but-2-ene

17.11.1.2 Chiral Compounds

Chiral centers are described by *R* or *S* in italics for atoms and by *P* or *M* in helical molecules.[2,3] In the tetrahedral models below, the atoms or groups are ranked as $a > b > c > d$. In the *R* and *P* configurations, *a*, *b*, and *c* trace a clockwise path when they are viewed as opposite *d*. In the *S* and *M* configurations, *a*, *b*, and *c* trace a counterclockwise path.

R configuration   *S* configuration

(2*S*,5*R*)-2,5-dimethylheptan-1-ol

The "*R*" and "*S*" symbols describe an absolute configuration. If the absolute configuration is not known, the relative configuration can be given by using the symbols "*R**" (spoken as "*R*-star") and "*S**." These are assigned in the same manner as "*R*" and "*S*" but in such a way that *R** is the first cited chiral center.

17.11.2 Inorganic Compounds

Inorganic compounds exhibit many more geometries than organic compounds. Consequently, descriptions of stereochemical configurations for inorganic compounds are more complex than those for organic compounds. Not only the geometry of polyhedra must be describedbut so too must the order of the attached ligating atoms. In addition,

Table 17.12. Inorganic Polyhedra

Coordination polyhedron	Coordination number	Polyhedral symbol
Linear	2	*L-2*
Angular	2	*A-2*
Trigonal plane	3	*TP-3*
Trigonal pyramid	4	*TPY-4*
Tetrahedron	4	*T-4*
Square plane	4	*SP-4*
Square pyramid	5	*SPY-5*
Trigonal bipyramid	5	*TBPY-5*
Octahedron	6	*OC-6*
Trigonal prism	6	*TPR-6*
Pentagonal bipyramid	7	*PBPY-7*
Monocapped octahedron	7	*OCF-7*
Square face monocapped trigonal prism	7	*TPRS-7*
Cube	8	*CU-8*
Square antiprism	8	*SAPR-8*
Dodecahedron	8	*DD-8*
Hexagonal bipyramid	8	*HBPY-8*
Trans bicapped octahedron	8	*OCT-8*
Triangular face bicapped trigonal prism	8	*TPRT-8*
Square face bicapped trigonal prism	8	*TPRS-8*
Square face tricapped trigonal prism	9	*TPRS-9*
Heptagonal bipyramid	9	*HBPY-9*

organic compounds having their own stereochemical configurations can be ligands in inorganic coordination compounds.

17.11.2.1 Geometrical Descriptors for Inorganic Polyhedra

The geometrical descriptors for the polyhedra of inorganic coordination compounds are listed in Table 17.12.[1]

17.11.2.2 Configuration Index

The configuration index portion of the inorganic stereochemical descriptor is a single digit or a pair of digits that identifies the positions of the ligating atoms on a polyhedron. Consult the International Union of Pure and Applied Chemistry's recommendations[1] for the detailed procedure for deriving the digits.

17.11.2.3 Chirality Symbol

The final component of the inorganic coordination stereochemical descriptor is the chirality symbol. This symbol is determined by orienting each polyhedron in a specific way and noting the direction, going from a lower-priority ligating atom to a higher-priority one along the edge of the polyhedron. If the direction is clockwise, the chirality symbol is an italic "*C*," and if it is counterclockwise, the chirality is "*A*."

(*OC*-6-32-*C*)-diaminedibromo(ethane-1,2-diamine-*N*,*N'*)cobalt(1+)

17.11.2.4 Other Descriptors for Inorganic Structural and Stereochemical Nomenclature

Other descriptors used in systematic inorganic stereochemical and geometric nomenclature include the following:

antiprismo-	8 atoms forming a rectangular antiprism
cis-	2 groups occupying adjacent positions in a coordination sphere
dodecahedro-	8 atoms forming a triangular dodecahedron
fac-	3 groups occupying the corners of the same face of an octahedron
hexahedro-	8 atoms forming a hexahedron (eg, a cube)
hexaprismo-	12 atoms forming a hexagonal prism
icosahedro-	12 atoms forming a triangular icosahedron
mer-	meridonal, 3 groups occupying the vertices of an octahedron in such relationship that one is *cis* to the other 2, which are both *trans*
octahedro-	6 atoms forming an octahedron
pentaprismo-	10 atoms forming a pentagonal prism
quadro-	4 atoms forming a quadrangle (eg, a square)
tetrahedro-	4 atoms forming a tetrahedron
trans-	2 groups directly across a central atom from each other (ie, in the polar positions of a sphere)
triangulo-	3 atoms forming a triangle
triprismo-	6 atoms forming a triangular prism

17.12 POLYMER NOMENCLATURE

Polymers are characterized by multiple repetitions of atoms or groups of atoms linked to each other in amounts sufficient to provide a set of properties that would not differ significantly with the addition or deletion of one or a few of the repeating units.

17.12.1 Source-Based Nomenclature

Polymers have long been named by attaching the prefix "poly" to the name of the real or assumed monomers from which they are derived. Enclose the monomer name in parentheses when it consists of 2 or more words or is prefixed by a descriptor.

> polystyrene
> poly(methyl acrylate)
> poly(L-lysine)

However, source-based nomenclature for polymers is misleading in that the chemical structures of monomers are different from the chemical structures of the monomeric units in polymers (eg, a monomer $CH_2=CHX$ versus the polymer unit $-CH_2-CHX-$).

"Copolymers" are polymers derived from more than one species of monomer. The series of italicized infixes listed below provides information about the structures of copolymers[5]:

-alt-	alternating, as in "poly(styrene-*alt*-maleic anhydride)"
-block-	block, as in "polystyrene-*block*-polybutadiene"
-co-	unspecified, as in "poly(styrene-*co*-methyl methacrylate)"
-graft-	graft, as in "polybutadiene-*graft*-polystyrene"
-per-	periodic, as in "poly[formaldehyde-*per*-(ethylene oxide)-*per*-(ethylene oxide)]"
-ran-	random, as in "poly[ethylene-*ran*-(vinyl acetate)]"
-stat-	statistical, as in "poly(styrene-*stat*-butadiene)"

Other infixes to describe polymer structures have been published by the International Union of Pure and Applied Chemistry.[11]

17.12.2 Structure-Based Nomenclature

For regular organic polymers (ie, polymers that have only one species of repeating group), names are formulated by using the pattern "poly(constitutional repeating unit)." The constitutional repeating unit, or "CRU," is the name of a bivalent organic group or a combination of bivalent groups.[5]

Structure-based name	Source-based name
poly(methylene)	polyethylene
poly(propylene)	polypropene
poly(1-acetoxyethylene)	poly(vinyl acetate)

Names for single-strand and quasi-single-strand inorganic and coordination polymers are formed in a similar way to names for organic polymers.[5] The prefix "*catena*" in italics is used before "poly" to emphasize the linear nature of these polymers.

catena-poly(dimethyl tin)   *catena*-poly(palladium-di-μ-chloro)

17.13 COMPOUNDS OF BIOCHEMICAL IMPORTANCE

Style, format, and nomenclature conventions for many compounds of biochemical importance are based on recommendations for natural products developed by the International Union of Pure and Applied Chemistry (IUPAC).[12] The major compound classes, including alkaloids, steroids, and terpenoids, and some relatively minor classes, such as carotenoids, corrinoids, and porphyrins, fit cleanly into these general rules (see Section 17.13.1, "Natural Products"). For other compound classes, specific guidelines and recommendations are described in Section 17.13.2, "Amino Acids, Peptides, and Polypeptides," through Section 17.13.19, "Vitamins."

17.13.1 Natural Products

The IUPAC rules for natural products[12] provide guidance for naming and numbering. Unless otherwise specified in those rules, methods of name construction and the principles of organic nomenclature, such as adding suffixes and prefixes, are applied to fundamental parent structures. For each class, a parent structure is selected that reflects the basic skeletal structure that is common to most compounds in the class. To the extent possible, the parent structure is given a name derived from one of the members of the class and numbered according to the numbering established for the class.

The following special operations, which are denoted by prefixes in roman type, can be applied to any fundamental parent structure:

abeo   migration of one end of a single bond from its original position in a fundamental parent structure to another position

Note: ("Abeo" is preceded by a descriptor $x(y{\rightarrow}z)$, in which "x" is the locant of the unchanged end of the bond, "y" is the locant of the original position of the moving bond, and "z" is the locant of the new position of the moving bond, as in "3aH-5(4→3)-abeopodocarpane."

cyclo   creation of an additional ring by means of a direct link between any 2 nonadjacent atoms of a fundamental structure

Note: "Cyclo" is preceded by locants of the positions joined (eg, "16βH-1,16-cyclocorynan.")

de   removal of an atom or group from a parent structure

Note: Examples include "deoxy," the removal of an oxygen atom, and "dehydro," the removal of a hydrogen atom. Sometimes when a hydrogen atom is removed, it is subsequently restored, as in the case of demethylmorphine.

des   removal of a terminal ring from a fundamental parent structure with the addition of an appropriate number of hydrogen atoms

Note: The prefix "des" is followed by a letter in italics that designates the removed ring (eg, "des-A-androstane")

homo   addition of a methylene group (ie, $-CH_2-$) between 2 skeletal atoms of a fundamental parent structure

Note: "Homo" is preceded by the locant of the added atom (eg, "10a-homotaxane" and "5(6)a-homoergoline").

nor   removal of an unsubstituted saturated or unsaturated skeletal atom from a ring or from an acyclic portion of a fundamental parent structure

Note: "Nor" is preceded by the locant of the atom removed (eg, "1,20-dinorprostane").

seco   cleavage of a saturated or unsaturated ring bond with the addition of an appropriate number of hydrogen atoms at each new terminus created

Note: "Seco" is preceded by the locants at the ends of the removed bond (eg, 2,3-secoyohimban).

Combinations of these prefixes can be used to describe more drastic changes to fundamental parent structures.

13(17)a-homo-12,18-dinor-5α-pregnane

In addition, skeletal replacement nomenclature can be used to describe the introduction of heteroatoms (eg, "1-thiaergoline"). It can also describe rings or ring systems added to the fundamental structures by adaptation of fusion nomenclature (eg, "benzo[2,3]-5α-androstane"), by bridging (eg, "3α,8-epidioxy-5α,8α-androstane"), and by functionalization (eg, "aspidospermidine-3α,4α-diyl carbonate").

Some compound classes, such as electron-transfer proteins and folates, have no unique style and format issues. For compounds not listed in the sections below, consult the IUPAC's website at https://iupac.org/.

17.13.2 Amino Acids, Peptides, and Polypeptides

17.13.2.1 Amino Acids

Amino acids are compounds containing both an amino group and a carboxylic acid group. Most naturally occurring amino acids are α-amino acids. An α-amino acid is a carboxylic acid in which an amino group has been substituted at the carbon atom immediately adjacent to the carboxylic acid group (ie, the α-carbon).

The simplest α-amino acid is glycine, whose chemical structure is NH_2-CH_2-COOH and whose systematic name is 2-aminoacetic acid. All other α-amino acids contain at least one chiral center (ie, the α-carbon atom), and they have 2 optically active forms, or enantiomers, that are termed "D" and "L." The D-amino acids are often found in microorganisms, and the L-amino acids are found in higher organisms. See Table 17.13 for a list of α-amino acids frequently found in proteins.

Semisystematic names of substituted α-amino acids are formed according to the general principles of organic nomenclature, including those for stereochemistry, by attaching to the trivial name of the amino acid the name of the substituent group along with the appropriate numerical locant.

> *trans*-4-hydroxy-L-proline

Trivial names for unmodified amino acids are preferred in straight text. Standard 3-letter symbols, consisting of a capital letter followed by 2 lowercase letters, are frequently used in tables, diagrams, peptide sequences, and terms indicating residue numbers (eg, "Tyr-110" for the tyrosine residue 110 in a protein sequence). The 3-letter symbols represent the L- configuration of the chiral amino acids unless the enantiomer label D- or DL- is placed before the symbols. Although not generally needed, the L- enantiomer may be added to the 3-letter symbols for emphasis. Trivial names also have 1-letter symbols, which should be used only in tables and sequences and never in text. (See Table 17.13 for examples of both 3-letter and 1-letter symbols for amino acids.)

Additional details for naming amino acids are published in the *CRC Handbook of Chemistry and Physics*.[6]

17.13.2.2 Peptides and Polypeptides

Peptides are compounds containing 2 or more amino acids joined through a peptide linkage (ie, the peptide bond $-CO-NH-$) between the carboxylic acid group of one amino acid and the amino group of another. Peptides with 10 to 20 amino acid residues (ie, specific monomers within the polymeric chain) are termed "oligopeptides," while those with more amino acid residues are called "polypeptides."[6] A systematic name of any peptide with one or more peptide linkages is formed using the acyl forms of the names of all the peptide's constituent amino acids (see Table 17.13). The acyl forms are listed in sequence from the amino acid at the N-terminal on the left to the carboxy at the C-terminal on the right to the amino acid on the C-terminal. The last amino acid retains its trivial name instead of using its acyl form.

> L-valyl-L-alanylglycyl-D-phenylalanyl-L-tryptophyl-D-proline

The example above is a hexapeptide, which is a peptide containing 6 amino acid residues. In this example, "L-alanylglycyl" is not hyphenated between "L-alanyl" and "glycyl" because the latter is the only amino acid without a chiral carbon atom.

Additional details for naming peptides can be found in the *CRC Handbook of Chemistry and Physics*.[6]

Table 17.13. Names and Symbols of Common α-Amino Acids[a]

| Trivial name | Symbols | | Systematic name | Acyl radical | |
	3-letter	1-letter[b]		Structure	Name
alanine	Ala	A	2-aminopropanoic acid	$CH_3CH(NH_2)CO-$	alanyl
arginine	Arg	R	2-amino-5-guanidinopentanoic acid and 2-amino-5-(carbamimidoylamino)pentanoic acid	$H_2NC(=NH)NH(CH_2)_3CH(NH_2)CO-$	arginyl
asparagine	Asn	N	2-amino-3-carbamoylpropanoic acid and 2,4-diamino-4-oxobutanoic acid	$H_2NCOCH_2CH(NH_2)CO-$	asparaginyl
aspartic acid	Asp	D	2-aminobutanedioic acid	$-COCH_2CH(NH_2)CO-$, $HOOCCH_2CH(NH_2)CO-$, and $HOOCCH(NH_2)CH_2CO-$	Aspartoyl, α-aspartyl, and β-aspartyl
cysteine	Cys	C	2-amino-3-sulfanylpropanoic acid	$HSCH_2CH(NH_2)CO-$	cysteinyl[c]
glutamic acid	Glu	E	2-aminopentanedioic acid	$-COCH_2CH_2CH(NH_2)CO-$, $HOOCCH_2CH_2CH(NH_2)CO-$, and $HOOCCH(NH_2)CH_2CH_2CO-$	Glutamoyl, α-glutamyl, and γ-glutamyl
glutamine	Gln	Q	2-amino-4-carbamoylbutanoic acid	$H_2NCOCH_2CH_2CH(NH_2)CO-$	glutaminyl
glycine	Gly	G	2,5-diamino-5-oxopentanoic acid and 2-aminoacetic acid	H_2NCH_2CO-	glycyl
histidine	His	H	2-amino-3-(1H-imidazol-4-yl)propanoic acid	$(N_2C_3H_3)CH_2CH(NH_2)CO-$	histidyl
isoleucine	Ile	I	2-amino-3-methylpentanoic acid	$CH_3CH_2CH(CH_3)CH(NH_2)CO-$	isoleucyl
leucine	Leu	L	2-amino-4-methylpentanoic acid	$(CH_3)_2CHCH_2CH(NH_2)CO-$	leucyl
lysine	Lys	K	2,6-diaminohexanoic acid	$H_2N(CH_2)_4CH(NH_2)CO-$	lysyl
methionine	Met	M	2-amino-4-(methylsulfanyl)butanoic acid	$CH_3SCH_2CH_2CH(NH_2)CO-$	methionyl
phenylalanine	Phe	F	2-amino-3-phenylpropanoic acid	$C_6H_5-CH_2CH(NH_2)CO-$	phenylalanyl
proline	Pro	P	pyrrolidine-2-carboxylic acid	$NHCH_2CH_2CH_2CHCO-$	prolyl
serine	Ser	S	2-amino-3-hydroxypropanoic acid	$HOCH_2CH(NH_2)CO-$	seryl
threonine	Thr	T	2-amino-3-hydroxybutanoic acid	$CH_3CH(OH)CH(NH_2)CO-$	threonyl
tryptophan	Trp	W	2-amino-3-(1H-indol-3-yl)propanoic acid	$(C_8H_6N)CH_2CH(NH_2)CO-$	tryptophyl
tyrosine	Tyr	Y	2-amino-3-(4-hydroxyphenyl)propanoic acid	$4-HOC_6H_4CH_2CH(NH_2)CO-$	tyrosyl
valine	Val	V	2-amino-3-methylbutanoic acid	$(CH_3)_2CHCH(NH_2)CO-$	valyl
unspecified	Xaa	X			

[a] The trivial name refers to the L, D, or DL amino acid. For those that are chiral, only the L form is used for protein biosynthesis.

[b] Use of 1-letter symbols should be restricted to comparisons of long sequences.

[c] An exception to the rule for naming an acyl radical to differentiate it from the cysteic acid radical.

17.13.2.3 Sequence Conventions for 3-Letter Symbols

When representing the sequences of polypeptides or short proteins, it is customary to use the 3-letter symbols for amino acids in text and tabular material. The hyphens between the configurational prefixes and the amino acid symbols may be omitted for brevity.

With hyphens after configurational prefixes	Without hyphens after configurational prefixes
L-Val-L-Ala-Gly-D-Phe-L-Trp-D-Pro	LVal-LAla-Gly-DPhe-LTrp-DPro

Other conventions for writing peptides are as follows:

Ala-Leu	a hyphen between amino acid abbreviations represents a peptide linkage
-Ala	a hyphen preceding an amino acid abbreviation represents loss of H from the NH$_2$ of the amino group
Ala-	a hyphen following an amino acid abbreviation represents loss of OH from the COOH of the amino acid
-Ala-	hyphens preceding and following an amino acid abbreviation represent loss of H from the NH$_2$ and loss of OH from the COOH
| Lys *or* Lys |	a vertical bar represents a linkage through the side chain functional group
|5 Lys	represents substitution at C-5 of lysine

17.13.2.4 Sequence Conventions for 1-Letter Symbols

The 1-letter symbols for amino acids are used most frequently to display polypeptide or protein sequences and to display the alignment of homologous sequences. In the latter case, the spacing of symbols and punctuation is critical. When using 1-letter symbols, use an equal-space font (eg, MonoLisa or Monoid) in which each letter and each punctuation mark occupy the same amount of horizontal space. When reporting a single sequence, insert a space after every 10 symbols.

GDVEKGKKIF IMKCSQCHTV EKGGKHKTGP N

When comparing sequences, insert hyphens instead of spaces, and vary the number of hyphens as needed to maintain alignment.

human	GDVEKGKKIFIMKCSQCHTV-------EKGGKHKTGPN
Neurospora crassa	GFSAGDSKKGAKLFKTRCAQCHTL----EEGGGNKIGPA
Euglena gracilis	GDAERGKKLFESRAAQCHSA--------QKG-VNSTGPS
Paracoccusi denitrificans	NEGDAAKGEKEFN-KCKACHMIQAPDGTDI-KGGKTGPN

17.13.2.5 Side Chains

The term "side chain" refers to C-3 and higher-numbered carbon atoms and their substituents. One way to present a substitution involving a side chain is to place the symbol of the substituent within parentheses immediately after the symbol of the amino acid being substituted.

Cys(Et) *for S*-ethylcysteine
Ser(Ac) *for O^3*-acetylserine
Ac-Glu(OEt)-OMe *for O^5*-ethyl methyl *N*-acetylglutamate *or O^5*-ethyl *O^1*-methyl *N*-acetylglutamate

17.13.2.6 Modification of Named Peptides

Amino acids in named peptides are numbered starting from the terminal amino acid with the free amino group. When one or more of these amino acids is replaced, the name or symbol of the replacement amino acid is preceded by the number of the amino acid that was replaced (ie, the "residue number"). This combination is then placed within square brackets in front of the name of the peptide. Commas separate multiple amino acids. In the abbreviated form, the residue numbers are in superscript.

> [8-citrulline]vasopressin *or* [Cit[8]]vasopressin
> [5-isoleucine,7-alanine]angiotensin II *or* [Ile[5],Ala[7]]angiotensin II

Indicate that a peptide bond was replaced by another group by placing the Greek uppercase letter psi (ie, "Ψ") between the 2 residue numbers as superscripts. This should be followed by a comma and the symbol for the group. The entire combination should be placed within square brackets before the name of the peptide. In the abbreviated form, however, use parentheses around the symbol for the group and hyphens to represent the bonds attaching the group to the rest of the peptide.

> [[3]Ψ[4], CH$_2$-S]oxytocin *or* . . . -Ψ (CH$_2$-S)- . . .

Extensions of the peptide chain are named for the amino terminal by placing the acyl form of the extending amino acid before the name of the peptide (see Table 17.13). In contrast, extensions are named for the carboxy terminal by adding "-yl" to the name of the peptide and attaching that "yl" form to the name of the extending amino acid.

> **Extension named for amino terminal:**   valylvasopressin *or* Val-vasopressin
> **Extension named for carboxy terminal:**   (angiotensin II)ylglycine *or* angiotensin II-Gly

Insertion and deletion of amino acid residues are indicated by the terms "*endo*" in italics and "des" in roman, respectively.

> **Insertion:**   *endo*-4a-tyrosine-angiotensin II *or* *endo*-Tyr[4a]-angiotensin II
> **Deletion:**   des-7-proline-oxytocin *or* des-Pro[7]-oxytocin

17.13.3 Antibiotics and Other Medications

Antibiotics and other drugs are usually classified or grouped by their chemical structures. In addition, many medications have trade names, which are developed by their manufacturers.

17.13.4 Carbohydrates

The principles for naming carbohydrates have been outlined by the International Union of Pure and Applied Chemistry and the International Union of Biochemistry and Molecular Biology. Monosaccharide classes are listed in Table 17.14.

17.13.4.1 Nomenclature for Monosaccharides

The systematic names for monosaccharides have 4 components:

(1) the configurational symbol D or L in small capital Roman letters that is linked by a hyphen to the name of a monosaccharide
(2) a configurational prefix in lowercase italic letters that represents the relative positions of hydroxy groups
(3) a stem name that indicates the number of carbon atoms
(4) a characteristic ending that indicates the type of monosaccharide (eg, "aldose" or "ketose")

> D-*gluco*-hexose   L-*glycero*-D-*manno*-heptose   D-*ribo*-pentose

Table 17.14. Monosaccharide Classes[a]

General formula	Y	Class	Characteristic ending
	CH$_2$–OH	aldose	ose
X	CHO	dialdose	dialdose
|	COOH	uronic acid	uronic acid
[CH–OH]^n	CH$_2$–OH	alditol	itol
|	CH$_2$–OH	aldonic acid	onic acid
Y	COOH	aldaric acid	aric acid
X			
|			
[CH–OH]^n	CH$_2$–OH	ketose	ulose
|			
CO	CH$_2$–OH	aldoketose	osulose
|			
[CH–OH]^m	CH$_2$–OH	ketoaldonic acid	ulosonic acid
|			
Y			

[a] Adapted from Fox RB, Powell WH, editors. *Nomenclature of Organic Compounds: Principles and Practice*, 2nd ed. Table 31.3. Copyright 2001 American Chemical Society.

17.13.4.2 Abbreviations

Because the systematic names of substituted carbohydrates are quite long, abbreviations are used extensively.

> D-Glc*p with a hyphen or* DGlc*p without a hyphen*
> ("Glc" is the symbol for glucose, and italic "*p*" is the symbol for a pyranose ring. The entire abbreviation stands for "D-glucopyranose.")

> L-Rib*f or* LRib*f*
> ("Rib" is the symbol for ribose, and italic "*f*" is the symbol for a furanose ring. The entire abbreviation stands for "L-ribofuranose.")

> β-D-Xyl*p*-(1→4)-α-D-Gal*p*-(1→6)-α-D-Man*p*-(1→2)-β-D-Fru*f*
> *or*
> DXyl*p*(β1−4)DGal*p*(α1−6)DMan*p*(α1−2β)DFru*f*
> (Both of these condensed forms represent the same compound, namely, *O*-D-xylopyranosyl-(1→4)-α-D-galactopyranosyl-(1→6)-α-D-mannopyranosyl-β-D-fructofuranoside. In the shorter of the 2 condensed forms, en dashes replace arrows between anomeric descriptors.)

17.13.4.3 Stereoisomers

Enantiomers (ie, pairs of stereoisomers whose images mirror each other) can rotate plane-polarized light in the positive direction (ie, clockwise) or in the negative direction (ie, counterclockwise), or they may not rotate plane-polarized light at all. These optical rotations or lack thereof may be indicated by adding "(+)-," "(−)-," or "(±)-" before the configurational prefix. Do not use the obsolete symbols "*d*-," "*l*-," and "*dl*-." Racemates (ie, mixtures containing equal amounts of left- and right-handed enantiomers) are indicated by the prefix "DL" in small capital letters. Optically inactive structures that have a plane of symmetry are indicated by the italicized prefix "*meso*."

> (+)-D-glucose   (±)-glucose   DL-glucose   *meso*-galactitol

17.13.5 Carotenoids

Many carotenoids are acyclic tetraterpenes, a class of hydrocarbons called "carotenes" and their oxygenated derivatives, which are called "xanthophylls." Acyclic tetraterpenes consist of 2 groups of 4 isoprenoid units, each joined at their terminal carbon atoms.[6] The methyl groups of one 4-unit group are in opposite configurational relationships to the methyl groups of the second 4-unit group.[6] Each group of isoprenoid units is numbered from 1 to 20 and from 1' to 20' as shown below.

Specific names are constructed by adding 2 Greek-letter prefixes to the stem name "carotene." The appropriate Greek letters depend on the type of C-9 end groups (ie, positions 1–6, 16–18, 1'–6', and 16'–18'). The prefixes are in Greek-letter alphabetical order. The prefixes are separated by commas, and they are connected to the stem names with hyphens. (See Table 17.15 for the C-9 end group designations.) Substituents on the carbon chains are named according to the general rules of organic nomenclature, including those that apply to natural products.

> 1-ethoxy-1,2,7',8'-tetrahydro-ω,ω-carotene 3-hydroxy-3'-oxo-β,ε-caroten-16-oic acid
> 2,2'-dinor-β,β-carotene 2,3-seco-ε,ε-carotene

Carotenoid derivatives that are shortened by removing fragments from one or both ends of the molecule are indicated with the roman prefix "apo" or "diapo," which follows locants to indicate the point of cleavage.

> 9-apo-β-caroten-9-one   6'-apo-β-carotene   4,4'-diapo-ζ-carotene

Carotenoid derivatives in which all single and double bonds of a conjugated polyene system are shifted by one position are denoted by a pair of locants followed by the italic descriptor "*retro*."

> 4',11-*retro*-β,ψ-carotene   3-hydroxy-6',7-*retro*-β,ε-caroten-3'-one

17.13.6 Cyclitols

Cyclitols are cycloalkanes that have one hydroxy group on each of 3 or more ring atoms. They usually consist of 5- and 6-membered rings. Cyclitols and related compounds have stereochemical features characteristic of their class, so special methods of designating that stereochemistry are used. In other respects, the nomenclature of cyclitols follows the general rules of organic nomenclature.[6]

Cyclitols with one hydroxy group on each atom of a cyclohexane ring (ie, 1,2,3,4,5,6-hexahydroxycyclohexane) are called "inositols." Individual inositol stereoisomers are distinguished either by italicized prefixes or by a series of numbers enclosed in parentheses representing the positions of the hydroxy groups. A slash separates the numbers indicating hydroxy groups above and those below the pseudoplane of the cyclohexane

Table 17.15. Carotenoid C-9 End Groups[a]

End group type	Formula	Structure	Greek letter prefix
acyclic	C_9H_{15}		Ψ (psi)
cyclohexene	C_9H_{15}		β (beta)
cyclohexene	C_9H_{15}		ε (epsilon)
cyclohexane	C_9H_{15}		γ (gamma)
aryl	C_9H_{11}		φ (phi)
aryl	C_9H_{11}		x (chi)
cyclopentane	C_9H_{17}		κ (kappa)

[a] Adapted from International Union of Pure and Applied Chemistry; International Union of Biochemistry. Nomenclature of carotenoids (rules approved 1974). Pure Appl Chem. 1975;41(3):405–431, Rule 3.2. https://doi.org/10.1351/pac197541030405.

ring. Numbers preceding the slash indicate the hydroxy groups that are above, and those following the slash indicate the hydroxy groups below.

Italicized prefixes	Number series
cis-	(1,2,3,4,5,6/0)
neo-	(1,2,3/4,5,6)
chiro-	(1,2,4/3,5,6)
epi-	(1,2,3,4,5/6)
myo-	(1,2,3,5/4,6)
scyllo-	(1,3,5/2,4,6)
allo-	(1,2,3,4/5,6)
muco-	(1,2,4,5/3,6)

Several features are unusual to cyclitol nomenclature, even though some of them are common to carbohydrate nomenclature. The following are among the most notable:

(1) Because the principal functional group in cyclitols is most commonly hydroxy, variations are often emphasized in naming cyclitols.

> (2,3,4/1(COOH),5)-2,3,4,5-tetrahydroxycyclopentane-1-carboxylic acid

(2) Because most carbon atoms in cyclitols have hydroxy substituents with hydrogen atoms that are replaceable, the names of cyclitols should emphasize the substitution of the hydrogen atoms directly attached to carbon atoms.

> 2-*C*-methyl-*myo*-inositol

(3) Configurational D and L symbols in small capitals often follow a locant. The absence of the prefixes D-, L-, and DL- indicates a *meso* configuration. Consequently, the prefixes should not be omitted unless the *meso* form is intended.

> 1L-1-*O*-methyl-6-mercapto-6-deoxy-*chiro*-inositol

Current rules for configurational prefixes and for locants deviate from those used for older, traditional names for cyclitols. Consequently, care should be taken to avoid errors when converting from the older to the current format.

17.13.7 Enzymes

Enzymes are proteins with catalytic activity. Multienzymes are proteins with 2 or more catalytic functions to which distinct parts called "domains" of a polypeptide chain or distinct subunits contribute.

Enzyme names are formulated according to the recommendations of the Nomenclature Committee of the International Union of Biochemistry and Molecular Biology (IUBMB).[13] Most enzyme names end in "-ase." However, the long-established names of peptidases, such as "chymotrypsin," "thrombin," "coagulation factor Xa," and "subtilisin," are exceptions.

Restriction endonucleases have a standardized trivial nomenclature of their own in addition to their IUBMB designations. Italicization of the names of restriction endonucleases was abandoned in 2003. Each name consists of a 3-letter abbreviation for the source organism followed by designations derived from the name of a strain or the number of enzymes in the source organism (see also Section 21.2.8, "Single-Nucleotide Variants").

> Bce1229   TaqI   XmaII

17.13.7.1 Classification

Enzymes are classified into 6 main classes based on the reactions they catalyze. In turn, each main class has subclasses, and subclasses have sub-subclasses. Based on these 3 tiers of classification, 4 code numbers separated by periods are assigned to each enzyme. Each of these codes is preceded by the abbreviation "EC," which stands for IUBMB's Enzyme Commission.

(1) The first number represents the main class to which an enzyme belongs.

- Class 1 consists of oxidoreductases, which catalyze oxidoreduction reactions. The substrate is the hydrogen donor, so these names often end with "dehydrogenase" or "reductase." Use "oxidase" only when O_2 is the acceptor. "Oxidoreductase" is normally used only in systematic names.

> EC 1.1.1.1 *for* alcohol dehydrogenase

- Class 2 is for transferases, which transfer a group from one compound to another. The names in this class often end with "transferase."

  EC 2.4.1.162 *for* aldose β-D-fructosyltransferase
  EC 2.1.1.1 *for* nicotinamide *N*-methyltransferase

- Class 3 is for hydrolases, which cleave C–O, C–N, C–C, and some other bonds. Names of hydrolases are often formed by appending "ase" to the substrate name.

  EC 3.2.1.15 *for* polygalacturonase

- Class 4 consists of lyases, which cleave C–O, C–N, C–C, and some other bonds by elimination, leaving double bonds or rings, or by adding groups to double bonds. Names often end in "decarboxylase," "aldolase," or "dehydratase."

  EC 4.1.2.13 *for* fructose–bisphosphate aldolase (note that the prefix for "bisphosphate" is "bis," not "bi")

- Class 5 is for isomerases, which catalyze geometric or structural changes within individual molecules. Names vary with the type of geometric or structural change.

  EC 5.3.1.9 *for* glucose-6-phosphate isomerase

- Class 6 consists of ligases, which catalyze the coupling of 2 molecules and the hydrolysis of a pyrophosphate bond in ATP or similar triphosphate. Most ligases have names in the form of "X–Y ligase."

  EC 6.3.1.1 *for* aspartate–ammonia ligase

(2) The second number represents the enzyme's subclass.
(3) The third number represents the sub-subclass.
(4) The last number is a serial number within the sub-subclass, assigned according to the order in which the enzymes were included in the sub-subclass.

The IUBMB has a complete list of the subclasses and sub-subclasses. Below are examples of common and systematic names for 2 enzymes:

Example No. 1
Common name: adenosinetriphosphatase, EC 3.6.1.3
Systematic name: ATP phosphohydrolase

(Adenosinetriphosphatase is a hydrolase, so it belongs to class 3. It catalyzes the hydrolytic cleavage of acid anhydrides, which places it in subclass 6 of hydrolases. It is also in sub-subclass 1, which is for phosphorus-containing hydrides. Adenosinetriphosphatase was the third enzyme thus categorized.)

Example No. 2
Common name: asparagine synthase (glutamine-hydrolyzing), EC 6.3.5.4
Systematic name: L-asparatate:L-glutamine amido–ligase (AMP-forming)

(Asparagine synthase [glutamine-hydrolyzing] is a ligase, which belongs to class 6. It catalyzes the formation of carbon–nitrogen bonds, which places it in subclass 3. This catalysis involves an amido–N donor, which places it in sub-subclass 5. This was the fourth enzyme thus categorized.)

17.13.7.2 Domain and Association Symbols

Catalytic domains, which are parts of polypeptide chains with catalytic functions, are indicated by capital letters from early in the alphabet (eg, "A," "B," and "C"), while substrate-carrier domains are indicated by capital letters from late in the alphabet (eg, "R," "S," and "T"). Regulatory domains are given lowercase letters from early in the alphabet (eg, "a," "b," and "c"). Domains in the same polypeptide chain are placed within the same pair of parentheses, and those from different chains are placed in separate sets of parentheses. Thus, "(ABC)" represents a multienzyme polypeptide, and "(A)(BC)" represents a multienzyme complex. Braces (ie, "{ }") may be used to indicate a stable association. For example, tryptophan synthase from *Escherichia coli* may be written as "{(A)2(B)2}," which indicates that the association is stable, or as "{(A)(B)}2," which indicates that each A chain binds one B chain tightly but that the 2 (A)(B) units are more loosely associated.

17.13.8 Lignans and Neolignans

Lignans and neolignans are large groups of natural products characterized by the coupling of 2 C_6C_3 units, each consisting of a propylbenzene structure. If the 2 C_6C_3 units are connected at the β position of each propyl group, the compound is lignane, which is the parent structure for all lignans. If the 2 C_6C_3 units are joined at any other position, the compound is neolignane, the parent structure of all neolignans. For a neolignane compound, the locants of the bond linking the 2 units are given in front of the name, and the second of these locants is primed.

lignane

8,4′-neolignane

Nomenclature for lignans and neolignans follows the general principles for naming natural products (see Section 17.13.1, "Natural Products"). Substituents are described by substitutive nomenclature principles. Several different numbering systems are in use, so the numbering system being used should be indicated with trivial, systematic, and semisystematic names and their modifications.

9-norlignane
4′,7-epoxy-8,3′-neoligna-7,8′-diene
7,9′:7′,9-diepoxylignane
3,3′,4,4′-tetramethoxy-7,9′-epoxylignan-7′-yl acetate
5′H-2V-oxa-8,3′-neolignane

Higher analogs of the lignans and neolignans composed of 3 or more C_6C_3 units are called "sesquineolignans," "dineolignans," etc, analogous with the terminology of terpenes. In addition, one of the terminal units for these higher analogs has unprimed numbers as locants, and locants for the other units are primed serially.

17.13.9 Lipids

17.13.9.1 Fatty Acids, Neutral Fats, Long-Chain Alcohols, and Long-Chain Bases

Fatty acids are the aliphatic monocarboxylic acids obtained from hydrolysis of naturally occurring fats and oils. Neutral fats are esters of glycerol and fatty acids and are termed "monoacylglycerol," "diacylglycerol," and "triacylglycerol." The terms "glyceride," "diglyceride," and "triglyceride" are discouraged for neutral fats. "Long-chain alcohol" refers to an alcohol whose carbon chain length is greater than 10. "Long-chain base" refers to any base containing a long-chain aliphatic radical. Sphinganine—(2S,3R)-2-aminooctane-1,3-diol—and its homologs, stereoisomers, and derivatives are called "sphingoids" or "sphingoid bases."

The prefix "*sn*," which stands for "stereospecifically numbered," designates as "C-1" the carbon atom at the top of a Fischer projection that shows a vertical carbon chain with the C-2 hydroxy group to the left. The lowercase italic symbol "*sn*" followed by a hyphen appears before the glycerol term.

The following examples are representative of these natural product classes:

4D-hydroxysphinganine	*also*	(2S,3S,4R)-2-aminooctadecane-1,3,5-triol *and* Phytosphingosine
sphingosine	*also*	(4E)-sphingenine, *trans*-4-sphingenine, *and* (2S,3R,4E)-2-aminooctadec-4-ene-1,3-diol
tristearoylglycerol	*also*	*tri-O*-stearoyl glycerol *and* glycerol tristearate
glycerol 2-phosphate	*also*	2-phosphoglycerol
sn-glycerol 1-phosphate	*also*	L-(glycerol 1-phosphate) *and* D-(glycerol 3-phosphate)
sn-glycerol 3-phosphate	*also*	D-(glycerol 1-phosphate) *and* L-(glycerol 3-phosphate)

Do not abbreviate the term "fatty acid." In tables and text in which several fatty acids are described, each fatty acid can be designated by the number of carbon atoms and the number of double bonds, separated by a colon. When 2 or more fatty acids have the same notation, they can be differentiated by adding the positions of the double bonds in parentheses. In contrast, radicals of fatty acids can be identified by adding "acyl."

16:0 *for* palmitic acid
16:0(acyl) *for* the acyl radical of palmitic acid
18:1 *for* oleic acid

18:3(6,9,12) *for* γ-linolenic acid
18:3(9,12,15) *for* α-linolenic acid

17.13.9.2 Phospholipids

A phospholipid is any lipid containing phosphoric acid as a monoester or diester. The prefix "phospho" may be used as an infix to designate phosphodiester bridges, as in "glycerophosphocholine." In addition, "phospho" can be used as an alternative to "*O*-phosphono" and "*N*-phosphono."

Phosphatidic acids are derivatives of glycerol phosphate in which both of the hydroxy groups of glycerol are esterified with fatty acids. In naming phosphatidic acids, stereospecific numbering may be used to emphasize the position of the phosphate group.

Trivial name: lecithin
Semisystematic name: 3-*sn*-phosphatidylcholine
Systematic name: 1,2-diacyl-*sn*-glycero-3-phosphocholine

17.13.9.3 Glycolipids

Glycolipids are compounds in which one or more monosaccharide residues are linked to a lipid group through a glycosyl group. Because the structures are often complex, abbreviations are prevalent in glycolipid names. However, "acyl" should not be abbreviated. To further shorten the abbreviated formulas, "D-" may be omitted unless needed to avoid ambiguity. In addition, hyphens may replace left-to-right arrows between anomeric descriptors.

> 1,2-diacyl-3-β-D-galactosyl-*sn*-glycerol *or* 1,2-diacyl-*sn*-glycerol
> 3-β-D-galactoside mucotriaosylceramide *or* McOse₃Cer *or* Gal(β1→4)Gal(β1→4)Glc(1→4)Cer

17.13.10 Nucleic Acids and Related Compounds
17.13.10.1 Nucleic Acids
Nucleic acids are compounds consisting of nucleotide units, purine bases, or pyrimidine bases attached to a ribosyl or deoxyribosyl group that is joined by phosphoric acid residues in ester linkages with the hydroxy groups of monosaccharides. The 5′-phosphates (ie, mono, di, tri) of the common nucleosides may be represented by commonly used abbreviations of the form (ie, "AMP," "ADP," and "ATP" for adenosine). For single nucleoside and nucleic acid representations, the 3-letter formulations in Table 17.16 are preferred (eg, "Ado-5′*PPP*"), whereas for chains, the single-letter symbols in Table 17.16 are used, with connecting hyphens or the lowercase letter "p" representing phosphate linkages.

Designate the purine and pyrimidine bases and their nucleosides by the 3-letter symbols in Table 17.16. Each 3-letter symbol has an initial capital letter followed by 2 lowercase letters. Use these symbols for single bases or short sequences of bases. Limit the corresponding 1-letter symbols in Table 17.16 to tables and sequences, and never use them in text.[6]

Table 17.16. Purines, Pyrimidines, and Nucleosides: Recommended Symbols[a]

Purine or pyridimine	3-letter symbol	Nucleoside	3-letter symbol	1-letter symbol
adenine	Ade	adenosine	Ado	A
		guanosine	Guo	G
guanine	Gua	inosine	Ino	I
		thioinosine	Sno	. . .
xanthine	Xan	xanthosine	Xao	X
hypoxanthine	Hyp	. . .	. . .	. . .
thymine	Thy	ribosylthymine	Thd	T
cytosine	Cyt	cytidine	Cyd	C
		uridine	Urd	U
uracil	Ura	thiouridine	Srd	S
		pseudouridine	Ψrd	Ψ[b]
orotate	Oro	orotidine	Ord	O
unknown purine	Pur	unknown purine nucleoside	Puo	R
unknown pyrimidine	Pyr	unknown pyrimidine nucleoside	Pyd	Y
unknown base	. . .	unknown nucleoside	Nuc	N

[a] Adapted from the International Union of Pure and Applied Chemistry and the International Union of Biochemistry.[14]
[b] Q may replace Ψ for computer work.[14]

17.13.10.2 General Principles for Describing Chains

One method for describing chains is based on chain direction. The conventional representation of a polynucleotide chain is from the 5′ terminus to the 3′ terminus. A nucleotide, which is the repeating unit of a polynucleotide chain, is composed of 3 parts: the D-ribose, or 2-deoxy-D-ribose monosaccharide ring; the phosphate group; and the purine or pyrimidine base. The monosaccharide ring and the phosphate form the backbone of the chain. A nucleotide unit is defined by the sequence of atoms from the phosphorus atom at the 5′ end to the oxygen atom at the 3′ end of the monosaccharide. The following are equivalent representations for a 6-nucleotide unit:

GAATTC *or* G-A-A-T-T-C *or* pGpApApTpTpC

Another method for describing chains is to designate specific nucleotide units by a letter or an italic number in parentheses, starting with the first nucleotide residue in the sequence.

A(1)   pU(5)   C(10)p   pG(*i*)p

With this method, atoms in each of the constituents follow the standard numbering practice.

C2, C2(5)   N3, N3(10) O^5 P(*i*+1)   H1   H2″O5′H

17.13.10.3 Functional Forms of DNA and RNA

When used as general terms or in reference to preparations of specific molecules, different types of deoxyribonucleic acid (DNA) and ribonucleic acid (RNA) and the methods that use them may be abbreviated:

cDNA *for* complementary DNA   RNAi *for* RNA interference
cRNA *for* complementary RNA   rRNA *for* ribosomal RNA
dsDNA *for* double-stranded DNA   siRNA *for* small interfering RNA
dsRNA *for* double-stranded RNA   snRNA *for* small nuclear RNA
mRNA *for* messenger RNA   snoRNA *for* small nucleolar RNA
miRNA *for* micro-RNA   ssDNA *for* single-stranded DNA
mtDNA *for* mitochondrial DNA   ssRNA *for* single-stranded RNA
nRNA *for* nuclear RNA   tRNA *for* transfer RNA

17.13.10.4 Transfer RNAs

Transfer RNAs that accept specific amino acids are named by attaching the name of the amino acid or its 3-letter symbol to the stem "tRNA" in the following format:

alanine tRNA *or* tRNA^{Ala} (for the nonacylated form)
alanyl-tRNA *or* Ala-tRNA *or* Ala-tRNA^{Ala} (for the aminoacylated form)

When 2 or more tRNAs accept the same amino acid, differentiate the isoacceptors by subscript numbers. Sources may be included in parentheses after the symbol.

tRNA$_1^{Ala}$ tRNA$_2^{Ala}$ alanyl-tRNA$_2^{Ala}$ (*E. coli*)

Write the codon triplet as a short sequence of symbols for the nucleic acids with no punctuation (eg, the codons for histidine are "CAU" and "CAC").

17.13.10.5 Conventions for 1-Letter Symbols

DNA and RNA sequences are usually presented from left to right, indicating the order from the 5′ to the 3′ end of the molecule. If another order is used, indicate the correct orientation. Because the spacing of symbols and punctuation is critical in the alignment of homologous sequences, use an equal-space font in which each letter and each punctuation mark occupy the same amount of horizontal space. When reporting a single sequence, insert a space after every 10 symbols.

> GAATTCCAUT CCTAGTAAGG GTCTCATCCT AGT

When comparing sequences, insert hyphens instead of spaces to maintain alignment. Additional hyphens may be needed before the last segments to keep the sequences aligned.

> | *Bacillus subtilis* | GAATTC-AUT-CCTAGTAAGG-GTCTCA-----CTAG |
> | *Staphylococcus aureus* | GAATTCCAUT-CCTAGTAAGG-GTCTGA--TCCTAGT |
> | *Tetrahymena caudatus* | GAAATCCAUT-CCCTAGTAAC-GGTCTCATCC-TAGT |

Indicate noncovalent associations between base pairs by a single raised dot (eg, "A·T base pair"), and indicate ratios between bases in the format "(A+T)/(G+C)," not "AT/GC" or "A+T/G+C."

17.13.11 Peptide Hormones

Many peptide hormones have well-established trivial names, some of which are so long that the hormones are known mainly by abbreviations. To preclude the proliferation of abbreviations and to decrease the use of existing abbreviations, IUPAC and IUBMB proposed alternative names based on 3 principles.

(1) New names for hormones of the adenohypophysis should end with "tropin."

> corticotropin *for* adrenocorticotropic hormone (*do not* abbreviate as "ACTH")

(2) Hypothalamic-releasing factors should end with "liberin."

> corticoliberin *for* corticotropin-releasing factor (*do not* abbreviate as "CRF")

(3) Hypothalamic release-inhibiting factors should end with "statin."

> somatostatin *for* somatotropin release-inhibiting factor

The peptide sequences of these hormones vary significantly between species. Therefore, the names of the hormones are essentially generic terms that do not denote distinct chemical entities. Consequently, the source of each hormone should be indicated in manuscripts when appropriate.

17.13.12 Phosphorus-Containing Compounds

In inorganic and organic chemistry, the nomenclature for phosphorus-containing compounds is well defined. However, strict application of those rules to biochemically important compounds usually results in complicated names that are inconvenient for most biochemists and biologists. The following principles are recommended for use in biochemical, biological, and medical fields[6]:

(1) Phosphoric esters, $RO-PO(OH)_2$, are named as O-substituted phosphoric acids or as substituted alcohols.

> glycerol 1-phosphate *or* glycero-1-phosphate (the second form is acceptable)
> choline *O*-(dihydrogen phosphate) *or* choline *O*-(phosphate)

O-phosphonocholine ("phosphono" may be contracted to "phospho"; "phosphoryl" is incorrect for this group)

(2) The prefixes "bis," "tris," and so on should be used to indicate 2 or more independent phosphoric acid residues. In contrast, the prefixes "di," "tri," and so on should be used to denote phosphate chains.

fructose 1,6-bisphosphate *not* fructose 1,6-diphosphate
adenosine diphosphate *not* adenosine bisphosphate

(3) Phosphodiesters, or $-PO(OH)-$, are named by using the infix "phospho."

glycerophosphocholine

(4) Nucleoside triphosphate analogs (ie, compounds in which methylene groups, imido groups, or sulfur atoms replace an oxygen atom bridging 2 phosphorus atoms) may be named by indicating within brackets the locants of the phosphorus atoms being bridged. The brackets indicate the name of the replaced group, and the segment in brackets should be followed by the name of the replacement group. Alternatively, the brackets may be omitted.

adenosine 5'-[α,β-methylene]triphosphate *or* adenosine 5'-α,β-μ-methylene-triphosphate

17.13.13 Prenols

Prenols are alcohols containing one or more isoprene units. The term "prenol" (from "iso*pren*oid alco*hol*") is recommended to describe the structure shown below.

$$\text{H--(CH}_2\text{--}\underset{3}{\overset{\overset{\displaystyle CH_3}{|}}{C}}\text{=}\underset{2}{CH}\text{--}\underset{1}{CH_2)_n}\text{--OH}$$

The carbon adjacent to the hydroxy group is numbered 1. The carbon of the methyl group attached to C-3 is numbered "31." This unusual representation occurs in many natural products containing rings and chains with fixed numbering systems. The repeating C_5H_8 unit that is enclosed within parentheses in the above structure is called an "isoprene unit" or an "isoprene residue." The compounds containing this unit are called "isoprenoids."

In addition, if the value of n is known, the compounds and their derivatives are named accordingly.

$n = 6$ *is* hexaprenol
$n = 7$ *is* heptaprenyl diphosphate (ester of diphosphoric acid)

17.13.14 Prostaglandins

Prostaglandins are icosanoids, naturally occurring compounds derived from the parent C_{20} acid, prostanoic acid. Systematically, prostaglandins are named on the basis of the parent hydride names "prostane" and "thromboxane." The structure-modifying prefixes given in Section 17.13.1, "Natural Products," may be applicable.

Trivial names for prostaglandins are divided into A, B, C, D, E, F, G, H, and I families. These families are distinguished by the presence and position of unsaturation groups and oxo and hydro groups, which result in trivial names such as "prostaglandin A_1," (PGA$_1$) and "prostaglandin A_2" (PGA$_2$). In addition, the F family is subdivided into α and β groups to distinguish between 2 stereochemical forms. "Prostaglandin $F_{2\alpha}$" and "prostaglandin $F_{2\beta}$" are typical designations. The subscript numbers used with family-letter designations indicate specific double-bond configurations.

Subscript number	Double-bond configuration
1	13E
2	5Z and 13E
3	5Z, 13E, and 17Z

Because of the complexity of prostaglandin compounds and their cumbersome systematic names derived from the rules for naming organic compounds, trivial names should be used when substituting or modifying common prostaglandin compounds.

Trivial name	Semisystematic prostane name (see Section 17.13.1, "Natural Products")
ent-prostanoic acid	(8α,12β)-prostan-1-oic acid
4,5,6-trinor-3,7-*inter*-*m*-phenylene-3-oxaprostaglandin A$_1$	(13E,15S)-15-hydroxy-9-oxo-3-oxa-4,5,6-trinor-3,7-*inter*-[1,3]phenyleneprosta-10,13-dien-1-oic acid (The infix *"inter"* has no official recognition for semisystematic prostane names.)
	2-[3-({{(1S,5R)-5-[(3S)-3-hydroxyoct-1-en-1-yl]cyclopent-3-en-1-yl}methyl)phenoxy]acetic acid (This is a substitutive name. See Section 17.6, "Substitutive Nomenclature.")
(19R)-19-hydroxyprostaglandin B$_1$	(13E,15S,19R)-15,19-dihydroxy-9-oxoprosta-8(12),13-dien-1-oic acid
(18Z)-18,19-didehydroprostaglandin E$_1$	(13E,15S,18Z)-11α,15-dihydroxy-9-oxoprosta-13,18-dien-1-oic acid
11,15-anhydro-11-*epi*-prostaglandin E$_1$	(13E,15S)-11β,15-epoxy-9-oxoprost-13-en-1-oic acid
2,3,4,5-tetranorprostaglandin E$_1$	(13E,15S)-11α,15-dihydroxy-9-oxo-4,5,6,7-tetranorprost-13-en-1-oic acid
8-*iso*-prostaglandin E$_1$	(13E,15S)-11α,15-dihydroxy-9-oxo-8α-prost-13-en-1-oic acid

The infixes *"ent," "epi," "inter,"* and *"iso"* in trivial names are italicized. *"Ent"* means inversion of all stereochemical features; *"epi,"* inversion of the normal configuration of a substituent at the numbered position; *"inter,"* replacement of the carbons at the numbered positions by the specified infix; and *"iso,"* inversion of the normal chirality at the numbered center.

17.13.15 Proteins

Proteins are polypeptides of specific sequences of numerous amino acids. Although authors disagree on the minimum number of amino acids required to classify a polypeptide as a protein, the consensus is that more than 50 are required.[6] In contrast to the standardized nomenclature for enzymes (see Section 17.13.7, "Enzymes"), no broadly applicable system exists for naming nonenzyme proteins. The following are some of the common systems.

17.13.15.1 Protein Designations Based on Open Reading Frames

Gene sequencing has led to identifying potential protein-coding regions that are referred to as "open reading frames" (ORFs). An identified protein may be designated according to the convention "protein ORFnumber," which includes the number of amino acids in the protein or the sequential numbering of the amino acids (eg, "protein ORF216" would indicate a protein that is 216 amino acids in length). This system of designation refers to both the DNA region containing the ORF and the protein resulting from the translation coded by the DNA region.

17.13.15.2 Protein Designations Based on Molecular Weight

Molecular weight may be used as the basis for the names of proteins. For example, "p21" and "p75" are proteins with molecular weights of 21 and 75 kilodaltons (kDa), respectively. If posttranslationally modified by proteolytic cleavage, these proteins may be referred to as "gp21" and "pp75," which would be a glycoprotein of 21 kDa and a phosphoprotein of 75 kDa, respectively.

Designations based on molecular weight may be further modified by including the gene symbol as a superscript. For example, $gp160^{env}$ designates the 160-kDa glycoprotein encoded by the *env* gene in the human immunodeficiency virus (HIV). This form, in turn, may be used to identify proteins that are processed to smaller molecular-weight forms encoded by a single gene, such as the processing of HIV's $gp160^{env}$ to $gp120^{env}$ and $gp41^{env}$.

Molecular-weight designations may also be used in combination with a protein's class, especially if the protein has multiple isoforms or multiple subunits. This is done by designating the class as an abbreviation in superscript after the molecular-weight designation (eg, "$p75^{NGFR}$" is a 75-kDa isoform of a nerve growth factor receptor).

17.13.15.3 Protein Designations Based on Gene Names

Many protein names are derived from the symbols for corresponding gene names. Variations of this system can be found throughout the literature. For example, when gene symbols are in lowercase italic letters, a protein can be designated by capitalizing the initial or all letters of the gene symbol and setting the resulting designation in Roman letters. When nonmutant gene symbols are fully capitalized (eg, the symbol for yeast), a protein is sometimes indicated by setting the gene symbol in roman characters, capitalizing all letters or just the initial letter, and adding "p" as a suffix. Essential to distinguishing these closely related pairs of symbols is using italic characters for the gene symbols and roman characters for the protein symbols.

Gene symbol	Encoded protein
ras	Ras
myc	Myc
NPL3	NPL3p or Npl3p

Names for proteins representing mutants characterized by the replacement of a single amino acid are also sometimes derived from the corresponding gene symbol. For example, if an amino acid is replaced with valine at position 12 of the Ras protein, the resulting protein may be indicated as "Ras^{Val12}."

17.13.15.4 Abbreviated Protein Sequences

In tables and other formats that use abbreviations, protein sequences can be abbreviated using the 3-letter or 1-letter symbols for the proteins' constituent amino acids (for

Table 17.17. Characteristic Groups of Retinoids

Characteristic group	Systematic name	Other names
$-CH_2OH$	retinol	vitamin A, vitamin A alcohol, vitamin A_1, vitamin A_1 alcohol, axerophthol, and axerol
$-CHO$	retinal	vitamin A aldehyde, vitamin A_1 aldehyde, retinene, and retinene$_1$
$-COOH$	retinoic acid	vitamin A acid, vitamin A_1 acid, and tretinoin
$-CH_3$	deoxyretinol (*not* retinene *or* retinane)	axerophthene
$-CH_2-O-CO-CH_3$	retinyl acetate	
$-CH_2-NH_2$	retinylamine	
$-CH_2=NOH$	retinal oxime	
$-CO-O-CH_2CH_3$	ethyl retinoate	

conventions, see Sections 17.13.2.3, "Sequence Conventions for 3-Letter Symbols," and 17.13.2.4, "Sequence Conventions for 1-Letter Symbols").

17.13.16 Retinoids

Retinoids are compounds consisting of 4 isoprenoid units[6] joined to form a cyclohexene ring attached to an acyclic chain with 4 double bonds terminated by a characteristic group. See Table 17.17 for the systematic and other names of characteristic groups of retinoids.

Basic retinoid structure

"Vitamin A" is a generic descriptor for retinoids having the biological activity of retinol, which is a retinoid with a terminal hydroxy functional group. Recommended names are based on 3 defined parent compounds: retinol, retinal, and retinoic acid.

Substituted derivatives are named using prefixes according to the general rules of organic nomenclature. The stereoparent name implies that the polyene chain has the *trans* configuration around all double bonds unless the contrary is indicated. Use the stereochemical prefixes "*Z*" and "*E*" for all double bonds whenever "*cis-*" and "*trans-*" might be ambiguous.

neovitamin A *or* 13-*cis*-retinol *or* (7*E*,9*E*,11*E*,13*Z*)-retinol

17.13.17 Steroids

Steroids are compounds having the skeletal structure of cyclopenta[*a*]phenanthrene, with or without methyl groups at positions C-10 and C-13. From steroids, derivatives can be created by bond scissions, ring expansions or contractions, and bond rearrangements. The names of steroids and their derivatives are based on the standard principles of the nomenclature of organic chemistry.

The basic structure of steroids with numbering is shown above. Atoms or groups projecting below the structure are indicated by broken lines (ie, ||||||||| or -----), while those projecting above the structure are designated by solid wedges (ie, ◄■) or thick lines (ie, ■■■).[6] In steroid names, "α" indicates atoms or groups projecting below the structure, while "β" indicates those above the structure.

The absolute stereochemistry of some chiral centers of steroids is defined by the name of the parent. The stereochemistry of other chiral centers is indicated with "α," "β," "R," and "S." When the configuration is not known, "ξ" (ie, xi) is used. Examples are given in Table 17.18.

17.13.18 Terpenes

Terpenes are hydrocarbons of biological origin with carbon skeletons formally derived from isoprene (ie, $CH_2=C(CH_3)CH=CH_2$). They are divided into classes on the basis of their number of isoprene units: Hemiterpenes have C_5; monoterpenes, C_{10}; sesquiterpenes, C_{15}; diterpenes, C_{20}; sesterterpenes, C_{30}; tetraterpenes (also known as "carotenes"), C_{40}; and polyterpenes, C_{5n}. Derivatives, especially oxygenated derivatives, are called "terpenoids." Terpenes may be structurally modified significantly by the prefixes described in Section 17.13.1, "Natural Products." These prefixes give rise to a wide variety of organic structures.

17.13.19 Vitamins

"Vitamin" is a generic term for many unrelated organic compounds with specific biological activities essential in minute quantities for normal metabolism and growth of organisms. Its use in phrases like "vitamin A activity" and "vitamin B_6 deficiency" is appropriate. Trivial names for individual vitamins are widely used, but they are often ambiguous. Therefore, trivial names should be used for vitamins only if there is no chance of misinterpretation.

The term "vitamin" is lowercased. In addition, numbers in vitamin names are often written as subscripts rather than on the line with a hyphen (eg, "vitamin B_{12}" instead of "vitamin B-12"). However, either form is acceptable as long as it is used consistently throughout a manuscript.

Table 17.18. Representative Steroids

CH$_3$ at C-19?	CH$_3$ at C-18?	Side chain at C-17	Implied configurations	Name
No	No	None	8β,9α,10β,13β,14α	gonane
No	Yes	None	8β,9α,10β,13β,14α	estrange
Yes	Yes	None	8β,9α,10β,13β,14α	androstane
Yes	Yes	$-CH_2CH_3$	8β,9α,10β,13β,14α	pregnane
Yes	Yes	Me⁄⁄⁄⁄ —20(H) —22—23—24 (attached to C-17)	8β,9α,10β,13β,14α,20R	cholane
Yes	Yes	Me⁄⁄⁄⁄ —20(H) —22—23—24—25—27, 26	8β,9α,10β,13β,14α,20R	cholestane
Yes	Yes	Me⁄⁄⁄⁄ —20(H) —22—23—24(H, Me)—25—27, 26	8β,9α,10β,13β,14α,20R,24S	ergostane
Yes	Yes	Me⁄⁄⁄⁄ —20(H) —22—23—24(H, CH$_2$—CH$_3$)—25—27, 26	8β,9α,10β,13β,14α,20R,24S	poriferastane and stigmastane[a]

[a] "Poriferastane" is the International Union of Pure and Applied Chemistry's name for this steroid, and "stigmastane" is the Chemical Abstracts Service's name.

Systematic names of vitamins and descriptions of their stereochemistry are determined by the rules of the individual classes of compounds to which the vitamins belong.

Vitamin	Class of compound
A	retinoids
B$_6$	pyridoxines
B$_{12}$	Corrinoids and tetrapyrroles
D	steroids
E	tocopherols
K	prenols

CITED REFERENCES

1. Connelly NG, Damhus T, Hartshorn RM, Hutton AT, editors. Nomenclature of inorganic chemistry, IUPAC recommendations 2005. RSC Publishing; 2005. Also known as the IUPAC Red Book. Also available at https://iupac.org/wp-content/uploads/2016/07/Red_Book_2005.pdf

2. Favre HA, Powell WH. Nomenclature of organic chemistry: IUPAC recommendations and preferred names 2013. RSC Publishing; 2014. Also known as the IUPAC Blue Book. https://doi.org/10.1039/9781849733069

3. Jones RG et al. Compendium of polymer terminology and nomenclature, IUPAC recommendations 2008. 2nd ed. RSC Publishing; 2009. Also known as the IUPAC Purple Book. https://doi.org/10.1039/978 1847559425

4. Liébecq C, editor. Biochemical nomenclature and related documents. 2nd ed. Portland Press; 1992. Also known as the IUPAC White Book.

5. Leigh J, editor. Principles of chemical nomenclature: a guide to IUPAC recommendations, 2011 ed. Royal Society of Chemistry; 2011.

6. Rumble J, editor. CRC handbook of chemistry and physics. 102nd ed. CRC Press; 2021.

7. Speight JA, editor. Lange's handbook of chemistry. 17th ed. McGraw-Hill; 2004.

8. Banik GM, Baysinger G, Kamat P, Pient N. The ACS guide to scholarly communication. American Chemical Society; 2020. https://doi.org/10.1021/acsguide

9. International Union of Pure and Applied Chemistry, Organic Chemistry Division, Commission on Nomenclature of Organic Chemistry. Revised nomenclature for radicals, ions, radical ions and related species (IUPAC Recommendations 1993). Pure Appl Chem. 1993;65(6):1357–1455. Prepared by Powell WH. https://publications.iupac.org/pac-2007/1993/pdf/6506x1357.pdf

10. Chemical Abstracts Service. Naming and indexing of chemical substances for chemical abstracts 2007. American Chemical Society; 2008.

11. International Union of Pure and Applied Chemistry, Macromolecular Division, Commission on Macromolecular Nomenclature. Source-based nomenclature for non-linear macromolecules and macromolecular assemblies. (IUPAC recommendations 1997). Pure Appl Chem. 1997;69(12):2511–2521. Prepared by Kahovec J et al. https://doi.org/10.1351/pac199769122511

12. International Union of Pure and Applied Chemistry, Organic Chemistry Division, Commission on Nomenclature of Organic Chemistry. Revised section F: natural products and related compounds. (IUPAC recommendations 1999). Pure Appl Chem. 1999;71(4):587–643. Prepared by Giles PM Jr. https://doi.org/10.1351/pac199971040587. Corrections and modifications (2004). Pure Appl Chem. 2004;76(6):1283–1292. Prepared by Favre HA et al. https://doi.org/10.1351/pac200476061283

13. Moss GP, preparer. Enzyme nomenclature: recommendations of the Nomenclature Committee of the International Union of Biochemistry and Molecular Biology on the nomenclature and classification of enzymes by the reactions they catalyse. International Union of Biochemistry and Molecular Biology; [updated 2021]. Prepared in consultation with the Joint Commission on Biochemical Nomenclature of the International Union of Biochemistry and Molecular Biology and the International Union of Pure and Applied Chemistry. https://iubmb.qmul.ac.uk/enzyme/

14. International Union of Pure and Applied Chemistry; International Union of Biochemistry. Abbreviations and symbols for nucleic acids, polynucleotides and their constituents. Rules approved 1974. Pure Appl Chem. 1974;40(3):277–290. https://doi.org/10.1351/pac197440030277

18 Chemical Kinetics and Thermodynamics

Editor: Raymond Lambert, MLS, ELS

18.1 INTRODUCTION

Kinetic studies are used in interpreting biochemical reactions and in extrapolating the results of in vitro and in vivo scientific studies. Because of the long history of accurate notation in the physical sciences, terminology is much more consistent in the field of kinetics than in the life sciences.[1–4] This can be attributed in part to multiple commissions in biothermodynamics working from the late 1970s to the mid-1980s to achieving greater systemization and consistency for presenting data and results across fields.[5–7]

18.2 UNITS IN BIOTHERMODYNAMICS

Examples of quantities used specifically in thermodynamics are listed in Table 18.1.

18.2.1 Mass

The term "specific" preceding the name of a quantity means "divided by mass." For this use of "mass," the Système International d'Unités (SI) uses "kilogram" for the unit of measurement. For a quantity represented by a capital letter (eg, "V" for volume), the symbol for the specific quantity is often the corresponding lowercase letter (eg, "v" for specific volume).

specific volume ($v = V/m$), with units $m^3 \cdot kg^{-1}$

The term "molar" preceding the name of a quantity means "divided by the amount of substance" or, occasionally, "divided by substance concentration." For this use, the

Table 18.1. Symbols and Units for Some Thermodynamic Quantities[a]

Quantity name	Quantity symbol	Name of SI unit	Symbol of SI unit
Volume	V	cubic meter	m^3
Force	F	newton	$N = m \cdot kg \cdot s^{-2}$
Density	ρ	kilogram per cubic meter	$kg \cdot m^{-3}$
Pressure	p	pascal	$Pa = N \cdot m^{-2}$
Viscosity (ie, dynamic viscosity)	n	pascal second	$Pa \cdot s$
Energy	E	joule	$J = N \cdot m$
Heat	q or Q	joule	J
Work	w or W	joule	J
Internal energy	U	joule	J
Enthalpy	H	joule	J
Gibbs energy	G	joule	J
Helmholtz energy	A	joule	J
Entropy	S	joule per kelvin	J/K or $J \cdot K^{-1}$
Power	P	watt	W
Heat capacity			
At constant pressure	C_p	joule per kelvin	J/K or $J \cdot K^{-1}$
At constant volume	C_V	joule per kelvin	J/K or $J \cdot K^{-1}$
Osmotic pressure	Π	pascal	Pa
Chemical potential of B	μ_B	joule per mole	J/mol or $J \cdot mol^{-1}$
Absolute activity of B	λ_B	(dimensionless)	
Activity coefficient of B			
With reference to Raoult's law	f_B	(dimensionless)	
With reference to Henry's law			
Molality basis	γ_m or $_B$	(dimensionless)	
Concentration basis	γ_c or $_B$	(dimensionless)	
Mole fraction basis	γ_x or $_B$	(dimensionless)	
Osmotic coefficient	φ	(dimensionless)	

SI, Système International d'Unités (International System of Units)

[a] Symbols for physical quantities represented by single Greek or Roman letters should be printed in italic type. Symbols for units should be in roman type.

amount of substance is in the SI unit "mole," whose symbol is "mol." "Molar" is typically represented by a lowercase subscript letter "m" when referring to a quantity and by a capital "M" when referring to an amount concentration.

     molar volume (V_m), with unit $m^3 \cdot mol^{-1}$ 0.5 M

Where possible, report thermodynamic quantities in molar quantities. However, for research with macromolecules and research in quantum chemistry, either molecular mass or molar mass may be used. The values of molecular mass and molar mass are numerically identical, but the units differ.

(1) For molecular mass, use either the dalton (Da) or the unified atomic mass unit (u), which are alternative names for the same unit: $1\ Da = 1\ u = m(^{12}C)/12$. The expression "$m(^{12}C)$" represents the mass of the carbon atom in its electronic ground state and at rest. Neither

the dalton nor the unified atomic mass unit is an SI unit, but they are appropriate units for the mass of an atom. Both terms have been accepted for use in conjunction with the SI.[7]

> The molecular mass of subunit A is 76,000 Da.
> The molecular mass of subunit A is 76,000 u.

(2) For molar mass, the coherent SI unit is kilogram per mole (kg·mol^{-1}). However, gram per mole (g·mol^{-1}) is commonly used.

> The molar mass of subunit A is 76,000 g·mol^{-1}.

Do not use "dalton" when referring to molecular weight. The term "molecular weight" is a historical synonym for "relative molecular mass" and is, therefore, a quantity of dimension 1 (ie, it is dimensionless).

> The molecular weight of subunit A is 76,000.

18.2.2 Volume, Temperature, Energy, and Density

The SI unit for volume is the cubic meter (m^3). However, in the expressions for most units, the use of submultiples, including "deci" and "centi," is acceptable. The cubic centimeter (cm^3, not cc) and the cubic decimeter (dm^3) are also acceptable. It is acceptable to use "liter" (L) as a specialized name for the cubic decimeter. The submultiples "milliliter" and "microliter" also can be used as specialized names.

In thermodynamic calculations and kinetic studies, express thermodynamic temperature in the SI base unit of kelvin (K). However, temperatures may also be expressed in degrees Celsius (°C).

The SI unit for energy is the joule (J). The use of "calorie" is generally discouraged because various different "calories" exist. In the field of nutrition, for example, "calorie" was used for "kilocalorie." What nutritionists formerly called "1 calorie" is now "4.184 kJ."

The SI unit for density is kilogram per cubic meter (kg·m^{-3}), but a more convenient unit is gram per cubic centimeter (g·cm^{-3}).

18.3 NOTATIONS FOR VARIABLES, STATES, AND PROCESSES

18.3.1 Variables

Symbols for quantities should generally be single letters of the Greek or Roman alphabet. This rule covers both variables and fixed quantities, such as the speed of light and the mass of a proton. Both capital and lowercase letters may be used. Single-letter variables should be printed in italic type (see also Section 4.2.2, "Type Styles for Symbols"). Where necessary, a symbol may be modified by a subscript or superscript that conveys a specified meaning, or a symbol may be qualified by information in parentheses.[3,7] For example, a symbol for a variable may be appended within parentheses to the symbol for a thermodynamic function.

> $\mu^{\circ}_{B}(T)$ (the chemical potential of substance B is modified by superscript "°," which indicates "standard state," and by "T" in parentheses, which indicates dependence on thermodynamic "temperature")

18.3.2 States

Particular states of matter are indicated by superscript symbols added to the symbols for thermodynamic functions (eg, "$G°$," which represents standard Gibbs energy). The 7 most frequently encountered states are represented by the following symbols, set superscript:

Activated complex	‡
Apparent	app
Excess	E
Ideal	id
Infinite dilution	∞
Pure substance	⋆
Standard	**o** *or* ⊖

States of aggregation are indicated by abbreviations that are set in roman type, enclosed in parentheses, and placed after the symbols for properties.[3] For example, "$V^*(cr)$" represents the volume of a substance in its pure crystalline state. The symbols below are used for states of aggregation.

Amorphous solid	am
Aqueous solution	aq
Crystalline	cr
Fluid	f
Gas	g
Liquid	l
Liquid crystal	lc
Solid	s
Solution	sln
Vitreous substance	vit

The following examples demonstrate how these symbols are used:

$C_V(f)$	constant volume heat capacity of a fluid
NaCl(sln)	sodium chloride in solution
$NH_3(aq)$	ammonia in aqueous solution

18.3.3 Processes

For state functions in which a total differential exists, use the Greek capital letter delta (Δ) before a thermodynamic quantity symbol to indicate a change in that thermodynamic quantity (ie, the final value minus the initial value). Use a subscript immediately after the Δ to indicate the type of change. For example, the subscript "f" in "$\Delta_f H$" indicates that the change is in enthalpy of formation. Symbols for the most frequently occurring processes are listed below[3]:

Combustion	c
Formation	f
Melting (fusion)	fus
Mixing of fluids	mix
Reaction (general)	r
Solution (dissolution)	sol
Sublimation (evaporation)	sub
Transition (between 2 phases)	trs
Vaporization (evaporation)	vap

The following examples demonstrate how these symbols are used:

Table 18.2. Selected Terms, Symbols, and Units Related to Microbial Processes[a]

Quantity	Quantity symbol	Symbol of SI unit	Symbol of customary unit
Vapor pressure	$p^\star$	Pa	atm *or* bar
Molar activation energy	E_A	J·mol^{-1}	J·mol^{-1}
Molar activation energy for specific growth rate	E_μ	. . .	kJ·mol^{-1}
Molar chemical energy[b]	U, H, G, A, and others	J·mol^{-1}	kJ·mol^{-1}
Area per volume	a	m^{-1}	cm^{-1}
Yield of cell biomass per amount of ATP produced in cells	Y_{ATP}	kg·mol^{-1}	g·mol^{-1}
Gas hold-up (volume of gas divided by volume of liquid)	ε	L *or* l	L *or* l
Doubling time, biomass	t_d	s	min *or* h *or* d
Growth rate, colony radial (rate of extension of biomass colony on a surface)	K_r	m·s^{-1}	m·h^{-1}
Specific mass metabolic rate	q	s^{-1}	h^{-1}
Mutation rate	w	s^{-1}	h^{-1}

SI, Système International d'Unités (International System of Units)
[a] Based on International Union of Pure and Applied Chemistry, Applied Chemistry Division, Commission on Biotechnology.[8]
[b] For types, see entries under "Energy" in Table 18.1.

$\Delta_f S^\circ(HgCl_2, cr, 298.15\ K)$ indicates the change in standard entropy due to the formation of crystalline mercuric chloride from its elements at a thermodynamic temperature of 298.15 K.
$\Delta_r G^\circ(1{,}000\ K)$ indicates the change in standard Gibbs energy due to a chemical reaction at standard pressure and a thermodynamic temperature of 1,000 K.

In contrast, reaction quantities, such as the heat of a reaction, are written as $\Delta_r H$, in which "Δ_r" means differentiation with respect to extent of reaction. The reaction quantities always use the unit "per mole."

18.4 MICROBIAL PROCESSES

The International Union of Pure and Applied Chemistry (IUPAC) published a detailed list of terms important in the biotechnology of microbial processes.[8] Many of these terms, definitions, and symbols are the same as those in other IUPAC documents (eg, the IUPAC Green Book[3]), or they have been brought into agreement with those IUPAC recommendations. These terms describe and define general quantities, intensive quantities, rate quantities, concentrations, and amount fractions. In addition, the IUPAC recommends symbols and units in common use, including units recommended by the Système International d'Unités (SI). See Table 18.2 for representative examples.

18.5 ION MOVEMENT

Research in physiology involves aspects of solute transport studied by convection, diffusion, and permeation across membranes (ie, ion movement and transport). Stan-

Table 18.3. Selected Symbols for Ion Mass Transport and Exchange[a]

Symbol	Definition or description	Symbol of unit (if applicable)
Principal symbol[b]		
a	relative activity, molar	
A	area of indicator concentration–time curve, excluding recirculation	$mol \cdot s \cdot L^{-1}$
C	concentration	$mol \cdot L^{-1}$
CV	coefficient of variation	(dimensionless)
D	diffusion coefficient	$cm^2 \cdot s^{-1}$
E	electrical potential	V
F	flow	$cm^3 \cdot s^{-1}$
F_B	blood flow to an organ	$cm^3 \cdot g^{-1} \cdot min^{-1}$
$h(t)$	transport function	s^{-1}
Hct	hematocrit, the fraction of the blood volume composed of erythrocytes	(dimensionless)
J	flux	$mol \cdot s^{-1} \cdot cm^{-2}$
P	permeability coefficient for a solute traversing a membrane	$cm \cdot s^{-1}$
RD	relative dispersion	(dimensionless)
Subscript symbol		
$_A$	arterial	
$_B$	blood	
$_C$ or $_{cap}$	capillary, or the region of blood–tissue exchange	
$_i$ and $_j$	indices in series, in summations, or in elements of arrays	
$_{in}$ or $_i$	into, inside, and inflow	
$_{ISF}$ or $_I$	interstitial fluid space (extravascular and extracellular fluid)	
$_M$	membrane	
$_{out}$ or $_o$	out of, outside, and outflow	
$_P$	plasma	
$_{RBC}$	red blood cell	
$_S$	solute	
$_T$	total	
$_V$	venous	

[a] Based on J.B. Bassingthwaighte and colleagues.[9]
[b] Single-letter principal symbols for physical quantities should be in italic type, and multiletter principal symbols should be in roman type.

dardized symbols and terms exist[9] (see Table 18.3 for some important examples of these symbols).

18.6 SOLUTIONS

A solution is a liquid phase or solid phase containing more than one component. The term "solvent" is generally used for the substance with the largest concentration in a solution, and the other components are called "solutes." However, in biological solutions, "solvent" is often used to refer to a fixed mixture of components (eg, water plus buffer components), and all other substances are considered to be solutes.

Use the following notations for the composition of solutions, recommended symbols, and appropriate units:

Composition	Symbol	Unit
Amount of substance B	n_B	mol
Concentration amount of solute substance B	c_B	mol·dm^{-3} *or* mol·L^{-1}
Mass concentration of substance B	P_B *or* ρ_B	g·dm^{-3}
Molality of solute substance B	b_B *or* m_B	mol·kg^{-1}
Mole fraction of substance B	x_B	(dimensionless)
Mass fraction of substance B	w_B	(dimensionless)
Volume fraction of substance B	φ_B	(dimensionless)

Concentration, which refers to the amount-of-substance concentration, is sometimes called "molarity." A solution of 1 mol·dm^{-3} may be referred to as a "1 molar solution" or a "1 M solution."[3] The symbol m_B can refer to both the mass of B and the molality of B.[3] To avoid confusion, the symbol b_B is recommended for molality instead of m_B.[3]

18.7 ENZYME KINETICS

The practices followed in the field of enzyme kinetics closely resemble those in chemical kinetics.[1,4] However, enzyme-catalyzed reactions most frequently occur in liquid phases at constant pressure. Depending on the sensitivity of experiments, enzyme-catalyzed reactions may not require the rigorous detail associated with all variables in chemical kinetics.

18.7.1 Kinetic Equations

The International Union of Biochemistry and Molecular Biology[1] (IUBMB) has developed definitions of consumption and formation rates, reaction rates, and elementary and composite reactions. Especially important are the IUBMB's kinetic equations describing Michaelis–Menten and non-Michaelis–Menten relationships.

(1) Michaelis–Menten relationships use the following form to express the relationship between the rate (v) of an enzyme-catalyzed reaction and the substrate concentration (A):

$$v = V[A]/(K_{mA} + [A])$$

(2) Non-Michaelis–Menten relationships use the following form:

$$v = V'[A]/(K'_{mA} + [A] + [A]^2/K_{iA})$$

See Table 18.4 for examples of symbols for kinetic reactions.

18.7.2 Enzyme Activity

The catalytic activity of an enzyme is measured as the increase in the rate of conversion of a specified chemical reaction that the enzyme produces in a specific assay system.[1] "Rate of conversion," in turn, is defined as the rate of reaction expressed as an extensive quantity, measured as the increase in the amount of substance per unit of time. Catalytic activity may be expressed using the unit "katal," which uses the symbol "kat." Katal is

Table 18.4. Selected Terms, Symbols, and Units Recommended for Enzyme Kinetics[a]

Term	Symbol	Symbol of customary unit[b]
Concentration of substrate A	[A]	$mol \cdot dm^{-3}$
Concentration of inhibitor I	[I]	$mol \cdot dm^{-3}$
Concentration of product Y	[Y]	$mol \cdot dm^{-3}$
Catalytic constant	k_0	s^{-1}
Rate constant of any order n	k	$(mol \cdot dm^{-3})^{1-n} \cdot s^{-1}$
Forward and reverse rate constants for ith step	k_i, k_{-i}	as k
Inhibition constant	K_i	$mol \cdot dm^{-3}$
Michaelis constant	K_m	$mol \cdot dm^{-3}$
Michaelis constant for substrate A	K_{mA}	$mol \cdot dm^{-3}$
Time	t	s
Rate, or velocity, of reaction	v	$mol \cdot dm^{-3} \cdot s^{-1}$
Initial rate of reaction	v_0	$mol \cdot dm^{-3} \cdot s^{-1}$
Rate of conversion	ξ	$kat = mol \cdot s^{-1}$

[a] Based on International Union of Biochemistry, Nomenclature Committee.[1] The union's name has since been changed to the International Union of Biochemistry and Molecular Biology.

[b] In all cases, "dm³" may be replaced with "L" or "l" for "liter."

equivalent to mole per second ($mol \cdot s^{-1}$). One katal is the catalytic activity that will raise the rate of reaction by 1 mole per second in a specified assay system.[2]

The katal was approved at the 1999 Conférence Générale des Poids et Mesures as a special name for the unit "mole per second" within the Système International d'Unités.[7]

CITED REFERENCES

1. International Union of Biochemistry, Nomenclature Committee. Symbolism and terminology in enzyme kinetics: recommendations 1981. Eur J Biochem. 1982;128(2–3):281–291. Prepared by Cornish-Bowden A et al. Note: The International Union of Biochemistry is now the International Union of Biochemistry and Molecular Biology. https://doi.org/10.1111/j.1432-1033.1982.tb06963.x. [Correction in Eur J Biochem. 1993;213(1):1. https://doi.org/10.1111/j.1432-1033.1993.tb17727.x_1]

2. International Union of Biochemistry, Nomenclature Committee. Units of enzyme activity: recommendations 1978. Eur J Biochem. 1979;97(1):319–320. Note: The International Union of Biochemistry is now the International Union of Biochemistry and Molecular Biology. https://doi.org/10.1111/j.1432-1033.1979.tb13116.x

3. International Union of Pure and Applied Chemistry, Physical and Biophysical Chemistry Division. Quantities, units and symbols in physical chemistry. 3rd ed. RSC Publishing; 2007. Prepared for publication by Cohen ER et al. Also known as the IUPAC Green Book. Also available at https://doi.org/10.1039/9781847557889

4. International Union of Pure and Applied Chemistry, Physical Chemistry Division, Subcommittee on Chemical Kinetics. Symbolism and terminology in chemical kinetics (appendix No. V to manual of symbols and terminology for physicochemical quantities and units). Pure Appl Chem. 1981;53(3):753–771. Prepared for publication by Laidler KJ. https://doi.org/10.1351/pac198153030753

5. Interunion Commission on Biothermodynamics. Recommendations for measurement and presentation of biochemical equilibrium data. J Biol Chem. 1976;251(22):6879–6885. https://doi.org/10.1016/S0021-9258(17)32917-4

6. International Union of Pure and Applied Chemistry, International Union of Pure and Applied Biophysics, and International Union of Biochemistry, Interunion Commission on Biothermodynamics. Recommendations for the presentation of thermodynamic and related data in biology (1985). Eur J Biochem. 1985;153(3):429–434. Prepared for publication by Wadsö I. https://doi.org/10.1111/j.1432-1033.1985.tb09320.x

7. Bureau International des Poids et Mesures. Le Système international d'unités (SI) (The International System of Units [SI]). 9th ed. Organisation Intergouvernementale de la Convention du Mètre; 2019. English translation available in PDF format at https://www.bipm.org/documents/20126/41483022/SI-Brochure-9-EN.pdf

8. International Union of Pure and Applied Chemistry, Applied Chemistry Division, Commission on Biotechnology. Selection of terms, symbols and units related to microbial processes. (IUPAC recommendations 1992). Pure Appl Chem. 1992;64(7):1047–1053. Prepared for publication by Eroshin VK. Also available at http://www.iupac.org/publications/pac/1992/pdf/6407x1047.pdf

9. Bassingthwaighte JB et al. Terminology for mass transport and exchange. Am J Physiol. 1986;250(4 Pt 2):H539–H545. https://doi.org/10.1152/ajpheart.1986.250.4.H539

ADDITIONAL REFERENCES

Banik G, Baysinger G, Kamat P, Pient N. The ACS guide to scholarly communication. American Chemical Society; 2020. https://doi.org/10.1021/acsguide

International Union of Pure and Applied Chemistry, Physical Chemistry Division. Abbreviated list of quantities, units and symbols in physical chemistry. The Union; 2000 [modified 2000 Jun 12]. Prepared for publication by Homann KH. http://old.iupac.org/reports/1993/homann/index.html

International Union of Pure and Applied Chemistry, Physical Chemistry Division, Commission on Chemical Kinestics. A glossary of terms used in chemical kinetics, including reaction dynamics (IUPAC recommendations 1996). Pure Appl Chem. 1996;68(1):149–192. Prepared for publication by Laidler KJ. https://doi.org/10.1351/pac199668010149

Speight JG, editor. Lange's handbook of chemistry. 17th ed. McGraw-Hill; 2017.

Ulicky L, Kemp T, editors. Comprehensive dictionary of physical chemistry. Prentice Hall; 1992.

19 Analytical Chemistry

Editor: Tina L. Fleischer

19.1 FUNDAMENTALS AND SOURCES

"Analytical chemistry," also known as "chemical analysis," is the science of obtaining, processing, and communicating information about the composition of matter. The information may be qualitative (eg, specifies what is in a sample) or quantitative (eg, specifies how much is in a sample). Accordingly, much of analytical chemistry is built around methods, which in turn can involve a variety of techniques.

In general, the form and style of presenting information in analytical chemistry are the same as those for any of the other physical sciences. ASTM International, formerly the American Society for Testing and Materials, offers relevant guidance in its style manual, *Form and Style for ASTM Standards*.[1] Especially helpful is Part G in that style manual. In addition, the American Chemical Society's style guide[2] provides detailed information on presenting chemical information, including the results of analytical studies.

When established methods have been used, it is sufficient to present the experimental details in summary form:

> . . . was determined by a Lineweaver–Burk evaluation.

However, if new techniques are used or if important modifications are introduced to basic methods, the experimental procedures should be described in sufficient detail to allow other experienced researchers to duplicate the procedures.

> . . . titanium was determined by precipitation of the hydrous oxide followed by filtration and ignition to TiO_2.

The International Union of Pure and Applied Chemistry (IUPAC) and other organizations have developed terminology covering the broad range of techniques used by analytical chemists. For example, *Quantities, Units, and Symbols in Physical Chemistry*,[3] known as the "IUPAC Green Book," is a collection of all IUPAC-recommended terms in the field of physical chemistry. In addition to defining a wide range of physical quantities and units, the IUPAC Green Book lists fundamental physical constants, provides conversion factors for units, and defines acronyms and other abbreviations used in physical chemistry.

The IUPAC has also published collections of all its terminology recommendations in the fields of analytical chemistry (ie, the "IUPAC Orange Book"[4]) and clinical laboratory sciences (ie, the "IUPAC Silver Book"[5]). In addition, a complete compilation of IUPAC terminology recommendations has been published as the "IUPAC Gold Book."[6]

19.2 ANALYTICAL TECHNIQUES AND TERMINOLOGY
19.2.1 Types of Techniques

The IUPAC recognizes 13 types of analytical techniques.[4] Some are qualitative, some are quantitative, and some can be both. The relevant terminology and the applicable units for each type of techniques are available in the IUPAC Orange Book.[4]

(1) Thermoanalytic and enthalpimetric methods encompass the following:
   - thermal analysis, in which a physical property of a substance is measured as a function of temperature
   - pyrometric analysis, in which the products of exposure to high temperatures are described
   - enthalpimetric analysis, in which the enthalpy change of a chemical reaction is measured

(2) In titrimetric analysis, the concentration of one substance is determined by adding known quantities of another substance with which it reacts. The result depends on the reaction's measurable end point, which is often visual. Visual end points include colors, precipitates, fluorescence, and turbidity. Examples of titrimetric analyses detailed in the Orange Book are acid–base, complexometric, chelatometric, oxidation–reduction, and precipitation titrations.

(3) Automatic analysis reduces human intervention through mechanization, instrumentation, or automation, each of which incorporates some degree of nonhuman decision-making.

(4) Electrochemical analysis is based on electrochemical processes and phenomena, including ion-selective and chemically modified electrodes. Such analyses use potentiometric, amperometric, and voltammetric techniques.

(5) Analytical separation methods encompass chromatography, ion exchange, liquid–liquid distribution (also known as "solvent extraction"), precipitation, electrophoresis, and centrifugation. In chromatography, for example, components in a mobile phase are separated from a stationary component. Chromatographic methods may be classified according to the shape of the chromatographic bed (eg, column or planar), the physical state of the mobile and stationary phases (eg, gas–liquid, liquid–liquid, gas–solid, and liquid–solid), the mechanism of separation (eg, adsorption, partition, ion exchange, exclusion, and affinity), and special methods (eg, gradient elution, isothermal chromatography, and reversed-phase chromatography). In addition, in precipitation, a substance in a solution is converted into a nonsoluble form. In electrophoresis, an electric current is applied through a matrix to separate molecules based on their size and electrical charge. In centrifugation, solutions of different densities are separated by spinning the sample at high speed around a central axis.

(6) Spectrochemical analysis uses techniques in which the components of a sample emit or

absorb radiation. These techniques include arc emission spectrometry, X-ray spectroscopy, flame emission spectrometry, and vapor generation methods.

(7) Other optical methods include photochemistry and light scattering. Photochemistry is the study of the chemical processes that occur when matter is exposed to light. In light scattering, small particles scatter light, and the resulting optical phenomena are observed.

(8) In mass spectroscopy, the masses of different molecules within a sample are measured. The results are used to identify the compounds within the sample.

(9) Diffraction methods (eg, X-ray, electron, and neutron diffraction) are used to investigate crystalline structures (see Section 25.6, "Crystals").

(10) Magnetic methods take advantage of the magnetic properties of substances to investigate their molecular structures. Magnetic resonance spectroscopy techniques include electron paramagnetic resonance (EPR) spectroscopy and nuclear magnetic resonance (NMR) spectroscopy.

(11) Kinetic methods use rates of reaction to determine the concentration of substances.

(12) In radioanalytical methods, the substance under investigation may be radioactive. Alternatively, a radioactive substance or ionizing radiation may be used to analyze the substance under investigation.

(13) Surface analysis is conducted primarily by means of spectroscopic methods, including several types of photoelectron emission spectroscopy, electron impact spectroscopies, Auger electron spectroscopies, and field emission electron spectroscopies. For each type of spectroscopy, the IUPAC Orange Book[4] lists the particles detected and their incidence.

19.2.2 Terminology and Abbreviations

The method of analysis is often abbreviated, but multiple abbreviations for the same term have arisen, which creates confusion. The IUPAC recommends a systematic method of generating abbreviations for technique names.[4] For each abbreviation, the first letter or letters represent descriptive adjectives, the second-to-last letter is for the type of probe or particle, and the last letter is for the type of technique. The following is a selection of abbreviations for those 3 elements.

Descriptive adjective	Probe or particle	Technique
A (absorption)	P (photon *or* potential)	S (spectroscopy *or* scattering)
E (emission *or* energy)	E (electron)	R (resonance)
R (reflection)	N (neutron)	D (diffraction)
T (transmission)	A (atom)	M (microscopy)
F (far)	I (ion)	I (ionization)
M (mid)	F (field)	
N (near)	R (radiation)	
H (high)		
L (low)		
S (scanning *or* surface)		
AR (angle-resolved)		
X *or* XR (X-ray)		
U *or* UV (ultraviolet)		

The following examples demonstrate how the IUPAC's systematic method creates abbreviations for methods of analysis.

LEED *for* low-energy electron diffraction
STEM *for* scanning transmission electron microscopy

However, some well-established abbreviations for techniques used in surface analysis do not follow this pattern. In such cases, use the established abbreviation.

NMR *for* nuclear magnetic resonance

See Table 17.5 in the IUPAC Orange Book[4] for an index of recommended abbreviations and techniques.

CITED REFERENCES

1. ASTM International. Form and style for ASTM standards. ASTM International; 2018. https://www.astm.org/FormStyle_for_ASTM_STDS.html

2. Banik GM, Baysinger G, Kamat P, Pient N. The ACS guide to scholarly communication. American Chemical Society; 2020. https://doi.org/10.1021/acsguide

3. International Union of Pure and Applied Chemistry, Physical Chemistry Division. Quantities, units, and symbols in physical chemistry. 3rd ed. RSC Publishing; 2007. Prepared for publication by Cohen ER et al. Also known as IUPAC Green Book. Also available at https://doi.org/10.1039/9781847557889

4. Hibbert DB, editor. Compendium of terminology in analytical chemistry. 4th ed. Royal Society of Chemistry, 2023. Prepared for International Union of Pure and Applied Chemistry. Also known as IUPAC Orange Book. Also available at https://doi.org/10.1039/9781788012881

5. International Union of Pure and Applied Chemistry, Clinical Chemistry Division. Compendium of terminology and nomenclature of properties in clinical laboratory sciences, recommendations 2016. RSC Publishing; 2017. Prepared for publication by Férard G, Dybkaer R, Fuentes-Arderiu X. Also known as IUPAC Silver Book. https://doi.org/10.1039/9781782622451

6. International Union of Pure and Applied Chemistry. Compendium of chemical terminology. 2nd ed. ICT Publishing House; 1997. Prepared for publication by McNaught AD, Wilkinson A. Updated 2019 July 1 by Nic M, Jirat J, Kosata B. Also known as IUPAC Gold Book. https://goldbook.iupac.org/

ADDITIONAL REFERENCES

ASTM dictionary of engineering science and technology. 10th ed. ASTM International; 2005.
International Organization for Standardization. International vocabulary of metrology—basic and general concepts and associated terms (VIM). 3rd ed. Bureau International des Poids et Mesures; 2012.
Joint Committee for Guides in Metrology, Working Group 1. Evaluation of measurement data—guide to the expression of uncertainty in measurement. Bureau International des Poids et Mesures; 2008. GUM 1995 with minor corrections. (JCGM 100:2008). Also available at https://www.bipm.org/en/committees/jc/jcgm/publications

20 Drugs and Pharmacokinetics

Editor: Beva Nall-Langdon

20.1 DRUG NOMENCLATURE

Because most drugs have complex molecular structures, their systematic chemical names are also complex (see Section 20.1.3, "Chemical Names"). Shorter, more convenient names are used in clinical medicine as well as in scientific literature that is not concerned specifically with the chemical characteristics of drugs. These shorter drug names fall into 2 main categories: nonproprietary names, also known as generic names, and proprietary names, which are used as trade names and brand names.

A detailed summary of various drug designations and their bases appears in the preface to the *USP Dictionary of USAN and International Drug Names*,[1] issued annually by the United States Pharmacopeial Convention (see Section 20.1.1.1, "United States Adopted Names").

20.1.1 Nonproprietary (Generic) Drug Names

Nonproprietary, or generic, names are used to refer to drugs as chemical entities independent of their commercial ownership. Using nonproprietary names increases the objectivity with which investigators conduct and report pharmacological and other medical research. Generic names also reduce bias on the part of others who read the published findings of such research.

20.1.1.1 United States Adopted Names

In the United States, drug developers propose nonproprietary names to the United States Adopted Names Council, which is sponsored by the American Medical Association, the United States Pharmacopeia Convention, and the American Pharmacists Association. Proposed names must conform to the criteria set forth in "Guiding Principles for Coining US Adopted Names for Drugs," which is published as an appendix to the *USP Dictionary of USAN and International Drug Names*.[1] Included in this document are guidelines for naming drugs based on the category to which a drug belongs, such as interferons, interleukins, somatotropins, colony-stimulating factors, erythropoietins, gene therapies, oligonucleotides, and monoclonal antibodies. Besides outlining rules for coining names, this document lists approved stem terms and terms that represent contractions for radicals and adducts.

Most nonproprietary names consist of a single word, but those representing chelates, complexes, esters, and salts consist of 2 terms.

rolitetracycline	rolitetracycline nitrate
remdesivir	fluocinolone acetonide
isotretinoin	magnesium salicylate

In contrast, radiopharmaceuticals are multiword terms (see Section 20.1.5, "Radiopharmaceuticals").

Once generic names are adopted, they are immediately listed online in the "New Names" section of *Pharmacopeial Forum*,[2] and they are published in the annually updated *USP Dictionary*[1]—both of which are products of the United States Pharmacopeia Convention. The *USP Dictionary* also includes information relevant to other designations for drugs (see Table 20.1 for a list and examples of the types of information in the dictionary). Other valuable sources of information on drugs are *The Merck Index Online*,[3] the combined *United States Pharmacopeia and the National Formulary*,[4] and the National Library of Medicine's PubChem.[5]

Although the nonproprietary drug names in such resources may be capitalized, generic names should not be capitalized, except when used at the beginning of sentences and in titles (see Sections 9.2.1, "First Word of a Sentence," and 9.3.1, "Titles").

20.1.1.2 International Nonproprietary Names (Recognized Worldwide)

Within the World Health Organization (WHO), the Expert Advisory Panel on the International Pharmacopoeia and Pharmaceutical Preparations serves a function akin to that of the United States Adopted Names Council by establishing the international nonproprietary name (INN) of each drug. Newly accepted INNs are published in the journal *WHO Drug Information*,[9] and they are identified in various other authoritative sources, including the WHO's web page "International Nonproprietary Names Programme and Classification of Medical Products"[10] and the *USP Dictionary of USAN and International Drug Names*.[1] Information about proposed and recommended INNs is also available through MedNet, the WHO's medicine information service at https://extranet.who.int/soinn.

Like the United States, many other countries rely on organizations other than the WHO

Table 20.1. Types of Information in Entries of the *USP Dictionary of USAN and International Drug Names*[a]

Type of information	Example, as presented in *USP Dictionary*[b]
US-adopted name (USAN)	Celecoxib
Year of publication of the USAN	1998
Pronunciation guide	(SEL-i-KOK-sib)
Official compendium	USP
Molecular formula and weight	$C_{17}H_{14}F_3N_3O_3S$ 381.373 g·mol^{-1}
Other systems in which nonproprietary name is entered	INN, BAN, DCF, MI
Chemical name	4-[5-(4-methylphenyl)-3-(trifluoromethyl)-1H-pyrazol-1-yl]benzenesulfonamide
Chemical Abstracts Service (CAS) registry number	C169590-42-5
Pharmacologic and/or therapeutic category	Nonsteroidal anti-inflammatory
Brand name(s), manufacturer(s)	Celebrex (Pfizer), Celebra (Pfizer, Germany), Celecox (Dejima Pharmacy Japan), Riva-Celecox (Laboratoire Riva, Inc, Canada)
Code designation	CP-870,893
Graphic formula	

USP, United States Pharmacopeia; INN, international nonproprietary name; BAN, British-approved name; DCF, Dénomination Commune Française; MI, Merck Index Online; CP, Charles Pfizer

[a] Based on *USP Dictionary of USAN and International Drug Names*.[1]

[b] The examples given all refer to the drug celecoxib. The categories of information are presented in the sequence and format in which they appear in the *USP Dictionary of USAN and International Drug Names*.[1] The *USP Dictionary* capitalizes chemical and nonproprietary drug names. However, this style manual recommends that chemical and nonproprietary drug names be lowercased except in titles and other places in which common nouns are capitalized (see Section 20.1.1.1, "United States Adopted Names").

to establish nonproprietary names, such as the British Pharmacopoeia Commission, the Chinese Pharmacopoeia Commission, and the Norwegian Pharmacopoeia Authority. The primary Canadian source for both nonproprietary and brand names is the *Compendium of Pharmaceuticals and Specialties*,[6] while comprehensive sources for drug names in European countries include *Martindale: The Complete Drug Reference*[7] and *Index Nominum*.[8]

In countries without national authorities for nonproprietary names, publications and authors should treat the WHO's INNs as properly established names.

20.1.1.3 Pharmacy Equivalent Names

To designate dosage combinations of 2 or more therapeutic substances, the United States Pharmacopeia Convention allows the use of pharmacy equivalent names (PENs). These names are derived from stems in the separate drug names, which are combined with the prefix "co-" to indicate a combination dosage.[1]

> co-trimoxazole *for* trimethoprim and sulfamethoxazole
> co-oxycode/APAP *for* oxycodone hydrochloride and acetaminophen

Using PENs reduces the proliferation of nonsystematic, unofficial, and colloquial names as well as ad hoc abbreviations for these dosage combinations. Although PENs are included in the *USP Dictionary of USAN and International Drug Names*,[1] they are not among the official names in *The United States Pharmacopeia and the National Formulary*.[4]

20.1.1.4 Abbreviations of Chemical Names and Multidrug Regimens

Some drugs are identified by abbreviations based on their chemical names.

> AZT or ZDV *for* zidovudine   5-FU *for* fluorouracil

However, because many of these abbreviations are not standardized, using them can result in confusion and error. Avoid such abbreviations by using United States adopted names (USANs), INNs, or the names recognized by the naming authority in the country of publication.

Similarly, abbreviations are used to identify some drug treatment regimens that combine 2 or more individual drugs administered together or sequentially, even though they may have PENs. These abbreviations usually are based on the initial letters of the drug names in the regimens.

> CMF *for* **c**yclophosphamide, **m**ethotrexate, and **f**luorouracil
> ABCP *for* **a**tezolizumab, **b**evacizumab, **c**arboplatin, and **p**aclitaxel

In some instances, an original abbreviation continues to be used even after the names of one or more of the drugs in the regimen change, such as when nonproprietary names replace proprietary names. For example, even though the trade name Oncovin has been replaced by the generic name vincristine, "MOPP" is still used as the abbreviation for the regimen of **m**echlorethamine, **O**ncovin, **p**rednisone, and **p**rocarbazine hydrochloride, and "CHOP" is still used for **c**yclophosphamide, **h**ydroxydaunorubicin, **O**ncovin, and **p**rednisone.

When such abbreviations are used in a journal article or other scientific document, the drug names of the full regimens should be explained either at first use of the abbreviations in both the abstract and the text or in a table of abbreviations on the first page of the text. However, these abbreviations can be used alone in the abstract and text if the abbreviations are well known in the publication's scientific discipline. In addition, these abbreviations can be used alone in a document's title regardless of how well known they are. For more on abbreviating drug names, see Section 11.3.2, "Scientific Usage."

20.1.1.5 Endocrinologic and Metabolic Drugs

Some USANs refer to hormones and vitamins that are used as endocrinologic and metabolic drugs for diagnosis or treatment in such disciplines as endocrinology and nutrition. When used for such purposes, these hormones and vitamins are referred to by names that are conventional in these disciplines instead of by their USANs. When these vitamins and hormones are used in other disciplines, their USANs should be used.

USAN	Conventional name in certain disciplines
calcitriol	1,25-dihydroxycholecalciferol
leucovorin calcium	calcium folinate

20.1.2 Proprietary Names (Trade Names, Brand Names, and Trademarks)

Proprietary names are established by pharmaceutical manufacturers and vendors for their products. Such names are proper nouns, so they should be capitalized (see Sections 9.4.1, "Proper Nouns," and 9.4.8, "Trademarks"). Although manufacturers and inventors are obligated to add a superscript symbol indicating trademarking or registration after proprietary names, authors and publications are not.

Proprietary names can be verified in sources such as the *USP Dictionary of USAN and International Drug Names*,[1] the *Compendium of Pharmaceuticals and Specialties*,[6] and *Prescribers' Digital Reference*[11] (formerly the *Physicians' Desk Reference*).

20.1.3 Chemical Names

Although most drugs are complex chemical compounds, their nonproprietary names are not intended to convey the details of their chemical structures. Those details are indicated in systematic chemical names that are based on the principles set forth by the International Union of Pure and Applied Chemistry (IUPAC) (see Chapter 17, "Chemical Formulas and Names"). The *USP Dictionary of USAN and International Drug Names*[1] provides 2 chemical names for each nonproprietary name: its uninverted IUPAC chemical name, which is preferred by the WHO, and its inverted name from the Chemical Abstracts Service,[12] which lends itself to electronic storage and retrieval.

Uninverted form:	2-[2-(2,6-dichlorophenoxy)ethyl]hydrazinecarboximidamide sulfate (2:1)
Inverted form:	hydrazinecarboximidamide, 2-[2-(2,6-dichlorophenoxy) ethyl]-, sulfate (2:1)

20.1.4 Chemical and Structural Formulas

Chemical molecular formulas for drugs indicate the relative proportions of elements that make up the compounds. Unlike chemical names, chemical formulas give little or no indication of chemical structure.

Chemical formula:	$C_{21}H_{26}ClNO$
Chemical name:	2-[(2R)-1-methylpyrrolidin-2-yl]ethanol
Nonproprietary name:	clemastine

In contrast, structural formulas provide information about the way atoms are connected in molecules. Structural formulas can be either line formulas, which are 2-dimensional representations of compounds, or stereoformulas, which indicate the 3-dimensional distribution of atoms in space.

Line formula	Stereoformula

$$[\text{Pt}(-\text{CH}_2-\text{CH}_2\text{NH}_2)_2]\text{Cl}_2$$

$$\begin{array}{cc} (\text{C}_6\text{H}_5)_3\text{P} & \text{Cl} \\ & \diagdown \quad \diagup \\ & \text{Pt} \\ & \diagup \quad \diagdown \\ (\text{C}_6\text{H}_5)_3\text{P} & \text{Cl} \end{array}$$

The structural formulas of drugs are illustrated in 2 major US reference sources on drugs: *The Merck Index Online*[3] and the *USP Dictionary of USAN and International Drug Names*.[1] For additional information and more examples, see Section 17.2.3, "Structural Formulas."

20.1.5 Radiopharmaceuticals

The nonproprietary names of pharmaceuticals with radioisotopes do not follow the conventions applied to other nonproprietary names (see Section 17.10, "Isotopically Modified Compounds"). When using the name of a radiopharmaceutical, place the symbol for the radioactive isotope after the carrier name, followed by the isotope's atomic weight on the same line rather than in a subscript.

> albumin, chromated Cr 51 serum   sodium pertechnetate technetium Tc 99m
> iodopyracet I 125

20.1.6 Chemical Abstracts Service (CAS) Registry Numbers

The American Chemical Society's Chemical Abstracts Service (CAS)[12] is a registry that randomly assigns unique numbers to drugs and other complex chemical entities to facilitate rapid handling. The file for each CAS number displays all the names for a single substance along with related bibliographic references in the chemical literature. The CAS registry is one of the databases available through STN International at https://www.stn-international.com. Other sources of CAS numbers are the MEDLINE database, *Martindale: The Complete Drug Reference*,[7] and the *USP Dictionary of USAN and International Drug Names*.[1]

Because CAS registry numbers are assigned to unique substances, some entries in the *USP Dictionary of USAN and International Drug Names*[1] carry more than one CAS number. For example, an anhydrous compound will have one number, and its hydrate will have a different number:

> CAS 58-55-9 *for* theophylline (anhydrous)   CAS 5967-84-0 *for* theophylline monohydrate

20.1.7 Code Designations

During the drug development process, pharmaceutical companies and other drug developers often assign alphanumeric designations to compounds. The alphabetical component is usually derived from the drug developers' names. For example, the alphabetical code for drugs developed by the pharmaceutical company Pfizer, Inc, is "CP," which is derived from the name of the company's founder, Charles Pfizer.

> CP-870,893 (the code designation for a CD40 agonist antibody developed by Pfizer, Inc)
> GSK 1349572 (the code designation for the brand version of dolutegravir developed by
>   GlaxoSmithKline, PLC)

Code designations are listed in an appendix to the *USP Dictionary of USAN and International Drug Names.*[1]

20.2 USE OF NONPROPRIETARY (GENERIC) OR PROPRIETARY (TRADE) NAMES

Whether to use nonproprietary or proprietary names depends in part on the audience for the publication and the context in which the drug names are used. Because of drug advertising in medical journals and direct-to-consumer advertising, many physicians and patients are more accustomed to trade names than to nonproprietary names, especially for relatively new drugs available only from their original manufacturers.

However, scientific publications generally prefer that authors use nonproprietary names for 3 reasons:

(1) Nonproprietary names are more likely to suggest the drugs' chemical characteristics.
(2) Trade names often differ greatly from one country to another, whereas nonproprietary names are more likely to be unique and recognized worldwide.
(3) Most bibliographic indexing services use nonproprietary names, not trade names, in indexing the drug-related content of publications.

However, when reporting research about a drug, authors should identify parenthetically the drug's trade name and its manufacturer at least in the manuscript's Methods. Only the nonproprietary name should be used in the manuscript's title.

> **Title:**   Etanercept Compared with Placebo in the Treatment of Patients with Psoriatic Arthritis
>
> **Methods:**   The treatment group received etanercept (Embrel, Amgen, Inc)

On the other hand, if research compares 2 or more drugs with the same nonproprietary name, authors should use the trade names along with the nonproprietary name in the manuscript's title, abstract, and Methods.

> **Title:**   Comparison of Oral Lisinopril Brands Prinivil and Zestril in the Treatment of Patients with Hypertension

20.3 SYMBOLS FOR RECEPTORS OF DRUGS AND OTHER HUMORAL MEDIATORS

The symbols for receptors of drugs and other humoral mediators have not been systematically standardized. The receptor symbols currently in wide use generally employ alphanumeric symbols: Roman or Greek characters followed by alphanumeric or numeric designators that are usually in subscript.

alpha adrenergic receptors	$á_{1A}$	$á_{1B}$	$á_{1C}$	$á_{1D}$	$á_{2A}$	$á_{2B}$	$á_{2C}$
adenosine receptors	A_1	A_{2b}					
bradykinin receptors	B_1	B_2					
cholecystokinin and gastrin receptors	CCK_A	CCK_B					
dopamine receptors	D_1	D_2	D_4	D_5			
purine nucleotide receptors	P_{2x}	P_{2u}	P_{2t}				

Table 20.2. Examples of Common Pharmacokinetic Symbols and Qualifiers[a]

Symbol	Term symbolized
\multicolumn{2}{l}{Measured or calculated variables (italic type for single-lettered symbols and roman type for multilettered symbols)[b]}	
A^{ss}	amount of drug in the body at steady state
AUC^t	area under plasma concentration–time curve from time zero to time t
AUC^∞	area under plasma concentration–time curve from time zero to infinity
C	drug concentration in plasma at any time t
CL	total body clearance of drug from plasma
CL_{CR}	creatinine clearance
C^{ss}	steady-state concentration of drug in plasma during infusion at a constant rate
E	organ extraction ratio (specify organ by subscript; see elsewhere in this table)
f, F	fraction of the dose systemically available; bioavailability
f_a	fraction of administered dose absorbed
f_u	fraction of drug unbound in plasma
k	first-order rate constant
K_m	Michaelis–Menten constant
Q_R	renal blood flow; qualifier used for plasma flow
R_0	constant infusion rate (zero-order)
t_{max}	time to reach peak or maximum concentration after administration of a drug
$t_{1/2}$	elimination half-life
$t_{1/2a}$	absorption half-life
τ	dosage interval
V_c	pharmacokinetic volume of central or plasma compartment
V_d	apparent volume of distribution
V^{ss}	apparent volume of distribution at steady state
V_{max}	maximum rate of metabolism by an enzyme-mediated reaction

Additional symbols and other related information can be found in the "*TiPS* Nomenclature Supplement" to *Trends in Pharmacological Sciences*[13] and in *Purinoceptor Nomenclature: A Status Report*.[14] Some of these receptor symbols might be similar to other symbols, such as those for some vitamins, but they are unlikely to be used in potentially confusing contexts.

20.4 UNITS FOR DRUG CONCENTRATIONS IN BIOLOGICAL FLUIDS

Drug concentrations in biological fluids are generally measured and reported in gravimetric units, such as milligrams per liter (mg/L) or nanograms per milliliter (ng/mL). Drug concentrations may also be reported in molar units, such as millimoles per liter (mmol/L) or nanomoles per liter (nmol/L).

Although gravimetric units are widely used in the United States, the Scientific Committee of the Association of Clinical Biochemists[15] in the United Kingdom recommends

Table 20.2. Examples of Common Pharmacokinetic Symbols and Qualifiers[a] (*continued*)

Symbol	Term symbolized
Modifiers representing sites of measurement (roman type and subscript)	
b	blood
p	plasma
s	serum
t	tissue
u	unbound species
ur	urine
Modifiers representing organs or elimination routes (roman type and subscript)	
e	excreted into urine
H	hepatic
m	metabolized
NR	nonrenal
R	renal
Modifiers representing routes of administration (roman type and subscript)	
im	intramuscular
ip	intraperitoneal
iv	intravenous
o	oral
po	oral
pr	rectal
sc	subcutaneous
sl	sublingual
top	topical

[a] Based on symbols preferred by the American College of Clinical Pharmacology's Committee for Pharmacokinetic Nomenclature[16] and *Clinical Pharmacokinetics*.[17] The first of these sources also lists applicable units, previously used symbols, and recommended pharmacokinetic equations.

[b] This table adheres to the international convention of italicizing single-letter symbols for variables and using roman type for multilettered symbols.

using molar units from the Système International d'Unités to measure the concentrations of drugs and metabolites (see Section 12.2.1, "Système International d'Unités (SI)"). Using molar units instead of gravimetric units has strong scientific rationale because molar units express drug concentration as drug molecules per unit volume, not mass per unit volume.

20.5 PHARMACOKINETICS

Symbols for variables in pharmacokinetic studies are combinations of uppercase and lowercase Roman and Greek letters[16,17] (see Table 20.2). Subscripts are used to denote modifiers specifying sites of measurement, organ sources, and routes of administration and elimination. Superscripts are used to denote modifiers describing general conditions, such as "SS" for "steady state." A number of publications, however, use subscripts for all such modifiers.

Single-letter symbols for variables should be italicized in accordance with the international style for symbols for quantities and chemical kinetics (see Table 4.5). Use roman type for multiletter symbols to distinguish them from multiplied single-letter variables.

rate of administration = $CL \cdot C^{ss}$, where "CL" = clearance, and "C^{ss}" = steady-state drug concentration

CL = dose/AUC, where "CL" = clearance, and "AUC" = area under the curve

CITED REFERENCES

1. USP dictionary of USAN and international drug names. United States Pharmacopeia Convention; 2014. Annual (subscription required). http://store.usp.org/

2. Pharmacopeial Forum. United States Pharmacopeia Convention. https://www.uspnf.com/pharmacopeial-forum

3. The Merck index online. Royal Society of Chemistry; 2013. https://www.rsc.org/merck-index

4. The United States Pharmacopeia and the National Formulary (USP-NF). United States Pharmacopeia Convention. https://www.uspnf.com

5. PubChem. National Library of Medicine, National Center for Biotechnology Information. https://pubchem.ncbi.nlm.nih.gov/

6. Repchinsky C, editor. Compendium of pharmaceuticals and specialties. Canadian Pharmacists Association; 2020. Annual. https://www.pharmacists.ca/products-services/compendium-of-pharmaceuticals-and-specialties/

7. Sweetman SC, editor. Martindale: the complete drug reference. 39th ed. Pharmaceutical Press; 2017. https://www.pharmpress.com/product/MC_MART/martindale-the-complete-drug-reference

8. Swiss Pharmaceutical Society, editor. Index nominum: international drug directory. 19th ed. Medpharm Scientific Publications; 2008.

9. WHO Drug Information. World Health Organization. [accessed 2023 Mar 2]. https://www.who.int/home/search?indexCatalogue=genericsearchindex1&searchQuery=WHO%20Drug%20Information&wordsMode=AnyWord

10. International Nonproprietary Names Programme and classification of medical products. World Health Organization. [accessed 2023 Mar 2]. https://www.who.int/medicines/services/inn/en/

11. Prescribers' Digital Reference. [accessed 2023 Mar 2]. https://www.pdr.net

12. Chemical Abstract Service. American Chemical Society. [accessed 2023 Mar 2]. https://cas.org/

13. Endothelin receptors. Trends in Pharmacological Sciences. 2001;22(Supp1):42–43. https://doi.org/10.1016/S0165-6147(01)80019-5

14. Abbracchio MP, Cattabeni F, Fredholm BB, Williams M. Purinoceptor nomenclature: a status report. Drug Dev Res. 1993;28(3):207–213. https://doi.org/10.1002/ddr.430280304

15. Ratcliffe JG, Worth HGJ. Recommended units for reporting drug concentrations in biological fluids. Lancet. 1986;327(8474):202–203.https://doi.org/10.1016/s0140-6736(86)90665-3

16. American College of Clinical Pharmacology, Committee for Pharmacokinetic Nomenclature. Manual of symbols, equations & definitions in pharmacokinetics. J Clin Pharmacol. 1982;22(7):18–23S. https://doi.org/10.1002/j.1552-4604.1982.tb02675.x

17. Clinical Pharmacokinetics preferred symbols. Clin Pharmacokinet. 2001;40(1):73–75.

21 Genes, Chromosomes, and Related Molecules

Editors: Patricia K. Baskin, MS, and Peter J. Freeman, PhD

21.1 GENERAL RULES AND PRINCIPLES OF GENETIC NOMENCLATURE

For some organisms, detailed systems for the symbolic representation of wild-type and variant genes and of normal and abnormal chromosomes are available. For other organisms, little or no work has been done to describe either the genes or the chromosomes. The conventions presented in this chapter represent taxa for which published, generally accepted rules and guidelines were available when the chapter was prepared. They do not cover all conventions for all organisms that have been studied with genetic or molecular methods, but they do cover the most-studied organisms. Many of these sets of conventions are similar, differing only in minor details of symbolization and notation.

21.1.1 Standardization Efforts

The first formal system of genetic nomenclature and symbolization appears to have been the rules on symbols for genes of the laboratory mouse.[1] A later set of recommendations[2] had wider application and influenced the systems developed for many organisms.

Although there is no one system for genetic nomenclature, considerable standardization exists, thanks to cooperation among the nomenclature committees of various scientific organizations. Efforts to standardize genetic nomenclature date back to 1957, with the International Committee on Gene Symbols and Nomenclature of the International Union of Biological Sciences (IUBS).[2] The IUBS's original recommendations are still in use today for most organisms covered in this style manual, although some recommendations have been modified and many have been expanded.[3-9] The general recommendations are summarized in Table 21.1.[2,3]

The Gene Ontology Consortium (GO Consortium) supplements gene nomenclature and the rules for different organisms. The GO Consortium's goal is to produce dynamic controlled vocabularies for molecular functions, biological processes, and cellular components as knowledge of gene and protein roles in cells accumulates and changes.[10]

21.1.2 Existing Standards and Unambiguous Identifiers

Standard gene names and symbols for major organisms facilitate communication among researchers. Although the scientific community is working toward a standard nomencla-

Table 21.1. General Rules and Principles: Recommendations for Symbolization[a]

Feature	Convention
Genes	
Gene name (the protein name is frequently the same as that of the gene that encoded the protein)	Use languages that are highly understood internationally. Traditionally, gene names consist of a concise word or phrase that describes the main diagnostic feature of a mutant phenotype, the protein encoded, the metabolic requirement, or the sensitivity or resistance of the gene to a drug or other agent. Only a few gene names are based on wild-type phenotypes. Some guidelines advocate naming genes on the basis of their gene family and the similarity of the sequences to each other (see Section 21.14.1, "Gene Families").
Gene symbol	Derive the symbol from the gene's full, original name by shortening the name, using the initials of a multiword term, or otherwise devising a symbol that is recognizable and, preferably, pronounceable. Although many older gene symbols are only 1 or 2 letters and some are as many as 5 letters, most guidelines now call for 3-letter symbols in italic type. Most gene symbols do not include Greek letters, roman numerals, superscripts, or subscripts. If superscripts or subscripts are included, they can be replaced with recognized substitutions. In addition, commas, colons, and semicolons are used minimally in gene symbols. Almost all guidelines call for gene symbols to be in italics. The unapproved symbols for oat are the lone exception outlined in this style manual.
Dominant trait	The name and symbol begin with capital letters, provided that the capital letters do not create ambiguity.
Recessive trait	The name and symbol begin with lowercase letters, provided that the lowercase letters do not create ambiguity.
Allele series	To represent the different members in the series, superscripts are added to the gene symbol.
Standard, wild-type alleles	Represented by either the gene symbol with a superscript plus symbol in roman type or a plus symbol with the gene symbol in superscript. In formulas, the plus symbol may be used alone.
Different loci with similar phenotypes	
Nonalleles (eg, mimics and polymeric genes)	Gene symbol followed by an additional letter or an arabic numeral either on the same line after a hyphen or as a subscript
Alleles of independent mutational origin	Gene symbol with a subscript
Enhancers, inhibitors, lethals, and suppressors	*En*, *I*, *L*, or *Su* for dominant traits, followed by a hyphen and the symbol of the allele affected. For recessive traits, substitute lowercase *en*, *i*, *l*, or *su*.
Sterility or incompatibility	*S* for dominant trait, followed by a hyphen and the symbol of the allele affected. For recessive trait, substitute lowercase *s*.
Genic formulas	Write as fractions with maternal alleles first or above. Each fraction corresponds to a single linkage group (LG). Order the LGs in numerical sequence separated by semicolons. Enclose unlocated genes in parentheses at the end of the formula. For euploids and aneuploids, repeat the gene symbol as many times as there are homologous loci.
Extrachromosomal factors	Enclose symbols in square brackets, and list them at the beginning of a genetic formula.

Table 21.1. General Rules and Principles: Recommendations for Symbolization[a] (*continued*)

Feature	Convention
Chromosomes	
Autosomes	Designate by arabic numerals in ascending order from longest to shortest.
Sex chromosomes	Animals in which the male is heterogametic have sex chromosomes designated X and Y. Therefore, males have XY chromosomes, and females have XX. Animals in which the female is heterogametic have sex chromosomes designated Z and W. Therefore, females have ZW chromosomes, and males have ZZ.
Designations	Use arabic numerals to designate individual chromosomes, and use roman numerals for linkage groups. The designations for the chromosomes, their bands, and other markers should be in roman type, not italics.
Karyotypes and ideograms	Arrange the autosomes vertically in horizontal rows on the page. Place them in decreasing order by length with the short arm on top and with the sex chromosomes at the end of the series.
Chromosomal aberrations	Use the following abbreviations for chromosomal aberrations: • "Df" or "Def" for deficiency • "Del" for deletion • "Dp" or "Dup" for duplication • "In" or "Inv" for inversion • "T" or "Tran" for translocation • "Tp" for transposition Rules vary on whether these abbreviations should be in italics and on which abbreviation in a pair should be used.
Chromosome count	Indicate the zygotic number of chromosomes by $2n$, the gametic number by n, and the basic number by x.

[a] Based on the report of the International Committee on Genetic Symbols and Nomenclature[2]; Milislav Demerec, PhD, and colleagues[3]; and the prevailing patterns outlined in this chapter.

ture for all organisms, this style manual recommends adhering to the existing guidelines set by the nomenclature committees for specific organisms. When comparing genes and alleles from different organisms, use the symbols, case, and type style recommended by the relevant nomenclature committees.

However, gene symbols do not serve as stable and persistent identifiers for genes. This is because gene symbols and gene names are "mutable" (ie, they may change over time).[11] For example, the name for the human gene *leucine proline-enriched proteoglycan (leprecan) 1* and its symbol, *LEPRE1*, were changed in the mid-2010s to *prolyl 3-hydroxylase 1* and *P3H1*, respectively, because the "leprechaun" moniker was deemed politically incorrect. Consequently, literature before the mid-2010s uses the symbol *LEPRE1* for this human gene, but later literature uses *P3H1*. To avoid this situation in the future, nomenclature entities like the Vertebrate Gene Nomenclature Committee[12] (VGNC) are recommending that researchers identify genes by stable gene identifiers in addition to gene symbols and names. In the case of *prolyl 3-hydroxylase 1*, the VGNC recommends using "HGNC:19316"[13] as a stable identifier. In addition, the National Center for Biotechnology Information (NCBI) in the United States has assigned "64175"[14] as the gene's stable identifier, and the Ensembl project has assigned "ENSG00000117385."[15]

Another source of stable identifiers is the UniProt Knowledgebase (UniProtKB), formerly SWISS–PROT. UniProtKB identifies species with mnemonics, which are used as suffixes to—but not part of—gene symbols. Along with species mnemonics and gene symbols, UniProtKB entries provide the names of the proteins the genes encode, which are frequently the same as the genes' full names. Each gene and protein listing in UniProtKB has a unique entry number.

> The UniProtKB entry name "*CHI_ARATH*" is for the gene and protein named "*endo-chitinase CHI*," which uses the gene symbol "*CHI*." The gene and protein are from an organism named *Arabidopsis thaliana*, which has the UniProtKB mnemonic suffix "*ARATH*." The gene's UniProtKB entry number is O24603.

Similarly, global genetics and genomics organizations, including the majority of diagnostic hospitals, have adopted the standardized location-based descriptions of DNA sequence variants developed by the Human Genome Variation Society (HGVS).[16] HGVS's terminology unambiguously describes the location, nature of change, and predicted coding effect of any variant in the human genome, and it is the de facto standard for reporting variation in clinical reports.

Although the nomenclature was established in the first decade of the 2000s,[17] at least as recently as 2021, prominent medical genomics journals were failing to verify that all DNA variants were described using HGVS's unambiguous sequence variation description (SVD) format. An international working group under the umbrella of the Human Genome Organisation (HUGO) drew attention to poor practice in reporting variants relevant to disease and highlighted how such poor practice is detrimental to diagnostics. Consisting of editors of genetics journals and experts in describing genome sequence variation, the HUGO Reporting of Sequence Variants Working Group[18] recommended that variant description validation software be used to facilitate robust and consistent reporting of variations and to eliminate erroneous reporting. The report advised adopting HGVS's SVD format as the global standard for journals reporting human variation.

Because HGVS's SVD format is also used in major genomics databases such as ClinVar, and the ClinGen Allele registry, the HUGO working group is attempting to establish workflows to ensure that journal entries containing genetics and genomics information are automatically exported into these databases, thus making the information much more findable. As more DNA variants are described in the SVD format, the diagnostic process will become faster, and diagnostic accuracy will improve.

Other naming schemes are being developed for human genomics, including the variant representation specification of the Global Alliance for Genomics and Health.[19]

For species other than humans, however, SVD formats have not been widely adopted, so sharing of variant information is even more difficult. However, HGVS's SVD format is not intended solely for human use, and it can be applied to any organism as long as valid and stable reference sequences are available.[20] RefSeq and the Ensembl project, for example, provide such reference sequences for a wide variety of organisms.

A major drawback to HGVS's SVD format is that it is more useful for describing small-scale variations than large-scale structural variations, to which a substantial proportion of human diseases are attributed. Publicly accessible databases such as

DECIPHER,[21,22] gnomAD,[23] and the NCBI's dbVar[24] are better able to collect data on large-scale structural variants. In general, these databases use formats that collect chromosome number, range, and an indication as to whether the region is deleted, duplicated, or inverted.

Where possible, researchers should report sequence variations at both the genome level and the transcript level to enable accurate mapping between transcript reference sequences. This particularly applies to variants identified via DNA sequencing rather than direct RNA sequencing.

To demonstrate the value of unambiguous stable identifiers, Sections 21.4, "Insertions and Transposons," through 21.15.7, "Human," list identifiers for species and genes from 2 or more nomenclature databases, such as the NCBI's database and UniProtKB.

21.1.3 Determining whether a Gene Is New

Researchers who believe they have discovered a new gene or allele should search relevant gene lists and consult with the appropriate nomenclature committee or gene registry before publishing a new gene name and symbol. The discovery may be a new member of an existing gene family, or the proposed symbol may already be in use for a different gene family. Before assigning a symbol, most nomenclature committees require researchers to provide statistically valid segregation or sequence data to support the existence of a "new" gene.

When the same symbol is inadvertently assigned to different genes or when 2 or more symbols are assigned to the same gene, priority in publication is usually the primary criterion for establishing the preferred symbol. However, symbols not given priority can be treated as synonyms. Such situations with potential for confusion reinforce the value of using stable gene identifiers to unambiguously distinguish genes.

Some publishers require that newly determined sequences be made available in public databases on or before the date of publication of the first article describing each sequence. When this is done, the sequences' citations in the databases should be included in each article.

21.1.4 Institute for Laboratory Animal Research

The symbols for transgenes, DNA loci, targeted and induced mutations, and chromosome anomalies include unique 1-to-5-letter codes for the investigators, laboratories, or institutions that identified these elements. Originally designed for designating substrains of mice, rats, and rabbits, these codes are assigned by an international registry at the Institute for Laboratory Animal Research (ILAR), which is part of the National Academies of Sciences, Engineering, and Medicine in the United States.

Although the codes can have as many as 5 letters, 2 or 3 are preferred. In addition, only the first of the letters in each code is capitalized. In this style manual, these codes are referred to as "laboratory codes." Information on the ILAR's International Laboratory Code Registry and the assignment of ILAR codes is available from the National Academies' website[25] (search for "Lab Code Database").

ILAR lab code	Registered to
Jah	National Institute of Animal Health (Japan)
Mck	Nissar Darmani, PhD
O	University of Oslo
Zan	Andrew Zannettino, PhD

21.1.5 Identifying Genes in Manuscripts

In titles, articles, and tables, use only gene names, symbols, and stable identifiers that are approved by the nomenclature committees appropriate for the organisms being researched. Avoid beginning titles or sentences with gene names or symbols that begin with a lowercase letter.

> **In titles:**
> Protein-Coding Gene chl1b in Both Swamp Eel and Great White Shark
> *not* chl1b Is a Protein-Coding Gene in Both Swamp Eel and Great White Shark

> **In text:**
> Insertion mutations in the *b* gene . . .
> *not b* gene insertion mutations . . .

In abstracts and introductions, introduce symbols and approved identifiers after an early mention of the gene names. Although symbols can be enclosed in parentheses immediately after gene names, identifiers should be introduced with a short explanation of what they are.

> Isopentenyl-diphosphate delta isomerase 1 (IDI1) has been assigned the stable identifier HGNC:5387.
> Laminin subunit beta 4, which uses the symbol LAMB4 and has the stable identifier ENSG00000091128, is a gene with 8 transcripts.

Similarly, well-known synonyms should be introduced after either gene names or symbols. Set the synonyms off with either parentheses or commas, and introduce the synonyms with an explanatory phrase such as "also known as." To avoid ambiguity, separate 2 synonyms with "and" and 3 or more with commas and "and," not slashes.

> *RPS14B*, also known as "*CRY2*," . . . *not RPS14B* (*CRY2*) . . .
> Cyclin dependent kinase inhibitor 1A (also known as "*CIP1*," "*WAF1*," and "*p21*") . . . *not* Cyclin dependent kinase inhibitor 1A (CIP1/WAF1/p21) . . .

After stable identifiers and synonyms are introduced, use only the approved names or symbols throughout the rest of the text.

21.2 GENETIC UNITS, MEASURES, AND TOOLS

21.2.1 Base Pairs

The length of a DNA sequence (ie, the distance between 2 loci on a contig map) can be designated by the number of base pairs encompassed. For double-stranded DNA molecules, use the symbol "bp," which stands for "base pairs," for relatively short sequences (eg, fewer than 1,000 bases). Use "kb," which stands for "kilobases," or "Mb," which stands for "megabases," for longer sequences.

> The total length of DNA sequenced from the clone is 255 bp.
> The sequence-tagged site sY80 is 1.075 kb long.
> The total amount of mapped sequence is now more than 10.3 Mb.

For single-stranded nucleotide sequences (eg, RNA), the symbols "bp," "kb," and "Mb" cannot be used. Instead, use "nt," which stands for "nucleotides."

21.2.2 Centimorgans

The genetic distance between 2 loci on a chromosome can be expressed in centimorgans (cM). The centimorgan numerically equals the statistically corrected recombination frequency expressed as a percentage. Thus, distances of more than 1 morgan (1M) cannot exist. For example, if the recombination frequency between 2 given markers is 15%, the markers are separated by 15 cM. The genetic distance in humans of 1 cM approximates the physical distance of 1 million base pairs (1 Mb), but this correlation varies throughout the genome, especially at telomeres, in which recombination is more frequent.

The "M" in "cM" is capitalized, the symbol having been named for geneticist Thomas Hunt Morgan, PhD, who studied the function of chromosomes in heredity.

21.2.3 Sequence-Tagged Sites

A "sequence-tagged site" (STS) is a short DNA segment that identifies a unique location on a chromosome. STSs can be located by polymerase chain reaction (PCR) techniques.

    sY163 *for sequence 163 on the Y chromosome*

Such DNA sequences of known map locations serve as markers for genetic and physical mapping of genes.[26] STSs include microsatellites, sequence-characterized amplified regions (SCARs), cleaved amplified polymorphic sequences (CAPSs), and inter-simple sequence repeats (ISSRs). The National Center for Biotechnology Information in the United States (https://www.ncbi.nlm.nih.gov) maintains databases of STSs.

The symbolization of STSs is not standardized within or across species, but many STSs are given D segment symbols (see Section 21.2.6, "Anonymous DNA Segments").

21.2.4 Contigs

A set of overlapping, contiguous cloned DNA sequences that constitute a continuous DNA segment of a genome is known as a "contig." Contigs may be clone based, or they may be generated from genome-sequencing projects.

If a contig is clone based, its components are generally represented by the symbols of the clones carrying the overlapping nucleotide sequences. Complicating the naming of contigs is that the symbolization of clones is not standardized. In some cases, clones are represented by the sequence database accession numbers of the components of contigs. A contig itself may also be given a name and accession number.

If a contig is based on a genome-sequencing project, the contig is usually designated with a 12-character database accession number composed of a 4-letter designation for the project, a 2-digit version number, and a 6-digit designation for the contig[27] (eg, CAAA01119629).

Related contigs can be aligned to form a scaffold, and related scaffolds can be aligned to represent a chromosome.

21.2.5 Loci and Markers

A "locus" is the location on any of the following: a chromosome, a plasmid, another genetic molecule, an anonymous DNA segment, a fragile site, a breakpoint, an insertion, and any other distinguishable sequence by which the location can be specified. For genes, the terms "gene" and "locus" are often used interchangeably. In contrast, only "locus" is used for sections of genetic code that do not specify gene products but serve regulatory roles, serve as attachment sites, or encompass a cluster of genes.

When possible, specify the type of locus (eg, "gene locus" or "restriction fragment length polymorphisms locus"). In a genome database, a locus entry is likely to include some genetic or cytogenetic location information, the official name and symbol of each locus, a stable identifier that unambiguously and persistently identifies the locus, the mapping methods used, and the homologous genes of a reference organism. Genes are represented by gene symbols. The locus tag (also called the "systematic name" and the "ORF name," for "open reading frame") is often used instead of a gene name, especially in descriptions of a whole genome or when the specific locus is known but the gene product has not yet been determined. The style of locus tags varies with organisms and organelles.

> YKR123C (Y = yeast, K = chromosome 11 converted to a letter, R = right arm, 123 = the OPF, and
> C = the Crick strand)
> *ycf42* (temporary designation for the hypothetical chloroplast ORF that was later renamed *bas1*)
> *C2orf1* (temporary designation for chromosome 2, OPF 1, in humans)

A "marker" can be a gene locus associated with a particular and usually readily identifiable phenotype, a subdivision of a gene, or an anonymous DNA segment.

21.2.6 Anonymous DNA Segments

An "anonymous DNA segment" is a segment with no known functional identity. Anonymous DNA segments have unique alphanumeric D-number symbols assigned either by genome-specific databases (eg, the Genome Reference Corsortium,[28] Mouse Genome Informatics,[29] and the Rat Genome Database[30]) or by the laboratories that perform the sequencing. All D segments are loci.

The alphanumeric D-number symbols for human DNA segments consist of the following:

- the letter "D" to designate an anonymous DNA segment
- a number between 1 and 22 or the letter "X" or "Y" to designate the chromosome on which the segment resides ("0" is used if the chromosome is unknown)
- a 1-letter code to designate the type of sequence (ie, "S," "Z," or "F")
- a unique sequential number

> D7S645 denotes the 645th single-copy sequence assigned to chromosome 7.

The letter "S" denotes a unique DNA segment, while "Z" denotes a repetitive DNA segment found at a single chromosome site. Both letters are followed by a sequential number unique for the chromosome on which the sequence resides.

> DYZ3 denotes the third repetitive sequence assigned to Y.

The letter "F" denotes a small, undefined "family" of homologous sequences found on multiple chromosomes. "F" is followed first by a sequential number unique in the genome for the sequence family and then by the letter "S" and a sequential number that is unique in the family to designate the specific member of the family.

> D14F23S1 denotes the first member of the 23rd sequence family assigned in the genome. This member of the family is on chromosome 14.
> D14F23S2 denotes the second member of the 23rd family, also on chromosome 14.
> D2F23S3 denotes the third member of the 23rd family. This member is on chromosome 2.

The letter "E" is added to the end of a sequence if the sequence is known to be "expressed." The fact that the sequence is expressed means that the sequence is likely to be a gene, but it cannot be given a gene symbol because its function is unknown.[6]

> D9S220E denotes the 220th single-copy sequence assigned to chromosome 9. The sequence is expressed.

The segment symbols for mice and rats differ slightly from those for humans. Instead of using the letter "S," "Z," or "F" to designate the type of sequence, mouse and rat symbols use laboratory codes.

> D1Mit1 denotes the first sequence for mouse chromosome 1 from the Massachusetts Institute of Technology, whose laboratory code is "Mit."

21.2.7 Probes

A probe is a DNA segment, RNA segment, protein, or other cellular constituent used to visualize a genetic target object or structure. Probe names, which are assigned by the originating laboratories, should be distinct from gene symbols and the names of loci. No standardized nomenclature exists for probes, although many probes are identified by D segment symbols (see Section 21.2.6, "Anonymous DNA Segments").

Superscripts and subscripts must not be used in probe names. The following lowercase-letter prefixes indicate vectors: "p" for "plasmid," "c" for "cosmid," "l" for "lambda phage," and "y" for "yeast."

21.2.8 Single-Nucleotide Variants

As the name suggests, a "single-nucleotide variant" (SNV) is a single-nucleotide change. SNVs are the most abundant form of genomic variation, and they are the most prevalent type of variation associated with disease. The term "single-nucleotide polymorphism" (SNP, pronounced as "snip") is commonly used instead of "SNV." However, "SNP" is no longer recommended in the context of human genetics because it implies a benign effect. Similarly, the term "mutation" is no longer preferred because it implies a pathogenic effect.[16] To prevent such misunderstandings, the term "SNV" is preferred, and SNVs are classified as "pathogenic," "benign," or "of uncertain significance."[31]

SNVs, which occur frequently throughout the genomes of all species, serve as useful biological markers. Because their physical locations remain identical, SNVs can be tracked. Consequently, SNVs can be used to construct genome maps showing the relative positions of known genes and markers with respect to a specified reference sequence (eg, a genome build).

As SNV-association studies increased and genome annotation improved, the need arose for standardized nomenclature to record and share the location of the SNVs identified in those studies. To facilitate identifying and classifying SNVs, the National Center for Biotechnology Information (NCBI) in the United States established the public-domain database called dbSNP, which can be accessed at NCBI's website (select "SNP" from the drop-down menu at https://www.ncbi.nlm.nih.gov). SNV records in the dbSNP are linked to other NCBI sources, including GenBank records, the Entrez gene database, and PubMed. The records are also linked to multiple external resources, including the genome browsers at Ensembl and the University of California, Santa Cruz. Beyond SNV records, dbSNP contains records on other variant types (eg, deletions, duplications, inversions, and deletion insertions with fewer than 50 base pairs).

Since September 2017, dbSNP has limited new submissions to data on human organisms. New submissions of data on human and other organisms can be made to the European Molecular Biology Laboratory's European Variation Archive (EVA) at https://www.ebi.ac.uk/eva/.

When researchers submit a new SNV or SNV cluster to dbSNP or the EVA, the SNV is assigned an "ss number," which is defined as a "submitted SNP number." Once the submitted SNV or SNV cluster is mapped to the correct location on the genome, the NCBI or the EVA assigns the SNV or cluster an "rs number," which the NCBI defines as a "RefSNP number" and the EVA defines as a "reference SNP." Many journals currently include ss or rs numbers in published articles, but this is not a sufficient way to describe SNVs because rs numbers refer to genomic coordinates, not individual variants. Therefore, many dbSNP records contain multiple variants identified by the same rs number. In practical terms, rs numbers do not actually describe SNVs. Instead, they confer information about map locations where these variations have been identified. For this reason, dbSNP records contain terms that unambiguously describe individual genetic variants (eg, sequence variants described using the Human Genome Variation Society's format).

21.2.9 Banding Patterns

Chromosomes can be stained by a variety of banding techniques, each of which produces a different type of banding pattern. "A band is defined as the part of a chromosome that is clearly distinguishable from its adjacent segments by appearing darker or lighter."[5] The results of staining, which can be photographed and compared, vary considerably among organisms. Most karyotypes and ideograms are based primarily on Giemsa banding, usually referred to as "G-banding." Supporting data for karyotypes and ideograms are based on other types of banding, such as constitutive (C), quinacrine (Q), reverse (R), telomeric (T), and nucleolus organizing regions (NORs) banding.

21.2.10 Restriction Endonucleases

Although restriction endonucleases are not genetic material themselves, they play a key role in genetic research. Restriction endonucleases have a standardized nomenclature that begins with a 3-letter italic abbreviation that represents the source organism. This abbreviation consists of an initial uppercase letter that is the first letter of the genus name. The remaining 2 letters are the first letters of the specific epithet. The abbreviation may be followed by the strain designation, which consists of arabic numerals. The strain designation should be in roman type, and it should be separated from the 3-letter abbreviation by a space. If a designation other than a strain designation is used, it is not separated from the 3-letter abbreviation.

> *Bce* 1229 (an enzyme from *Bacillus cereus* of the strain IAM 1229)
> *Xma*II (the second enzyme from *Xanthomonas malvacearum*)

See Richard J. Roberts, PhD,[32] for more details, and see REBASE[33] for a catalog of restriction enzymes.

21.3 TRANSCRIPTION-RELATED PROTEINS

Initiation factors needed to start protein synthesis are designated with the symbol "IF" in roman type followed by a hyphen and a numerical identifier. The symbols for bacterial factors (eg, those for *Escherichia coli*) have no prefix, but those for eukaryotic factors use the lowercase prefix "e" in roman type. The hyphenated number indicates the group assignment. To create a symbol for a new initiation factor in a group, add the next uppercase letter available from among the other symbols in the group. To create subunits of initiation factors, add Greek letters (eg, α, β, and γ).

> IF-2   eIF-2   eIF-4B   eIF-3β

Elongation factors (EFs) have similar representations.

> EF-Ts   EF-Tu   eEF-1α   eEF-1βγ

The symbols for prokaryotic factors differ in minor ways from those of EFs. For a prokaryote, the first letter of a protein symbol is capitalized (eg, "RecA").

For further details, consult Brian Safer's recommendations.[34]

21.4 INSERTIONS AND TRANSPOSONS

(1) Simple bacterial insertion elements are represented by the symbol "IS" (for "insertion sequences") followed by an arabic numeral. The symbol IS should be in roman type, and the numeral should be in italics. Bacterial transposons are represented by the symbol "Tn" in roman type followed by specific alphanumeric designators in italics.[4,35]

The symbols for transposons in eukaryotes vary with the organism, but most are in roman type. Transposons in filamentous fungi are given gene names that begin with an uppercase letter and are displayed in italics.[36]

> IS*2* and IS*4* (bacterial insertion elements)
> Tn*1* and Tn*5 lac* (bacterial transposons)
> Tcr1 and Tcr2 (*Chlamydomonas reinhardtii* transposons)
> Ty1 and Ty2 (yeast transposons)
> *Hidaway*, *Ans1*, and *MGR586* (filamentous fungal transposons)

(2) An example of a mnemonic designation for a transposon in the UniProt Knowledgebase is "TY1A_YEASX," for the transposon *TyH3 Gag polyprotein* in *Saccharomyces cerevisiae*. This transposon's UniProtKB entry number is P08405, and its identifier with the National Center for Biotechnology Information in the United States is P08405.2.

Table 21.2. Cytochrome P450 Gene Superfamily, Families, and Subfamilies: Symbols[a]

Feature	Convention	Examples
Gene root symbol for the superfamily	Italicized symbol in all uppercase for human and some other species, and initial capital only for other species	*CYP* for cytochrome P450 (for human) *Cyp* for cytochrome P450 (for mouse and genus *Drosophila*)
Cytochrome P450 family	Root symbol with an italic arabic numeral for the family	*CYP2* *Cyp7*
Subfamily	Family symbol plus an italic capital letter for the subfamily. Lowercase letter for mouse subfamily.	*CYP11B* *Cyp7a*
Individual gene	Subfamily symbol plus an arabic numeral for the gene	*Cyp2b2* *Cyp7a1*
Pseudogene	Closing italic capital letter *P*, but *ps* for mouse and *Drosophila*	*CYP2B7P*
Gene product	Gene symbol in uppercase characters in roman	CYP1A2 (from *CYP1A2*) CYP24 (from *Cyp24*)
Trivial names	Genes known by their trivial names before it is discovered that they encode cytochrome P450 enzymes may continue under their original names, and their *CYP* names will serve as their official synonyms. If the original names continue to be used in publications, include the official names in the title, the summary, or a footnote. Be consistent within a publication.	*P450-OLF1 = Cyp2g1* in *Rattus norvegicus* *P450 IIC16 = CYP2C16* in *Oryctolagus cuniculus*

[a] Based on David R. Nelson, PhD, and colleagues.[38]

21.5 TRANSSPECIES GENE FAMILIES

In contrast to the traditional mutant phenotype-based method of naming genes, many gene nomenclature committees have recommended that gene names and symbols be based on function whenever possible. Other approaches to gene families are designed to transcend taxonomic and functional boundaries by basing names and symbols on sequence similarities and the presence of various motifs, also known as "domains." Although motifs may imply function, they are not conclusive.

An example of such a system of nomenclature and symbolization is one proposed for the cytochrome P450 enzymes in eukaryotes and prokaryotes.[37,38] Table 21.2[38] summarizes this system. See Section 21.14.1, "Gene Families," for details of the system in plants.

21.6 ONCOGENES

(1) Oncogenes were originally assigned 3-letter italic symbols based on the retrovirus in which each oncogene was first identified. In this nomenclature system, single lowercase-letter prefixes in roman type specify the immediate origin of the gene, and uppercase-letter prefixes in roman type indicate the cellular localization of the protein in eukaryotic cells.

v-*fes* (viral gene, from feline sarcoma virus)
c-*myc* (cellular, from myelocytomatosis)
N-*erb A* (nucleus localization, from avian erythroblastosis)
S-*sis* (secreted, from simian sarcoma virus)

(2) Nomenclature for human and mouse cellular oncogene sequences now follows the standard nomenclature for genes of those species. When referring specifically to the human locus, use the name of the homologous retroviral oncogene, but omit the "v-" or "c-" prefix. For example, use "*JUN*" instead of "v-*jun*" for sarcoma virus 17 oncogene homolog (avian). Name mouse loci in a similar fashion, but capitalize only the initial letter. For example, use "*Hras1*," not "*c-HRAS1*," for Harvey rat sarcoma-1 oncogene.

(3) The names and symbols of oncogenes should be regarded as provisional until the true functions of the genes become known and the genes are renamed. For example, "*Erbb*" became the symbol "*Egfr*" for "epidermal growth factor receptor," and "*Sis*" became "*Pdgfb*" for "platelet-derived growth factor, beta polypeptide."

(4) An example of a mnemonic designation for an oncogene in the UniProt Knowledgebase is "JUN_HUMAN," which is for the *V-jun avian sarcoma virus 17 oncogene homolog* in *Homo sapiens*. This oncogene's UniProtKB entry number is P05412, and its stable gene identifier with the National Center for Biotechnology Information (NCBI) is 3725. In addition, the HUGO Gene Nomenclature Committee (HGNC) assigned the identifier HGNC:6204 to *JUN*, and the Ensembl project assigned the identifier ENSG00000177606 to it.

Another example of a mnemonic UniProtKB designation is "PDGFB_FELCA," which is used for *platelet-derived growth factor subunit B* in *Felis catus*. This oncogene's UniProtKB entry number is P12919, and its stable NCBI gene identifier is 100135774. In addition, the Vertebrate Gene Nomenclature Committee's identifier for *PDGFB* is VGNC:68770.

21.7 PLASMIDS

The notations for plasmid loci, genes, and alleles generally follow the rules set forth for bacteria by Milislav Demerec, PhD, and colleagues[3] (see Section 21.10, "Bacteria"). Richard P. Novick and colleagues[39] developed detailed recommendations along with definitions of relevant terms. Naturally occurring plasmids often have descriptive names (eg, "circular plasmid" in *Saccharomyces cerevisiae*, 2 μm).

In general, each newly described plasmid and newly isolated genotypic modification of a known plasmid is given a unique alphanumeric designation with the form "pXY1234." In this designation, "p" stands for "plasmid," "XY" are the initials for the laboratory or the reporting scientist, and "1234" stands for the laboratory's numerical designation. Indicate a deletion by an uppercase Greek delta (Δ) followed by unique serial numbers and a list of deleted genes. Indicate insertions, transpositions, and translocations by an uppercase Greek omega (Ω) followed by unique serial numbers and a list of translocated genes.

The phenotypic notation for a plasmid gene consists of 1 uppercase letter and 1 or 2 lowercase letters. In addition, the notation should reflect the phenotypic trait for which the gene is responsible. The genotype is given in the form recommended for bacteria by Dr Demerec and colleagues[3] and the American Society for Microbiology[4] (see Section 21.10, "Bacteria"). The complete genotypic identification includes as a prefix the name of the plasmid on which the gene was found.

21.8 VIRUSES

21.8.1 Bacteriophages (Phages)

(1) The rules for the symbolization of bacteriophage genes vary with the phage. In general, specify the phage itself with a prefix (eg, "λ" for "phage lambda," "Mu" for "phage Mu," and "P1" for "phage P1"). The symbols for genes themselves are 1, 2, or 3 italic letters. In some cases, a space separates the prefix from the gene, and in others, they are unspaced. Because phages do not have metabolism outside bacterial cells, genotype and phenotype are not differentiated. Symbols may be combined, spaced or unspaced, to represent mutations. Superscripts indicate hybrid genomes.

> λ*che*22
> Mu dII345 (phage Mu)
> P1 *vir* (phage P1)
> λ*cI*857*int*2*red*114*susA*11 (mutations in genes *cI*, *int*, and *red* and a suppressible [*sus*] mutation in gene *A*)
> λ *att*^{434} *imm*^{21} (hybrid of phage λ carrying the attachment [*att*] region of phage 434 and the immunity [*imm*] region of phage 21)

For more details and resources for individual phages, see the "Nomenclature" section of the "Writing Your Paper" instructions for the American Society for Microbiology's journals.[40]

(2) An example of a mnemonic suffix for a bacteriophage in the UniProt Knowledgebase is "BPT4," which UniProtKB uses for enterobacteria phage T4. The stable taxon identifier for the same bacteriophage is 10665 with the National Center for Biotechnology Information (NCBI). Similarly, "CAPSP_BPT4" is UniProtKB's designation for this bacteriophage's gene *protein capsid vertex protein*, which uses the symbol *24*. The NCBI's stable gene identifier for gene *24* is 1258587.

21.8.2 Human Retroviruses

(1) The genes of the human immunodeficiency viruses (ie, HIV-1 and HIV-2) and the human T-cell leukemia viruses (ie, HTLV-1 and HTLV-2) have been represented by various 1-, 2-, and 3-letter and alphanumeric symbols, all of which are set in italics.

> *tel*   *sor*   *X*   *tat-3*   *p40x*

Robert Gallo, MD, and colleagues[41] recommended consistently using 3-letter italic symbols, with the letters reflecting gene functions.

> *tax*$_1$ (transactivator)   *rev* (regulator of expression of virion proteins)
> *nef* (negative factor)

(2) An example of a mnemonic suffix for a human retrovirus in the UniProt Knowledgebase is "HV1H2," which UniProtKB uses for HIV-1. The stable taxon identifier for the same virus is 11706 with the National Center for Biotechnology Information (NCBI). Similarly, "NEF_HV1H2" is UniProtKB's designation for this retrovirus's gene *protein Nef*, which uses the symbol *nef*. The NCBI's stable gene identifier for *nef* is 156110.

21.9 ARCHAEA

(1) No rules for genetic nomenclature have been published for archaea as a group separate from bacteria. Until such guidelines are established, the gene names for archaea are the same as those for bacteria.

(2) An example of a mnemonic suffix for an archaeon in the UniProt Knowledgebase is

"SULAC," which UniProtKB uses for one strain of *Sulfolobus acidocaldarius*. The stable taxon identifier for the same strain of *S. acidocaldarius* with the National Center for Biotechnology Information (NCBI) is 330779. Similarly, "Q4J9J7_SULAC" is UniProtKB's designation for the *S. acidocaldarius* gene *acetyl-coenzyme A synthetas*, which uses the symbol *Saci_1184*. The NCBI's stable gene identifier for *Saci_1184* is 33336239.

21.10 BACTERIA

(1) A unified system for genetic nomenclature for bacterial genes and phenotypes was first proposed in 1966,[3] and it has been updated and expanded since.[40,42] Table 21.3[3,35,39,40,43,44] summarizes the major conventions for bacterial genes and phenotypes.

(2) An example of a mnemonic suffix for a bacterium is "ECOLI," which is the UniProt Knowledgebase's suffix for one strain of *Escherichia coli*. The stable taxon identifier for the same strain of *E. coli* with the National Center for Biotechnology Information (NCBI) is 83333. Similarly, "DCDA_ECOLI" is UniProtKB's designation for the *E. coli* gene *diaminopimelate decarboxylase*, which uses the symbol *lysA*. The NCBI's stable gene identifier for *lysA* is 947313.

21.11 YEASTS

21.11.1 *Saccharomyces cerevisiae*

(1) *Saccharomyces cerevisiae*'s stable taxon identifiers with the National Center for Biotechnology Information (NCBI) are 559292 and 4932, and its mnemonic suffixes in the UniProt Knowledgebase are "YEAST" and "YEASX," respectively. An example of a UniProtKB gene designation for *S. cerevisiae* is "ARGI_YEAST" for *arginase*, which uses the gene symbol *CAR1*. *CAR1*'s UniProtKB entry number is P00812, and its NBCI stable gene identifier is 855993. In addition, the Saccharomyces Genome Database (SGD) assigned *CAR1* the identifier SGD:S000006032.

(2) The gene symbols for *S. cerevisiae* have been updated over time.[3,45,46] Table 21.4[45–47,74] summarizes the major conventions. Authors should search the SGD[46] and refer to the SGD Gene Naming Guidelines before naming newly discovered genes and designating symbols for them. Gene names and symbols can be reserved while manuscripts are being prepared for publication.

The open reading frame (ORF) name (also called the "locus tag" and the "systematic name") is often used instead of the gene name, especially in descriptions of the whole genome or when the specific locus is known but the gene product has not yet been determined.

Gene systematic names for *S. cerevisiae* consist of 3 parts in the order below[46]:

- 3 uppercase Roman letters (The first is "Y" for "yeast," the second for the chromosome number converted to letters A through P, and the last for the arm designation of "L" or "R.")
- a 3-digit number for the ORF, counting from the centromere
- "C" or "W" for the "Crick strand," which is the sense or reverse complementary strand, or the Watson strand, which is the antisense, coding, or template strand

YJL191W (The second letter, "J," indicates that this gene is on chromosome 10, and the number 191 is the ORF.)

(3) The chromosomes of *S. cerevisiae* are designated by the roman numerals I to XVI, and the chromosomes' arms are labeled "L" for "left" and "R" for "right."

Table 21.3. Bacteria: Symbols for Genes and Phenotypes[a]

Feature	Convention	Examples
Gene or locus symbol	3-letter, lowercase italic symbol followed by a capital italic letter. For bacteria, the terms "gene" and "locus" are used interchangeably. However, only the term "locus" is used for sections of genetic code that are: • regions that do not specify a gene product but serve regulatory roles • attachment sites • a cluster of genes	*araA* and *araB* (genes or loci controlling arabinose utilization) *oriC* locus (origin of replication)
Mutation (ie, allele)	Locus symbol plus italic serial isolation number (ie, allele number).	*araB1*
	If the exact locus is unknown, the capital letter before the serial isolation number is replaced with a hyphen.	*Ara-2*
	When written as part of a genotype, the 3-letter gene symbol and capital locus letter (with or without a superscript minus symbol) imply an unnumbered mutation. Although such notations are frequently encountered, they are incorrect. Instead, authors should assign allele numbers or use phenotypic notations (see below).	*araB*
Wild-type gene or allele	Gene or locus symbol with superscript plus symbol. Do not use a superscript minus symbol to designate a mutation.	*Ara*$^+$ *araB*$^+$
Promoter site	Locus symbol with italic *p* added. Subscript numerals in roman type may be added to distinguish multiple promoters.	*lacZp* *glnAp*$_1$ *glnAp*$_2$
Terminator site	Locus symbol with italic *t* added	*lacAt*
Operator site	Locus symbol with italic *o* added	*lacZo*
Attenuator site	Locus symbol with italic *a* added	*trpAa*
Operon	Gene symbol followed by more than one locus letter	*trpEDCBA* or *trp* operon
Phenotypes	Generally, 3 or fewer characters in roman type with the first letter capitalized. With rare exceptions, a superscript character is required to indicate the nature of the phenotype.	Lac$^+$ Ara$^-$ Ts
Wild-type, or positive, phenotype	Phenotype symbol with superscript plus symbol. Locus letters are not used.	Lac$^+$
Mutant, or negative, phenotype	Phenotype symbol with superscript minus symbol	Lac$^-$
Drug-resistance trait carried on host chromosome	3-letter symbol for drug in roman type with superscript "s" for "sensitive" or superscript "r" for "resistant." Superscript is mandatory.	Amp^r
Drug-resistance trait carried on plasmid	2-letter symbol for drug in roman type with superscript "s" for "sensitive" or superscript "r" for "resistant." The superscript "s" is mandatory, but the "r" is optional. A 3-letter symbol is permissible if necessary for clarity.	Ap^s or Ap or Ap^r Asa (arsenate resistance) Asi (arsenite resistance)
Phenotypic property or gene characteristic	Phenotypic symbol in parentheses added to the end of the gene symbol for the trait described	*araA230*(Am) *hisD21*(Ts) *rpsL20*(Str^r)

Table 21.3. Bacteria: Symbols for Genes and Phenotypes[a] (*continued*)

Feature	Convention	Examples
Plasmids and episomes	Generally, 3 or 4 letters in roman type followed by numbers. The first letter is a lowercase "p." When part of a strain name or bacterial genotype, plasmids and episomes are enclosed in parentheses. Names of older, naturally occurring plasmids may have various names.	pUC19 pBR322 *E. coli* (pXY1234) Col E1 F[+] F128
Deletions	The uppercase Greek delta is placed before the symbol for the gene or region being deleted. Parentheses around the region are in roman type, but the gene or region within the parentheses and the number after them are in italics.	Δ*trpA432* Δ(*aroP-aceE*)*419*
Inversions	The uppercase symbol "IN" in roman type is placed before the symbols for the genes being inverted. Parentheses around the region are in roman type, but the gene symbols within the parentheses and the number after them are in italics.	IN(*rrnD-rrnE*)*1*
Insertions	The symbol for the locus affected is followed by a double colon, which in turn is followed by the symbol for the gene or other element being inserted.	*galT236*::Tn5
	For a complex insertion into a plasmid, the plasmid name precedes the uppercase Greek omega, which is followed by a description in parentheses.	pSC101 Ω(0kb::K-12*hisB*)*4* (This symbol indicates that the *E. coli his* gene is inserted into pSC101 at 0 kb.)
Fused genes (ie, fusions)	The uppercase Greek phi is placed before the symbols of the fused genes. Parentheses around the region are in roman type, but the symbols of the fused genes within the parentheses and the number after them are in italics. An apostrophe indicates a truncated gene in a fusion. Fused genes are sometimes indicated by a double colon (see the entries for "Insertions" and "Promoter Fusions").	Φ(*ara-lac*)*95* Φ(*ara'-lac*)*96*
Promoter fusions	The symbol for the promoter site, including the lowercase "*p*," is followed by a double colon and the symbol for the unrelated structural gene fused to the promoter.[b] The symbol is in italics except for the double colon.	*lacZp*::*trp*
Transposons	The 2-letter symbol "Tn" in roman type is followed by italic numbers or letters.	Tn5 and Tn5-1 Tn*A* and Tn*phoA* zef-123::Tn5
Foreign genes	The source or origin of the gene is indicated by a subscript mnemonic of the scientific name or the name and strain designation. The scientific name should be used if ambiguity exists.	*lacZYA*$_{Eco}$ (This symbol is for the *lac* operon of *E. coli*.) *lacZ*$_{EcoO157}$

E. coli, Escherichia coli

[a] Based on Milislav Demerec, PhD, and colleagues[3]; K. Brooks Low[44]; Richard P. Novick and colleagues[39]; Forrest G. Chumley and colleagues[43]; A. Campbell and colleagues[35]; "Nomenclature" section of the "Writing Your Paper" instructions for the journals of the American Society for Microbiology[40]; Mary Berlyn, Yale University, personal communications, 11 Jan 2001, 25 Jan 2001, and 3 Sep 2003; and Donald P. Nierlich, University of California, Los Angeles, personal communications, 27 Mar 2002, 29 Mar 2002, and 20 Oct 2003.

[b] The fusion of a promoter to an unrelated gene is rarely written in a conventional style. The American Society for Microbiology discourages the form widely used (eg, *Plac-trp* and *Ptac-araBAD-ftsQ*) because it can easily be mistaken for a plasmid or protein name. Regardless of which form is used, the mnemonic used for the promoter should be explained (American Society of Microbiology[4]; Mary Berlyn, Yale University, personal communication, 3 Sep 2003; and Donald P. Nierlich, University of California, Los Angeles, personal communication, 20 Oct 2003).

Table 21.4. *Saccharomyces cerevisiae*: Symbols for Genes and Phenotypes[a]

Feature	Convention	Examples
Gene name	3 italic letters	*ARG* and *arg*
Gene symbol	Abbreviation of gene name followed by an arabic numeral. For *S. cerevisiae*, the terms "gene" and "locus" are often used interchangeably. However, only the term "locus" is used for sections of genetic code that are: • open reading frames • regions that do not specify a gene product but serve regulatory roles • attachment sites • a cluster of genes	*ARG2* and *arg2* for acetyl-CoA:L-glutamate *N*-acetyltransferase, whose locus tag is *YJL071W* *ADE12* for adenylosuccinate synthase, whose locus tag is *YNL220W*
Allele designation	The gene symbol followed by a hyphen and an arabic numeral in italics. Although gene numbers are those originally assigned to genes, allele numbers may be specific to laboratories.	*arg2-14*
Gene cluster[b]	The gene symbol followed by an uppercase italic letter	*hIS4A* and *hIS4B*
Dominant allele	The gene symbol in uppercase italic letters	*ARG2*
Recessive allele	The gene symbol in lowercase italic letters	*arg2*
Wild-type gene	The gene symbol followed by a superscript plus symbol. Wild-type genes may be designated simply as +, provided that doing so will not create confusion.	$ARG2^+$
Gene conferring resistance or susceptibility	The gene symbol followed by a superscript capital "R" or "S" in roman type	CUP^R1 CAN^S1
Mating type	Wild-type mating types are designated "*MAT*a" and "*MAT*α" in genotype descriptions. The phenotypes are simply a and α.	. . .
Homothallic genes	Wild-type homothallic genes at the *HMR* and *HML* loci are "*HMR*a," "*HMR*α," "*HML*a," and "*HML*α." Mutations at these loci are denoted "*hmr*a-1," "*hml*α-1," and so forth.	. . .
Informational suppressors	The symbol "*SUP*" or "*sup*" followed by a locus designation. Frameshifts take the symbols "*SUF*" and "*suf*."	*SUP4* and *sup35* *SUF1* and *suf11*
Deletions	Insert an uppercase Greek delta after the hyphen in the symbol for the allele being deleted.	*Arg2-Δ1*
Insertions	The symbol for the gene or locus affected, followed by a double colon and by the symbol for the gene or other element being inserted	*ARG2::LEU2* *arg2-10::LEU2* *yjl031c::KanMX4*
Transposons (also called "Ty elements")	The symbol "Ty" in roman type and a number	Ty1 and Ty5 YERCTy1-1
Non-Mendelian and mitochondrial genes	Enclose the symbols in square brackets when necessary to distinguish them from chromosomal genes. Avoid Greek letters for newly discovered genes, but retain the symbols "ρ^+," "ρ^-," "Ψ^+," and "Ψ^-," as well as the symbols for their transliterations, which are "*rho*$^+$," "*rho*$^-$," "*psi*$^+$," and "*psi*$^-$."	[*COX1*] and [*cox1*] [ρ^+] and [*rho*$^+$]
Phenotypes	The same characters as the gene symbols in roman type. Phenotype symbols have initial capital letters, and the symbols are followed by a superscript plus symbol to indicate the independence of a substance or by a superscript minus symbol to indicate a substance is required.	Arg$^+$ (independent of arginine) Arg$^-$ (requires arginine)

[a] Based on Fred Sherman, PhD[45]; *Saccharomyces* Genome Database[46]; J. Michael Cherry, PhD[47]; and the UniProt Knowledgebase.[73]

[b] This convention also is used for complementation groups within a gene and for domains within a gene having different properties.

21.11.2 *Schizosaccharomyces pombe*

(1) *Schizosaccharomyces pombe*'s NCBI taxon identifier is 284812, and its UniProtKB mnemonic suffix is "SCHPO." An example of a UniProtKB gene designation for a *S. pombe* gene is "TPIS_SCHPO" for *triosephosphate isomerase*, which uses the gene symbol *tpi1*. The UniProtKB entry number for *tpi1* is P07669, and the NBCI stable gene identifier is 2538729. In addition, PomBase assigned to *tpi1* the identifier SPCC24B10.21.

(2) Symbols for *S. pombe* genes are based on the proposals of Milislav Demerec, PhD, and colleagues.[3] Jürg Kohli[48] explained the deviations seen in practice for *S. pombe* and bacteria. Table 21.5[48,49] summarizes the major conventions. For more details and examples, see Peter Fantes and Jürg Kohli.[49]

(3) Chromosomes of *S. pombe* are designated by roman numerals I, II, and III, and the chromosomes' arms are labeled "L" for "left" and "R" for "right."

21.12 FILAMENTOUS FUNGI

21.12.1 Ascomycete Mating-Type Genes

Nomenclature for mating-type genes of filamentous ascomycetes is based on the type of protein encoded.[50] Mating-type alleles are designated *MAT1-1* (for the idiomorph encoding a protein with an α box motif) and *MAT1-2* (for the idiomorph encoding a protein with a high-mobility group motif). Each gene is indicated by the symbol followed by a hyphen and a number (eg, *MAT1-2-1*). The number assigned to any particular gene within an idiomorph should correspond to that of its homolog in other fungi that have characterized *MAT* genes. Conventions for *Neurospora* and *Podospora* are specifically excluded from this nomenclature convention. For details and examples, see B. Gillian Turgeon and Olen C. Yoder.[50]

21.12.2 *Aspergillus nidulans*

(1) *Aspergillus nidulans*'s stable taxon identifiers are 227321 and 286162 with the National Center for Biotechnology Information (NCBI) in the United States, and its UniProt Knowledgebase mnemonic suffixes are "EMENI" and "EMEND," respectively. An example of a UniProtKB gene designation for *A. nidulans* is "BIMC_EMENI" for the gene symbol *bimC* and for kinesin-like protein bimC. This gene has the UniProtKB entry number P17120 and the NBCI stable gene identifier 2874347.

(2) Table 21.6[51,73] summarizes the major conventions for the genes of *A. nidulans*. For more details and examples, see A.J. Clutterbuck.[52] A linkage map[53] and other genome data[54] are available online.

(3) The 8 chromosomes/mitotic linkage groups of *A. nidulans* are numbered I to VIII in the order of their discovery.[55] A chromosomal aberration is indicated by the symbol for the aberration (eg, "T" for "translocation" and "Inv" for "inversion"), a serial number, and the relevant chromosomes in parentheses in numerical order, separated by a semicolon. Indicate a mutation associated with an aberration by attaching the mutation's symbol with a hyphen.

Aberration:	T1(III;VIII)
Mutation:	Inv2(I)-*areB405*

21.12.3 *Neurospora crassa*

(1) *Neurospora crassa*'s stable NCBI taxon identifier is 367110, and its UniProtKB mnemonic suffix is "NEUCR." An example of a UniProtKB gene designation for a *N. crass* gene is

Table 21.5. *Schizosaccharomyces pombe*: Symbols for Genes[a]

Feature	Convention	Examples
Gene symbol	3 lowercase italic letters	*arg*
Gene locus	Abbreviation of gene name followed by an arabic numeral	*arg1* for acetylornithine aminotransferase
Allele	The locus symbol followed by a hyphen and an arabic numeral or by a hyphen and a combination of letters and numerals. All of the characters are set in italics. No distinction is made between dominant and recessive alleles.	*Arg1-230* *ade6-M210*
Complementation group	The locus symbol followed by an italic uppercase letter	*trp1A* and *trp1B*
Wild-type gene	The locus symbol followed by a superscript plus symbol. Wild-type genes may be designated simply as "+," provided that doing so will not create confusion.	*Arg1*⁺ *mcm2*⁺
Gene conferring resistance or susceptibility	The allele symbol followed by the italic superscript "*r*" or "*s*" for "resistance" or "susceptibility," respectively	*can1-1r*
Gene conferring temperature sensitivity	The allele symbol followed by the italic superscript "*ts*" for "heat sensitive" or "*cs*" for "cold sensitive," respectively	*cdc2-5ts*
Phenotype	The same characters as the gene symbol, but neither italicized nor capitalized. Superscript plus and minus symbols indicate the wild type and the mutant, respectively.	Arg1⁺ Arg1⁻
Mating type	Homothallic (ie, wild-type) strains are designated "*h*,"[90] while heterothallic strains are "*h*⁺" and "*h*⁻."	
Mitochondrial and other non-Mendelian genes	The same rules apply for nuclear and non-Mendelian genes. Enclose the symbols in square brackets when necessary to distinguish them from chromosomal genes.	*Ana1* or *[ana1]*
Plasmid designations	Letters and numerals in roman type, beginning with lowercase "p"	pFL20
Transposons (also called "retro-transposons")	The symbol "Tf" in roman type and a number	Tf1-107 Tf2
Open reading frames	When an open reading frame is known but its biological function is unassigned, it may be named "*orf*" until a function is assigned.	*orf42*
Ribosomal RNA	The symbols for ribosomal protein genes begin with either "*rpl*" or "*rps*," which stands for "ribosomal protein, large subunit" or "ribosomal protein, small subunit," respectively.	*rps6*
Gene products	The gene or allele designation in roman type followed by a lowercase "p" for "protein"	*Mcm2*p

[a] Based on Jürg Kohli[48] and on Peter A. Fantes and Kohli.[49]

"TBA2_NEUCR" for *tubulin alpha-B chain*, which uses the gene symbol *tba-2*. The UniProtKB entry number for *tba-2* is P38669, and the NBCI stable gene identifier is 3879371.

(2) The conventions for *Neurospora* genes[56] are similar to those for *Drosophila* (see Section 21.15.2, "*Drosophila melanogaster*"), but *Neurospora* genes predate and differ significantly from those of many other organisms. Table 21.7[56] summarizes the major conventions. Researchers should consult the Fungal Genetics Stock Center[57] before assigning

Table 21.6. *Aspergillus nidulans*: Symbols for Genes and Phenotypes[a]

Feature	Convention	Examples
Gene symbol	Italic 3-letter symbol for new mutants. Older symbols remain unchanged and may be longer or shorter.[b]	*cre* *msc*
Nonallelic locus with the same primary symbol	The gene symbol with capital letter added	*creA* *creB*
Allele	The locus symbol with a serial number added	*creA1*, *creA2*, and *creB3*
Undetermined allelic relationships of a mutant	A hyphen replaces the capital letter in the allele symbol.	*cre-99*
Wild-type allele	The locus symbol with a superscript plus symbol	*creA⁺*
Dominant mutant	Described in written text, not as part of the symbol[c]	. . .
Multimarked strains	The allele symbols separated by spaces. Linkage groups in the order of I to VIII and separated by semicolons. Diploids shown as fractions, preferably one linkage group per fraction.	*proA1 biA1* *proAI BiA1* ; $\frac{+}{+}$ $\frac{+}{+}$ $\frac{}{pyroA4}$ *proA1/+ biA1/+*; *pyroA4/+*
Specific property of mutants	An optional, temporary superscript letter is added to the mutant gene symbol. The superscript is not to be regarded as a permanent or essential part of the symbol.	*amdR6ʳ*
Suppressor	The symbol "*su*" with locus letters and mutant number followed by the symbol of the mutant being suppressed. Alternatively, standard gene symbols may be used.	*suA1creA1* *snx4A*
Phenotype	The unabbreviated word or phrase from which the gene symbol is derived is preferred. An alternative is the gene symbol in roman type with an initial capital letter.	Chitin deacetylase Cda
Proteins	The gene symbol in roman type with initial capital letter[d]	CreA
Mitochondrial gene symbol	Use the same gene symbol that is used for a nuclear gene. Enclose the symbol in square brackets when writing it as part of a genotype.	*camA* *proA1* *[camA]*
Transposon name	No specific convention for symbols exists for *Aspergillus*.[d] Standard fungal nomenclature calls for a symbol in italic letters and numbers with an initial capital.[36]	*Afut1* *F2PO8*

[a] Based on A. John Clutterbuck[51] and the UniProt Knowledgebase.[73]

[b] Where appropriate, use the symbols proposed by Milislav Demerec, PhD, and colleagues[3] or by Giuseppe Sermonti.[60] Pronounceable symbols are preferred. Older, unchanged symbols are 1 to 5 italic letters.

[c] Attention can be drawn to the dominance of a mutant by using a superscript or an initial italic capital letter. However, "*IvoA*" and "*ivoA*" refer to the same mutation.

[d] Recommendation from A. John Clutterbuck, University of Glasgow, personal communication, 4 Aug 2003.

new names, symbols, locus numbers, or allele-number prefixes to avoid creating duplications.

Most laboratories have adopted the Oak Ridge strains OR23-1A and ORS-6a as the standard wild-type strains of *Neurospora*.[56]

(3) Chromosome linkage groups (LGs) are designated by roman numerals I to VII.[56] In contrast, chromosomes, which are more difficult to identify than LGs, are designated by arabic numerals 1 to 7 and by the abbreviations "L" and "R" for "left" and "right" arms.

Table 21.7. *Neurospora crassa*: Symbols for Genes and Phenotypes[a]

Feature	Convention	Examples
Gene name	The name of the mutant phenotype in italics	*arginine-1* *invertase*
Gene symbol	1 to 4 italic letters or arabic numerals, with 3-letter symbols preferred. All lowercase letters unless the name is based on a dominant allele.	*arg-1* *inv* *Asm-1* *tom22*
Different loci with similar phenotypes	The gene symbol followed by a hyphen and an arabic numeral. When the name applies to only one locus, the number 1 is optional.	*arg-1*, *arg-2*, and *arg-3* *hsp70-1* and *hsp70-2* *inv* or *inv-1*
Suppressors	The symbol "*su*" followed by the symbol of the suppressed gene in parentheses and the locus number, if it is known. The symbol "*su+*" is used for the wild type.	*su(met-7)* *su(met-7)-2*
Enhancers	The symbol "*en*" followed by the symbol of the enhanced gene in parentheses and the locus number	*en(am)-1*
Mating types	Either the format "*mat A*" or "*mat a*." The symbol may be abbreviated to "*A*" or "*a*" if the context is clear. The symbol is usually printed at the end of a genotype.	*cr-1 A* *al-2 A* *cot-1 a*
Transposons	The symbol in italic letters with an initial capital	*Tad* *Pogo*
Recessive mutant alleles	All lowercase italic letters for the name and symbol	*albino* and *al*
Dominant mutant allele	Italic letters with an initial capital for the name and symbol	*Banana* and *Ban*
Codominant alleles	All lowercase italic letters for the name and symbol	*heterokaryon incompatibility* and *het*
Anonymous genes	The symbol "*anon*" followed by a unique isolation designation in parentheses	*anon(NP6C9)*
Ectopic genes	The gene symbol followed by the abbreviation "*EC*" in parentheses	*am+(EC)*
Gene fusions	A double colon placed between the symbols of the genes being fused	*mtr::Asm-1+(EC)*
Genotypes	The gene symbols in linear order for each linkage group separated by spaces with the linkage groups separated by semicolons and spaces. Diploids are shown as fractions, with each fraction representing one linkage group.	*cr-1 al-2; am inl; nic-3* $\frac{al\text{-}2\ \ arg\text{-}6}{al\text{-}2^+\ +}$; $\frac{am\ \ inl}{am\ \ inl}$
Wild-type allele	The locus symbol with a superscript plus symbol	*Bml+*
Mutant allele	The locus symbol without a superscript	*Bml*
Multiple alleles at one locus	The locus symbol with italic superscript serial numbers. Use italic capital letters "*R*" and "*S*" in superscript to indicate resistance and sensitivity, respectively, to a toxic agent. Use square brackets in italics if superscripts are not available.	frq^1 and frq^2 $cyh\text{-}2^R$ and $cyh\text{-}2^S$ *frq[1]*, *frq[2]*, and *frq[3]* *cyh-2[R]* and *cyh-2[S]*
Deletion of part of a gene	Treat as an allele at the gene locus.	. . .
Deletion of an entire gene	An uppercase Greek delta precedes the gene symbol.	Δ*am*
Pseudogenes	The gene symbol followed by superscript "*ps*." Use square brackets in italics when superscripts are not available.	$Fsr63^{ps}$ *Fsr63[ps]*

Table 21.7. *Neurospora crassa*: Symbols for Genes and Phenotypes[a] (*continued*)

Feature	Convention	Examples
Isolation numbers (ie, allele numbers)	The gene or locus symbol followed by the laboratory code in roman type enclosed in parentheses. Use the laboratory code only when necessary to distinguish between alleles.	*pyr-3*(KS43)
Heterokaryon	The symbols for component nuclei separated by a plus symbol. Enclose the whole genotype in parentheses.	(*col-20 A* + *ad-3B cyh1 A*)
Phenotype	The gene symbol in roman type with an initial capital letter and a superscript plus, minus, or other allele designation	Al⁻ Arg⁺ Cplˢ
Gene product	The symbols for gene products are the same characters as the gene symbols, except the gene product symbols are in all capital letters and in roman type. When a gene product name is spelled out, capital letters are not necessary.	INV *and* invertase
Centromeres and telomeres	The symbols "*Cen-*" and "*Tel-*" are followed by the roman numeral for the linkage group. All characters are in italics.	*Cen-III* *Tel-VIR*
Mitochondrial genes	Lowercase letters and arabic numerals in italics without hyphen or space	*cox* *atp7*
Mutant mitochondrial genomes	Italicized names or symbols enclosed in nonitalicized square brackets	[*mi-2*] [*poky*] [*stp*]

[a] Based on David D. Perkins, PhD.[56]

Aberrations on the same chromosome are indicated by the italic symbols "*Df*," "*Dp*," "*In*," "*T*," and "*Tp*" for "deficiency," "duplication," "inversion," "translocation," and "transposition," respectively. In addition to beginning with the symbol for the aberration, each description includes the LGs with arm designation and an identification number. Semicolons separate LGs. The entire symbol is italicized with no intervening spaces. Arrows indicate the direction of action.

> *T(IIIR;VR)P1226* (a translocation of the right arm of LGIII and the right arm of LGV)
> *T(IL→IIR)39311* (a translocation of the left arm of LGI to the right arm of LGII)
> *Dp(IL→IIR)39311* (a duplication of the left arm of LGI in the right arm of LGII)

If a rearrangement breakpoint is inseparable from the mutant phenotype of an associated gene, the gene symbol follows the rearrangement symbol and is separated from it by a space with no comma.

> *T(IR;IIR)4637 al-1* (a translocation between the right arms of LGI and LGII at the *al-1* locus)

21.12.4 *Phycomyces* spp.

(1) The species of the *Phycomyces* genus have such stable NCBI taxon identifiers as 4837 and 763407, and their UniProtKB mnemonic suffixes include "PHYBL" and "PHYB8." An example of a UniProtKB gene designation for the *Phycomyces* spp. is "PYRF_PHYB8" for *orotidine 5′-phosphate decarboxylase*, which uses the gene symbol *pyr*. The UniProtKB entry number for *pyrG* is P21593, and the NBCI stable gene identifier is 29000633.

(2) The genetic nomenclature of *Phycomyces*[58] follows the recommendations of Milislav Demerec, PhD, and colleagues.[3] The genes of the *Phycomyces* species are named with 3

lowercase italic letters that refer to the gene's general function. To specify a gene, an uppercase letter is added to the 3-letter designation without a space.

> *car* (genes related to carotene)
> *carB* (the gene for phytoene dehydrogenase)

Mutant alleles are designated with the name of the gene and a number. When the specific gene is unknown, the uppercase letter is replaced by a hyphen. The genotype is a list of alleles that differ from the wild type. Mating types are indicated by plus and minus symbols within parentheses.

> *carB10* (–) (one of the alleles of *carB* with mating type minus)

Heterokaryons are indicated by using the names or genotypes of the components separated by an asterisk.

> C115*NRRL1555 (heterokaryon of strains C115 and NRRL1555)

(3) *Phycomyces* has at least 11 chromosome LGs, which are designated by roman numerals I to XI.[59] The total number of LGs and chromosomes is unknown. The standard wild-type strain of *Phycomyces* used for genetic studies is NRRL1555.

21.12.5 Plant-Pathogenic Fungi

(1) To standardize genetic nomenclature and symbolization for plant-pathogenic fungi, the Genetics Committee of the American Phytopathological Society generated recommendations[60] based mainly on the conventions that were current for yeast at the time.[45] The recommendations for plant-pathogenic fungi call for designating gene loci with unique 3-letter italic symbols that are based on the names of the mutant phenotypes they represent. The first letter in each symbol is an uppercase letter, and the other 2 are lowercase.

> *Met* (for methionine auxotrophy)

A dominant allele is indicated by 3 uppercase letters, and a recessive allele is indicated by 3 lowercase letters. Wild-type alleles are identified by a plus symbol following the locus number, while mutant alleles are identified by a minus symbol.

> *MET* (dominant allele)   *Met+* (wild-type allele with methionine-independent growth)
> *met* (recessive allele)

Mutations at different loci yielding similar phenotypes can be identified with identical letter symbols followed by a number unique to the locus. Resistance and sensitivity associated with a particular allele are indicated by adding italic "*R*" and "*S*," respectively, to the gene symbol.

> *Met1*, *Met2*, and *Met3* (different loci with similar phenotypes)
> *Cyh1S* (locus 1 confers sensitivity to cyclohexamide)

Mating-type alleles are designated *MAT1-1* and *MAT1-2*. All letters in the symbols for both alleles are uppercase because both alleles are needed for activity.

Symbols for cytoplasmically inherited genes follow the same rules as for nuclear genes. They are enclosed in brackets when it is necessary to distinguish them from nuclear genes, as in the genotype of a strain.

The conventions for phenotypes are similar to those for cytoplasmically inherited and nuclear genes. However, the symbols for phenotypes are in roman type.

> Met+ (methionine-independent growth)   Met– (methionine auxotrophy)

(2) Chromosome LGs are indicated by roman numerals, preferably ordered by size from largest to smallest chromosome as determined by gel electrophoresis. However, the LGs can be listed in the order in which they are recognized. Linked genes in a genotype are

separated by spaces, and unlinked genes are separated by semicolons. See Turgeon and Yoder[50] and Olen C. Yoder and colleagues[61] for more details.

21.13 PROTISTS

21.13.1 *Chlamydomonas reinhardtii*

(1) *Chlamydomonas reinhardtii*'s stable taxon identifier is 3055 with the National Center for Biotechnology Information (NCBI) in the United States, and its UniProt Knowledgebase mnemonic suffix is "CHLRE." An example of a UniProtKB gene designation for *C. reinhardtii* is "RBL_CHLRE" for ribulose bisphosphate carboxylase large chain, which uses the gene symbol *rbcL*. The UniProtKB entry number for rbcL is P00877, and the NBCI stable gene identifier is 2717040.

(2) Table 21.8[62,63] summarizes the major conventions for the genes of *C. reinhardtii*.

(3) The 17 chromosomes of *Chlamydomonas* are designated by roman numerals I to XIX. Two pairs of chromosomes have been consolidated: XII/XIII and XVI/XVII. Detailed gene maps are available at the website of the Chlamydomonas Resource Center.[64]

21.13.2 *Dictyostelium discoideum*

(1) The stable NCBI taxon identifier for *Dictyostelium discoideum* is 3055, and the UniProtKB mnemonic suffix is "DICDI." An example of a UniProtKB gene designation for *D. discoideum* is "AMT2_DICDI" for *ammonium transporter 2*, which uses the gene symbol *amtB*. The UniProtKB entry number for *amtB* is Q9BLG3, and the NBCI stable gene identifier is 8621332. In addition, the gene has the dictyBase identifier DDB_G0277889.

(2) Gene names and symbols for *D. discoideum* follow the rules recommended by Milislav Demerec, PhD, and colleagues[3] for bacterial genes, which are summarized in Table 21.3.[3,35,39,40,43,44] Robert J. Kay, PhD, and colleagues[65] summarized the rules as adapted for the genus *Dictyostelium*. The basic symbol is 3 lowercase italic letters, followed by uppercase letters to indicate different loci for the same phenotype and then by arabic numerals to indicate different alleles at the same locus. Until the exact locus is known, a hyphen is used instead of an uppercase letter, but the isolation number is retained. Wild-type and mutant alleles can be distinguished by superscript plus and minus symbols. There is no established system for distinguishing between allele types.

 acpA   *acpB*   *aggA2*   *aga*⁺   *aga*⁻

(3) *D. discoideum* has 6 chromosomes, which are designated by arabic numerals 1 through 6.

21.13.3 *Leishmania* spp. and *Trypanosoma* spp.

(1) *Leishmania* species have such stable NCBI taxon identifiers as 5664 and 5671, and their UniProtKB mnemonic suffixes include "LEIMA" and "LEIIN," respectively. An example of a UniProtKB gene designation within *Leishmania* spp. is "GP63_LEIDO" for *leishmanolysin*, which uses the gene symbol *gp63*. The UniProtKB entry number for *gp63* is P23223, and the NBCI stable gene identifier is 13389283.

 Trypanosoma species have such stable NCBI taxon identifiers as 5692 and 5702, and their UniProtKB mnemonic suffixes include "TRYCO" and "TRYBB." An example of a UniProtKB gene designation within the *Trypanosoma* spp. is "A0A1X0NHW0_9TRYP," for *Trypanosoma vivax*, which uses the gene symbol *TM35_000601010*. TM35_000601010 has the UniProtKB entry number A0A1X0NHW0 and the NBCI stable gene identifier 39990714.

(2) Gene nomenclature for *Leishmania* spp. and *Trypanosoma* spp. is based on the rules for

Table 21.8. *Chlamydomonas reinhardtii*: Symbols for Genes and Phenotypes[a]

Feature	Convention	Examples
Gene name	The name of mutant phenotype but in roman type	Carbonic anhydrase
Nuclear gene symbol	Uppercase italic letters based on gene name without hyphens, superscripts, or subscripts	*CAH*
Different loci with similar phenotypes	The gene symbol with arabic numerals added	*CAH1*, *CAH2*, and *CAH3*
Wild-type allele	Same as the locus symbol but with no plus symbol	*CAH1*
Mutant allele	The locus symbol in lowercase italics followed by a hyphen and an arabic numeral	*cah1-1*
Allele made by insertion of transforming DNA	The symbol for the initial allele followed by a double colon and the symbol for the introduced DNA	*abc1-1::NIT1*
Allele made by insertion of a transposon	The symbol for the initial allele followed by a double colon and the symbol for the transposon	*abc1-1::Tcr1*
Gene product	The gene symbol without italics	CAH1, CAH2, and CAH3
Phenotype	The gene symbol without italics and with only the initial letter capitalized	Cah1
Chloroplast gene	3 lowercase italic letters followed by either an arabic numeral or an uppercase letter	*rps7* *psaA*
Mitochondrial gene	3 lowercase letters followed by an arabic numeral, all in italics	*atp1*
Mating-type locus	Mating types are designated as "MT" on maps and "*mt+*" or "*mt–*" in genotypes. This is an exception to the uppercase rule for gene names.	MT *mt+* *mt–*
Genotypes		
Linked genes	The locus symbols separated by spaces	*ac17 nit2 tua1*
Unlinked genes	The locus symbols separated by semicolons and spaces	*arg7*; *ac17*; *tar1*
Diploids	The symbols for linked genes followed by a slash and the symbols for the homologous genes. Unlinked sets are separated by semicolons.	*ac17 NIT2/AC17 nit2* *tar1/TAR1*; *ac17/AC17*
Transposons	If the symbol can be pronounced as a word, set it in italics. If the name begins with the symbol "Tcr," set it in roman type when it stands alone and in italics when it is part of an allele name.	*Gulliver* *Pioneer* *TOC1* Tcr1 and Tcr3 *abc1-1::Tcr1*

[a] Based on Susan K. Dutcher, PhD, and Elizabeth H. Harris[63] and on Chlamydomonas Resource Center.[62]

Saccharomyces cerevisiae, which were proposed by Christine E. Clayton, PhD, and colleagues[66] in 1998. Table 21.9[66] summarizes the major conventions. Genes are assumed to be wild type unless a mutation has caused a significant loss of function.

21.13.4 *Paramecium* spp.

(1) *Paramecium* species have such stable NCBI taxon identifiers as 5888 and 5671, and their UniProtKB mnemonic suffixes include "PARTE" and "PAROT," respectively. An example of a UniProtKB gene designation within the *Paramecium* spp. is "D8L7U1_PARCA" for *NADH dehydrogenase subunit 1*, which uses the gene symbol *nad1_b*. The UniProtKB entry number for *nad1_b* is D8L7U1, and the NBCI stable gene identifier is 9384815.

Table 21.9. *Leishmania* spp. and *Trypanosoma* spp.: Symbols for Genes and Phenotypes[a]

Feature	Convention	Examples
Wild-type gene name	3 to 6 uppercase italic letters with no hyphens	*AAH* *TRYR*
Mutant gene	The wild-type gene name in lower-case, even when the mutant gene is dominant	*aah* *tryr*
Members of a gene family	The gene name with numbers or letters added	*RPL11* and *RPL18*
Different alleles of the same gene	The gene name followed by a hyphen and a number	*SMP-1*, *SMP-2*, and *SMP-3*
Deletions	The gene name preceded by the uppercase Greek delta	Δ *tryr*
Replacements	The uppercase Greek delta followed by the name of the gene being replaced, a double colon, and the gene being inserted	Δ *tryr::AAH*
Fused genes	Gene names in 5′→3′ order with double colon between the names	*DHFRTS::GFP*
Insertional inactivation	A caret (ie, "^," which is created by Unicode 005E) followed by the name of the gene being inactivated, a double colon, and the name of gene being inserted	^Δ *tryr::AAH*
Genes from different taxa	The gene name with a prefix consisting of a code for the taxon. Use only when needed to compare the genes of different taxa.	*Lmjgp63* (for *Leishmania major*) *Lmxgp63-C1* (for *L. mexicana*)
Diploids and triploids	The gene names for homologous genes separated by a slash	*dhfrts-3/DHFRTS* Δ*dhfrts/*Δ*dhfrts/DHFRTS*
Extrachromosomal elements	Names and symbols in square brackets, all of which are in italics. A high copy number may be added as a subscript in italics.	*[pX HYG GFP]*$_{85}$
Wild-type protein	The gene name in roman type. If a species prefix is used, the entire protein name should be in italics.	AAH TRYR *Lmj GP63*
Mutant protein	The gene name in roman type with an initial capital letter	Smp-1

[a] Based on Christine E. Clayton, PhD, and colleagues.[66]

(2)   Table 21.10[67] summarizes the major conventions for the genes of *Paramecium* spp. The strains from which genes are derived should be identified for each gene by an alphanumeric designation within square brackets after the locus name of "1."

   nd2-1[d4-2] (mutant ND2 locus from inbred strain d4-2)

   The macronucleus contains about 800 copies of each gene. If all copies of a particular gene in the macronucleus are the same, the gene name is listed only once in the genotype. If there are 2 or more alleles, they are separated by slashes. Such use of the slash does not indicate the number of copies of each allele in the macronucleus and is not meant to imply that the macronucleus is diploid.

(3)   Each of the 2 micronuclei (ie, the germline nuclei) of *Paramecium* has approximately

Table 21.10. *Paramecium* spp. and *Tetrahymena* spp.: Symbols for Genes and Phenotypes[a]

Feature	Convention	Examples
Micronuclear genes		
Gene symbol	3 italic letters. For wild type, use all uppercase letters.	*BTU*
Locus name	The gene symbol followed by an arabic numeral	*BTU2*
Induced mutants	The allele name in all lower-case followed by a hyphen and a number	*chx1-1*
Insertions	The mutant allele name followed by a double colon and the name of the inserted element	*btu2-1::neo2*
Deletions	The mutant allele name followed by the uppercase Greek delta and the range of deleted amino acids	*btu2-3Δ1-40*
Mutant alleles with known modifica-tions	The mutant allele name followed by appropriate designations	*btu2-2A251K* (allele of *BTU2* in which alanine at position 251 is replaced by lysine)
Genotypes	The symbols for alleles separated by a slash. Linked genes may be grouped to-gether, separated by commas on the same side of a slash. Symbols for unlinked genes are separated by semicolons and spaces.	*BTU2/BTU2*; *CHX1, EST1/CHX1, EST1*
Variations for Paramecium spp.		
Locus name	The gene symbol followed by a 3-digit arabic number, the first digit of which indicates the gene family. The entire symbol is set in italics.	*TUB101* and *TUB102* (β-tubulin loci in the same family) *TUB201* (β-tubulin locus in a different family)
Strain of origin	The allele names followed by the designation of the strain of origin in square brackets	*CHX[d4-2]*
Variations for Tetrahymena spp.		
Strain of origin	Assumed to be strain B unless otherwise indicated. For other strains, allele names are followed by the designa-tion of the strain of origin in square brackets.	*CHX* (no strain designation needed when strain B is the strain of origin) *CHX1[C3]*
Randomly applied polymor-phic DNAs (RAPDs)	A numeric code for the labora-tory of origin followed by the 2 initials of the person who discovered the polymorphism and a sequential number, all in italics (see Sally Lyman Allen, PhD, and colleagues[67] for additional details and examples)	*1JB11* and *1JB12*

Table 21.10. *Paramecium* spp. and *Tetrahymena* spp.: Symbols for Genes and Phenotypes[a] (*continued*)

Feature	Convention	Examples
Macronuclear genes and phenotypes		
Placement within genotype	The designation follows the micronuclear genotype. The designation, which is in italics, is enclosed in parentheses, which are in roman type.	*BTU2/BTU2*; *CHX1, EST1/CHX1, EST1* (*CHX1/chx1-1*)
Gene names	The symbols conform with those of micronuclear genes. In *Paramecium*, an "O" or "E" is used to indicate macronuclear mating-type alleles.	*CHX* *chx1-1[C3]* *MPR1* *tam6-1[d4-2]* *mtA-l/mtA-1* (E, o), in which "E" refers to the macronuclear genotype and "o" to the phenotype
Genotypes	If more than one allele is present for a gene, the symbols for alleles are separated by a slash. Linked genes may be grouped together, separated by a comma and a space on the same side of a slash. Symbols for unlinked genes are separated by a semicolon and a space. If all copies of the gene are the same, only one allele need be indicated.	(*CHX1/chx1-1*) (*CHX1, PJB1/chx1-1, PJB1*; *MPR1*)
Phenotypes	2- or 3-letter designations in roman type. The designations are listed after the genotype. Multiple phenotype designations are separated by commas. Drug phenotypes are followed by a hyphen and either "s" or "r" for "sensitive" or "resistant," respectively.	(exo⁻) (exo⁺) (cy-s, pm-r)
Mating types	In *Tetrahymena*, mating types are in uppercase numerals. In *Paramecium*, mating types are either "o" for "odd" or "e" for "even." In both cases, the mating types are listed in roman type after the genotype, phenotype, or both.	(*CHX1, PJB1/chx1-1, PJB1*; *MPR1/MPR1*; cy-r, mp-s, IV) (*tam6-1[d4-2]*; *O*; exo⁻, o)

[a] Based on Sally Lyman Allen, PhD, and colleagues.[67]

50 pairs of chromosomes and is, therefore, diploid. The single macronucleus (ie, the somatic nucleus) is derived from the micronuclei during conjugation and is not diploid.

21.13.5 *Tetrahymena* spp.

(1) *Tetrahymena* species have such stable NCBI taxon identifiers as 5911 and 5898, and their UniProtKB mnemonic suffixes include "TETTH" and "TETFU," respectively. An example of a UniProtKB gene designation within *Tetrahymena* spp. is "Q9XMS1_TETPY" for

apocytochrome b, which uses the gene symbol *cob*. The UniProtKB entry number for *cob* is Q9XMS1, and the NBCI stable gene identifier is 800774.

(2) Table 21.10[67] summarizes the major conventions for the genes of *Tetrahymena* spp. Unless otherwise indicated, these genes are assumed to be from inbred strain B, the reference strain for *Tetrahymena thermophila*. If any other strain is used, the designation for each gene derived from the strain should identify the strain by an alphanumeric designation within square brackets after the locus name.[67]

> *CHX1[C3]* (wild-type *CHX1* locus from inbred strain C3)

The macronucleus contains approximately 50 copies of each gene. If all copies of a particular gene in the macronucleus are the same, list the gene name only once in the genotype. If there are 2 or more alleles, separate them by slashes. Such use of the slash does not indicate the number of copies of each allele in the macronucleus and is not meant to imply that the macronucleus is diploid.

(3) The single micronucleus (ie, the germline nucleus) of *Tetrahymena* has 5 pairs of chromosomes and is, therefore, diploid. The single macronucleus (ie, the somatic nucleus) is derived from the micronuclei during conjugation[67] and is not diploid.

21.14 PLANTS

The following guidelines for plants are based on published rules for genes and chromosomes. Most of these guidelines are adaptations of the rules that the International Committee on Genetic Symbols and Nomenclature[2] proposed in 1957. Others are based on gene families. Tables 21.11[9,68–72,75,79,81,83,84,88,89,91,94,99,100,102,103,106] and 21.12[69,74–76,80,81,84,88,89,93,94,97,99,100,103,106,117,118] summarize and index the recommendations from Section 21.14.1, "Gene Families," through Section 21.14.7, "Other Plants."

21.14.1 Gene Families

A common nomenclature for sequenced plant genes was developed by the Commission on Plant Gene Nomenclature (CPGN) of the former International Society for Plant Molecular Biology. This nomenclature is based on the principle that plant genes that encode similar products and have similar coding sequences are members of the same gene family.[9] Consequently, these plant genes are assigned the same gene family number and the same gene family name.[9]

CPGN recommendations were used to build the Mendel database, which has since been replaced by other databases. To avoid creating a duplicate name, authors should consult recognized genome databases such as the UniProt Knowledgebase,[57] which is maintained by the European consortium UniProt; xGDBvm,[77] which is supported by the National Science Foundation in the United States; and the Plant Genome DataBase Japan,[78] which is supported by Japan's National Bioscience Database Center.

Gene family names of nuclear, endomorph, and viral genes are in the form "*XyzN*." Sets of genes that encode products with similar functions but whose sequences contain distinct motifs are numbered sequentially (eg, "*Xyz1*," "*Xyz2*," and "*Xyz3*"). Organelle gene names begin with lowercase letters (eg, "*xyzA*," "*xyzB*," and "*xyzC*").

The "*Xyz*" and "*xyz*" portions of gene family names are usually mnemonics that reflect

Table 21.11. Plants: Gene Nomenclature

Plant	Section or table[a]	NCBI taxon identifier[b]	UniProtKB species code[c]	Model	Reference
Alfalfa (*Medicago sativa*)	Section 21.14.5.1, "Alfalfa"	3879	MEDSA	CPGN and ICGSN	Carl A. Price and Ellen M. Reardon[9]
Arabidopsis (*Arabidopsis thaliana*)	Section 21.14.3, "*Arabidopsis thaliana* and Other Crucifers," and Table 21.17	3702	ARATH	…	David W. Meinke and Maarten Koornneef[70]
Barley (*Hordeum vulgare*)	Section 21.14.2.1, "Barley"	4513	HORVU	…	Jerome Franckowiak, PhD, and colleagues[79]
Cotton (*Gossypium* spp.)	Section 21.14.7.1, "Cotton"	Varies with species. Includes 3635, 29729, and 3634.	Varies with species.[c] Includes GOSHI, GOSAR, and GOSBA.	ICGSN	Russell J. Kohel, PhD[106]
Crucifers (*Brassica* spp.)	Section 21.14.3, "*Arabidopsis thaliana* and Other Crucifers"	Varies with species. Includes 3708 and 3713.	Varies with species.[c] Includes BRANA and BRAOV.	*Arabidopsis*	Crucifer Genetics Cooperative[91]
Cucumber (*Cucumis sativus*)	Section 21.14.4, "Cucurbits" and Table 21.18	3659	CUCSA	cucurbits	Cucurbit Genetics Cooperative[94]
Cucurbits (*Citrullus, Cucumis,* and *Cucurbita* spp.)	Section 21.14.4, "Cucurbits," and Table 21.18	Varies with species. Includes 3654, 61887, and 3660.	Varies with species.[c] Includes CITLA, CUCZE, and 9ROSI.	ICGSN and tomato	Cucurbit Genetics Cooperative[94]
Lettuce (*Lactuca sativa*)	Section 21.14.7.2, "Lettuce"	4236	LACSA	ICGSN	R.W. Robinson and colleagues[75]
Maize (*Zea mays*)	Section 21.14.2.2, "Maize (Corn)," and Table 21.13	4577	MAIZE	CPGN	A standard for maize genetics nomenclature[81]
Melon (*Cucumis melo*)	Section 21.14.4, "Cucurbits," and Table 21.18	3656	CUCME	cucurbits	Cucurbit Genetics Cooperative[94]
Oat (*Avena* spp.)	Section 21.14.2.3, "Oat," and Table 21.14	4498	AVESA	ICGSN	M.D. Simons and colleagues[72] and H.G. Marshall and Gregory E. Shaner[83]

Plant	Section reference[a]	NCBI identifier[b]	UniProtKB code	Genetic nomenclature	Authority
Pea (*Pisum* spp.)	Section 21.14.5.2, "Pea," and Table 21.19	Varies with species. Includes 3888.	Varies with species.[c] Includes PEA.	...	Thomas Henry Noel Ellis, PhD, and Mike Ambrose, MPhil[199]
Pepper (*Capsicum* spp.)	Section 21.14.6.1, "Pepper"	Varies with species. Includes 4072 and 80379.	Varies with species.[c] Includes CAPAN and CAPCH.	...	Deyuan Wang and Paul W. Bosland, PhD[102]
Pumpkin (*Cucurbita* spp.)	Section 21.14.4, "Cucurbits," and Table 21.18	Varies with species. Includes 3661.	Varies with species.[c] Includes CUCMA.	cucurbits	Cucurbit Genetics Cooperative[94]
Rice (*Oryza sativa*)	Section 21.14.2.4, "Rice," and Table 21.15	4530	ORYSA	...	Toshiro Kinoshita[84]
Rye (*Secale cereale*)	Section 21.14.2.5, "Rye"	4550	SECCE	wheat	Jacob Sybenga[88]
Soybean (*Glycine* spp.)	Section 21.14.5.3, "Soybean," and Table 21.20	Varies with species. Includes 3847.	Varies with species.[c] Includes SOYBN.	...	Soybean Genetics Committee[100]
Squash (*Cucurbita* spp.)	Section 21.14.4, "Cucurbits," and Table 21.18	Varies with species. Includes 1094860, 3661, and 3663.	Varies with species,[c] Includes 9ROSI, CUCMA, and CUCPE.	cucurbits	Cucurbit Genetics Cooperative[94]
Tomato (*Lycopersicon* spp.)	Section 21.14.6.2, "Tomato," and Table 21.21	Varies with species. Includes 4081, 62890, and 62890.	Varies with species.[c] Includes SOLLC, SOLHA, and SOLHA.	...	Tomato Genetics Cooperative[103] and Kevin D. Livingstone and colleagues[69]
Watermelon (*Citrullus lanatus*)	Section 21.14.4, "Cucurbits," and Table 21.18	3654	CITLA	cucurbits	Cucurbit Genetics Cooperative[94]
Wheat (*Triticum aestivum*)	Section 21.14.2.6, "Wheat," and Table 21.16	4565	WHEAT	...	Robert A. McIntosh, PhD, and colleagues[89]; W. Jon Raupp and colleagues[71]; and GrainGenes[68]

NBCI, National Center for Biotechnology Information of the National Library of Medicine in the United States; UniProtKB, UniProt Knowledgebase (formerly SWISS–PROT); CPGN, Commission on Plant Gene Nomenclature of the International Society for Plant Molecular Biology; ICGSN, International Committee on Genetic Symbols and Nomenclature

a Entries refer to the section or table for the model organism when there are no separate rules for the plant named in the first column.

b The National Center for Biotechnology Information's unique stable identifiers are numerical. The identifiers in this table are for taxa. The center also assigns numerical identifiers to genes.

c When the common name of the organism in English is 5 characters or fewer, the UniProt Knowledgebase's identification code is usually the common name (eg, "MAIZE"). When the common name is longer, the prefix is usually the first 3 letters of the genus name and the first 2 letters of the specific epithet (eg, "ARATH" for "*Arabidopsis thaliana*"). For details, see the ExPASy Proteomics Server at https://www.expasy.org/search/proteomics%20server.

Table 21.12. Plants: Cytogenetic Nomenclature

Plant	Section	Chromosome (Chr) or linkage group (LG)[a]	Reference
Alfalfa (*Medicago sativa*)	Section 21.14.5.1, "Alfalfa"	$n = 8$; autotetraploid with 4 nearly identical chromosomes	Gary R. Bauchan and M. Azhar Hossain[97]
Arabidopsis (*Arabidopsis thaliana*)	Section 21.14.3, "*Arabidopsis thaliana* and Other Crucifers"	$n = 5$	Paul Fransz and colleagues[93] and David W. Meinke and colleagues[74]
Barley (*Hordeum* spp.)	Section 21.14.2.1, "Barley"	1H to 7H; $n = 7$	Ib Linde-Laursen[80]
Cotton (*Gossypium* spp.)	Section 21.14.7.1, "Cotton"	A1 to A13 and D1 to D13; $n = 13$; tetraploid	Russell J. Kohel, PhD[106]
Cucumber (*Cucumis sativus*)	Section 21.14.4, "Cucurbits"	$n = 7$	Cucurbit Genetics Cooperative[94]
Cucurbits (see cucumber, melon, pumpkin, squash, and watermelon in this list for basic numbers)	Section 21.14.4, "Cucurbits"	(varies with species)	Cucurbit Genetics Cooperative[94]
Lettuce (*Lactuca sativa*)	Section 21.14.7.2, "Lettuce"	$n = 9$	R.W. Robinson and colleagues[75]
Maize (*Zea mays*)	Section 21.14.2.2, "Maize (Corn)"	$n = 10$	A standard for maize genetics nomenclature[81]
Melon (*Cucumis melo*)	Section 21.14.4, "Cucurbits"	$n = 12$	Cucurbit Genetics Cooperative[94]
Oat (*Avena* spp.)	Section 21.14.2.3, "Oat"	$n = 7$; genomes A, B, C, and D	H. Thomas[76]
Onion (*Allium cepa*)	Section 21.14.7.3, "Onion"	1C to 8C; $n = 8$	E.R. Kalkman[118] and J.N. de Vries[117]
Pea (*Pisum* spp.)	Section 21.14.5.2, "Pea"	Chr 1 to Chr 7; LGI to LGVII; $n = 7$[b]	Thomas Henry Noel Ellis, PhD, and Mike Ambrose, MPhil[99]
Pepper (*Capsicum* spp.)	Section 21.14.6.1, "Pepper"	$n = 12$	Kevin D. Livingstone and colleagues[69]
Potato (*Solanum tuberosum*)	. . .	$n = 12$	Kevin D. Livingstone and colleagues[69]
Pumpkin (*Cucurbita* spp.)	Section 21.14.4, "Cucurbits"	$n = 20$	Cucurbit Genetics Cooperative[94]
Rice (*Oryza sativa*)	Section 21.14.2.4, "Rice"	$n = 12$	Toshiro Kinoshita[84]
Rye (*Secale cereale*)	Section 21.14.2.5, "Rye"	1R to 7R; $n = 7$	Jacob Sybenga[88]
Soybean (*Glycine* spp.)	Section 21.14.5.3, "Soybean"	$n = 20$	Soybean Genetics Committee[100]
Squash (*Cucurbita* spp.)	Section 21.14.4, "Cucurbits"	$n = 20$	Cucurbit Genetics Cooperative[94]
Tomato (*Lycopersicon* spp.)	Section 21.14.6.2, "Tomato"	$n = 12$	Tomato Genetics Cooperative[103] and Kevin D. Livingstone and colleagues[69]
Watermelon (*Citrullus lanatus*)	Section 21.14.4, "Cucurbits"	$n = 11$	Cucurbit Genetics Cooperative[94]
Wheat (*Triticum* spp.)	Section 21.14.2.6, "Wheat"	1A to 7D; $n = 7$; genomes A, B, and D	Robert A. McIntosh, PhD, and colleagues[89]

[a] Chromosomes and linkage groups are designated by sequential arabic numerals unless specified otherwise.

[b] The numbers of the chromosomes for the *Pisum* spp. do not correspond to those of the linkage groups.

the functions of the gene products. These mnemonics consist of up to 8 characters, and they are set in italics.

Designations for alleles follow the conventions of the relevant plant species. Mutant alleles are usually represented in lowercase and are separated from the gene name by a hyphen.

> Adh1-Fm335 (*Fm335* allele of the wild-type gene for maize alcohol dehydrogenase)

21.14.2 Cereals

21.14.2.1 Barley

(1) *Hordeum vulgare* has such stable taxon identifiers as 4513 and 112509 with the National Center for Biotechnology Information (NCBI) in the United States, and its UniProtKB mnemonic suffixes include "HORVU" and "HORVV," respectively. An example of a UniProtKB gene designation for *H. vulgare* is "Q5ITT0_HORVV" for *hordoindo-line-a*, which uses the gene symbol *hina*. The UniProtKB entry number for *hina* is Q5ITT0, and the NBCI stable gene identifier is 123395916.

(2) The rules for names of barley genes were outlined by Jerome Franckowiak, PhD, and colleagues.[79] Gene names should be as descriptive as possible of the phenotype. The names and symbols of dominant genes begin with uppercase letters, while those of recessive genes begin with lowercase letters. A barley gene symbol consists of 3 letters that designate the gene's character and a number that represents a particular locus.

> *Mdh1* (a malate dehydrogenase locus)

A particular allele or mutational event at a locus is represented by one or more lowercase letters. A period separates the locus from the allele symbol. The entire symbol is written on the text line in italics. Wild-type alleles are not designated with a plus symbol as they are in many other species.

> *Mdh1.a* and *Mdh1.b* (alleles of the *Mdh1* locus)

Inhibitors, suppressors, and enhancers are designated by "*I*," "*Su*," and "*En*," respectively, if they are dominant. If they are recessive, they are designed by "*i*," "*su*," and "*en*," respectively. In either case, they are followed by a hyphen and the symbol of the allele affected.

Genic formulas are written as fractions with the maternal alleles listed first or above. Each fraction corresponds to a single linkage group, and the groups are listed in numerical order separated by semicolons. In euploids and aneuploids, the gene symbols are repeated as many times as there are homologous loci. Symbols for extrachromosomal factors are placed within square brackets before the genic formula.

(3) Barley chromosomes are designated "1H" to "7H" in accord with those of other species of the tribe Triticeae, and the short and long arms are designated by "S" and "L," respectively, without a space after the chromosomes' alphanumeric designations.[80]

> 2HS   6HL

Indicate chromosomal aberrations with the roman symbols "Df," "Dp," "In," "T," and "Tp," which stand for "deficiency," "duplication," "inversion," "translocation," and "transposition," respectively.

The genome of barley cultivar Betzes is the reference genome in the tribe Triticeae, and the definitions of translocations and short arm and long arm reversals are standardized in all Triticeae species.[80]

21.14.2.2 Maize (Corn)

(1) The NCBI stable taxon identifier for *Zea mays* is 4577, and the UniProtKB mnemonic suffix is "MAIZE." An example of a UniProtKB gene designation for *Z. mays* is "Q00LN8_MAIZE" for *16 kDa gamma zein*, which uses the gene symbol *gz16*. The UniProtKB entry number for *gz16* is Q00LN8, and the NBCI stable gene identifier is 542262.

(2) The definitions and standards for gene nomenclature and symbolization for maize are published periodically in the *Maize Genetics Cooperation Newsletter*[81] (*MNL*). A catalog of symbols for named genes is published annually in the *MNL*. Table 21.13[81,82] summarizes the major conventions. Because no specific wild-type strain has been designated for maize, naturally occurring variants are not properly termed "mutants."

Although it is strongly recommended that all newly identified genes be given 3-letter symbols, it is also recommended that older names and symbols be retained, even though many have just 1 or 2 letters. When naming newly detected genes previously known in other species, reference the list of gene families compiled by the Maize Genetics and Genomics Database[81] (see also Section 21.14.1, "Gene Families"). Give a newly identified locus a unique name, and designate it as "1," even though there is no second locus yet.

(3) The basic complement of A maize chromosomes is designated by arabic numerals 1 to 10 in decreasing order by length, and the chromosomes' short and long arms are designated by "S" and "L," respectively. Chromosome linkage groups (LGs) are represented beginning with position zero at the distal end of the short arm, which is termed the "left," "top," or "north" end on the linkage map.

>   4S   9L

Unless otherwise specified, translocations and other aberrations are assumed to affect the A chromosomes. Designate reciprocal translocations by "T," followed by the numbers of the rearranged chromosomes separated by a hyphen and followed by a letter or an isolate number. If an isolate number is used, encompass it within parentheses.

>   T1-2a   T1-2b   T1-9(4995)

The supernumerary B chromosomes are highly variable, unnumbered, and devoid of structural genes. Designate translocations of A chromosomes with B chromosomes by "TB" followed by a hyphen, the number of the A chromosome, the arm translocated, and a letter or an isolate number.

>   TB-1La   TB-5Sc

Designate inversions, deletions, and duplications with the abbreviations "Inv," "Del," and "Dup," respectively, followed by the chromosome number and a letter or an isolate number. For additional details, see *A Standard for Maize Genetics Nomenclature*.[81]

>   Inv1c   Inv2(8865)

21.14.2.3 Oat

(1) The NCBI stable taxon identifier for *Avena sativa* is 4498, and the UniProtKB mnemonic suffix is "AVESA." An example of a UniProtKB gene designation for oat is "H2B515_AVESA" for *translation initiation factor 1*, which uses the gene symbol *infA*. The UniProtKB entry number for *infA* is H2B515, and the NBCI stable gene identifier is 25016663.

(2) Tables 21.12 and 21.14[68,72,83] summarize the major conventions for oat. The proper display of oat gene symbols is determined by the International Oat Nomenclature Committee. Approved symbols should be set in italics, while yet-to-be-approved symbols should be in roman type.[83]

Table 21.13. Maize (Corn): Symbols for Genes and Phenotypes[a]

Feature	Convention	Examples
Gene name	Lowercase italic characters	*defective kernel*
Gene symbol	3-letter italic symbols based on the gene name are recommended, but older, shorter names are retained.	*Dek*
Different loci with similar phenotypes	The gene symbol with an arabic number added[b]	*dek12*
Dominant allele	Italicized locus symbol with initial capital letter	*Dek12* *Cg1*
Codominant alleles	Italicized locus symbol with initial capital letter followed by a hyphen and an arabic allele number	*Pgm2-5* and *Pgm2-7*
Recessive allele	Italicized locus symbol in all lowercase letters	*dek12* *cg1*
Reference allele	The locus symbol followed by "-*Ref*" or "-*R*"	*bz1-Ref* or *bz1-R*
Newly identified allele	The allele symbol followed by a hyphen and a laboratory number[c]	*sh2-6801*
Provisional symbol for new mutation at given phenotype	The symbol for the known gene followed by an asterisk and the laboratory number[c]	*bt*-8711*
Gene product and phenotype	The symbol in all capital letters. The name of the gene product is neither capitalized nor italicized.	ADH1 for alcohol dehydrogenase 1
Nonmutant alleles	Part of the allele designation may include the name of the inbred strain.	*Bz1-W22*
Mutations as result of deletion	The symbol for a deleted gene is enclosed in parentheses, preceded by the abbreviation "def," indicating a "deficiency." If 2 genes were deleted, their symbols are separated by double periods.	*def(wx1)-C34* *def(an1.. bz2)-6923* *def(bz1..sh1)-X2*
Mutation from transposable element insertion	The mutation is indicated with double colons or single apostrophe.	*wx-m1::Ds1* *Bz1'-7801* *dnap2094::Ac*
Genotypes		
Homozygous unlinked genes	Semicolons separate the gene symbols.	*a1; a2; c1; c2; r*
Homozygous linked genes	Spaces separate the gene symbols.	*C1 sh1 bz1 Wx1*
Heterozygous unlinked genes	Slashes separate homologous genes, and semicolons separate sets of unlinked gene symbols.	*Sh2/sh2; Bt2/bt2*
Heterozygous linked genes	Slashes separate sets of linked gene symbols.	*C1 Bz1/cl1 bz1*
Restriction fragment length polymorphisms (RFLPs) and random amplified polymorphic DNAs (RAPDs)	Lowercase 3- or 4-letter laboratory code followed by a laboratory number. cDNA or subclones of a gene are enclosed in parentheses after the RFLP locus designation.	*umc000(a1)*
Transposons	The names of transposons are in italics. The names of defective transposons begin with a lowercase italic *d*.	*Ac* *Ds* *dSpm(1)*

[a] Based on "A Standard for Maize Genetics Nomenclature"[81] and MaizeGDB: Maize Genetics and Genomics Database.[82]

[b] No hyphen separates the gene name from the numerical suffix. For symbols for mutant alleles, the hyphen is reserved for separating the allele designation from a suffix specifying the particular allele.

[c] See MaizeGDB[82] for a list of laboratory codes in current use. The laboratory number could indicate the date of identification. In the example shown, "6801" would represent "January 1968."

Table 21.14. Oat: Symbols for Genes and Chromosomes[a]

Feature	Convention	Examples
Gene symbol	Approved gene symbols are in italic letters, and unapproved symbols are letters in roman type with an initial capital letter. Except for genes for reactions to living organisms, the symbols stand for the dominant allele in genes with only 2 known alleles.	*Sg-1* Dw-7a
Dominant allele	Same as the gene symbol	Cda-4
Recessive allele	Same as the gene symbol, except first letter is lowercase	cda-1
Alleles	The gene symbol followed by an arabic numeral. The first locus is numbered 1, the second 2, and so forth. The members of allelic series are distinguished by lowercase letters after the locus number.	*Av-1*, *Av-2*, and *Av-3* *Px-1a* and *Px-1b*
Reactions to living organisms	The allele symbol is used with an uppercase initial letter. Alleles ending in the lowercase letter "a" indicate resistance, and those ending in "b" indicate susceptibility.	Ts-1a and Ts-1b
Genotypes		
Homozygous unlinked genes	Semicolons separate the gene symbols, and linkage groups are in numerical order.	*Cda-6*; *Cda-5*; *Cda-4*
Homozygous linked genes	Spaces separate the gene symbols.	*cda-1 ma-1 L-1*
Heterozygous unlinked genes	Slashes separate homologous gene symbols, with the maternal gene written first. Semicolons separate sets of unlinked gene symbols. Linkage groups are in numerical order.	. . .
Heterozygous linked genes	Slashes separate sets of linked gene symbols, with the maternal genes written first.	*A-1/Ba-1*
Unlocated genes	The gene names are enclosed in parentheses at the end of genic formulas.	. . .
Euploids and aneuploids	The gene symbols are repeated in genic formulas as many times as there are homologous loci.	. . .
Extrachromosomal factors	The factors are enclosed in square brackets preceding genic formulas.	. . .
Aberrations	Use the abbreviations "Df" for "deficiency," "Dp" for "duplication," "In" for "inversion," "T" for "translocation," and "Tp" for "transposition."	. . .

[a] Based on M.D. Simons and colleagues[72]; H.G. Marshall and Gregory E. Shaner[83]; and the GrainGenes database[68] of the US Department of Agriculture's Agricultural Research Service.

21.14.2.4 Rice

(1) The NCBI stable taxon identifiers for *Oryza sativa* (ie, Asian rice) include 4530 and 39947, and the UniProtKB mnemonic suffix is "ORYSJ." In addition, the NCBI stable identifier for *Oryza glaberrima* (ie, African rice) is 4538, and the UniProtKB mnemonic suffix is "ORYGL."

An example of a UniProtKB gene designation for *O. sativa* is "FAS1_ORYSJ" for *chromatin assembly factor 1 subunit FSM-like*, which uses the gene symbol *FSM*. *FSM*'s UniProtKB entry number is B2ZX90, and its NBCI stable gene identifier is 4324973.

(2) The rules for symbolizing rice genes were described by Toshiro Kinoshita[84] and updated by Kinoshita and G. Rothschild.[85] Table 21.15[73,84–87] summarizes the major conventions.

Table 21.15. Rice: Symbols for Genes and Phenotypes[a]

Feature	Convention	Examples
Gene name	Use an international language, preferably English, to describe the trait.	cytochrome c biogenesis B granule-bound starch synthase 1, chloroplastic/amyloplastic acetyl-CoA carboxylase carboxyltransferase beta subunit
Gene symbol	Abbreviation of the gene name in italics. Commonly used symbols should be retained even if they do not fit the rule.	*ccmB* *WAXY* *accD*
Dominant alleles	The name and symbol begin with a capital letter.	Chitin elicitor-binding protein and *CEBIP* Lactoylglutathione lyase and *GLYI-11*
Recessive alleles	The name and symbol use all lowercase letters.	ribulose bisphosphate carboxylase large chain and *rbcL*
Wild-type alleles	The gene symbol with a superscript plus symbol	*AL1⁺*
Multiple alleles	The gene symbol with letter or number superscripts	*PHRʰ*, *PHRⁱ*, and *PHRʲ* *FZP¹*, *FZP²*, and *FZP³*
Nonalleles with similar phenotypes	The gene symbols are differentiated by number or letter suffixes, which may be written on the line or as superscripts.	*psbD1*, *psbD2*, and *psbD3* *psbD¹*, *psbD²*, and *psbD³*
Enhancers, inhibitors, modifiers, and suppressors	If dominant, use "*En*," "*I*," "*M*," and "*Su*," respectively. If recessive, use "*en*," "*i*," "*m*," and "*su*." Follow abbreviation with a hyphen and the symbol of the gene affected.	*En-RAB5A* *I-PAO5* *M-RAD* *Su-ERG1*
Cytoplasmic genes	The names and symbols follow the same rules as nuclear genes.	. . .
Uncertain allelic relationships	The gene symbol followed by a lowercase italic "t" for "tentative" in parentheses. Used until the allelic relationship is determined.	*Piz(t)*
Quantitative trait loci (QTLs)	A lowercase "q" is followed by a 2- to 5-letter name in all uppercase that describes the trait being measured. This is followed by a hyphen and the number of the chromosome to which the trait has been mapped. The formula ends with another hyphen and a number indicating the individual QTL.	qDTH-2-3 (third QTL for the days-to-heading trait reported on chromosome 2)

[a] Based on Toshiro Kinoshita[84,87]; Kinoshita and G. Rothschild[85]; Susan R. McCouch, PhD, and colleagues[86]; the UniProt Knowledgebase[73]; and the National Center for Biotechnology Information's dbSNP database at https://www.ncbi.nlm.nih.gov.

The standard strain may be any strain that researchers select as long as it is named and its genic formula is made explicit.

(3) Nomenclature for rice chromosomes and chromosome LGs has been described by Kinoshita and Rothschild.[84,85,87] Chromosomes of rice are numbered 1 through 12 from longest to shortest, according to the length at the pachytene stage. The short arm of each chromosome is designated as the "left arm," and the zero position is the distal end of the short arm.

Aberrations and other structural changes are represented by the italic symbols "*Dp*,"

"*In*," "*T*," and "*Tp*," which stand for "duplication," "inversion," "translocation," and "transposition," respectively. Each symbol also includes a chromosome number in roman type. To distinguish similar aberrations involving the same chromosome, lowercase letters in roman type should follow the chromosome number.

In(2)a   *T*(1–2)b

Monosomics and trisomics are designated by the number of the affected chromosome.

Mono-1   Triplo-2

21.14.2.5 Rye

(1) *Secale cereale*'s NCBI stable taxon identifier is 4550, and the UniProtKB mnemonic suffix is "SECCE." An example of a UniProtKB gene designation for rye is "S4Z8W2_SECCE" for *cytochrome b6-f complex subunit 8*, which uses the gene symbol *petN*. The UniProtKB entry number for *petN* is S4Z8W2, and the NBCI stable gene identifier is 16792764.

(2) The recommendations for symbolizing genes of rye are based on those for wheat as reported by Jacob Sybenga.[88] Rye symbols are usually 2 letters, and they correspond to wheat symbols, provided that the rye symbols cannot be misinterpreted as wheat symbols. When a proposed rye symbol might be confused with a wheat symbol, a different symbol should be used for the rye gene.

Rye gene symbols, which are derived from gene names, are in italic letters. When they are unambiguous, the name and symbol of a dominant allele begin with an uppercase letter and those of a recessive allele with a lowercase letter. All letters and numerals in the symbol should be printed on the same text line.

Two or more genes producing similar phenotypes are designated by the same basic symbol. Nonallelic loci, such as mimics and polymeric genes. are designated by arabic numerals added to the locus symbol or by a hyphen and the genome symbol.

Inhibitors, suppressors, and enhancers are designated by "*I*," *Su*," and "*En*," respectively, if they are dominant and by "*i*," "*su*," and "*en*" if they are recessive. Each of these is followed by a space and the symbol of the allele affected.

(3) The rules for rye chromosome nomenclature were published by Sybenga.[88] Chromosomes are designated "1R" to "7R," and their short and long arms are designated as "S" and "L," respectively.

To designate the *Secale* species in the derivation of a chromosome arm, a 3-letter superscript is added to the chromosome symbol (eg, "2R[cer] L" represents the long arm of chromosome 2R of *S. cereale*).

The standard rye chromosome set is that of the chromosomes of the cultivar Imperial added to those of the wheat cultivar Chinese Spring.

21.14.2.6 Wheat

(1) *Triticum aestivum*'s primary NCBI stable taxon identifier is 4565, and its UniProtKB mnemonic suffix is "WHEAT." An example of a UniProtKB gene designation for *T. aestivum* is "XIP1_WHEAT" for xylanase inhibitor protein 1, which has the gene symbol *XIPI*. *XIPI*'s UniProtKB entry number is Q8L5C6, and its NBCI stable gene identifier is 542997.

(2) Table 21.16[68,89] summarizes the major conventions for wheat genes.

(3) Chromosome LGs and the corresponding chromosomes of wheat are designated by arabic numerals 1 to 7 followed by an uppercase letter in roman type representing genome A, B, or D.[89] Short and long chromosome arms are designated by "S" and "L." The designation of a particular chromosomal band requires the following 5 symbol elements[90]: a chromosome number, a genome designation, an arm symbol, a region number, and the band number within the region.

Table 21.16. Wheat and Related Species: Symbols for Genes and Phenotypes[a]

Feature	Convention	Examples
Gene symbol	The gene name abbreviated to 2, 3, or 4 italic letters. All letters are on the line, and none are superscripts or subscripts.	*Pc* (purple culm) *Co1* (corroded) *Adh* (alcohol dehydrogenase)
Dominant alleles	The name and symbol begin with a capital letter.	*Hs* and *Pln*
Recessive alleles	The name and symbol in all lowercase, except for reactions to pathogens	*bg* and *ki*
Gene pairs with similar effects	Use same gene symbol.	. . .
Mimics and poly-meric genes	Sequential arabic numerals are added to the gene symbols. Different alleles of independent origin are indicated by lowercase letters that follow the locus designation.	*Sr9* *Sr9a*
Orthologous sets of genes	The gene symbol followed by a hyphen, the accepted genome symbol, and an arabic numeral representing the set number. Different alleles of independent origin are indicated by a lowercase letter after the locus designation.	*Adh-A1* and *Adh-B1* *Adh-A1a*
Temporary symbols	The gene symbol followed by an abbreviation for the line and an arabic numeral referring to the gene	*SrFr1* and *SrFr2* (2 genes for reaction to *Puccinia graminis* in cultivar Federation)
Enhancers, inhibitors, and suppressors	If dominant, use "*En*," "*I*," and "*Su*," respectively. If recessive, use "*en*," "*i*," and "*su*." Follow with a space and the symbol of the affected allele.	. . .
Genetic formulas	Formulas are written as fractions, with maternal alleles given first or on top. Each fraction corresponds to a single linkage group.	. . .
Extrachromo-somal factors	The symbols enclosed in parentheses before the genetic formulas	. . .
Locus symbol	Italicized 2-, 3-, or 4-letter abbreviation of the trivial name of the macromolecule affected, with initial letter capitalized	*Adh*
Phenotypes	Same as the symbol for the locus, except in all capital letters in roman type	ADH-1
Pathogen-reaction genes	Symbols begin with a capital letter, even if they behave as recessive alleles.	*Sr17*

[a] Based on Robert A. McIntosh, PhD, and colleagues[89] and the GrainGenes database[68] of the US Department of Agriculture's Agricultural Research Service.

1BL21 (chromosome 1, genome B, long arm, region 2, band 1)

The cultivar Chinese Spring has the standard chromosome arrangement. Chromosomal aberrations are indicated by the roman abbreviations "Df," "Dp," "In, "T," and "Tp," which stand for "deficiency," "duplication," "inversion," "translocation," and "transposition," respectively. For a gene not in the standard chromosome position, insert the new chromosome designation within parentheses following the gene designation.

Hp (Tp 6D) (the introgressed *hairy neck* gene on chromosome 6D instead of the standard chromosome 4A)

21.14.3 *Arabidopsis thaliana* and Other Crucifers

(1) *Arabidopsis thaliana*'s NCBI stable taxon identifier is 3702, and its UniProtKB mnemonic suffix is "ARATH." An example of a UniProtKB gene designation for *A. thaliana* is

Table 21.17. *Arabidopsis thaliana* and Other Crucifers: Symbols for Genes and Phenotypes[a]

Feature	Convention	Examples
Gene name	The wild type in uppercase italics and mutant alleles in lowercase italics	*ARMADILLO BTB ARABIDOPSIS PROTEIN 1* *histidine kinase 2*
Gene symbols	The wild-type symbol in uppercase italics and mutant alleles in lowercase italics. Well-known symbols may have only 2 letters.	ABAP and *ahk*
Different loci with the same symbol	The gene symbol followed by an arabic numeral	*ABAP1* and *ahk2*
Different alleles of the same locus	The gene symbol is followed by a hyphen and an arabic numeral.	*ABAP1-1* and *ABAP1-2*
	If only one allele is known, the hyphen and number are not needed.	*fps1* = *fps1-1*
	If an allele shows dominance relative to the wild type, add an italic "*D*" to the allele number.	*ABAP1-2D* (allele 2 is dominant to the wild type)
Phenotype	The gene symbol in roman type with an initial capital letter. A superscript plus symbol is added for the wild type, and a superscript minus symbol is added for a mutant.	Abap1[+] (wild type) Abap1[-] (mutant)
Symbol for gene's protein product	The gene symbol in all uppercase letters in roman type	ABAP

[a] Based on David W. Meinke and Maarten Koornneef,[70] the Crucifer Genetics Cooperative,[91] and the UniProt KnowledgeBase.[73]

"KNATM_ARATH" for *KNOX meinox protein*, which uses the gene symbol *KNATM*. *KNATM*'s UniProtKB entry number is F4HXU3, and its NBCI stable gene identifier is 838041.

(2) Table 21.17[70,91] summarizes the major conventions for *A. thaliana*. When a locus has long been known by the mutant phenotype rather than by gene function, retain and use the original gene symbol even after the gene is cloned. When the mutant gene symbol is vague, misleading, or not widely known, contact the *Arabidopsis* Information Resource, known as TAIR, to change the symbol to one that does reflect the gene function or to provide information about the function or expression pattern of a new gene. Further information is available from the TAIR website.[92]

(3) The Crucifer Genetics Cooperative has adopted the guidelines used by the *Arabidopsis* research community for designating genotypes and phenotypes of *Brassica* spp.[91] Symbols for phenotypic traits may be followed by a number from 0 to 9 in parentheses to indicate the level of expression, with 0 indicating that the trait is absent and 9 indicating that the trait has very high expression.

    Ant(0) (no expression of anthocyanin pigment)
    Ant(8) (high expression of the anthocyanin pigment)

(4) At the pachytene stage, *Arabidopsis* has 5 bivalent chromosomes that conform to known LGs.[93] These chromosomes are designated by arabic numerals 1 to 5. Unlike the ideograms for most other organisms, the ideogram for *Arabidopsis* does not arrange chromosomes in descending order by length.

21.14.4 Cucurbits

(1) An example of a mnemonic UniProtKB suffix for a cucurbit is "CITLA," which UniProtKB uses for *Citrullus lanatus*. The stable NCBI taxon identifier for *C. lanatus* is 3654. Similarly, "A0A343A8E1_CITLA" is UniProtKB's designation for *C. lanatus*'s gene *cytochrome b6-f complex subunit 4*, which uses the symbol *petD*. The NCBI's stable gene identifier for *petD* is 30436321.

(2) The rules for cucurbit gene names and symbols are developed by the Cucurbit Genetics Cooperative (CGC).[94] These rules are published in 5 sets of CGC reports, which are periodically updated.[94] Table 21.18[94] summarizes the major conventions. Because of the diversity of cultivars, one cannot be selected to represent the norm. So researchers must distinguish between the normal and mutant alleles of genes being named, and researchers need to choose appropriate gene and allele symbols.[95]

(3) The chromosomes of cucurbits are small and not easily differentiated, so their morphology is not well characterized. See Table 21.12 for the haploid numbers of each group.

21.14.5 Legumes
21.14.5.1 Alfalfa

(1) The NCBI stable taxon identifiers for *Medicago sativa* include 3879 and 36902, and the UniProtKB mnemonic suffixes include "MEDSA" and "MEDSV." An example of a Uni-

Table 21.18. Cucurbits[a]: Symbols for Genes and Phenotypes[b]

Feature	Convention	Examples
Gene name	A short description in italics of the characteristic feature of the mutant	*bitter fruit* *short internode*
Gene symbol	The symbol consists of a minimum number of italic letters, the first letter being the same as that of the name of the gene. The symbol applies to allelic genes in compatible species.	*Bt* *Si*
Dominant alleles	The gene symbol with the initial letter capitalized	*Si*
Recessive alleles	The gene symbol in all lowercase letters	*fa*
Wild-type normal alleles	The gene symbol with initial capital letter. Alternatively, a superscript plus symbol may be used.	*Si* *Si*⁺
Multiple alleles	The gene symbol followed by a distinguishing superscript Roman letter or arabic numeral	*yg*ʷ
Mimics and polymeric genes	The names and symbols may be distinct from each other, or they may be assigned the same gene symbol followed by a hyphen and arabic numerals or Roman letters. The suffix "-1" may be used for the original gene in the series, or it may be omitted because it is understood.	*ms-1* and *ms-2*
Indistinguishable alleles	Identical phenotypes should have identical symbols. If distinctive symbols are assigned to recurrences of the same mutation, use the same symbol with distinguishing numbers or letters in parentheses as superscripts.	*ms-2*⁽ᴾˢ⁾
Intensifiers, inhibitors, and suppressors	The gene symbol is followed by a hyphen and the symbol of the affected allele. Alternatively, assign a distinctive name without the symbol of the gene affected.	*In-F* *Hi* *bi* (*bitterfree* suppresses *Bt* gene for bitterness)

[a] Cucurbits include cucumber (*Cucumis sativus*), melon (*Cucumis melo*), pumpkin (*Cucurbita* spp.), squash (*Cucurbita* spp.), watermelon (*Citrullus lanatus*), and other genera of the Cucurbitaceae.
[b] Based on Cucurbit Genetics Cooperative Gene List Committee.[94]

ProtKB gene designation for alfalfa is "PSBA_MEDSA" for *photosystem II protein D1*, which uses the gene symbol *psbA*. The UniProtKB entry number for *psbA* is P04998, and the NBCI stable gene identifier is 40503355.

(2) In naming the genes and symbols for alfalfa, authors should use the guidelines developed by the Commission on Plant Gene Nomenclature.[9]

(3) Cultivated alfalfa, *M. sativa*, has purple flowers and is tetraploid ($2n = 4x = 32$). Alfalfa has diploid progenitors ($2n = 2x = 16$): *M. sativa* ssp. *caerulea*, which has purple flowers, and *M. sativa* ssp. *falcata*, which has yellow flowers.[96] The standard karyotype of cultivated alfalfa shows 32 tiny chromosomes that are similar in length and morphology. C-banding studies have identified all of the chromosomes on the basis of their unique patterns. The chromosomes are paired in 8 sets, which are numbered 1 to 8. The first 7 pairs are numbered in descending order by length, and the eighth pair is the satellite chromosome.[97,98]

21.14.5.2 Pea

(1) The NCBI stable taxon identifiers for *Pisum sativum* include 3888 and 208194, and the UniProtKB mnemonic suffix is "PEA." An example of a UniProtKB gene designation for *P. sativum* is "RK33_PEA" for *50S ribosomal protein L33, chloroplastic*, which uses the gene symbol *rpl33*. The UniProtKB entry number for *rpl33* is P51416, and the NBCI stable gene identifier is 9073126.

(2) Table 21.19[73,99] summarizes the major conventions for pea. The rules presented in this table are by Thomas Henry Noel Ellis, PhD, and Mike Ambrose, MPhil.[99]

(3) Pea chromosomes are numbered 1 to 7, and their short and long arms are designated as "S" and "L," respectively. The pea LGs, which do not correspond by number to the chromosomes, are designated by roman numerals I to VII. The standard karyotype is that of line JI4. Symbols for chromosome aberrations consist of a prefix for the type of aberration, the numbers of the chromosomes affected in parentheses, and a unique lowercase suffix.

T(3L-7S)a (the first reported translocation between the long arm of chromosome 3 and the short arm of chromosome 7)

21.14.5.3 Soybean

(1) The NCBI stable taxon identifier for *Glycine max* is 3847, and the UniProtKB mnemonic suffix is "SOYBN." An example of a UniProtKB gene designation for *G. max* is "SLE2_SOYBN" for protein SLE2, which uses *SLE2* for its symbol. The UniProtKB entry number for *SLE2* is I1JLC8, and the NBCI stable gene identifier is 547493.

(2) The rules for gene symbols for soybean are updated by the Soybean Genetics Committee (SGC),[100] and online resources on soybean genetics are available on the US Department of Agriculture's website SoyBase at https://soybase.org. Table 21.20[73,100] summarizes the major conventions. The standard strain may be any strain selected by researchers as long as the strain and its genic formula are made explicit. The distinction between traits that are to be symbolized with identical, similar, or unrelated base letters is not necessarily clear-cut. The decision for intermediate cases is at the discretion of the researchers, but the intermediate cases should agree with previous practice for the trait.

(3) The rules for soybean cytogenetics are also updated by the SGC.[100] Classic LGs and the corresponding chromosomes are represented by arabic numerals 1 for the longest

Table 21.19. Pea: Symbols for Genes and Phenotypes[a]

Feature	Convention	Examples
Gene name	For dominant traits, use italics and an initial uppercase letter. For recessive traits, use all lowercase letters.	*Dihydrolipoyl dehydrogenase, mitochondrial* *Ribulose bisphosphate carboxylase large chain*
Gene symbol	The symbol is usually 3 italic letters derived from the gene name, but some symbols are shorter or longer.	*LPD* *SPP* *rbcL* *psbA*
Dominant allele	The gene symbol with initial uppercase letter	*LPD* *SPP*
Recessive allele	The gene symbol with all lowercase letters	*rbcL* *psbA*
Multiple alleles	The gene symbol with superscript lowercase letters or arabic numerals. Alternatively, the gene symbol is followed by a hyphen and the allele-specific designation.	*Cry^s* *D^{co}* *rug3-a* *lv-1* cry2-1
Phenotype	The gene name or gene symbol with or without italics	*translocase of chloroplast 34* or translocase of chloroplast 34
Different loci with similar phenotypes	The gene symbol followed by an arabic number	*TOC34* and *TOC75*
Isozyme loci	The italic abbreviation of the isozyme name with an initial uppercase letter. Follow with a hyphen and the code for the cellular compartment when appropriate.	*Mdh* *Aat-p* (*-p* indicates the plastid)
Genes of known function but no known phenotype	The gene symbol with an initial uppercase letter. When more than one locus is involved, each symbol is followed by an arabic numeral.	*Rbcs1* and *Rbcs2*
Genes identified by homology with those in other organisms	The prefix "*Ps*" for "*Pisum sativum*" is followed by the symbol for the homolog.	*PsENOD40*
Pseudogenes	The uppercase Greek letter psi followed by the gene symbol	Ψ*legD*

[a] Based on Thomas Henry Noel Ellis, PhD, and Mike Ambrose, MPhil,[99] and the UniProt Knowledgebase.[73]

through 20 for the shortest. In contrast, molecular LGs are represented by uppercase letters in roman type.[101]

Each symbol for chromosomal aberrations consists of an abbreviation for the aberration followed by the chromosome number or numbers and a letter for each additional aberration on a chromosome. The abbreviations for "deficiency," "inversion," "translocation," and "primary trisomics" are written in roman type as "Def," "Inv," "Tran," and "Tri," respectively.

Tran 1-2a (the first case of reciprocal translocations between chromosomes 1 and 2)
Tran 1-2b (the second case of such translocations)

Cytoplasmic factors are represented by the hyphenated prefix "*cyt-*" followed by one or more italic letters.

cyt-G (cytoplasmic factor for maternal green cotyledons)

Table 21.20. Soybean: Symbols for Genes, Alleles, and Cytoplasmic Factors[a]

Feature	Convention	Examples
Gene symbol	Italic letters to which superscripts, subscripts, or both may be appended	*Ep* for peroxidase *NARK* for nodule auto-regulation receptor kinase
Allele group base symbol	Each group of alleles has the same base letters as a symbol.	. . .
Gene pairs with similar effects	Sets of duplicate, complementary, and polymeric genes have the same base letter symbol, differentiated by numerical suffixes assigned in order of publication. The suffixes may be printed as subscripts or on the same line as the base. This is the only use of numerals in soybean symbols.	*Rps1*, *Rps2*, and *Rps2* *APX¹* and *APX²*
Alleles	Of the first published pair, the dominant gene has an initial capital letter, and the recessive is all lowercase. Subsequent alleles are designated by 1 or 2 lowercase superscript letters. This is the only use of superscripts in soybean symbols. As an alternative to using a superscript, the entire symbol may be written on the same line if a hyphen is inserted.	*Man4* and *man4* *NR*ᵇ and *NR*ᶜ *Rps1-b* and *Rps1-k*
Dominant alleles	An initial capital letter is used for the dominant or a partially dominant allele.	*Ab12* dominant over *ab12* *Br1*, *br2*
Codominant alleles	Capitalized symbol assigned at the authors' discretion	
Tentative alleles	Italicized gene symbol followed by an identifying designation in roman type enclosed in parentheses	*ms1* (Tonica) *ms1* (Ames 2)
Any allele	An underscore in place of a gene symbol represents any allele at the indicated locus.	*A_*
Unknown alleles	A question mark is used in place of a symbol. Following the question mark, place in parentheses the name of the line in which the gene was identified.	*Rps?* (Harosoy) (an allele at an unknown locus)
Standard strain	The plus symbol used in place of the assigned gene symbols of a designated homozygous strain	. . .
Isozyme variants	A gene symbol is used for the enzyme. Include the Enzyme Commission name and number for the specific enzyme activity investigated.	*HIUH* (*hydroxyisourate hydrolase* and EC: 3.5.2.17)
Genotypes	Precede the linked genes with the linkage group number, and list the gene symbols in the order with which they occur on the chromosome. Separate linkage groups with a semicolon and space.	*5E3Dt1*; *15Pgm1ms2*
Probe-detected loci	Locus designations are in roman type, and they consist of the following: • a prefix that indicates the origin of the probe • a hyphen • a string of letters or whole numbers that identify the probe used • a code for the restriction endonuclease used If needed, use a hyphen followed by a whole number to distinguish between loci detected by the same probe.	IaSU-B317I-1 and IaSU-B317T-2 (2 loci detected at Iowa State University in the United States, the first digested with *Eco*RI [I] and the second with *Taq*I [T])
Random amplified polymorphic DNA (RAPD) loci	The locus designations consist of an uppercase letter identifying the primer, a code for the primer name, and the fragment size in base pairs as a subscript. The designations are in roman type.	OA14$_{800}$ (an 800-bp fragment amplified with Operon Technologies primer14 from kit A)

Table 21.20. Soybean: Symbols for Genes, Alleles, and Cytoplasmic Factors[a] (*continued*)

Feature	Convention	Examples
Simple sequence repeat (SSR) loci	The locus designations consist of a prefix indicating the origin of the probe, a hyphen, a string of letters identifying the core nucleotide repeat of the SSR, and an identifying number. The code after the hyphen should not exceed 8 characters. The designations are in roman type.	IaSU-at275 BARC-gata3412
Cytoplasmic factors	The symbol "*cyt-*" is followed by one or more letters for the trait that are consistent with the letters for nuclear gene traits.	*cyt-G* (maternal green cotyledons) *cyt-Y* (maternal yellow cotyledons)

[a] Based on the Soybean Genetics Committee,[100] the National Center for Biotechnology Information's dbSNP database at https://www.ncbi.nlm.nih.gov, and the UniProt Knowledgebase.[73]

21.14.6 Solanaceous Plants

21.14.6.1 Pepper

(1) The NCBI stable taxon identifiers for species of the *Capsicum* genus include 4072, 4073, and 80379, and the UniProtKB mnemonic suffixes include "CAPAN," "CAPFR," and "CAPCH," respectively. An example of a UniProtKB gene designation for *Capsicum annuum* is "J7H3W8_CAPAN" for *ATP-dependent Clp protease proteolytic subunit*, which uses the gene symbol *clpP*. The UniProtKB entry number for *clpP* is J7H3W8, and the NBCI stable gene identifier is 13540245.

(2) The rules for the gene nomenclature of *Capsicum* were developed by Deyuan Wang and Paul W. Bosland, PhD.[102] Gene names concisely describe a characteristic feature of the mutant type in English or Latin. If the mutant is dominant, the first letter is capitalized in both the name and the symbol. If the mutant is recessive, the name and symbol are in lowercase. Mimics (ie, mutants with similar phenotypes that are controlled by different genes) may have distinctive names and symbols, or they may be assigned the same symbol followed by a unique arabic numeral or Roman letter. The original gene in the series is numbered *1*. Multiple alleles have the same symbol, followed by a superscript letter or arabic numeral.

> *Bzt* (mutant is dominant)　　　*ms-1 and ms-2* (mimics)　　　*et^a and et^f* (multiple alleles)

Modifying *Capsicum* genes may have a symbol for an appropriate name (eg, intensifier, suppressor, or inhibitor), followed by a hyphen and the symbol of the allele affected. Alternatively, the gene may be given a distinctive name that does not include the symbol of the gene being modified.

> *Mo-A* (modifier of *A*)
> *t* (interacts with *B* to produce high levels of β-carotene)
> *B* (interacts with *t* to produce high levels of β-carotene)

21.14.6.2 Tomato

(1) *Solanum lycopersicum*'s NCBI stable taxon identifier is 4081, and its UniProtKB mnemonic suffix is "SOLLC." An example of a UniProtKB gene designation for *S. lycopersicum* is "GAI_SOLLC" for *DELLA protein GAI*, which uses the gene symbol *GAI*. The UniProtKB entry number for *GAI* is Q7Y1B6, and the NBCI stable gene identifier is 54388.

(2) The Tomato Genetics Cooperative (TGC) outlined the rules for tomato gene nomenclature.[103] Those rules are summarized in Table 21.21.[104] The cultivar Marglobe is the stan-

Table 21.21. Tomato: Symbols for Genes and Phenotypes[a]

Feature	Convention	Examples
Mutant name	An appropriate descriptive name in italics, preferably in Latin or English	*abscisic stress-ripening protein 1* phytoene desaturase
Mutant gene symbol	An abbreviation of the mutant name in one or more italic letters or numbers[b]	*ASR1* *pds*
Standard, normal allele	The mutant gene symbol with a superscript plus symbol. When the context is clear, the plus symbol alone is sufficient.	*SPPF⁺* *MDHAR⁺* +
Dominant alleles	The name and symbol beginning with a capital letter	*Alternaria stem canker resistance protein* and *Asc-1*
Recessive alleles	The name and symbol in all lowercase letters	*endonuclease precursor-like* and *tbn1*
Additional alleles	A superscript uppercase "*D*" designates a dominant allele of a normally recessive gene. Other alleles are designated with appropriate number or letter superscripts, with numbers preferred. For the first member of a series, the number "*1*" is understood but not used.	*rnalx^D* (dominant allele appearing later) *mcpi*, *mcpi^2*, and *mcpi^3*
Indistinguishable alleles of independent origin (supposed recurrences)	Avoid unique symbols. Instead, use the existing mutant gene symbol with a series symbol as a superscript. The series symbol consists of an arabic numeral and a laboratory code assigned by the Tomato Genetics Cooperative's now-defunct Gene List Committee. Enclose the superscript series symbol in parentheses.	*ag1$^{(1K)}$* and *ag1$^{(2K)}$* (first and second recurrences of *ag*, with superscript "*K*" as the laboratory code)
Enhancers, modifiers, and suppressors	The symbol of the affected gene enclosed in parentheses and preceded by "*Enh*" for enhancer, "*Mo*" for a dominant modifier, "*mo*" for a recessive modifier, or "*Sup*" for a suppressor. Add a numerical suffix for subsequent nonallelic modifiers.	*Enh(wus)* *Mo(wus)* and *Mo(wus)2* *Sup(wus)2*

[a] Based on the Tomato Genetics Cooperative[104] and the National Center for Biotechnology Information's dbSNP database at https://www.ncbi.nlm.nih.gov.
[b] Because of their long-term, widespread use, the symbols "*c*," "*r*," "*s*," and "*y*" are retained for the gene *compound inflorescence*, even though the symbols were derived from the normal standard lines, not mutant lines.

dard, normal type. The TGC also proposed the rules for the gene nomenclature of *Arabidopsis*[104] (see Section 21.14.3, "*Arabidopsis thaliana* and Other Crucifers," and Table 21.17[70,73,91]), which have major differences with the guidelines for tomato genes.

(3) The rules for tomato cytogenetic nomenclature are intended for all members of the paraphyletic genus *Lycopersicon*.[104] Chromosomes are designated by arabic numerals in descending order by length, with 1 being the longest and 12 the shortest. The chromosome arms are designated short and long with the abbreviations "S" and "L," respectively. In addition, the short arm of each chromosome is designated the left arm, and the zero locus position is the distal, or left, end of the short arm.

Aberrations are designated by the roman symbols "Df," "In," and "T," which stand for "deficiency," "inversion," and "translocation," respectively. To distinguish aberrations involving the same chromosomes, lowercase letters are added after the chromosome numbers.

T (1–2)a   T (1–2)b   T (1–2)c

Aneuploids are designated according to the missing or extra chromosome. Those with whole-arm interchanges are symbolized by the component arms, with a connective dot representing the centromere.

triplo-1 (primary trisomic of chromosome 1)
haplo-12 (monosome for chromosome 12)
1S·12L (interchange between the short arm of chromosome 1 and the long arm of chromosome 12)

21.14.7 Other Plants

21.14.7.1 Cotton

(1) The NCBI stable taxon identifiers for species of the *Gossypium* genus include 3634, 3635, and 34284, and the UniProtKB mnemonic suffixes include "GOSBA," "GOSHI," and "GOSTU," respectively. An example of a UniProtKB gene designation for cotton is "H2B915_GOSRA" for *photosystem II protein D1*, which uses the gene symbol *psbA*. The UniProtKB entry number for *psbA* is H2B915, and the NBCI stable gene identifier is 11538581.

(2) The recommendations for gene names and symbols for cotton are based on those by Richard G. Percy, PhD, and Russell J. Kohel, PhD.[105] The nomenclature is based on classic genetics that deal with morphological mutants and chromosome aberrations. The names of mutant genes describe the main diagnostic feature of each mutant with a concise word or short phrase. The names and symbols of dominant mutants begin with an uppercase letter, and those of recessive mutants begin with a lowercase letter. The first letter of the symbol is the same as that of the mutant's name.

Newly discovered alleles at a given locus are assigned the original locus's symbol, which is followed by a superscript lowercase letter. New mutants with phenotypes similar to previously described mutants are designated with the symbol of the original mutant followed by a numeric subscript. The original mutant is considered number 1, and subsequent mutants are numbered serially. Gene symbols are in italic letters.

as_1 and as_2 (asynapsis) B_1 (Blight resistance)   *ml* (mosaic leaf) Sm_3 (Smooth leaf)

(3) The chromosome nomenclature of cotton was outlined by Dr Kohel.[106] Chromosomes are designated by arabic numbers 1 to 26, and chromosome LGs are designated by roman numerals. The genomes are designated by uppercase letters A and D. Aberrations are designated by the italic symbols "*Del*," "*Df*," "*Dp*," "*In*," "*T*," and "*Tp*," which stand for "deletion," "deficiency," "duplication," "inversion," "translocation," and "transposition," respectively. These abbreviations are followed by the symbols for the involved chromosomes. Different inversions and translocations in the same chromosomes are distinguished with lowercase letters.

In1a and *In1b* (inversions *a* and *b* on chromosome 1)
T1–2a and *T1–2b* (translocations *a* and *b* on chromosome 2)

Working designations for primary monosomics and monotelodisomics are "M#" and "Te#." Primary monosomics and trisomics are designated by the italic terms "*mono*" and "*triplo*" followed by the number of the chromosome.

mono-6   *triplo-6*

The standard for *Gossypium hirsutum* shall be a type similar to cultivar Texas Marker 1, and the standard for *G. barbadense* shall be a type similar to cultivar Pima S-4 for the pima cottons.[107] The research community has established a doubled haploid from the cultivar 3-79 as the standard for the cultivated extra-long staple *G. barbadense*. See the qualitative genetic summary by Drs Percy and Kohel.[105]

Cotton genetics changed with the advent of DNA markers. The first DNA marker map used restriction–fragment-length polymorphisms, or RFLPs, in an interspecific hybrid population.[108] Because DNA markers revealed a low level of polymorphism within *G. hirsutum*, mapping populations have been constructed primarily from hybrid populations between *G. hirsutum* and *G. barbadense*. As additional types of DNA markers were developed, they were applied to cotton. However, the low level of polymorphism in cotton slowed the development of genetic maps and restricted their development to interspecific populations.[107,109–114] Finally, with the use of DNA markers and cytological markers, the chromosomes and LGs were aligned and identified with the original nomenclature.[115]

Jing Yu and colleagues[116] applied a mathematical algorithm to construct a genome-wide comprehensive reference map based on the relative location of markers.

21.14.7.2 Lettuce

(1) The NCBI stable taxon identifiers for species of the *Lactuca* genus include 4236 and 75943, for which the UniProtKB mnemonic suffixes are "LACSA" and "9ASTR," respectively. An example of a UniProtKB gene designation for lettuce is "C7BL2_LACSA" for *costunolide synthase*, which uses the gene symbol *CYP71BL2* (also known as *LOC111890503*). The UniProtKB entry number for *CYP71BL2* is F8S1I0, and the NBCI stable gene identifier is 111890503.

(2) The gene nomenclature of lettuce and closely related species is adapted from recommendations of the International Committee on Genetic Symbols and Nomenclature[2] to meet the needs of lettuce researchers.[75] Although the names of newly discovered genes should conform to the basic guidelines, gene names and symbols already in widespread use should not be changed. In addition, genes with capitalized German names remain in use despite the apparent contradiction between the allele names and the allele symbols. Gene names for a normal, or wild-type, phenotype rather than a mutant phenotype are also retained. Where there is diversity in *Lactuca sativa* and the primitive form cannot be readily identified, the predominant form in *L. serriola* will represent the wild-type (+) allele.

21.14.7.3 Onion

(1) *Allium cepa*'s NCBI stable taxon identifier is 4679, and its UniProtKB mnemonic suffix is "ALLCE." An example of a UniProtKB gene designation for onion is "D2XT95_ALLCE" for *cytochrome b*, which uses the gene symbol *cob*. The UniProtKB entry number for *cob* is D2XT95, and the NBCI stable gene identifier is 27815641.

(2) The nomenclature and notation for the chromosomes of *A. cepa* also constitute the standard for *Allium* species that cross-fertilize with *A. cepa* to yield F_1 hybrids.[117,118] In an ideogram, the longest chromosome is represented on the left, the shortest is on the right, the centromeres are at the same level, and the short arms are uppermost. Beginning at the left of the ideogram, the chromosomes are numbered with arabic numerals, which are followed immediately by uppercase letters representing specific epithets. Thus, the chromosomes for *A. cepa* are numbered from 1C to 8C.

21.15 ANIMALS

The following guidelines are based on published rules for genes and chromosomes. Most conform to human or mouse rules, but there are several significant exceptions. Some organisms have elaborate, detailed sets of rules. Others have only minor differences from the model. Tables 21.22[29,119–125,140,141,144,146,153,156,168,170] and 21.23[5,29,126–138,140,141,143,146,152,153,159–161,166,167,169,170] summarize and index Sections 21.15.1, "*Caenorhabditis elegans*," to 21.15.8, "Transgenic Animals."

21.15.1 *Caenorhabditis elegans*

(1) *Caenorhabditis elegans*'s stable taxon identifier is 6239 with the National Center for Biotechnology Information (NCBI) in the United States, and its UniProt Knowledgebase mnemonic suffix is "CAEEL." An example of a UniProtKB gene designation for *C. elegans* is "DCTN2_CAEEL" for *putative dynactin subunit 2*, which uses the gene symbol *dnc-2*. The UniProtKB entry number for *dnc-2* is Q09248, the NBCI stable gene identifier is 175837, and the Ensembl identifier is WBGene00001018.

(2) The original recommendations for genetic nomenclature of *C. elegans* by H. Robert Horvitz, PhD, and colleagues[139] have been expanded. Table 21.24[140] summarizes the major conventions. The wild-type, standard strain is Bristol N2. Authors should consult the *Caenorhabditis* Genetics Center[140] when proposing new gene or allele names or when registering new laboratory codes.

(3) *C. elegans* has 5 pairs of autosomes (ie, I, II, III, IV, and V) and 1 pair of sex chromosomes (ie, X). Hermaphrodites are XX, and males, which are rare, are XO. Aberrations are indicated by the italic symbols "*Df*," "*Dp*," "*In*," and "*T*," which stand for "deficiency," "duplication," "inversion," and "translocation," respectively. Each name consists of a prefix in lowercase italic letters for a laboratory mutation, the abbreviation for the aberration, an arabic number, and, optionally, the numbers of the affected linkage groups.[140]

   mnDp2   *sDf3(I)*   *mnDp1(X;V)*

21.15.2 *Drosophila melanogaster*

(1) *Drosophila melanogaster*'s NCBI stable taxon identifier is 7227, and its UniProtKB mnemonic suffix is "DROME." However, "DROME" is not used in FlyBase, which is a database of *Drosophila* genes and genomes.

   An example of a UniProtKB gene designation for *D. melanogaster* is "ZEST_DROME" for the *regulatory protein zeste*, which uses the gene symbol z. The UniProtKB entry number for z is P09956, the NCBI gene identifier is 31230, and the FlyBase identifier is FBgn0004050.

(2) The conventions for *D. melanogaster* genes are similar to those for *Neurospora crassa* genes (see Section 21.12.3, "*Neurospora crassa*"). However, the *D. melanogaster* conventions differ significantly in some ways from those of many other organisms. The gene names for *D. melanogaster* must be italicized, concise, descriptive, unique, and inoffensive. Symbols derived from them should contain no spaces or subscripts, and all the characters should be letters in roman type. However, superscripts are required for alleles, including the superscript plus symbol for the wild-type allele.[141,142] Table 21.25[143] summarizes the major conventions.

   When necessary to distinguish a *D. melanogaster* gene from one of another organism that would otherwise have the same symbol, use *Dmel* as a prefix. Other species have similar 4-letter symbols to distinguish them.

(3) The symbols representing chromosomal aberrations in *Drosophila* spp. and the rules for their use were summarized by Dan L. Lindsley, PhD, and Georgianna G. Zimm.[143] Updates are made online as needed in FlyBase,[142] which includes a catalog of genes, chromosomal aberrations and their names, the inducing agents, alternative symbols and names, and specific instructions for genetic nomenclature for *D. melanogaster* and related species.

Table 21.22. Animals: Gene Nomenclature

Animal	Section or table[a]	Prefix named in rules	NCBI taxon identifier[b]	UniProtKB species code[c]	Model	Reference
Caenorhabditis elegans	Section 21.15.1, "*Caenorhabditis elegans*," and Table 21.24	*Ce-*	6239	CAEEL	...	*Caenorhabditis elegans* website[140]
Cat (*Felis catus*)	Table 21.30	*FCA*	9685	FELCA	mouse	Committee on Standardized Genetic Nomenclature for Cats[120] and Committee on Standardized Genetic Nomenclature for Mice[121]
Cattle (*Bos* spp.)	Section 21.15.5.1, "Cattle," and Table 21.29	*BBO*	9913	BOVIN	ruminants	Thomas E. Broad and colleagues[156]
Chicken (*Gallus domesticus*) (see also entry below for Poultry and other avian species)	Table 21.28	...	9031	CHICK	...	...
Deer mouse (*Peromyscus* spp.)	Section 21.15.6.1, "Deer Mouse," and Table 21.30	...	Varies with species. Includes 57102, 1587529, and 57104.	Varies with species. Includes PERAZ, PERMX, and 9RODE	mouse	Mary F. Lyon, PhD[168]
Dog (*Canis familiaris*)	Section 21.15.5, "Domestic Mammals," and Table 21.29	*CFA*	9615	CANLF	domestic animals	Anatoly Ruvinsky, PhD, DSc, and Jeff Sampson[125]
Domestic animals: ruminants, cattle, sheep, goats, deer, camelids, pigs, and horses	Section 21.15.5, "Domestic Mammals," and Table 21.29	...	Varies with species. Includes 9837, 59525, and 9894.	Varies with species. Includes CAMBA, ANTCE, and GIRCA.	ruminants	Thomas E. Broad and colleagues[156]
Drosophila (*Drosophila melanogaster*)	Section 21.15.2, "*Drosophila melanogaster*," and Table 21.25	*Dmel*	7227	DROME	Neurospora	FlyBase[141]
Fish (excluding zebra danio)	Section 21.15.3.1, "Fish Proteins," and Tables 21.26 and 21.32	...	Varies with species. Includes 311080, 7913, and 7892.	Varies with species. Includes HUSHU, POLSP, and NEOFS.	human	James B. Shaklee and colleagues[344]

Organism	Section/Table	Code	ID	UniProt	Category	Reference
Goat (*Capra hircus*)	Table 21.29	*CHI*	9925	CAPHI	ruminants	Thomas E. Broad and colleagues[156]
Horse (*Equus caballus*)	Section 21.15.5.2, "Horse and Other Equines," and Table 21.29	*ECA*	9796	HORSE	ruminants	Thomas E. Broad and colleagues[156] and Charles H.S. Dolling[123]
Human (*Homo sapiens*)	Section 21.15.7, "Human," and Table 21.32	*HSA*	9606	HUMAN	human	Human Gene Nomenclature Committee
Mouse (*Mus* spp.)	Section 21.15.6.2, "Mouse and Rat," and Table 21.30	*MMU*	Varies with species. Includes 10090, 41269, and 1916.	Varies with species. Includes MOUSE, 9MURI, and MUSPA.	mouse	Mouse Genome Informatics[29]
Pig (*Sus scrofa*)	Table 21.29	*SSC*	9825	PIG	ruminants	Thomas E. Broad and colleagues[156]
Poultry and other avian species	Section 21.15.4, "Poultry, Domestic Fowl, and Other Avian Species," and Table 21.28	...	Varies with species. Includes 33616, 9103, and 8839.	Varies with species.[b] Includes GYMCA, MELGA, and ANAPL.	human	Lyman B. Crittenden and colleagues[153] and *Trends in Genetics*[119]
Rabbit (*Oryctolagus cuniculus*)	Table 21.30	*OCU*	Varies with species. Includes 9986 and 48083.	Varies with species. Includes RABIT and BRAID.	mouse	R.R. Fox[124]
Rat (*Rattus norvegicus*)	Table 21.30	*RNO*	10116	RAT	mouse	G. Levan and colleagues[170]
Sheep (*Ovis aries*)	Table 21.29	*OOV*	9940	SHEEP	ruminants	Charles H.S. Dolling[122]
Zebra danio (*Danio rerio*, also known as *Brachydanio rerio*)	Section 21.15.3.2, "Zebra Danio (Zebrafish)," and Table 21.27	...	7955	DANRE	...	ZFIN[146]

[a] Entries refer to the section or table for the model organism. In a number of cases, the rules are not unique to the animal named in the first column.

[b] The National Center for Biotechnology Information's unique stable identifiers are numerical. The identifiers in this table are for taxa. The center also assigns numerical identifiers to genes.

[c] When an organism's common name in English is 5 characters or fewer, the UniProt Knowledgebase's identification code is usually the common name (eg, "HUMAN" and "PIG"). When the common name is longer than 5 characters, the prefix is usually the first 3 letters of the genus name and the first 2 letters of the specific epithet's name (eg, "DROME" for "*Drosophila melanogaster*"). Exceptions are made when a short common name can be applied to more than one species (eg, "goat") or when the name can be shortened without ambiguity (eg, "BOVIN" for "bovine"). For details and a list of organisms, see ExPASy at https://www.expasy.org/.

Table 21.23. Animals: Cytogenetic Nomenclature

Animal	Section or table[a]	Chromosomes or linkage groups (LG)	Model	Reference
Caenorhabditis elegans	Section 21.15.1, "*Caenorhabditis elegans*"	I, II, III, IV, V, and X	. . .	*Caenorhabditis elegans* website[140]
Cat (*Felis catus*)	. . .	A1, A2, A3, B1, B2, B3, B4, C1, C2, D1, D2, D3, D4, E1, E2, E3, F1, F2, X, and Y	human	K.W. Cho and colleagues,[132] ISCN 1985,[137] and ISCN 1995,[138] and ISCN 2020[5]
Cattle (*Bos* spp.)	Section 21.15.5.1, "Cattle"	1 to 29, plus X and Y	domestic animals	ISCNDA 1989[159] and C. Paul Popescu and colleagues[160]
Deer mouse (*Peromyscus* spp.)	Section 21.15.6.1, "Deer Mouse"	1 to 23, plus X and Y	human	Ira F. Greenbaum, PhD, and colleagues[169]
Dog (*Canis familiaris*)	. . .	1 to 38, plus X and Y	. . .	Marek Świtoński and colleagues,[129] N. Reimann and colleagues,[128] and A. Ruvinsky and colleagues[131]
Domestic animals (see also specific animal)	Section 21.15.5, "Domestic Mammals"	. . .	human	ISCNDA 1989[159]
Drosophila (*Drosophila melanogaster*)	Section 21.15.2, "*Drosophila melanogaster*"	1 to 4, plus X and Y	. . .	Dan L. Lindsley, PhD, and Georgianna G. Zimm[143] and FlyBase[141]
Fox (*Vulpes fulvus*)	. . .	1 to 18, plus X and Y (including 2 pairs of microsomes)	. . .	Committee for the Standard Karyotype of *Vulpes fulvus* Desm[126]
Goat (*Capra hircus*)	. . .	1 to 29, plus X and Y	ruminants	C. Paul Popescu and colleagues[160]
Hamster (*Cricetulus griseus*)	. . .	1 to 10, plus X and Y	. . .	Committee on Chromosome Markers[127]
Horse (*Equus caballus*)	Section 21.15.5.2, "Horse and Other Equines"	1 to 31, plus X and Y	human	C. Richer and colleagues,[166] Ann T. Bowling, PhD, and colleagues,[167] and ISCN 2020[5]
Human (*Homo sapiens*)	Section 21.15.7, "Human," and Tables 21.33 and 21.34	1 to 22, plus X and Y	human	ISCN 2020[5]
Mouse (*Mus* spp.)	Section 21.15.6.2, "Mouse and Rat," and Table 21.31	1 to 19, plus X and Y	mouse	Mouse Genome Informatics[29]
Pig (*Sus scrofa*)	. . .	1 to 18, plus X and Y	human	Committee for the Standardized Karyotype of the Domestic Pig[135] and Larry D. Young[130]
Poultry and other avian species (includes chicken [*Gallus domesticus*])	Section 21.15.4, "Poultry, Domestic Fowl, and Other Avian Species"	1 to 38, plus Z and W (those smaller than chromosome 8 are considered microchromosomes)	human	T.B. Shows and colleagues[152] and Lyman B. Crittenden and colleagues[153]
Rabbit (*Oryctolagus cuniculus*)	. . .	1 to 21, plus X and Y	human	Committee for Standardized Karyotype of *Oryctolagus cuniculus*[134]
Rat (*Rattus norvegicus*)	. . .	1 to 20, plus X and Y	human	Committee for a Standardized Karyotype of *Rattus norvegicus*[133] and G. Levan and colleagues[170]

Table 21.23. Animals: Cytogenetic Nomenclature (*continued*)

Animal	Section or table[a]	Chromosomes or linkage groups (LG)	Model	Reference
River buffalo (*Bubalus bubalis*)	. . .	1 to 24, plus X and Y	domestic animals and human	ISCN 2020,[5] ISCNDA 1989,[159] and Leopoldo Iannuzzi[136]
Sheep (*Ovis aries*)	. . .	1 to 26, plus X and Y	human	ISCN 2020[5] and H.A. Ansari and colleagues[161]
Zebra danio (*Danio rerio*, also known as *Brachydanio rerio*)	Section 21.15.3.2, "Zebra Danio (Zebrafish)"	LG01 to LG25		ZFIN[146]

ISCN, International System for Human Cytogenomic Nomenclature; ISCNDA, International System for Cytogenetic Nomenclature of Domestic Animals; ZFIN, Zebrafish Information Network

[a] Entries refer to the sections and tables for the model organisms. In a number of cases, the rules are not unique to the animals named in the first column.

Table 21.24. *Caenorhabditis elegans*: Symbols for Genes and Phenotypes[a]

Feature	Convention	Examples
Gene symbol	3 lowercase italic letters derived from the description of the mutant phenotype, followed by a hyphen and an arabic numeral in italics. May include a roman numeral in italics to indicate the linkage group of the gene. The gene symbol and the linkage group are separated by a space.	*dpy-18* *mlc-2* *dpy-18 III* *lon-2 X* *wrn-1* (corresponds to *WRN1* in human)
Homologous genes in related species	The gene name preceded by 2 italic letters referring to the species, with a hyphen in between	*Cb-tra-1* (homologue of the *Caenorhabditis elegans* gene *tra-1* in *C. briggsae*) *Ce-snt-1* (specifies the *C. elegans* gene)
Homologous set of genes related to a single gene from another species	The gene name followed by decimal numbers	*sir-2.1, sir-2.2, sir-2.3,* and *sir-2.4* (all correspond to *SIR2* in *Saccharomyces cerevisiae*)
Pseudogenes	The gene name followed by "*ps*"	*msp-38ps*
Different loci conferring similar phenotypes	The gene name prefix followed by a hyphen and serial arabic numerals in italics	*mlc-1, mlc-2,* and *mlc-3*
Alleles and mutations	Mutation symbols consist of a laboratory code and an arabic numeral for the specific mutation. Both are in italics. The symbols may be followed by a lowercase symbol in roman type for the description of the mutation.	*e1348*rl ("*e*" for "MRC Lab in England," and "rl" for "recessive lethal")
	When the gene name and mutation symbol are used together, the mutation symbol is enclosed in parentheses.	*dpy-5(e61)*
Wild-type allele	The gene name followed by a plus symbol or by a plus symbol in italicized parentheses	*sma-2*+ or *sma-2(+)*
Chromosomal duplication	The symbol for the chromosomal aberration, which may be followed by the name of one or more of the duplicated genes in square brackets	*mnDp1(X;V)* or *mnDp1(X;V) [unc-3(+)]*

Table 21.24. *Caenorhabditis elegans*: Symbols for Genes and Phenotypes[a] (*continued*)

Feature	Convention	Examples
Extrachromosomal array	The laboratory code, the letters "*Ex*," and a number, all in italics. May be followed by the description of the transgene in square brackets.	*eEx3* *stEx5* *stEx5[sup-7(st5) unc-22(+)]*
Integrated transgenes	The laboratory code, the letters "*Is*," and a number, all in italics	*eIs2*
Gene fusions	The gene name followed by a double colon and the name of the reporter gene	*pes-1::lacZ* *mab-9::GFP*
Genotypes		
Homozygotes	The gene symbols in order of chromosome number. The chromosomes are separated by a semicolon and space.	*dpy-5(e61) I; bli-2(e768) II*
Heterozygotes	The same as for homozygotes, except slashes separate homologous chromosomes, not semicolons and spaces. Use plus symbol for wild-type alleles.	*dpy-5/unc-13* *dpy-5/+ I; +/bli-2 II*
Transposons	Not italicized except when included in a genotype. Indicate insertions by adding "::*Tc*" to the relevant mutation symbol.	Tc1 *and* Tc2 *unc-54(r293::Tc1)*
Phenotypes	Description in words in all lowercase roman type or a 3-letter gene symbol with initial letter capitalized	dumpy animal *or* Dpy uncoordinated *or* Unc
Wild-type phenotype	In comparisons of the wild type to the mutant, use the prefix "non-" before the mutant phenotype.	non-Dpy (wild type is not dumpy)
Protein product	The gene symbol in roman capital letters	UNC-13 (protein encoded by the *unc-13* gene)
Mitochondria	The gene name using the standard nomenclature, followed by the nuclear genes in a genotype with "*M*" as the abbreviation for the linkage group	*dpy-5/+ I; +/bli-2 II; cyt-1 M*

[a] Based on the *Caenorhabditis* Genetics Center at https://cgc.umn.edu/.[140]

The autosomes of *D. melanogaster* are represented by arabic numerals 1 to 4, and the sex chromosomes are represented by X and Y. The symbols "*X*" and "*1*" for the X chromosome are synonymous, but "*1*" is preferred.

Aberrations on the same chromosome are indicated by the italic prefixes "*Df*," "*Dp*," "*In*," "*R*," "*T*," and "*Tp*," which stand for "deficiency," "duplication," "inversion," "ring chromosomes," "translocation," and "transposition," respectively. These prefixes are followed by the chromosome designation within parentheses and a specific designation that identifies the rearrangement. Superscripts are allowed only in names for synthetic inversions and in aberration names with allele symbols. No spaces are used.

    Dp(1;1)   *In(2R)C72*   *T(1;2;3)OR14*

21.15.3 Fish

21.15.3.1 Fish Proteins

(1) An example of a mnemonic UniProtKB suffix for a fish species is "ORYLA," which UniProtKB uses for *Oryzias latipes*. The stable NCBI taxon identifier for *O. latipes* is 8090. Similarly, "TERT_ORYLA" is UniProtKB's designation for *O. latipes*'s gene *telomerase re-*

Table 21.25. *Drosophila melanogaster*: Genes and Phenotypes[a]

Feature	Convention	Examples
Gene name	A description of the gene's function or mutant phenotype in italics	*white*, *Deformed* "Flies were scored for *white* mutations."
Gene symbol	Italic letters derived from the gene name	*w* (symbol for *white*) *Dfd* (symbol for *Deformed*)
Different loci with similar function	The gene symbol followed by arabic numerals and capital letters in italics. Hyphens are not used, except to separate numbers or letters that would lose their descriptive content otherwise.	*Act5C* and *Act42A*
Gene named for an allele recessive to wild type	The name and symbol begin with a lowercase letter.	*awd* (for *abnormal wing discs*)
Gene named for an allele dominant to wild type	The name and symbol begin with an uppercase letter.	*R* (for *Roughened*)
Gene named for a protein	The symbol begins with an uppercase letter.	*Adh* (for alcohol dehydrogenase)
Gene named for a tRNA	The symbol begins with "*tRNA:*"	*tRNA:Ser-AGA-2-4*
Gene named for small-nuclear RNA	The symbol begins with "*snRNA:*"	*snRNA:U6:96Aa*
Mitochondrial gene	The symbol begins with "*mt:*"	*mt:ND4* and *mt:tRNA:Leu:TAA*
Transposons	The symbol consists of the designations for *ends{genes=construct-symbol}*, all in italics.	$P\{w^{+mC}ovo^{D1-18}=ovoD\text{-}18\}$ $H\{Lw2\}dpp^{151H}$
Alleles	The gene symbol followed by superscript characters. Square brackets may be used instead of superscript characters.	sc^1, sc^2 or *sc[1]*, *sc[2]*
Recessive allele of a gene named for a dominant mutation	The gene symbol with superscript or bracketed "*r*" for "recessive"	Hn^{r2} or *Hn[r2]*
Dominant allele of a gene named for a recessive mutation	The gene symbol with superscript or bracketed "*D*" for "dominant"	ci^D or *ci[D]*
Genotypes	Genes on the same chromosome are separated with spaces. Genes on different chromosomes are separated with semicolons and spaces. Homologues are separated with slashes.	$Y^1\,w^1\,f^1$ $y^1; bw^1; e^4; ci^1\,ey^R$ $y^1\,w^1\,f^1\,eor^{16\text{-}3\text{-}119}/Basc$
Genes identified during a genomic sequencing project	The symbol consists of "BG:" or "EG:,"[b] the clone name, a decimal, and a serial number, all in roman type.	BG:DS07851.5 EG:152A3.3
Genes identified during annotation of the genomic sequencing projects	The symbol consists of "CG" and a unique number in roman type.	CG10809
Phenotypes	Description in words in roman type	"Flies with white eyes were selected."
Protein products	The gene symbol in capital letters in roman type	HH (protein encoded by the *hh hedgehog* gene *hh*)

[a] Based on FlyBase[143] at https://flybase.org/.

[b] "BG" stands for the "Berkeley Drosophila Genome Project" and "EG" for the "European Drosophila Genome Project." The clone name is the name or number of the clone on which the gene was first identified. The integer after the decimal is a serial number, and the serial number does not imply gene order on the clone.

Table 21.26. Fish: Symbols for Genes and Phenotypes[a]

Feature	Convention	Examples
Gene symbol	Uppercase letters[b] in italics. The symbol preferably ends with "*." All characters are on the baseline, and none is a superscript or subscript.	*IDHA** *MPI** *YAP1**
Multiple loci	For unknown relationships, use a hyphen and arabic numeral in italics. For established orthologies, use a hyphen and uppercase letter in italics.	*Idha-1** *LDH-A** and *LDH-B**
Isoloci	The gene symbol with a comma between the 2 locus numbers	*GPI-B1,2**
Regulatory locus suffix	The gene symbol with a lowercase "*r*" in italics	*LDH-Ar**
Subcellular prefix	The gene symbol with a lowercase letter as a prefix[c]	*sMDH-B**
Allele	The gene symbol followed by an asterisk, followed in turn by an arabic numeral, a lowercase letter, or relative electrophoretic mobility in italics. The preferred convention is to sequentially assign number codes to designate alleles.	*MDH-B*1* *ADA-1*a* *EST-2*75*
Enzyme	IUBMB-specified name and number[b]	L-lactate dehydrogenase, EC 1.1.1.27
Enzyme symbol	The uppercase letters in roman type. Numbers, Greek-letter stereochemical isomer symbols, or hyphens are used as necessary.	LDH
	A lowercase prefix is used for subcellular localization.[c]	mIDHP

IUBMB, International Union of Biochemistry and Molecular Biology

[a] Based on James B. Shaklee and colleagues[144] and the UniProt Knowledgebase.[73] See Section 21.15.3.2, "Zebra Danio (Zebrafish)," and Table 21.27 for discussion and symbols for zebra danio (*Danio rerio*), which has its own gene nomenclature.

[b] These are the same alphanumeric symbols as the abbreviations for the coded proteins, which are derived from the names assigned by the Nomenclature Committee of the International Union of Biochemistry and Molecular Biology (see https://www.enzyme-database.org/). Italicization and asterisks distinguish the symbols from those for enzymes and proteins. The asterisk follows the locus symbol and precedes the allele designation.

[c] The prefix indicates the subcellular location of the enzyme coded by the gene: The letter "l" is used for "lysosomal," "m" for "mitochondrial," "p" for "peroxisomal," and "s" for "cytosolic (ie, "supernatantor soluble"). These prefixes should be ignored when alphabetizing a list.

verse transcriptase, which uses the symbol *tert*. The NCBI's stable gene identifier for *tert* is 100049443.

(2) A formal system developed by James B. Shaklee and colleagues[144] established the nomenclature and symbolization for gene coding for enzymes and other proteins of fish. The system is similar to that for human genes, and draws on the enzyme nomenclature of the International Union of Biochemistry and Molecular Biology.[145] Table 21.26[73,144] summarizes the major conventions. This system is quite different from the one adopted for zebra danio (see Section 21.15.3.2, "Zebra Danio (Zebrafish)," and Table 21.27[146]).

21.15.3.2 Zebra Danio (Zebrafish)

Many researchers who work with this species as a laboratory animal prefer the name "zebrafish" to "zebra danio." However, the term "zebrafish" is ambiguous because it has been applied to *Melambaphes zebra*, which is now known as "*Girella zebra*," and to *Danio rerio*, which is now also known as "*Brachydanio rerio*." This style manual uses

Table 21.27. Zebra Danio: Symbols for Genes and Phenotypes[a]

Feature	Convention	Examples
Gene name	The descriptive name is in lowercase italics. Genes identified as orthologs of mammalian genes take the mammalian name and symbol.	*engrailed 1a* and *engrailed 2b*
Gene symbol	Short alphanumeric string derived from the gene name set in lowercase italics. When a zebrafish gene has been renamed to the mammalian ortholog from an older zebrafish name, append the previous name in parentheses.	*eng1a* and *eng2b* *shha (syu)*
Duplicate genes	Use the approved name of the mammalian ortholog, followed by *a* and *b*.	*hoxa13a* and *hoxa13b*
Mutants and genes identified by mutation	Gene nomenclature for mutants is the same as that for genes identified by mutation.	. . .
Names of mutants and genes identified by mutation	Genes identified by mutation are typically named to reflect the mutant phenotype. Numbers should generally not be used in naming mutants.	*touchy feely*
Symbols for mutants and genes identified by mutation	The symbol should be derived from the full name.	*tuf*
Wild-type allele	The symbol with a superscript plus symbol	*lof⁺*, *ndr2⁺*, and *brs⁺*
Mutant allele (recessive allele)	The symbol for the mutant with superscript laboratory code and allele designation	*ndr2^m101* (laboratory code "m" is for Massachusetts General Hospital in Boston in the United States)
Dominant alleles	The symbol for the mutant or allele with a "*d*" in the first position of the superscript	*lof^dt2* (laboratory code "t" is for the Max Planck Institute for Biological Cybernetics in Tübingen, Germany)
Transgenes	The symbol is in the format "*Tg(xxx),*" in which "*Tg*" is the symbol for "transgene" and "*(xxx)*" describes the salient features of the transgene. If a transgene has a regulatory sequence derived from either enhancers or promoters, use a colon to separate the regulatory sequence from the coding sequence. In general, the regulatory sequence is named for the gene from which it was derived or the gene or transcript that it regulates. Several other configurations are used, depending on variations such as transgenes with transcripts, with fusions, with promoter elements of differing sizes, or with a gene from a different species.	*Tg(pitx2-002:GFP)*—transgene with transcript *pitx2-002* *Tg(actb2:stk11-mCherry)*—transgene with fused protein stk11 *Tg(-3.5hhex:sptb-GFP)*—transgene with promoter elements of differing sizes *Tg(Hsa.FGF8:GFP)*—transgene with the human FGF8 gene as the promoter

Genotypes

Feature	Convention	Examples
Homozygotes	For homozygotes with a single locus, separate each allele with a forward slash ("/"). For homozygotes with multiple loci, list the genotype at each locus in order according to chromosome number, and use semicolons to separate loci on different chromosomes.	*ednrb1a^b140/ednrb1a^b140*— homozygote with a single locus *ednrb1a^b140; slc24a5^b16⁺*— homozygote with multiple loci
Heterozygotes	For heterozygotes with a single locus, separate each allele with a forward slash ("/"). For heterozygotes with multiple loci, separate loci on homologous chromosomes with slashes.	*ednrb1a^b140/ednrb1a⁺*— heterozygote with a single locus *fgf3^t21142/fgf3^t24149; slc24a5^b16/ slc24a5^m592*—heterozygote with multiple loci, of which 2 pair are on homologous chromosomes

Table 21.27. Zebra Danio: Symbols for Genes and Phenotypes[a] (*continued*)

Feature	Convention	Examples
Unmapped loci	The genotypes of unmapped loci follow the genotypes of mapped loci on different chromosomes. The genotypes of the unmapped loci are listed alphabetically within braces.	*ednrb1a^{b140}; mycbp2^{tj236} {edi^{tl35}}*
Poorly resolved loci on the same chromosome	List gene symbols in alphabetical order within braces.	*{abc^{b000} def^{m000}}*—poorly resolved loci on same chromosome
Protein products	The gene symbol with an initial capital letter in roman type	Ndrw, Brs, Eng1a, Eng2b, and Ntl
Phenotypes	Mutant or gene name and symbol in italics	*touchy feely* and *tuf*

[a] Based on ZFIN Zebrafish Nomenclature Conventions.[146]

"zebra danio," which is the name for "*Danio rerio*" approved by the American Fisheries Society and other authorities.

(1) Zebra danio's stable NCBI taxon identifier is 7955, and its UniProtKB mnemonic suffix is "DANRE." An example of a UniProtKB gene designation for zebra danio is "OPSD_DANRE" for rhodopsin, which uses the gene symbobl *rho*. The UniProtKB entry number for *rho* is P35359, the NBCI stable gene identifier is 30295, and the Ensembl identifier is ENSDARG00000002193. In addition, the identifier that The Zebrafish Information Network (ZFIN) assigned to *rho* is ZDB-GENE-990415-271.

(2) The rules for the gene nomenclature for zebra danio were published by Mary C. Mullins, PhD, in 1995.[147] They are updated as needed by ZFIN (ZFIN).[146,148–150] Table 21.27[146] summarizes the major conventions. Members of gene families follow existing naming conventions. Symbols should not begin with "*Z*" or "*Zf*." Authors should register with ZFIN new locus and allele names, their symbols, and codes for the laboratories researching the genes.

When different laboratories give different names to the same new gene or mutant allele of a previously known gene, the first name in print is given priority. When a mutation is found in a previously cloned zebra danio gene whose name has been published, the mutant takes the name of the cloned gene. If both the cloned gene and the mutation are already known by different names but are later found to be the same gene, the name of the mutation usually takes priority. The unique name of the mutation may be more relevant to the function of the gene than the cloned gene's name, and this unique mutant name may be easier to remember or use than the gene name. Exceptions include cases in which orthologous relationships with genes of other species are well established and these orthologs have commonly used names.

(3) The 25 original linkage groups, designated "LG01" to "LG25," are now designated as chromosomes "(Chr)01" to "(Chr)25." For the first 9 chromosomes, the number begins with zero so that computer programs order them correctly. The upper and long arms are indicated by "U" and "L" without a space after the chromosome number.[148] Chromosome rearrangements are indicated by the prefixes "*Df*," "*Dp*," "*In*," "*Is*," "*T*," and "*Tg*," which stand for "deficiency," "duplication," "inversion," "insertion," "translocation," and "transgene," respectively. Chromosomal differences have not been observed between males and females. Detailed instructions for describing chromosomes for zebra danio and their aberrations are available at ZFIN.[148]

21.15.4 Poultry, Domestic Fowl, and Other Avian Species

(1)   The rules for poultry gene nomenclature were developed by the Poultry Science Association and described by Ralph G. Somes Jr.[151] A later review prompted the Poultry Gene Nomenclature Committee of the International Society for Animal Genetics to adopt a system based on human locus and allele nomenclature.[152] This system was adapted to fit the needs of the poultry research community. Table 21.28[153] summarizes the major conventions. See Lyman B. Crittenden and colleagues[153] for additional details and examples. Another resource is the Chicken Gene Nomenclature Consortium, which is an international group of researchers interested in providing standardized gene nomenclature for chicken genes.[154]

Although allele symbols should be brief and not attempt to summarize all known information about a gene, they may convey information about morphological characteristics, subcellular locations, control properties, and amino acid substitutions. Use an

Table 21.28. Poultry, Domestic Fowl, and Other Avian Species: Symbols for Genes and Phenotypes[a]

Feature	Convention	Examples
Gene symbol (locus designation)	No more than 5 capital italic letters or a combination of letters and arabic numerals derived from the gene name. No superscripts, subscripts, Greek letters, or roman numerals are used in gene or allele symbols.	*PA* (pre-albumin) *MM7* (micromelia VII) *GPDA* (α-glycerol phosphate dehydrogenase-liver) *HBA* (α-hemoglobin) *HBB* (β-hemoglobin)
Different loci with similar phenotypes	The same symbols as used for genes with arabic numerals added	*PA2* and *PA3*
Mutation site	Mutation indicated by an italic letter or number representing the characteristic	*HBB*6V*
Alleles	A capital letter or arabic numeral in italics is added to the locus designation. Allele characters are separated from the locus characters by an asterisk. After initial full identification, alleles may be designated without the locus characters.	*OV*A* and *OV*B* *EAA*1* and *EAA*7* **1* and **7* (for subsequent use after initial identification)
Heterozygotes		
at a single locus	Allele symbols are separated by a horizontal line or, if written on the same line, by a slash.	<u>*EAA*1*</u> *EAA*2* or *EAA*1/EAA*2* or *EAA*1/*2*
at multiple loci	Linked alleles are separated by spaces.	<u>*EAJ*1 SE*N*</u> *EAJ*2 SE*S* or *EAJ*1 SE*N/EAJ*2 SE*S*
	Unlinked alleles are separated by semicolons.	<u>*EAA*1*</u> ; <u>*EAB*1*</u> ; <u>*EAP*1*</u> *EAA*2 EAB*1 EAP*2* or *EAA*1/EAA*2; EAB*1/EAB*1; EAP*1/EAP*2* or *EAA*1/*2; EAB*1/*1; EAP*1/ *2*
for sex-linked traits	Genotypes distinguish between males, which are heterozygous, and females, which are hemizygous.	male: <u>*GHR*A*</u> or *GHR*A/GHR*B* *GHR*B* female: <u>*GHR*A*</u> or *GHR*A/W* *W*

[a] Based on Lyman B. Crittenden and colleagues.[153]

italic uppercase letter "*N*" to indicate the normal allele. Indicate lack of activity of an allele with an italic "*O*." Present complex data on dominance, recessiveness, and wild type in tables.

The laboratory that first conducts the genetic segregation analysis or assigns a gene to a specific chromosomal location also names the loci and alleles. Follow the naming conventions presented by Crittenden and colleagues,[153] and include the population in which the alleles were found. When a poultry gene is homologous to a human gene or when there is strong evidence for homology, use the same full gene name as that of the human gene listed by the Gene Nomenclature Committee of the Human Genome Organisation (HUGO) (see Section 21.15.7, "Human.").

(2) Karyotypes for domestic fowl (ie, *Gallus domesticus*) and other avian species follow the system developed by K. Ladjali-Mohammedi and colleagues.[155] Chickens have 38 pairs of autosomes, which are numbered 1 to 38, and a pair of sex chromosomes, which are Z and W. The female is the heterogametic sex. The 8 largest autosomes have been differentiated by morphology and banding patterns, but the others are too small to be distinguished microscopically. In karyotypes, the autosomes are arranged in descending order by length with the short arm on top, and the Z and W chromosomes are shown at the end. G-band landmarks remain the standard. R-, Q-, and C-bands are useful for some studies, but they are too variable for mapping.

21.15.5 Domestic Mammals

(1) Genetic loci in ruminants and other domestic mammals are represented by conventions developed by the Committee on Genetic Nomenclature of Sheep and Goats (COGNOSAG).[156] The rules are recommended for all ruminants (including cattle, sheep, goats, deer, and camelids), other farmed mammals (such as pigs and horses), and dogs. The guidelines are generally consistent with those of human and mouse. Table 21.29[12,156] summarizes the major conventions. In addition, Online Mendelian Inheritance in Animals[157] is a database of inherited disorders and traits in many animal species, excluding human and mouse. The database contains textual information, references, and links to PubMed and gene records at the NCBI.

New locus and allele names and symbols should be as brief as possible, but they should not be a single letter. In addition, these names and symbols should accurately convey the character affected or the function by which the locus is recognized, and they should reflect interspecies homology but not duplicate or confuse existing nomenclature. For keratins and keratin-associated proteins, use the names and symbols proposed by Barry C. Powell, PhD, and George E. Rogers, PhD.[158]

Ensure that the names of new alleles at a recognized locus conform to the established nomenclature for that locus. In loci detected by biochemical, serological, or nucleotide methods, alleles' names and symbols may be identical. Do not use either the plus or minus symbol for alleles detected by those methods. The plus symbol identifies the standard wild type for alleles having visible effects. Designate null alleles by the number zero.

(2) Karyotypes of domestic animals were standardized for cattle, goats, and sheep with both G- and R-bands.[159] More specific recommendations have been published for cattle[160] and sheep.[161] Autosomes are numbered beginning with chromosome 1, and the sex chromosomes are designated as "X" and "Y" (see Table 21.23[126–138]).

21.15.5.1 Cattle

(1) The NCBI's stable taxon identifier for cattle (ie, *Bos taurus*) is 9913, and the UniProtKB mnemonic suffix for cattle is "BOVIN." An example of a UniProtKB designation for

Table 21.29. Domestic Mammals (Ruminants and Other Farmed Mammals): Symbols for Genes and Phenotypes[a]

Feature	Convention	Examples
Locus name	The name is in italic letters or in a combination of letters and arabic numerals. The initial letters of all words are capitalized. Greek letters are spelled out and moved to the end of the name.	*Agouti Signaling Protein* *Casein Beta* *Growth Hormone–Releasing Hormone* *Leptin* *Prolactin*
Locus symbol	The symbol is in italic letters. In some cases, the letters are combined with arabic or roman numerals. The initial letter is capitalized and is the same as the initial letter of the locus name. The locus symbol does not use superscripts, subscripts, Greek letters, or roman numerals.	*ASIP* for *Agouti Signaling Protein* *CSN2* for *Casein Beta* *GHRH* for *Growth Hormone–Releasing Hormone* *LEP* for *LeptinPRL* for *Prolactin*
Allele name	The allele name is in italics. An initial lowercase letter is preferred, including for dominant alleles.	*hornless* (recessive) *polled* (dominant)
Allele symbol	The symbol is in italic letters, arabic numerals, or both. It is an abbreviation of the allele name and begins with the same letter. The letters are all lowercase unless the allele is dominant or codominant. No Greek letters or roman numerals are used. The allele symbol is always written with the locus symbol, either as a superscript or following an asterisk on the same line. No spaces are placed between the locus and allele symbols. For loci other than coat color and other visible traits, the symbol is in all uppercase letters or a combination of uppercase letters and arabic numerals.	*Ho^{hl}* or *Ho*hl* *Ho^P* or *Ho*P* *ALOP^{alop1}* or *ALOP*alop1*
Wild-type allele symbol	For alleles with visible effects, a plus symbol is used as a superscript to the symbol, or the plus symbol is used alone. Null alleles are designated by the number zero in superscript.	*E$^+$* or *+* *E^0*

Genotypes

Homologues	The symbols are separated by slashes.	*Ho^P/Ho^P* *Ho*hl/Ho*hl* *ALOP1^{alop1}/ALOP1^{alop1}*
Unlinked loci	The symbols are separated by semicolons.	*INHBB; INFW*
Linked loci	The symbols are separated by spaces and listed in alphabetical order when gene order or phase are not known.	. . .
X-linked loci	The symbol for the X-linked locus is followed by "/X."	*FecX*1/X*
Y-linked loci	The symbol for the Y-linked locus is followed by "/Y."	. . .
Phenotype symbols	The same characters as are used for locus and allele symbols. No italics or underlining. Insert a space instead of an asterisk between locus and allele designations if all characters are on the line. Alternatively, enclose the symbols in square brackets, place the allele symbol in superscript, and remove the space between the locus and allele designations.	Ho P *or* [Ho^P] ALOP1 Alop1, *or* [ALOP1^{Alop1}]

[a] Based on Thomas E. Broad and colleagues[156] and the Vertebrate Gene Nomenclature Committee.[12] The Committee on Genetic Nomenclature of Sheep and Goats recommends that its rules be adopted not only for cattle, sheep, goats, deer, and camelids but also for other farmed animals, such as pigs and horses. Anatoly Ruvinsky, PhD, DSc, and Jeff Sampson[125] recommend that the same rules apply to dogs.

B. taurus is "CATH_BOVIN" for *pro-cathepsin H*, which uses the gene symbol *CTSH*. *CTSH*'s UniProtKB entry number is Q3TOI2, the NBCI stable gene identifier is 510524, and the Ensembl identifier is ENSBTAG00000010992. In addition, the Vertebrate Gene Nomenclature Committee (VGNC) assigned to *CTSH* the stable identifier VGNC:27816 to *CTSH*.

(2)   Guidelines for genetic loci in cattle are essentially those developed by COGNOSAG for all ruminants.[156] Table 21.29[156] summarizes the major conventions for cattle and other domestic animals. When necessary to distinguish a bovine locus from one of another organism that would otherwise have the same symbol, use *BBO* as a prefix.

The nomenclature for DNA segments was proposed by C.H.S. Dolling.[162] Whenever possible, the symbols for unmapped DNA segments with no known homologs, official names, or symbols will be the same as those first reported, except that all the letters in the symbols will be in uppercase. Whenever possible, the symbols for mapped DNA segments with no known homologs, official names, or symbols will consist of the same symbols assigned to the unmapped components, followed by D numbers assigned by The Bovine Genome Database.[163] When unnamed newly mapped loci are identified, compare them with the database maintained by the HUGO Gene Nomenclature Committee.[6]

RM095 (unmapped)   *RM095(D1S13)* (the same marker after it was mapped)

Loci are divided into 4 categories: coat color, visible trait other than coat color, blood and milk polymorphisms, and mapped loci and other genetic systems. Locus symbols for blood and milk polymorphisms are written in all uppercase letters, but recessive alleles are in all lowercase.[160]

(3)   A standing committee of the International Society for Animal Genetics compiled the nomenclature for bovine leukocyte antigens, which is based on the human leukocyte antigen system.[164] The names of alleles are based on amino acid sequences and consist of 4 or 5 digits: The first and second digits indicate the major type; the third and fourth indicate the subtype; and if the fifth is used, it indicates any unexpressed variations. Class I alleles of a single major type differ by no more than 4 amino acids in the first and second domains. Class II alleles differ by no more than 4 amino acids in the first domain. If a name is based on a partial sequence, the first full-length sequence that includes the original partial sequence will assume the allele name. Minor sequence errors are corrected when identified.[165]

(4)   The karyotypes of cattle, sheep, and goats were standardized by the second International Conference on Standardization of Domestic Animal Karyotypes.[157] The rules for cattle were revised in 1995 as the "Texas nomenclature."[160] All ideograms were retained from the International System for Cytogenetic Nomenclature of Domestic Animals 1989,[159] but the chromosomes were renumbered in a different order. The nomenclature for chromosomal abnormalities and markers follows that for human chromosomes.[157,158] Autosomes are numbered 1 to 29, and the sex chromosomes are X and Y.

21.15.5.2 Horse and Other Equines

(1)   The NCBI's stable taxon identifier for *Equus caballus* is 9796, and the UniProtKB mnemonic suffix is "HORSE." An example of a UniProtKB gene designation for *E. caballus* is "CYC_HORSE" for *cytochrome c*, which uses the gene symbol *CYCS*. *CYCS*'s UniProtKB entry number is P00004, its NBCI stable gene identifier is 100053958, its Ensembl identifier is ENSECAG00000014330, and its VGNC identfier is VGNC:50468.

(2)   Tables 21.22[119-125] and 21.29[156] summarize the major conventions for the genes of *E. caballus* and other equines.

(3)   The cytogenetic nomenclature for the domestic horse and other equines is based on the Paris Standard[166] as endorsed by the International Committee for the Standardization of the Domestic Horse Karyotype in 1996.[167] Karyotypes arrange 13 pairs of metacentric/sub-

metacentric autosomes in one group and 18 pairs of acrocentrics in another group. Within each group, the autosomes are arranged in descending order by length. The sex chromosomes are placed in the middle after the smallest bi-armed autosomes. The scheme for numbering regions and bands corresponds to the International System for Human Cytogenetic Nomenclature (2009).[5]

21.15.6 Rodents

21.15.6.1 Deer Mouse

(1) The various deer mouse species have several UniProtKB mnemonic suffixes, including "PERSJ" for *Peromyscus sejugis*, "PERGO" for *Peromyscus gossypinus*, and "PERPL" for *Peromyscus polionotus*. Similarly, all *Peromyscus* spp. have their own NCBI stable taxon identifiers, such as 44241, 42411, and 42413.

   An example of a UniProtKB gene designation for deer mouse is "CP17A_PERLE" for steroid 17-alpha-hydroxylase/17,20 lyase in *Peromyscus leucopus*. This steroid uses the gene symbol *Cyp17a1*, which has UniProtKB entry number Q91Z85 and the NBCI stable gene identifier 114696710.

(2) The gene nomenclature conventions described by Mary F. Lyon, PhD,[168] for the mouse are applied to the deer mouse species with a few exceptions.

(3) The chromosomes of the deer mouse are described and symbolized to the extent possible by the same conventions used for *Homo sapiens*. However, differences between the karyotypes of the single species *H. sapiens* and the many species of *Peromyscus* account for various departures from the rules for humans and from some rules specific to mice.[169]

   Autosomes are numbered Chromosome 1 (Chr 1) to Chr 23 in descending order by size, and the sex chromosomes are X and Y. The ideogram of *P. boylii* is the reference for establishing G-band nomenclature. Bands are grouped into regions, which are designated alphabetically. Within regions, bands are numbered sequentially distal to the centromere. Decimal points denote subdivisions within bands. For details, see Ira F. Greenbaum, PhD, and colleagues.[169]

   Chr 20   band 3C1   band 2C1.4

21.15.6.2 Mouse and Rat

In 2003, the International Committee on Standardized Genetic Nomenclature for Mice and the Rat Genome and Nomenclature Committee unified the rules and guidelines for gene, allele, and mutation nomenclature in mouse and rat. Those rules were updated in 2011.[7]

21.15.6.2.1 MOUSE

(1) The UniProtKB mnemonic suffix for *Mus musculus* is "MOUSE," and the corresponding NCBI stable taxon identifier is 10090. An example of a UniProtKB gene designation for *M. musculus* is "MUP6_MOUSE" for major urinary protein 6, which uses the abbreviation *Mup6*. *Mup6*'s UniProtKB entry number is P02762, its NBCI stable gene identifier is 620807, and its Ensembl identifier is ENSMUSG00000078689.

(2) Mouse Genome Informatics (MGI) provides information on genes, mapping, multispecies homology, probes, clones, sequences, strain characteristics, and nomenclature rules, as well as links to other databases on mouse genetics. The complete guidelines, instructions for registering a new locus symbol, and advice on nomenclature are available from the MGI[15] at http://www.informatics.jax.org.

   The symbols for unique mouse genes and loci are composed of italic letters and arabic numerals with no punctuation. However, hyphens are used in gene symbols when needed to separate 2 numbers that would be in adjacent positions, to separate related sequence and pseudogene designations from gene symbols, to separate characters for loci in a complex from the complex symbol, and to separate components of mutant allele symbols. Table 21.30[7] summarizes the major conventions.

Laboratory codes are included in the gene symbols. These codes have initial upper-case letters. Genes encoded by the opposite, or antisense, strand have their own symbols. Alternative transcripts from the same gene do not have different symbols. In cross-hybridizing DNA segments, the species code is uppercase. Homologous genes in different vertebrate species (ie, orthologs) should be given the same gene nomenclature, if possible.

To distinguish mRNA, genomic DNA, and cDNA, the relevant prefix should be placed in parentheses, as in "(mRNA) *Rbp1*."

When necessary to distinguish a mouse gene from a gene of another organism that would otherwise have the same symbol, use "*MMU*" as a prefix.

(3) The rules and guidelines for mouse cytogenetic nomenclature[8] are summarized here and in Table 21.31.[8] The 19 pairs of autosomes of the mouse are designated by arabic numerals in roman type, and the sex chromosomes are X and Y. The short and long arms are designated "p" and "q," respectively. Chromosome bands are numbered from the centromere toward the telomeres and are designated by the chromosome number, the p or q arm, the region number, and the band-specific letter. However, reference to the q arm may be omitted when the meaning is clear. Subdivisions are designated by numbers following a decimal point. Symbols for chromosome anomalies are not italicized.

Distinguish each successive anomaly in a series by a symbol consisting of a number followed by the laboratory code of the researcher or laboratory that discovered the anomaly (ie, the same code used for the designation of inbred substrains or sublines). Indicate the chromosomes involved in the anomaly by adding the appropriate arabic numerals in parentheses between the initial letter and the series symbol (see examples in Table 21.31).

21.15.6.2.2 RAT

(1) The NCBI's stable taxon identifier for *Rattus norvegicus* is 10116, and the UniProtKB mnemonic suffix is "RAT." An example of a UniProtKB gene designation for *R. norvegicus* is "PERI_RAT" for peripherin, which uses the gene symbol *Prph*. *Prph*'s UniProtKB entry number is P21807, its NBCI stable gene identifier is 24688, and its Ensembl identifier is ENSRNOG00000052880. In addition, the Rat Genome Database assigned 3414 as its identifier for *Prph*.

(2) Guidelines for rat gene nomenclature follow those for mice.[170,171]Anonymous DNA segment symbols are assigned by the Rat Genome Database[171] at https://rgd.mcw.edu.

(3) Rat chromosome nomenclature[21] follows that for humans (see Section 21.15.7, "Human"). *R. norvegicus*'s 20 pairs of autosomes are designated by arabic numerals in roman type, and its sex chromosomes are X and Y.[21]

21.15.7 Human

The NCBI's stable taxon identifier for *Homo sapiens* is 9606, and the UniProtKB mnemonic suffix is "HUMAN." An example of a UniProtKB gene designation for *H. sapiens* is "STX5_HUMAN" for *syntaxin-5*, which uses the gene symbol *STX5*. The UniProtKB entry number for *STX5* is Q13190, the NBCI stable gene identifier is 6811, and the Ensembl

Table 21.30. Mouse: Symbols for Genes and Phenotypes[a]

Feature	Convention	Examples
Gene name	A unique, brief description in roman type of the character by which the gene is recognized or of the protein encoded	dwarf hemoglobin β-chain
Gene symbol	The symbol is a 2-, 3-, or 4-character abbreviation[b] of the gene name in italics. Arabic numerals are used when needed. The symbol must begin with the same letter as the gene name. Hyphens are used only to separate numbers that would be in adjacent positions or to separate components of mutant alleles.	*Dw* (dwarf) *Hbb* (hemoglobin β-chain complex) *G6pdx* (glucose-6-phosphate dehydrogenase X-linked) *Hba-ps3* (hemoglobin alpha, pseudogene 3)
Gene families	The same gene symbol is used for all members. An identifying numeral is added to each symbol.	*H1* and *H2* *Es2* and *Es3* *Adam8* and *Adam9*
Allele	The gene symbol with superscript[c] italic letters. No more than 10 characters.	Hbb^d $Mitf^{wh}$ Tyr^c (albino) $Tyr^{c\text{-}ch}$ (chinchilla allele of Tyr^c)
Recessive allele	The gene symbol with a lowercase initial letter for the mutant gene	*a* (nonagouti) $gusb^{mps}$ (recessive mutation in the *Gusb* gene)
Viral expression or immune response	The gene symbol with superscript italic "*a*" to indicate the presence of virus expression or immune response. A superscript italic "*b*" is used for the absence of both traits.	$Mtv1^a$ $Mtv1^b$
Resistance or susceptibility	A superscript italic "*r*" is added to the gene symbol to indicate "resistance" to infectious organisms or other agents. A superscript italic "*s*" indicates "susceptibility" to an agent.	$Pla2g2a^{Mom1\text{-}r}$ $Pla2g2a^{Mom1\text{-}s}$
Cloned gene	When a mutant gene is cloned, the symbol for the structural gene becomes the gene symbol, and the former mutation symbol is hyphenated as an allele symbol.	*Mbpshi-mld* (*shi* is the structural gene for myelin basic protein [Mbp].)
Targeted, or knock-out, mutation	The gene or allele symbol plus a superscript consisting of the abbreviation "*tm*," the number of the knockout, and the laboratory code, all in italics	$Cftr^{tm1Unc}$ (the first targeted mutation in this gene made at The University of North Carolina in the United States)
Transgene	The symbol follows the convention "*Tg(YYY)###Zzz*," in which "*Tg*" is the mode, "*(YYY)*" is the insert designation, "*###*" is the laboratory assigned number, and "*Zzz*" is the laboratory code. After the full symbol has been used in a manuscript, it may be abbreviated by omitting the insert designation.	*Tg(GPD1)A1Bir* (insertion of the *GPD1* gene at the A1 transgenic line, produced by geneticist Edward H. Birkenmeier, MD. The symbol may be abbreviated as *TgA1Bir* when mentioned later in the same manuscript.)
Wild type	A plus symbol in roman type with a superscript gene symbol in italics. A plus symbol may be used alone when the context leaves no doubt as to which locus is represented.	$+^{Ap3b1}$ $+^{Myo5a\text{-}d}$ $+^{Tyr\text{-}c}$ +
Reversion to wild type	The symbolization is the reverse of the wild-type symbol, with the gene symbol on the line in italics and a plus symbol in roman type as part of the superscript.	$Ap3b1^+$ $Myo5a^{d\text{-}+2J}$ Tyr^{c+}

Table 21.30. Mouse: Symbols for Genes and Phenotypes[a] (*continued*)

Feature	Convention	Examples
Mutant genes with definite wild type	The gene symbol for the first discovered mutant allele is used for both the gene and the allele.	Krt71 (the first mutant) Krt71^Ca-J (a remutation)
Mitochondrial genes	The prefix "*mt-*" with the gene symbol	*mt-Cytb* *mt-Nd2*
Pseudogenes	The gene symbol followed by the abbreviation "*ps*" and an appropriate serial number	*Hba-ps3* and *Hba-ps4* (pseudogenes of α-globin located away from the *Hba* complex) *Pgk1-ps2* (the second pseudogene to the functional Pgk1 locus)
Related loci	The gene symbol followed by the abbreviation "*rs*" and an appropriate serial number	*Ela1-rs1* (the first related locus identified by the elastase-1 gene probe)
Expressed sequence tags (ESTs)	Symbolized by a GenBank sequence identification number	AA066038
Sequence tagged sites (STSs)	No symbols assigned other than those given by the laboratory that produced or used the STSs	. . .
Loci recognized by anonymous DNA probes	D-number symbols composed of "D" for "DNA," the chromosome designation, the laboratory code, and a unique serial number, all in italics	*D1Pas5* (the fifth D-locus developed and mapped on Chromosome 1 at the Pasteur Institute in Paris, France)
Phenotype (antigen loci and enzyme loci)	The same elements as for the gene symbol but with capital letters in roman type and superscripts lowered to the type line	GPI1A GPI1B

[a] Based on the International Committee on Standardized Genetic Nomenclature for Mice[7] at http://www.informatics.jax .org/mgihome/nomen/gene.shtml.

[b] In general, the initial letter is uppercase and the following letters are lowercase. However, for genes recognized only by a recessive mutation, the symbol begins with a lowercase letter. Numerals may be used in the symbol if they are in the name or the abbreviation on which the symbol is based. Roman numerals and Greek letters are not used. The complete gene or allele symbol, including superscripts, must not exceed 10 characters.

[c] When superscript symbols are not available, as in computer printouts, the superscript letters may be replaced by an asterisk followed by the letters or by angle brackets in which the letters are enclosed (eg, *Hbb*d* and *Hbb<d>* rather than *Hbb^d*).

identifier is ENSG00000162236. Most important for *H. sapiens* genes, the HUGO Gene Nomenclature Committee[172] (HGNC) assigned HGNC:11440 as *STX5*'s identifier.

21.15.7.1 Genes

The guidelines for human gene nomenclature were developed by the HGNC.[172] Table 21.32[145,173–175] summarizes the major conventions. Researchers should consult the HUGO committee before publishing any new human gene symbol to ensure that the symbol is unique and suitable and to reserve the symbol for the gene in question.

Because gene and protein names can change over time, researchers and scientific journals should ensure that published descriptions of DNA sequence variants include standardized location-based descriptions, such as the sequence variation description of the Human Genome Variation Society[16] and the variant representation specification of the Global Alliance for Genomics and Health.[19] (See Section 21.1.2, "Existing Standards and Unambiguous Identifiers.")

Table 21.31. Mouse Chromosome Terminology: Conventions and Examples[a]

Symbol	Feature	Convention	Examples
. . .	chromosome bands	Chromosome bands are numbered from the centromere toward the telomeres. The bands are designated by the chromosome number, the region number, and the band-specific letter. Further subdivisions are numbered following a decimal point. Because all mouse chromosomes but Chromosome Y are acrocentric, they do not have short arms other than a telomere proximal to the centromere. Consequently, short and long arms (ie, "p" and "q" arms, respectively) are designated only on Chromosome Y.	17B 17B1 17B1.1
		When the positions of the chromosomal breakpoints relative to G-bands are known, add the band numbers after the appropriate chromosome numbers.	T(2H1;8A4)26H (reciprocal translocation, with breaks in band H1 of Chr 2 and band A4 of Chr 8)
. . .	series symbol	Successive anomalies discovered at the same laboratory are designated by adding the series number and laboratory code to the end of the description of the anomalies. The laboratory code may be based on either the name of the laboratory or a person conducting research at the laboratory.	Dp(1)6H (the sixth duplication found in Chr 1 by the Mary Lyon Centre at MRC Harwell in Oxfordshire, United Kingdom) In(2)5Rk (the fifth inversion in Chr 2 found by Thomas H. Roderick, PhD, at the Jackson Laboratory in Bar Harbor, Maine, in the United States)
Cen	centromere	Use the symbol "Cen" when referring to the centromere itself. In Robertsonian translocations, indicate the centromere with a period.	Rb(9.19)163H
Chr	chromosome	Begin the word "chromosome" with a capital letter when referring to a specific chromosome. Abbreviate it to "Chr" after the first use in a manuscript.	Chromosome 2 *and* Chr 2
Del	deletion	Use the symbol "Del" to define interstitial losses that are often, but not always, visible. Do not use "Del" for single-gene deletions. Treat those deletions as alleles instead.	Del(7E1) Tyr8Rl (deletion of band 7E1, which manifests as a mutation to the albino allele, Tyr^c, and which was found by zoologist William L. Russell, PhD, and geneticist Liane B. Russell, PhD, at the Oak Ridge National Laboratory in Tennessee in the United States)
Dp	duplication	Follow "Dp" with the number of the chromosome involved in parentheses.	Dp(1)6H (the sixth duplication of Chr 1 found by the Mary Lyon Centre at MRC Harwell in Oxfordshire, United Kingdom)
Hc	pericentric heterochromatin	Follow the symbol "Hc" with the number of the chromosome involved.	Hc14 (pericentric heterochromatin on Chr 14)

Table 21.31. Mouse Chromosome Terminology: Conventions and Examples[a] (*continued*)

Symbol	Feature	Convention	Examples
In	inversion	Follow the symbol "In" with the chromosome number in parentheses. Separate bands with semicolons.	In(2)5Rk (the fifth inversion in Chr 2 found by Thomas H. Roderick, PhD, at the Jackson Laboratory in Bar Harbor, Maine, United States) In(5C2;15E1)Rb3Bnr 1Ct (this inversion involves bands 5C2 and 15E1, and it is the first inversion in Rb3Bnr found by Bruce M. Cattanach, PhD, at the Mammalian Genetics Unit at MRC Harwell in Oxfordshire, United Kingdom)
Is	insertion	Right after the symbol "Is," encompass in parentheses the number of the chromosome that donated the inserted portion, followed by a semicolon and the number of the chromosome in which the portion was inserted.	Is(7;1)40H (an insertion of part of Chr 7 into Chr 1) Is(In7F1-7C;XF1)1Ct (an inverted insertion of a segment of band F1-C of Chr 7 into the X Chromosome at band F1)
Ms	monosomy	Follow the symbol "Ms" with the number of the chromosome involved.	Ms12 (an animal with only one copy of Chr 12)
. . .	multiple anomalies	When an animal has 2 or more anomalies that are potentially separable by recombination, list the symbols for all of them.	Rb(16.17)Bnr T(1;17)190Ca /++ (an animal with a Robertsonian translocation and a reciprocal translocation, both of which involve Chr 17) Rb(5.15)3Bnr +/+ In(5)9Rk (an animal with a Robertsonian translocation and an inversion involving Chr 5 in repulsion)
		When one anomaly is contained within another or is inseparable from it, combine the symbols.	T(In1;5)44H (translocation between Chrs 1 and 5, with the Chr 1 segment inverted)
Ns	nullisomy	Follow the symbol "Ns" with the number of the chromosome involved.	Ns2 (an animal with no copies of Chr 2)
p	short arm	Breaks in the short arm must be designated with a "p."	T(Yp;5)## (translocation involving a break in the short arm of Chr Y and the long arm of Chr 5)
pq	pericentric inversion	Use the symbol "pq," the appropriate band numbers, or both.	In(Ypq) (pericentric inversion involving the Y chromosome) In(8pqA2) (pericentric inversion of the region between the short arm and band A2 of the long arm of Chr 8)
. . .	polymorphic loci	Designate polymorphic loci within the ribosomal DNA region with the symbol "Rnr" and the chromosome number.	Rnr12 (ribosomal DNA region on Chr 12)
q	long arm	Rearrangements in mouse chromosomes are assumed to be in the long arm unless stated otherwise.	. . .
Rb	Robertsonian translocation	Indicate the centromere with a period.	Rb(9.19)163H
T	translocation	Separate the chromosomes involved with a semicolon.	T(4;X)37H T(2H1;8A4)26H (reciprocal translocation, with breaks in band H1 of Chr 2 and band A4 of Chr 8)

Table 21.31. Mouse Chromosome Terminology: Conventions and Examples[a] (*continued*)

Symbol	Feature	Convention	Examples
Tel	telomere	Use the symbol "Tel" when referring to the telomere itself. Follow the symbol with the number of the chromosome involved. Use roman type in karyotype descriptions, and use italics when the telomere is treated as a locus.	Tel14 Tel14p1 (the first telomere sequence mapped at the centromeric end of Chr 14) Tel19q2 (the second telomere sequence mapped to the distal end of Chr 19)
Tet	tetrasomy	Follow the symbol "Tet" with the number of the chromosome involved.	Tet14 (an animal with 4 copies of Chr 14)
Tp	transposition	Follow the symbol "Tp" with the number of the chromosome involved in parentheses.	Tp(Y)1Ct (transposition of Chr Y)
Ts	trisomy	Follow the symbol "Ts" with the number of the chromosome involved.	Ts16 (an animal with 3 copies of Chr 16)

[a] Based on the International Committee on Standardized Genetic Nomenclature for Mice.[8]

As an additional precaution, when it is necessary to distinguish a human gene from one of another organism that would otherwise have the same symbol, use the roman designation "HUMAN" in parentheses before the gene symbol, as in "(HUMAN) *ABCA1*."

Another general consideration is that the designations and orthography for messenger RNA (mRNA) and complementary DNA (cDNA) generally follow those for the corresponding genes.

21.15.7.2 Human Leukocyte Antigen

To provide guidance to the scientific community on the human major histocompatibility complex, the World Health Organization established its Nomenclature Committee for Factors of the HLA System.[176] In addition, the IPD-IMGT/HLA database is devoted to the complex.[165,177]

Gene names for humans begin with "*HLA-*," which stands for "human leukocyte antigen." This prefix is followed by a locus symbol and a specific allele designation.

   HLA-DRB1*13   *HLA-DRB1*1301*   *HLA-DRB1*13012*

21.15.7.3 Chromosomes

Human chromosome nomenclature is based on the recommendations of the International Standing Committee on Human Cytogenetic Nomenclature (ISCN). The most recent report[5] incorporates and supersedes all previous ISCN recommendations.

21.15.7.4 Chromosome Number and Morphology

The human autosomes are numbered from 1 to 22 in decreasing order of length. The sex chromosomes are referred to as "X" and "Y." When the chromosomes are stained by methods that do not produce bands, they can be arranged into 7 readily distinguishable groups (ie, A through G), based on descending order by size and the position of the centromere.

Table 21.32. Human Genes: Symbols for Genes and Phenotypes[a]

Feature	Convention	Examples
Gene name	The name is a brief description of the character or function of the gene in American spelling and in roman type. Begin with a lowercase letter unless the gene is named for an eponymous disease, syndrome, or phenotype.	alcohol dehydrogenase 4 (class II), pi polypeptide aconitase 1 Down syndrome critical region 10
Gene symbol	Uppercase italic letters[b] or a combination of uppercase letters and arabic numerals based on the gene name. The first letter of the gene name is also the first letter of the symbol. All elements are set on the line.	*ADH4* *ACO1* *DSCR10*
	Greek letters, roman numerals, superscripts, subscripts, and punctuation are not used in gene symbols.	β GAL-1 *becomes GLB1*
Allele	The gene symbol followed by an asterisk and then the allele characters, which are 4 or fewer[c]	*ADA*1* and *HBB*6V*
Gene cluster	The character "@" at the end of the symbol represents a gene cluster.	*PCDHG@* for *protocadherin gamma cluster*
Oncogenes	The symbols correspond to the homologous retroviral oncogenes but do not use the "v-" or "c-" prefixes. When referring specifically to the human locus, omit the prefix and capitalize the symbol.	*MYC* and *HRAS*
Putative genes	Genes are regarded as putative if they are predicted from expressed sequence tag (EST) clusters or genomic sequence alone. Putative genes are designated[c] by the chromosome of origin and an arbitrary number in the format "*C#orf#*" in italics.	*C2orf15* (Chromosome 2 open reading frame 15)
DNA segment	Arbitrary DNA fragments and loci can be designated with a capital letter "*D*" followed by the chromosome number, an uppercase letter indicating sequence type, and arbitrary numbers, all in italics. (For details, see Section 21.2.6, "Anonymous DNA Segments.") When known to be an expressed sequence, the suffix "*E*" can be added.	*D9S14* *D1F10S1* *DXS1004E*
Pseudogenes	Pseudogenes are identified with the gene symbols of the structural gene followed by a "*P*" for "pseudogene" and, if necessary, a number.	*PGK1P1* *RPL32P1* *CALCP* *DDX12P*
Transposable elements	The HUGO Gene Nomenclature Committee no longer names transposable elements.	. . .

[a] Based on the HUGO Gene Nomenclature Committee at https://www.genenames.org/about/guidelines/.

[b] Base the names and symbols for genes on the names recommended by the International Union of Biochemistry and Molecular Biology.[145] Allele terminology is the responsibility of the Human Genome Variation Society.[175] The system has been described in Stylianos E. Antonarakis, MD, and the Nomenclature Working Group,[173] and in Johan T. den Dunnen, PhD, and Antonarakis.[174]

[c] Obtain *C#orf* numbers from the HUGO Gene Nomenclature Committee, which keeps records of all assigned numbers at https://www.genenames.org/.

After chemical treatment and staining, each chromosome displays a continuous series of light and dark bands. Each chromosome band is clearly distinguishable from its adjacent segments, which appear darker or lighter. As a consequence, there are no interbands. The bands are allocated to various regions along the chromosome arms. A region is an area of a chromosome lying between 2 adjacent landmarks. The landmarks, in turn, are consistent and distinct morphological features, including the ends of the chromosome arms, the centromere, and certain bands.

Regions and bands are numbered consecutively from the centromere outward along each chromosome arm. The symbols "p" and "q" designate each chromosome's short and long arms, respectively. The centromere (cen) is designated as "10," while the region facing the short arm is "p10" and the region facing the long arm is "q10." The 2 regions in each arm adjacent to the centromere are labeled as "1." Each successive region between the centromere and p10 and q10 is numbered sequentially.

In designating a particular band, 4 items are required: the chromosome number, the arm symbol, the region number, and the band number within that region. These items are listed in the order above without spacing or punctuation.

> 1p31 (chromosome 1, short arm, region 3, band 1)

Whenever an existing band is divided, a decimal point and the numbers assigned to the subbands are placed after the original band designation. The subbands are numbered sequentially from the centromere outward. For example, if the original band 1p31 is divided, the subbands are labeled "1p31.1," "1p31.2," and so on. If a subband is divided, additional digits are used without further punctuation. For example, subband 1p31.1 might be divided into 1p31.11 and 1p31.12.

21.15.7.5 Karyotype Designation

In the description of a karyotype, the first item recorded is the total number of chromosomes, including the sex chromosomes. This number is followed by a comma and the symbols for sex chromosome constitution. The autosomes are specified only when an abnormality is present.

> 46,XX (normal female)   46,XY (normal male)

In descriptions of chromosome abnormalities, sex chromosome aberrations are listed first, followed by abnormalities of the autosomes in numerical order, irrespective of aberration type. Each abnormality is separated from the next by a comma. The only exception to this rule is that constitutional numerical sex chromosome abnormalities are designated by listing all sex chromosomes after the chromosome number (eg, "48,XXXY").

Letter designations are used to specify rearranged chromosomes (ie, structurally altered chromosomes). All symbols and abbreviated forms used to designate chromosome abnormalities in humans are listed in Table 21.33,[5] and examples of karyotypic descriptions are listed in Table 21.34.

In single-chromosome rearrangements, the chromosome involved in the change is specified within parentheses immediately following the symbol identifying the type of rearrangement.

> inv(2)   del(4)   r(18)

Table 21.33. Human Chromosomes: Symbols and Abbreviations[a]

Symbol	Description
AI	first meiotic anaphase
AII	second meiotic anaphase
ace	acentric fragment
add	additional material of unknown origin
approximate sign (~)	denotes intervals and boundaries of a chromosome segment
arrow (→ or -›)	indicates "from–to"; separates the unaltered karyotype from the altered karyotype; the direction of the arrow indicates the direction of change
b	break
brackets, angle (‹ ›)	surround the ploidy level
brackets, square ([ ])	surround number of cells
c	constitutional anomaly
Cen	centromere
chi	chimera
Chr	chromosome
cht	chromatid
colon, single (:)	break
colon, double (::)	break and reunion
comma (,)	separates chromosome numbers, sex chromosomes, and chromosome abnormalities
cp	composite karyotype
cx	complex chromatid interchanges
decimal point (.)	denotes a subband
del	deletion
de novo	designates a chromosome abnormality that has not been inherited
der	derivative chromosome
dia	diakinesis
dic	dicentric
dip	diplotene
dir	direct
dis	distal
dit	dictyotene
dmin	double minute
dup	duplication
e	exchange
end	endoreduplication
equal sign (=)	number of chiasmata
fem	female
fis	fission at the centromere
fra	fragile site
g	gap
h	constitutive heterochromatin
hsr	homogeneously staining region
i	isochromosome
idem	denotes the stemline karyotype in subclones

Table 21.33. Human Chromosomes: Symbols and Abbreviations[a] (*continued*)

Symbol	Description
ider	isoderivative chromosome
idic	isodicentric chromosome
inc	incomplete karyotype
ins	insertion
inv	inversion or inverted
lep	leptotene
MI	first meiotic metaphase
MII	second meiotic metaphase
mal	male
mar	marker chromosome
mat	maternal origin
med	medial
min	minute acentric fragment
minus symbol (–)	loss
ml	mainline
mn	modal number
mos	mosaic
multiplication symbol (×)	multiple copies of rearranged chromosomes
oom	oogonial metaphase
or	alternative interpretation
p	short arm of chromosome
PI	first meiotic prophase
pac	pachytene
parentheses ( )	surround structurally altered chromosomes and breakpoints
pat	paternal origin
pcc	premature chromosome condensation
pcd	premature centromere division
Ph	Philadelphia chromosome
plus symbol (+)	gain
prx	proximal
psu	pseudo
pvz	pulverization
q	long arm of chromosome
qdp	quadruplication
qr	quadriradial
question mark (?)	questionable identification of a chromosome or chromosome structure
r	ring chromosome
rcp	reciprocal
rea	rearrangement
rec	recombinant chromosome
rob	Robertsonian translocation
roman numerals I–IV	indicate univalent, bivalent, trivalent, and quadrivalent structures, respectively

Table 21.33. Human Chromosomes: Symbols and Abbreviations[a] (*continued*)

Symbol	Description
s	satellite
sce	sister chromatid exchange
sct	secondary constriction
sdl	sideline
semicolon (;)	separates altered chromosomes and breakpoints in structural rearrangements involving more than one chromosome
Sl	stemline
slant line (/)	separates clones
spm	spermatogonial metaphase
stk	satellite stalk
t	translocation
tan	tandem
tas	telomeric association
tel	telomere
ter	terminal (ie, end of chromosome)
tr	triradial
trc	tricentric chromosome
trp	triplication
underlining (single)	used to distinguish homologous chromosomes
upd	uniparental disomy
v	variant or variable region
xma	chiasma and chiasmata
zyg	zygotene

[a] Adapted from ISCN 2020.[5]

If 2 or more chromosomes have been altered, a semicolon is used to separate their designations. If one of the rearranged chromosomes is a sex chromosome, then it is listed first. Otherwise, the chromosome having the lowest number is always specified first.

t(X;3)   t(2;5)

An exception to this rule involves certain 3-break rearrangements in which part of one chromosome is inserted at a point of breakage in another chromosome. In this circumstance, the recipient chromosome is specified first, regardless of whether it is a sex chromosome and regardless of whether it is an autosome with a higher or lower number than that of the donor chromosome.

ins(5;2)

For balanced translocations involving 3 separate chromosomes with one breakpoint in each chromosome, the rule is still that the sex chromosome or the autosome with the lowest number is specified first. The chromosome listed next is the one that receives a segment from the first chromosome, and the chromosome specified last is the one that donates a segment to the first listed chromosome. The same rule is followed in complex balanced translocations with 4 or more breaks. To distinguish homologous chromosomes, one of the numerals may be underlined.

Table 21.34. Human Chromosomes: Examples of Karyotypic Descriptions[a]

Karyotype	Explanation
45,X	karyotype with one X chromosome (ie, Turner syndrome)
45,XX,−7	female karyotype with monosomy 7
47,XXY	karyotype with 2 X chromosomes and 1 Y chromosome (ie, Klinefelter syndrome)
47,XY,+21	male karyotype with trisomy 21
47,XY,+mar	an additional marker chromosome
46,XX,inv(3)(q21q26)	inversion of one Chr 3 with breakpoints in bands 3q21 and 3q26
46,XX,del(5)(q13)	terminal deletion with a break in band 5q13
46,XX,del(5)(q13q33)	interstitial deletion with breakage and reunion of bands 5q13 and 5q33
46,XX,del(5)(q?)	deletion of the long arm of Chr 5; but it is unclear whether it is a terminal or an interstitial deletion, and the breakpoints are unknown
46,XX,dup(1)(q22q25)	direct duplication of the segment between bands 1q22 and 1q25
46,XY,add(19)(p13)	additional material of unknown origin attached to band 19p13
46,XX,ins(5;2)(p14;q22q32)	direct insertion. The long-arm segment between bands 2q22 and 2q32 has been inserted into the short arm of Chr 5 at band 5p14. The original orientation of the inserted segment has been maintained in its new position (ie, 2q22 remains more proximal to the centromere than 2q32). Note that the recipient chromosome is specified first.
46,XY,t(2;5)(q21;q31)	reciprocal translocation between Chr 2 and Chr 5. Breakage and reunion have occurred at bands 2q21 and 5q31. The segments distal to these bands have been exchanged.
46,XX,t(2;7;5)(p21;q22;q23)	3 translocated segments. The segment on Chr 2 distal to 2p21 has been translocated onto Chr 7 at band 7q22, the segment on Chr 7 distal to 7q22 has been translocated onto Chr 5 at 5q23, and the segment of Chr 5 distal to 5q23 has been translocated onto Chr 2 at 2p21.
46,XX,der(1)t(1;3)(p22;q13)	derivative caused by translocation. A derivative Chr 1 has resulted from a translocation of the Chr 3 segment distal to 3q13 to the short arm of Chr 1 at band 1p22. Der(1) replaces a normal Chr 1. There is no need to indicate the missing chromosome. There are 2 normal Chrs 3. The karyotype is unbalanced with loss of the segment 1p22-pter and the gain of 3q13-qter.
46,XY,del(8)(q21) or i(8) (p10)	deletion or isochromosome. A deletion of the long arm of Chr 8 with a breakpoint in 8q21 or an isochromosome for the short arm of Chr 8. Note that there should be a space before and after the symbol "or."

Chr, Chromosome
[a] Table courtesy of Felix Mitelman, MD, PhD.

Breakpoints are specified within parentheses immediately after the designation of the type of rearrangement and the chromosomes involved. The breakpoints are identified by band designations and are listed in the same order as the chromosomes involved, again separated by semicolons. No semicolon is used between breakpoints in single-chromosome rearrangements.

> **Breakpoints in multiple-chromosome rearrangements:** t(2;5)(q22;p14)
> **Breakpoints in single-chromosome rearrangements:** inv(2)(p21q31)

A derivative chromosome (der) is a structurally rearranged chromosome generated either by a rearrangement involving 2 or more chromosomes or by multiple aberrations within a single chromosome. The term always refers to chromosomes with intact centromeres. Derivative chromosomes are specified in parentheses, followed by all aberrations involved in the generation of the derivative chromosome. The term "Philadelphia chromosome" is retained for historical reasons to describe the derivative chromosome 22, which is generated by the translocation t(9;22)(q34;q11). The abbreviation "Ph," formerly named "Ph1," may be used in text, but for the description of the karyotype, "der(22)t(9;22)(q34;q11)" is recommended.

A plus or minus symbol is placed before a chromosome or an abnormality designation to indicate additional or missing normal or abnormal chromosomes.

> +21, −7, +der(2)

When normal chromosomes are replaced by structurally altered chromosomes, the normal ones should not be recorded as missing. A plus or minus symbol placed after a chromosome arm symbol (ie, "p" or "q") may be used in text to indicate an increase or decrease in the length of a chromosome arm (eg, "4p+, 5q−"). However, a plus or minus symbol should not be used in the description of karyotypes. The multiplication symbol (×) can be used to describe multiple copies of a rearranged chromosome, but the symbol should not be used to denote multiple copies of normal chromosomes.

Uncertainty in chromosome or band designation may be indicated by the question mark or the approximate sign (ie, "~"). The term "or" is used to indicate alternative interpretations of an aberration.

The rules for designating chromosome aberrations are also used for describing constitutional and acquired chromosome aberrations. When an acquired chromosome abnormality is found in an individual with a constitutional chromosome anomaly, the latter is indicated by the lowercase letter "c" in roman type after the abnormality designation.

In addition, the karyotype designations of different clones are separated by a slash. The absolute number of cells in each clone is enclosed in square brackets and placed after the karyotype description.

21.15.7.6 In Situ Hybridization

The results obtained by in situ hybridization (ISH) are described using the symbol "ish." If a standard cytogenetic observation has been made, it may be given followed by a decimal point, the abbreviation "ish," and the ISH results. If a standard cytogenetic observation has not been made, only the ISH observations are given. For details, see ISCN 2020.[5]

Observations on a structurally abnormal chromosome are expressed by the symbol "ish" followed by the symbol for the structural abnormality, regardless of whether the structural abnormality is seen by standard techniques and ISH or by ISH alone. Following the symbol for the structural abnormality in separate parentheses are the chromosomes, the breakpoints, and the loci for which probes were used, designated according to the Human Genome Database[28] (GDB). These last 3 elements are ordered from pter to qter (ie, from the terminal of the p arm to the terminal of the q arm). If no GDB locus is available, the probe name can be used instead. The locus designations are set in uppercase letters in roman type, and they are separated by commas. The status of each locus is given immediately after the locus designation.

> 46,XY.ish del(22)(q11.2q11.2)(D22S75−)

Observations on normal chromosomes are expressed by the symbol "ish" followed by the chromosome, region, band, or subband designation of the loci tested. These are separated from the symbol "ish" by a space, but they are not in parentheses. The rest of the expression consists of the following in parentheses: the loci tested, a multiplication symbol (×), and the number of signals observed.

> 46,XY.ish 22q11.2(D22S75×2)

ISCN 2020[5] offers additional guidance, and it provides examples of nomenclature based on newer ISH technologies, such as the following:

- interphase, or nuclear, ISH (nuc ish)
- extended chromatin, or fiber, ISH (fib ish)
- reverse ISH (rev ish)
- comparative genomic hybridization (cgh)
- chromosome painting
- microarray techniques, including multiple ligation-dependent probe amplification (MLPA)

21.15.8 Transgenic Animals

The International Committee on Standardized Genetic Nomenclature for Mice approved the symbolic representation of transgenic animal lines. The guidelines are posted on the Mouse Genome Informatics' website.[7]

Use the symbol "*Tg*" to designate genetically engineered transgenic events that result from random insertion of DNA into the genome. The full symbol for the transgenic line consists of 4 parts in italic characters in the form "*Tg(YYY)###Zzz*," in which "*Tg*" indicates a "transgenic insertion by microinjection," "*(YYY)*" indicates the inserted sequence in the official nomenclature for that gene, "*###*" is the laboratory's transgene line designation or a serial number, and "*Zzz*" is the laboratory code assigned by the Institute for Laboratory Animal Research[25] at the National Academies of Sciences, Engineering, and Medicine in the United States.

The strain background is independent of the nomenclature but should precede the nomenclature whenever possible.

> C57BL/6J-*Tg(CD8)1Jwg* (the human CD8 sequence inserted into C57BL/6 mice from the Jackson Laboratory [J])

Whether the transgene was inserted "benignly" into the genome or disrupted, a locus is irrelevant to the nomenclature, but that information should be reported elsewhere. However, if a disrupted locus is identified, the transgene becomes an allele of that locus.

> *Tg(Crya1)2Ove*

The nomenclature in the example above indicates that the mouse α1 crystallin gene was inserted by microinjection and reported by Paul Overbeek, PhD. This was Dr Overbeek's second transgenic line. In a later publication, Dr Overbeek reported that the insertion was allelic with the downless (*dl*) mutant mouse. The symbol for the allele became *dl*[Tg(Crya1)2Ove]. In yet another publication, the *dl* locus and the locus disrupted by the transgene insertion were identified as the gene encoding the ectodysplasin-A receptor (*Edar*). The symbol changed to *Edar*[dl-Tg(Crya1)2Ove].

When a construct is composed of roughly equal parts of 2 genes (ie, a fusion gene), the symbols of the genes are separated by a slash.

> *Tg(TCFE2A/HLF)1Mlc* (the human transcription factor E2Aα and the hepatic leukemia factor were expressed as a chimeric cDNA)

A "reporter construct" provides a way to study a gene's function and localization of its product. "Promoter reporter constructs" allow a protein to be expressed under the control of a target gene. When the construct is a reporter (eg, GFP or LacZ) or a recombinase (ie, a catalyst for site-specific recombination events, such as Cre), the promoter is an essential construct element. The promoter is included in parentheses, separated from the inserted element by a hyphen.

> *Tg(Zp3-Cre)3Mrt* (the *Zp3* promoter plus *Cre* in the third transgenic line)

Transgene symbols can be abbreviated by omitting the insert. The full symbol should be used the first time the transgene is mentioned in the abstract and text of a publication, and the abbreviation should be used thereafter.

Full transgene symbol:	*TgN(GPDHIm)1Bir*
Abbreviated symbol:	*TgN1Bir*

CITED REFERENCES

1. Dunn LC, Grünberg H, Snell GD. Report of the Committee on Mouse Genetics Nomenclature. J Hered. 1940;31:505–506.

2. International Committee on Genetic Symbols and Nomenclature. Report of the International Committee on Genetic Symbols and Nomenclature. Union Int Sci Biol, Ser B. 1957;(30):1–6.

3. Demerec M, Adelberg EA, Clark AJ, Hartman PE. A proposal for a uniform nomenclature in bacterial genetics. Genetics. 1966;54(1):61–76. https://doi.org/10.1093/genetics/54.1.61

4. ASM style manual for journals and books. American Society for Microbiology; 1991.

5. McGowan-Jordan J, Hastings RJ, Moore S, editors. ISCN 2020: an international system for human cytogenetic nomenclature. Karger; 2020.

6. HGNC Guidelines. Human Genome Organisation, Gene Nomenclature Committee; [updated 2022]. http://www.genenames.org/

7. International Committee on Standardized Genetic Nomenclature for Mice. Mouse nomenclature home page; [updated 2021 Dec]. http://www.informatics.jax.org/mgihome/nomen/

8. International Committee on Standardized Genetic Nomenclature for Mice. Rules for nomenclature of mouse chromosome aberrations; [updated 2021 Dec]. http://www.informatics.jax.org/mgihome/nomen/anomalies.shtml

9. Price CA, Reardon EM. Mendel, a database of nomenclature for sequenced plant genes. Nucleic Acids Res. 2001;29(1):118–119. https://doi.org/10.1093/nar/29.1.118

10. The Gene Ontology Resource. Gene Ontology Consortium; [updated 2021 Dec 15]. http://www.geneontology.org

11. Ford J. Frustration with Excel has caused geneticists to rename some genes. NewScientist. 2020 Aug 19. https://www.newscientist.com/article/mg24732961-400-frustration-with-excel-has-caused-geneticists-to-rename-some-genes/

12. Vertebrate Gene Nomenclature Committee. European Bioinformatics Institute; [accessed 2023 Mar 7]. https://vertebrate.genenames.org/

13. Symbol report for P3H1. Human Genome Organisation, Gene Nomenclature Committee; [accessed 2023 Mar 7]. https://www.genenames.org/data/gene-symbol-report/#!/hgnc_id/HGNC:19316

14. P3H1 prolyl 3-hydroxylase 1 [Homo sapiens (human)]. National Library of Medicine (US), National Center for Biotechnology Information; [accessed 2023 Mar 7]. https://www.ncbi.nlm.nih.gov/gene/64175

15. Gene: P3H1 ENSG00000117385. Ensembl (release 107). 2022 Jul. https://www.ensembl.org/Homo _sapiens/Gene/Summary?g=ENSG00000117385;r=1:42746335-42767084

16. den Dunnen et al. HGVS recommendations for the description of sequence variants: 2016 update. Hum Mutat. 2016;37(6):564–569. https://doi.org/10.1002/humu.22981

17. den Dunnen J, Antonarakis S. Nomenclature for the description of human sequence variations. Hum Genet. 2001;109:121–124. https://doi.org/10.1007/s004390100505

18. Higgins J et al. Verifying nomenclature of DNA variants in submitted manuscripts: guidance for journal. Hum Mutat. 2020;42(1);3–7. https://doi.org/10.1002/humu.24144

19. Wagner AH et al. The GA4GH variation representation specification (VRS): a computational framework for variation representation and federated identification. Cell Genomics. 2021;1(2):100027. https:// doi.org/10.1016/j.xgen.2021.100027

20. Reference sequences definition. Human Genome Variation Society; [accessed 2023 Mar 7]. https:// varnomen.hgvs.org/bg-material/refseq/

21. Bragin E et al. DECIPHER: database for the interpretation of phenotype-linked plausibly pathogenic sequence and copy-number variation. Nucleic Acids Res. 2014;42(D1):D993–D1000. https://doi.org/10 .1093/nar/gkt937

22. DECIPHER database. Wellcome Sanger Institute, Genome Research, Ltd; [accessed 2023 Mar 7]. https://www.deciphergenomics.org/

23. The Genome Aggregation Database (gnomAD). Springer Nature, Ltd; 2020. https://www.nature.com /immersive/d42859-020-00002-x/index.html

24. Database of Genomic Structural Variation (dbVar). National Library of Medicine (US), National Center for Biotechnology Center; [accessed 2023 Mar 7]. https://www.ncbi.nlm.nih.gov/dbvar

25. Institute for Laboratory Animal Research. National Academies of Sciences, Engineering, and Medicine (US); c2022. https://www.nationalacademies.org/ilar/institute-for-laboratory-animal-research

26. Olson M, Hood L, Cantor C, Botstein D. A common language for physical mapping of the human genome. Science. 1989;245(4925):1434–1435. https://doi.org/10.1126/science.2781285

27. Whole genome shotgun submissions. National Library of Medicine (US), National Center for Biotechnology Information; [updated 2020]. http://www.ncbi.nlm.nih.gov/Genbank/wgs.html

28. The Genome Reference Consortium at National Institutes of Health (US). 2022. https://www.ncbi .nlm.nih.gov/grc

29. Mouse Genome Informatics (MGI). The Jackson Laboratory; 2021. http://www.informatics.jax.org/

30. Rat Genome and Nomenclature Committee. Rat Genome Database. Medical College of Wisconsin (US); [accessed 2023 Mar 7]. https://rgd.mcw.edu/nomen/nomen.shtml

31. Richards S et al. Standards and guidelines for the interpretation of sequence variants: a joint consensus recommendation of the American College of Medical Genetics and Genomics and the Association for Molecular Pathology. Genet Med. 2015;17(5):405–424. https://doi.org/10.1038/gim.2015.30

32. Roberts RJ. Restriction enzymes and their isoschizomers. Nucleic Acids Res. 1989;17(Suppl):347–387. https://doi.org/10.1093/nar/17.suppl.r347

33. re3data.org: REBASE. Registry of Research Data Repositories; [updated 2022 Dec 16]. https://www .re3data.org/repository/r3d100012171

34. International Union of Biochemistry, Nomenclature Committee. Nomenclature of initiation, elongation and termination factors for translation in eukaryotes: recommendations 1988. Eur J Biochem. 1989;186(1–2):1–3. Prepared by Safer B. https://doi.org/10.1111/j.1432-1033.1989.tb15169.x

35. Campbell A et al. Nomenclature of transposable elements in prokaryotes. Gene. 1979;5(3):197–206. https://doi.org/10.1016/0378-1119(79)90078-7

36. Kempken F, Kück U. Transposons in filamentous fungi—facts and perspectives. BioEssays. 1998; 20(8):652–659. https://doi.org/10.1002/(SICI)1521-1878(199808)20:8<652::AID-BIES8>3.0.CO;2-K

37. Nebert DW et al. The P450 gene superfamily: recommended nomenclature. DNA. 1987;6(1):1–11. https://doi.org/10.1089/dna.1987.6.1

38. Nelson DR et al. P450 superfamily: update on new sequences, gene mapping, accession numbers and nomenclature. Pharmacogenetics. 1996;6(1):1–42. https://doi.10.1097/00008571-199602000-00002

39. Novick RP et al. Uniform nomenclature for bacterial plasmids: a proposal. Bacteriol Rev. 1976;40(1): 168–189. https://doi.10.1128/br.40.1.168-189.1976

40. Nomenclature. American Society for Microbiology Journals; [accessed 2023 Mar 7]. https://journals.asm.org/nomenclature

41. Gallo R, Wong-Stahl F, Montagnier L, Haseltine WA, Yoshida M. HIV/HTLV gene nomenclature [letter]. Nature. 1988;333(6173):504. https://doi.org/10.1038/333504a0

42. Berlyn M, Rudd K, Chater K. Bacteria. Trends in Genetics. 1998;14(11 Suppl 1):S1–S4. https://doi.org/10.1016/S0168-9525(98)80003-6

43. Chumley FG, Menzel R, Roth JR. Hfr formation directed by Tn10. Genetics. 1979;91(4):639–655. https://doi.org/10.1093/genetics/91.4.639

44. Low KB. *Escherichia coli* K-12 F-prime factors, old and new. Bacteriol Rev. 1972;36(4):587–607. https://doi.10.1128/br.36.4.587-607.1972

45. Sherman F. An introduction to the genetics and molecular biology of the yeast *Saccharomyces cerevisiae*. In: Meyers, RA, editor. Encyclopedia of molecular biology and molecular medicine. Vol 6. VCH Publisher. 1998 [updated 2000 Sep 19]. https://www.uni-ulm.de/fileadmin/website_uni_ulm/nawi.inst.155/Praktikumsskripte/Genetische_Grunduebungen/F.Sherman_An_Introduction_to_the_Genetics_and_Molecular_Biology_of_the_Yeast.pdf

46. SGD: *Saccharomyces* Genome Database. Stanford University; c1997–2011. https://www.yeastgenome.org

47. Cherry JM. *Saccharomyces cerevisiae*. Trends Genet. 1998;14(11 Suppl 1):S10–S11. https://doi.org/10.1016/S0168-9525(98)80006-1

48. Kohli J. Genetic nomenclature and gene list of the fission yeast *Schizosaccharomyces pombe*. Curr Genet. 1987;11(8):575–589. https://doi.org/10.1007/BF00393919

49. Fantes P, Kohli J. *Schizosaccharomyces pombe*. Trends Genet. 1998:14(11 Suppl 1):S7–S9. https://doi.org/10.1016/S0168-9525(98)80005-X

50. Turgeon BG, Yoder OC. Proposed nomenclature for mating type genes of filamentous ascomycetes [commentary]. Fung Genet Biol. 2000;31(1):1–5. https://doi.org/10.1006/fgbi.2000.1227

51. Clutterbuck AJ. Gene symbols in *Aspergillus nidulans*. Genet Res. 1973;21(3):291–296. https://doi.org/10.1017/S0016672300013483

52. Clutterbuck AJ. *Aspergillus nidulans*. Trends Genet. 1998:14(11 Suppl 1):S12–S13. https://doi.org/10.1016/S0168-9525(98)80007-3

53. Clutterbuck J, compiler. The *Aspergillus nidulans* linkage map. University of Glasgow; [updated 2008 Nov]. http://www.fgsc.net/Aspergillus/gene_list/index.htm

54. Aspergillus nidulans FGSC A4. National Library of Medicine (US), National Center for Biotechnology Information; [accessed 2023 March 3]. https://www.ncbi.nlm.nih.gov/bioproject/?term=130

55. Clutterbuck AJ. The validity of the *Aspergillus nidulans* linkage map. Fungal Genet Biol. 1997;21(3):267–277. https://doi.org/10.1006/fgbi.1997.0984

56. Perkins DD. *Neurospora* genetic nomenclature. Fungal Genetics Newsl. 46:31–41. http://www.fgsc.net/fgn46/perkins.htm

57. Fungal Genetics Stock Center. University of Missouri–Kansas City; c2005, c2009–2020. http://www.fgsc.net

58. Cerda-Olmedo E. Carotene mutants of *Phycomyces*. Methods Enzymol. 1985;105:220–243. https://doi.org/10.1016/S0076-6879(85)10080-7

59. Eslava AP, Alvarez MI. Genetics of *Phycomyces*. In: Bos CJ, editor. Fungal genetics: principles and practice. Marcel Dekker; 1996. p 385–406.

60. Sermonti G. Genetics of antibiotic-producing micro-organisms (techniques in pure and applied microbiology). Wiley-Interscience; 1969.

61. Yoder OC, Valent B, Chumley F. Genetic nomenclature and practice for plant pathogenic fungi [letter]. Phytopathology. 1986;76(4):383–385. https://www.apsnet.org/publications/phytopathology/backissues /Documents/1986Articles/Phyto76n04_383.PDF

62. Chlamydomonas Resource Center. University of Minnesota (US); [accessed 2023 Mar 7]. https:// www.chlamycollection.org/

63. Dutcher S, Harris E. *Chlamydomonas reinhardtii*. Trends Genet. 1998;14(11 Suppl 1):S18–S19. https:// doi.org/10.1016/S0168-9525(98)80010-3

64. Lefebvre P. Molecular and genetic maps of the nuclear genome. University of Minnesota (US), Chlamydomonas Resource Center. https://www.chlamycollection.org/products/plasmids/

65. Kay R, Loomis B, Devreotes P. *Dictyostelium discoideum*. Trends Genet. 1998:14(11 Suppl 1):S5–S6. https://doi.org/10.1016/S0168-9525(98)80004-8

66. Clayton C et al. Genetic nomenclature for *Trypanosoma* and *Leishmania*. Mol Biochem Parasitol. 1998;97(1–2):221–224. https://doi.org/10.1016/S0166-6851(98)00115-7

67. Allen SL et al. Proposed genetic nomenclature rules for *Tetrahymena thermophila*, *Paramecium primaurelia* and *Paramecium tetraurelia*. Genetics. 1998;149(5):459–462. https://doi.org/10.1093/genetics /149.1.459

68. GrainGenes: a database for *Triticeae* and *Avena*. Department of Agriculture (US); 2019. https://wheat .pw.usda.gov/GG3/

69. Livingstone DL, Lackney VK, Blauth JR, van Wijk R, Jahn MK. Genome mapping in *Capsicum* and the evolution of genome structure in the *Solanaceae*. Genetics. 1999;152(3):1183–1202. https://doi.org/10 .1093/genetics/152.3.1183

70. Meinke D, Koornneef M. Community standards for *Arabidopsis* genetics. Plant J. 1997;12(2):247–253. https://doi.org/10.1046/j.1365-313X.1997.12020247.x

71. Raupp WJ, Friebe B, Gill BS. Suggested guidelines for the nomenclature and abbreviation of the genetic stocks of wheat, *Triticum aestivum* L. em Thell., and its relatives. Wheat Inf Serv. 1995;81:51–55. Also available at http://wheat.pw.usda.gov/ggpages/nomenclature.html

72. Simons MD et al. Oats: a standardized system of nomenclature for genes and chromosomes and catalog of genes governing characters. Agriculture Handbook Number 509. Department of Agriculture (US); 1978.

73. UniProtKB. UniProt consortium. [updated 2022 Apr 28]. https://www.uniprot.org/

74. Meinke DW, Cherry JM, Dean C, Rounsley SD, Koornneef M. *Arabidopsis thaliana*: a model plant for genome analysis. Science. 1998;282(5389):662–682. Includes a foldout genome map. https://doi.10.1126 /science.282.5389.662

75. Robinson RW, McCreight JD, Ryder EJ. The genes of lettuce and closely related species. In: Janick J, editor. Plant Breeding Reviews. Vol 1. Springer; 1983. p 267–293. Also available at https://link.springer .com/chapter/10.1007/978-1-4684-8896-8_9

76. Thomas H. Cytogenetics of *Avena*. In: Marshal HG, Sorrells ME, editors. Oat science and technology. American Society of Agronomy; 1992. p 473–507. Agronomy; No. 33.

77. Project home page for xGDBvm. xGDBvm; [accessed 2023 Mar 7]. http://goblinx.soic.indiana.edu/

78. Plant Gene DataBase Japan. [accessed 2023 Mar 7]. http://pgdbj.jp/?ln=en

79. Franckowiak JD, Lundqvist U, Konishi T. New and revised names for barley genes. Barley Genet Newsl. 1997 May;26:4–8.

80. Linde-Laursen I. Recommendations for the designation of the barley chromosomes and their arms. Barley Genet Newsl. 1997 May;26:1–3.

81. A standard for maize genetics nomenclature. Maize Genet Coop Newsl. 1995;69:182–184 [updated 2015]. Also available at https://www.maizegdb.org/nomenclature

82. MaizeGDB: Maize Genetics and Genomics Database. Department of Agriculture (US), Agricultural Research Service. 2005 [updated 2020]. http://www.maizegdb.org

83. Marshall HG, Shaner GE. Genetics and inheritance in oat. In: Marshal HG, Sorrells ME, editors. Oat science and technology. American Society of Agronomy; 1992. p 509–571. Agronomy; No. 33. https://doi .org/10.2134/agronmonogr33.c15

84. Kinoshita T, convener. Report of the Committee on Gene Symbolization, Nomenclature and Linkage Groups. Rice Genet Newsl. 1986;3:4–5. https://shigen.nig.ac.jp/rice/oryzabase/asset/rgn/vol3/v3C.html

85. Kinoshita T, Rothschild G, chairmen. Report of the Coordinating Committee of Rice Genetics Cooperative. Rice Genet Newsl. 1995;12:5. https://shigen.nig.ac.jp/rice/oryzabase/asset/rgn/vol12/v12p5.html

86. McCouch SR et al, coordinators. Report on QTL nomenclature. Rice Genet Newsl. 1997;14:11. https:// shigen.nig.ac.jp/rice/oryzabase/asset/rgn/vol14/v14p11.html

87. Kinoshita T. Linkage mapping using mutant genes in rice. Rice Genet Newsl. 1998;15:13–74. https:// shigen.nig.ac.jp/rice/oryzabase/asset/rgn/vol15/v15p13II.html

88. Sybenga J. Rye chromosome nomenclature and homoeology relationships: workshop report. Z Pflanzenzüchtg. 1983;90(4):297–304.

89. Part I, introduction: 1. Recommended rules for gene symbolization in wheat. In: McIntosh RA et al. Catalogue of gene symbols for wheat: 2003 supplement. Proceedings of the 10th International Wheat Genetics Symposium; 2003 Sept 1–6; Paestum, Italy. Department of Agriculture (US); 2003. http://wheat.pw .usda.gov/ggpages/wgc/2003/Catalogue.pdf

90. Gill B. Chromosome banding methods, standard chromosome band nomenclature, and applications in cytogenetic analysis. In: Heyne E. Wheat and wheat improvement. American Society of Agronomy, Crop Science Society of America, and Soil Science Society of America; 1987;243–254.

91. Crucifer Genetics Cooperative. CrGC Information Catalogue. University of Wisconsin–Madison. 2003. p 9–10.

92. TAIR: The *Arabidopsis* information resource. Phoenix Bioinformatics Corporation [accessed 2003 Mar 7]. http://www.arabidopsis.org/

93. Fransz P et al. Cytogenetics of the model system *Arabidopsis thaliana*. Plant J. 1998;13(6):867–876. https://doi.org/10.1046/j.1365-313X.1998.00086.x

94. Cucurbit Genetics Cooperative, Gene List Committee. Gene nomenclature for the *Cucurbitaceae*. Cucurbit Genetics Cooperative; [updated 2022 Mar 29]. Created by Wehner TC, Ng T. https://cucurbit.info /home/gene-lists/

95. Hutton MG, Robinson RW. 1992 gene list for *Cucurbita* spp. Cucurbit Genet Coop Rep. 1992;15:102–109. https://cucurbit.info/1992/07/1992-gene-list-for-cucurbita-spp/

96. Quiros CF, Bauchan GR. The genus *Medicago* and the origin of the *Medicago* sativa complex. In: Hanson AA, Barnes DK, Hill RR Jr. Alfalfa and alfalfa improvement. Monograph Series 29. American Society of Agronomy, Crop Science Society of America, and Soil Science Society of America; 1988;93–124. https:// doi.org/10.2134/agronmonogr29.c3

97. Bauchan GR, Hossain MA. Advances in alfalfa cytogenetics. In: One hundred years of alfalfa genetics, the alfalfa genome. Proceedings of the Alfalfa Genome Conference; 1999 Aug 1–3; North American Alfalfa Improvement Conference; 1999. Department of Agriculture (US), Agriculture Research Service, Soybean & Alfalfa Research Lab. http://www.naaic.org/TAG/TAGpapers/Bauchan/advcytog.html

98. Bauchan GR, Hossain MA. Karyotypic analysis of C-banded chromosomes of diploid alfalfa: *Medicago sativa* ssp. *caerulea* and ssp. *falcata* and their hybrid. J Hered. 1997;88(6):533–537. https://doi.org/10.1093 /oxfordjournals.jhered.a023152

99. Ellis TN, Ambrose M. Pea. Trends Genet; 1998;14(11 Suppl 1):S22–S23. https://doi.org/10.1016 /S0168-9525(98)80012-7

100. Soybean Genetics Committee. Rules for genetic symbols. Soybean Genet Newsl. 1996;23:19–23. https://archive.org/details/CAT75654695023/page/18/mode/2up

101. Cregan PB et al. An integrated genetic linkage map of the soybean genome. Crop Sci. 1999;39(5):1464–1490. https://doi.org/10.2135/cropsci1999.3951464x

102. Wang D, Bosland PW. The genes of *Capsicum*. HortScience; 2006;41(5):1169–1187. https://doi.org /10.21273/HORTSCI.41.5.1169

103. Tomato Genetics Cooperative. Rules for nomenclature in tomato genetics [editorial]. TGC Rep. 1992;42:6–7. https://tgc.ifas.ufl.edu/vol42/vol42.pdf

104. Tomato Genetics Cooperative. *Arabidopsis* rules proposed for tomato [editorial]. TGC Rep. 1992;42:5. https://tgc.ifas.ufl.edu/vol42/vol42.pdf

105. Percy RG, Kohel RJ. Qualitative genetics. In: Smith CW, Cothren JT, editors. Cotton: origin, history, technology, and production. John Wiley & Sons; 1999.

106. Kohel RJ. Genetic nomenclature in cotton. J Hered. 1973;64(5):291–295. https://doi.org/10.1093/oxfordjournals.jhered.a108415

107. Frelichowski JE Jr et al. Cotton genome mapping with new microsatellites from Acala "Maxxa" BAC-ends. Mol Genet Genomics. 2006;275(5):479–491. https://doi.org/10.1007/s00438-006-0106-z

108. Reinisch MJ et al. A detailed RFLP map of cotton, *Gossypium hirsutum* × *Gossypium barbadense*: chromosome organization and evolution in a disomic polyploidy genome. Genetics. 1994;138:(3)829–847. https://doi.org/10.1093/genetics/138.3.829

109. Guo W et al. A microsatellite-based, gene-rich linkage map reveals genome structure, function and evolution in *Gossypium*. Genetics. 2007;176(1):527–541. https://doi.org/10.1534/genetics.107.070375

110. Lacape JM et al. A new interspecific, *Gossypium hirsutum* × *G. barbadense*, RIL population: towards a unified consensus linkage map of tetraploid cotton. Theor Appl Genet. 2009;119:281–292. https://doi.org/10.1007/s00122-009-1037-y

111. Lacape JM et al. A combined RFLP–SSR–AFLP map of tetraploid cotton based on a *Gossypium hirsutum* × *Gossypium barbadense* backcross population. Genome. 2003;46(4):612–626. https://doi.org/10.1139/g03-050

112. Nguyen TB, Giband M, Brottier P, Risterucci AM, Lacape JM. Wide coverage of the tetraploid cotton genome using newly developed microsatellite markers. Theor Appl Genet. 2004;109(1):167–175. https://doi.org/10.1007/s00122-004-1612-1

113. Yu JZ et al. A high-density simple sequence repeat and single nucleotide polymorphism genetic map of the tetraploid cotton genome. Genes Genomes Genet. 2012;2(1):43–58. https://doi.org/10.1534/g3.111.001552

114. Van Deynze A et al. Sampling nucleotide diversity in cotton. BMC Plant Biol. 2009;9:125. https://doi.org/10.1186/1471-2229-9-125

115. Wang K et al. Complete assignment of the chromosomes of *Gossypium hirsutum* L. by translocation and fluorescence in situ hybridization mapping. Theor Appl Genet. 2006;113(1):73–80. https://doi.org/10.1007/s00122-006-0273-7

116. Yu J, Kohel RJ, Smith CW. The construction of a tetraploid cotton genome wide comprehensive reference map. Genomics. 2010;95(4):230–240. https://doi.org/10.1016/j.ygeno.2010.02.001

117. de Vries JN. Onion chromosome nomenclature and homoeology relationships: workshop report. Euphytica. 1990;49(1):1–3. https://doi.org/10.1007/BF00024124

118. Kalkman ER. Analysis of the C-banded karyotype of *Allium cepa* L.: standard system of nomenclature and polymorphism. Genetica. 1984;65(2):141–148. https://doi.org/10.1007/BF00135278

119. Chick. Trends Genet. 1998:14(11 Suppl 1);S34–S36. https://doi.org/10.1016/S0168-9525(98)80017-6

120. Committee on Standardized Genetic Nomenclature for Cats. Standardized genetic nomenclature for the domestic cat. J Hered. 1968;59(1):39–40. https://doi.org/10.1093/oxfordjournals.jhered.a107638

121. Committee on Standardized Genetic Nomenclature for Mice. A revision of the standardized genetic nomenclature for mice. J Hered. 1963;54(2):159–162. https://doi.org/10.1093/jhered/54.4.159

122. Dolling CHS. Standardized genetic nomenclature for sheep. In: Piper L, Ruvinsky A, editors. The genetics of sheep. CABI Publishing. 1997;593–601.

123. Dolling CHS. Standardized genetic nomenclature for the horse. In: Bowling AT, Ruvinsky A, editors. The genetics of the horse. CABI Publishing; 2000. p 499–506. https://doi.org/10.1079/9780851994291.0499

124. Fox RR. Taxonomy and genetics. In: Manning PJ, Ringler DH, Newcomer CE, editors. The biology of the laboratory rabbit. Academic Press; 1974. p 1–22.

125. Ruvinsky A, Sampson J. Standardized genetic nomenclature for the dog. In: The genetics of the dog. CABI Publishing; 2001. p 537–540.

126. Committee for the Standard Karyotype of *Vulpes fulvus* Desm. The standard karyotype of the silver fox (*Vulpes fulvus* Desm). Mäkinen A, coordinator. Hereditas. 1985;103(2):171–176. https://onlinelibrary .wiley.com/doi/pdf/10.1111/j.1601-5223.1985.tb00498.x

127. Committee on Chromosome Markers. Report of the Committee on Chromosome Markers: proposed banding nomenclature for the Chinese hamster chromosomes (*Cricetulus griseus*). Cytogenet Cell Genet. 1976;16(1–5):83–91. Prepared by Ray M, Mohandas T. https://doi.org/10.1159/000130559

128. Reimann N, Bartnitzke S, Nolte I, Bullerdick J. Working with canine chromosomes: current recommendations for karyotype description. J Hered. 1999;90(1):31–34. https://doi.org/10.1093/jhered/90.1.31

129. Świtoński M et al. Report on the progress of standardization of the G-banded canine (*Canis familiaris*) karyotype. Chromosome Res. 1996;4(4):306–309. https://doi.org/10.1007/BF02263682

130. Young LD. Standard nomenclature and pig genetic glossary. In: Rothschild MF, Ruvinsky A, editors. The genetics of the pig. 2nd ed. CABI Publishing; 2011.

131. Ruvinsky A, Ostrander E, Sampson J, editors. The genetics of the dog. 2nd ed. CABI Publishing; 2012.

132. Cho KW et al. A proposed nomenclature of the domestic cat karyotype. Cytogenet Cell Genet. 1997;79(1–2):71–78. https://doi.org/10.1159/000134686

133. Committee for a Standardized Karyotype of *Rattus norvegicus*. Standard karyotype of the Norway rat, *Rattus norvegicus*. Cytogenet Cell Genet. 1973;12(3):199–205. https://doi.org/10.1159/000130455

134. Committee for Standardized Karyotype of *Oryctolagus cuniculus*. Standard karyotype of the laboratory rabbit, *Oryctolagus cuniculus*. Cytogenet Cell Genet. 1981;31(4):240–248. https://doi.org/10.1159 /000131653

135. Committee for the Standardized Karyotype of the Domestic Pig. Standard karyotype of the domestic pig. Hereditas. 1988;109(2):151–157. Gustavsson I, coordinator. https://doi.org/10.1111/j.1601-5223.1988 .tb00351.x

136. Iannuzzi I. Standard karyotype of the river buffalo (*Bubalus bubalis* L., 2n = 50). Report of the committee for the standardization of banded karyotypes of the river buffalo. Cytogenet Cell Genet. 1994;67(2):102–113. https://doi.org/10.1159/000133808

137. Harnden DG, Klinger HP, editors. ISCN 1985. An international system for human cytogenetic nomenclature. Birth Defects: Original Article Series, vol 21(1). March of Dimes Birth Defects Foundation; 1985.

138. Mitelman F, editor. ISCN 1995: an international system for human cytogenetic nomenclature. Recommendations of the International Standing Committee on Human Cytogenetic Nomenclature, Memphis, Tennessee, USA, 1994 Oct 9-13. Karger AG; 1995. Published in collaboration with Cytogenetics and Cell Genetics.

139. Horvitz HR, Brenner S, Hodgkin J, Herman RK. A uniform genetic nomenclature for the nematode *Caenorhabditis elegans*. Mol Gen Genet. 1979;175(2):129–133.

140. Caenorhabditis Genetics Center (CGC). University of Minnesota; [accessed 2023 Mar 7]. https:// cgc.umn.edu/

141. FlyBase: a database of *Drosophila* genes and genomes. Indiana University; [update released 2023 Feb 15]. http://flybase.org/

142. FlyBase Consortium. *Drosophila melanogaster*. Trends Genet. 1998;14(11 Suppl 1):S28–S31. https:// doi.org/10.1016/S0168-9525(98)80015-2

143. Lindsley DL, Zimm GG. The genome of *Drosophila melanogaster*. Academic Press; 2012.

144. Shaklee JB, Allendorf FW, Morizot DC, Whitt GS. Gene nomenclature for protein-coding loci in fish. Trans Am Fish Soc. 1990;119(1):2–15. Published erratum in Trans Am Fish Soc. 1990;119(4):unnumbered page after 790. https://doi.org/10.1577/1548-8659(1990)119<0002:GNFPLI>2.3.CO;2

145. International Union of Biochemistry and Molecular Biology, Nomenclature Committee (NC-IUBMB). Enzyme nomenclature: recommendations of the Nomenclature Committee of the International Union of Biochemistry and Molecular Biology on the nomenclature and classification of enzymes by the reactions they catalyse. [updated 19 Dec 2022]. Developed in consultation with the Joint Commission on Biochemical

Nomenclature of the International Union of Pure and Applied Chemistry and the International Union of Biochemistry and Molecular Biology. https://iubmb.qmul.ac.uk/enzyme/

146. ZFIN Zebrafish Nomenclature Conventions. University of Oregon. c1994 [updated 2022 May 18]. https://zfin.atlassian.net/wiki/spaces/general/pages/1818394635/ZFIN+Zebrafish+Nomenclature+Conventions

147. Mullins M. Genetic methods: conventions for naming zebrafish genes. In: Westerfield M, editor. The zebrafish book: a guide for the laboratory use of zebrafish *Danio* (*Brachydanio*) *rerio*. 4th ed. University of Oregon Press; c1994. https://zfin.org/zf_info/zfbook/chapt7/7.1.html

148. Westerfield M. The zebrafish book, a guide for the laboratory use of zebrafish (*Danio rerio*). 5th ed. University of Oregon Press; 2007.

149. Bradford Y et al. ZFIN: enhancements and updates to the zebrafish model organism database. Nucleic Acids Res. 2011;39(Suppl 1):D822–D829. https://doi.org/10.1093/nar/gkq1077

150. Sprague J et al. The Zebrafish Information Network: the zebrafish model organism database. Nucleic Acids Res. 2006;34:D581–585. https://doi.org/10.1093/nar/gkj086

151. Somes RG Jr. Alphabetical list of the genes of domestic fowl. J Hered. 1980;71(3):168–174. https://doi.org/10.1093/oxfordjournals.jhered.a109341

152. Shows TB et al. Guidelines for human gene nomenclature: an international system for human gene nomenclature (ISGN, 1987). Cytogenet Cell Genet. 1987;46(1–4):11–28. https://doi.org/10.1159/000132471

153. Crittenden LB, Bitgood JJ, Burt DW, Ponce de Leon FA, Tixier-Boichard M. Nomenclature for naming loci, alleles, linkage groups, and chromosomes to be used in poultry genome publications and databases. Genet Sel Evol. 1996;28(3):289–297. https://doi.org/10.1186/1297-9686-28-3-289

154. Burt DW et al. The Chicken Gene Nomenclature Committee report. BMC Genomics. 2009;10(Suppl 2):S5. https://doi.org/10.1186/1471-2164-10-S2-S5

155. Ladjali-Mohammedi K, Bitgood JJ, Tixier-Boichard M, Ponce de Leon FA. International system for standardized avian karyotypes (ISSAK): standardized banded karyotypes of the domestic fowl (*Gallus domesticus*). Cytogenet Cell Genet. 1999;86(3–4):271–276. https://doi.org/10.1159/000015318

156. Broad TE, Dolling CHS, Lauvergne JJ, Miller P. Revised COGNOSAG guidelines for gene nomenclature in ruminants 1998. Genet Sel Evol. 1999;31(3):263–268. https://doi.org/10.1186/1297-9686-31-3-263

157. OMIA—Online Mendelian Inheritance in Animals. The University of Sydney (AU), Sydney School of Veterinary Science. https://www.omia.org/home/

158. Powell BC, Rogers GE. Differentiation in hard keratin tissues: hair and related structures. In: Leigh IM, Lane EB, Watt FM. The keratinocyte handbook. Cambridge University Press; 1994. p 401–436.

159. Di Berardino D, Hayes H, Fries R, Long S, editors. ISCNDA 1989. International system for cytogenetic nomenclature of domestic animals (1989). Cytogenet Cell Genet. 1990;53(2–3):65–79. https://doi.org/10.1159/000132898

160. Popescu CP et al. Standardization of cattle karyotype nomenclature: report of the committee for the standardization of the cattle karyotype. Cytogenet Cell Genet. 1996;74(4):259–261. https://doi.org/10.1159/000134429

161. Ansari HA et al. Standard G-, Q-, and R-banded ideograms of the domestic sheep (*Ovis aries*): homology with cattle (*Bos taurus*). Report of the Committee for the Standardization of the Sheep Karyotype. Cytogenet Cell Genet. 1999;87(1–2):134–142. https://doi.org/10.1159/000015380

162. Dolling CHS. Standardized genetic nomenclature for cattle. In: Fries R, Ruvinsky A, editors. The genetics of cattle. CABI Publishing; 1999. p 657–666.

163. The Bovine Genome Database. University of Missouri (US); [accessed 2023 Mar 8]. https://bovinegenome.elsiklab.missouri.edu/

164. Bodmer JG et al. Nomenclature for factors of the HLA system, 1995. Tissue Antigens. 1995;46(1):1–18. https://doi.org/10.1111/j.1399-0039.1995.tb02470.x

165. IPD-IMGT/HLA Database. Release 3.51 (2022-01). EMBL's European Bioinformatics Institute. https://www.ebi.ac.uk/ipd/imgt/hla/

166. Richer CL, Power MM, Klunder LR, McFeely RA, Kent MG. Standard of the domestic horse (*Equus caballus*). Hereditas. 1990;112(3):289–293. Decisions made at the second International Conference for

Standardization of Domestic Animal Karyotypes, held 1989 22–26 May in Jouy-en-Josas, France. https://doi.org/10.1111/j.1601-5223.1990.tb00069.x

167. Bowling AT et al. International System for Cytogenetic Nomenclature of the Domestic Horse. Report of the third International Committee for the Standardization of the Domestic Horse Karyotype, Davis, CA, USA, 1996. Chromosome Res. 1997;5(7):433–443. Also called: ISCNH (1997). https://doi.org/10.1023/A:1018408811881

168. Lyon MF. Rules and guidelines for gene nomenclature. In: Lyon MF, Searle AG, editors. Genetic variants and strains of the laboratory mouse. 2nd ed. Oxford University Press; 1989. p 1–11.

169. Greenbaum IF et al. Cytogenetic nomenclature of deer mice, *Peromyscus* (Rodentia): revision and review of the standardized karyotype. Cytogenet Cell Genet. 1994;66(3):181–195. https://doi.org/10.1159/000133696

170. Levan G, Hedrich HJ, Remmers EF, Serikawa T, Yoshida MC. Standardized rat genetic nomenclature. Mamm Genome. 1995;6(7):447–448. https://doi.org/10.1007/BF00360651

171. RGD: Rat Genome Database. Medical College of Wisconsin, Bioinformatics Program; [accessed 2023 Mar 8]. http://rgd.mcw.edu/

172. Seal RL, Gordon SM, Lush MJ, Wright MW, Bruford EA. Genenames.org: the HGNC resources in 2011. Nucleic Acids Res. 2011;39(Suppl 1):D514–519. https://doi.org/10.1093/nar/gkq892

173. Antonarakis SE, Nomenclature Working Group. Recommendations for a nomenclature system for human gene mutations. Hum Mutat. 1998;11(1):1–3. https://doi.org/10.1002/(SICI)1098-1004(1998)11:1<1::AID-HUMU1>3.0.CO;2-0

174. den Dunnen JT, Antonarakis SE. Mutation nomenclature extensions and suggestions to describe complex mutations: a discussion. Hum Mutat. 2000;15(1):7–12. https://onlinelibrary.wiley.com/doi/pdf/10.1002/%28SICI%291098-1004%28200001%2915%3A1%3C7%3A%3AAID-HUMU4%3E3.0.CO%3B2-N

175. Human Genome Variation Society's website. Human Genome Variation Society; [updated 2021 Mar 1]. https://www.hgvs.org/

176. Marsh SGE et al. Nomenclature for factors of the HLA system. Tissue Antigens. 2002;60:407–464. https://onlinelibrary.wiley.com/doi/pdf/10.1034/j.1399-0039.2002.600509.x

177. Robinson J et al. IMGT/HLA and IMGT/MHC: sequence databases for the study of the major histocompatibility complex. Nucleic Acids Res. 2003;31(1):311–314. https://doi.org/10.1093/nar/gkg070

ADDITIONAL REFERENCES

Aspergillus Genome Projects. Massachusetts Institute of Technology, Broad Institute, Center for Genome Research; 2022. https://www.broadinstitute.org/fungal-genome-initiative/aspergillus-genome-projects

Fokkema IF et al. The LOVD3 platform: efficient genome-wide sharing of genetic variants. Eur J Hum Genet. 2021;29:1796–1803. https://doi.org/10.1038/s41431-021-00959-x

King RC, Stansfield WD. A dictionary of genetics. 8th ed. Oxford University Press; 2012.

Landrum MJ et al. ClinVar: improving access to variant interpretations and supporting evidence. Nucleic Acids Res. 2018:46(D1);D1062–D1067. https://doi.org/10.1093/nar/gkx1153

Nussbaum RL, McInnes RR, Willard HF. Thompson & Thompson genetics in medicine. 8th ed. WB Saunders; 2016.

O'Brien SJ, editor. Genetic maps: locus maps of complex genomes. 6th ed. Cold Spring Harbor Laboratory Press; 1993.

Rieger R, Michaelis A, Green MM. Glossary of genetics: classical and molecular. 5th ed. Springer-Verlag; 1991.

Singh RJ. Plant cytogenetics. 3rd ed. CRC Press; 2016. https://doi.org/10.1201/9781315374611

Singleton P, Sainsbury D. Dictionary of microbiology and molecular biology. 3rd ed. John Wiley & Sons; 2001.

Wood R, editor. Genetic nomenclature guide. Trends Genet. 1998;14(11 Suppl 1). https://www.sciencedirect.com/journal/trends-in-genetics/vol/14/issue/11/suppl/S1

22 Taxonomy and Nomenclature

Editors: Beth E. Hazen, PhD, and Amy McPherson

22.1 OVERVIEW OF TAXONOMY AND NOMENCLATURE

The concepts of systematics, taxonomy, and nomenclature are often confused, and the terms are often used incorrectly. "Systematics" is the study of biodiversity and evolutionary relationships among organisms. Molecular systematists often use nucleotide sequences from genes or entire genomes from a group of organisms to reconstruct evolutionary history or phylogenetic relationships among organisms, as well as to construct phylogenetic trees of life.

"Taxonomy," a subdivision of systematics, is the science of describing, identifying, naming, and classifying living organisms and nonliving viruses into hierarchical systems of taxa based on similarities and differences with other organisms.[1,2] In a formal description of an unknown organism, features are assigned, and attributes are delimited. These features are used to assign the unknown organism to a known taxon or to establish a taxon that is new to science. A new organism is assigned genus and species names according to the rules of a codified system of nomenclature and classified and assigned to a family and higher ranks as appropriate.

"Nomenclature" is the formal naming of taxa in Latin. It is governed by nomenclatural regulations, or codes, that are standardized for organisms in specific kingdoms and for viruses and virus-like entities based on traditional taxonomic hierarchies (see Section 22.2, "Taxonomy and Nomenclature: Commonalities across Hierarchical Systems"). Many assignments to traditional kingdoms have become outdated as the discovery of evolutionary relationships has led to new classifications, to proposals for new classification systems, and even to uncertainty in higher rankings and naming (see Section 22.2.1, "Kingdoms and Other High Taxonomic Ranks"). For example, blue-green "algae" were originally named according to the code for plants, but they are now known to be cyanobacteria.

Among the most reclassified organisms are protists. These were originally classified based on whether they most closely resembled animals, plants, or fungi, and then they were named using the nomenclatural code for that group. However, the diverse group of organisms clumped together as protists include amoebas, slime molds, apicomplexans, diatoms, and many other organisms that could not be placed as animals, bacteria, fungi, or plants. The simplest, most inclusive definition of protists is that they are unicellular, eukaryotic organisms that lack any cellular differentiation into tissue.[3] Molecular phylogenetic work on the protists has revealed key evolutionary relationships that have disrupted traditional classification systems. New classification systems continue to be developed and expanded as phylogenomic work advances. The protists are now classified among different kingdoms and higher ranks, but no internationally accepted nomenclatural code has been developed specifically for the new classifications. Generally, the naming of these organisms is governed by the most appropriate international nomenclatural codes.

22.2 TAXONOMY AND NOMENCLATURE: COMMONALITIES ACROSS HIERARCHICAL SYSTEMS

The hierarchical taxonomic systems and the nomenclatures that have been developed for groups of organisms have some common features. As noted in Section 22.2.1, "Kingdoms and Other High Taxonomic Ranks," the formal naming of organisms is governed by specific nomenclatural codes that were developed for organisms in traditional kingdoms and for nonliving viruses. This section focuses on the commonalities across these systems. (For features that are specific to each nomenclatural code, see Sections 22.4, "Nomenclature and Conventions for Plants, Fungi, and Associated Nontaxonomic Groups," 22.5, "Nomenclature and Conventions for Animals," 22.6, "Nomenclature and Conventions for Prokaryotes," and 22.7, "Taxonomy, Nomenclature, and Conventions for Viruses and Virus-like Mobile Genetic Elements.")

The following are the 7 basic taxonomic ranks in their hierarchy: kingdom, phylum (or division), class, order, family, genus, and species. Note that the plural of the term "phylum" is "phyla" and the plural of "genus" is "genera," while "species" is both singular and plural.

"Species" is generally considered the basic unit. Every species is a member of a genus, every genus is a member of a family, every family is a member of an order, and so forth. Botanists sometimes use the term "division" rather than "phylum" for the second-highest rank. Any of these ranks may be prefixed with "sub" or "super" to provide additional taxonomic categories. Other supplementary ranks (eg, "tribe," a rank below subfamily) are sometimes used to further extend the structure of the hierarchy.

Except for animals, taxa in all taxonomic ranks are written in italic type. For animals, italics are used only for taxa at the level of genus and below, including subgenus, species, subspecies, variety, and forma. Roman type is often incorrectly applied to taxa in ranks of family and above for plants, fungi, and other organisms governed by the *International Code for Algae, Fungi and Plants* (see Section 22.4, "Nomenclature and Conventions for Plants, Fungi, and Associated Nontaxonomic Groups"). For important variations for other organisms, see Sections 22.6, "Nomenclature and Conventions for Prokaryotes," and 22.7, "Taxonomy, Nomenclature, and Conventions for Viruses and Virus-like Mobile Genetic Elements."

The terms for rank are written in all lowercase letters, while an initial capital letter is used for the names of all taxa down through genus (eg, "phylum Protozoa" and "family Terebratulidae"). The names of species are lowercased. While the names of taxa from kingdom to genus may stand alone, species names need to follow the names of their genera. In addition, the names of families and higher-order taxa are treated as plural, while the names of genera and species are treated as singular.

> The family *Orchidaceae* include more than 700 genera.
> *Poa* is a widespread genus in temperate North America.
> *Carchardon carcharias* is a species of mackerel shark.

22.2.1 Kingdoms and Other High Taxonomic Ranks

As noted in Section 22.1, "Overview of Taxonomy and Nomenclature," new kingdoms and hierarchies have been proposed based on the evolutionary relationships revealed by molecular phylogenetics. New classifications are continually contested and evolving as more genomes are sequenced. For example, the system used by the online database Catalogue of Life has 7 kingdoms: Animalia; *Archaea*; *Bacteria Cavalier-Smith, 2002*; *Chromista*; *Fungi*; *Plantae*; and *Protozoa*.[4,5] In addition, the Catalogue of Life treats viruses as an unranked group. Another classification for living organisms consists of 2 superkingdoms (ie, *Prokarya* and *Eukarya*), 5 kingdoms (ie, *Bacteria*, *Proctotista*, Animalia, *Plantae*, and *Fungi*), and 2 subkingdoms within the kingdom *Bacteria* (ie, *Archaea* and *Eubacteria*).[6]

After the bacterium-like *Archaea* were discovered and classified, another system was proposed that consists of 3 domains, which are equivalent to superkingdoms. Two of the domains are for the prokaryotes *Archaea* and *Bacteria*, and one is for the eukaryotes *Eucarya*, also known as "*Eukaryota*."[7] Although the superkingdom and the domain systems are currently used, the domain system is not formally recognized by international codes of nomenclature. In addition, parts of the 3-domain system have been challenged,[8-10] and work in the mid-2010s proposed a new 2-domain system.[11]

Also proposed were eukaryotic trees of life with 5 to 8 monophyletic "supergroups."[3] A monophyletic group, also called a "clade," consists of all organisms descended from a recent common ancestor. The members within a supergroup share a conspicuous biological property. In 2005, a classification for higher ranks of the eukaryotes emphasized the taxonomy of protists by incorporating molecular phylogenetics.[3] This classification recognizes 6 supergroups, or clusters, of eukaryotes that are similar to traditional kingdoms. In a departure from hierarchies with named ranks, this system does not assign names to ranks. Instead, each supergroup is followed by first, second, and third ranks and so on. This system is considered more flexible for incorporating new data and easier to modify.

In 2020, a new tree of life for eukaryotes[12] was proposed. Based on phylogenomic analyses, this tree incorporates key lineages of protists and other organisms that were revealed during the previous 15 years. In this tree, most of the previous supergroups were subsumed into new taxa or were excluded altogether. Unlike the earlier supergroups, the new taxa are not based on shared, derived biological features. Instead, they reflect clades in phylogenetic trees. Genomic data for this tree of life are still missing for many organisms, and many questions remain.

Given the ongoing reclassifications, the introduction of new systems, and the lack of consensus within the discipline of taxonomy, researchers and editors should ensure that the system chosen for a publication at least recognizes the differences among the higher groups and that the system is used consistently throughout the work.

22.2.2 Scientific Binomial System for Species and Lower Ranks

Much of the material in this section applies to the scientific names of organisms, not viruses. For detailed information on virus names, see Section 22.7, "Taxonomy, Nomenclature, and Conventions for Viruses and Virus-like Mobile Genetic Elements."

22.2.2.1 Binomial System for Species Names

In formal taxonomy, a binomial system is used for the names of species.[13] Species names consist of 2 Latin words: The first word is a "genus taxon," and the second is called a "specific epithet" in botany and bacteriology and a "specific name" in zoology. Such a species name is called a "scientific binomial," also known as a "binomen" and as a "scientific name." Specific epithets, specific names, and the names of infraspecific taxa are written in all lowercase letters, even if they are derived from proper nouns to honor individuals or to reflect geographic names. Unlike a genus name, the second name in a scientific binomial cannot stand alone.

Species name	Genus name	Specific epithet or name
Quercus alba	*Quercus*	*alba*
Equus caballus	*Equus*	*caballus*
Shigella boydii	*Shigella*	*boydii* (honors bacteriologist Sir John Boyd, DPH, MD, FRS)

Genus names, specific epithets, and specific names may not actually be Latin words, but they are treated as if they are, and they use Latin endings. Both words in scientific binomials are italicized. These conventions apply wherever a species name is used, including in running text, titles, tables, figures, and indexes.

Running text:	*Homo sapiens* and *Homo neanderthalensis* emerged 300,000 to 200,000 years ago.
Title:	The Metabolism of *Drosophila melanogaster*
Index entry:	*Musca domestica* 25–37

When a species name is part of a title that is italicized, such as a book title in running text, write the species name in roman type to distinguish between the title and the species name. Unfortunately, this convention may confuse readers as to the full title when a species name appears at the beginning or end of a title.

His major monograph was *The Metabolism of* Drosophila melanogaster: *A New Treatise.*
David E. Saunders wrote the book *The Tiger*—Panthera tigris.

In scientific writing, organisms are referred to by their scientific binomials. However, if an organism is widely known by a vernacular name, the vernacular may be used if it is clearly associated with its Latin name at the first mention of the organism in both the abstract and the running text.

The plains zebra (ie, *Equus quagga*) has broader stripes than the Grévy's zebra (ie, *Equus grevyi*) and the mountain zebra (ie, *Equus zebra*).
The polar bear, or *Ursus maritimus*, adapted to cold temperatures and open water.

When scientific binomials are first mentioned in running text, it is customary to append the names of the authors who validly published the binomials' names. These authors are considered to be the "authorities" for the Latin species names. For the different rules for appending author names, see Sections 22.4.1.1, "Authors of Plant Taxa," 22.5.2, "Authors of Animal Taxa," and 22.6.2, "Authors of Prokaryotic Taxa." However, some scientific journals no longer require that the authors of Latin names be mentioned unless the subject of an article is taxonomy or nomenclature.

22.2.2.2 Monomials

The names of genera and higher ranks may stand alone as monomials.

strains or species of *Rhizobium*	the genus *Conus*	a member of the *Poaceae*

The same applies to some of the subordinate categories of genus, namely subgenus, section, and series. However, monomials are not permitted for the rank of species and below because these ranks are not meaningful on their own.

If one or more species of a genus have not been identified, the genus name may be followed by the term "species" or the abbreviation "sp." for a single species or by the term "species" or the abbreviation "spp." for more than one species. Do not italicize "species" or the abbreviations "sp." and "spp." (See also Section 22.2.2.4, "Proper Use of 'sp.' and 'spp.'")

> *Phaseolus* species (refers to one or more unknown species of bean plants of the genus *Phaseolus*)
> *Bifidobacterium* sp. (refers to a single unidentified species of probiotic of the genus *Bifidobacterium*)
> *Mirabilis* spp. (refers to 2 or more species of herb plants of the genus *Mirabilis*)

22.2.2.3 Abbreviation of Scientific Binomials

When a scientific binomial is first used in a document, spell out the genus taxon. Subsequently, the genus name may be abbreviated but only if paired with the full specific epithet or specific name. The abbreviation for a genus name should consist of the genus's first letter and a period. This use of the period is an exception to this style manual's general recommendation to use abbreviations without periods (see Section 11.2, "Punctuation and Typography"). As it does for other abbreviations, this style manual allows abbreviated genus names to be used at the start of sentences if doing so does not create confusion.

> *Electrolux addisoni* lives on reefs and is endemic to the coast of South Africa. *E. addisoni* was first recorded in 1984, but it was not described until 17 years later.
> We included several species of *Festuca* in our study. *not* We included several species of *F.* in our study.

If 2 or more species in the same genus are mentioned in the same document, the genus name of the additional species may be abbreviated the first time they are mentioned, provided that they will not be confused with any other genus already mentioned.

> *Salmo salar* is larger and more aggressive than *S. macedonicus* and *S. lumi.*
> *Impatiens acaulis* and *Io ambondrombeensis* are flowering plants, as is *Impatiens walleriana. not Impatiens acaulis* and *Io ambondrombeensis* are flowering plants, as is *I. walleriana.*

When 2 or more genera with the same first letter are mentioned in the same document, this style manual recommends spelling out each organism's full name on first mention in the abstract and running text. On subsequent references, each genus can be abbreviated to its first letter because the specific name or epithet will serve to uniquely identify each species.

> Although *Mesocricetus auratus* and *Meriones unguiculatus* both belong to the same order,
> *M. auratus* is of the family Cricetidae while *M. unguiculatus* is of the family Muridae.

To avoid variation from one document to another, this style manual discourages the use of multilettered abbreviations to distinguish genus names that begin with the same letter. However, Recommendation 25A of the *International Code of Zoological Nomenclature*[14] ("Zoological Code") allows multilettered abbreviations of animal genus names in this situation. Under the Zoological Code, the genus of the first organism mentioned would still be abbreviated with its first letter, while the genus of other organisms with the same first letter would use multilettered abbreviations. So in the example above, "*Mesocricetus auratus*" would continue to use "*M. auratus*," but "*Meriones unguiculatus*" would be abbreviated as "*Mer. unguiculatus*." Although the *International Code of Nomenclature of Prokaryotes*[15] ("Prokaryotic Code") mirrors the recommendations in this style manual, relevant taxonomic subcommittees of the International Committee on Systematics of Prokaryotes[16,17] officially sanctioned 3-letter abbreviations for the genera of *Halobacteriaceae* and *Halomonadaceae* and for phototrophic bacteria. In contrast, the *International Code of Nomenclature for Algae, Fungi, and Plants*[18] ("Shenzhen Code") makes no recommendation on this matter, even though its examples for plant names use only single-letter abbreviations.

Given the various options recommended by different authorities, it is never wrong to write out full genus names at each mention. This approach is preferable to creating a unique abbreviation for each genus.

22.2.2.4 Proper Use of "sp." and "spp."

When unable or unwilling to identify the specific name or epithet of a known genus, authors can substitute the abbreviation "sp." for one species or "spp." for more than one. Alternatively, the word "species" can be used to indicate a single unknown species or more than one unknown species. Unlike specific names or epithets, "sp.," "spp.," and "species" are set in roman type rather than in italics (eg, "*Angelica* sp.," "*Angelica* spp.," and "*Angelica* species").

Do not use "spp." when referring to the genus as a whole. Instead, use the genus name alone (eg, *Ceriodaphnia*) to avoid confusion. For example, "*Canis*" refers to all species in the same genus, while "*Canis* spp." applies to some but not necessarily all species in the genus.

22.2.2.5 Plurals of Genus Names

Names of genera do not have plural forms. When referring to more than one species in the same genus, write out each scientific binomial, use the genus name with the abbreviation "spp.," or write "species of" before the genus.

Iris junonia, *I. pumila*, and *I. sambucina*	*Poa abbreviata* and *P. flexuosa*
species of *Iris*	species of *Poa*
Iris spp. (*not Irises*)	*Poa* spp. (*not Poas*)

Note that in the examples above, "*Iris*" is a genus name, not the common name "iris."

Like genus names, binomial names are always singular, but they can be used as adjectives with either singular or plural nouns. Although the use of Latin binomials as adjectives is increasing, some publishers discourage this practice.

Escherichia coli was . . .	*Escherichia coli* strains were . . .

22.2.2.6 Diacritical Marks

The nomenclatural codes for plants and their allies, animals, bacteria, and viruses[12–17] prohibit using diacritical marks in Latin binomials. If Latin binomials contain such marks, the names are modified to eliminate the diacritics.

ä, ö, and ü *become* ae, oe, and ue, respectively
é, è, and ê *become* e
ø, æ, and å *become* oe, ae, and aa, respectively
ñ *becomes* n

As an exception to this rule, the Shenzhen Code[18] allows a diaeresis to be used to indicate that a vowel is to be pronounced separately from the vowel just before it (eg, *Cephaëlis* and *Isoëtes*). Using diaereses, however, is optional. (Vowels with diaereses are created with Unicode 00E4 for ä, Unicode 00EB for ë, Unicode 00EF for ï, Unicode 00F6 for ö, and Unicode 00FC for ü.)

22.2.2.7 Validity of Names for Taxa

Only the legitimate name of a taxon should be used in scientific writing. For the name of a binomial or any other taxonomic rank to be legitimate, it must be published according to the rules set forth in the appropriate governing code (ie, the Shenzhen Code,[18] the Zoological Code,[14] the Prokaryotic Code,[15] or the *International Code of Virus Classification and Nomenclature*[19] ["Virology Code"]). Because the same organism may have more than one name, these codes generally maintain that the legitimate name is the first one to be validly published. Fossils, for example, often acquire multiple names based on fragmentary material, resulting in names applied to different parts, life-history stages, and preservational states.

When a new taxon is being described, the convention in many journals is to place the author's surname after the first mention of the taxon's name. The author's name is followed by a comma and an abbreviated designation indicating the taxon's rank, such as "sp. nov." (*species novum*) for a new species, "gen. nov." (*genus nova*) for a new genus, and "fam. nov." (*familia nova*) for a new family. In botany, the abbreviations "nothogen." (nothogenus) and "nothosp." (nothospecies) may be used with a multiplication symbol to designate hybrids (for more on naming hybrids, see Section 22.4.1.4, "Names of Hybrids, or Nothotaxa").

> *Piper allardii* Yuncker, sp. nov.
> *Enterobacter huaxiensis*, sp. nov.
> *Dictyocoprotus* Krug & Khan, gen. nov. (fungus)
> *Strotheria gypsophila* B.L. Turner, gen. et sp. nov.
> *Moraxellaceae* fam. nov. (bacterial family names are italicized)
> Anatidae (animal family names are not italicized)
> ×*Dryostichum singulare* W.H. Wagner, nothogen. et nothosp. nov.

22.2.2.8 Nomenclatural Types

Nomenclatural codes for plants and their allies, animals, and bacteria all use the concept of nomenclatural type. As of March 2021, however, the Virology Code[19] does not require a type.

The concept of type applies to the taxonomic ranks of up to at least family, depending on the nomenclatural code. When a new species of a plant or animal is named, a physical, preserved specimen is designated as the "nomenclatural type," and the type name remains permanently attached to this specimen. In contrast, the type specimen for a prokaryote is a living, pure culture descended from a strain designated as the "nomenclatural type."[15] Before January 2000, a description, illustration, or preserved specimen could suffice for a prokaryote's nomenclatural type.[15]

The nomenclatural type for a genus is the "type species," and the nomenclatural type for a family is the "type genus." Other kinds of type specimens, such as "holotype," "neotype," and "lectotype," are described in the various nomenclatural codes.

22.2.2.9 Grammar of Latin Names for Taxa

Latin binomials should conform to the rules of Latin grammar. Nevertheless, a name that is validly published cannot be rejected for violating Latin grammar rules. Grammar is an issue primarily for articles in which a new name is proposed. A diagnosis, also known as a description, written in Latin or English is required for valid publication of a new plant, fungal, or algal name, and the diagnosis must conform to the rules of Latin or English grammar. (For writing diagnoses, English has been allowed since January 2012.)

Scientific Latin is more formalized in style and vocabulary than classical Latin. However, as in classical Latin, all adjectives used in nomenclature must agree in gender, number, and case with the nouns they modify. Latin and Greek words adopted as genus names generally retain their classical gender unless exceptions are specified within the various nomenclatural codes. Some of the challenges in determining gender are as follows[20]:

(1) Latin names ending in "us" are masculine, unless they are trees recognized by ancient Romans, which are feminine (eg, "*Helianthus annuus*" is masculine, but "*Pinus resinosa*" is feminine even though "*Pinus*" has a masculine ending).

(2) Genus names ending in "on" are masculine (eg, "*Rhododendron canadense*"), and genus names ending in "ma" are neuter (eg, "*Alisma subcordatum*").

(3) Except for genus names ending in "ma," those ending in "a" are feminine (eg, "*Cicuta maculata*").

(4) Genus names ending in "is" are either feminine or masculine but are treated as feminine (eg, "*Amerorchis rotundifolia*").

(5) Some other feminine endings are "ago" (eg, "*Plantago cordata*"), "es" (eg, "*Prenanthes aspera*"), "ix" (eg, "*Larix laricina*"), "odes" (eg, "*Erythrodes querceticola*"), and "oides" (eg, "*Typhoides arundinacea*").

(6) Genus names ending in "um" or "e" are neuter (eg, "*Viburnum dentatum*" and "*Daphne mezereum*").

Latin grammar and gender are covered in detail in other resources,[21-22] as are the meanings for many Latin names applied to animals.[23] The genders of bacterial genera and the meanings of many words used in the names of prokaryotes are given in the online LPSN—List of Prokaryotic Names with Standing in Nomenclature.[24]

22.2.3 Vernacular Names

"Vernacular names," also known as "common names," are names in the language commonly spoken in a country or region. The discussion here applies only to English names.

22.2.3.1 Vernacular Names for Species

Many species have vernacular English names in addition to their Latin names. Some of these names are widely known (eg, "chimpanzee" and "dandelion"), while others are known mostly to people with special interests (eg, the bird "plain chachalaca" and the plant "Utah bladderpod"). Once the scientific name has been clearly associated with its vernacular equivalent in the text, the vernacular name may then be used by itself. This practice is especially common with vernacular names that are widely recognized.

> *Puma concolor*, or cougar, is native to North and South America.
> Lady tulip (ie, *Tulipa clusiana*) is pink on the outside and white on the inside.

The vernacular names of nearly all mammals, fish, plants, and invertebrates, including insects and crustaceans, are generally written in lowercase, except for components that are proper names. This style is reflected in the lists of vernacular names developed by professional societies (eg, the American Fisheries Society's lists for fish, mollusks, decapod crustaceans, Cnidaria, and Ctenophora) and in common dictionaries (eg, *Webster's New World Dictionary of American English*).

> | | | |
> |---|---|---|
> | common juniper | confused flour beetle | golden retriever |
> | Vancouver groundcone | Japanese beetle | Chesapeake Bay retriever |

A few professional societies, however, recommend capitalizing the vernacular names of birds, reptiles, and amphibians (eg, "Blue-winged Teal" and "Diamondback Terrapin"). The Society for the Study of Amphibians and Reptiles, the Herpetologists' League, and the American Ornithological Union advocate capitalizing these vernacular names to distinguish between compound names and names modified by adjectives. For example, does the name "white heron" mean a heron that is the color white, or is it a compound vernacular name? Although capitalizing vernacular names usually eliminates ambiguity, it does not always. For example, does the name "Arctic Tern" refer to a tern living in the Arctic, or is it the full vernacular name?

Until a cross-disciplinary authority such as the International Union of Biological Sciences makes a style recommendation for all the biological sciences, this style manual recommends lowercasing vernacular names. Ambiguity can be reduced if on first mention, scientific binomials are used with vernacular names and adjectives for race, strain, or stock are avoided. After the first mention, such adjectives can be used before the vernacular name.

> populations of striped bass (ie, *Morone saxatilis*) from the Chesapeake *not* populations of Chesapeake striped bass (ie, *Morone saxatilis*)

If a species within a genus is unidentified, the vernacular name should reflect that, or it should be omitted.

> larval salmon or trout (ie, *Oncorhynchus* sp.) *not* larval trout (ie, *Oncorhynchus* sp.)
> larval salmon and trout (ie, *Oncorhynchus* spp.) *not* larval salmon (ie, *Oncorhynchus* spp.)
> *Oncorhynchus* spp. larvae

22.2.3.2 Problems with Vernacular Names

The use of vernacular names presents several problems for precise scientific communication:

- The same species may have more than one vernacular name.
- The same vernacular name may be applied to quite different species.
- Vernacular names are specific to a language and often to a region.
- Many species do not have vernacular names.
- Vernacular names may be misleading.

These problems reinforce the need to use scientific binomial names with vernacular names on first mention in abstracts and running text. For more information on vernacular names, see Sections 22.4.1.6, "Vernacular Names for Plants," 22.5.7, "Vernacular Names for Animals," 22.6.3, "Vernacular Names for Prokaryotes," and 22.7.3, "Vernacular Names for Viruses."

22.2.3.3 Vernacular Names of Genus and Family Names
Many genus names are used as vernacular names. Unlike genus names, vernacular names are neither italicized nor capitalized.

Vernacular name	Genus name
aster	*Aster*
gorilla	*Gorilla*
octopus	*Octopus*
python	*Python*

However, genera often have 2 or more vernacular names, some of which are mutually exclusive. For example, the genus *Oncorhynchus* includes some "salmon" and some "trout." In such cases, the vernacular and scientific names need to be conveyed in a format that is unambiguous.

The genus *Oncorhynchus* (ie, some salmon and some trout) . . .
not Pacific salmon (*Oncorhynchus*) . . .

The first example above indicates that the genus *Oncorhynchus* includes more than salmon, while the second example might lead some readers to conclude that the genus is limited to Pacific salmon.

Similar to genera, families often have vernacular names that reflect the vernacular names of only some species within the families. For example, "sunfish" is the vernacular name for the family *Centrarchidae*, which consists of 8 genera, of which only some have sunfish species. Given that *Centrarchidae* also include species of black bass, crappies, and other fish, the vernacular family name "sunfish" is ambiguous.

Consequently, widespread practices evolved that convert the taxa for animal and plant families to quasi-vernacular names. The "idae" ending for the scientific names of animal families is shortened to "id" for singular usage and "ids" for plural, and the initial capital letter is lowercased (eg, the singular of "*Centrarchidae*" becomes "centrarchid," and the plural becomes "centrarchids"). Similarly, for plant families, the "aceae" ending is replaced by "id" or "iad" for singular usage and "ids" or "iads" for plural (eg, "*Orchidaceae*" becomes "orchid" or "orchids," and "*Bromeliaceae*" becomes "bromeliad" or "bromeliads"). Plant family names are sometimes converted to quasi-vernacular adjectives by replacing the "aceae" ending with "ous" and the initial capital with a lowercase letter (eg, "plants in the *Solanaceae*" become "solanaceous plants").

Such quasi-vernacular family names should refer to valid family names, and their Latin taxa should be used at least once early in the running text.

22.3 INTERNATIONAL NOMENCLATURAL CODES AND OTHER NAMING CONVENTIONS

International codes of nomenclature and other naming conventions are used in various disciplines. The following sections of this chapter summarize those codes in the order in which they were developed. The first of those international codes addressed names for plants and plant "allies," such as algae, fungi, and plant-like protists. The second international code covered animals and animal-like protists, and the third addressed bacteria initially and expanded later to prokaryotes. Developed last, the code for viruses and related nonliving entities differs in several respects from those for organisms.

22.4 NOMENCLATURE AND CONVENTIONS FOR PLANTS, FUNGI, AND ASSOCIATED NONTAXONOMIC GROUPS

22.4.1 Naming Plant Taxa

For the name of a species or any other taxonomic rank of plants to be legitimate, it must be validly published according to the requirements set forth in the *International Code of Nomenclature for Algae, Fungi, and Plants*[18] or its predecessor, the *International Code of Botanical Nomenclature* ("Botanical Code"). The code was revised and ratified in 2017 at the International Botanical Congress in Shenzhen, China, and that new edition was published in 2018.[18] Also known as the "Shenzhen Code," the 2017 edition was published in English, and it was translated into several other languages. When any differences are perceived among the translations, the English version is considered official. In botanical fields, the words "algae, fungi, and plants" are frequently lowercased in the official name of this international code because these terms are not formal names of clades, which are groups of organisms with common ancestors (ie, *International Code of Nomenclature for algae, fungi, and plants*).

The ultimate authority for the code rests with the International Botanical Congress, which is traditionally held every 6 years. The next congress is scheduled for 2024 in Madrid, Spain.

The latest edition of the code supersedes all previous editions and is known by the name of the city where the code was ratified. Consequently, the Shenzhen Code will be replaced with the Madrid Code once it is ratified. The code's rules and recommendations are intended to bring uniformity and clarity to future nomenclature. They apply to the names of all fossil and nonfossil organisms historically treated as plants, including fungi, algae, and lichen-forming fungi. The rules of nomenclature for fossil plants are generally the same as for nonfossil plants, except for some modifications of the rule for priority that relate to applying names to fragmentary material (see Article 11.7 in the Shenzhen Code).

Family, genus, and species names found to be incorrect according to the current code can be retained, or conserved, if the updated names would cause confusion and considerable inconvenience, especially for scientists in different disciplines. Conserving names requires specific action by the International Botanical Congress. Such names are categorized as "nomina conservanda" (nom. cons.). In the Shenzhen Code, those names are listed in Appendixes II–IV.[25]

> *Maclura* Nutt., nom. cons.

22.4.1.1 Authors of Plant Taxa

Authors who validly publish the scientific taxa under the provisions of the then-current code are the authors, or authorities, of those names. Under the Shenzhen Code, for example, citing author names is optional, and author names are not part of the official name of a taxon. When citing authors after a taxon, full surnames or unambiguous abbreviations may be used (see Section 22.4.1.1.2, "Abbreviation of Author Names").

> *Myriophyllum tenellum* Bigelow   *Proserpinaca palustris* L.

If an author's name is written in a non-Roman alphabet, the name should be romanized.

22.4.1.1.1 TRANSFER OF A SPECIES TO ANOTHER GENUS

As more is learned about taxonomic relationships, it is sometimes necessary to move a species from its originally designated genus to another. When such a change is made for plants and other organisms covered by the Shenzhen Code, the name of the author who first validly described the species is enclosed in parentheses, followed by the name of the author who moved the species to another genus. The same rule applies to infraspecific taxa.

> Edward Lee Greene, BPhil, validly described *Tetraneuris herbacea*, so its name was originally "*Tetraneuris herbacea* Greene." Arthur Cronquist, PhD, moved this species to the genus *Hymenoxys*, so the name became "*Hymenoxys herbacea* (Greene) Cronquist."
> Variety *rydbergii* St-Yves of *Festuca ovina* L. was reclassified by Arthur Cronquist, PhD, to the genus *Festuca*, so it became *Festuca brachyphylla* Schult. var. *rydbergii* (St-Yves) Cronquist.

Note that when a taxon with 2 sets of authors is used with a phrase or sentence that is enclosed in fences, the parentheses that enclose the original author's name are retained, and square brackets are used to enclose the entire phrase or sentence, as in "[see the entry for *Hymenoxys herbacea* (Greene) Cronquist on the Global Biodiversity Information Facility's website]." This practice is an exception to Section 5.3.7.1, "General Uses," which calls for using square brackets within parentheses. See also Section 22.4.1.1.2, "Abbreviation of Author Names."

For up-to-date taxonomic classifications and synonyms, consult online plant and fungal databases, such as the following:

- World Flora Online[26]
- Tropicos[27]
- World Checklist of Vascular Plants[28]
- International Plant Names Index[29]
- Index Fungorum[30]
- Catalogue of Life[4]

22.4.1.1.2 ABBREVIATION OF AUTHOR NAMES

For scientific names published by 18th-century botanist and physician Carl Linnaeus, who introduced the first nomenclatural system,[31] the author abbreviation is simply "L." The surnames of many other early botanists are also abbreviated, especially the names of authors responsible for many scientific names. The surnames of more recent authors are usually written out in full. Abbreviations and full names can be found online in the International Plant Names Index[29] and Index Fungorum,[30] both of which are hosted by the Royal Botanical Gardens, Kew, in the United Kingdom. These names are also in Tropicos,[27] which is maintained by the Missouri Botanical Garden in the United States. Author abbreviations and surnames are also listed in appendixes of some plant-identification manuals.

22.4.1.1.3 MULTIPLE AUTHORS

When a scientific name has been published by 2 or more authors, the surnames of both are cited, and they are linked by the Latin word "et" or by an ampersand.

> *Panicum ravenelii* Scribn. et Merr. *or Panicum ravenelii* Scribn. & Merr.

In literature written before 1995, "in" between 2 names meant that the first-named person described the plant in a work written or published by the second person.

> *Solanum sarrachoides* Sendtn. in Mart.

Now, only the author of the taxon is used. The name of the person who published the taxon is no longer included as part of the authority, but it is considered part of the bibliographic citation (see Article 46.2, Note 1, of the Shenzhen Code[18]).

> *Solanum sarrachoides* Sendtn.

The Latin word "ex" between 2 names means that the first author proposed the scientific name but did not validly publish it and that the second author validly published the scientific name.

> *Cypripedium candidum* Muhl. ex Willd.

The author citation is omitted from a scientific name used in the title of a journal article unless the article is about different authors of the same taxon or about a taxon being described for the first time.

22.4.1.1.4 AUTHORS OF INFRASPECIFIC TAXA

Authors are used for infraspecific ranks, such as subspecies (subsp.), variety (var.), subvariety (subvar.), and form (f.). Author names may be used after both the species taxon and the infraspecific taxon or after just the infraspecific taxon.

> *Medicago sativa* L. subsp. *falcata* (L.) Arcangeli *or Medicago sativa* subsp. *falcata* (L.) Arcangeli
> *Clematis flammula* L. var. *maritima* (L.) D.C. *or Clematis flammula* var. *maritima* (L.) D.C.
> *Salix candida* Flüggé f. *denudata* (Andersson ex D.C.) Rouleau *or Salix candida* f. *denudata* (Andersson ex D.C.) Rouleau

However, the name of the nomenclatural type of an infraspecific taxon does not include the author of that taxon. Only the author of the species epithet is named in the nomenclatural type. Similar to the nomenclatural types of family, genus, and species taxa, the nomenclatural type of an infraspecific taxon of a plant is a physical specimen of dried plant preserved in an herbarium to which the type name is permanently attached (see also Section 22.2.2.8, "Nomenclatural Types"). The name of a nomenclatural type of an infraspecific taxon automatically receives an infraspecific epithet that is identical to its specific epithet. The 3-part name of genus name, specific epithet, and infraspecific epithet is termed an "autonym."

> *Rorippa palustris* (L.) Besser var. *palustris*

22.4.1.2 Infrageneric and Infraspecific Taxa

Under the Shenzhen Code, an epithet referring to an infrageneric or infraspecific rank must be preceded by a word indicating the taxonomic rank. This word is often abbreviated. The word or abbreviation indicating the rank is not italicized, and it is written without an initial capital letter (eg, ser.).

> *Costus* subg. *Metacostus* ("subg." is the abbreviation for "subgenus," and *Metacostus* is the subgenus epithet)
>
> *Ricinocarpos* sect. *Anomodiscus* ("sect." is the abbreviation for "section," and *Anomodiscus* is the section epithet)
>
> *Desmodium* ser. *Stipulata* Schub. ("ser." is the abbreviation for "series," and *Stipulata* is the series name)
>
> *Rorippa palustris* (L.) Besser var. *fernaldiana* (Butters & Abbe) Stuckey ("var." is the abbreviation for "variety," and *fernaldiana* is the varietal name)

The trinomial names of organisms below the rank of species can be shortened by abbreviating the genus name. However, the specific epithet should not be abbreviated, even though it precedes the lower rank. In addition, the Shenzhen Code does not allow the infraspecific rank to be omitted from trinomial names (ie, "undesignated trinomials"). In contrast, the *International Code of Zoological Nomenclature*[12] allows undesignated trinomials.

> *R. palustris* var. *fernaldiana* (Butters & Abbe) Stuckey *not R. p. fernaldiana* (Butters & Abbe) Stuckey

The abbreviations for infrageneric or infraspecific ranks use periods, which is an exception to this style manual's general recommendation against using periods with abbreviations (see Section 11.2, "Punctuation and Typography").

22.4.1.3 Names of Families and Higher Taxa

The names of plant families end in "aceae." This ending is appended to the type genus name (eg, "*Cyperaceae*" for the family name for the type genus *Cyperus*). The Shenzhen Code[18] permits alternative names to be used for 8 families. Either name is acceptable as long as it is consistently used within a given work.

Family name based on type genus name	Alternative family name
Apiaceae	*Umbelliferae*
Arecaceae	*Palmae*
Asteraceae	*Compositae*
Brassicaceae	*Cruciferae*
Clusiaceae	*Guttiferae*
Fabaceae	*Leguminosae*
Lamiaceae	*Labiatae*
Poaceae	*Gramineae*

Similar to plant family names ending in "aceae," order names end in "ales," class names end in "opsida," and subclass names end in "idea."

The names at the rank of family and above are plural in form and, therefore, require plural verbs and pronouns.

> *Polypodiopsida* include the royal ferns.
> The *Rosales* comprise an estimated 6,600 species.
> The *Liliaceae* are very diverse and have been separated into numerous smaller families by some botanists.

22.4.1.4 Names of Hybrids, or Nothotaxa

The rules for naming hybrids, or "nothotaxa," are addressed in Chapter H of the Shenzhen Code.[18] These rules apply to naturally occurring hybrids, as well as to agronomic and horticultural ones. Hybrids are designated in 2 ways: by formula and by name.

In a formula, the names of the known parents or putative parents (ie, assumed parents) are separated by a multiplication symbol (ie, "×," which is created by Unicode U-00D7). A space is used before and after the multiplication symbol, and the symbol is not italicized. Because a formula consists of existing names, it requires none of the formalities for new names. The names of the parents should be listed in alphabetical order. Alternatively, the seed parent (ie, the female parent) can be listed first if this convention is explained in the manuscript.

> *Digitalis lutea* L. × *D. purpurea* L. (for the hybrid between *D. lutea* and *D. purpurea*)
> *Fatsia* × *Hedera* (for the hybrid between an unnamed species of the genus *Fatsia* and an unnamed species of *Hedera*)

The multiplication symbol is also used to designate a new name as a hybrid. In this case, the symbol is placed just before the new hybrid name without a following space.

> *Mentha* ×*piperita* (for a hybrid of 2 species, both of which are within the genus *Mentha*)
> ×*Stiporyzopsis* (for a hybrid of the genera *Stipa* and *Oryzopsis*)
> ×*Mahoberberis* sp. (for a hybrid between a species of the genus *Mahonia* and a species of the genus *Berberis*)

If the multiplication symbol (×) is not available, the lowercase letter "x" may be used, but in this situation, a space is needed between "x" and the hybrid name to avoid ambiguity.

> A naturally occurring nothospecies is *Quercus* x *asheana*, a hybrid between *Quercus cinerea* and *Q. laevis*.

The multiplication symbol is not used for infraspecific ranks, but the abbreviations denoting their ranks carry the prefix "notho" (eg, *Polypodium vulgare* L. nothosubsp. *mantoniae* Schidlay).

When applied to a new hybrid, a name is subject to essentially the same rules for publication that the Shenzhen Code or other appropriate code applies to any other new name of the same rank.

22.4.1.5 Names of Cultivated Plants

The *International Code of Nomenclature for Cultivated Plants*[32] (ICNCP), which supplements the Shenzhen Code,[18] promotes uniformity, accuracy, and stability in the naming of agricultural, horticultural, and silvicultural cultivated varieties, including cultivars, groups, and chimeras. First published in 1953 and revised most recently in 2016, the

ICNCP is maintained and modified by the International Commission for the Nomenclature of Cultivated Plants of the International Union of Biological Sciences.

Although a plant brought from the wild into cultivation retains the name that was applied to it when it was growing in nature, designations for additional varieties are based on rules established by the ICNCP. However, nothing precludes using the names for these additional varieties that were published in accordance with the requirements of the Shenzhen Code.[18]

According to Article 21 of the ICNCP,[32] a "cultivar" is an assemblage of plants that has been selected for a particular attribute or combination of attributes and that is clearly distinct, uniform, and stable in these characteristics. Additionally, the characteristics are retained when a cultivar is propagated by appropriate means.

The term "cultivar" encompasses clones derived vegetatively from a single parent, lines of selfed or inbred individuals, and collections of individuals that are resynthesized only by crossbreeding (eg, F_1 hybrids). A cultivar name can include a forestry provenance, which is the geographic origin of seed or trees. A cultivar name also can include a particular growth-habit form that can be retained by appropriate methods of propagation, such as clonal reproduction.

To ensure that a cultivar has only one correct name, the ICNCP requires that each cultivar be named by means of priority acts, which consist of publication and registration of the name in documents that are dated and distributed to the public. Appendixes I, II, and III of the ICNCP contain a list of international cultivar registration authorities, a directory of statutory plant registration authorities, and a list of places maintaining nomenclatural standards.

Cultivar names, sometimes referred to as "fancy names," are placed within single quotation marks after Latin species names. Cultivar names use initial capital letters, and they are written in roman type.

> *Rubus flagellaris* Willd. 'American Dewberry'
> *Triticum aestivum* L. 'Era'
> *Hordeum vulgare* L. 'Proctor'
> *Juniperus communis* L. var. *depressa* Pursh. 'Plumosa'

Article 14.1 in the ICNCP[32] prohibits using double quotation marks around a cultivar name. This article also prohibits using the abbreviations "cv." and "var." between the Latin species name and the cultivar name.

> *Triticum aestivum* L. 'Era' *not Triticum aestivum* L. cv. "Era"

Single quotation marks are not necessary when a cultivar name is written without the species name, unless omitting the quotation marks would create confusion.

> Era is a widely grown cultivar of wheat.
> *Hedera helix* L. 'Chicago' is a popular ivy because of its fast-growing habit and good keeping qualities. It resembles *H. helix* 'Pittsburgh' except that plants of 'Chicago' grow more slowly than those of 'Pittsburgh.'

A "group" is a formal category for assembling cultivars, individual plants, or combinations thereof on the basis of a defined similarity, according to Article 3 of ICNCP.[32] Each element of a group name, including the word "Group," begins with a capital letter.

> *Allium cepa* L. Shallot Group
> *Rosa* L. Polyantha Group

If the group name is followed by a cultivar name, enclose the group name in parentheses or square brackets.

> . . . a shrub like *Hydrangea macrophylla* Ser. (Hortensia Group) 'Ami Pasquier.'
> *or*
> . . . a shrub (eg, *H. macrophylla* [Hortensia Group] 'Ami Pasquier').

A "chimera" is an individual plant or organ consisting of tissues of different genetic constitution resulting from a graft union (ie, a "graft chimera"). A chimera name can be coined from elements of the names of the genera of the grafted species. When formed this way, a chimera name is preceded by a plus symbol without a space after the symbol. Alternatively, a graft chimera can be designated by writing the genus names of its component taxa in alphabetical order with a plus symbol between them and a space on either side of the symbol. The plus symbol is used instead of the multiplication symbol because a graft chimera is not a true hybrid.

> +*Crataegomespilus* 'Dardarii' (graft chimera between *Crataegus monogyna* and *Mespilus germanica*)
> *Crataegus* + *Mespilus*

When component taxa belong to the same genus, the graft chimera can be designated by the genus name and a cultivar name without a plus symbol.[32]

> *Camellia* 'Daisy Eagleson'   *Syringa* 'Correlata'

Because the ICNCP has no legal status, the commercial interests of plant breeders are protected by national and international law. The Council of the International Union for the Protection of New Varieties of Plants (https://www.upov.int) is an international organization that coordinates laws that protect plant cultivars in the council's member countries.

In addition to having no legal status, the ICNCP does not cover trademarks because they are not considered to be names.

22.4.1.6 Vernacular Names for Plants

Because no authoritative source exists for writing vernacular names for plants, vernacular plant names are used inconsistently. This style manual recommends that authors and editors consult an authoritative dictionary or plant-identification manual for guidance on vernacular names, including recommendations for using hyphens and joining vernacular names. Because different sources are likely to disagree on the same names, a general approach to vernacular names is outlined below.

In general terms, vernacular names may consist of a group name corresponding to a genus or family. Simple group names consist of a single word.

ash	fern	lily	mustard	tulip
aster	grass	mallow	orchid	willow

A modifier that corresponds to a species may precede a simple group name (eg, "soapwort gentian," "Indian grass," and "three-birds orchid"). In contrast, some vernacular names do not include group names (eg, "adder's-mouth" for the orchid *Pogonia ophioglossoides*).

Common vernacular names are not capitalized unless they are derived from proper nouns.

Blue Ridge gayfeather	field pansy	pinwheel zinnia
Dutchman's-pipe	Good King Henry	Texas sage
English ivy	Jerusalem artichoke	yellow cosmos

A genus name used as a vernacular name is neither capitalized nor italicized, and its plural is formed by the same rules for plurals in English.

Singular	Plural
aster for the genus *Aster*	asters
camellia for *Camellia*	camellias
iris for *Iris*	irises
crocus for *Crocus*	crocuses (*not* croci)

A common inconsistency arises from whether to designate misapplied group names by inserting a hyphen before the group names (eg, "Douglas-fir" for *Pseudotsuga menziesii*, which is not a true fir). Although entomologists are consistent on this matter (see Section 22.5.7, "Vernacular Names for Animals"), botanists and authoritative dictionaries are not. For example, the vernacular name for *Xerophyllum* (a member of the lily family that is not a true grass) is given as "beargrass" in one manual,[33] as "bear-grass" in another,[34] and as "bear grass" in a compilation of plant names.[35]

Even when a vernacular name reflects the true taxonomic position of a plant, the elements of the name may be separated by a space, joined with a hyphen, or joined into a single word. For example, the following vernacular names for various true grasses are given in the *Manual of Vascular Plants of Northeastern United States and Adjacent Canada*.[36]

bluegrass for *Poa* spp.	needle-and-thread grass for *Stipa comata*
Indian ricegrass for *Oryzopsis hymenoides*	porcupine-grass for *Stipa spartea*
meadow-fescue for *Festuca pratensis*	western fescue for *Festuca occidentalis*

Were botanists to follow the same guidelines as entomologists, the words "grass" and "fescue" in the examples above would all be separated from the preceding words without hyphens. In contrast, the American Joint Committee on Horticultural Plant Nomenclature[37] and the Composite List of Weeds[38,39] advocate joining words to standardize vernacular names of plants in horticultural, forestry, and agricultural industries.

Despite these challenges, authors and editors should strive for consistent style. This style manual recommends adhering to the following guidelines for vernacular plant names, which are derived from long-established sources[40–42]:

(1) Join words of 1 or 2 syllables that refer to plants in general (eg, "plant," "grass," "tree," "wort," and "weed") or to part of a plant (eg, "vine," "nut," "berry," and "flower") to the preceding word, unless the resulting name would be long and unwieldy. Examples are "blackberry," "bluegrass," "bladderwort," "duckweed," and "sunflower."

(2) If the second part of a vernacular name correctly refers to a specific taxonomic group instead of a general term, separate it from the preceding word with a space (eg, "orange lily" and "tumbling mustard").

(3) If the second part of a vernacular name is misapplied according to taxonomic principles, hyphenate the name (eg, "Douglas-fir," "poison-oak," and "rue-anemone"). However, do not hyphenate vernacular names that begin with "false" because this prefix already indicates that the last word in the term has been misapplied (eg, "false indigo"). Vernacular names beginning with the modifier "Indian" also are not hyphenated (eg, "Indian bean").

(4) If the second part of the name has nothing to do with plants or plant parts, hyphenate the name (eg, "blazing-star" and "adder's-mouth").

(5) For names composed of more than 2 words, hyphenate each word (eg, "Jack-in-the-pulpit," "Joe-Pye-weed," "Star-of-Bethlehem," and "Queen-Anne's-lace").

(6) Do not use the word "common" in vernacular names.

(7) For vernacular names of plants that refer to animals, use the guidelines of John T. Kartesz and John W. Thieret.[42]

(8) For capitalization of vernacular plant names, follow the recommendations of an authoritative dictionary.

Another important consideration when using vernacular names is to distinguish the vernacular names of crops from those of harvest products.

Crop	Harvest product
apple	apples
bean	beans
beet	beets
grape	grapes
oat	oats
pea	peas

22.4.2 Naming Fungi

In addition to addressing plants, the *International Code of Nomenclature for Algae, Fungi, and Plants*[18] ("Shenzhen Code") applies to fossil and nonfossil fungi, as well as to chytrids, oomycetes, and slime molds except *Microsporidia*.[17] However, although Chapter F in the Shenzhen Code addresses fungi, that chapter is superseded by decisions made at the 11th International Mycological Congress, held in 2018 in San Juan, Puerto Rico.[43] The congress's modifications to Chapter F can be accessed via a hyperlink at the beginning of Chapter F in the online version of the Shenzhen Code.

As noted in Section 22.2, "Taxonomy and Nomenclature: Commonalities across Hierarchical Systems," fungi were included in the plant kingdom in early 2-kingdom classifications. Now, they are their own kingdom, which is named "*Fungi*." Fungi are nongreen, eukaryotic organisms with cell walls that differ chemically from those of green plants. In addition, fungi have different modes of nutrition than do green plants. The kingdom *Fungi* consists of 3 divisions: *Zygomycota*, *Basidiomycota*, and *Ascomycota*, also known as "zygomycetes," "basidiomycetes," and "ascomycetes," respectively. (The Latin names of the 3 divisions are capitalized, but the alternative names ending in "mycetes" are lowercased.)

Fungi are named using the same binomial method as used for plants. The genus and specific epithet are written in Latin and are italicized, and the surname or abbreviation of the taxon's author is written in roman type (eg, "*Amanita phalloides* Fr.").

One difference between fungal and plant nomenclature is that the colon has replaced the Latin word "ex" to indicate that the first author proposed the scientific name but did not validly publish it and that the second author validly published the scientific name (see Section 22.4.1.1.3, "Multiple Authors"). In such a situation, the binomial scientific name of the fungus is followed by the originating author's surname or abbreviation, a colon with a space before and after it, and the sanctioning author's surname or abbreviation. For example, the taxon "*Boletus piperatus* Bull. : Fr." indicates that botanist and physician Pierre Bulliard first applied the name but mycologist Elias Magnus Fries, FRS, was the first to validly publish the taxon.

After first mention of the complete species name, the genus can be abbreviated to a single capital letter followed by a period, unless there is potential for ambiguity as to the genus. The genus cannot be abbreviated unless it is followed by the specific epithet.

The names of fungal genera do not have plural forms. When referring to a group of species in the same genus, the genus name should be followed by "spp.," the abbreviation for the plural word "species." (See Section 22.2.2.4, "Proper Use of 'sp.' and 'spp.'")

Acaulopage spp. *not* acaulopages	*Helminthosporium* spp. *not* helminthosporia
Calocera spp. *not* calocerae	*Pythium* spp. *not* pythia

Do not apply adjectival endings to the names of fungal genera.

Fusarium head blight *not* Fusarial head blight *and not* fusarial head blight

22.4.2.1 Anamorph, Teleomorph, and Holomorph Names

Before 2013, a fungus in the *Ascomycota* or *Basidiomycota* division with a pleiomorphic life cycle that does not form lichens had separate names as an anamorph (ie, asexual morph) and a teleomorph (ie, sexual morph) at the genus and species levels. Although both the anamorph and teleomorph names could be used, the teleomorph name was considered the correct name for the holomorph (ie, the organism in all of its life stages). This rule changed with the 2011 version of the *Code of Nomenclature for Algae, Fungi, and Plants,*[44] which took effect 1 January 2013. Known as the "Melbourne Code," the 2011 version dictated that separate names cannot be used for an anamorph and a teleomorph. Instead, the valid name for a fungus is established as the name of the morph with the earlier date of publication. No priority is given to the teleomorph.

Often the anamorph stage is the first form discovered and the first validly named. Therefore, under the Melbourne Code's one-name rule, the valid holomorph names of numerous fungi were changed from their teleomorph to their anamorph names. For example, the holomorph name *Gibberella zeae* (Schwein.) Petch (1936) was changed to *Fusarium graminearum* Schwabe (1839) because the former was a teleomorph name published later than the fungus's anamorph name.

However, the Melbourne Code also established rules that allowed widely used names to be protected en masse and names of uncertain application to be rejected en masse. Conserved and rejected names can be found in the online version of the Shenzhen Code's Appendixes II–IV.[25] Other good sources for searching for names and publication dates for fungi are the online database Index Fungorum[30] and the searchable CABI database based on the *Dictionary of the Fungi.*[45]

22.4.2.2 Sanctioned Names

Fungal taxa are considered "sanctioned" names if they were adopted by mycologist Christiaan Hendrik Persoon in his works on the class *Gasteromycetes* and the orders *Pucciniales* and *Ustilaginales* in the early 1800s and by mycologist Elias Magnus Fries, FRS, in his works a few decades later on other fungi.[46,47] "Sanctioned" names are privileged with special priority, and they are based on taxa with typification status.

The abbreviation "Pers." is used in the names of taxa originally published by Persoon, and "Fr." is used in the names of taxa originally published by Fries.

> *Lycoperdon perlatum* Pers.   *Cortinarius cyanites* Fr.

When Persoon or Fries published names previously applied by others, the names recognize both the originating and sanctioning authors. A colon with a space before and after it separates the surnames or abbreviations of the originating authors from the abbreviation "Pers." or "Fr." (see Section 22.4.2, "Naming Fungi").

22.4.2.3 Infraspecific Names for Fungi

Fungi have several infraspecific ranks, including subspecies, variety, and form (abbreviated as "subsp.," "var.," and "f.," respectively). "Form" is the lowest rank in formal taxonomy that is governed by the Shenzhen Code.[18] These infraspecific taxa are usually recognized based on morphological characteristics.

Another infraspecific rank used with plant-pathogenic fungi is the "special form," or "forma specialis" (abbreviated as "f. sp." for the singular and "ff. sp." for the plural). Instead of being recognized by morphological characteristics, special forms are characterized by physiological criteria, such as host adaptation. The Shenzhen Code does not govern nomenclature for this rank, so Latin diagnoses and author citations are not required.

> *Blumeria graminis* f. sp. *hordei* (the special form epithet *hordei* designates a form of the species
> *B. graminis* that parasitizes barley)

Another category used for fungal pathogens of plants is variously referred to as "race," "physiologic race," "pathogenic race," "pathotype," and "biotype." Because the nomenclature for this category is also not governed by the Shenzhen Code, scientists working with specific groups of fungi establish the rules and methods for race nomenclature, and considerable variation exists among species in different races. Races are usually identified by the disease reactions of "differential" host cultivars, each of which carries a different gene or genes for resistance to the pathogen species after the cultivar is inoculated with the fungus. Isolates of fungi are placed in races according to which differential cultivars are susceptible and which are resistant to the fungi. In contrast, pathotype and biotype names are based on the specific hosts infected.

In some instances, races are named according to the host-resistance genes that they "overcome." The genes are designated by numbers (eg, "race 1" and "race 2" or "*R1*" and "*R2*"). For example, R_1R_2 of *Phytophthora infestans* can infect potato cultivars that have either the *R1* or *R2* gene for resistance but not cultivars that carry other *R* genes. Races for many other pathogens are simply numbered consecutively as they are discovered, starting with *R1*.

In contrast, a special system was developed to deal with the large number of host-resistance genes used to differentiate races of wheat stem rust.[48] This system uses letter codes to designate patterns of reaction on subsets of differentials. Thus, a race of the stem rust fungus *Puccinia graminis* f. sp. *tritici* might be designated TNM, which indicates a pattern of reaction on each of 3 sets of 4 differential cultivars. A similar system exists for leaf rust of wheat.[49]

22.4.2.4 Fossil Fungi

The Shenzhen Code[18] also deals with fossil fungi. The names of fossil fungi often have the suffix "ites," and sometimes, that suffix is added to the names of living fungi.

> *Graphiolites* Fritel (this genus name is based on the living fungus *Graphiola* Poit.)
> *Pleuricellaesporites* v.d. Hammen and *Fungites* Hallier (these 2 names for fossil fungi are not derived from living fungi)

Alternatively, fossil fungi may be given new form-generic names just like living fungi and fossil plants.

> *Grilletia* Renault & Betrand   *Plectosclerotia* Stach & Pickh

22.4.3 Nontaxonomic Groups of "Fungi" and "Plants"

Certain terms are used broadly to refer to growth forms or types of organisms, but they have no taxonomic standing. Except for "lichens," these words should not be used in technical writing. The following sections discuss these terms and their meanings.

22.4.3.1 Yeasts

A yeast is a unicellular fungus that reproduces by budding or fission. Rather than being a taxonomic term, "yeast" refers to these 2 growth forms. Yeasts that produce spores are classified in the divisions *Ascomycotina* and *Basidiomycotina*, while those without spores are placed in the artificial class *Hyphomycetes* (formerly fungi imperfecti or *Deuteromycetes*).[50] The nomenclature and typification of yeasts follow the guidelines for other fungi in the Shenzhen Code.[18]

> *Saccharomyces cerevisiae* Meyen ex E. Hansen var. *ellipsoideus* (Hansen) Dekker is used as baker's or brewer's yeast.

Budding yeasts include such genera as *Candida* and *Saccharomyces* in class *Saccharomycetes* of the division *Ascomycota* and *Cryptococcus* in class *Tremellomycetes* in the division *Basidiomycota*. Fission yeasts include *Schizosaccharomyces* in class *Schizosaccharomycetes* of division *Ascomycota*. Some of these are pathogenic yeasts.

The term "yeast" is sometimes applied to fungi in genera such as *Aspergillus* and *Neurospora* because these fungi have a yeast-like phase with budding cells at some stage in their life cycles. However, these fungi are not unicellular. In addition, "black yeast" is jargon referring to yeast-like states of *Aureobasidium*, *Cladosporium*, *Moniliella*, and other nonyeast fungi. Industrial yeasts, such as those used in distilling, brewing, and wine making, are often not identified by genus or species.

If the word "yeast" is used, identify the organism by its Latin name at least once in the abstract and once early in the text. Although the word "yeast" is a noun, it may be used as an adjective (eg, "yeast cells," "yeast DNA," and "yeast culture") if only one kind of yeast is described in a manuscript.

22.4.3.2 Molds

The term "mold" usually refers to any fungal growth that does not produce a visible fruiting body. However, the term may be used in nontechnical writing for fungi that grow in damp areas of dwellings and other structures. It can also be used to refer to fungi in the class *Hyphomycetes* that produce asexually via conidiation of hyphae (ie, without a sexual fruiting body).

"Mold" should not be used in technical writing to refer to an organism unless it is part of an accepted vernacular name for an organism or group (eg, "slime molds" and "sooty molds"). Originally considered to be fungi,[50] slime molds were reclassified as amoeboid protists *Mycetozoa* (*Myxomycota*).[12]

22.4.3.3 Algae

"Algae" and its singular form, "alga," are nontaxonomic words. An alga is any plant-like organism that carries on photosynthesis and differs structurally from ordinary land plants, such as mosses, ferns, and seed plants. Algae were originally classified as plants, so they were named using the same rules for nomenclatural style and format as plants. For current classifications and names of algae, consult online plant and fungal databases.[4,26-30]

Algae are now regarded as having diverse origins. They have been reclassified in different kingdoms and other ranks in various ways, depending on taxonomic systems. For example, those known as "blue-green algae" are cyanobacteria, and they are classified as prokaryotes. Like the nomenclatures of other organisms that have been reclassified or that have contested classifications, the nomenclature of blue-green algae is governed according to their traditional classification (see also Section 22.1, "Overview of Taxonomy and Nomenclature").

22.4.3.4 Lichens

Lichens are a biological, not a systematic, group. A lichen is a stable, self-supporting association of a fungus and an alga or cyanobacterium. In a lichen, the association is called a "consortium," the fungus is called a "mycobiont," and the alga or cyanobacterium is called a "photobiont."

For nomenclatural purposes, the names given to lichens apply to their fungal components, and lichens are classified mainly in the order *Lecanorales* of the division *Ascomycota*. The algae in lichen consortia are usually green algae or occasionally blue-green algae in the division *Cyanobacteria*. The algal components of lichens have separate names.

> *Cladonia cristatella* Tuck. is a lichen in the order *Lecanorales* of the kingdom *Fungi*. Its algal component is *Trebouxia* sp., a green alga.
> The algal component of the lichen *Peltula polysora* (Tuck.) Wet. is the cyanobacterium *Anacystis montana* (Lightf.) Dr. et Daily.

For information on lichen biology, keys to identification, and descriptions of species, see *Lichens of North America*.[51]

22.5 NOMENCLATURE AND CONVENTIONS FOR ANIMALS

22.5.1 Naming Animal Taxa

The rules and recommendations for naming a taxonomic group of animals, including fossil animals, are found in the *International Code of Zoological Nomenclature* ("Zoological Code")[14] and its amendments. The criteria for valid publication of scientific names of animals are set out in Articles 7–9 in Chapter 3 of the Zoological Code. The code has several articles dealing with fossils. The key concepts and essential features of the Zoological Code, nomenclature, and taxonomy are explained in an article on best practices for editors of technical journals in the 1 December 2011 issue of the *Bulletin of Zoological Nomenclature*.[52] Current developments are reported in *Systematic Biology*.[53] Applications to the International Commission on Zoological Nomenclature for approval of scientific names and comments on those applications are published in the *Bulletin of Zoological Nomenclature*.[54] The commission's official decisions regarding names and works (ie, publications) are also published in the *Bulletin*. In addition, the commission produces official lists of approved names and official indexes of rejected names and works.[55,56]

Because the Zoological Code is independent of codes for organisms in other kingdoms, a valid name for an animal can be the same as a valid genus in another kingdom. For example, *Sida* is the valid name of both a genus of the crustacean waterflea and a genus of plants. However, the Zoological Code recommends that before authors propose new genus names for animals, they check lists of approved names for plants and bacteria to avoid using names for animals that are included in those lists.

22.5.2 Authors of Animal Taxa

Zoological nomenclature dates to the mid-1700s with the publication of *Systema Naturae*[57] by taxonomist and physician Carl Linnaeus and *Svenska Spindlar*[58] by arachnologist Carl Alexander Clerck. Articles 50 and 51 in Chapter 11 of the Zoological Code[14] cover the rules and recommendations for authorship of animal names.

Although the author of a genus or species is not part of an animal's scientific name, including the author's name is optional. Nevertheless, Appendix B of the Zoological Code recommends that when a genus name or taxon of lower rank is mentioned in a document, the taxon's name should include the surname of the author and the year of naming. The surname and year are often separated by a comma and space (eg, *Aphis gossypii* Clover, 1877), but the comma is not prescribed in the code. The surname and date should be in roman type. The surname should not be abbreviated, but when a taxon has more than 3 authors, "et al" can be used after the first surname (see also Section 22.5.4, "Abbreviation of Author Names").

The author's surname and the year need appear only once in an article, preferably at first mention of the taxon in the text. The surname and year should not appear in either the article's title or abstract. In a taxonomic work, a full reference to the publication in which a taxon's name was published should be included in the manuscript's list of references.

> In-text reference in the citation–sequence or citation–name format:
> *Taenia diminuta* Rudolphi, 1819[353]

> In-text reference in the name–year format:
> *Taenia diminuta* Rudolphi, 1819 (Rudolphi 1819)

22.5.3 Transfer of a Species to Another Genus

When an animal species is transferred from its originally designated genus to another on the basis of new information or a new interpretation of its characteristics, the surname of the original author is placed within parentheses. Unlike the practice for plants (see Section 22.4.1.1.1, "Transfer of a Species to Another Genus") and the practice for bacteria (see Section 22.6.2, "Authors of Prokaryotic Taxa"), the surname of the author of the revised animal name is not included in the new combination, according to Article 51.3 in the Zoological Code.[14]

> *Taenia dunubyta* Rudolphi was reclassified as *Hymenolepis diminuta* (Rudolphi).

If the new genus name for the organism differs in gender from that of the previous genus name, the ending of the specific name changes to match the gender of the new genus name.

> *Taeniothrips albus* (masculine) was reclassified as *Frankliniella alba* (feminine).

22.5.4 Abbreviation of Author Names

Appendix B of the Zoological Code[14] recommends that surnames of authors of scientific names be spelled out in full. Although the surname "Linnaeus" is commonly abbreviated as either "L." or "Linn.," this practice is inconsistent with the code's recommendation.

When 3 or more authors publish a scientific name, all authors must be cited in full on first mention in the text or in the references. The surname of the first author may then be followed by "et al" (see Recommendation 51C in the Zoological Code[14]).

For a species name or other taxon that was published anonymously, the abbreviation "Anon." is used in place of an author's surname (see Recommendation 51D in the Zoological Code[14]). However, if the author of an anonymously published taxon is known or inferred from reliable information, the author's surname is given in square brackets after the taxon. For this reason, parentheses or square brackets around author surnames should not be converted to other fences, as opposed to the common practice when parenthetical statements are nested within other parenthetical statements.

22.5.5 Diacritical Signs and Other Punctuation Marks

The Zoological Code[14] does not permit diacritical symbols, apostrophes, or hyphens (see Section 22.2.2.6, "Diacritical Marks"). An exception to that rule is a hyphen used to set off an individual letter in a compound species-group name.

> The *Polygonia c-album* Linnaeus, 1758 is a butterfly with a white mark on its wings that resembles the letter "c."

Article 32.5.2 of the Zoological Code[14] states that a name published with a diacritical symbol, ligature, apostrophe, or hyphen is to be corrected.

> *nuñezi* corrected to *nunezi* as in *Haplocochlias nunezi* Espinosa, Ortea, and Fernandez-Garces, 2004
> *mjøbergi* corrected to *mjobergi* as in *Leptobrachella mjobergi* Smith, 1925

For names published before 1985 that contain a vowel with an umlaut, this diacritical mark should be omitted from the vowel, and the letter "e" should be inserted after it.

> *mülleri* corrected to *muelleri* as in *Phalacrognathus muelleri* Macleay, 1885

22.5.6 Infraspecific Taxa

The Zoological Code[14] regulates subspecies names but not infrasubspecific names (eg, "variety," "form," "aberration," and "morph" names). However, an infrasubspecific name may be regarded as a subspecific name if the name was first published before 1961 and if the author was ambiguous as to the animal's designation as an infrasubspecific name (see Articles 45.6.3 and 45.6.4 in Chapter 10 of the Zoological Code[14]).

Infraspecific terms such as "subspecies," "variety," "form," "aberration," and "morph" do not need to be included in the trinomial names for animals, contrary to the practice for plants (see Section 22.4.1.2, "Infrageneric and Infraspecific Taxa") and bacteria. If a full trinomen is mentioned, both its genus and species names can be abbreviated on subsequent use to their initial letters followed by periods and spaces.

> *Panthera tigris altaica* is the scientific name for the Siberian tiger. *P. t. altaica* is the most power-ful subspecies of tigers.
> The Sumatran tiger's scientific name is *Panthera tigris sumatrae*. The smallest subspecies of tiger, *P. t. sumatrae* exists only on the Indonesian island of Sumatra.

If a subspecies contains the species type, it has the same specific name, author, and date as the species. This is known as a "nominotypical subspecies" (see Article 47.1 in Chapter 10 of the Zoological Code[14]).

> *Tamias amoenus amoenus* Allen, 1890

22.5.7 Vernacular Names for Animals

Lists exist of approved vernacular names for species in different phyla in North America[59-66] and for European and world herpetofauna.[67,68] In addition, recognized authorities have published illustrated checklists of mammals and birds of the world with both scientific and common names.[69,70] However, the common names are not necessarily approved by authoritative bodies.

The format for common names varies among authorities. The American Fisheries Society capitalizes common names of species and subspecies.[64] Similarly, the American Ornithological Union recommends capitalizing the first letter of all words in bird names (eg, "Brown Shrike" for *Lanius cristatus* and "White-collared Manikan" for *Manacus candei*). In contrast, only proper nouns are capitalized in the approved lists of common names for insects on the Entomological Society of America's website.[65] Common names of other animals are usually not capitalized.

Because of outstanding disagreements on approved names, especially for birds, authors should consult the guidelines of the publishers to which they intend to submit their work and follow those specifications. (See also Section 22.2.3, "Vernacular Names.")

In addition to differing on capitalization styles, recommendations may differ on how to use 2-part common names. For 2-part common names of insects, for example, the second part is used as a separate word when it is a systematically correct name but is combined with the preceding modifier when it is not.

> bed bug (a true bug)   house fly (a true fly)   butterfly (not a true fly)

In contrast, in the common names for insect larvae, the term "worm" is always combined with the preceding modifier because larvae are not true worms (ie, annelids).

> silkworm *not* silk worm   striped cutworm *not* striped cut worm

22.5.8 Laboratory Strains of Animals

The Committee on Standardized Genetic Nomenclature for Mice[71] (CSGNM) established nomenclatural rules and guidelines for inbred strains of mice, as well as a symbolic system for designating these strains. In addition, the Mouse Genome Database follows the nomenclatural guidelines of the International Committee on Standardized Genetic Nomenclature for Mice and the Rat Genome and Nomenclature Committee.[72] The Mouse Genome Database details codes for inbred strains and sublines and designates recombinant inbred, coisogenic, congenic, and segregating strains of mice and rats. The nomenclature for outbred laboratory animals is based on a separate set of guidelines published in 1972.[73] Guidelines for nomenclature of transgenic animals can be found online in the *ILAR Journal*.[74]

22.6 NOMENCLATURE AND CONVENTIONS FOR PROKARYOTES

The *International Code of Nomenclature of Prokaryotes*[15] ("Prokaryotic Code") governs the naming of prokaryotes, which include eubacteria (ie, "true" bacteria) and archaea (also called "archaebacteria"). Archaea are morphologically similar but genetically distinct from eubacteria.

In the 7-kingdom system, bacteria are placed in kingdom *Bacteria Cavalier and Smith, 2002*, and archaea are placed in the kingdom *Archaea*. In the 5-kingdom system,[6] bacteria and archaea are placed in the kingdom *Bacteria*, which was formerly referred to as the kingdom *Monera* or *Prokaryotae*. In contrast, in the 3-domain system, bacteria constitute the *Bacteria* domain, also called the *Eubacteria* domain, and archaea constitute the *Archaea* domain.[7] (For more on these systems, see Section 22.2.1, "Kingdoms and Other High Taxonomic Ranks").

The singular for "bacteria" is "bacterium," and the singular for "archaea" is "archaeon." Among the better-known bacteria are rickettsiae (singular, "rickettsia"), chlamydiae (singular, "chlamydia"), mycoplasmas (singular, "mycoplasma"), cyanobacteria (singular, "cyanobacterium"), and actinomycetes (singular, "actinomycete"). Examples of archaea are methanogens, halophiles, thermophiles, and psychrophiles (the singular of which are all formed by removing the letter "s").

22.6.1 Naming Prokaryotic Taxa

The taxonomic ranks for prokaryotes are similar to those of plants. The Prokaryotic Code deals only with taxonomic ranks of class through subspecies. Taxa from the rank of order to the rank of subtribe are formed by adding suffixes to the names of the type genera. The names of all taxonomic ranks for prokaryotes are italicized, as is the case with plants but not animals (see Section 22.2, "Taxonomy and Nomenclature: Commonalities across Hierarchical Systems").

Under the provisions of the Prokaryotic Code, taxa must be both effectively and validly published to be recognized. It is the names, not the organisms to which they refer, that are validly published.

To be effectively published, the description of a taxon must appear in English in a printed publication that is distributed to the public or to bacteriologic institutions and that is recognized as a permanent record. Valid publication, in turn, requires that the name be published in the *International Journal of Systematic and Evolutionary Microbiology* (*IJSEM*) or its predecessor, the *International Journal of Systematic Bacteriology* (*IJSB*). In addition, the article in *IJSEM* or *IJSB* in which the name is published must either include a description of the organism or reference a description "effectively" published elsewhere.

Validly published scientific names appear in the *List of Prokaryotic Names with Standing in Nomenclature*.[24] This list is updated after the publication of each issue of *IJSEM*.

Names published before 1980 are valid only if they were included in the *Approved Lists of Bacterial Names (Amended)*[75] and the *Index of the Bacterial and Yeast Nomenclatural Changes*.[76] It is the *Approved Lists* that is approved, not the names. Thus, there can be more than one valid name for the same entity, all of which appear on the *Approved Lists*. Names often carry a superscript "AL," meaning that the names are on the *Approved Lists*. The *Approved Lists*[75] also provides nomenclatural changes for bacteria and yeast from 1 January 1980 to 1 January 1992.

> *Methanococcus vannielii* Stadtman and Barker 1951[AL] is the type species for the genus.

A name that was effectively published before the publication of the *Approved Lists* but was not included in the lists should be referred to as a "nomen revictum," or "revived name," if the name is subsequently validly published. In such a case, the name ends with the abbreviation "nom. rev." after the designation for the taxon's rank (eg, "sp. nov." and "gen. nov."). In addition, the authors of the original effective publication can be acknowledged, but the format differs from that used in zoology and botany. For prokaryotic names, the surnames or abbreviations of the original authors are placed in parentheses before the surnames or abbreviations of the validating authors, and "*ex*" in italics precedes the original authors' surnames.

> *Pseudomonas cannabina* (*ex* Sutic and Dowson 1959) Gardan et al 1999 sp. nov., nom. rev.

One useful online reference for bacterial taxonomy and classification is *Bergey's Manual of Systematics of Archaea and Bacteria*[77] (previously *Bergey's Manual of Systematic Bacteriology*[78–81]). However, because *Bergey's* has no standing in nomenclature, the manual cannot be used to validly publish names, and not all the names in the manual are validly published elsewhere. A related resource, *Bergey's Manual of Determinative*

Bacteriology,[82] is useful for identifying unknown prokaryotes that have been cultured and described.

Names that have not been validly published should be enclosed in single roman quotation marks. This format even applies to names that have been effectively but not validly published and to names that are coined, such as those deposited in a sequence database.

> '*Actinomadura spinosa*' Saitoh et al

In a case in which the original authors misclassified an organism, part or all of the taxon may be enclosed in square brackets, such as "[*Cytophaga*] *latercula*." The name *Cytophaga latercula* was included in the *Approved Lists* so it is validly published, but the species has been recognized for some time as not belonging to the genus *Cytophaga* and has been reclassified as *Stanierella latercula*.

Since 1999, orthographic and grammatical errors in the names of prokaryotes cannot be corrected. The only exception is that adjectival epithets that are in the wrong gender must be corrected if a species is reclassified to a new genus. However, the gender of the epithet remains uncorrected in the original genus name.

22.6.2 Authors of Prokaryotic Taxa

In referring to a prokaryotic species, the surname of the author should be placed after the specific epithet along with the date of valid publication of the taxon, not the year of effective publication if the 2 dates differ. Unlike the recommendation for animal taxa, no punctuation is used between the specific epithet and the author's surname. Another difference from the nomenclature for animal taxa is that when a prokaryotic taxon has 2 authors, the surnames are separated by the conjunction "and" instead of "et" or an ampersand.

> *Erwinia quercina* Hildebrand and Schroth 1967

If a taxon has more than 2 authors, all of the surnames can be listed, or "et al" can be used after the first surname.

> *Legionella pittsburghensis* Pasculle, Feeley, Gibson, Cordes, Myerowitz, Patton, Gorman, Carmack, Ezzel, and Dowling 1980
>
> *or*
>
> *Legionella pittsburghensis* Pasculle et al 1980

If a species has been reclassified using the same specific epithet, the original author's surname appears within parentheses after the specific epithet, followed by the surname of the person who reclassified it.

> *Enterobacter agglomerans* Ewing and Fife 1972 was reclassified 17 years later, at which time the name changed to *Pantoea agglomerans* (Ewing and Fife 1972) Gavini et al 1989.

If the same species is transferred again, the author of the basonym (ie, the taxon's first validly published name) should be given in parentheses, followed by the surname of the latest person to reclassify the taxon.

> *Hydrogenomonas facilis* Schatz and Bovell 1952 was subsequently reclassified as *Pseudomonas facilis* (Schatz and Bovell 1952) Davis 1969 and again as *Acidovorax facilis* (Schatz and Bovell 1952) Willems et al 1990, *not as Acidovorax facilis* (Davis 1969) Willems et al 1990

22.6.3 Vernacular Names for Prokaryotes

Vernacular names for prokaryotes are always set in lowercase Roman letters. The plurals for some of these vernacular names take Latin form, while others use English form.

bacillus, bacilli	pseudomonad, pseudomonads
brucella, brucellae	vibrio, vibrios

Avoid using "bacillus" or "bacilli" if doing so would create confusion as to whether the genus *Bacillus* or a nonspecific, rod-shaped bacterium is meant. In addition, when italicized and capitalized, the word *Bacilli* is never correct.

Organisms in the genus *Treponema* can be designated as singular by using "treponema" and plural by using "treponemas" or "treponemata." The following are some additional examples of the singular and plural forms of vernacular names.

citrobacter, citrobacters	mycobacterium, mycobacteria
klebsiella, klebsiellae	streptomycete, streptomycetes

Additional vernacular names and their plural forms can be found in medical dictionaries and some general dictionaries.

Although international rules do not cover adjectives derived from genera, some guidelines are widely followed. Adjectival forms usually end in "al," but noun forms can also serve as adjectives for prokaryotes. When a formal genus name is used as a vernacular adjective, it should be in lowercase Roman letters.

corynebacterial multiplication	streptococcal infections
corynebacterium multiplication	streptococcus infection

A formal genus name should be used as a modifier only if the name is meant to refer to all species of the genus. In contrast, a binomial can be used as a modifier[83] when it only refers to that species.

> The patient's symptoms were due to *Pseudomonas aeruginosa* bacteremia. (*not Pseudomonas* bacteremia)
> The researchers studied all *Pseudomonas* species.
> A nonproductive cough is one symptom of *Pneumocystis carinii* pneumonia. (*not Pneumocystis* pneumonia)

Adjectival forms must not be derived from specific epithets alone. For example, "coli" from *Escherichia coli* and "amylovora" from *Erwinia amylovora* should not be used as adjectives.

22.6.4 Nomenclatural Types

Although the taxon for a family, genus, or species of plant or animal remains permanently attached to a preserved specimen, the taxon of a prokaryote remains permanently attached to a living specimen descended from the type strain of the taxon[15] (see Section 22.2.2.8, "Nomenclatural Types"). The type strain of a prokaryote is designated by the author in the original, validly published description of the species or by a subsequent author in a validly published description that reclassifies the species. As of 2001, a description must include both the designation of the type strain and the designations for viable cultures of the type strain.[15] The author must deposit the cultures in at least 2 public culture collections in different countries.[15] From these deposits, subcultures can be requested for additional research. For publication of a new or reclassified species in the *International Journal of Systematic and Evolutionary Microbiology*, the species' type strain must be designated with a superscript "T" at each occurrence in the text, tables, and figures.

> We obtained strain NRS 966 of *Bacillus thuringensis* (ATCC 10792^T) from . . .

The type strain serves as the nomenclatural type for a prokaryotic species or subspecies. For the genus or subgenus of a prokaryote, however, the nomenclatural type is the type species that was identified when the species name was originally validly published. The nomenclatural type for an order, suborder, family, subfamily, tribe, or subtribe is the nomenclatural type of the genus on which the names of the order through subtribe are based. For ranks higher than order, the nomenclatural type is the type of one of the orders contained in those higher ranks. The rules for types are explained in Section 4 of the Prokaryotic Code.[15]

> *Xanthomonas fragariae* Kennedy and King 1962, type strain NCPPB 1469

Until 31 December 2000, if a species or subspecies could not be maintained in culture or if its type did not exist, an original description, a nonviable preserved specimen, or an illustration could serve as the type. Starting in 2001, the original published description of a taxon had to include a viable specimen. An illustration or image is no longer sufficient.

If an organism is a symbiont, the type culture can be maintained as a defined co-culture with another named strain.

> The type of *Syntrophobacter wolinii* Boone and Bryant 1984 is strain DB, which is maintained in the DSMZ culture collection as a co-culture with *Desulfovibrio* sp. strain G-11 as DSM 2805 and as strain DSM 2805M, a monoculture isolated from the co-culture DSM 2805.

Ideally, the strain designation should be listed with the full binomial.

> *Xanthomonas campestris* (Pammel) Dowson pathovar *campestris* NCPPB 528

If the specific epithet is missing, insert the word "strain."

> *Pseudomonas* strain B13 *or Pseudomonas* sp. strain B13

A subspecies that contains the type specimen of the species must bear the same epithet as the species, according to the Prokaryotic Code.[15] If the first subspecies of a species does not contain the type specimen, a subspecies is automatically created that contains the type specimen, and the epithet of that subspecies becomes the same as the epithet of the species. In addition, the author of the species name, not the author of the subspecies name, is cited as the author of the automatically created subspecies name.

> The publication of *Bacillus subtilis* subsp. *viscosus* Chester 1904 automatically created a new subspecies, *Bacillus subtilis* subsp. *subtilis* (Ehrenberg 1835) Cohn 1872.

22.6.5 Organisms That Cannot Be Cultured

If a prokaryotic organism cannot be isolated in culture, its name can no longer be validly published. However, the Judicial Commission of the International Committee on Systematics of Prokaryotes[84,85] recommends giving "*Candidatus*" status to organisms that are well characterized but that cannot be maintained in culture. Appendix 11 of the Prokaryotic Code[15] describes the *Candidatus* concept and the information to be included in the description of the prokaryotic organism. A taxon with *Candidatus* status is not considered to be validly published and thus is not included in the *Approved Lists of Bacterial Names (Amended)*.[75] In addition, the name is not considered to be a "new" taxon, and it should not be accompanied by designations such as "sp. nov." Instead, on first reference, the word "*Candidatus*" should be placed before the name, and this combined designation should be placed within quotation marks. "*Candidatus*" should be in italics, and the taxon should be in roman. On subsequent uses of the taxon, the word *Candidatus* can be abbreviated to "*Ca.*," and the genus name can be abbreviated to its capitalized initial letter.

> **First mention:** "*Candidatus* Arsenophonus triatominarum" Hypsa and Dale 1997
> **Subsequent mentions:** "*Ca.* A. triatominarum"

22.6.6 Infraspecific Taxa

Some prokaryotes that cannot be differentiated taxonomically at the level of subspecies are given infraspecific designations that are not governed by the Prokaryotic Code[15] and are, therefore, excluded from the *Approved Lists*[75] (see Section 22.6.1, "Naming Prokaryotic Taxa"). These infraspecific designations include "biovar" (bv.), or "biotype"; "chemovar," also known as "chemotype" and "chemoform"; "cultivar"; "*forma specialis*" (f.sp.), or "special form"; "morphovar," or "morphotype"; "pathovar" (pv.), or "pathotype"; "state"; "phagovar," or "phagotype"; "phase"; and "serovar" (sv.), or serotype (see Appendix 10 B of the Prokaryotic Code[15]). To avoid confusion with nomenclatural types, the designations with the suffix "var" are preferred to those with the suffix "type." In addition, the designation "variety" (var.) is not used for prokaryotes.

The term "pathovar" is derived from "pathovarietas." Pathovars are distinguished primarily but not solely by their pathogenic characteristics. The pathovar designation is written in italics, but the abbreviation "pv." is written in roman.

> *Corynebacterium michiganense* (Smith) Jensen 1977 pv. *tritici* causes spike blight in wheat.

Because the Prokaryotic Code does not govern the nomenclature of pathovars, a committee of the International Society for Plant Pathology developed a set of international standards for naming pathovars of phytopathogenic bacteria and compiled lists of pathovars and their representative pathotype cultures.[86–89]

These international standards recommend that a new pathovar include the designation "pv. nov.," which stands for "pathovarietas nova." In addition, the pathovar should bear the surname of the author of the publication in which the pathovar epithet was formally proposed, followed by the date of publication and the page of the publication on which the proposed name can be found. No punctuation should be used between the author's surname and the date of publication, but a comma should be placed between the publication date and the page number. The page number should also be included in the full citation of the publication. The page number should be the location in the text where the name was proposed, not the location in the summary or abstract.

> *Xanthomonas campestris* (Pammel 1895) Dowson 1939 pv. *cannabina* Severin 1978, 13

Although the names of some serovars have been validly published as specific epithets (eg, "*Salmonella typhi*" and "*Salmonella paratyphi*"), most serovar names should not be used as if they are species names in Latin binomials. In general, the serovar name is placed in roman after the subspecies name. After the first use of the full name, the name can be shortened to the genus followed by the serovar in roman.

> **First mention:** *Salmonella enterica* subsp. *enterica* serovar Dublin
> **Subsequent mentions:** *Salmonella* Dublin *not Salmonella dublin and not S.* Dublin

More information on serovar designations and *Salmonella* nomenclature is available online.[90–92]

In addition to using latinized words to designate infrasubspecific taxa, authors can use vernacular names, numbers, letters, and formulae (see Appendix 10 C of the Prokaryotic Code[15]).

22.7 TAXONOMY, NOMENCLATURE, AND CONVENTIONS FOR VIRUSES AND VIRUS-LIKE MOBILE GENETIC ELEMENTS

Viruses and virus-like mobile genetic elements that are considered part of the virosphere are neither prokaryotes nor eukaryotes. Instead, they are a type of mobile genetic element. Technically called "viruses *sensu stricto*," viruses are nucleic acid molecules that can enter cells, replicate in them, and encode at least one protein that is a major component of the outer protein shell of the virus particle (ie, virion). Because viruses are not cellular, do not grow, can only organize their replication, and function only within living cells, they are not living organisms.[93] Viruses appear to be partial or degenerate forms of living systems, but they are not placed in any classification system of living things.

Virus-like mobile genetic elements include viroids, satellite nucleic acids, and viriforms. When these elements can be demonstrated to be members of an evolutionary lineage from a shared virion–protein-encoding ancestor, they are classified into a group of viruses.[19] Section 3.3 of the *ICTV Code: The International Code of Virus Classification and Nomenclature*[19] (ICVCN), also known as the "Virology Code," provides definitions of the various types of mobile genetic elements.

22.7.1 Classification of Viruses

The International Committee on Taxonomy of Viruses (ICTV) is responsible for classifying members of the virosphere. Virosphere members are mobile and selfish genetic elements. Because they are replicons subject to selective pressures and because they are largely independent of other replicons, these genetic elements have distinct evolutionary histories. However, they need cellular hosts for energy and chemical building blocks.

The ICTV[19] governs the taxonomy and nomenclature of viruses *sensu stricto* and the remaining replicator space of the virosphere. As of 2020, the ICTV[19] allows viruses and virus-like mobile genetic elements to be classified into 15 taxonomic ranks. Although 15 ranks are possible, the ICTV does not require a virus species to be described above the rank of genus. In addition, the ICTV does not deal with taxa below species. Instead, specialist groups deal with serotypes, strains, and other taxa below species.

See Table 22.1[95] for the taxonomic ranks of viruses and virus-like mobile genetic

Table 22.1. Suffixes for the 15 Taxonomic Ranks of Viruses, Satellite Nucleic Acids, Viroids, and Viriforms[a]

Rank	Virus suffix	Example[b]	Satellite suffix	Example[c]
realm	*viria*	*Duplodnaviria*	*satellitia*	...
subrealm	*vira*	...	*satellita*	...
kingdom	*virae*	*Heunggongvirae*	*satellitae*	...
subkingdom	*virites*	...	*satellitites*	...
phylum	*viricota*	*Uroviricota*	*satelliticota*	...
subphylum	*viricotina*	*Haploviricotina*	*satelliticotina*	...
class	*viricetes*	*Caudoviricetes*	*satelliticetes*	...
subclass	*viricetidae*	...	*satelliticetidae*	...
order	*virales*	*Kirjokansivirales*	*satellitales*	...
suborder	*virineae*	*Abnidovirineae*	*satellitineae*	...
family	*viridae*	*Crevaviridae*	*satellitidae*	*Alphasatellitidae*
subfamily	*virinae*	*Doltivirinae*	*satellitinae*	*Petromoalphasatellitinae*
genus	*virus*	*Bertelyvirus*	*satellite*	*Kobbarisatellite*
subgenus	*virus*	*Aplyccavirus*	*satellite*	...
species[e]	(varies)	*Bertelyvirus BL9*	(varies)	*Capsicum India alphasatellite*

Rank	Viroid suffix	Example[d]	Viriform suffix	Example[d]
family	*viroidae*	*Avsunviroidae*	*viriformidae*	*Polydnaviriformidae*
subfamily	*viroiditinae*	...	*viriformitinae*	...
genus	*viroid*	*Avsunviroid*	*viriform*	*Bracoviriform*
subgenus	*viroid*	...	*viriform*	...
species[e]	*viroid* as separate word	*Avocado sunblotch viroid*	(varies) *Bracoviriform curvimaculati*	*Bracoviriform curvimaculati*

[a] Based on the International Committee on the Taxonomy of Viruses' Virus Taxonomy: 2021 Release.[95]

[b] As of 2021, the International Committee on the Taxonomy of Viruses has not approved names for virus subrealms, subkingdoms, and subclasses.

[c] As of 2021, the International Committee on the Taxonomy of Viruses has not approved names in the ranks above family or in the rank of subgenus for satellites.

[d] As of 2021, the International Committee on the Taxonomy of Viruses has not approved names in the ranks above family or in the ranks of subfamily and subgenus for viroids and viriforms.

[e] Unlike scientific binomials, the species names of viruses, satellite nucleic acids, viroids, and viriforms do not include genus names.

elements, the formal suffix endings of taxa in those ranks, and examples of how those endings are used. Although the ICTV has established suffixes for most taxonomic ranks, the ICTV has not yet classified viruses and virus-like elements to all of those ranks.

22.7.2 Naming Viral Taxa

Virologists are only required to assign a new virus to a genus and species.[94] Other ranks are optional and should be assigned only if scientifically justified.[94] To be valid, scientific names of virus species have to be approved by the ICTV and published in either the ninth print edition of *Virus Taxonomy: The ICTV Report on Virus Classification and Taxon Nomenclature*[96] or ongoing online updates to the ICTV's report.[94] The ICTV maintains an authoritative classification database[95,97] on virus taxonomy.

The orthography of virus names differs from that used for organisms. Although virus species do not have Latin names, virus names that have been approved by the ICTV are written in italics when used in a formal taxonomic sense. In this context, the first word of the species name is capitalized. Capitalize other parts of a virus name only if they are proper nouns.

> *Avian leukosis virus*   *Sandfly fever Naples virus*
> *Maize dwarf mosaic virus*   *Yellow fever virus*
> *Murray River encephalitis virus*

However, when writing about a physical viral entity (ie, the actual viral particles as opposed to a virus's formal taxonomic sense), the species name is written in roman type with all lowercase letters.

> The virus causing the symptoms is cucumber mosaic virus.

Similarly, tentative virus names are written in roman type until formally approved by the ICTV.

Although virus species are placed in genera, genus names are generally not part of the species names. However, a genus name may be an acronym of a species name, termed a "sigla" in the Virology Code[19] (see Section 22.7.4, "Abbreviated Virus Designations").

> *Alfalfa mosaic virus* is in the genus *Alfamovirus*.
> *Tobacco mosaic virus* is in the genus *Tobamovirus*.

The formal name of a viral taxon should be preceded by its actual rank.

> the family *Caulimoviridae*   the species *Cryphonectria hypovirus virus 1*
> the genus *Betavirus*

This system of orthography is irregularly applied by virologists, and adoption varies from discipline to discipline. Consequently, authors should ascertain the practices followed by the publications to which they intend to submit manuscripts.

22.7.3 Vernacular Names for Viruses

As mentioned in Section 22.7.2, "Naming Viral Taxa," the species names of physical viral entities are written as vernacular names in roman script and are not capitalized unless English grammar requires it.

> RNA of barley stripe mosaic virus   Fraser Point virus virions

Genus names used in a vernacular sense are also written in roman script and not capitalized.

> potexvirus   rhabdovirus   rhinovirus

22.7.4 Abbreviated Viral Designations

Most viruses are designated by their siglas,[19] or acronyms, after their full names are first mentioned in scientific manuscripts.

First mention	Subsequent mentions
Porcine adenovirus B (PAdV-B)	PAdV-B
Lymphocystis disease virus 2 (LCDV-2)	LCDV-2
Tobacco mosaic virus (TMV)	TMV

The Virology Code makes a distinction between "sigla" and "acronyms," a distinction that is not recognized by standard dictionaries, such as *Webster's New World Dictionary of*

American English. In the Virology Code, "sigla" is defined as "names comprising letters and/or letter combinations taken from words in a compound term."[19]

> The genus *Comovirus* has the sigla stem "Co" taken from "cowpea" and "mo" taken from "mosaic."
>
> The genus *Reovirus* has the sigla stem "R" from "respiratory," "e" from "enteric," and "o" from "orphan."

22.7.5 Strain Designations

As mentioned in Section 22.7.1, "Classification of Viruses," the ICTV regulates virus names only down to the level of species and leaves the classification and naming of serotypes, genotypes, strains, variants, and isolates to acknowledged international specialist working groups.[19]

Details of strain designations may be placed within the virus names.

> influenza virus A/WS/33 *or* influenza A/WS/33 virus

In some cases, more detailed designations may be needed.

> A/Equine/Prague/1/56 (H7N7)
> "A" is the virus type. "Equine" is the host of origin (for animal influenza isolates). "Prague" is the geographic origin of the name. "1" is the strain number. "56" stands for the year the strain was isolated, 1956. "(H7N7)" is the antigenic description.

Alternately, a strain name may follow the virus name or its sigla, with a hyphen between the virus name and sigla.

First mention	Subsequent mentions
Barley yellow dwarf virus-PAV	BYDV-PAV
Barley yellow dwarf virus-MAV	BYDV-MAV

22.7.6 Databases of Viral Names and Strain Designations

The ICTV maintains an updated, downloadable list of validly published virus names and valid synonyms.[19,95] The list has more than 10,000 species names.

CITED REFERENCES

1. Simpson MG. Plant systematics. 3rd ed. Academic Press; 2019.

2. Winston JE. Twenty-first century biological nomenclature—the enduring power of names. Integr Comp Biol. 2018;58(6):1122–1131. https://doi.org/10.1093/icb/icy060

3. Adl SM et al. The new higher level classification of eukaryotes with emphasis on the taxonomy of protists. J Eukaryot Microbiol. 2005;52(5):399–451. https://doi.org/10.1111/j.1550-7408.2005.00053.x

4. Bánki O et al. Catalogue of life checklist. Catalogue of Life; 2021. https://doi.org/10.48580/d4sb

5. ITIS. The Integrated Taxonomic Information System; 2021. https://itis.gov

6. Margulis L, Chapman MJ. Kingdoms and domains: an illustrated guide to the phyla of life on Earth. 4th ed. Academic Press; 2009.

7. Woese CR, Kandler O, Wheelis ML. Towards a natural system of organisms: proposal for the domains Archaea, Bacteria, and Eucarya. Proc Natl Acad Sci USA. 1990;87(12):4576–4579. https://doi.org/10.1073/pnas.87.12.4576

8. Gupta RS. Life's third domain (Archaea): an established fact or an endangered paradigm?: a new proposal for classification of organisms based on protein sequences and cell structure. Theor Pop Biol. 1998;54(2):91–104. https://doi.org/10.1006/tpbi.1998.1376

9. Mayr E. Two empires or three? Proc Natl Acad Sci USA. 1998;95(17):9720–9723. https://doi.org/10.1073/pnas.95.17.9720

10. Cavalier-Smith T. The neomuran origin of archaebacteria, the negibacterial root of the universal tree and bacterial megaclassification. Int J Syst Evol Microbiol. 2002;52(1):7–76. https://doi.org/10.1099/00207713-52-1-7

11. Spang A. Complex archaea that bridge the gap between prokaryotes and eukaryotes. Nature. 2015;521(7551):173–179. https://doi.org/10.1038/nature14447

12. Burki F et al. The new tree of eukaryotes. Trends Ecol Evol. 2020;35(1):43–55. https://doi.org/10.1016/j.tree.2019.08.008

13. Winston JE. Describing species. Columbia University Press; 1999.

14. International Commission on Zoological Nomenclature. International code of zoological nomenclature. 4th ed. The Natural History Museum, International Trust for Zoological Nomenclature; 1999. https://www.iczn.org/the-code/the-code-online/

15. Parker CT, Tindall BJ, Garrity GM. International code of nomenclature of prokaryotes: prokaryotic code (2008 revision). Int J Syst Evol Microbiol. 69(1A):S1–S111. https://doi.org/10.1099/ijsem.0.000778

16. Oren A, Ventosa A. Subcommittee on the Taxonomy of Halobacteriaceae and Subcommittee on the Taxonomy of Halomonadaceae: minutes of the joint open meeting, 2013 June 24, Storrs, Connecticut, USA. Int J Syst Evol Microbiol. 2013:63(Pt 9):3540–3544. https://doi.org/10.1099/ijs.0.055988-0

17. Imhoff JF, Madigan MT. International Committee on Systematics of Prokaryotes Subcommittee on the Taxonomy of Phototrophic Bacteria: minutes of the meetings, 27 August 2003, Tokyo, Japan. Int J Syst Evol Microbiol. 54(3):1001–1003. Erratum Int J Syst Evol Microbiol. 54(5):1907.

18. Turland NJ et al, editors. International code of nomenclature for algae, fungi, and plants (Shenzhen Code) adopted by the Nineteenth International Botanical Congress Shenzhen, China, July 2017. Regnum Vegetabile 159. Koeltz Botanical Books; 2018. https://doi.org/10.12705/Code.2018

19. International Committee on Taxonomy of Viruses. The international code of virus classification and nomenclature (ICVCN), March 2021. Available at https://talk.ictvonline.org/information/w/ictv-information/383/ictv-code

20. Manara B. Some guidelines on the use of gender in generic names and species epithets. Taxon. 1991;40(2):301–308. https://doi.org/10.2307/1222983

21. Nybakken OE. Greek and Latin in scientific terminology. Iowa State University Press; 1959.

22. Stearn WT. Botanical Latin: history, grammar, syntax, terminology and vocabulary. 4th ed. Timber Press Inc; 1995. Available at https://archive.org/details/botanicallatin00will

23. Gotch AF. Latin names explained: a guide to the scientific classification of reptiles, birds, and mammals. Facts on File; 1996.

24. LPSN—list of prokaryotic names with standing in nomenclature. Leibniz Institute, DSMZ-German Collection of Microorganisms and Cell Cultures GmbH; [accessed 2023 Mar 17]. https://LPSN.dsmz.de

25. Wiersema JH et al, editors. 2018+ [continuously updated]: International code of nomenclature for algae, fungi, and plants (Shenzhen Code) adopted by the Nineteenth International Botanical Congress, Shenzhen, China, July 2017: Appendixes I–VII. https://naturalhistory2.si.edu/botany/codes-proposals/

26. WFO. 2021. World Flora Online; [accessed 2023 Mar 17]. http://www.worldfloraonline.org

27. Tropicos.org. Release 3.4.0. Missouri Botanical Garden; [accessed 2023 Mar 17]. https://tropicos.org

28. WCVP. World checklist of vascular plants. Version 2.0. Facilitated by the Royal Botanic Gardens, Kew; 2021. http://wcvp.science.kew.org/

29. IPNI. 2021. International plant names index. The Royal Botanic Gardens, Kew, Harvard University Herbaria & Libraries, and Australian National Botanic Gardens. http://www.ipni.org

30. Index Fungorum. Royal Botanical Garden, Kew [accessed 2023 March 17]. http://www.indexfungorum.org/names/names.asp

31. Linnaeus C. Species plantarum: a facsimile of the first edition. Vols 1–2. The Ray Society; 1957 and 1959.

32. Brickell CD et al, editors. International code of nomenclature for cultivated plants. 9th ed. Scripta Horticulturae 18. International Society of Horticultural Science; 2016.

33. Dorn RD. Vascular plants of Wyoming. 3rd ed. Mountain West Publishing; 2001.

34. Peck ME. A manual of the higher plants of Oregon. 2nd ed. Oregon State University Press; 1961.

35. Mabberley DJ. The plant-book: a portable dictionary of higher plants. 2nd ed. Cambridge University Press; 1997.

36. Gleason HA, Cronquist A. Manual of vascular plants of northeastern United States and adjacent Canada. 2nd ed. New York Botanical Garden; 1991.

37. American Joint Committee on Horticultural Plant Nomenclature. Standardized plant names. J. Horace McFarland Co; 1942.

38. Patterson DT. Composite list of weeds. Weed Science Society of America; 1989.

39. Composite list of weeds (online version). Weed Science Society of America; [accessed 2023 Mar 17]. https://wssa.net/wssa/weed/composite-list-of-weeds/

40. Rickett HW. The English names of plants. Bull Torrey Bot Club. 1965;92(2):137–139.

41. Brako L, Rossman AY, Farr DF. Scientific and common names of 7,000 vascular plants in the United States. APS Press; 1995.

42. Kartesz JT, Thieret JW. Common names for vascular plants: guidelines for use and application. Sida Contrib Bot. 1991;14(3):421–434. https://www.jstor.org/stable/41966904

43. May TW et al. Chapter F of the *International Code of Nomenclature for algae, fungi, and plants* as approved by the 11th International Mycological Congress, San Juan, Puerto Rico, July 2018. IMA Fungus. 2019;10:21. https://doi.org/10.1186/s43008-019-0019-1

44. McNeill J et al. International code of nomenclature for algae, fungi, and plants (Melbourne Code). International Association of Plant Taxonomy; 2012. http://www.iapt-taxon.org/nomen/main.php

45. Kirk PM, Cannon OF, Stalpers JA. Dictionary of the Fungi. 10th ed. CAB International; 2011. http://speciesfungorum.org/Names/fundic.asp

46. Korf RP. Citation of authors' names and the typification of names of fungal taxa published between 1753 and 1832 under the changes in the code of nomenclature enacted in 1981. Mycologia. 1982;74(2):250–255. https://doi.org/10.2307/3792891

47. Korf RP. Simplified author citations for fungi and some old traps and new complications. Mycologia. 1996;88(1):146–150. https://doi.org/10.2307/3760796

48. Roelfs AP, Martens JW. An international system of nomenclature for *Puccinia graminis* f. sp. *tritici*. Phytopathology. 1988;78(5):526–533. https://www.ars.usda.gov/ARSUserFiles/50620500/Cerealrusts/Pgt/Pgt_code_Roelfs_Martens%201988.pdf

49. Long DL, Kolmer JA. A North American system of nomenclature for *Puccinia recondita* f. sp. *tritici*. Phytopathology. 1989;79(5):525–529. https://www.apsnet.org/publications/phytopathology/backissues/Documents/1989Articles/Phyto79n05_525.PDF

50. Alexopoulos CJ, Mims CW, Blackwell M. Introductory mycology. 4th ed. John Wiley & Sons; 1996.

51. Brodo IM, Sharnoff SD, Sharnoff S. Lichens of North America. Yale University Press; 2001.

52. Notton DG, Michel E, Dale-Skey N, Nikolaeva S, Tracey S. Best practice in the use of the scientific names of animals: support for editors of technical journals. Bulletin of Zoological Nomenclature. 2011;68(4):313–322. http://doi.org/10.21805/bzn.v68i4.a15

53. Page R, editor. Systematic Biology. A quarterly of the Society of Systematic Biologists. Taylor & Francis. 1992;41(1). Continues: Systematic Zoology. https://academic.oup.com/sysbio

54. Bulletin of Zoological Nomenclature. British Museum, International Trust for Zoological Nomenclature. 1943;1. https://bioone.org/journals/the-bulletin-of-zoological-nomenclature

55. Melville RV, Smith JDD, editors. Official lists and indexes of names and works in zoology. International Trust for Zoological Nomenclature; 1987.

56. Smith JDD, editor. Official lists and indexes of names and work of zoology. Supplement 1986–2000. International Trust for Zoological Nomenclature; 2001. Also available at https://web.archive.org/web/20070325000337/http:/www.nhm.ac.uk/hosted_sites/iczn/names_works_supplement.pdf

57. Linne C-von, Salvius L. Systema naturae per regna tria naturae: secundum classes, ordines, genera, species, cum characteribus, differentiis, synonymis, locis. Holmiae: Impensis Direct. Laurentii Salvii,

1758–1759. Available at https://www.biodiversitylibrary.org/item/10277#page/3/mode/1up; https://doi .org/10.5962/bhl.title.542

58. Clerck C, Bergquist C, Borg E, Gottman L, Salvisu L. Svenska spindlar: uti sina hufvud-slägter indelte samt under några och sextio särskildte arter beskrefne: och med illuminerade figurer uplyste. Stockholmiae: Literis Laur. Salvii, 1757. https://doi.org/10.5962/bhl.title.119890

59. Crother BI, committee chair. Scientific and standard English names of amphibians and reptiles of North America north of Mexico, with comments regarding confidence in our understanding. 8th ed. Society for the Study of Amphibians and Reptiles, Committee on Standard English and Scientific Names; 2017. (Herpetological Circular no. 43). https://ssarherps.org/wp-content/uploads/2017/10/8th-Ed-2017-Scientific -and-Standard-English-Names.pdf

60. Cairns SD et al. Common and scientific names of aquatic invertebrates from the United States and Canada: Cnidaria and Ctenophora. 2nd ed. American Fisheries Society; 2002. (Special publication 28).

61. Turgeon DD et al. Common and scientific names of aquatic invertebrates from the United States and Canada: mollusks. 2nd ed. American Fisheries Society; 1998. (Special publication 26).

62. McLaughlin PA et al. Decapoda. In: McLaughlin PA, Camp DK, Angel MV, editors. Common and scientific names of aquatic invertebrates from the United States and Canada: crustaceans. American Fisheries Society; 2005. (Special publication 31).

63. Williams AB et al. Common and scientific names of aquatic invertebrates from the United States and Canada: decapod crustaceans. American Fisheries Society; 1989. (Special publication 17).

64. Page LM et al. Common and scientific names of fishes from the United States, Canada, and Mexico. 7th ed. American Fisheries Society; 2013. (Special publication 34).

65. Entomological Society of America, Committee on the Common Names of Insects. Common names of insects database. Entomological Society of America; [accessed 2023 Mar 17]. http://www.entsoc.org /common-names

66. Chesser RT et al. Check-list of North American Birds (online). American Ornithological Society; 2020. http://checklist.aou.org/taxa

67. Speybroeck J et al. Species list of the European herpetofauna—2020 update by the Taxonomic Committee of the Societas Europaea Herpetologica. Amphibia-Reptilia. 41;(2):139–189. https://doi.org/10 .1163/15685381-bja10010

68. Werner Y, Frank N, Ramus E. A complete guide to scientific and common names of reptiles and amphibians of the world. National Guard Bureau Publications & Forms Library; 1995. Also available at https://archive.org/details/completeguidetos00fran

69. Burgin CJ et al. Illustrated checklist of the mammals of the world. Lynx Edicions; 2020.

70. Hoyo J del et al. HBW and BirdLife International Illustrated checklist of the birds of the world. Vol 1, Non-passerines; vol 2, Passerines. Lynx Edicions; 2021.

71. Lyon MF. Rules for nomenclature of inbred strains. In: Lyon MF, Searle AG, editors. Genetic variants and strains of the laboratory mouse. 2nd ed. Oxford University Press; 1989. p 632–635.

72. International Committee on Standardized Genetic Nomenclature for Mice and Rat Genome and Nomenclature Committee. Guidelines for nomenclature of mouse and rat strains. Rat Genome Database; 2010 [revised 2021]. http://www.informatics.jax.org/mgihome/nomen/strains.shtml

73. Festing M, Kondo K, Loosli R, Poiley SM, Spiegel A. International standardized nomenclature for outbred stocks of laboratory animals. Z Versuchstierkd. 1972;14(4):215–224.

74. National Research Council (US), Commission on Life Sciences, Institute of Laboratory Animal Resources, Committee on Transgenic Nomenclature. Standardized nomenclature for transgenic animals. ILAR J. 1992;34(4):45–52. https://doi.org/10.1093/ilar.34.4.45

75. Skerman VBD, McGowan V, Sneath PHA, editors. Approved lists of bacterial names (amended). American Society for Microbiology; 1989. Available at https://www.ncbi.nlm.nih.gov/books/NBK814/

76. Moore WEC, Moore LVH. Index of the bacterial and yeast nomenclatural changes. American Society for Microbiology; 1992. Available at https://www.ncbi.nlm.nih.gov/books/NBK815/

77. Bergey's manual of systematics of archaea and bacteria. Bergey's Manual Trust; [accessed 2023 Mar 18]. https://onlinelibrary.wiley.com/doi/book/10.1002/9781118960608

78. Boone DR, Castenholz RW, Garrity GM, editors. Bergey's manual of systematic bacteriology. Vol 1, The Archaea and the deeply branching and phototropic bacteria. 2nd ed. Springer; 2001. https://doi.org/10.1007/978-0-387-21609-6

79. Brenner et al, editors. Bergey's manual of systematic bacteriology. Vol. 2, The *Proteobacteria*, Part A Introductory essays and Part B: The *Gammaproteobacteria*. 2nd ed. Springer; 2005. https://doi.org/10.1007/0-387-28021-9 and https://doi.org/10.1007/0-387-28022-7

80. Vos P et al, editors. Bergey's manual of systematic bacteriology. Vol 3, The *Firmicutes*. 2nd ed. Springer; 2009. https://doi.org/10.1007/978-0-387-68489-5

81. Krieg NR et al, editors. Bergey's manual of systematic bacteriology. Vol 4, The *Bacteroidetes*, *Spirochaetes*, *Tenericutes* (*Mollicutes*), *Acidobacteria*, *Fibrobacteres*, *Fusobacteria*, *Dictyoglomi*, *Gemmatimonadetes*, *Lentisphaerae*, *Verrucomicrobia*, *Chlamydiae*, and *Planctomycetes*. 2nd ed. Springer; 2010. https://doi.org/10.1007/978-0-387-68572-4

82. Holt JG, Krieg NR, Sneath PHA, Staley JT, Williams ST. Bergey's manual of determinative bacteriology. 9th ed. Lippincott, Williams & Wilkins; 1994.

83. Huth EJ. Style notes: taxonomic names in microbiology and their adjectival derivatives. Ann Intern Med. 1989;110(6):419–420. https://doi.org/10.7326/0003-4819-110-6-419

84. Murray RG, Schleifer KH. Taxonomic notes: a proposal for recording the properties of putative taxa of procaryotes. Int J Syst Bacteriol. 1994;44(1):174–176. https://doi.org/10.1099/00207713-44-1-174

85. Murray RG, Stackebrandt E. Taxonomic note: implementation of the provisional status *Candidatus* for incompletely described procaryotes. Int J Syst Bacteriol. 1995;45(1):186–187. https://doi.org/10.1099/00207713-45-1-186

86. Dye DW et al. International standards for naming pathovars of phytopathogenic bacteria and a list of pathovar names and pathotype strains. Rev Plant Pathol. 1980;59(4):153–168.

87. Young JM et al. Nomenclatural revisions of plant pathogenic bacteria and list of names 1980–1988. Rev Plant Pathol. 1991;70:211–221.

88. Young JM et al. Names of plant pathogenic bacteria 1864–1995. Rev Plant Pathol. 1996;75:721–763. https://www.isppweb.org/names_bacterial.asp

89. Young JM et al. Committee on the Taxonomy of Plant Pathogenic Bacteria: international standards for naming pathovars of phytopathogenic bacteria. International Society of Plant Pathology, 2002. https://www.isppweb.org/about_tppb_naming.asp

90. Brenner FW, Villar RG, Angulo FJ, Tauxe R, Swaminathan B. *Salmonella* nomenclature. J Clin Microbiol. 2000;38(7):2465–2467. https://doi.org/10.1128/JCM.38.7.2465-2467.2000

91. Tindall BJ, Grimont PAD, Garrity GM, JP Euzéby JP. Nomenclature and taxonomy of *Salmonella*. Int J Syst Evol Microbiol. 2005;55:521–524. https://doi.org/10.1099/ijs.0.63580-0

92. Grimont PAD, Weill FX. Antigenic formulae of the *Salmonella* serovars. 9th ed. WHO Collaborating Centre for Reference and Research on *Salmonella*, Institut Pasteur; 2007. https://www.researchgate.net/publication/283428414_Antigenic_Formulae_of_the_Salmonella_serovars_9th_ed_Paris_WHO_Collaborating_Centre_for_Reference_and_Research_on_Salmonella

93. Hull R. Matthews' plant virology. 5th ed. Academic Press; 2013.

94. International Committee on Taxonomy of Viruses, Executive Committee. The new scope of virus taxonomy: partitioning the virosphere into 15 hierarchical ranks. Nat Microbiol. 2020;5(5):668–674. https://doi.org/10.1038/s41564-020-0709-x

95. Current ICTV taxonomy release. International Committee on the Taxonomy of Viruses; [accessed 18 Mar 2023]. https://talk.ictvonline.org/taxonomy

96. Virus taxonomy: the ICTV report on virus classification and taxon nomenclature. International Committee on Taxonomy of Viruses; [accessed 18 Mar 2023]. https://talk.ictvonline.org/ictv-reports/ictv_online_report/

97. Walker PJ et al. Changes to virus taxonomy and to the International Code of Virus Classification and Nomenclature ratified by the International Committee on Taxonomy of Viruses (2021). Arch Virol. 2021(Jul);166, 2633–2648. https://doi.org/10.1007/s00705-021-05156-1

23 Anatomical and Physiological Descriptions for Macroorganisms

Editors: Beth E. Hazen, PhD, and Amy McPherson

23.1 CHAPTER FOCUS AND REFERENCES ON TERMINOLOGY

This chapter deals with style and format matters pertaining to anatomical and physiological descriptions of macroorganisms when they differ from standard use. Many reference works are available on the specialized terminology for describing the structure and function of viruses,[1-4] microorganisms,[5,6] plants,[7,8] and animals, including humans.[9] Specialized dictionaries and glossaries[10-15] provide guidance on terminology and proper usage in particular disciplines, as do some journals.

23.2 PLANT ANATOMY AND PHYSIOLOGY

Terms to describe the orientation of parts on plants and plant allies differ from those used for animals, including humans. These terms can be found by consulting appropriate references.[7,8,10,14]

Table 23.1. Selected Conditions and Units of Measure for Controlled Environments for Plant Studies[a]

Condition	Unit
Radiation	
Photosynthetic photon flux (PPF), 400–700 nm	μmol m^{-2}s^{-1}
Energy flux (irradiance)	W m^{-2} (nm waveband)
Spectral photon flux (bandwidths $\leq$ 2 nm)	μmol m^{-2}s^{-1}mm^{-1}
Spectral energy flux (spectral irradiance)	W m^{-2}
Temperature (air, soil, and liquid)	°C
Atmospheric moisture	
Relative humidity (RH)	% RH
Dew point temperature	°C
Water vapor density	g m^{-2}
Air velocity	m s^{-1}
Watering	L, l
Nutrients	mol m^{-3} or mol kg^{-1}

[a] Based on Tibbitts TW, Sager JC, Krizek DT. Guidelines for measuring and reporting environmental parameters in growth chambers. Biotronics. 2000;29(1):9–16. https://catalog.lib.kyushu-u.ac.jp/opac_download_md/8260/KJ00004506925.pdf.

Table 23.2. Terms for Light Measurements and Photosynthesis[a]

Term	Abbreviation	Unit
Apparent photosynthesis	AP	none
CO_2 exchange rate[b]	CER	μmol cm^{-2}s^{-1}
Photosynthetic irradiance	PI	W m^{-2}
Photosynthetic photon flux density	PPFD	μmol m^{-2}s^{-1}
Photosynthetically active radiation[c]	PAR	PPFD *or* PI

[a] Terms originally defined by Shibles.[18] Updated by the American Society of Agronomy, the Crop Science Society of America, and the Soil Science Society of America.[19]
[b] Use this term instead of "net CO_2 exchange."
[c] For reporting PAR, photon units (PPFD) are preferred to energy units (PI), but both are acceptable. The term "light intensity" is no longer used.

For clarity, leaf surfaces are described in relation to the stem axis. Use "adaxial," not "upper," to specify the surface facing or toward the stem and "abaxial," not "lower," for the surface facing away from the stem.

Most plant biology journals use the units of measurement of the Système International d'Unités, or the International System of Units (see Section 12.2.1, "Système International d'Unités (SI)"). These journals also follow the standards for terminology and data presentation recommended in *Units, Symbols, and Terminology for Plant Physiology*.[16]

Guidelines for measuring and reporting conditions in controlled environments were updated in 2015.[17] Some typical units of measure appear in Table 23.1. Terms and units for light measurement and photosynthesis are in Table 23.2.[18,19]

Growth stages of agricultural crops are described using decimal-based growth-stage codes that are specific for each crop.[20] Those codes range from 00 for "dry seed" to 99 for "harvested product."

23.3 FISH: AGES

23.3.1 Age Notation for Salmonids and Nonsalmonids

In general, fish age is expressed with an arabic numeral and without a plus symbol. One year is counted from 1 January after a fish hatches to 31 December. This first year is described as "age 0" and "young of the year," not "YOY."[21]

A special notation system is used to designate the age of anadromous fish (eg, salmon), which begin life in freshwater streams, spend time in the ocean, and return to their natal streams to spawn and die. The freshwater phase and the ocean, or seawater, phase of anadromous life history are commonly represented as "$x.y$" by the "European system," in which "x" is the number of winters the fish spent in freshwater and "y" the number of winters spent at sea. The x and y numbers are separated by a period without a space. Use arabic numerals to designate the age of adult and juvenile salmon without indicating units of measurement. According to this system, an adult salmon aged 1.2 lived 1 year in freshwater and 2 years at sea. The first winter spent during embryonic development is not counted as part of either phase.

Often, only one phase of the salmon life history is described. In European notation, when only one phase is described, the x freshwater age is written as a numeral followed by a period, and the y ocean age is written as a numeral preceded by a period.

> Salmon age 2. emigrated from Black Lake. (freshwater phase only)
> At age .3, sockeye salmon return to their natal stream. (seawater phase only)

Because of potential ambiguity, do not use a plus symbol after the age (eg, "2.3+") to indicate either an additional summer of growth after the last winter counted in the age or to indicate that a group of salmon is composed of fish at least as old as the given age.[21,22]

Textual descriptions and systems other than the European system are also used to designate salmonid ages. No standard sets of terminology for these descriptions have been adopted, so word descriptions of age can be awkward and ambiguous. For example, fish designated as age 2. in the European system may be described as "2-freshwater fish" or "2-years-in-the-lake type" fish. In turn, a salmon designated as age .3 may be described with terms such as "3-ocean fish." Although the construction used in the second example is common, it is potentially confusing to those outside the discipline, who might interpret "3-ocean" to mean that the salmon lived in 3 different oceans.

23.3.2 Age in Years

As indicated in Section 23.3.1, "Age Notation for Salmonids and Nonsalmonids," the birth date for salmonids is conceptually standardized as 1 January of the year following the brood year, regardless of when a brood actually hatched.[22] The "brood year" is the year in which adult salmonid return to freshwater to spawn, which may not be the same as the year in which the embryos produced by those spawning adults hatch. A fish less than 1 year of age is age 0.

Ages in years are determined differently for Atlantic salmon (*Salmo salar*) and Pacific salmon (*Oncorhynchus* spp.). Atlantic salmon ages are either the total number of winters a fish has lived since hatching from its egg or the number of annuli on the fish's scales.[23] An Atlantic salmon hatched in the spring of 2023 would be 3 years old in 2026. In contrast, Pacific salmon ages are based on their brood year. Eggs spawned by Pacific salmon in 2023 (ie, the 2023 brood) would be 4 years old in 2026.

23.4 BIRDS, WILDLIFE, AND DOMESTIC MAMMALS

The conventions for symbolization for the structure and function of nonhuman animals were developed mainly within zoology. Some of these conventions are formally documented, but many have become established simply through usage. Where specific conventions for the animal sciences are lacking, veterinarians and veterinary medical researchers have adopted many of the style conventions established by the medical and basic science communities. The information provided in this section covers topics for which specific conventions exist for nonhuman animals. Otherwise, consult the relevant parts of Section 23.5, "Humans," or the *AMA Manual of Style: A Guide for Authors and Editors.*[24]

23.4.1 Anatomical Description

For the official list of anatomical terms for most nonhuman animals, consult the *Nomina Anatomica Veterinaria*, published by the World Association of Veterinary Anatomists.[25] Nonhuman animal anatomy incorporates a distinct set of terms indicating orientation of a body part, which differ from those used for humans. For example, the anatomical term "caudal" means "toward the tail."

Useful resources for anatomical descriptions of birds include *Avian Anatomy: Textbook and Colour Atlas*,[26] the *Handbook of Avian Anatomy: Nomina Anatomica Avium*,[27] and the *Handbook of Bird Biology.*[28]

23.4.2 Clinical Chemistry

Information on the clinical chemistry of laboratory animals has been compiled in *The Clinical Chemistry of Laboratory Animals.*[29] This work covers the mouse, rat, guinea pig, hamster, rabbit, dog, and nonhuman primates. In addition, it includes extensive lists of reference values for various analyses, arranged by substance analyzed and species. A more general reference is *Clinical Biochemistry of Domestic Animals,*[30] which focuses on domesticated animals, such as the dog, cat, horse, cow, and pig.

23.4.3 Blood Groups

Blood typing is one of the most significant areas in which human medical conventions cannot be applied in veterinary medicine.

The red blood cell antigens and blood group systems of a number of domesticated and livestock species have been defined by specific nomenclature. The history and specific details of these systems are described in *Schalm's Veterinary Hematology.*[31] Several of the systems covered by *Schalm's* are briefly described next.

Most of the more than 12 known canine blood groups are designated by the initialism "DEA," which stands for "dog erythrocyte antigen." This initialism is followed by a number that designates the locus (eg, "DEA 1" or "DEA 3").[31] The DEA 1 locus is sometimes followed by a decimal point and another number to indicate a specific allele (eg, "DEA 1.1," "DEA 1.2," and "DEA 1.1-negative"). However, recent work has shown that the alleles for DEA 1 have the same epitope, so they should instead be designated as "DEA 1-positive" or "DEA 1-negative." Some of the rarer blood group types are Dal and Kai 1 and 2.

In cats, blood group phenotypes include types A, B, and AB.[31]

Horses have 7 blood group systems: A, C, D, K, P, Q, and U. Distributed among these blood groups are 34 blood factors.[31]

For livestock, the traditional blood groups may be preceded by "EA" to designate "erythrocyte antigen."

Cattle have 11 blood group systems, which have more than 70 blood group factors.[31] The blood groups are A, B, C, F, J, L, M, R, S, T, and Z, which may be followed by a numerical subscript to designate serologically related subtypes (eg, A_1 and A_2). In addition, the symbol prime or double prime after the blood group letter (eg, A' or A'') can be used to designate a second or third use of the same letter, respectively. Multiple blood group factors cosegregate in specific combinations that define allelic variants called "phenogroups."

Pigs have 16 blood group systems, ranging from A to P.[31] Some of these are complex and diverse, as are the B and C systems of cattle. In addition, the A system in pigs is related to the cattle J, human A, and sheep R systems.

23.4.4 Immunologic Systems

The cluster of differentiation nomenclature developed for human leukocyte antigens is widely accepted for nonhuman animals, including the mouse[32] and the rat[33] (see Section 23.5.6.3, "Lymphocytes and Surface Antigens of Immune Cells," which focuses on humans). The major histocompatibility complex antigens of many mammals and birds have been characterized.[34]

23.5 HUMANS

23.5.1 Anatomical Description

The official list of human anatomical terms is *Terminologia Anatomica*,[35] prepared by the Federative International Programme for Anatomical Terminology of the International Federation of Associations of Anatomists.

The official terms for human anatomy take Latin forms, but English equivalents that are either direct translations of Latin terms or idiomatic substitutes may be used in most scientific texts. For example, "brachial plexus" may be used instead of the Latin term "plexus brachialis," and "stomach" may be used instead of "ventriculus." Some terms have to be used in their Latin forms because they do not have formal equivalents in English (eg, "cisterna chyli"). Anatomical terms derived from Latin are common nouns, and therefore, they are not capitalized or italicized in either their Latin or English forms.

Terms describing the relative orientation of body parts in humans differ from those for nonhuman animals and birds because adult humans can take the upright position.

23.5.2 Blood

23.5.2.1 Blood Groups

Human blood groups have various classification systems (eg, the ABO, Rh, Lewis, and MNS systems). Antigens and factors are usually represented by capital and lowercase letters. Some of these designations are arbitrary (eg, the A, B, and O blood groups of the ABO system), and others stand for the names of the persons who were the sources of the initially identified components (eg, "Le" for "Lewis" and "Fy" for "Duffy").

Antigens, phenotypes, and genes for blood groups are all represented by the same letter or letters for a particular group, with the gene designations set in italic. However, the specifics of how each system represents subclasses differ.

Antigen	Phenotype	Gene
K	K+	*K*
A_1	A_1	*A^1*
Fy^a	Fy(a+)	*Fy^a*

Some systems, including the Kell, Lutheran, and Duffy systems, have both letter and number designations (eg, "K" and "K1" for the same antigen, "K−" and "K:−1" for the same phenotype, and "*K*" and "*K^1*" for the same gene).

One of the most complex symbolizations is that for the Rh system, which is named after the rhesus monkey. Three systems of symbols have been applied: the Rh-Hr notation, the CDE notation, and the numeric system.

Rh-Hr notation	CDE notation	Numeric system
Rh_0	D	1
rh″	E	3
hr″	e	5
rh^G	G	12
Hr^B	Bas	34

These examples illustrate some of the uses of capital, lowercase, and superscript letters to symbolize the Rh classification system. Additional details can be found in Chapter 5, "The Rh Blood Group System," of *Mollison's Blood Transfusion in Clinical Medicine*.[36]

The Committee on Terminology for Red Cell Surface Antigens of the International Society of Blood Transfusion (ISBT) described a system of 6-digit numerical codes for established serologically determined blood groups.[37] The first 3 numbers represent the system, and the last 3 numbers represent antigenic specificity.

The ISBT Working Party for Red Cell Immunogenetics and Blood Group Terminology maintains an official, public record of the blood group systems.[38] The ISBT codes are intended primarily for computer applications. For publication purposes, phenotype designations may consist of the system symbol in all capital letters, a colon, and the appropriate numbers for the specificities. In text, any leading zeros in the 3 digits designating specificity are dropped. For example, in the ABO system, the A antigen becomes "ABO1," not "ABO001," and in the Lutheran system, Lua becomes "LU1," not "LU001."

Antigens that are missing are preceded by a minus symbol. Genotypes consist of the system symbols, an asterisk, and the alleles or haplotypes separated by a forward slash, all in italics (eg, "*KEL*2,3/2,4*").

System	Traditional	ISBT
Colton	Co(a+b−)	CO:1, −2
	Co^a/Co	*CO*1/0*
Kell	K− k+ Kp(a−b+)	KEL:−1,2,−3,4
	k,Kp^b,Js^b/K⁰	*KEL*2,4,7/0*

Within the ISBT system, other antigens and specificities are grouped under "collections" of specificities that have serological, biochemical, or genetic connections. These collections have 6-digit designations beginning with "2xx" and followed by 3 digits for each member of the collection. For example, the Cost collection is number "205," and the codes for the 2 antigens in this collection are" 205001" and "205002." Antigens not assigned to systems or collections are put into 1 of 2 series: the 700 series for low-incidence antigens and the 901 series for high-incidence antigens.

23.5.2.2 Hemoglobins

Human hemoglobins were conventionally designated by the symbol "Hb" followed by a space and a letter (ie, A, C through Q, or S). For more recently identified hemoglobin variants, the letter has been replaced by a word indicating the location or laboratory of the discovery.

Hb A   Hb S   Hb Providence   Hb Hopkins-2

The normal variants of human hemoglobin are Hb A, Hb A_2, and Hb F. Each complete molecule of normal hemoglobin is a tetramer containing a total of 4 globin polypeptide chains. Each chain, in turn, comprises 2 pairs of 4 possible globin subunits, which are designated by the lowercase Greek letters alpha (α), beta (β), gamma (γ), and delta (δ). In addition, globin chains designated epsilon (ε) and zeta (ζ) are found in embryonal hemoglobins.

The composition of the normal major fraction of adult human hemoglobin, Hb A, is 2 alpha and 2 beta chains, which are designated as "$\alpha_2\beta_2$." Hb A_2 contains 2 alpha and 2 delta chains, which are designated as "$\alpha_2\delta_2$." The fetal human hemoglobin, Hb F, consists of 2 alpha and 2 gamma chains, which are designated as "$\alpha_2\gamma_2$." Some forms of hemoglobin contain only nonalpha chains.

Hb H: β_4 Hb Portland: $\gamma_2\zeta_2$

23.5.2.3 Platelet Antigens

A nomenclature for platelet antigens was developed by the ISBT's Working Party on Platelet Serology and by the International Council for Standardization in Haematology.[39] The nomenclature consists of the initialism "HPA" for "human platelet antigen" and numbers indicating the order of date of publication (eg, "HPA-1" and "HPA-2"). Allelic antigens are designated alphabetically in order of their frequency in the population from high to low (eg, "HPA-1a" is more frequent than "HPA-1b").

23.5.2.4 Coagulation Factors

Human clotting factors are designated by the term "factor" followed by a roman numeral from "I" to "XIII," excluding "VI" (eg, "factor III" and "factor XI"). A lowercase "a" designates a factor's activated form (eg, "XIa"). Factor VIII and von Willebrand factor have special terminology.[40]

23.5.3 Bone Histomorphometry

The system of nomenclature, symbols, and units for bone histomorphometry includes a set of 123 initialisms that are used in combination to produce standard representations of the variables reported in histomorphometry (see the report of the American Society for Bone and Mineral Research's Histomorphometry Nomenclature Committee[41]). Symbols are used for the primary measurements in both 2-dimensional and 3-dimensional reporting systems.

With one exception, the initialisms consist of 1 or 2 letters. Both capital and lowercase letters are used for 1-letter initialisms, while 2-letter initialisms consist of an initial capital letter followed by a lowercase letter. The sole 3-letterinitialism, "BMU," is all capitals. When used in combination, single-letter initialisms are run together, and multiletter initialisms are separated by a period from adjacent multiletter and single-letter abbreviations.

> BS *for* bone surface   Ct.Th *for* cortical thickness
> B.Pm *for* bone perimeter   Ob.S *for* osteoblast surface

23.5.4 Circulation

23.5.4.1 Hemodynamics

Symbols for hemodynamic quantities should, as far as possible, follow the international conventions for symbolizing quantities (see Sections 4.2, "Symbols and Signs," and 10.1.1.2, "Scientific Uses"). In those conventions, italic type is used for single-letter symbols representing quantities, and roman type is used for subscript or superscript symbols modifying the main symbol, unless the modifier is itself a symbol for a quantity. See Table 23.3 for examples of symbols, their modifiers, and their applications for specific measurements.

Multiletter abbreviations have been widely used to represent quantities (eg, "LVEDP" for "left ventricular end-diastolic pressure"), but these should be replaced whenever possible by standard symbols (see Table 23.3).

23.5.4.2 Clinical Notations

Leads (ie, recording electrodes and connecting wires) for recording electrocardiographic tracings are designated by alphanumeric characters. The standard leads are designated with roman numerals. In contrast, in the unipolar lead system, the central terminal is designated by "V" for "voltage" followed by the letter "R" for "right arm," "L" for "left arm," or "F" for "foot." In the augmented lead system, the central terminal is designated as "aV."

Table 23.3. Examples of Variables Related to Hemodynamics

Variable	Symbol	SI unit
Main symbols		
Area	A	mm^2
Quantity (volume)	Q	mL *and* L
Flow rate (volume/unit time)	$Q\ (V/t)$	mL s^{-1} *and* L min^{-1}
Flow velocity	v	cm s^{-1}
Flow acceleration	v/t	cm s^{-2}
Tissue or organ mass	M	G *and* kg
Perfusion	Q/M	mL min^{-1}g^{-1} *and* mL min^{-1} (100 g)$^{-1}$
Pressure	P	kPa (also mm Hg *or* cm H$_2$O for clinical use)
Resistance	R	kPa L^{-1}
Velocity	v	mm s^{-1} *and* cm s^{-1}
Volume	V	mL *and* L
Modifying symbols		
Arterial	a	. . .
Venous	ven	. . .
Capillary	cap	. . .
Atrium	A	. . .
Ventricle	V	. . .
Systemic	S	. . .
Systolic	syst	. . .
Diastolic	diast	. . .
Examples of modifying symbols		
Pressure, arterial	P_a	kPa (also mm Hg)
Systolic	$P_{a,\ syst}$	. . .
Diastolic	$P_{a,\ diast}$	. . .
Aortic	P_{aor}	. . .
Pressure, venous	P_{ven}	. . .
Pressure, left ventricular	P_{LV}	kPa (also cm H$_2$O)

SI, Système International d'Unités (International System of Units)

Standard leads:	lead I	lead II	lead III
Unipolar lead:	VR	VL	VF
Augmented lead:	aVR	aVL	aVF

Chest leads are designated as "V" for the central terminal and a subscript arabic numeral for the chest electrode. "R" refers to the right side of the chest and roman numerals to intercostal-space locations above the standard locations. Esophageal leads have the designation of "E" with a number representing the distance in centimeters from the nares to the esophageal electrode.

lead V$_1$ lead V$_2$ lead V$_1$R lead V$_3$III lead VE28

Electrocardiographic waves (ie, deflections from the tracing baseline) are designated with the capital letters "P," "Q," "R," "S," "T," and "U." Wave complexes are represented by capital letters without spaces, and a minor wave in a complex is represented by a lowercase letter. Prime signs indicate waves located after their usual position. An interval between 2 waves is represented by an en dash or hyphen between the wave symbols.

P wave   QRS complex   rS   rRV   P–R interval

The severity of manifestations of cardiac and vascular disease is commonly represented by roman or arabic numerals.

grade IV hypertensive retinopathy   mitral stenosis grade 2/6 (2 on a scale of 6)

In contrast, the New York Heart Association uses a 2-number system: The first number is a roman numeral between I and IV, which represents functional capacity, and the second is a capital letter between A and D, which represents objective assessment.[42] Class I functional capacity indicates the least impairment, and class IV indicates the greatest. Objective assessment A indicates no objective evidence of cardiovascular disease, while objective assessment D represents objective evidence of severe cardiovascular disease.

23.5.5 Clinical Chemistry

In many countries, the measurement units of the Système International d'Unités (SI) are used for reporting clinical chemistry findings. These units have not been widely adopted in the United States, so US journals may require reporting clinical findings in older metric units or require that findings be reported in both SI and older metric units. Chemical pathologist Donald S. Young, MB, PhD,[43] compiled tables for converting the older metric units to SI units and offered recommendations on significant digits and minimum increments for clinical hematology and clinical chemistry measurements.

23.5.6 Immunologic Systems

The human leukocyte antigen (HLA) system of the human major histocompatibility complex determines a person's tissue type. HLA genetic "haplotypes," which are a set of closely linked genes inherited together on a chromosome, are important for determining tissue compatibility for tissue and organ transplantation. The major histocompatibility complex genomic region contains several subregions with genes that encode the various HLA antigens and other immune elements.

The gene loci for the class I antigens, which are located on the surface of most nucleated cells and on platelets, are lettered A, B, C, E, F, G, H, J, K, L, N, S, X, and Z and are symbolized as follows:

HLA-A   HLA-C   HLA-E   HLA-H

The D region contains the gene loci for class II antigens and is divided into subregions, which are distinguished by a second capital letter. A third capital letter indicates an alpha (A) or beta (B) chain. If needed, a number distinguishes among several chains.

HLA-DRA   HLA-DOB   HLA-DQA1   HLA-DRB4   HLA-DPA2

The class III region of the major histocompatibility complex contains various genes, including those for some complement components.

Alleles of the HLA genes are designated by the letters of the locus, followed by an asterisk and at least 4 numerals. Sometimes, 6 or 8 numerals are used in "fields" to indicate various nucleotide variations. Each field consists of 2 numerals, and fields are separated by colons. In addition, the allele designations are italicized. In contrast, the expressed protein products of the HLA genes are designated by the gene letters and the first 2 fields, all of which are set in roman type.

Genes:	*Cw*02:03*	*DPB1*01:02*	*A*02:01:01*	*DRB1*08:02:03*
Proteins:	Cw*02:03	DPB1*01:02	A*02:01	DRB1*08:02

The nomenclature for the HLA system is provided in reports from the World Health Organization's Nomenclature Committee for Factors of the HLA System. For example, the 2010 report[44] describes the system of symbols and the variant information provided for each field, and it has complete lists of all officially recognized HLA class I and class II genes and alleles. The 2010 report also introduced "P" and "G" designations to describe ambiguities in polypeptide and nucleotide sequences. Monthly updates are published in *HLA: Immune Response Genetics,*[45] *Human Immunology,*[46] and the *International Journal of Immunogenetics.*[47] Resources related to HLA nomenclature are accessible through the National Marrow Donor Program[48] in the United States.

The major histocompatibility complexes have been described for other species (eg, RT1 in the rat,[49,50] DLA in the dog,[51] *H-2* genes in the mouse,[52,53] and RhLA in the rhesus monkey). Some discussion has taken place on establishing a uniform system of symbols for major histocompatibility complexes across species, except for the human and the mouse.[54] *The HLA System in Clinical Transplantation*[55] and other sources[56] provide additional guidance on major histocompatibility complexes in domestic animals.

23.5.6.1 Immunoglobulins

The 5 classes of human immunoglobulins are represented by the symbol "Ig" followed by a capital letter without a space (ie, "IgA," "IgD," "IgE," "IgG," and "IgM"). Subclasses are indicated by an arabic numeral after the class letter (eg, "IgA2" and "IgG1").

Immunoglobulin molecules are composed of heavy and light polypeptide chains. The heavy chains are different for each immunoglobulin class. These chains are symbolized by the Greek letter that corresponds to the roman capital letter representing the class (ie, "α" for "A," "δ" for "D," "ε" for "E," "γ" for "G," and "μ" for "M"). The heavy chains of IgG1 and IgA2 would be designated as "γ1" and "α2," respectively. The light chains of all classes are either kappa (ie, κ) or lambda (ie, λ) chains.

The basic structure of an immunoglobulin molecule consists of 4 chains: 2 heavy chains, which are specific to the class, and 2 light chains, which are either kappa or lambda chains but never both. Thus, an IgG molecule may contain 2 γ chains and either 2 κ or 2 λ chains. IgA may exist as a dimer, as in "$(\alpha_2\lambda_2)_2$," and IgM may exist as a pentamer, as in "$(\mu_2\lambda_2)_5$." Because each molecule has only one type of light chain, the subscripts indicating number can be omitted in a shorthand representation of immunoglobulins (eg, "IgG-κ" and "IgM-λ").

An immunoglobulin molecule can be fragmented into portions, which are designated with a capital "F" and lowercase letters without spacing.

Fab   F(ab')2   Fd   Fd′   Fv   Fc

Each heavy and light chain contains variable, joining, and constant regions designated by the capital letters "V," "J," and "C," respectively. The variables are modified by the subscript "H" for "heavy chain" or "L" for "light chain." Subgroups are designated by arabic numerals attached to the symbol. Roman numerals are used for the 4 subgroups of the heavy chain variable region, the 4 subgroups of the kappa chain variable region, and the 6 subgroups of the lambda chain variable region.

Variable region	Generic	Specific
Heavy chain	V_H	$V_\alpha III$
Light chain	V_L	$V_\kappa I$

Variable region	Generic	Specific
Heavy chain	C_H	$C_\gamma 1$
Light chain	C_L	C_λ

The surface membrane molecules that specifically bind immunoglobulins via the Fc portion are called "Fc receptors," symbolized by "FcR."[57] Receptors specific to an immunoglobulin class are designated by a subscript Greek letter between the "c" and the "R" in the symbol "FcR." These subscript Greek letters correspond to the class of immunoglobulin. Subtypes are designated by roman numerals without spacing. Species abbreviations consist of prefixes (eg, "hu" for human, "mo" for mouse, "rt" for rat, and "rb" for rabbit).

$Fc_\alpha R$ $moFc_\gamma RII$

Paul's Fundamental Immunology[58] is among the resources that provide an overview of immunoglobulins.

23.5.6.2 Complement

"Complement" is the system of proteins involved in antigen–antibody reactions in all vertebrates. The components of the classical complement pathway are designated by the capital letter "C," to which arabic numerals are attached.[59] Note that the numerals do not follow the exact order of the classical pathway's reaction sequence. Specifically, the second component is C4, and the third and fourth are C2 and C3. The complete sequence of components is C1, C4, C2, C3, C5, C6, C7, C8, and C9. Component C1 has subcomponents C1q, C1r, and C1s.

Other complement factors take part in the alternative complement pathway, and those factors are represented by the letters "B," "D," "H," "I," and "P."[60]

Fragments of a complement molecule are designated by adding lowercase letters, generally with the larger fragment designated by the letter "b" and the smaller by "a" (eg, C3b and C3a). Inactive fragments are designated by the letter "i" as a suffix (eg, C4bi). In contrast, a letter "i" as a prefix indicates the loss of activity from a protein through peptide hydrolysis without fragmentation (eg, iC3b).

An overview of the complement pathways can be found in *Paul's Fundamental Immunology*.[58]

23.5.6.3 Lymphocytes and Surface Antigens of Immune Cells

Using monoclonal antibodies to identify lymphocyte cell–surface antigens led to the development of the system of nomenclature for clusters of differentiation (CDs).[61] A cluster of antibodies that have the same cellular reactivity are assigned numbers after the prefix "CD."

CD4   CD11a   CD11b   CD45 RB

Lowercase and uppercase letters after the CD number indicate molecular characteristics specific to the CD. A lowercase "w" indicates that a provisional designation was made at one of the workshops at which decisions are made about designations (eg, CDw186). The CD system encompasses more than 400 surface molecules.[62]

23.5.6.4 Interleukins and Interferons

The nomenclature for interleukins is established by the Nomenclature Subcommittee on Interleukin Designation of the World Health Organization (WHO) and the International Union of Immunological Societies (IUIS).[63] Interleukins are designated numerically.

Full name:	interleukin-1α	interleukin-1β	interleukin-5	interleukin-12
Abbreviation:	IL-1α	IL-1β	IL-5	IL-12

The symbol for interferon is "IFN."[64] The prefixes "Hu" and "Mu" indicate "human" and "murine," respectively, and a Greek letter preceded by a hyphen specifies the type of interferon. Nonallelic variants can be indicated by an arabic numeral or a capital letter immediately following the Greek letter.

IFN-β   HuIFN-α3   MuIFN-γB

23.5.6.5 Allergens

The IUIS and the WHO[65] have recommended a formal nomenclature system for allergens. Highly purified allergens are designated as follows: the first 3 letters of the genus, a space, the first letter of the species name, a space, and an arabic numeral, all in roman type. For example, pollen from perennial rye grass, *Lolium perenne*, would be designated Lol p 1. An allergen from the western honey bee *Apis mellifera* would be Api m 4.

When this system produces identical designations, either the genus or the species designation is extended by one or more letters. For example, allergens from *Canis domesticus* are designated "Can d," while those from *Candida albicans* are designated "Cand a" by using 4 instead of 3 letters from the genus.

For allergens, roman type is used instead of italics. The accepted practices for genetic nomenclature reserve italics to represent genes.

23.5.7 Renin–Angiotensin System

Systematic names and abbreviations for the renin–angiotensin system were established by a nomenclature committee of the International Society of Hypertension.[66] The conventions include using abbreviations for components of the system and superscript numbers indicating amino acid substitutions and their positions.

ANG II, ANG-(1–8) *for* angiotensin-(1–8)octapeptide
[Sar^1, Val^5, Ala^8]ANG II *for* [Sar^1,Val^5,Ala^8]angiotensin-(1–8)octapeptide

23.5.8 Kidney and Renal Function

The International Union of Physiological Sciences[67] developed a detailed standardized nomenclature for the components of renal structure, which is essentially the same for humans and nonhuman mammals.

Although symbols representing renal functions are not standardized, many of the symbols in use are generally accepted. Italicize single-letter symbols representing quantities (see Sections 4.2, "Symbols and Signs," and 10.1.1.2, "Scientific Uses"), but use roman type for modifiers that do not represent quantities.

C_{cr} *for* creatinine clearance

23.5.9 Respiration

The symbols and abbreviations used in respiratory physiology that have been developed to describe human pulmonary function in studies are also applied to the pulmonary function of other animals.

Use italic type for single capital letters representing the major symbols in respiratory physiology (see Section 4.2, "Symbols and Signs," Table 4.5, and Section 10.1.1.2, "Scientific Uses"), but do not italicize multiletter symbols representing such variables. Use small capital or lowercase letters in roman type for modifiers (ie, specifications) of the main symbols. Depending on the modifier, it is set on the line with the main symbol or it is subscripted. Multiple subscripts are separated by commas. If a main symbol has several modifiers, the modifiers should be placed in the following order: anatomic location (where), time (when), and condition or quality (what or how). The same letter may be used for different meanings (eg, "*C*" for both "concentration" and "compliance"). Similarly, the same letter may be used as a small or subscripted capital letter and as a lowercase letter, which carry different meanings (eg, "A" for "alveolar" and "a" for "arterial").

V_{CO_2} *for* CO_2 production per unit time

$FEF_{200-1,200}$ *for* forced expiratory flow between 200 and 1,200 mL of forced vital capacity

*P*aw *for* pressure at any point along the airways

See Table 23.4[68] for the main symbols and examples of modifying symbols. More detailed descriptions of nomenclature and symbols are available in *Fishman's Pulmonary Diseases and Disorders*[68] and the *AMA Manual of Style: A Guide for Authors and Editors*.[24]

Table 23.4. Examples of Variables for Respiratory Function[a]

Variable	Symbol[b]	Variable	Symbol[b]
Main symbols		Forced vital capacity	FVC
Concentration, compliance	*C*	Residual volume	RV
Fractional concentration of a gas	*F*	Total lung capacity	TLC
Pressure, partial pressure	*P*	**Modifiers (on the line)**	
Gas volume	*V*	Alveolar (small capital)	A
Gas flow	*V̇*	Expired (small capital)	E
Blood volume	*Q*	Inspired (small capital)	I
Blood flow	*Q̇*	Lung (small capital)	L
Abbreviations		Arterial (lowercase)	a
Inspiratory capacity	IC	Blood (lowercase)	b
Forced expiratory volume	FEV	Capillary (lowercase)	c

[a] See *Fishman's Pulmonary Diseases and Disorders*[68] for a complete list of symbols.

[b] Main symbols are combined with modifiers and with gas abbreviations as subscripts as appropriate (eg, "PA_{CO_2}" for alveolar pressure of carbon dioxide and "Ca_{O_2}" for arterial concentration of oxygen).

23.5.10 Thermal Physiology

The symbols for thermal physiology[69] represent physical quantities. They may include modifying subscripts to define physical and physiological specificities. The *Glossary of Terms for Thermal Physiology*[70] provides a list of relevant symbols, abbreviations, and SI units. However, some of these symbols have been given more than one meaning. Table 23.5[70] lists many of the quantities and their symbols with the proper SI units. Standard 80000-5:2019[71] of the International Organization for Standardization requires that symbols for thermal physiology be italicized, even though other sources may not conform to that convention.

23.5.11 Reproduction: Inheritance and Pedigrees

The symbols used in genetics to designate generations are listed in Table 23.6. Each of these symbols consists of a single roman capital letter and a subscript indicating the number of the generation.

Figure 23.1 illustrates the symbols for pedigree diagrams recommended by the National Society of Genetic Counselors[72] (NSGC) in the United States. Generations are assigned roman numerals in consecutive order, beginning with the uppermost (ie, first)

Table 23.5. Examples of Variables for Thermal Physiology[a]

Quantity	Symbol	SI unit
Area, total body	A_b	m^2
Metabolic rate, basal	BMR	W, W m^{-2}, *and* W kg^{-1}
Metabolic heat production	H	W m^{-2}
Radiant intensity, spectral	$I\lambda$	W sr^{-1} nm^{-1}
Humidity, absolute	γ	kg m^{-3}
Pressure, vapor (saturated) at temperature T	$P_{s,T}$	Pa
Pressure, water vapor	P_w	Pa
Temperature, ambient	T_a	°C

SI, Système International d'Unités (International System of Units)
[a] A complete list of symbols and modifiers is available in the *Glossary of Terms for Thermal Physiology*.[70]

Table 23.6. Symbols for Designating Generations

Symbol	Meaning	Example
F_1, F_2, . . .	Filial generations	$P_1 \times P_1 \rightarrow F_1$ $F_1 \times F_1 \rightarrow F_2$
P_1, P_2, . . .	Parental generations	P_1 = parents of F_1 P_2 = grandparents of F_1
B_1, B_2, . . .	Backcross generations	$F_1 \times P_1 \rightarrow B_1$ $B_1 \times P_1 \rightarrow B_2$
S_1, S_2, . . .	Self-fertilized generations (only for plants)	Parental self-fertilization $\rightarrow S_1$ S_1 self-fertilization $\rightarrow S_2$
I_1, I_2, . . .	Inbred generations	. . .
E_1, E_2, . . .	Generations after experimental manipulation	. . .
X_2	Offspring of F_1 testcross	. . .

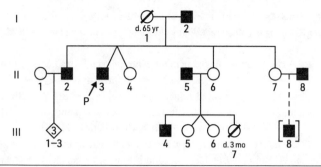

Figure 23.1. Symbols for pedigree diagrams. Squares represent males, and circles represent females. Diamonds are used for individuals of unknown or unspecified gender, transgender individuals, and those with congenital disorders of sex development. A slash through the symbol for an individual means that the individual is deceased; the age at death is given below the symbol. Vertical square brackets around the symbol for an individual indicate that the individual was adopted; a dashed vertical line connects the individual to the adoptive parents. Members of the same generation may be grouped in a single symbol, which should contain a numeral for the number of individuals. Symbols for individuals carrying the trait tracked by a diagram are solid. All other symbols are open. Symbols may be subdivided to indicate multiple traits.

generation in the diagram. Individuals on the same horizontal generation line are identified by arabic numerals assigned from left to right.

Symbols for individuals carrying the trait tracked by a diagram are solid. In addition, a symbol for an individual identified by the letter P and an arrow represents a proband, which is "an affected individual coming to medical attention independent of other family members."[73]

Parents are connected by a horizontal line in such diagrams. Offspring are arrayed below on a parallel line in birth order from left to right. Identical (ie, monozygotic) twins are connected to the horizontal sibling line by a single line. Fraternal (ie, dizygotic) twins have separate lines that join the horizontal sibling line at the same point. For individuals with no offspring, a vertical line exits the symbol downward. This line terminates either in a short horizontal line, which indicates no offspring by choice or reason unknown, or in a double line, which indicates infertility. A slash through a symbol indicates that the individual has died.

Additional symbols can be found in the NSGC's recommendations.[72] For nomenclature and symbols for chromosomes and genes, see Chapter 21, "Genes, Chromosomes, and Related Molecules."

CITED REFERENCES

1. Flint J, Racaniello VR, Rall GF, Hatziioannou T, Skalka AM. Principles of virology. 5th ed. ASM Press; Wiley; 2020. 2 vols.

2. Hull R. Plant virology. 5th ed. Elsevier; 2013.

3. Maclachlan N, Dubovi EJ, editors. Fenner's veterinary virology. 5th ed. Academic Press; 2016.

4. Collier L, Kellam P, Oxford J. Human virology. 5th ed. Oxford University Press; 2016.

5. Lee JJ, Leedale GF, Bradbury PC. An illustrated guide to the protozoa. 2nd ed. Society of Protozoologists; 2000. 2 vols.

6. Pommerville JC. Fundamentals of microbiology. 12th ed. Jones & Bartlett Learning; 2021.

7. Evert RF, Eichhorn S. Esau's plant anatomy. 3rd ed. Wiley; 2006.

8. Taiz L, Zeiger E, Møller IM, Murphy A. Plant physiology and development. 6th ed. Sinauer Sunderland; 2014.

9. Tortora GJ, Derrickson BH. Principles of anatomy & physiology. 16th ed. Wiley; 2020.

10. Allaby M. Dictionary of plant sciences. 4th ed. Oxford University Press; 2019.

11. Gordh G, Headrick DH, compilers. A dictionary of entomology. 2nd ed. CABI; 2011.

12. D'Arcy CJ, Eastburn DM, Schumann GL. Illustrated glossary of plant pathology. The Plant Health Instructor; [accessed 2023 Mar 19]. https://doi.org/10.1094/PHI-I-2001-0219-01

13. Holliday P. A dictionary of plant pathology. 2nd ed. Cambridge University Press; 2001.

14. Kirk PM, Cannon PF, Minter DW, Stalpers JA, editors. Ainsworth and Bisby's dictionary of the Fungi. 10th ed. CABI; 2011.

15. Lackie JM. The dictionary of cell and molecular biology. 5th ed. Academic Press; 2013.

16. Salisbury FB, editor. Units, symbols, and terminology for plant physiology: a reference for presentation of research results in the plant sciences. Oxford University Press; 1996.

17. Both AJ et al. Guidelines for measuring and reporting environmental parameters for experiments in greenhouses. Plant Methods. 2015;11:43. https://doi.org/10.1186/s13007-015-0083-5

18. Shibles R. Terminology pertaining to photosynthesis. Crop Sci. 1976;16(3):437–439. Report by the Crop Science Society of America's Committee on Crop Terminology. https://doi.org/10.2135/cropsci1976.0011183X001600030033x

19. American Society of Agronomy. Publications handbook and style manual. American Society of Agronomy; [updated 2021 Nov]. Jointly published with the Crop Science Society of America and the Soil Science Society of America. https://www.agronomy.org/publications/journals/author-resources/style-manual/

20. Meier U, editor. Growth stages of mono- and dicotyledonous plants. 2nd ed. Engl ed. Federal Biological Research Centre for Agriculture and Forestry (DE); 2018. https://doi.org/10.5073/20180906-074619

21. Transactions of the American Fisheries Society: guide for authors. Trans Am Fish Soc. 2011;140(1):201–206. https://doi.org/10.1080/00028487.2011.566490

22. Alaska Department of Fish and Game, Interdivisional Publications Committee. Writer's guide. 3rd ed. Alaska Department of Fish and Game; 2010. Also available at http://www.adfg.alaska.gov/static/home/library/PDFs/writersguide_full.pdf

23. Koo TSY. Age designation in salmon. In: Koo TSY, editor. Studies of Alaska red salmon. University of Washington Press; 1962. p 34–48. University of Washington publications in fisheries, new series, vol 1.

24. JAMA Network Editors. AMA manual of style: a guide for authors and editors. 11th ed. Oxford University Press; 2020. Also available at https://www.amamanualofstyle.com

25. World Association of Veterinary Anatomists, International Committee on Veterinary Gross Anatomical Nomenclature. Nomina anatomica veterinaria. 6th ed. World Association of Veterinary Anatomists, Editorial Committee; 2017. 178 p. Available at http://www.wava-amav.org/wava-documents.html

26. Koenig HE, Korbel R, Liebich HHG, editors. Avian anatomy: textbook and colour atlas. 2nd ed. Klupiec C, translator. 5m Publishing; 2016.

27. Baumel JJ, King AS, Breazile JE, Evans HE, Van den Berge JC, editors. Handbook of avian anatomy: nomina anatomica avium. 2nd ed. Nuttall Ornithological Society; 1993. Nuttall Ornithological Club publication; 23. Prepared by the World Association of Veterinary Anatomists, International Committee on Avian Anatomical Nomenclature. Also available at https://www.academia.edu/20947312/HANDBOOKOF_AVIAN_ANATOMY_NOMINA_ANATOMICA_AVIUM_Second_Edition

28. Lovette IJ, Fitzpatrick JW, editors. Handbook of bird biology. 3rd ed. Wiley; 2016. Prepared for Cornell Lab of Ornithology.

29. Kurtz DM, Travlos GS, editors. The clinical chemistry of laboratory animals. 3rd ed. Taylor & Francis; 2017.

30. Kaneko JJ, Harvey JW, Bruss ML, editors. Clinical biochemistry of domestic animals. 6th ed. Academic Press; 2008.

31. Blais MC, Penedo MCT. Erythrocyte antigens and blood groups. In: Brooks M, Harr KE, Seelig D, Wardrop KJ, Weiss DJ, editors. Schalm's veterinary hematology. 7th ed. Wiley-Blackwell; 2022. p 877–890.

32. Alaverdi N. Monoclonal antibodies to mouse cell-surface antigens. Curr Protoc Immunol. 2004;62(1):Appendix 4B. https://doi.org/10.1002/0471142735.ima04bs47

33. Puklavec JM, Barclay AN. Monoclonal antibodies to rat leukocyte surface antigens, MHC antigens, and immunoglobulins. Curr Protoc Immunol. 2001 May; Appendix 4C. https://doi.org/10.1002/0471142735.ima04cs31

34. Hess P. Major histocompatibility complex antigens. In: Brooks MB, Harr KE, Seelig DM, Wardrop KJ, Weiss DJ, editors. Schalm's veterinary hematology. 7th ed. Wiley-Blackwell; 2022. p 48–62.

35. International Federation of Associations of Anatomists, Federative International Programme for Anatomical Terminology. Terminologia anatomica. 2nd ed. Federative International Programme for Anatomical Terminology; 2019; approved 2020. https://fipat.library.dal.ca/TA2/

36. Klein H, Anstee D. Mollison's blood transfusion in clinical medicine. 12th ed. Wiley-Blackwell; 2014.

37. Daniels GL et al. Blood group terminology 2004: from the International Society of Blood Transfusion committee on terminology for red cell surface antigens. Vox Sang [The International Journal of Transfusion Medicine]. 2004;87(4):304–316. https://doi.org/10.1111/j.1423-0410.2004.00564.x

38. International Society for Blood Transfusion, Working Party for Red Cell Immunogenetics and Blood Group Terminology. Red cell immunogenetics and blood group terminology. International Society for Blood Transfusion. https://www.isbtweb.org/working-parties/red-cell-immunogenetics-and-blood-group-terminology

39. von dem Borne AEG, Decary F. ICSH/ISBT Working Party on Platelet Serology. Vox Sang [The International Journal of Transfusion Medicine]. 1990;58(2):176. https://doi.org/10.1111/j.1423-0410.1990.tb02085.x

40. Marder VJ, Mannucci PM, Firkin BG, Hoyer LW, Meyer D. Standard nomenclature for factor VIII and von Willebrand factor: a recommendation by the International Committee on Thrombosis and Haemostasis. Thromb Haemost. 1985;54(4):871–872.

41. Dempster DW et al. Standardized nomenclature, symbols, and units for bone histomorphometry: a 2012 update of the report of the ASBMR Histomorphometry Nomenclature Committee. J Bone Miner Res. 2013;28(1):1–16. https://doi.org/10.1002/jbmr.1805

42. New York Heart Association Criteria Committee. Nomenclature and criteria for diagnosis of diseases of the heart and great vessels. 9th ed. Little, Brown; 1994.

43. Young DS. Implementation of SI units for clinical laboratory data: style specifications and conversion tables. Ann Intern Med. 1987;106(1):114–129. Errata in Ann Intern Med. 1987;107(2):265; Ann Intern Med. 1989;110(4):328; Ann Intern Med. 1991;114(2):172. https://doi.org/10.7326/0003-4819-106-1-114

44. Marsh SGE et al. Nomenclature for factors of the HLA system, 2010. Tissue Antigens. 2010;75(4):291–455. https://doi.org/10.1111/j.1399-0039.2010.01466.x

45. HLA: immune response genetics. Vol 1. 1971. Continues: Tissue Antigens.

46. Human immunology. Elsevier. Vol 1. 1980.

47. International Journal of Immunogenetics. Vol 32. 2005. Continues: European Journal of Immunogenetics, 1991.

48. National Marrow Donor Program (US). HLA education; [accessed 2023 Mar 20]. http://bioinformatics.nmdp.org/Education/HLA_Educational_Resources.aspx

49. Gunther E, Walter L. The major histocompatibility complex of the rat (*Rattus norvegicus*). Immunogenetics. 2001;53(7):520–542. https://doi.org/10.1007/s002510100361

50. Walter L. Nomenclature report on the major histocompatibility complex genes and alleles of the laboratory rat (*Rattus norvegicus*). Immunogenetics. 2020;72(1–2):5–8. http://doi.org/10.1007/s00251-019-01131-y

51. Wagner JL. Organization of the canine major histocompatibility complex. J Hered. 2003;94(1):23–26. https://doi.org/10.1093/jhered/esg002

52. Klein J et al. Revised nomenclature of mouse H-2 genes. Immunogenetics. 1990;32(3):147–149. https://doi.org/10.1007/BF02114967

53. Stuart PM. Major histocompatibility complex (MHC): mouse. In: eLS. John Wiley & Sons, Ltd (Ed); 2015. https://doi.org/10.1002/9780470015902.a0000921.pub4

54. Klein J et al. Nomenclature for the major histocompatibility complexes of different species: a proposal. In: Solheim BG, Ferrone S, Möller E, editors. The HLA system in clinical transplantation: basic concepts and importance. Springer; 1993. p 407–411. https://doi.org/10.1007/978-3-642-77506-2_32

55. Ellis SA. Minireview: MHC studies in domestic animals. Int J Immunogenet. 1994;21(3):209–215. https://doi.org/10.1111/j.1744-313X.1994.tb00194.x

56. Schook LB, Lamont SJ. The major histocompatibility complex region of domestic animal species. CRC Press; 1996.

57. International Union of Immunological Societies and World Health Organization, IUIS–WHO Subcommittee on Nomenclature of Fc receptors. Nomenclature of Fc receptors. Bull World Health Organ. 1994;72(5):809–810. https://www.ncbi.nlm.nih.gov/pmc/articles/PMC2486555/

58. Flajnik MF, Singh NJ, Holland SM, editors. Paul's fundamental immunology. 8th ed. Lippincott Williams & Wilkins; 2022.

59. Kemper C, Pangburn MK, Fishelson Z. Complement nomenclature 2014. Mol Immunol. 2014;61(2):56–58. https://doi.org/10.1016/j.molimm.2014.07.004

60. International Union of Immunological Societies and World Health Organization, subcommittee of the WHO–IUIS Nomenclature Committee. Nomenclature of the alternative activating pathway of complement. Bull World Health Organ. 1981;59(3):489–491. https://www.ncbi.nlm.nih.gov/pmc/articles/PMC2396063/

61. Beare A, Stockinger H, Zola H, Nicholson I. Monoclonal antibodies to human cell surface antigens. Curr Protoc Immunol. 2008;80(1):Appendix 4A. https://doi.org/10.1002/0471142735.ima04as80

62. Engel P et al. CD nomenclature 2015: human leukocyte differentiation antigen workshops as a driving force in immunology. J Immunol. 2015;195(10):4555–4563. https://doi.org/10.4049/jimmunol.1502033

63. World Health Organization and International Union of Immunological Societies, WHO–IUIS Nomenclature Subcommittee on Interleukin Designation. Nomenclature for secreted regulatory proteins of the immune system (interleukins). Bull World Health Organ. 1991;69(4):483–484. https://www.ncbi.nlm.nih.gov/pmc/articles/PMC2393236/

64. Interferon nomenclature. Arch Virol. 1983;77(2–4):283–285. https://doi.org/10.1007/BF01309277

65. International Union of Immunological Societies and World Health Organization, IUIS–WHO Allergen Nomenclature Subcommittee. Allergen nomenclature. Bull World Health Organ. 1994;72(5):797–800. https://www.ncbi.nlm.nih.gov/pmc/articles/PMC2486548/

66. International Society of Hypertension, Nomenclature Committee. Nomenclature of the renin-angiotensin system. Report of the Nomenclature Committee of the International Society of Hypertension, Hypertension. 1979;1(6):654–656. https://doi.org/10.1161/01.HYP.1.6.654

67. International Union of Physiological Sciences, Renal Commission. A standard nomenclature for structures of the kidney [editorial review]. Kidney Int. 1988;33(1):1–7. https://doi.org/10.1038/ki.1988.1

68. Grippi MA et al, editors. Fishman's pulmonary diseases and disorders. 3rd ed. McGraw-Hill; 2015. 2 vols.

69. Proposed standard system of symbols for thermal physiology. J Appl Physiol. 1969:27(3):439–446. https://doi.org/10.1152/jappl.1969.27.3.439

70. International Union of Physiological Sciences, Commission for Thermal Physiology. Glossary of terms for thermal physiology. 3rd ed. Jpn J Physiol. 2001:51(2):245–280.

71. International Organization for Standardization. Quantities and units—part 5: thermodynamics. International Organization for Standardization. (ISO 80000-5:2019). https://www.iso.org/obp/ui/#iso:std:iso:80000:-5:ed-2:v1:en

72. Bennett RL, French KS, Resta RG, Doyle DL. Standardized human pedigree nomenclature: update and assessment of the recommendations of the National Society of Genetic Counselors. J Genet Couns. 2008;17(5):424–433. https://doi.org/10.1007/s10897-008-9169-9

73. Bennett RL et al. Reply to Mazarita and Curtis [letter]. Am J Hum Genet. 1995;57(4):983–984.

ADDITIONAL REFERENCES

Boden E, Andrews A, editors. Black's student veterinary dictionary. 22th ed. Bloomsbury Information; 2017.

Floyd MR. The modified triadan system: nomenclature for veterinary dentistry. In: DeForge DH, Colmery BH, editors. An atlas of veterinary dental radiology. Iowa State University Press; 2000. p 265–266.

Studdert VP, Gay CC, Hinchcliff KW, editors. Saunders comprehensive veterinary dictionary. 5th ed. WB Saunders; 2020.

24 Disease Names

**Editors: Ellen Lazarus, MD, ELS; Stephanie Mowat, MA;
and Soo-Hee Chang, MS, ELS**

24.1 PLANT DISEASES

24.1.1 Standard Nomenclature for Plant Diseases

Plant disease names are not governed by formal rules of taxonomy and nomenclature. A plant disease is usually named for its major symptom or sign or for the pathogen responsible for the disease. A few names refer to geographic locations, but eponyms are rare in plant pathology (see Table 24.1).

The same name may be used for diseases affecting several host species, even though each disease is caused by a different pathogen. For example, scab of wheat and scab of barley are caused by several species of *Fusarium*, whereas common scab of potato is caused by *Streptomyces scabies*. Therefore, depending on the context, it may be important to include the host name in the disease name.

> scab of wheat   *Phytophthora* rot of soybean
> potato scab   southern blight of red clover

A pathogen species may be moved to another genus for taxonomic reasons. Although the original disease names are often retained in such situations because they are familiar to lay people, it is preferable to use the names that reflect the current nomenclatural status of the pathogens.

> *Stenocarpella* stalk and ear rot of maize caused by *Stenocarpella maydis*
> (formerly *Diplodia* stalk and ear rot of maize caused by *Diplodia maydis*)

A Dictionary of Plant Pathology[1] suggests 2 principles for naming plant diseases without regard to pathogens, but they are not always followed.

(1) A name should be readily usable by plant growers, who are more concerned with the disease than with the pathogen.
(2) A name should mention the host's and the disease's most conspicuous abnormality instead of the name of the pathogen.

Table 24.1. Sources of Plant Disease Names

Plant disease	Source of name
anther smut	plant tissue affected
aster yellows	name of the pathogen
bacterial soft rot	general pathogen group and a symptom of the disease
black root rot	a symptom of the disease and the tissue affected
Dwarf Cavendish tip rot	cultivar name and the tissue affected
Gibberella ear rot	genus name of the pathogen and the part of the host affected
Granville wilt	county in North Carolina where this tobacco disease was first reported
gray mold rot	appearance of the fungal pathogen
Karnal bunt	region in India where this wheat disease was first found
northern corn leaf blight	each word refers to a different characteristic of the disease, specifically the general geographic area where the disease occurs, the host, the tissue affected, and the disease's major symptom
Pierce disease	named after plant pathologist Newton B. Pierce, who first described the disease in 1892
powdery mildew	appearance of the fungal pathogen on the host plant's surface
Stewart wilt	named after F.C. Stewart, who first isolated the pathogen

> cotton wilt *for Verticillium* wilt
> eucalyptus canker *for Cytospora* canker
> soybean root rot *for Phytophthora* root rot
> Dutch elm wilt *for* Dutch elm disease (wilt is the primary symptom, and the word "disease" should not be part of a plant disease name)

These principles do have limitations. If 2 or more pathogens produce similar symptoms in the same host, it may be difficult to coin distinct disease names based only on the host and the symptoms. For example, *Stagonospora nodorum* and *Septoria tritici* both cause a leaf blotch on wheat with symptoms that can be difficult to distinguish. Therefore, the names "*Stagonospora* blotch" and "*Septoria* blotch" are used rather than "wheat blotch." Similarly, various stalk rots of corn are difficult to distinguish symptomatically, so pathogen names are included in the names of these stalk rots (eg, *Fusarium* stalk rot and *Gibberella* stalk rot).

Diseases and injuries caused by pest nematodes are generally referred to by the common names of the nematode pathogens.

> root knot of tomato (caused by root knot nematode)
> lesion nematode on corn
> soybean cyst nematode

The American Phytopathological Society maintains a list of recommended common names for plant diseases.[2] While these names do not carry the force of nomenclatural codes for organisms, their use is encouraged. Besides *A Dictionary of Plant Pathology*[1] and the American Phytopathological Society's *Common Names of Plant Diseases*,[2] reliable sources for plant disease names include the *Thesaurus of Agricultural Organisms: Pests, Weeds, and Diseases*[3] and *Westcott's Plant Disease Handbook*.[4]

24.1.2 Capitalization and Italicization in Plant Disease Names

Journals that report on plant diseases differ on whether to capitalize and italicize the scientific Latin names of pathogens that are part of disease names. Some journals use both an initial capital and italics for scientific names of pathogens, some use an initial capital but no italics, and some use neither. For consistency and accuracy, this style manual recommends italicizing the scientific names of plant-pathogenic organisms and using an initial capital with genus names (see Section 22.2.2.1, "Binomial System for Species Names"). In addition, when a genus name is used without a specific epithet in plant names, the genus's noun form, not its adjectival form, should be used. The noun form should be capitalized and italicized.

> *Fusarium* head blight *not* Fusarium head blight, *not* fusarium head blight, *and not* fusarial head blight

24.2 DISEASES OF HUMANS AND OTHER ANIMALS

24.2.1 Standard Nomenclature for Diseases of Humans and Other Animals

Various standard nomenclatures for diseases and for histological classifications of tumors exist worldwide. Generally accepted names and synonyms can be found in standard medical, dental, and veterinary dictionaries, as well as vocabulary and statistical databases. Health care journals and other scientific publications will typically designate a preferred scientific dictionary or database to ensure that disease names and other biomedical terms are used consistently.

One such authoritative source is the Unified Medical Language System (UMLS),[5] which is maintained by the National Library of Medicine (NLM) in the United States. This large multipurpose vocabulary database contains biomedical and other health-related terminology, as well as various names for health concepts and the relations among them. As a metathesaurus, the UMLS is built from the electronic versions of more than 100 thesauri, classifications, code sets, and lists of terms used in patient care; health services billing; public health statistics; indexing and cataloging of biomedical literature; and basic, clinical, and health services research.

The Medical Subject Headings (MeSH) vocabulary thesaurus[6] is another source for verifying disease terminology used in human and veterinary medicine. The NLM uses MeSH terms for indexing its databases, and many other medical indexers use MeSH to ensure that indexes in biomedical books and journals are compatible with the NLM databases.

The widespread adoption of electronic health records has increased the need for standard code sets and terminologies for clinical data, such as those found in the UMLS. One comprehensive compilation that has gained international recognition is the Systematized Nomenclature of Medicine–Clinical Terms[7] (SNOMED CT). SNOMED CT covers the nomenclature of diseases, disorders, syndromes, and their manifestations in humans and other animals, as well as such related topics as clinical procedures, devices, and pharmaceutical products.

For epidemiological and health management purposes, the World Health Organization

(WHO) maintains the International Statistical Classification of Diseases and Related Health Problems[8] (ICD) and its extensions, which focus on specific disciplines. For example, the extension titled International Classification of Diseases for Oncology (ICD-O)[9] is the international standard for the classification and nomenclature of cancer based on tissue type and histopathology.

Standard nomenclatures for narrower fields have been published. Some of them are included in the UMLS metathesaurus, as is the American Psychiatric Association's *Diagnostic and Statistical Manual of Mental Disorders*,[10] which contains terms and classifications for psychiatric diagnoses. Among other specialized nomenclatures is a comprehensive series that the WHO publishes to classify human tumors.[11-23] Similarly, the Union for International Cancer Control (UICC) publishes a widely used classification system for tumors, nodes, and metastases (TNM) that stages tumors based on characteristics of the tumor site of origin, involvement of regional lymph nodes, and the absence or presence of distant metastases.[24,25] In addition, the International Society for Human and Animal Mycology has recommendations on naming fungal diseases.[26]

Among the resources for finding standardized names for diseases in animals other than humans is the Standardized Nomenclature of Animal Parasitic Diseases, better known as SNOAPAD.[27,28] Standardized nomenclatures for animal tumors are also available.[29-31]

24.2.2 Eponymic Disease Names

Most journal and textbook editors recommend using descriptive disease names rather than eponymic names. Medical dictionaries and other specialized dictionaries[32-34] provide synonyms for the eponymic names of diseases.

> amyotrophic lateral sclerosis *not* Lou Gehrig disease
> *Bartonella* infection (systemic or cutaneous) *not* Carrión disease

If an eponymic name is necessary, use the proper noun in the term in its nonpossessive form rather than its possessive form (see Section 6.5.6, "Eponymic Terms"). In addition, the adjectival and derivative forms of proper nouns in disease names should be lowercased (see Section 9.4.3, "Eponymic Terms").

> Alzheimer disease *not* Alzheimer's disease   addisonian anemia *from* Thomas Addison
> Graves disease *not* Graves' disease   parkinsonism *from* James Parkinson
> Sjögren syndrome *not* Sjögren's syndrome

CITED REFERENCES

1. Holliday P. A dictionary of plant pathology. 2nd ed. Cambridge University Press; 2001.

2. American Phytopathological Society, Committee on Standardization of Common Names for Plant Diseases. Common names of plant diseases. American Phytopathological Society; c2022. http://www.apsnet.org/publications/commonnames/Pages/default.aspx

3. Derwent Publications. Thesaurus of agricultural organisms: pests, weeds, and diseases. Chapman & Hall; 1990. 2 vols.

4. Horst RK. Westcott's plant disease handbook. 8th ed. Springer Dordrecht; 2013.

5. Unified Medical Language System (UMLS). National Library of Medicine (US); 2019. http://www.nlm.nih.gov/research/umls/

6. Medical subject headings. National Library of Medicine (US); 2020. http://www.nlm.nih.gov/mesh/

7. SNOMED CT. SNOMED International; 2021. http://www.snomed.org/

8. World Health Organization. International statistical classification of diseases and related health problems (ICD). 11th rev. World Health Organization; 2018. http://www.who.int/classifications/icd/en/

9. Fritz A et al, editors. International classification of diseases for oncology. 3rd ed. World Health Organization; 2000. https://www.who.int/standards/classifications/other-classifications/international-classification-of-diseases-for-oncology

10. American Psychiatric Association. Diagnostic and statistical manual of mental disorders. 5th ed. American Psychiatric Association; 2013.

11. El-Naggar AK, Chan JKC, Grandis JR, Takata T, Slootweg PJ, editors. WHO classification of head and neck tumours. 4th ed. World Health Organization; 2017.

12. WHO Classification of Tumours Editorial Board, editors. Digestive system tumours. 5th ed. World Health Organization; 2019.

13. Lloyd RV, Osamura RY, Klöppel G, Rosai J, editors. WHO classification of tumours of endocrine organs. 4th ed. World Health Organization; 2017.

14. Moch H, Humphrey PA, Ulbright TM, Reuter VE, editors. WHO classification of tumours of the urinary system and male genital organs. 4th ed. World Health Organization; 2016.

15. WHO Classification of Tumours Editorial Board, editors. Soft tissue and bone tumours. 5th ed. World Health Organization; 2020.

16. Elder DE, Massi D, Scolyer RA, Willemze R, editors. WHO classification of skin tumours. 4th ed. World Health Organization; 2018.

17. Louis DN, Ohgaki H, Wiestler OD, Cavenee WK, editors. WHO classification of tumours of the central nervous system. 4th ed. World Health Organization; 2016.

18. Swerdlow SH et al. WHO classification of tumours of haematopoietic and lymphoid tissues. 4th rev ed. World Health Organization; 2017.

19. WHO Classification of Tumours Editorial Board, editors. Breast tumours. 5th ed. World Health Organization; 2019.

20. WHO Classification of Tumours Editorial Board, editors. Female genital tumours. 5th ed. World Health Organization; 2020.

21. Travis WD, Brambilla E, Burke AP, Marx A, Nicholson AG, editors. WHO classification of tumours of the lung, pleura, thymus and heart. 4th ed. World Health Organization; 2015.

22. WHO Classification of Tumours Editorial Board, editors. Thoracic tumours. 5th ed. World Health Organization; 2021.

23. Grossniklaus HE, Eberhart CG, Kivelä TT, editors. WHO classification of tumours of the eye. 4th ed. World Health Organization; 2018.

24. Brierley JD, Gospodarowicz MK, Wittekind C, editors. TNM classification of malignant tumours. 8th ed. Wiley-Blackwell; 2016.

25. Union for International Cancer Control. What is the TNM cancer staging system? Union for International Cancer Control; 2021. [approximately 6 screens]. https://www.uicc.org/resources/tnm

26. Odds FC et al. Nomenclature of fungal diseases: a report and recommendations from a sub-committee of the International Society for Human and Animal Mycology (ISHAM). J Med Vet Mycol. 1992;30(1):1–10. https://doi.org/10.1080/02681219280000021

27. Kassai T, Burt MDB. A plea for consistency. Parasitol Today. 1994;10(4):127–128. https://doi.org/10.1016/0169-4758(94)90257-7

28. Kassai T et al. Standardized nomenclature of animal parasitic diseases (SNOAPAD). Vet Parasitol. 1988;29(4):299–326. https://doi.org/10.1016/0304-4017(88)90148-3

29. International histological classification of tumours of domestic animals. Bull World Health Organ. 1974;50(1–2):1–142. Also available at https://www.ncbi.nlm.nih.gov/pmc/issues/169501/

30. International histological classification of tumours of domestic animals. Part 2. Bull World Health Organ. 1976;53(2–3):137–304. Also available at https://www.ncbi.nlm.nih.gov/pmc/issues/165698/

31. Meuten DJ, editor. Tumors in domestic animals. 5th ed. Wiley-Blackwell; 2017.

32. Jablonski S. Jablonski's dictionary of syndromes and eponymic diseases. 2nd ed. Krieger Publishing Company; 1991.

33. Magalini SI, Magalini SC, editors. Dictionary of medical syndromes. 4th ed. Lippincott; 1997.

34. Dorland's illustrated medical dictionary. 33rd ed. Elsevier; 2020. Also available at https://www.dorlandsonline.com/

25 Earth

Editors: David M. Schultz, PhD, FAMS, FRMetS, FGS, SFHEA, and Kaitlyn Aman Ramm

25.1 SCOPE

The atmosphere, rocks, soil, and bodies of water are studied intensively in the disciplines of meteorology, geology, soil science, limnology, oceanography, and related fields. This chapter covers the conventions used to report research on these subjects.

25.2 GEOLOGICAL TIME UNITS

Formal definitions of geological time divisions have been ratified through the International Union of Geological Sciences. Terms for international geological divisions of time can be found in the *International Stratigraphic Guide*[1] and the *International Stratigraphic Guide–Abridged Version*.[2] These terms are listed in order of increasing geological age in the International Chronostratigraphic Chart,[3] which uses different typography to distinguish formal names from informal ones.

25.2.1 Classification Systems

The 3 most common classification systems used for geological time units are chronostratigraphy, geochronology, and geochronometry. Chronostratigraphy is a time-rock, or time-stratigraphic, system in which units of time are based on all rocks formed during the same span of time. In geochronology, each unit of time is an interval during which a chronostratigraphic unit was deposited. Chronostratigraphic units delineate the relative positions of Earth's strata, whereas geochronologic units delineate the age of those strata. For example, rocks in the chronostratigraphic unit known as the Jurassic System were formed during the geochronologic unit known as the Jurassic Period. In contrast, in geochronometry, unit boundaries are based on absolute time measured in years.

Chronostratigraphic units are organized in order of decreasing rank as follows: eonothem, erathem, system, series, stage, and chronozone. Geochronologic units are organized in order of decreasing rank as follows: eon, era, period, epoch, age, and chron. Chronozones and chrons, however, are nonhierarchical, and they are usually smaller than the other units.[4]

Geochronologic units should be used to designate the age of material formed since 540 Ma ("Ma" stands for "mega-annum," or 1 million years). On the other hand, rocks and other material older than 540 Ma should be divided into geochronometric units with defined time boundaries.

> Archean Eon (older than 2,500 Ma)
> Proterozoic Eon (between 2,500 to 538.8 Ma)

The most common geochronometric units with agreed-upon boundaries are eons and eras. For example, the International Commission on Stratigraphy (ICS) divides the Proterozoic Eon into the Paleoproterozoic, Mesoproterozoic, and Neoproterozoic eras and the Archean Eon into the Eoarchean, Paleoarchean, Mesoarchean, and Neoarchean eras. The ICS[3] also divides the 3 eras of the Proterozoic into periods.

The North American Commission on Stratigraphic Nomenclature (NACSN) uses a similar naming system for eras within the Proterozoic: The Early, Middle, and Late

Proterozoic eras are equivalent to the Paleoproterozoic, Mesoproterozoic, and Neoproterozoic. However, the NACSN divides the Archean into 3, not 4, eras, namely the Early, Middle, and Late Archean eras.[4]

The terms for geochronologic rank should only be used for geochronometric units that have been formally accepted. Geochronometric units may also correspond to chronostratigraphic units, though they are not defined by them.[1]

Because geochronometric units are essentially arbitrary time boundaries that have been internationally agreed upon, type localities (ie, deposits that define the characteristics of a particular geological period) do not exist.

25.2.2 Capitalization

Capitalization style differs for formal and informal geological time terms. Because the same words (eg, "early," "late," "epoch," and "era") can be used in both formal and informal names, authoritative bodies within Earth science disciplines have established the following capitalization conventions to reduce ambiguity:

(1) Capitalize each word in formal names, including modifiers such as "Early" and "Late."[4-6] Note that formal names can be used with or without their rank terms.

> Phanerozoic Eon   Late Jurassic   Maastrichtian Stage
> Cretaceous Period   Mesozoic Era   Early Ordovician

(2) When formal names are modified by informal modifiers, capitalize the formal names, but lowercase the modifiers.

> early Mesozoic Era (here, "early" is an informal modifier of the formal name "Mesozoic Era")

(3) Lowercase rank terms when used alone and when they are used in plural form after 2 or more formal names.

> An eon divides into eras, and those eras divide into periods.
> Dinosaurs lived during the Triassic, Jurassic, and Cretaceous periods.

(4) Series, which are time units for rocks, are often termed "Upper" or "Lower," while the epochs they are in are termed "Late" or "Early." The paired terms "Upper and Lower" and "Late and Early" are not interchangeable. "Middle," however, may be used in both series and epoch names, and it can be either a formal or informal part of those names.

> Upper Triassic strata (position) were deposited in the Late Triassic (age).
> Lower Cambrian rocks (position) contain Early Cambrian fossils (age).
> Middle Devonian strata (position) formed before the middle Mississippian (informal age division) but after the Middle Ordovician (formal age division).

(5) In map explanations, in illustrations, and in tables, always use capitalized formal geological time terms instead of shortened terms. However, generic rank terms should be omitted, except in a column heading in a table.

> Permian and Pennsylvanian *not* Permo-Penn
> Cambrian and Ordovician *not* Cambro-Ordovician
> Middle Cambrian *not* Mid-Cambrian

Authors and editors should consult an authoritative source to determine whether a term is formal or informal rather than assume that common usage is correct.

25.2.3 Tentative Time Divisions

A question mark can be used to indicate that a time division is tentative or provisional. The question mark should be placed immediately after the part of the name whose accuracy is in doubt. In the discipline of geology, such question marks are considered more informative than less specific terms such as "probably."

> Late? Devonian (accuracy of epoch is in doubt)
> Late Devonian? (accuracy of period is in doubt)

To avoid confusion about the meaning of the question mark, avoid ending a sentence with a tentative or provisional name followed by a question mark.

25.2.4 Abbreviations for Geological Time

Terms for geological time may be abbreviated on maps (eg, "K" for Cretaceous) (see Section 25.3.2, "Map Units"), but do not abbreviate these terms in map explanations or in running text. Express absolute dates in geological time with the units giga-annum (Ga), which is equal to 10^9 years; mega-annum (Ma), which is 10^6 years; and kiloannum (ka), which is 10^3 years. In contrast, express duration of time in terms of gigayear (Gy), megayear (My), and kiloyear (ky).[4] Spell out units of time in running text when they are not preceded by numerical values.

> The Cretaceous Period lasted 80 My (duration) from 144 Ma to 65 Ma (absolute dates).

25.2.5 Magnetic Polarity

Throughout geological time, Earth's polarity has repeatedly reversed. Magnetic polarity imposes a record on rock as it is deposited. The following polarity-chronostratigraphic units distinguish primary magnetic polarity during the past 66 million years in order of decreasing age: Gilbert, Gauss, Matuyama, and Brunhes.

> Matuyama Reversed-Polarity Chronozone

25.3 STRATIGRAPHIC AND MAP UNITS

Stratigraphic unit names are given to formally defined mappable units of lithostratigraphic and lithodemic rock. Stratigraphic units may also have informal local names. Each of these mappable units is identified by its lithologic content and its boundaries. Each is assigned a place within the geological age sequence, and each has a stratigraphic rank.

25.3.1 Naming and Capitalization Conventions for Stratigraphic and Map Units

Rock units are usually discussed chronologically in text from oldest, known as "bottommost," to youngest, known as "topmost." A formally named unit will have a compound name consisting of a geographical name along with a descriptive lithic term (eg, shale, sandstone, or limestone), a rank term (eg, group, formation, member, or bed), or both lithic and rank terms. Capitalize each word in a formally named stratigraphic unit except for the species name in a biostratigraphic unit.

> Stony Point Tuff   Baltimore Gneiss
> Codell Sandstone Member   St Louis Limestone
> *Cordylodus intermedius* Zone

Refer to the *North American Stratigraphic Code*[4] and the *Lexicon of Canadian Stratigraphy*[7-9] for the recommended procedures for classifying and naming formal stratigraphic and related units.

For informal stratigraphic units, capitalize the place names, but lowercase the rank terms or descriptive lithics.

> St Louis coal   formation of Madeira Canyon   J sandstone

On first use, define and describe informal terms, including local stratigraphic units and commercial names, such as those used in the mining and oil industries. A modifier such as "the informal" may be used with the first mention of a name to alert readers that the name is for an informal stratigraphic unit.

> the informal St Louis coal (on first use) *but* St Louis coal (in subsequent uses)

Do not capitalize a stratigraphic rank term when it is used alone in text as a generic reference to a stratigraphic unit.

> the formation *for* Nevada Formation
> group in the West *for* Uinta Mountain Group

Stratigraphic rank terms may be abbreviated and capitalized in illustrations and tables but not in text.

> Gp *for* group   Fm *for* family   Mbr *for* member

In text, the full names of stratigraphic units need not be repeated after their first usage. On subsequent references, the descriptive lithic terms may be dropped, or the lithic and rank terms may be dropped together. In fact, in names with both lithic and rank terms, the lithic term cannot be retained if the rank term is dropped. The capitalization rules for formal and informal names of complete stratigraphic units also apply to their shortened names. Follow the recommendations in the *International Stratigraphic Guide*[1] for shortened forms of stratigraphic names.

> Cozzette Sandstone Member *shortens to* Cozzette Member *and to* Cozzette *but not to* Cozzette Sandstone

Unlike other stratigraphic units, biostratigraphic units are often named after the predominant plant or animal species that inhabited the biozones (eg, *Benueites benuensis* Zone). The taxonomy rules and formats for using species names as part of biozone names are the same as those for the names of plant and animal species themselves (see Chapter 22, "Taxonomy and Nomenclature").

The North American Commission on Stratigraphic Nomenclature[4] provides additional guidance on naming and capitalizing formal and informal stratigraphic names.

25.3.2 Map Units

Units on a geological map and within a map's stratigraphic sections are explained in the map's legend. Each map unit in the legend consists of a symbol and a description of the unit. Letter symbols for map units consist of uppercase and lowercase letters and sometimes subscripts. The symbols convey information about the chronostratigraphic, geochronologic, or geochronometric units, as well as the names of the map units.

A single capitalized first letter in a map symbol represents the geological period (eg, "K" for "Cretaceous" and "J" for "Jurassic"). Two capital letters indicate a range of 2 or more systems, with the younger system listed first (eg, "JK"). The lowercase letters that follow indicate either the name of the map unit or the type of rock. Special characters may be used to distinguish the names of systems or periods that begin with the same letter. For example, the symbol "P," which stands for "Permian," is modified to "PG" for "Paleogene," "Pz" for "Paleozoic," "[P" for "Pennsylvanian," "p€" for "Precambrian," and "P̲" for "Proterozoic."

Do not use map symbols in text as stand-alone abbreviations. However, map symbols may be placed in parentheses after unit names in text, especially when readers are referred to a map that uses the symbols.

> Locations of Jurassic Entrada Sandstone (Je) are shown on the map.
> *not* Locations of Je are shown on the map.

Letter symbols for map units are unique to each geological map, but some parts of symbols are standardized for all maps. For more information on map symbols, see *Suggestions to Authors of the Reports of the United States Geological Survey.*[6]

In the description of a map unit, the name of the stratigraphic unit is followed in parentheses by its position (usually the series term) or its age. Units should be described in order of increasing age (ie, upper members of a formation before middle members and middle members before lower members). The description of the unit is usually a brief account of the lithology, color, thickness, and other distinguishing characteristics of the rocks in each unit. The order in which these characteristics are described should be consistent within each unit description in the legend. Some map units are simply combinations of units that are described elsewhere in the legend.

> Tb   Bishop Conglomerate (Oligocene)—Light gray to pinkish gray, very poorly sorted, loosely cemented, pebbly, cobbly to bouldery conglomerate and sandstone. Thickness highly varied; may locally exceed 60 m.

25.3.3 Correlation Charts and Columnar Sections

A correlation chart is used to show the chart author's interpretation of rock units and their ages as they relate to units that others have identified. Relative or radiometric time is usually designated on the left side of these charts. Units are listed from youngest, known as "topmost," to oldest, known as "bottommost." If space permits in a chart, rock units should be identified by formal or informal names without abbreviations.

Stratigraphic columnar sections are graphic illustrations that describe in a vertical column the sequence and relations of rock or soil in a defined area. Colors or symbols used to distinguish units are unique to the particular illustration. Color terms used to describe rocks should be as specific as possible. Refer to the *Munsell's Rock Color Book.*[10] Standard patterns are used to indicate rock types (see pages 376–377 in *Geology in the Field*[11]).

For more details on the construction of correlation charts, columnar sections, and other illustrations, consult *Suggestions to Authors of the Reports of the United States Geological Survey.*[6]

25.4 PHYSICAL DIVISIONS

Physiographic divisions, provinces, and sections are specific physiographic entities that should be capitalized as proper nouns.

Appalachian Plateaus	Hudson Valley	New England Province
Floridian Section	Laurentian Upland	Osage Plains

See the *US Government Publishing Office Style Manual*[12] or a reputable atlas such as the *National Geographic Atlas of the World*[13] for a comprehensive listing of the names of physiographic divisions, provinces, and sections.

25.5 ROCKS AND MINERALS

Although the nomenclature for igneous, metamorphic, and sedimentary rocks has not been standardized, some proposed classifications are widely used.[6,14] In addition, *Fleischer's Glossary of Mineral Species*[15] lists the correct spellings and chemical formulas for approximately 5,000 mineral species. *Fleischer's* is periodically updated and corrected.

The International Mineralogical Association's Commission on New Minerals and Mineral Names has procedures and criteria for proposing new mineral species, as well as guidelines for mineral nomenclature.[16] Journals such as the *American Mineralogist* and *The Canadian Mineralogist* periodically publish the names and characterizations of newly designated minerals.

In general, avoid using colloquial, outdated, nonspecific, and varietal names unless the parent mineral species is referenced. Do not abbreviate names of minerals in text.

For charts, tables, and illustrations, the following 4 guidelines should be followed for devising mineral-phase and mineral-component abbreviations to use as symbols, subscripts, and superscripts[6]:

(1) Abbreviations should consist of 2 or 3 letters. Capitalize the first letter of abbreviations used as mineral-phase symbols, and use all lowercase letters for mineral-component symbols.
(2) The first letter of an abbreviation should be the first letter of the mineral's name. The other letters should be taken from the rest of the name, preferably from the consonants.
(3) Symbols should not be identical with those of chemical elements (see Table 16.3).
(4) Symbols should not form words likely to be used in scientific writing in any language.

Below are examples of abbreviations formed in accordance with these recommendations.

Di	the mineral phase "diopside"
di	the diopside component
Agt	aegirine-augite
Ak	åkermanite
Hl	halite
Rds	rhodochrosite
Ft	ferrotschermakite
Usp	ulvöspinel

Table 25.1. Symbols for Basic Symmetry Operations of a Space Group in Crystallography[a]

Symbol	Description
$\{E/0\}$	Identity operation; no rotation and no translation
$\{C_n/0\}$	n-Fold rotation
$\{\sigma/0\}$	Reflection
$\{I/0\}$	Inversion
$\{S_n/0\}$	n-Fold rotation followed by a fractional translation in a direction parallel to the plane
$\{E/\tau\}$	Translation
$\{\sigma/\tau/m\}$	Reflection followed by a fractional translation in a direction parallel to the plane
$\{C_n/\tau/m\}$	Rotation followed by a fractional translation parallel to rotation axis

[a] Based on Bennett and colleagues.[19]

Additional mineral-phase and mineral-component abbreviations can be found in *Suggestions to Authors of the Reports of the United States Geological Survey*[6] and *Symbols for Rock-Forming Minerals.*[17]

25.6 CRYSTALS

Crystals are distinguished by class, family, and system. Crystals fall into 32 classes determined either by such operations as inversion and rotation (see Table 25.1[19]) or by combinations that leave crystals invariant. Crystal families are determined by 17 plane groups and 230 space groups. Crystals are also categorized into 7 systems that are classified either by the unit-cell shapes of the crystals' Bravais lattices or by the main symmetry elements of the crystals' classes (see Table 25.2).

25.6.1 Symmetry, Planes, and Axes

A hallmark characteristic of crystals is symmetry, which is caused by atoms or a group of atoms periodically repeating at equal intervals throughout crystals and creating a 3-dimensional lattice. A lattice is an arrangement of points in space in which each point has identical surroundings in the same orientation. Crystals have 14 types of 3-dimensional lattices, which are called "Bravais lattices."

Crystals are described by the Miller indices of a plane, which are numbers derived from the reciprocals of the intercepts of a plane with the coordinate axes multiplied by the smallest number that will cause all numbers in the set of reciprocals to be integers. Crystals have 3 principal axes. Principal axes are chosen parallel to translation directions such that the axes are aligned with symmetry elements. The axes are characteristically labeled "a," "b," and "c." If the intercepts are all equal, the reciprocals of the intercepts are in the ratio 1:1:1, and the Miller indices of such a plane would be written as "(111)"—within parentheses and with no spaces between the numbers. When each of the 3 indices is less than 10, the numbers should be written without commas, but when one or more of the numbers is greater than 10, separate all the numbers with commas to prevent confusion, as in "(11,2,1)." To indicate a negative intercept, a bar should be

Table 25.2. Symbols in Crystallography

Symbol	Description
′	Prime applied to the symbol for any movement; the movement is accompanied by change of color
′	Prime applied to the prefix; change of color with translation
1	Not a diad
$\bar{1}$	Pure inversion
2	Diad
$\bar{2}$	Reflection in a plane perpendicular to the axis of rotation
A	Centered in yz plane
a	Glide reflection in the x direction
B	Centered in zx plane
b	Glide reflection in the y direction
C	Centered in xy plane
c	Centered net; glide rotation in the z direction
d	Diamond glide rotation
F	Centered in all 3 planes (xy, yz, and zx)
g	Glide reflection; for plane groups (ie, layer patterns), glide reflection in both x and y directions
I	Body-centered
i or I	Inversion
l′	Reflection of line l
l″	Reflection line of l′
m	Reflection lines; in parallel mirror lines, reflections, and glide reflections
n	Glide reflection in a diagonal direction for plane groups (ie, layer patterns)
$\bar{n}$	Rotation through $360°/n$ $(2\pi/n)$ combined with inversion
n_r	Rotation of $2\pi/n$ combined with an axial movement of r/n units
(n)	Number in parentheses as a superscript; number of colors in a polychromatic crystal
O	Center of inversion
P	Primitive plane
p	Primitive net
r	Row
σ	Reflection
T	Translation

placed above the number for that intercept, as in "$(11\bar{1})$." This example is read, "1, 1, bar 1." (The overbar can be created by inserting Unicode 0305 after the number.)

To refer to a group of planes, replace the parentheses with braces, also known as "curly brackets." Equal and plus symbols can be used to indicate which planes are in a group.

$$\{abc\} = (abc) + (acb) + (bac) + (bca) + (cab) + (cba)$$
$$\{100\} = (001) + (010) + (100)$$

Denote the Miller indices for direction with square brackets, as in "[abc]." Denote the set of directions that is equivalent by symmetry with angle brackets, as in "⟨abc⟩."

A crystal is said to be right-handed or left-handed depending on the direction in which the crystal rotates polarized light. Some crystals are twinned, which means that adjacent crystals share some crystal lattice points in a symmetrical manner. Japan, Dauphiné, and Brazil are 3 kinds of twinning. References to the faces of both twins should be in lowercase italic letters, separated by a comma, but references to the faces of the opposing twin should also have a bar underneath, created by inserting Unicode 0332 after a letter.

$a, \underline{a}$ $z, \underline{z}$

25.6.2 Rotation, Reflection, Inversion, and Translation

Rotation leaves unchanged a crystal's right-hand or left-hand coordinate system, whereas inversion changes a right-hand coordinate system to a left-hand one and vice versa. An integer subscript denotes the number of types of rotation around a fixed point 0. For example, "D_5" indicates a dihedral crystal with pentad rotation and 5 mirror lines, also known as "reflection lines." However, more often than not, the numbers are used alone (eg, 3, 4, and 2).

Mirror lines pass through the rotation center of crystals. In international notation, numbers indicating an odd number of rotations are followed by the letter "m" for the "mirror lines," and numbers indicating an even number of rotations are followed by "mm" (eg, 3m and 4mm). For even-numbered types, the lines are divided into 2 sets: those along the arms of the cross and those bisecting the angles formed by the cross. For odd-numbered types, the 2 sets of lines are the same.

Translation means moving from one point to another point in a crystal in a way in which the environment around the second point is exactly the same as the environment around the first. (For example, a point on the lower right side of a square in a cyclone fence is identical to all the other such points on the lower right sides of all the other squares in a cyclone fence.)

25.6.3 International Symbols (Hermann–Mauguin Symbols)

In both its short and full versions, the Hermann–Mauguin[19] symbol for a crystal consists of 2 parts: a letter indicating the centering type of the conventional cell and a set of characters indicating symmetry elements of the space group to which the crystal's family is a part (ie, "a modified point-group symbol"). Use capital italicized letters for 3-dimensional lattices, and use lowercase italic letters for 2-dimensional nets.

Full	Short
$C1m1$ or $C11m$	Cm
$P2_1/n2_1/m2_1/a$	$Pnma$
$P6_3/m\ 2/m\ 2/c$	$P6_3/mmc$

25.6.4 Patterson Symmetry

Patterson symmetry gives the space group of the Patterson function $P(x,y,z)$. The Patterson function represents the convolution of a structure with its inverse or the pair-correlation function of a structure.

25.6.5 Laue Class and Cell

The space-group determination starts with the assignment of the Laue class to the weighted reciprocal lattice and the determination of the cell geometry. The Laue class determines the crystal system.

The axial system should be thought of as right-handed. For crystal systems with symmetry higher than orthorhombic, the symmetry directions and the convention that the cell should be taken as small as possible determine the axes uniquely. Three directions are fixed by symmetry for orthorhombic crystals, but any of the 3 may be called "a," "b," or "c." The convention is for $c < a < b$. Monoclinic crystals have one unique direction. If there are no special reasons to decide otherwise, the standard choice "b" is preferred. For triclinic crystals, usually the reduced cell is taken, but the labeling of the axes remains a matter of choice.

25.7 SOILS

Soil information presented in technical publications may include pedon descriptions, laboratory data, soil classifications, and information from soil surveys.

25.7.1 Pedon Descriptions

The starting point for many soil investigations is a pedon description. A pedon is a 3-dimensional soil body about 1 m² at the surface and extending to the bottom of the soil, usually 1.5 to 2 m deep. (In contrast, a soil profile is 2 dimensional and cannot be sampled.) To prepare a pedon description, a pedon is divided into horizons that appear homogeneous. Typically, 6 to 8 horizons are identified (fewer for shallow soils and more for both deeper soils and detailed studies). The depth, horizon designation, and other morphological properties are recorded for each horizon.

Pedon descriptions usually are presented in narrative form in soil surveys and in tabular form in journal articles (see Figure 25.1). Narrative pedon descriptions are generally written as a series of short descriptive phrases separated by semicolons. A narrative description includes more information than a table. In both short narratives and tables, authors provide what they believe are the most important details. Guidelines for preparing pedon descriptions are available in the *Soil Survey Manual*[20] and the *Field Book for Describing and Sampling Soils*.[21] (See Sections 25.7.1.2, "Horizon Designation," and 25.7.1.3, "Morphological Properties," for descriptions of the horizons and morphological properties and the abbreviations used.)

25.7.1.1 Use of Pedon Descriptions

As the starting point for many soil investigations, pedon descriptions have several applications, including the following:

(1) as the basis for classifying the soil in a natural system
(2) to divide pedons into units that can be sampled for laboratory analysis
(3) as the basis for defining map units for soil surveys

Narrative descriptions for parts of pedons

Ap—0 to 25 cm; dark grayish brown (10YR 4/2) silt loam, pale brown (10YR 6/3) dry; moderate fine granular structure; friable; many fine and very fine roots; neutral; abrupt smooth boundary (15 to 25 cm thick)

E—25 to 33 cm; grayish brown (10YR 5/2) silt loam; weak fine subangular blocky structure; friable; common fine and very fine roots; common medium prominent yellowish brown (10YR 5/6) masses of iron accumulation in the matrix; moderately acid; clear smooth boundary (0 to 15 cm thick)

Bt1—33 to 53 cm; yellowish brown (10YR 5/4) silty clay loam; moderate medium subangular blocky structure; firm; few fine and common very fine roots; common distinct dark grayish brown (10YR 4/2) clay films on faces of peds; many medium distinct light brownish gray (10YR 6/2) iron depletions in the matrix; moderately acid; clear wavy boundary (10 to 40 cm thick)

Tabular descriptions of same 3 parts

Horizon	Depth (cm)	Color[a] (moist)	Texture[b]	Structure[c]	Consistence[d]	Ped coats
Ap	0–25	10YR 4/2	SiL	2 f gr	fr	
E	25–33	10YR 5/2 c 10YR 5/6	SiL	1 f sbk	fr	
Bt1	33–53	10YR 5/4 m 10YR 6/2	SiCL	2 m sbk	fi	clay, 10YR 4/2

[a] Munsell designation; abundance of mottles: c = common, m = many
[b] C = clay, L = loam, Si = silt(y)
[c] Grade: 1 = weak, 2 = moderate; size: f = fine, m = medium; shape: gr = granular, sbk = subangular blocky
[d] Consistence: fi = firm, fr = friable

Figure 25.1. Examples of narrative and tabular pedon descriptions

(4) to group individual soils into classes for specific uses, such as crop production and engineering

25.7.1.2 Horizon Designation

Horizon designations are shorthand labels that indicate the major properties and relations of the horizons. The sequence of horizons from the surface down to 1.5 to 2 m might be written as follows:

A, E, EB, 2Bt1, 2Bt2, 2Cd

An individual horizon designation may have up to 4 main parts.

(1) A numerical prefix indicates the horizon's parent material, which is the geological material from which the soil formed. For horizons from the upper parent material, the numeral 1 is implied. Subsequent horizons are preceded by numbers. Thus, in the horizon sequence above, the soil in the A, E, and EB horizons formed in one parent material, perhaps loess, and the soil in the 2Bt1 and lower horizons formed in a second parent material, perhaps glacial till.

(2) Master horizons, which are the most obvious horizons in a sample, are represented by single capital letters. The letters A, E, B, and C in the example above denote master horizons (see Table 25.3). Less obvious horizons, such as those that are transitional between master horizons, are assigned 2 letters (eg, EB).

(3) A major process or property is indicated by adding a lowercase letter as a suffix. In the 2Bt1 and 2Bt2 horizons, the suffix "t" (for the German "ton," meaning "clay") indicates

Table 25.3. Brief Definitions of Master Horizons

Horizon type	Description
O	Horizons that consist mainly of organic material, not mineral material like other horizons
A	Mineral horizons at the surface or below an O horizon that are relatively high in organic matter
E	Mineral horizons from which silicate clay, iron, aluminum, or some combination of these have been leached, usually to lower horizons
B	Horizons below an A, E, or O horizon that show one or more of the following: (1) accumulation (ie, movement in from above) of silicate clay, iron, aluminum, humus, or silica, alone or in combination (2) removal of carbonates (3) concentration of sesquioxides (eg, iron and aluminum) because other material, such as silica, weathered out faster (4) coatings of sesquioxides that make the horizon conspicuously darker, higher in chroma, or redder in hue than overlying and underlying horizons (5) formation of granular, blocky, or prismatic structure (6) brittleness
C	Horizons or layers, excluding hard bedrock, that are little affected by soil-forming processes and lack properties of O, A, E, or B horizons; most are mineral layers
L	Limnic soil materials; organic and inorganic materials deposited in water, including coprogenous earth (eg, sedimentary peat), diatomaceous earth, and marl
R	Hard bedrock
W	A layer of ice within a soil (ie, permafrost)

that clay has moved downward in the pedon and has become immobilized in the Bt horizon.

(4) A numerical suffix indicates a minor subdivision. The "2Bt1, 2Bt2" segment in the example above indicates that the thick 2Bt horizon was split into 2 subhorizons for sampling the soil.

25.7.1.3 Morphological Properties

The main morphological properties of soils are color, texture, structure, consistence, and ped coats.

Color is represented using designators described in the Munsell system.[22] For example, in the designation "10YR 5/2," the Munsell designator "10YR" is the hue, "5" is the value, and "2" is the chroma. When more than one color is described, the dominant color is listed first, followed by minor colors. The designation for the abundance of mottles in a soil sample is usually a lowercase letter that follows a Munsell designator (eg, "c" for common, "f" for few, or "m" for many).

Descriptive words, such as "grayish brown," may also be used. In soil surveys, soil color is traditionally written with the word description of the color appearing first, followed by the Munsell designations in parentheses.[23] In many scientific journals, however, only the Munsell designators are given, and they are used without parentheses. If a word description of the color is used in a journal article, it follows the dominant Munsell designator and is placed in parentheses, the opposite of the convention in soil

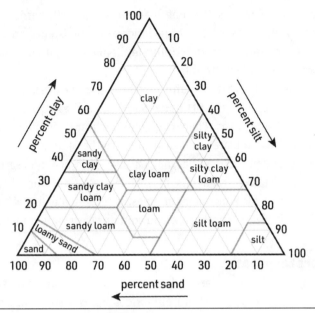

Figure 25.2. Example of a soil texture triangle. Reprinted from the US Department of Agriculture.

surveys.[23] In addition, the word description is included only the first time the color designation is used in a journal article's abstract and text.

> grayish brown (10YR 5/2) (10YR 5/6 c) for a soil sample
> 10YR 5/2 (grayish brown) 10YR 5/6 c for the first mention in a journal article's abstract and text

Texture describes the amount and size of mineral particles in the soil. Texture classes are based on defined upper and lower limits on the percentages by weight of each component particle less than 2 mm in diameter (eg, sand, silt, and loam). The upper and lower limits are defined in texture triangles, such as those in Figure 25.2, as well as those in the *Soil Survey Manual*[20] and the *Field Book for Describing and Sampling Soils*.[21] These texture classes are also used to describe sediment composition (see Section 25.10.2, "Texture").

> clay   sand   loam   silty clay loam   loamy sand

Texture modifiers are used to denote the presence of conditions or components other than sand, silt, and clay. A texture modifier can be used when the content of the component exceeds 15% by volume.

> gravelly loam   mucky loam

Structure refers to how soil particles are grouped together, ranging from particles that form larger units, or peds, to those that form soil that easily breaks into pieces. Three features of structure are part of a pedon description: grade, size, and type. Grade, also known as "distinctness," is represented in a pedon description by a number. Size is represented by a lowercase letter. Type, also known as "shape," is represented by lowercase letters. The following example describes a soil structure that has a moderate grade (2), fine-sized peds (f), and a granular shape (gr):

> 2 f gr

Consistence is a measure of soil strength. It is described with terms such as "loose," "soft," "friable," "firm," and "hard." Abbreviations for consistence are usually formed from the first letter or letters of the descriptive terms and are written in lowercase letters (eg, "l" for "loose" and "fr" for "friable").

Coatings on peds, such as clay, organic matter, and iron oxides, reflect processes of soil formation. In Figure 25.1, the Bt1 horizon is described as dark grayish-brown clay coats on subangular blocky peds. These clay coats, or "films," formed when clay moved down the profile and accumulated on ped surfaces in the Bt horizon.

Other morphological properties that may be covered in pedon descriptions include roots, pores, pH, cracks, rock fragments, and concentrations of iron or other materials. Pedon descriptions may also document effervescence, which is tested for by using hydrochloric acid to determine whether the soil contains carbonate minerals.

25.7.2 Laboratory Data

Generally 2 kinds of soil samples are collected from soil horizons: bulk samples, which are dried, crushed, and sieved, and natural-fabric samples, also known as "undisturbed samples." These samples are used to measure physical properties and to study soil morphology under a microscope.

The standard practice is to conduct laboratory analyses on bulk samples that have been oven dried at 105 °C until the samples reach a constant mass. Then the samples are passed through a 2-mm sieve so that all the material is less than 2 mm in size. However, the method is highly dependent on the purpose of the soil test. Consequently, for some purposes, the samples are sieved before being oven dried. Laboratory data from bulk samples are used to classify a soil, to study soil formation processes, and to help decide how best to use a soil.

Pedon descriptions and laboratory data for many soils are available online through the Web Soil Survey[24] of the US Department of Agriculture and through the FAO Soils Portal[25] of the Food and Agriculture Organization of the United Nations.

25.7.3 Soil Classification Systems

Soils are classified into groups with similar properties to organize knowledge about soils and to facilitate understanding the relations among soils. Natural classification systems group soils according to natural relations, such as how soils formed. Technical classification systems group soils according to specific purposes (eg, crop production, engineering applications, and erodibility). Generally, several technical classifications can be derived from a single natural classification.

Many countries have soil classification systems for soils within their boundaries. For example, Canada has *The Canadian System of Soil Classification*.[26] Soils from around the world are included in *Soil Taxonomy: A Basic System of Soil Classification for Making and Interpreting Soil Surveys*[27] and the *World Reference Base for Soil Resources*.[28]

25.7.3.1 Soil Taxonomy for the United States

Soil Taxonomy: A Basic System of Soil Classification for Making and Interpreting Soil Surveys[27] is a comprehensive classification system for soils in the United States, and it is nearly comprehensive for soils in the rest of the world. It is supported by the US National Cooperative Soil Survey, which is a coalition of the US Natural Resources Conservation Service, state universities, and other federal and state entities involved with soil resources.

To use *Soil Taxonomy* to classify soil, the diagnostic horizons and materials of the soil must first be identified. Most diagnostic horizons are subsoil horizons, but some are surface horizons. Diagnostic horizons are formed by broad-scale, soil-forming processes. For example, the argillic horizon formed as a result of clay migrating from upper to lower horizons in humid, originally forested areas of the world. Diagnostic materials have distinctive properties that are often related to the origin of the material, such as volcanic activity.

25.7.3.1.1 CATEGORIES

Soil Taxonomy[27] has 6 classification categories, or levels. From highest to lowest, the categories are called "order," "suborder," "great group," "subgroup," "family," and "series."

Soil Taxonomy has 12 soil orders, most of which are defined on the basis of diagnostic horizons or other features that reflect soil-forming processes (see Table 25.4[27]). In several of the soil orders, suborders are based on soil moisture regime, whereas in other orders, different criteria are used. For example, suborders in the Entisol order are based on the reasons why distinctive soil horizons have not formed, such as the soil being too sandy, too stony, or too steep.

Great groups are based on the presence or absence of certain kinds of soil horizons

Table 25.4. Brief Descriptions of the Orders of *Soil Taxonomy*[a]

Soil orders	Description
Alfisols	Soils with a subsoil accumulation of silicate clay that are moderately weathered (ie, soils that have a high base saturation)
Andisols	Soils formed from volcanic materials
Aridisols	Soils of arid environments
Entisols	Very weakly developed soils, including many sandy soils
Gelisols	Soils with permafrost
Histosols	Soils formed from organic materials
Inceptisols	Weakly developed soils, excluding sandy soils
Mollisols	Soils with thick, dark surface horizons that are high in organic matter content
Oxisols	Very highly weathered soils of tropical areas that are high in iron- and aluminum-oxide minerals
Spodosols	Soils with a subsoil accumulation of aluminum, organic matter, and usually iron
Ultisols	Soils with a subsoil accumulation of clay that are highly weathered (ie, soils that have a low base saturation)
Vertisols	Soils that undergo much shrinking and swelling

[a] Based on US Department of Agriculture, Natural Resources Conservation Service, Soil Survey Staff.[27]

and other soil properties. Subgroups indicate whether the soil is in the central range of its great group or whether the soil has marginal properties that tend toward those of soils in another great group. Families are defined by soil texture, mineralogy, temperature regime, and a few other properties.

Soil series are named for the communities where they were first identified. The map units in most published soil surveys are named after soil series. In addition, series names are used to identify soils in most scientific journals that publish soil investigations. All series descriptions in *Soil Taxonomy* provide such details as typical pedon descriptions and the series' classification at the family level.[29]

25.7.3.1.2 CLASSIFICATION OF SOILS

Classifying soils to their series involves the following 6 steps:

(1) identifying diagnostic soil horizons and diagnostic properties
(2) determining the soil order following a key to soil orders (eg, Chapter 8 of *Soil Taxonomy*[27])
(3) identifying the soil moisture regime
(4) determining the subgroup following a key for one of the orders (eg, Chapters 9 to 20 in *Soil Taxonomy*[27] or *Keys to Soil Taxonomy*[30])
(5) determining the family (eg, Chapter 21 in *Soil Taxonomy*[27])
(6) matching the description and data with official series descriptions (usually requiring input from local experts)

25.7.3.1.3 NOMENCLATURE

Soil Taxonomy[27] uses nomenclature based largely on Greek and Latin formative elements. For example, names for wet soils contain the formative element "aqu" (from the Latin "aqua" for "water"). Names for soils with high base saturation may contain the element "eutr" (from the Greek "eutrophos" for "fertile"). This formative element is also used in the term "eutrophication," which describes lakes when they become enriched with plant nutrients.

Soil Taxonomy's main rules for nomenclature are as follows:

(1) Orders have names with the suffix "sol" (eg, Alfisol and Mollisol [see Table 25.4]). In the order "Alfisol," the soils have a subsoil clay accumulation and high base saturation.
(2) Suborder names have 2 syllables, the second of which is based on the formative element of the order to which the suborder belongs. For example, "Aqualfs" are wet Alfisols, and "Udolls" are Mollisols in humid climates ("ud" stands for "udic," which refers to the soil moisture regime as being moist but not waterlogged).
(3) Great group names consist of prefixes added to suborder names (eg, "Ochraqualf" and "Argiudoll"). In the great group name "Fragiundalf," the prefix "Fragi" is from "fragilis," which means brittle. Soils in this great group have a "fragipan," a subsurface layer that resists root penetration and water movement.
(4) Subgroup names are formed by adding adjectives to great group names (eg, "Aeric Ochraqualf" and "Typic Argiudoll"). In the subgroup name "Aquic Fragiudalf," the prefix "aqu" means the soil is on the moist side of the udic moisture regime.
(5) Family names are designated by modifiers that describe particle-size distribution, mineralogy, clay activity, and temperature. In the family name "fine-silty, mixed, superac-

tive, mesic Aquic Fragiudalfs," the modifiers indicate that the subsoil is high in silt and moderately high in clay; that the soil is mixed and no single mineral dominates the mineralogy of the whole soil; that the soil is superactive in that it has a high ratio of cation exchange capacity to clay; and that the soil falls into the mesic, or medium, temperature regime.

(6) Unlike the higher categories, series are named for localities (eg, Cincinnati), and they are defined by a typical pedon and a range of characteristics, such as boundary conditions.

The second through fifth rules above are illustrated in the following example of a family-level name.

Element	fine-silty, mixed, superactive, mesic Aquic Fragiudalfs			
Rule	5	4	3 2	2

Except for family-level modifiers, all names used in classifying soils should be capitalized.

25.7.3.2 World Reference Base for Soil Resources

The *World Reference Base for Soil Resources*[28] is an international system that evolved from the Soil Map of the World. The *World Reference Base* is not a national soil classification system like *Soil Taxonomy*[27] but, rather, a database that correlates many national systems. The *World Reference Base* has many similarities to *Soil Taxonomy*. For example, many of the diagnostic horizons and materials defined in the *World Reference Base* are similar to those in *Soil Taxonomy*.

The 2 main categories in the *World Reference Base* are an upper level of 30 reference soil groups, similar to the order and suborder levels in *Soil Taxonomy*, and a lower level of soil units, each defined and named by adding an adjective to the reference soil group name. For example, "Luvisol" is the upper-level name of a reference soil group similar to "Alfisol" in *Soil Taxonomy*. The adjective "Fragic" may be added to "Luvisol" to describe the lower-level Luvisols with fragic horizons. The Fragic Luvisols are essentially the Fragiudalfs of *Soil Taxonomy*.

The *World Reference Base* defines 121 formative elements, but only a subset of 10 to 30 formative elements is used with each soil group.

25.7.4 Soil Surveys

Soil surveys contain maps, text, and tables about soils and their uses. These surveys also include pedon descriptions and soil names, and they may provide the survey locations using a coordinate system. Traditionally, soil surveys were published in print form, but electronic versions of the maps, text, and tables are now available for numerous areas.

Detailed soil surveys are usually published at a scale of 1:12,000 to 1:24,000 on aerial photo backgrounds. The map units, or delineated units, are generally named for soil series. Map units also list soil series other than the one used in the map unit name. For example, a soil map unit designated "Miami silt loam, 6% to 12% slopes" would be named for the major soil series in it, Miami, but the unit may include areas of the Crosby series, the Russell series, and other series.

Most soil surveys conducted before the year 2000 were done for legal entities such as counties. Copies of these surveys are generally available in county agriculture offices, in many libraries, and through state universities. Since the year 2000, soil surveys in the United States are more commonly conducted for major land resource areas, which are natural geographic areas characterized by similar soils and other similar characteristics. These major land surveys can be found in the Soil Survey Geographic Database,[31] and more general soil association units can be found in the US General Soil Map.[32] Most other countries have similar major land resource surveys.

25.7.5 Locations

Soil surveys should provide site locations based on a commonly used coordinate system, such as latitude–longitude, Universal Transverse Mercator, state plane coordinate, or US rectangular survey (township-range-section) (see Section 14.2, "Geographic Coordinates"). If soil surveys use the US rectangular survey, locations of specific points should be given relative to their distances from 1 of the 4 corners or from the center of a section.

100 m north and 56 m east of the southwest corner of section 4, T12N, R9E, Boise Meridian

Note that the site location in the example above is more specific than an area description (see Section 25.9, "Tracts of Land").

25.7.6 Soil Interpretations

Provided in soil surveys and other documents, soil interpretations predict how a soil might respond to various uses. These interpretations are based in part on research on and experience with a few soils. The interpretations are extended to similar soils in the area based on soil descriptions, laboratory data, soil classification systems, and soil surveys.

Soil interpretations may be used in agricultural production (eg, land use capability, crop yield predictions, tillage systems, and conservation practices); construction (eg, buildings, highways, and dams); forestry (eg, productivity and tree planting); onsite sewage disposal (eg, septic systems and lagoons); recreation (eg, playgrounds, paths, and camp areas); wildlife habitats; and other uses.

25.7.7 Classification of Land Capability

The US Natural Resources Conservation Service recognizes 8 classes of land capability.[27] Class I designates soils with the greatest capability of response to management (eg, intensive crop farming, pasture, range, and wildlife preserve) and the fewest limitations on use (eg, well drained, level, and fertile). Soils with the fewest capabilities (eg, not suitable for commercial plant production) and greatest limitations (eg, restricted to recreation, wildlife, water supply, or aesthetic uses) are in Class VIII. The land classification system in Canada is similar.[33]

25.7.8 Terminology and Additional Information

Definitions of additional soil science terms can be found in glossaries on geology and soil,[34,35] and more information about soil science is available in encyclopedias and handbooks of soil science.[36,37]

Names for soil parent material (ie, geological material) and landforms should follow the conventions in the *Glossary of Geology*.[34]

25.8 AQUIFERS

An "aquifer" is "a body of rock that is sufficiently permeable to conduct ground water and to yield economically significant quantities of water to wells and springs."[34] An aquifer may encompass thousands of square hectares.

The following terms are not capitalized even when used with proper nouns: "aquifer," "aquifer system," "zone," and "confining unit." Terms such as "sand and gravel aquifer" and "limestone aquifer" are neither capitalized nor hyphenated. Do not capitalize adjectival modifiers and relative-position terms unless they are part of formal geographic names. In addition, do not use quotation marks for aquifer names except in cases when the term "aquifer" is a misnomer.

> Nubian Sandstone aquifer system   Mississippi River alluvial aquifer
> Upper Canada aquifer

Although the term "aquifer" may be imprecise, it is widely accepted and used. Coining new terms will only add to the confusion, so do not use terms intended to be synonymous with "aquifer" and "aquifer system," such as "aquigroup."

Distinguish hydrologic from geological terms.

> water from the Madison aquifer *not* Madison water
> wells completed in Madison Limestone *or* wells completed in the Madison aquifer *not* Madison
>   wells

25.9 TRACTS OF LAND

With such exceptions as 14 Eastern states, Texas, and Hawaii, much of the United States was surveyed in a grid system called the Public Land Survey System. Under this system, tracts of US public land were divided into rows designated as townships (each 6 miles square) and columns designated as ranges. These grids were established in relation to east–west baselines and named north–south meridians. Land descriptions in these grids are designated with symbol groups.

> SE1/4 NW1/4 sec 4, T 12 S, R 15 E, Boise Meridian
> (designates the southeast quarter of the northwest quarter of section 4 in the 12th townships
>   south and 15th range east of the Boise Meridian)
>
> north half sec 20, T 7 N, R 2 W, 6th principal meridian
> (designates the northern half of section 20 in the 7th townships north and 2nd range west of the
>   6th principal meridian)

In land-description symbol groups, omit periods from abbreviations, and do not use a space between the compass directions and the fractions. If fractions are spelled out, use "half" and "quarter," not "one-half" and "one-quarter."

The abbreviation for the plural "townships" is "Tps," and the abbreviation for "ranges" is "Rs." Multiple township and range numbers in the same description should be separated by commas.

> Tps 9, 10, 11, and 12 S, Rs 12 and 13 W

If possible, do not break a land-description symbol group at the end of a line of text. If that is unavoidable, break the symbol group after a fraction and do so without a hyphen. For example, break the land description "NE1/4 SE1/4 sec 4" between "NE1/4 SE1/4" and "sec 4" or between "NE1/4" and "SE1/4 sec 4."

For more information on writing descriptions of US public lands, see *Specifications for Descriptions of Land: For Use in Land Orders, Executive Orders, Proclamations*, Federal Register *Documents, and Land Description Databases*.[38] Land in Canada is divided using similar systems: the Canadian Lands Survey System for most of the country and the Dominion Land Survey System for western Canada.[39]

25.10 SEDIMENT

Sediment is material deposited by water, wind, or glaciers. Sediment consists of rock and mineral fragments, organic material (eg, shell, bone, and other animal and plant material), loess (ie, loosely compacted yellowish-gray windblown sediment), desert sand, volcanic ash, precipitates from water, settled dust, and small atmospheric and cosmic particles.

The study of sediments provides information about the origin, age, and transport of material in riverbeds, lake beds, and ocean basins. Sediments are classified according to their source, texture, and color.

25.10.1 Source

Lithogenous and terrigenous sediments consist of materials originating from weathered rocks, whereas cosmogenous sediments consist of materials from the atmosphere and space. Hydrogenous sediments, in turn, result from chemical reactions in water. Biogenous sediments consist largely of plant and animal remains, and sediment composed of more than 30% biogenous material is called an "ooze."

25.10.2 Texture

Sediment texture is classified by grain size and sorting. The geometric Wentworth scale[40] and modifications of the Wentworth scale provide standardized grade classifications for particles based on maximum particle diameter, as does the arithmetic phi scale (see Table 25.5). In the phi scale, particle diameter (d), which is measured in millimeters, is converted from the geometric Wentworth scale to an arithmetic scale: $\Phi = -\log_2 d$, in which Φ is a measurement without a dimension unit such as millimeters.

The texture characteristic of sphericity is a measure of particle shape. Specifically, sphericity is the ratio of the surface area of a particle to the surface area of a perfect sphere. A perfectly spherical particle would have a sphericity of 1. The lower a particle's sphericity is, the less round the particle is. Because sphericity is a ratio, its measurements do not include a dimension unit.

Sorting is a texture measure of the similarity of particle sizes within sediment. The degree to which particles have been sorted can indicate how sediment was transported.

Table 25.5. Sediment Classification according to Grain Size[a]

Size class	Diameter (mm)	Phi (Φ)
Boulder	>256	less than −8
Cobble	256 to 64	−8 to −6
Pebble	64 to 4	−6 to −2
Gravel	4 to 2	−2 to −1
Sand		
Very coarse sand	2 to 1	−1 to 0
Coarse sand	1 to 0.5	0 to 1
Medium sand	0.5 to 0.25	1 to 2
Fine sand	0.25 to 0.125	2 to 3
Very fine sand	0.125 to 0.0625	3 to 4
Silt		
Coarse silt	0.0625 to 0.0310	4 to 5
Medium silt	0.0310 to 0.0156	5
Fine silt	0.0156 to 0.0078	6
Very fine silt	0.0078 to 0.0039	7
Clay	<0.0039	≥8

[a] Adapted from Wentworth.[40]

Generally, uniformly sorted sediments (ie, a mix of similarly sized particles) would have been transported by wind or waves, whereas poorly sorted sediments (ie, a random mix of different-sized particles) may have been transported by sea ice, floods, or strong currents.

Nomenclature Based on Sand-Silt-Clay Ratios[41] classifies sediments according to the relative composition of different-sized particles. Employing a triangle similar to that used for categorizing soil texture (see Section 25.7.1.3, "Morphological Properties"), *nomenclature* categorizes sediments according to their relative percentages of sand, silt, and clay particles.

25.11 WATER

25.11.1 Descriptive Terms

Descriptive terms for water types usually consist of the word "water" and a modifier. The same terms are often used as both nouns and adjectives, but the construction of noun and adjective forms may vary. For some water types, the name is a one-word combination of "water" and its modifier. In other cases, the modifier is separated from "water" by a space or a hyphen. Because the rules for forming these words are not consistent, check spellings and punctuation against a standard dictionary, preferably the one adopted by the intended publisher of the work.

> seawater (used as both noun and adjective)
> fresh water (noun) *but* freshwater (adjective)
> deep water (noun) *but* deepwater (adjective)
> open water (noun) *but* open-water (adjective)

25.11.2 Physical Properties of Water

This section outlines recommendations adopted by the International Association for the Physical Sciences of the Ocean (IAPSO).[42]

25.11.2.1 Salinity of Seawater

Salinity (S) is defined, measured, and reported differently depending on the chemical composition of the water being studied. In addition, the interpretation of salinity varies among oceanographers and their freshwater equivalents, limnologists.

Salinity can be defined as a measure of the total amount of dissolved material in water. Salinity also can be defined as the total ion concentration in water.[43,44] Under either definition, pure water would have a salinity of zero.

IAPSO recommends determining the salinity of seawater by measuring its electrical conductivity. This IAPSO-recommended method replaced a slower method in which ocean salinity was determined indirectly from chlorinity. The newer method, called "practical salinity," is defined as the ratio K_{15}, in which the antecedent is the electrical conductivity of a seawater sample at the temperature of 15 °C and the pressure of one standard atmosphere and the consequent is a potassium chloride (KCl) solution in which the mass fraction of KCl is 32.4356×10^{-3}, also at 15 °C and one standard atmosphere.[42] Because practical salinity is a ratio of 2 electrical conductivities, it is a "dimensionless unit" (ie, a unit without a dimension modifier such as parts per thousand).

Because the symbol "S" is used for "practical salinity" and older methods of determining salinity, "practical salinity" should be written out in full to avoid confusion with the other methods.

Before practical salinity became the standard measurement for salinity, oceanographers relied principally on indirect measurements of absolute salinity (S_A). Absolute salinity is defined as the total mass of solid material dissolved in a sample of water divided by the mass of the sample when all of the carbonate has been converted into oxide, all of the bromine and all of the iodine have been replaced by chlorine, and all of the organic matter has been oxidized.[45]

Absolute salinity is difficult to measure directly. Consequently, a variety of indirect measures were commonly used, and these measures varied according to the chemical composition of the water being examined. Among the most commonly used indirect methods was determination of chlorinity, which is based on using silver nitrate to titrate the chlorine, bromine, and iodine ions in seawater.[45] The chlorinity of seawater (Cl) represents the mass of pure silver necessary to precipitate the chlorine, bromine, and iodine contained in 0.3285234 kg of seawater. The following equation describes the relation between salinity and chlorinity:

$$S = 1.80655 \times Cl$$

25.11.2.2 Salinity of Fresh and Diluted Water

Whereas oceans have stable ionic compositions, fresh and diluted inland water, coastal water, and salt lakes are highly variable; consequently, chlorinity was never considered an accurate measure of salinity in the latter waterways. Other measures of the colligative, or binding, properties of molecules in water, such as osmotic pressure, boiling point, freezing point, density, conductivity, and total dissolved solids, are effective methods for determining the number of molecules in solutions of fresh and diluted water. These methods produce salinity values that are precise enough for most limnological purposes.

25.11.2.3 Salinity Units

The IAPSO recommends that salinity be expressed as a factor of 10 raised to the appropriate power and used without a dimension unit.[42] This convention replaced expressing salinity in units of parts per thousand (ppt or ‰), parts per million (ppm or ppM), and parts per billion (ppb).

Salinity is reported as a mass-to-mass ratio (g/kg) in some disciplines and as a mass-to-volume ratio (g/L) in other disciplines. If using a dimensionless unit (eg, 10^{-3}) creates confusion as to which ratio is being reported, use "g/kg" or "g/L" for clarification instead of "ppt," "ppM," or "ppb." Also discouraged is using the abbreviation "PSU" or "psu" for "practical salinity units."[46]

> Salinity decreased to a practical salinity of 26 near the mouth of the estuary.
> *not* Salinity decreased to 26 PSU near the mouth of the estuary.

25.11.2.4 Density

The density of water is measured by mass divided by volume (ie, kg/m^3). Density in the open ocean ranges from approximately 1,021 kg/m^3 at the surface to approximately 1,070 kg/m^3 at 10,000 m.[45] In oceanography, density measurements are calculated by the formula "$\sigma = \rho - 1{,}000$." In that formula, the Greek letter sigma (σ) represents a density anomaly from 1,000 kg/m^3, and the Greek letter rho (ρ) represents density, which, as noted above, is calculated by dividing mass by volume. Usually density measurements are written with 4 digits, 2 or 3 of which are after the decimal.

> If $\rho = 1{,}035.25$ kg/m^3, then 35.25 $kg/m^3 = 1{,}035.25$ $kg/m^3 - 1{,}000$

Because density varies with salinity (S), temperature (t), and pressure (p), density can also be expressed by the notation "$\rho(S, t, p)$."[42] The values for the 3 parameters enclosed in parentheses must be reported in the following order: salinity, temperature, and pressure. Within parentheses, values and symbols are separated by commas. The values should not be placed in subscript type.

> $\rho(34.85, 3.17\ °C, 17.20\ MPa)$ *not* $\rho_{(34.85,\ 3.17\ °C,\ 17.20\ MPa)}$

Parameters may also consist of a mixture of symbols and values.

> $\rho(S, 3.17\ °C, 17.20\ MPa)$

The recommended units for temperature and pressure are degrees Celsius and megapascals, respectively. Practical salinity, on the other hand, is not modified by a unit because it is dimensionless (see Section 25.11.2.1, "Salinity of Seawater"). When the values are written in the proper order and expressed in the recommended units, the unit symbols for temperature and pressure may be omitted within the parenthetical statement.

> $\rho(34.85, 3.17, 17.20)$

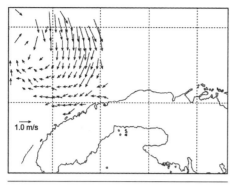

Figure 25.3. Example of a vector diagram

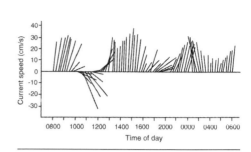

Figure 25.4. Example of a stick plot

Pressure's effect on density can be ignored in many applications (eg, for water at sea surface). When pressure is 1 atmosphere of atmospheric pressure, seawater density is commonly represented by σ_t, pronounced "sigma tee." Although IAPSO strongly discourages using this unit because it is not part of the Système International d'Unités,[42] it is widely used in oceanography.

The notation $\rho(S, t, p)$ is generally not used for pure water because salinity would be zero.

These rules for the notation $\rho(S, t, p)$ also apply to quantities other than density that are functions of the same 3 parameters.

25.11.3 Currents and Streamflow

Currents are generally horizontal movements of water. Currents may be temporary or permanent, small or large in scale, and variable or continuously circulating. Currents are described by the velocity and direction of water flow, and current flow is expressed as a rate.

Currents are often depicted in vector diagrams or stick plots. In vector diagrams, currents are represented by arrows that point in the direction of the current flow (see Figure 25.3). The length of each arrow is proportional to the current speed.

In stick plots, lines originate from a horizontal axis and extend in the direction the current is flowing (see Figure 25.4). The orientation of each line in relation to the axis indicates current direction. By convention, a line pointing to the top of the plot is pointing north, and a line pointing to the right side of the plot is pointing east. The position of each line along the horizontal axis indicates the time when the current measurement was made. The length of a line is proportional to current speed. Current speed is determined by the length of the sticks as if they were parallel to the vertical, or y, axis, not by drawing a horizontal line from the top of the angled sticks to the y axis.

Vector diagrams and stick plots are also used to depict the direction and speed of wind, but a different convention is used for describing wind direction (see Section 25.12.2, "Wind").

25.11.3.1 Names of Currents

Large-scale, permanent oceanic and coastal currents are generally named for where they exist. Capitalize the names of currents.

> South Equatorial Current   Benguela Current   Alaska Gyre

Currents off the coast of Japan are commonly referred to by their Japanese names. These names and the names of other well-known currents can be used with or without the word "current."

> Kuroshio *or* Kuroshio Current *or* Japan Current   Gulf Stream *or* Gulf Stream Current
> Oyashio *or* Oyashio Current   Cape Horn *or* Cape Horn Current

25.11.3.2 Current Direction

The direction of an oceanic or coastal current is the direction in which the water is flowing. The convention for naming current direction is the opposite of the one for naming wind direction (see Section 25.12.2, "Wind").

> A northerly ocean current flows from south to north. *but* A northerly wind blows from north to south.

Oceanic and coastal currents may circulate clockwise or counterclockwise (also known as "anticlockwise"). These directions may also be referred to as "cyclonic" and "anticyclonic." Cyclonic currents are counterclockwise in the Northern Hemisphere and clockwise in the Southern Hemisphere, whereas anticyclonic currents are clockwise in the Northern Hemisphere and counterclockwise in the Southern Hemisphere.

25.11.3.3 Streamflow

Streamflow is the volume of water moving past a location in a given period or the volume of water a stream or river discharges over a given period. Streamflow is also expressed as a rate, commonly in cubic meters per second (m^3/s) or megaliters per day (ML/d).

25.11.4 Tides

25.11.4.1 Tidal Datum

Tides are the periodic and alternating rise and fall of the surface of coastal ocean water and other marine water. Tides are caused by gravitational forces of the Sun and the Moon on Earth.

The height of the ocean surface at various stages of a tide is determined relative to a vertical tidal datum, which is a base elevation against which the heights and depths of phases of a tide are measured. The plural form of "tidal datum" is "tidal datums," not "tidal data."

Tidal datums are local, and no international standard exists. For example, the National Tidal Datum Convention of 1980[47] established a continuous tidal datum system for all marine waters of the United States and US territories. The convention created a uniform tidal datum system independent of whether tides are diurnal or semidiurnal. In addition, the system established "mean lower low water," or "MLLW," as the datum for nautical charts for both the Pacific and the Atlantic coasts.

In contrast, Canadian coastal charts use the tidal datum known as "lower low water, large tide," also known as "LLWLT," which is the average of the lowest low water level per year predicted for a tidal datum epoch, a period of 19 years. The United Kingdom and Australia use the "lowest astronomical tide" ("LAT") for the chart or sounding datum. The LAT is the lowest water level predicted for a tidal epoch.

Different tidal datums may be used in other countries. Other common tidal datums include "mean higher high water," or "MHHW"; "mean low water," or "MLW"; and "mean sea level," or "MSL."

Similar to tidal datums, nontidal datums provide the water height of upland rivers, lakes, inland waterways, and reservoirs. The International Great Lakes Datum for the Great Lakes and the Low Water Reference Plan for the middle Mississippi River are 2 examples of nontidal datums.

25.11.4.2 Tidal Height

Tidal height for a specific time is measured in meters relative to the local datum. Report a tide as a negative number if the tidal height is below the datum and as a positive number if the height is above the datum. Although a plus symbol can be used in reporting tidal heights above a datum, its use is not recommended unless required for clarity.

> This morning's low tide of −13 m was the lowest of the spring tide series.
> Mussels are exposed in the intertidal zone during tides of 1.5 m and lower.

Whereas tidal heights are reported in meters in scientific writing, feet are usually used in some nontechnical writing in the United States.

25.11.5 Depth

Depth soundings on nautical charts represent water depth below mean sea level, the average level measured over a tidal datum epoch. Depth soundings are recorded in meters on the nautical charts of all countries but the United States. The United States has traditionally recorded depth soundings in fathoms (see Table 12.8).

25.11.6 Seafloor Features

Capitalize the names of ocean floor features and their associated descriptive terms, such as "ridge," "trench," "bank," and "seamount."

Marianas Trench	Great Barrier Reef	Grand Banks
Patton Seamount	Mid-Ocean Ridge	Kuril-Kamchatka Trench

In contrast, general features on the seafloor are not capitalized.

continental shelf	deep-sea trench	fringing reef

25.12 AIR

25.12.1 Scales of Motion

Meteorology is the study of atmospheric motion, in particular how such motion creates weather patterns on Earth. Atmospheric motions are classified according to the scale on which they occur. The largest of these are on the planetary scale, which are on the order of thousands of kilometers. These atmospheric motions include large-scale circulations, such as the El Niño–Southern Oscillation, and the largest waves in the jet stream, which influence weather patterns. On the order of hundreds of kilometers, the synoptic scale refers to the highs, lows, fronts, and other weather patterns that are typically shown on weather maps, also referred to as synoptic maps. Mesoscale phenomena, which are on the order of a few kilometers to tens of kilometers, are regional in scale, with circulations ranging from as large as hurricanes down to the size of thunderstorms. Microscale phenomena, which are on the order of a kilometer or smaller, range from the size of clouds down to such small circulations as the flow of air past a building and even airflow within a few micrometers of the surface of a leaf.

The division between one scale and the next is intentionally imprecise because different aspects of many phenomena span adjacent scales. Despite their imprecision, the divisions provide a useful framework for discussing atmospheric phenomena. Computer simulation models are often labeled according to the scales they are intended to reproduce, such as global circulation models and mesoscale numerical models.

25.12.2 Wind

In technical writing, report winds in terms of the direction from which they are blowing. Use the suffix "erly" to convey the same sense.

> north wind *or* northerly wind (a wind blowing from the north)
> south wind *or* southerly wind (a wind blowing from the south)

Note that the convention for reporting the direction of winds is the opposite of the convention for reporting the direction of water currents (see Section 25.11.3.2, "Current Direction").

Wind direction can also be expressed in degrees with reference to the direction from which the wind is blowing. According to this convention, 0° and 360° winds are from the north, a 90° wind is from the east, a 180° wind is from the south, and a 270° wind is from the west.

In some nontechnical writing, wind may be described in terms of the direction toward which it is blowing. For this purpose, use the suffix "ward," not "erly."

> northward (a wind blowing toward the north)
> southward (a wind blowing toward the south)
> poleward (a wind blowing toward a pole)

25.12.3 Specialized Terminology

The American Meteorological Society publishes a glossary of terms specific to meteorology.[48] A few common specialized meteorological terms are explained in the next 2 sections.

25.12.3.1 Oscillations and Teleconnections

Complicated interactions between the atmosphere and the oceans can cause wind and current circulations called "oscillations." An oscillation shifts in a more or less periodic manner between 2 basic patterns, sometimes called "phases" or "states." In addition, oscillations are coincident with seasonal weather anomalies in other parts of the world (ie, a location might be warmer, colder, wetter, or drier than normal). These correlations are referred to as "teleconnections."

Several oscillations affect ocean currents and seasonal weather patterns for specific regions of the globe. The most well known is the El Niño–Southern Oscillation, whose oceanic component is the opposing El Niño and La Niña climate patterns and whose atmospheric component is the Southern Oscillation. Similar oscillations include the Arctic Oscillation, the North Atlantic Oscillation, and the Pacific Decadal Oscillation.

The formal names of oscillations should be capitalized.

25.12.3.2 Skill

In meteorology, the term "skill" has a precise meaning that differs from its common usage. In meteorological usage, "skill" is a statistical evaluation of a forecast or class of forecasts in comparison with some other forecast or with a climatological average. When used in this manner, the word "skill" should not be replaced by a synonym such as "aptitude."

> The skill of the 5-day forecast has improved dramatically, and such forecasts are now as skillful as the 3-day forecast was 15 years ago.

25.12.4 Names of Environmental Satellites

Environmental satellites have several purposes, including the following:

* tracking weather systems
* measuring winds, temperature, and the humidity structure of the atmosphere
* measuring chemical constituents, such as ozone
* observing oceans and land surfaces

Most satellites obtain their names and abbreviations from the series to which they belong. For example, the European Organisation for the Exploitation of Meteorological Satellites' series of polar-orbiting satellites is named the Meteorological Operational satellite program (MetOp). The satellites in a series are distinguished from each other by adding a number, a letter, or a combination of both (eg, "*MetOp-B*," "*NOAA-19*," and "*Elektro-L No. 2*").

The National Oceanic and Atmospheric Administration (NOAA) in the United States uses letters to designate specific satellites in a series during the satellites' design, construction, and launch. Once satellites reach geostationary orbit, their letters are changed to numbers. At that point, the satellites' names are changed from roman type to italics. For example, in the NOAA's Geostationary Orbiting Environmental Satellite (GOES) series, a satellite launched in 2018 was named as follows:

> GOES-S (during design, construction, and launch)
> *GOES-17* (after reaching geostationary orbit)

Although satellites are commonly referred to by their acronyms or other abbreviations, spell out the full names of satellites on first mention followed by the abbreviations in parentheses. Use all uppercase letters if the abbreviated name of a satellite or satellite series is an acronym. If an abbreviated name of a satellite name or series is not an acronym, follow the capitalization established by the organization that owns the satellite.

> The Indian National Satellite System (INSAT) includes the meteorologic satellites *INSAT-2E*, *INSAT-3D*, and *INSAT-3DR*.
> The People's Republic of China's *Fēngyún-3D (FY-3D)* is in a polar sun-synchronous orbit.

25.13 NON-SI UNITS

Although the metric units defined in the Système International d'Unités (SI) are generally recommended for scientific writing (see Section 12.2.1, "Système International d'Unités (SI)"), some non-SI units are still in common usage in Earth, oceanographic, and meteorological sciences, particularly in the United States and Canada.

For crop and soil scientists, acres are more convenient than their metric equivalent, hectares, when referring to the 160-acre parcels of land that the United States and Canadian governments surveyed in the 19th century. The unit called "board foot" (bd ft) is still used in measurements of lumber (see Table 12.8). Geological time is expressed in decimal multiples of the units annum and years (see Section 25.2, "Geological Time Units").

In meteorology and oceanography, a few non-SI units are still pervasive in the technical literature. The most common non-SI unit in meteorological literature is "millibar" (mb). Although the term "millibar" has a long tradition, its use is generally discouraged. The SI equivalent of "millibar" is "hectopascal" (ie, 1 hPa = 1 mb), which is the SI term for 100 pascals.

Similarly, "nautical mile" (n mi) and "knot" (kt) are strongly discouraged even though they still sometimes appear in the oceanographic literature (see Table 12.5). The metric equivalents for "nautical mile" and "knot" are "kilometer" and "meters per second" (m/s), respectively. In addition, the SI-derived unit "cubic hectometer per second" (hm/s) should be used instead of the "Sverdrup unit" ($Sv = 10^6$ m^3/s) for volume transport in physical oceanography, and the options outlined in Section 25.11.2.4, "Density," should be used instead of "sigma-t" (σ_t) for seawater density.[42]

CITED REFERENCES

1. Salvador A, editor. International stratigraphic guide: a guide to stratigraphic classification, terminology, and procedure. 2nd ed. Geological Society of America Publications; 1994.

2. Murphy MA, Salvador S, editors. International stratigraphic guide—abridged version. International Subcommission on Stratigraphic Classification of IUGS International Commission on Stratigraphy. Episodes. 1994;22(4):255–271.

3. International Commission on Stratigraphy. International chronostratigraphic chart. 2022. https://stratigraphy.org/ICSchart/ChronostratChart2022-02.pdf

4. North American Commission on Stratigraphic Nomenclature. North American stratigraphic code. Am Assoc Pet Geol Bull. 2005;89(11):1547–1591. Also available at http://www.agiweb.org/nacsn/code2.html#anchor514748

5. Gradstein FM, Ogg JG. Geologic time scale 2004—why, how, and where next! Lethaia. 2004;(37):175–181. https://doi.org/10.1080/00241160410006483

6. Hansen WR, editor. Suggestions to authors of the reports of the United States Geological Survey. 7th ed. United States Geological Survey; 1991. https://doi.org/10.3133/7000088

7. Lexique stratigraphique canadien (Lexicon of Canadian stratigraphy). Service Géologique du Québec; 1993. Vol 5B, Région des Appalaches, des Basses-Terres du Saint-Laurent et des îles de la Madeleine.

8. Glass DJ, editor. Lexicon of Canadian stratigraphy [CD-ROM]. Vol. 4, Western Canada. Canadian Society of Petroleum Geologists; 1990.

9. Lexicon of Canadian geological names on-line. Government of Canada; [accessed 2023 Mar 22]. https://weblex.nrcan.gc.ca/weblexnet4/weblex_e.aspx

10. Munsell's rock color book. Munsell Color Company. 2009. Also available at https://munsell.com/color-products/color-communications-products/environmental-color-communication/munsell-rock-color-chart/

11. Compton RR. Geology in the field. John Wiley & Sons; 1985.

12. US Government Publishing Office. Style manual: an official guide to the form and style of federal government publishing. US Government Publishing Office; 2016. https://www.govinfo.gov/collection/gpo-style-manual

13. National Geographic atlas of the world. 11th ed. National Geographic Society; 2019.

14. Le Maitre RW, editor. Igneous rocks: a classification of igneous rocks and glossary of terms: recommendations of the International Union of Geological Sciences Subcommission on the Systematics of Igneous Rocks. 2nd ed. Le Maitre RW et al, contributors. Cambridge University Press; 2002.

15. Back ME. Fleischer's glossary of mineral species. 12th ed. Mineralogical Record; 2018.

16. Nickel EH, Grice JD. The IMA Commission on New Minerals and Mineral Names: procedures and guidelines on mineral nomenclature, 1998. Can Mineral. 1998;36(3):913–926. https://pubs.geoscienceworld.org/canmin/article-abstract/36/3/913/12980/The-IMA-Commission-on-New-Minerals-and-Mineral?redirectedFrom=fulltext

17. Kretz R. Symbols for rock-forming minerals. Am Mineralog. 1983;68(1–2):277–279. https://pubs.geoscienceworld.org/msa/ammin/article-abstract/68/1-2/277/41456/Symbols-for-rock-forming-minerals

18. Aroyo MI, editor. International tables for crystallography. Vol A, space-group symmetry. 6th ed. Wiley; 2016. Published for the International Union of Crystallography.

19. Bennett A, Hamilton D, Maradudin A, Miller R, Murphy J. Crystals perfect and imperfect, by scientists of the Westinghouse Research Laboratories. Banigan S, editor. Walker; 1965.

20. Department of Agriculture (US), Soil Survey Division Staff. Soil survey manual. 4th ed. Ditzler C, Scheffe K, Monger HC, editors. Department of Agriculture (US), Natural Resources Conservation Service; 2017. https://www.nrcs.usda.gov/wps/portal/nrcs/detail/soils/scientists/?cid=nrcs142p2_054262

21. Schoeneberger PJ, Wysocki DA, Benham EC, Broderson WD, Soil Survey Staff. Field book for describing and sampling soils. Ver 3.0. Department of Agriculture (US), Natural Resources Conservation Service, National Soil Survey Center; 2012. https://www.academia.edu/30381925/Field_Book_for_Describing_and_Sampling_Soils

22. Department of Agriculture (US), Natural Resources Conservation Service. The color of soil. Department of Agriculture (US), Natural Resources Conservation Service. https://www.nrcs.usda.gov/wps/portal/nrcs/detail/soils/edu/?cid=nrcs142p2_054286. Adapted from Lynn WC, Pearson MJ. The color of soil. Sci Teach. 2000;67(5):20–23.

23. American Society of Agronomy. Publications handbook and style manual. American Society of Agronomy; [updated 2021 November]. Jointly published with the Crop Science Society of America and the Soil Science Society of America. https://www.agronomy.org/publications/journals/author-resources/style-manual/

24. Web soil survey. Department of Agriculture (US). https://websoilsurvey.sc.egov.usda.gov/

25. FAO soils portal. Food and Agriculture Organization of the United Nations. https://www.fao.org/soils-portal/en/

26. National Research Council Canada, Canadian Agricultural Services Coordinating Committee, Soil Classification Working Group. The Canadian system of soil classification. 3rd ed. NRC Research Press; 1998. https://sis.agr.gc.ca/cansis/publications/manuals/1998-cssc-ed3/cssc3_manual.pdf

27. Department of Agriculture (US), Natural Resources Conservation Service, Soil Survey Staff. Soil tax-

onomy: a basic system of soil classification for making and interpreting soil surveys. 2nd ed. Department of Agriculture (US), Natural Resources Conservation Service; 1999. https://www.nrcs.usda.gov/Internet /FSE_DOCUMENTS/nrcs142p2_051232.pdf

28. International Society of Soil Science; International Soil Reference and Information Centre; Food and Agriculture Organization of the United Nations. World reference base for soil resources 2014: international soil classification system for naming soils and creating legends for soil maps. Food and Agriculture Organization of the United Nations; 2014 [2015 update]. (World soil resources reports No. 106.) https:// www.fao.org/3/i3794en/I3794en.pdf

29. Soil series classification database (SC). Department of Agriculture (US), Natural Resources Conservation Service, Soil Survey Staff; [accessed 2023 Mar 22]. https://data.nal.usda.gov/dataset/soil-series -classification-database-sc

30. Department of Agriculture (US), National Resources Conservation Service, Soil Survey Staff. Keys to soil taxonomy. 13th ed. Department of Agriculture (US), National Resources Conservation Service; 2022. https://www.nrcs.usda.gov/resources/guides-and-instructions/keys-to-soil-taxonomy

31. Soil Survey Geographic database (SSURGO). Department of Agriculture (US), National Resources Conservation Service, Soil Survey Staff; [accessed 2023 Mar 22]. https://data.nal.usda.gov/dataset/soil -survey-geographic-database-ssurgo

32. United States General Soil Map (STATSGO2). Department of Agriculture (US), Natural Resources Conservation Service, Soil Survey Staff; [accessed 2023 Mar 22]. https://data.nal.usda.gov/dataset/united -states-general-soil-map-statsgo2

33. National ecological framework for Canada. Agriculture and Agri-Food Canada; 2021. https://open .canada.ca/data/en/dataset/3ef8e8a9-8d05-4fea-a8bf-7f5023d2b6e1

34. Klaus KEN, Mehl JP Jr, Jackson JA, editors. Glossary of geology. 5th ed (revised). American Geological Institute; 2011.

35. Glossary of soil science terms. Soil Science Society of America; 2021. https://www.soils.org /publications/soils-glossary

36. Lal R, editor. Encyclopedia of soil science. 3rd ed. Marcel Dekker; 2016.

37. Huang PM, Li Y, Sumner ME, editors. Handbook of soil science: properties and processes. 2nd ed. CRC Press; 2011.

38. Department of the Interior (US), Bureau of Land Management, Cadastral Survey. Specifications for descriptions of land: for use in land orders, executive orders, proclamations, Federal Register documents, and land description databases. Department of the Interior (US), Bureau of Land Management; 2015 [revised 2017]. https://www.blm.gov/sites/blm.gov/files/SpecificationsForDescriptionsOfLand.pdf

39. Natural Resources Canada. Manual of instructions for the survey of Canada lands. 3rd ed. Vol 2, Administrative requirements, general instructions and appendices. Geomatics Canada Legal Surveys Division; 1996. http://clss.nrcan.gc.ca/standards-normes/data/Manual%20of%20Instructions-3rd%20Edition -Vol%202E.pdf

40. Wentworth CK. A scale of grade and class terms for clastic sediments. J Geol. 1922;30(5):377–392. https://www.jstor.org/stable/30063207?seq=1

41. Shepard FP. Nomenclature based on sand-silt-clay ratios. J Sed Petrol. 1954;24:151–158. https://doi .org/10.1306/D4269774-2B26-11D7-8648000102C1865D

42. International Association for the Physical Sciences of the Ocean, Working Group on Symbols, Units and Nomenclature in Physical Oceanography. The international system of units (SI) in oceanography: report of IAPSO Working Group on Symbols, Units and Nomenclature in Physical Oceanography (SUN). United Nations Educational, Scientific and Cultural Organization; 1985. (UNESCO technical papers in marine science; 45); (IAPSO publication scientifique; No. 32). http://unesdoc.unesco.org/images/0006/000650 /065031eb.pdf

43. Hutchinson GE. A treatise on limnology. John Wiley & Sons; 1957.

44. Williams WD, Sherwood JE. Definition and measurement of salinity in salt lakes. Int J Salt Lake Res. 1994;3:53–63. https://doi.org/10.1007/BF01990642

45. Talley LD, Pickard GL, Emery WJ, Swift JH. Descriptive physical oceanography: an introduction. 6th ed. Academic Press; 2011. Also available at https://www.sciencedirect.com/book/9780750645522 /descriptive-physical-oceanography

46. Millero FJ. What is PSU? Oceanography. 1993;6(3):67. http://www.jstor.org/stable/43924646

47. Hicks SD. National Tidal Datum Convention of 1980. National Oceanic and Atmospheric Administration (US), National Ocean Survey; 1980. https://tidesandcurrents.noaa.gov/publications/National_Tidal _Datum_Convention_1980.pdf

48. Seguin W, Desai A, editors. Glossary of meteorology. American Meteorological Society. https:// glossary.ametsoc.org/wiki/Welcome

ADDITIONAL REFERENCES

Ashworth W, Little CE. Encyclopedia of environmental studies. Facts on File; 2001.

Bates RL, Adkins-Heljeson MD, Buchanan RC, editors. Geowriting: a guide to writing, editing, and printing in Earth science. 5th rev ed. American Geological Institute; 2004.

Canarache A, Vintila I, Munteanu I. Elsevier's dictionary of soil science. Elsevier; 2006.

Clark AM. Hey's mineral index: mineral species, varieties, and synonyms. 3rd ed. Chapman & Hall; 1993.

Fanning DS, Fanning MCB. Soil: morphology, genesis, and classification. John Wiley & Sons; 1989

GID Editorial Board. Guide to authors: a guide for the preparation of Geological Survey of Canada maps and reports. Geological Survey of Canada; 1998. Also available at https://doi.org/10.4095/209973

Gradstein FM, Ogg JG, Smith AG, editors. A geologic time scale 2004. Cambridge University Press; 2005.

Grant B. Geoscience reporting guidelines. [Publisher unknown]; 2003. (Includes geologic time scale.) Available at https://gac.ca/product/geoscience-reporting-guidelines/

Haq BU, van Eynsinga FWB. Geological time table [chart]. 6th rev ed. Elsevier Science; 2007.

Klein C, Dutrow B. Manual of mineral science. 23rd ed. John Wiley & Sons; 2008.

Ogg JG. Status of divisions of the international geologic time scale. Lethaia. 2004;37(2):138–139. https:// doi.org/10.1080/00241160410006492

Porteous A. Dictionary of environmental science and technology. 4th ed. John Wiley & Sons; 2008.

Price E. International tables for crystallography. Vol C, mathematical, physical, and chemical tables. 3rd ed. Springer Dordrecht; 2004. Published for the International Union of Crystallography.

Schultz DM. Eloquent science: a practical guide to becoming a better writer, speaker, and atmospheric scientist. Appendix B, Commonly misused scientific words and expressions. American Meteorological Society; 2009. p 351–365.

Thomas DSG, editor. The dictionary of physical geography. 4th ed. John Wiley & Sons; 2016. https://online library.wiley.com/doi/book/10.1002/9781118782323

Walsh SL, Gradstein FM, Ogg JG. History, philosophy, and application of the Global Stratotype Section and Point (GSSP). Lethaia. 2004;37(2):201–218. https://doi.org/10.1080/00241160410006500

Weil RR, Brady NC. Elements of the nature and properties of soils. 4th ed. Pearson; 2019. https://www .pearson.com/store/p/elements-of-the-nature-and-properties-of-soils/P100000860305

Witty JE, Arnold RW. Soil taxonomy: an overview. Outlook Agric. 1987;(16):8–13.

Wood DN, Hardy JE, Hardy AP, editors. Information sources in the Earth sciences. 2nd ed. Bowker-Saur; 1989.

26 Astronomical Objects and Time Systems

Editors: Craig Damlo, MSc; David M. Schultz, PhD, FAMS, FRMetS, FGS, SFHEA; Sherri Damlo, MS, ELS; and Julie Steffen, CAE

26.1 NAMES AND ABBREVIATIONS

During the past several decades, new instruments and methods that exceed the capabilities of traditional optical telescopes have rapidly expanded our knowledge of what lies beyond Earth's atmosphere. Simultaneously expanding is the nomenclature for identifying newly discovered astronomical objects and phenomena. The leading authority on this specialized nomenclature is the International Astronomical Union.[1,2]

An astronomical object may be designated by a proper name, a catalog number, or a composite name that indicates the type of object and its position. In general, newly designated objects receive composite names.

Proper names may be traditional names established long before telescopic observation. Some examples are planets and smaller bodies in the Solar System that are visible to the naked eye, as well as constellations (see Table 26.1) and prominent stars.

Catalog numbers are assigned by compilers of catalogs. Each of these numbers is preceded by an abbreviation for the name of the catalog in which the object is listed. For examples, see Section 26.5, "Objects outside the Solar System." For a list of some

Table 26.1. Constellation Names and Their Abbreviations

Name of constellation	Abbreviation	Name of constellation	Abbreviation
Andromeda	And	Lacerta	Lac
Antlia	Ant	Leo	Leo
Apus	Aps	Leo Minor	LMi
Aquarius	Aqr	Lepus	Lep
Aquila	Aql	Libra	Lib
Ara	Ara	Lupus	Lup
Aries	Ari	Lynx	Lyn
Auriga	Aur	Lyra	Lyr
Boötes	Boo	Mensa	Men
Caelum	Cae	Microscopium	Mic
Camelopardalis	Cam	Monoceros	Mon
Cancer	Cnc	Musca	Mus
Canes Venatici	CVn	Norma	Nor
Canis Major	CMa	Octans	Oct
Canis Minor	CMi	Ophiuchus	Oph
Capricornus	Cap	Orion	Ori
Carina	Car	Pavo	Pav
Cassiopeia	Cas	Pegasus	Peg
Centaurus	Cen	Perseus	Per
Cepheus	Cep	Phoenix	Phe
Cetus	Cet	Pictor	Pic
Chamaeleon	Cha	Pisces	Psc
Circinus	Cir	Piscis Austrinus	PsA
Columba	Col	Puppis	Pup
Coma Berenices	Com	Pyxis	Pyx
Corona Australis	CrA	Reticulum	Ret
Corona Borealis	CrB	Sagitta	Sge
Corvus	Crv	Sagittarius	Sgr
Crater	Crt	Scorpius	Sco
Crux	Cru	Sculptor	Scl
Cygnus	Cyg	Scutum	Sct
Delphinus	Del	Serpens[a]	Ser
Dorado	Dor	Sextans	Sex
Draco	Dra	Taurus	Tau
Equuleus	Equ	Telescopium	Tel
Eridanus	Eri	Triangulum	Tri
Fornax	For	Triangulum Australe	TrA
Gemini	Gem	Tucana	Tuc
Grus	Gru	Ursa Major	UMa
Hercules	Her	Ursa Minor	UMi
Horologium	Hor	Vela	Vel
Hydra	Hya	Virgo	Vir
Hydrus	Hyi	Volans	Vol
Indus	Ind	Vulpecula	Vul

[a] Serpens may be divided into Serpens Caput and Serpens Cauda.

Table 26.2. Abbreviations for the Titles of Selected Astronomical Catalogs[a]

Abbreviation	Catalog
AGK *n*	*Astronomische Gesellschaft Katalog Nummer n*
BD	*Bonner Durchmusterung*
BS *or* BSC	*Bright Star Catalogue*
CD	*Cordoba Durchmusterung*
CPD	*Cape Photographic Durchmusterung*
FK *n*	*Fundamental Katalog Nummer n*
GCVS	*General Catalogue of Variable Stars*
HD	*Henry Draper Catalogue*
HDE	*Henry Draper Extension*
HR	*Revised Harvard Photometry Catalogue*
IC	*Index Catalog*
M	*The Messier Catalog*
NGC	*New General Catalog of Nebulae and Clusters of Stars*
SAO	*Smithsonian Astrophysical Observatory Star Catalog*
3C	*Third Cambridge Catalogue*

[a] For additional titles and their abbreviations, see Ridpath.[3]

catalog names and their abbreviations, see Table 26.2. For a more comprehensive list, see *Norton's Star Atlas and Reference Handbook*.[3]

Some objects have 2 or more names of equal standing.

> Ring Nebula = M57 (Messier catalog number) = NGC 6720 (New General Catalog number)
> the Pleiades = Seven Sisters = M45

A temporary name may be given to an object when it is first observed. A permanent name is assigned when the object's properties are established.

> Originally designated as "S/1989 N 1," this moon was later given the designation "Neptune VIII" and the name "Proteus."

Because proper names and catalog numbers may be inadequate to unambiguously identify many objects, composite names provide essential characteristics of objects (by using alphanumeric code names) and their positions.

26.2 CAPITALIZATION AND LOWERCASE

Capitalize the proper names of galaxies, constellations (see Table 26.1), stars, planets and their satellites, asteroids, comets, other unique celestial objects, and space programs.

> Milky Way   Crab Nebula   Solar System   Comet Biela
> Orion   Coalsack   North Star   Artemis Program

To designate astronomical objects by catalog numbers, capitalize the catalog's full name and its abbreviation.

> NGC 6165   Bond 619   M81

Capitalize the word "Earth" when used in an astronomical context, particularly when used with the names of other planets. In this situation, the definite article (ie, "the") is usually omitted. Similarly, capitalize "Moon" and "Sun" when referring to Earth's natural satellite and Earth's star. Lowercase "earth" only when referring to soil, and lowercase "moon" and "sun" when referring to a satellite of a planet other than Earth and to a star in another solar system. (Note that many astronomy, atmospheric science, and Earth science publications always capitalize "Earth.")

> The first 3 planets from the Sun are Mercury, Venus, and Earth.
> Jupiter has many moons, but Earth has only one.
> Human exploration of the surface of the Moon was followed by robotic exploration of the surface of Mars.
> Gliese 876 is the sun to at least 4 planets.
> Archaeologists carefully remove earth at excavation sites.

Capitalize the names of spacecraft and artificial satellites. Once a spacecraft is launched into orbit, it becomes a satellite. Satellite names are set in roman type during design and construction phases but in italics after they reach orbit (see Section 25.12.4, "Names of Environmental Satellites").

> *Astra 3A*, which the Société Européenne des Satellites launched in 2002
> *Centauri-1*, which the Indian Space Research Organisation launched in 2018
> *Tianlian II-03*, which the China Aerospace Science and Technology Corporation launched in 2022
> PLATO, which the European Space Agency plans to launch in 2026

Use lowercase for 3 main classes for astronomical objects:

(1) descriptive terms applied to celestial phenomena, except for terms that are proper nouns

> the gegenschein   the Cassini division   transit of Venus
> star clusters   the rings of Saturn   the Laplacian plane of Saturn's rings

(2) terms applied to meteorologic and other atmospheric phenomena, except for terms that are proper nouns

> aurora borealis   sun dog   meteor shower   Tunguska fireball

(3) adjectival forms of proper nouns

> jovian *for* Jupiter   lunar *for* the Moon   mercurial *for* Mercury   solar *for* the Sun

26.3 CELESTIAL COORDINATES

Right ascension (designated "α") is given in hours, minutes, and seconds of sidereal time, with no spaces between numbers and units. The units for hours, minutes, and seconds can be designated in superscript "h," "m" and "s," respectively. In less precise designations, the general convention for units may be acceptable (see Section 12.2.1, "Système International d'Unités (SI)").

> $14^h6^m7^s$ ("14 h 6 min" in a less precise designation)

Declination (designated "δ") is given in degrees, minutes, and seconds of arc north or south of the celestial equator. Declinations north of the equator are marked with the + symbol or they are unmarked. Those south of the equator are marked with the – symbol.

> $+49°8'11''$ *or* $49°8'11''$ $-87°41'08''$

When decimal fractions are part of the units used in either right ascension or declination, the decimal point is placed between the unit symbol and the decimal value.

> $26^h6^m7^s.2$ $+34°.26$

Celestial coordinate systems are defined by the mean equator and the equinoxes. The equinoxes are the 2 points where Earth's equator projected into space crosses the ecliptic, which is the plane of Earth's orbit around the Sun. Celestial coordinate systems ignore small, short-period variations in the motion of the celestial equator, but they are affected by precession, the slow continuous westward motion of the equinoxes around the ecliptic that results from the simultaneous westward motion of Earth's axis. The positions of the mean equator and equinoxes at certain epochs are used to define standard reference systems for time, such as those for B1950.0 and J2000.0 (see Section 26.6.1.2, "Julian and Besselian Epochs").

See Table 26.3 for a list of abbreviations and symbols used frequently in astronomical literature, including those related to celestial coordinates.

Table 26.3. Abbreviations and Symbols Frequently Used in the Astronomical Literature

Abbreviation or symbol	Meaning
α	right ascension (*also used for* the brightest star in a constellation)
AU	astronomical unit, the mean distance between Earth and the Sun (UA in French)
B	Besselian
β	second brightest star in a constellation
c	speed of light
CMBR	cosmic microwave background radiation
ΔT	increment to be added to universal time to obtain terrestrial time[a]
δ	declination (*also used for* fourth brightest star in a constellation)
E	color excess
ET	ephemeris time, a measure of time for which a constant rate was defined (used from 1958 to 1983)
G	gravitation constant
γ	third brightest star in a constellation
H_0	Hubble constant
H II region	volume of hydrogen photoionized into protons and electrons by the ultraviolet radiation from a central, hot object
HA	hour angle
k	curvature index of space (*also used for* Gaussian gravitational constant)
kpc	kiloparsec
L	stellar luminosity
$L_\odot$	solar luminosity
λ	celestial longitude
LAST	local apparent sidereal time
LMST	local mean sidereal time
LST	local sidereal time
M	absolute magnitude
$M_\oplus$	mass of Earth
$M_\odot$	solar mass
MHD	magnetohydrodynamics, also known as "hydromagnetics"
mJy	milli-Jansky, a unit of luminous flux

Table 26.3. Abbreviations and Symbols Frequently Used in the Astronomical Literature (*continued*)

Abbreviation or symbol	Meaning
Mpc	megaparsec, a million parsecs, or 10^6 pc
μg	microgauss
pc	parsec
PZT	photographic zenith telescope (out of date)
Q	aphelion, the point in solar orbit farthest from the Sun
QSO	quasistellar object, or quasar
R	cosmic scale factor, a measure of the size of the universe as a function of time
RA	right ascension
R_i	radius of Earth
rv	radial velocity
SRS	southern reference system
t	hour angle
TAI	international atomic time ("TAI" is from the French term "temps atomique international")
TCB	coordinated barycentric time ("TCB" is from the French term "temps-coordonnée barycentrique")
TDB	barycentric dynamical time ("TDB" is from the French term "temps dynamique barycentrique")
TT	terrestrial time (formerly "terrestrial dynamical time [TDT]")
UT1	universal time 1
UTC	coordinated universal time ("UTC" is from the alternative term "universal time coordinated")
v	visual magnitude
VLBI	very-long-baseline radio interferometry
z	red-shift parameter
ZAMS	zero-age main sequence
ZHR	zenithal hourly rate

[a] Terrestrial time can also be obtained by adding 32.184 s to international atomic time.

26.4 OBJECTS IN THE SOLAR SYSTEM

26.4.1 Planets and Satellites

In August 2006, the International Astronomical Union (IAU) defined a planet in the Solar System as "a celestial body that (a) is in orbit around the Sun, (b) has sufficient mass for its self-gravity to overcome rigid body forces so that it assumes a hydrostatic equilibrium (nearly round) shape, and (c) has cleared the neighborhood around its orbit."[4] By this definition, the Solar System has 8 planets: Mercury, Venus, Earth, Mars, Jupiter, Saturn, Uranus, and Neptune.

In scientific publications, use the established full names for planets, not the symbols that are common in astrological and other informal literature. In general, when listing planets or natural satellites, place them in order of mean distance from the Sun or their planet, respectively.

A reliable source for names for bodies within the Solar System is the IAU's Working

Group for Planetary System Nomenclature. This IAU working group is responsible for naming newly discovered members of the Solar System other than small bodies, as well as naming surface features of planets and satellites. In addition to having names, satellites are assigned numbers in their chronological sequence of discovery. The decisions made by the IAU's Working Group for Planetary System Nomenclature are available in the *Gazetteer of Planetary Nomenclature*.[5] In addition, these decisions are reported in the *Proceedings of the International Astronomical Union*,[6] formerly titled *Transactions of the International Astronomical Union*. The Jet Propulsion Laboratory's Solar System Dynamics Group also maintains a list of numbers and names for known natural satellites.[7]

26.4.2 Dwarf Planets and Small Solar System Bodies (Asteroids and Comets)

In August 2006, the IAU created a new class of objects called "dwarf planets." Like a planet, a dwarf planet must be in orbit in the Solar System and be sufficiently massive to have become rounded by its own gravity. Unlike a planet, a dwarf planet may share its orbit with other objects, such as asteroids. Dwarf planets are generally smaller than Mercury. The first 5 dwarf planets designated by the IAU are Ceres, Pluto, Eris, Makemake, and Haumea.

With the exception of Ceres, these dwarf planets also belong to a new class called "Trans-Neptunian Objects." The IAU has designated these objects "plutoids," which are dwarf planets that orbit the Sun at a semimajor axis greater than that of Neptune.

The IAU uses the term "small Solar System bodies" to refer to objects orbiting the Sun that are insufficiently massive to be called either planets or dwarf planets. (The term "minor planet" is still in use, but the new term is preferred in most contexts.) This class includes asteroids and comets.[4]

Asteroids are given numbers serially along with proper names after their orbital elements have been reliably defined.

> 288 Glauke   719 Albert   878 Mildred   1627 Ivar

Since 1995, newly discovered comets are assigned designations reflecting their date of discovery.[8] The designation consists of the year of observation, an uppercase code letter identifying the half-month of the year in which the comet was observed, and a consecutive numeral indicating the order of discovery during that half-month. To indicate the nature of a comet, one of the following prefixes is added to the designation: "P/" for a periodic comet, "C/" for a comet that is not periodic, "X/" for a comet whose orbit cannot be computed, "D/" for a periodic comet that no longer exists, "I/" for an interstellar object, and "A/" for a minor planet that was mistakenly identified as a comet.

> C/2021 A1, a nonperiodic comet that was the first comet observed in the first half-month of 2021
> D/1770 L1, redesignation of a lost or broken-up comet that was the first comet observed the 12th half-month of 1770

The prefixes "I/" and "A/" were not used until 2017. As of 2022, "I/" had been used only twice, and in those cases, the cardinal numbers "1" and "2" preceded the prefix.

> 2I/2019 Q4, the second interstellar comet, which was the fourth comet observed in the 17th half-month of 2019

Prior to 1995, confirmed comets were designated by the names of their discoverers, their year of discovery, and capital Roman numerals. The Roman numerals indicated the chronological sequence during the year of discovery in which the comets passed through perihelion, the point in an astronomical body's path at which the body is closest to the Sun.

> Comet Arend–Roland 1957 III, the third comet to pass through perihelion in 1957, which was discovered by Belgian astronomers Sylvain Arend and Georges Roland (redesignated C/1956 R1)
> Comet Kohoutek 1973 XII, the 12th comet to pass through perihelion in 1973, which was discovered by Czech astronomer Luboš Kohoutek (redesignated C/1973 E1)

26.4.3 Meteor Showers and Meteorites

Meteor showers are usually named for the constellations in which their radiant points appear or according to the name of the comet with which they are associated.

> the Arietids   the Lyrids   the Eta Aquarids   the Quadrantids

A large meteorite is usually identified by the name of the place near where it and its crater were found. Although a large meteorite is named according to its fixed geographic location, a fireball is usually identified by the date on which it was seen.

> Seymour meteorite (in Indiana, United States)   Brent crater (in Canada)
> Hoba West meteorite (in Namibia)

Comprehensive sources for meteorite names include the *Catalogue of Meteorites: With Special Reference to Those Represented in the Collection of the Natural History Museum, London.*[9]

26.5 OBJECTS OUTSIDE THE SOLAR SYSTEM

26.5.1 Bright Stars

A "bright star" is any star that is visible from Earth by the naked eye. In general, bright stars have a magnitude of 6.5 or greater.

The Bayer designation of a bright star consists of a Greek letter followed by the name or standard 3-letter abbreviation for the constellation in which the bright star appears (see Table 26.4). The Greek letters are usually assigned in order of brightness (eg, α, β, γ, δ). The names of bright stars should be in roman type. If the Greek character at the beginning of a bright star's name is not available in the font being used, spell out the English name of the Greek letter.

> α Centauri A *or* α Cen A *or* alpha Centauri A *or* alpha Cen A

Table 26.4. Examples of Alternative Designations for Bright Stars[a]

Traditional name	Bayer system	Flamsteed system
Betelgeuse	α Ori	58 Ori
Sirius	α CMa	9 CMa
Ras-Alhague	α Oph	55 Oph
Polaris	α UMi	14 UMi

[a] For other designations in the Bayer and Flamsteed systems, see *The Astronomical Almanac for the Year 2023.* United States Naval Observatory, Nautical Almanac Office; 2023.

Table 26.5. Luminosity Classes of Stars

Class	Luminosity type	Class	Luminosity type
Ia and Ib	Supergiants	IV	Subgiants
II	Bright giants	V	Main sequence (ie, dwarfs)
III	Giants		

In contrast to the Bayer system, the Flamsteed system designates each star by a number followed by the 3-letter abbreviation for the star's constellation (see Table 26.4).

In addition to Bayer and Flamsteed designations, approximately 1,000 stars have proper names derived from early Arabic names or from peculiar characteristics. When using these traditional names, the stars should be identified elsewhere in the manuscript with their more precise Bayer or Flamsteed designations, especially if the traditional names differ in other languages.

The Bright Star Catalogue[10] lists approximately 900 traditional names and gives information on more than 9,000 stars that are brighter than magnitude 6.5. The catalog also lists names based on the Bayer system and the Flamsteed system (see Table 26.4). In addition, approximately 1,500 bright stars are listed in the annual *Astronomical Almanac*.[11] This almanac provides the Bayer, Flamsteed, and traditional bright star designations, as well as other information, such as each star's mean position.

Stars can be classified into 9 groups according to spectral temperature: O (for the hottest stars), B, A, F, G, K, M, L, and T (for the coolest stars, which are very red). In turn, these groups can be divided into divisions, which are designated by appending the numerals 0 to 9 to the alphabetical group designations (eg, A4, F8, G0). However, there are no K8 or K9 stars.

Stars may also be classified by luminosity by using the roman numerals I (for supergiants), II, III, IV, and V (for main-sequence, or dwarf, stars). In turn, the luminosity classes may be divided by adding lowercase letters to the roman numerals (see Table 26.5).

Although roman type should be used for spectral and luminosity classifications and similar symbols and abbreviations for stars, italics should be used for Roman and Greek symbols for quantity (see Section 4.2.2, "Type Styles for Symbols," and Table 4.5).

26.5.2 Faint Stars

For a star not listed in any star catalog and for which a designation has not otherwise been assigned, use an abbreviation along with the star's celestial coordinates in the standard form (see Section 26.5.7, "Radiation Sources"). The abbreviation should refer in some way to the star's discoverer or to a literature citation in which the star is reported.

26.5.3 Variable Stars and Binary Stars

The name of a variable star consists of the name of its constellation (see Table 26.1) preceded by either a 1- or 2-letter capitalized code or the capital letter V followed by a number. When a variable star has a classical name, use that name in titles of manuscripts. For the classical names of variable stars, consult the *General Catalogue of Variable Stars*[12] or the SIMBAD database[13] of the Centre de Données Astronomiques de Strasbourg.

W Vir	RV Tau	V356 Sgr	V827 Her 1987

Many binary, or double, stars are named in the same manner as variable stars (eg, RY Sct). Alternatively, a binary star system may take the name of the catalog star with which the binary stars are associated (eg, β Cet, α Her, Sirius A).

Some binary stars are cataclysmic variables (eg, WZ Sge, Her X-1). The stars in a cataclysmic variable are relatively close to each other, and they exhibit violent behavior.[14] One member of the pair is a compact, degenerate star, such as a white dwarf, a neutron star, or a black hole. The behavior of cataclysmic variables varies greatly, depending on the nature of the degenerate star in each pair.

26.5.4 Novae

A nova is designated by the word "Nova" followed by the constellation name or its abbreviation (see Table 26.1) and the year of discovery. These elements are separated by spaces without punctuation. If more than one nova is discovered in a constellation in the same year, the year is followed by a number.

Nova Sge 1987	Nova Sgr 1987 No. 1	Nova Per 1901	Nova Sge 1783

26.5.5 Supernovae

A supernova is designated with the initialism "SN" followed by a space, the year of discovery, and an alphabetical designation that indicates the order of discovery. At the start of each year, the alphabetical designations are single capital letters from A to Z. Once 26 supernovae are discovered in a year, the alphabetical designation switches to paired lowercase letters (eg, aa, ab, ac) and then to tripled lowercase letters (eg, dgk, dgl, dgm). No punctuation or space is used between the year and the alphabetical designation or between multiple lowercase-letter designations.

SN 1985A	SN 2005gl	SN 2018zd	SN 2022hrs

Newly discovered supernova remnants are designated as "SNR" followed by the position of the supernova's explosion with respect to the estimated date in the Julian system (see Section 26.6.1.2, "Julian and Besselian Epochs").

26.5.6 Nebulae, Galaxies, Clusters, Pulsars, and Quasars

One of the 2 most widely used catalogs of nebulae, galaxies, and clusters is *The Messier Catalog*.[15] Initially compiled by French astronomer Charles Messier, FRS, between 1771 and 1784, this catalog consists of 110 entries on the brightest nebulae, galaxies, and star clusters visible from France. Each object is designated by Messier's initial immediately followed by the serial number he assigned to the object.

M1 *for* the Crab Nebula   M45 *for* the Pleiades (the Seven Sisters)

The other widely used catalog is the *New General Catalog* (NGC),[16,17] originally compiled in 1888 by Danish astronomer John Louis Emil Dreyer, PhD. The objects in this catalog and its supplements (the *Index Catalog* [IC]) are designated by the catalog's abbreviation, NGC, or the supplements' abbreviation, IC. The abbreviation is followed by a space and the catalog number of each object.

NGC 1952 *for* the Crab Nebula   NGC 869 *for* the cluster h Persei
IC 443 *for* the Jellyfish Nebula   IC 819 and IC 820 *for* the Mice Galaxies

Similarly, objects identified in other catalogs are usually referred to by the catalogs' or compilers' abbreviations and the objects' serial numbers. Spell out the names of catalogs on first use with the abbreviations provided immediately afterward and enclosed in parentheses (see Table 26.2 for the names and abbreviations of a few catalogs). While the catalogs' full names should be in italic type, the object names, including the abbreviations, should be in roman type.

Another means of designating objects is to refer to the constellations in which they reside (see Table 26.1 for constellations and their official abbreviations). The term "constellation" now refers to the whole of a specific region of the sky in which a pattern formed by the bright stars is seen. However, the ancient constellation patterns are still used as a convenient guide to parts of the sky, and they are used for naming individual objects in those patterns. For example, the first object discovered by radio source in the Southern Hemisphere constellation of Centaurus was named "Centaurus A."

Similarly, many famous galaxies, nebulae, and clusters are still known by individual names, often derived from their appearance and their more or less fanciful resemblance to terrestrial objects.

Coalsack   Whirlpool Galaxy

26.5.6.1 Planetary Nebulae

Several planetary nebulae have proper names. Some have Messier numbers, many have NGC or IC numbers, and many have catalog numbers based on their discoverers' names. All of them have catalog numbers based on their galactic coordinates.

Owl Nebula *also designated* M97, NGC 3587, and PN G148.4+57.0 (a bright planetary nebula)
LoTr 5 *also designated* PN G339.9+88.4 (a faint planetary nebula discovered by Andrew J. Long-more and S.B. Tritton)

26.5.6.2 Galaxies

Some galaxies have proper names. When identifying galaxies by their proper names, also provide their Messier numbers and their NGC or IC numbers.

Pinwheel Galaxy, *also known as* M101 and NGC 5457
Sombrero Galaxy, *also known as* M104 and NGC 4594

26.5.6.3 Radio Sources

Since 1985, the International Astronomical Union has recommended designating radio sources with catalog acronyms followed by the sources' celestial coordinates for right ascension and declination (see Section 26.3, "Celestial Coordinates") with respect to the reference system of J2000.0 (see Section 26.6.1.2, "Julian and Besselian Epochs").

Radio sources include pulsars, which are designated by the abbreviation "PSR" before the condensed notations for their positions. Older pulsar names are obsolete, such as "CP1919" for the first pulsar ever discovered.

> PSR 1937 + 21 (the numerical designation after the abbreviation "PSR" refers to right ascension of 19^h37^m and declination of 21°.4)

26.5.6.4 Star Clusters

Although some star clusters have proper names, many are identified only by their numbers in the NGC, the IC, or the Messier catalog.

> NGC 4755 *also* Jewel Box   M11 *also* Wild Duck Cluster   NGC 6656 *also* M22

26.5.6.5 Quasistellar Objects, or Quasars

No standard form has been assigned to the designation of quasistellar objects, also known as "quasars." However, quasistellar objects do have catalog numbers.

26.5.7 Radiation Sources

Designations of radiation sources should be taken from catalogs of satellite surveys. When designations are used in listings, they should not be altered. To avoid ambiguity, each source listing should also contain positional information, a second designation, or both.

The designation of an astronomical source consists of the following 3 parts: acronym, sequence, and specifier, with a space between each element but no commas. The acronym and sequence are necessary in the designation. However, the specifier is optional, and when provided, the specifier must be placed in parentheses.

> NGC 205   PKS 1817–43   CO J0326.0+3041.0   H2O G123.4+57.6 (VSLR=–185)

Formerly called "origin," the acronym specifies a catalog or collection of sources. Acronyms may refer to catalog names (eg, NGC, *Bonner Durchmusterung* [BD]), the names of authors (eg, RCW), observatories (eg, PKS, VLA, IRAS, 3C, 51W), instruments, and so forth. Acronyms should consist of at least 2 letters or numerals, but they should have no special characters.

The alphanumeric sequence, also called "numbering," specifies a unique source within a catalog or collection. This element of a radiation source may consist of a sequence number within a catalog (eg, HD 224801) or a combination of fields, or it may be based on coordinates. If coordinates are used, they are preceded by one of the following codes for reference frames, also called "flag letters": "G" for galactic coordinates, "B" for Besselian 1950, and "J" for Julian 2000 equatorial coordinates.

Coordinates used in designations are considered to be names. Therefore, they do not change even if radio sources' positions become more accurately known (eg, "BD –25 765" remains a designation even though that source's declination is now –26). Subcomponents and multiples of objects are designated with letters or numerals added to the sequence after a colon (eg, "NGC 1818:B12").

The optional specifier indicates that a source is associated with larger radiating sources (eg, M31, W 3) or other object parameters. The reason that specifiers are enclosed within parentheses is that they are optional.[18]

Designation	Position in right ascension and declination	
[Acronym sequence (specifier)]	[α (2000)]	[δ (2000)]
BD −3 5750	00 02 02.4	−02 45 59
H20 B0446.6+7253.7	04 46 37.3	+72 53 47
AC 211 (= 1E2127+119; in M15)	21 30 15.54	+11 43 39.0
PN G001.2−00.3	17 49 36.9	−28 03 59
R 136:a3	05 38 42.4	−69 06 03

The following examples are incorrect:

BD 4°14 (The degree symbol should not be used, and the declination plus or minus symbol is missing.)

N221 (The acronym is only one letter instead of at least 2. Consequently, the source is unclear: "N" could refer to NGC or to the *Magellanic Catalogue of Stars*. In addition, a space is missing between the acronym and sequence.)

P 43578 (The single-letter acronym is ambiguous.)

26.6 ASTRONOMICAL DATES AND TIME

26.6.1 Dates

The basis of any date and time system used for astronomical purposes must be clearly stated because no system is free from ambiguity. For example, in the calendar system, the year 1 BC is followed by the year AD 1, but for astronomical purposes, the calendar year 1 BC is denoted as year 0. The formulas for calculating calendar dates for astronomical purposes are "n BC = $-(n-1)$" and "n AD = n." As a consequence, 350 BC is the year −349 and 350 AD is the year 350 without reference to BC or AD.

In addition to clearly stating the basis of astronomical dates, ambiguity should be reduced further by specifying corresponding coordinated universal time (UTC) (Section 26.6.2.2, "Coordinated Universal Time")

26.6.1.1 Julian Day

The date on which an astronomical event occurred is usually best represented by its Julian day (JD) accompanied by an appropriate conventional form, such as UTC. The JD of any event corresponds to the number of solar days between the event and noon on 1 January in the year 4713 BC in the Julian calendar (ie, 24 November 4714 BC in the Gregorian calendar). For example, midnight on 1 January 2023 was JD 2459945.5, which corresponds to 0000 UTC on 1 January 2023.

In converting calendar dates and times (see Sections 13.3, "Dates," and 13.2, "Time") to JDs, it is important to remember that JDs begin at 1200 UT (noon on the International Reference Meridian), whereas current calendar days begin at 0000 UT (midnight on the International Reference Meridian). Before 1925, Greenwich mean time (GMT) often, but not always, referred to days that began at noon, not midnight, on the International Reference Meridian. Because the precise meaning of GMT remains somewhat ambiguous, it is preferable to use UTC (see Section 26.6.2.2, "Coordinated Universal Time") instead of GMT for reporting astronomical observations in conjunction with their JD dates.

The Julian system may be used in conjunction with other time systems besides UTC, such as ephemeris time and international atomic time (TAI). In fact, the name "Julian ephemeris day" (JED) was formerly used to indicate the combination of JDs with ephemeris time. It is now more appropriate to combine JD with another time scale.

In text:
A total solar eclipse began on JD 2459198.05071, which corresponds to 13:12:28 TAI on 14 December 2020. Another began JD 2459552.72847, or 05:28:28 TAI on 4 December 2021.

In column heading in table on total solar eclipses:
JD (TAI)
2459198.05071 (13:12:28 on 14 December 2020)
2459564.72848 (05:28:28 on 4 December 2021)

The modified Julian date (MJD) is sometimes used for current dates because it is a shorter number than JD. The MJD counts solar days from JD 2400000.5, or 0000 UT on 17 November 1858. The MJD is calculated by the formula "JD – 2400000.5." For example, JD 2500000.5 will be MJD 100000.0, or 0000 UT on 1 September 2132.

26.6.1.2 Julian and Besselian Epochs

In astronomy, epochs are instantaneous moments in time, so they have no duration, only dates of occurrence.

Since 1984, an astronomical epoch has been expressed in a JD. Julian epochs are denoted by the letter J. For example, the date J2000.0 is JD 2451545.0, and it occurred at noon on 1 January 2000 (expressed as "2000 January 1.5"). JDs may be used with 3 different time scales: UTC; TAI; and ephemeris time (T_{eph}), or dynamical time. The choice of time scale should be indicated as follows: J2000.0 for UTC, J2000.0 (TAI) for TAI, and JED 2000.0 for T_{eph}.

Prior to 1984, Besselian years were used to designate epochs. A Besselian year is the period of Earth's orbit measured between 2 successive passages of the apparent right ascension of the mean Sun through the value of 18^h40^{min}. Besselian epochs are denoted by the letter "B" (eg, B1950.0).

Besselian epoch	Julian day	Calendar day
2023.272822	2460044.750000	10 April 2023 at 6 AM UTC
2016.85111	2457699.666667	7 November 2016 at 4 PM UTC

26.6.2 Time Systems

The 7 major time scales described in this section are currently used in astronomy, although in reality, universal time 1 (UT1) and Greenwich mean sidereal time (GMST) are more properly considered angles.

26.6.2.1 International Atomic Time

TAI, or temps atomique international, is a statistical time scale based on an ensemble of atomic clocks around the world measuring the "SI second." The Système International d'Unités, or the International System of Units (SI), defines the SI second as the duration of 9,192,631,770 periods of the radiation corresponding to the transition between the 2 hyperfine levels of the ground state of the cesium-133 atom. A less exacting description is 1/86,400 the length of the mean solar day. Broadcast by radio, TAI is used by scientists throughout the world.

26.6.2.2 Coordinated Universal Time

UTC, also known as "universal time coordinated," is used for civil timekeeping. UTC corresponds to time at Greenwich, United Kingdom, along the International Reference Meridian, also known as the "prime meridian." Although UTC normally runs at the same rate as TAI, UTC is occasionally adjusted to match UT1 so that UTC keeps approximate pace with Earth's rotation, which is generally slowing down. These adjustments usually take the form of insertions of "leap seconds" into UTC, which when needed are made either at the end of June or the end of December. Thus, UTC departs from TAI slowly by a series of leap seconds. A table of the leap seconds and other adjustments that reflect the difference between TAI and UTC is available from the International Earth Rotation and Reference System Service's Earth Orientation Center[19] at the Observatoire de Paris.

26.6.2.3 Universal Time 1

UT1 is a refined version of universal time. While both are based on Earth's rotation with respect to mean Sun, UT1 also accounts for the wandering of Earth's rotational pole with respect to its surface.

UT1 should be thought of as an angle, as should GMST. UT1 relates to GMST by a strict formula and is determined directly by measurements of sidereal time.

It is to UT1 that UTC is adjusted by the use of leap seconds. UTC is adjusted if the difference between UT1 and UTC is more than 0.9 s. Consequently, when the mean Sun at Greenwich is overhead at UT1 noontime, it may not be exactly UTC noontime.

26.6.2.4 Terrestrial Time

For historical continuity with previous uses of ephemeris time, terrestrial time (TT) is defined as TAI plus 32.184 s. Formerly known as terrestrial dynamical time, TT is used for the printed listings of crude astronomical ephemerides and as a rough approximation to T_{eph} (see Section 26.6.2.5, "Ephemeris Time Argument T_{eph}").

26.6.2.5 Ephemeris Time Argument T_{eph}

Ephemeris time argument T_{eph} was defined to properly express the "free-flowing time" used in the ephemerides. An independent variable in equations of motion for bodies in the Solar System, T_{eph} is what its predecessors ephemeris time and barycentric dynamical time, or "temps dynamique barycentrique" (TDB), were intended to be, namely rigorously relativistic coordinate time.

Through a formula based in relativity, T_{eph} is related to TT and, therefore, to TAI and UTC. The formula is $|T_{eph} - TT| < 0.002$ s. In this formula, T_{eph} never departs from TT by more than 0.002 s.

26.6.2.6 Coordinated Barycentric Time

Defined by the International Astronomical Union in 1991, coordinated barycentric time, or temps-coordonnée barycentrique (TCB), is a relativistic coordinate time, like T_{eph}. Based strictly on the SI second, TCB has the advantage that all quantities in a TCB-based system are expressed in SI units. Its disadvantage is that TCB runs at a different rate than

TT and, therefore, departs from TT at the rate of approximately 0.5 s per year. Thus, TT cannot be used to approximate TCB. The 2 time scales were defined as equal at the start of 1977, and they have been departing from each other ever since.

26.6.2.7 Greenwich Mean Sidereal Time

Like UT1, sidereal time should be thought of as an angle. It is, effectively, the spin of the planet Earth in space. Before the advent of atomic time, the Earth's rotation was used for timekeeping. This involved determining the planet's orientation in space to determine sidereal time. Now, the accuracy of atomic clocks allows the process to be reversed, and sidereal time is used to measure the variation of Earth's rotation rather than time.

Mean sidereal time and apparent sidereal time are given by the hour angle of the mean equinox and true equinox, respectively. Local sidereal time, in turn, is the hour angle at the location. For example, GMST is the hour angle at Greenwich, United Kingdom.

CITED REFERENCES

1. International Astronomical Union. IAU Secretariat;[accessed 2023 Mar 24]. http://www.iau.org/

2. Wilkins GA. IAU style manual (1989): the preparation of astronomical papers and reports. International Astronomical Union; 1989. https://www.iau.org/static/publications/stylemanual1989.pdf

3. Ridpath I, editor. Norton's star atlas and reference handbook. 20th ed. Pi Press; 2004.

4. International Astronomical Union. Pluto and the developing landscape of our Solar System. IAU Secretariat; [accessed 2023 Mar 24]. http://www.iau.org/public/pluto/

5. United States Geological Survey, Astrogeology Science Center. Gazetteer of planetary nomenclature. Astrogeology Science Center; [accessed 2023 Mar 24]. http://planetarynames.wr.usgs.gov/

6. Proceedings of the International Astronomical Union. Cambridge University Press;[accessed 2023 Mar 24]. https://www.cambridge.org/core/journals/proceedings-of-the-international-astronomical-union

7. Planetary satellite discovery circumstances. National Aeronautics and Space Administration (US), Jet Propulsion Laboratory, Solar System Dynamics Group; [modified 2011 Nov 9]. http://ssd.jpl.nasa.gov/?sat_discovery

8. International Astronomical Union, Committee on Small Body Nomenclature. Cometary names. International Astronomical Union; 1997 [updated 2011 Feb 8].

9. Grady MM. Catalogue of meteorites: with special reference to those represented in the collection of the Natural History Museum, London. 5th ed. Cambridge University Press; 2000.

10. Hoffleit D, Warren WH Jr. The Bright Star Catalogue. 5th ed. Yale University Observatory; 1991.

11. United States Naval Observatory, Nautical Almanac Office. Astronomical almanac for the year 2012 and its companion, the astronomical almanac online. US Government Printing Office; 2011.

12. Samus NN, Kazarovets EV, Durlevich OV, Kireeva NN, Pastukhova EN. General catalogue of variable stars: Version GCVS 5.1. Astronomy Reports; 2017. http://www.sai.msu.su/gcvs/gcvs/index.htm

13. Wegner M et al. SIMBAD Astronomical Database—CDS (Strasbourg). Release 4. Centre de Données Astronomiques de Strasbourg; [updated 2000]. http://simbad.u-strasbg.fr/simbad

14. Shore SN. Binary stars. In: Encyclopedia of Physical Science and Technology. 3rd ed. Elsevier Science Ltd; 2003. p 77–92. https://doi.org/10.1016/B0-12-227410-5/00052-1

15. The Messier catalog. Students for the Development and Exploration of Space; [modified 2008 Feb 25]. http://www.messier.seds.org/

16. Sinnott RW, editor. NGC2000–NGC 2000.0: The complete new general catalogue and index catalogue. Sky Publishing Corporation; [modified 2022 Nov 23]. https://heasarc.gsfc.nasa.gov/W3Browse/all/ngc2000.html

17. Frommert H. The interactive NGC catalog online. Students for the Development and Exploration of Space; [accessed 2023 Mar 24]. http://spider.seds.org/ngc/ngc.html

18. International Astronomical Union. Specifications concerning designations for astronomical radiation sources outside the Solar System. International Astronomical Union; [updated 2018 Aug]. http://cdsweb .u-strasbg.fr/Dic/iau-spec.htx

19. Earth Orientation Center. The leap second: when should we introduce leap second in UTC? On 30 June 2012, the last minute of the day has lasted 61 seconds. Why? Observatoire de Paris; [accessed 2023 Mar 24]. Table 1, relationship between TAI and UTC. http://hpiers.obspm.fr/eop-pc/earthor/utc/leapsecond.html

ADDITIONAL REFERENCES

The astronomical almanac for the year 2023. United States Naval Observatory, Nautical Almanac Office; 2023.

Astrophysics data system. Smithsonian Astrophysical Observatory; [accessed 2023 Mar 24]. https://ads .harvard.edu/

Cox AN, editor. Allen's astrophysical quantities. 4th ed. Springer; 2013.

Dasch EJ. A dictionary of space exploration. 3rd ed. Oxford University Press; 2018. https://doi.org/10.1093 /acref/9780191842764.001.0001

Dictionary of nomenclature of celestial objects. Université de Strasbourg/CNRS; [updated 2023 Mar 17]. http://cdsweb.u-strasbg.fr/cgi-bin/Dic

Heydari-Malayeri M. An etymological dictionary of astronomy and astrophysics. l'Observatoire de Paris— PSL; 2005–2022. http://dictionary.obspm.fr

Hoffleit D. Supplement to the bright star catalogue. Yale University, Department of Astronomy; 1984.

Lang KR. The Cambridge guide to the Solar System. 2nd ed. Cambridge University Press; 2011.

Maran SP, editor. The astronomy and astrophysics encyclopedia. Van Nostrand Reinhold; 1991.

Mason BD, Wycoff GL, Hartkopf WI. The Washington double star catalog. US Naval Observatory, Astrometry Department; [accessed 2023 Mar 24]. http://www.astro.gsu.edu/wds/

Schmadel LD. Dictionary of minor planet names. 6th ed. Springer; 2012.

Steinicke W. Revised new general catalogue and index catalogue. Burton J, software developer. 2022 Sep 1. http://www.klima-luft.de/steinicke/ngcic/rev2000/Explan.htm

Thompson A, Barry N. Taylor BN. Guide for the use of the International System of Units (SI). Department of Commerce (US), National Institute of Standards and Technology; 2008. NIST special publication 811. https://nvlpubs.nist.gov/nistpubs/Legacy/SP/nistspecialpublication811e2008.pdf

Unified astronomy thesaurus. Version 5.0.0. American Astronomical Society. 2022 Jun 6. https://astro thesaurus.org

Urban SE, editor. The explanatory supplement to the Astronomical Almanac. 3rd ed. University Science Books; 2012.

4

Technical Elements of Publications

27 Journal Style and Format

Editors: Janaynne do Amaral, PhD; Simona Fernandes, MSc, ELS; and Jae Hwa Chang, MA, ELS

27.1 DEFINITIONS

Journals, known as "serials" to librarians and bibliographers, are "publications appearing or intended to appear at regular or stated intervals, generally more frequently than annually, each issue of which normally contains separate articles or papers."[1]

The usefulness and ease of use of a journal are determined not only by its content but also by its style and format. Decisions on these characteristics should aim to serve the needs of readers, librarians, and bibliographers. Priority should be given to clarity, accuracy, and adequacy of information, which can be served by carefully attending to the many details of style and format described in this chapter. The recommendations in this chapter are derived mainly from the standards of the National Information Standards Organization[1-7] of the United States and the International Organization for Standardization[8-15] (see their websites at https://www.niso.org and https://www.iso.org, respectively).

Although this chapter primarily addresses style and format issues concerning print journals, many of the recommendations are relevant in principle to nonprint journals. In addition, Section 27.3.1.2, "Online Publication," addresses some special considerations regarding electronic journals.

27.2 FORMAT

During the past few decades, nearly all established journals have created online editions that complement their print editions. In addition, many established and new journals have decided to forgo print editions completely.

Journals published in print are generally static in nature. Their articles are fixed at the time of publication and do not change. Articles in online journals can be updated to correct errors and make similar changes. In addition, many journals publish early online versions of peer-reviewed and edited articles, making them available sooner than the traditional print production process permits. Many journals also post "preprints," which are versions of articles before they are peer reviewed and copyedited, and many allow authors to submit articles for consideration that were posted to preprint services before being submitted to journals.

Because most journals are issued in more than one format, publishers should designate one of these as the "version of record," which is the authoritative version in which errata and other changes are recorded. Although it may no longer be necessary to differentiate which format is being cited in a reference list, authors should be sure that the version they cite contains the information the authors reference in their manuscript. If an early online version of an article was used, for example, authors should not cite the final version of that article without first ensuring that no significant differences exist between the 2 versions. (See Chapter 29, "References," for details on citation.)

27.3 PUBLICATION
27.3.1 Publishers and Publishing
27.3.1.1 Overall Process
A journal publisher provides the following services: peer review, editorial, production, sales, marketing, subscription fulfillment, distribution, and finance. Managing editors, also called "acquisitions editors," represent the publisher to authors and serve as liaisons to journal editors, who are usually scientists in the disciplines covered by their journals. In addition, managing editors oversee manuscript editors and editorial assistants and provide financial and production management.[16]

The methods by which journal editors choose to accept or reject manuscripts vary widely. Most journal editors apply the following criteria[17]:

(1) relevance to the journal's scope and audience
(2) importance of the message to most of the journal's audience
(3) newness and innovative nature of the message
(4) scientific validity of the evidence supporting the article's conclusions
(5) usefulness to the journal in contributing to a suitable range of topics

In reaching decisions on submitted articles, most journal editors are assisted by peer reviewers, also known as "referees" and "assessors," who are invited to provide expert evaluations of articles (see Section 1.2, "Components of Scientific Publications"). Usually, a minimum of 2 reviewers are assigned to each article. Journals provide reviewers with guidelines outlining the type of information the journals are seeking on the originality of the work, the accuracy of the methods, and the soundness of the conclusions.[16]

Although most journals maintain the anonymity of their peer reviewers by using a single-blind or double-blind process, some journals allow authors to know who peer reviewed their articles, either as a general policy or with the permission of individual peer reviewers. This is termed "open peer review." Some journals also ask authors to suggest suitable peer reviewers for their manuscripts, but journals do not guarantee that they will use the suggested reviewers. In addition, some journals conceal the authors' identities from peer reviewers and vice versa, which is known as "double-blind" or "anonymous" review. Regardless of the practices they follow, journals should publish the names of their editors and their policies on peer review.

Based on the peer reviewers' observations and recommendations, journal editors decide whether to accept manuscripts, request that they be revised and resubmitted, or reject them.

Once a manuscript is accepted, manuscript editors adapt the manuscript to the journal's house style, check the manuscript for accuracy and consistency, and look for discrepancies and omissions. Copy editors, in turn, read the proofs and work with the authors and the journal editor on needed corrections.

Many books detail the publishing and editing process, including *How to Write and Publish a Scientific Paper*,[18] *How to Write a Paper*,[16] *Writing and Publishing in Medicine*,[17] and *Essentials of Writing Biomedical Research Papers*.[19] Others may be found in this manual's Additional Resources.

27.3.1.2 Online Publication
27.3.1.2.1 ADVANTAGES AND DISADVANTAGES

Online publication provides many advantages over print, including the following[20]:

(1) The size limitations of print do not apply to online publications. Some journals that publish in both print and online formats place longer versions of their print articles on their websites, and some publish additional articles online only.
(2) Supplemental materials that would be impossible or impractical to have in print can be made available online. Such materials include data files, complex tables and graphs, video and audio clips, and animation.
(3) Hyperlinks make articles more usable by allowing readers to quickly navigate to other places within documents and to other documents. For example, a hyperlink in an in-text reference can take readers to the full citation in the reference list, and a hyperlink embedded in the citation's digital object identifier can take readers to the full text of the referenced document.
(4) Color can be added to online documents without incurring appreciably higher production costs, unlike print documents.
(5) Overall production is faster and less expensive because printing, binding, and mailing are eliminated.
(6) Articles or entire issues can be posted ahead of the print publication to reach readers sooner.
(7) Online distribution creates a much wider audience for journals.
(8) Large archives of previous volumes can be made available online.
(9) Inventory costs are greatly reduced.

Online publishing also has unique challenges, including the following:

(1) Online journals offer a bewildering array of options for access. Some offer unrestricted access to all of their content. Some journals restrict access to just recent issues, while others restrict all but recent issues. Professional associations typically offer free online access to all of their journals as a benefit of membership, while offering nonmembers subscription options ranging from password-protected access to pay per view. Many publishers use open-access models that vary from permitting free access to just the articles for which authors paid open-access fees to permitting free access to all articles. In addition, many educational institutions pay for licenses that enable faculty, other employees, and students to gain access to numerous scholarly titles. One consistency with regard to access is that articles based on research funded by the National Institutes of Health (NIH) in the United States must adhere to NIH guidelines for access (see https://publicaccess.nih.gov).
(2) Protecting copyright is a major concern because online articles are easily downloaded, copied, altered, and plagiarized.

(3) No standards currently exist for archiving electronic material. In addition, because of the high cost of maintaining archives, many publishers have had to contract with archiving organizations, such as PubMed Central (see https://www.ncbi.nlm.nih.gov/pmc), HighWire (see https://www.highwirepress.com), ScienceDirect (see https://sciencedirect .com), Ovid Synthesis Clinical Evidence Manager (see https://synthesis.ovid.com), and Silverchair (see https://www.silverchair.com).

(4) Data degradation and media degradation are concerns, as is migration as newer formats are created.

(5) Online-only journals use a wide variety of schemes for numbering issues, ranging from adhering to the traditional print practice of numbering both volumes and issues to numbering issues but not collecting issues in volumes to providing date of publication without volume or issue numbers. In addition, online-only journals may not have page numbers, and journals that publish online before print may use their web-based article numbering system as their print pagination system. These numbering issues make bibliographic citations more difficult (see Section 29.4.11, "Websites").

For a discussion of the costs and related business issues in online publishing, see *Electronic Scientific, Technical, and Medical Journal Publishing and Its Implications: Report of a Symposium.*[21]

27.3.1.2.2 PRODUCTION ISSUES

Although the easiest way to prepare articles for online publication is to convert them to Portable Document Format (PDF) using any of the numerous conversion programs,[22,23] PDF files are simply digital copies of the print pages. To take advantage of the full range of features and capabilities that electronic publishing provides, journal articles need to use a markup language, such as HyperText Markup Language (HTML), Standard Generalized Markup Language (SGML), Extensible Markup Language (XML), or Extensible HyperText Markup Language (XHTML). For more on markup languages, see Section 31.5, "Markup Languages," and Chapter 33, "Digital Standards of Scholarly Journal Publishing."

Critical to enabling the newer features of markup languages are document type definitions (DTDs). A DTD organizes the contents of a document. It defines the parts of a document, which are called "elements"; describes the elements by means of attributes; and enables units of code to be created, which are called "entities."[20] In journal articles, examples of elements are article titles and author names.

Searching and retrieving online documents is made possible by metadata, which are standardized, descriptive pieces of information in a machine-readable, structured form. Online search engines, for example, rely on metadata. Metadata assist in the following[24]:

(1) Identifying, locating, and organizing resources
(2) Promoting interoperability across systems
(3) Providing for digital identification using such identifiers as International Standard Serial Numbers (ISSNs) and uniform resource locators (URLs)
(4) Archiving and preservation

Two major metadata vocabularies used in online publishing are the Dublin Core Metadata Initiative (DCMI) (see https://dublincore.org) and the Online Information Exchange (ONIX) (see https://www.editeur.org). DCMI describes 15 metadata elements, while ONIX uses more than 200 metadata elements. Two useful introductions to metadata are *Understanding Metadata*[24] and *Metadata Demystified: A Guide for Publishers*.[25]

In recent years, digital object identifiers (DOIs) have surpassed URLs as a means of directing readers to online articles. Developed by the International DOI Foundation (see https://www.doi.org), DOIs provide documents with persistent identification across networks. A DOI consists of a prefix and a suffix. The prefix contains the name of a DOI directory and the content owner's identifier that the publisher registered with the International DOI Foundation. The suffix is a numerical or alphanumeric string that the publisher supplies. To register a DOI, the publisher submits the document's suffix to a DOI registration agency along with the document's URL and appropriate metadata.[7] Although DOIs can be applied to any level of web-based objects, most publishers assign DOIs at the article level. For additional information, see the National Information Standards Organization's standard Z39.84, *Syntax for the Digital Object Identifier*.[7]

The Columbia Guide to Digital Publishing[20] provides a comprehensive discussion of online and other electronic documents. For a detailed list of publications on the topic, see *Scholarly Electronic Publishing Bibliography 2010*.[26] See also Chapter 31, "Typography and Manuscript Preparation," and the Additional Resources at the end of this manual. For information on citing online journal articles, see Section 29.4.1, "Journals."

27.3.2 Copyright

Traditionally, authors were required to assign copyright for their work to the scientific journal or other publisher that published the work. This was done in consideration for the costs the publisher incurred, such as editing, typesetting, printing, mailing, and posting the work. Exceptions were fairly limited, usually based on conditions for complying with research funding agreements. For example, works produced by employees of the US government cannot be copyrighted for publication in the United States, and they are in the public domain (see Section 3.1.5.3, "Government Employees and Contractors").

Today, copyright status varies widely. Whereas many journals continue to require that copyright be assigned to them, a newer trend is to permit authors to retain copyright. Open-access journals, in particular, do this because they charge authors fees to publish their work, provided the work is accepted for publication through the peer-review process. Even when authors retain copyright, publishers are likely to require authors to grant them the exclusive right to publish their work both in print and in electronic form.[16]

Journals and publishers should make their policy on copyright clear[27] and include this policy in their information for authors (see Section 27.6.2.5, "Information for Authors"). For more on copyright, see Chapter 3, "The Basics of Copyright."

27.3.3 Major Changes

In general, editors and publishers should not change the major characteristics of their publications (eg, title, frequency, and design) without giving careful thought to the consequences. They also should anticipate that the changes will stay in effect for an indefinitely long period. Frequent and hastily adopted changes can injure a publication's reputation among authors, readers, and librarians and hamper the ability of these users to find the publication and its articles in literature searches.

27.3.3.1 Change in Frequency of Issues

Changing the frequency with which issues are published (eg, switching from monthly to weekly) should begin with a new volume. Readers, librarians, bibliographers, and subscription agencies should be alerted to the change well in advance. In addition, the change should be announced prominently on the cover and in the table of contents of the last issue before the change and the first issue with the change.

In addition, when issues are combined or separated into parts, that change should be clearly indicated wherever the issue number is listed (eg, the cover, table of contents, and masthead). For example, 2 combined issues could be designated as "Volume 17, Issues 1 and 2," and if an issue is split into 2 parts, the first part could be designated as "Volume 17, Issue 5, Part 1," and the second as "Volume 17, Issue 5, Part 2."

27.3.3.2 Change in Design

Major design changes (eg, cover redesign, new page layouts, and change in trim size) should be made only with the first issue of a volume. However, the former format should be applied to any parts of the preceding volume (eg, volume index and supplements) produced after the first issue of the new volume.

27.3.3.3 Other Changes

Significant changes made between volumes should be called to the attention of readers in the first affected issue on the cover, at the head of the table of contents, and on the masthead page. The changes should be described to readers, and their expected duration should be noted.

27.4 JOURNAL TITLES

27.4.1 Function and Length

A journal's title should be chosen with great care, and it should not be changed except for extremely important reasons, such as a change in the journal's scope. The title should represent the best compromise between a title long enough to adequately convey the journal's scope and content and short enough to make it easy to reference and remember. The title should also be unique so that it can be registered as a trademark and sufficiently different from the titles of other journals so as not to be confused with them.

A title such as *Current Research* may be too short and vague. It raises questions such as whether the journal publishes all current research in all fields, whether it publishes research in a specific discipline, or whether it publishes only research conducted by members of its sponsoring society. In contrast, the title *International Archives of Botany, Plant Physiology, Zoology, and Related Fields* may be too long and overly detailed. The journal might be better represented by a title that encompasses all of the relevant disciplines, such as the *International Archives of Biology*.

A single-word title can be amplified with a compound title or subtitle that details the journal's scope.

> *Atoms: Scientific Annals of Matter*
> *Methuselah: The Multidisciplinary Journal of Anti-aging Medicine*

Such a title, however, might be listed in databases and indexes by the initial term only.

In general, acronymic titles, such as *"PNZHA"* for the *Proceedings of the New Zealand Hydrospheric Association*, can cause confusion. Although they may be readily recognized by members of the journals' professional associations, such titles are cryptic to others, and they fail to indicate the scope and content of the journals. For some journals, however, acronymic titles became so much more recognized than the full names that the acronyms replaced the full names as the journals' official names (eg, *JAMA* and *BMJ*).

27.4.2 Uniqueness

Care should be taken to ensure that a journal's title does not duplicate another title that is used or owned by another publisher. Adopting an existing title might constitute infringement on the legal right of a publisher to that title. At the very least, duplicate names could confuse readers, librarians, and bibliographers. Even a title very similar to an existing one can cause confusion and lead to conflict with the first title's owner.

Existing titles can be identified from many different sources, including database listings and print catalogs of serial titles, such as the *List of Serials Indexed for Online Users*[28] of the US National Library of Medicine, *BIOSIS Serial Sources*[29] for the BIOSIS Previews database, and *Ulrich's Periodicals Directory*.[30] (For more sources on scientific journal titles, see Appendix 29.1 and consult science and other academic libraries.) The National Serials Data Program will search such sources when responding to requests for information on the uniqueness of proposed titles and to requests for International Standard Serial Numbers (see Section 27.4.4, "ISSN and CODEN").

27.4.3 Locations and Form

Ideally, a journal's entire title should be used wherever that title is used in the journal, such as the cover, spine, table of contents, masthead, officers' page, running heads, footers, and text. If the title must be abbreviated to save space, the abbreviation should adhere to the form specified by the International Organization for Standardization's standard titled *Information and Documentation—Rules for the Abbreviation of Title Words and Titles of Publications*.[15] The database for this standard, maintained by the ISSN International Centre in Paris at https://www.issn.org, is available by subscription. Other sources of abbreviated titles and title-word abbreviations prepared in accordance with this standard can be found in Appendix 29.1.

To avoid ambiguity, a journal title's location and typography should sufficiently distinguish the title from other text near it. In addition, the acronym of a professional society's or publisher's name should not be placed near the journal's title if it might be misinterpreted as the acronym of the journal's title.

27.4.4 ISSN and CODEN

Two widely used numbering systems for identifying serials are the International Standard Serial Number (ISSN)[4,10] and CODEN. The ISSN is an 8-digit code that uniquely identifies a journal title. The eighth character is a check digit. When the check digit would be 10, the letter "X" is used instead.

Journal of South American Earth Sciences	ISSN: 0895-9811
Madagascar Conservation & Development	ISSN: 1662-2510
ScienceAsia	ISSN: 1513-1874
Clinical and Translational Neuroscience	ISSN: 2514-183X

Journals with both print and online editions should have a separate print ISSN (pISSN) and electronic ISSN (eISSN).

Canadian Geotechnical Journal	pISSN: 0008-3674, eISSN: 1208-6010
Crop and Pasture Science	pISSN: 1836-0947, eISSN: 1836-5795
Nigerian Journal of Basic and Applied Sciences	pISSN: 0794-5698, eISSN: 2756-4843

Publishers should request ISSNs before launching journals so that the ISSNs can appear in the first issues of those new titles. In approximately 100 countries, publishers can obtain ISSNs from their respective national ISSN centers. These centers belong to the ISSN International Centre, which coordinates the activities of member countries and assigns ISSNs to international journals and to journals in countries without national ISSN centers. Journals can locate national centers or request ISSN numbers on the ISSN International Centre's website at https://www.issn.org. The following is a select list of national ISSN centers:

Country	ISSN center	Website
United States	Library of Congress	https://www.loc.gov (search with term "ISSN Basics")
United Kingdom	British Library	https://www.bl.uk (search with phrase "Get an ISBN or ISSN")
Canada	Library and Archives Canada	https://www.bac-lac.gc.ca (search with term "About ISSNs")
Australia	National Library of Australia	https://www.nla.gov.au (search with term "ISSN")
China	National Library of China's Chinese Acquisitions and Cataloging Department	http://www.nlc.cn/newissn
India	ISSN National Centre, India, at the National Science Library	http://www.niscair.res.in/ (search with term "ISSN")
Nigeria	National Library of Nigeria's International Standards and Programmes Department	https://www.nln.gov.ng (select "ISBN/ISSN" from home page's top navigation bar)
Brazil	Instituto Brasileiro de Informação em Ciência e Tecnologia	https://cbissn.ibict.br (select ISSN icon on home page)

Whereas an ISSN is an 8-character identifier, a CODEN is a unique 6-character alphabetical code. CODENs are assigned by the International CODEN Service, which is maintained by the Chemical Abstracts Service (https://www.cas.org). CODENs are assigned to journals in all subject areas, not just in chemistry.

Indian Journal of Forestry	CODEN: IJFODJ
Journal of the American Chemical Society	CODEN: JACSAT
The Lancet	CODEN: LANCAO

27.4.5 Change of Title

A journal's title should be changed only for important, carefully considered reasons, such as a change in topical scope or sponsorship or the need for a clearer, more specific title. A title change should be made at the beginning of a volume, preferably with the first issue at the start of a calendar year. The same precautions should be taken in selecting a new title for an existing journal as are taken for choosing the title of a new journal (see Sections 27.4.1, "Function and Length," and 27.4.2, "Uniqueness").

The title will have to be changed in all of its locations in the journal (see Section 27.4.3, "Locations and Form"). For the first year after the change, the former title should appear subordinate in position and size to the new title in all of the title locations. The former title should be preceded by the term "Formerly."

The same recommendations apply when 2 journals are combined under the title of one of the journals. In such a case, the combined journal should continue the volume-number sequence of the retained title. In contrast, when a new title is given to combined journals, the recommendations for a new journal apply. For one year, the new title should be accompanied in all of its locations by a clarifying statement such as, "[*New Title*] represents the combined [*Previous Title 1*] and [*Previous Title 2*]."

If a journal is divided into 2 or more journals and its title is continued for one of the journals, that journal should continue its ISSN and volume numbering. The other journal or journals should be treated as new journals, except that the covers, tables of contents, and mastheads should carry for one year a statement such as, "Continues in part [*Original Title*, ISSN]."

Any changes in title should be called to the attention of subscribers well in advance with a notice mailed separately from the journal or journals that are being changed. Librarians, database producers, and subscription agencies should also be alerted in advance.

27.5 JOURNAL VOLUMES AND THEIR PARTS

The division of journals into units and subunits, usually designated as "volumes" and "issues," is a long-standing convention. Volumes and issues should be organized and formatted with a view to efficiently serving the needs of readers for a variety of information.

27.5.1 Period of Publication

Each volume should represent no more than one full calendar year of issues. If a year's worth of issues is divided into 2 or more volumes, all volumes should include the same number of issues.

27.5.2 Numbering

Volumes should be numbered sequentially with arabic, not roman, numerals starting with 1. The numeral should follow the term "Volume." If the journal's title changes, common practice is to continue the original title's volume numbering sequence with the new title, but numbering may begin again with 1.

27.5.3 Pagination

In print editions of scientific journals, page numbers, also known as "folios," begin at 1 and run sequentially through all text pages of the issues making up the volume, rather than restarting at 1 for each issue as is usually done in popular magazines.

The preliminary pages of an issue (eg, title page, table of contents, and information pages) may be numbered to distinguish them from text pages. Common practices include using lowercase roman numerals (eg, i, ii, and iii) or alphanumeric designators (eg, P1, P2, and P3) for the preliminary pages. Other print pages not used for articles, such as advertising pages, may also be numbered alphanumerically (eg, A1, A2, and A3) to distinguish them from text pages.

If a journal includes a volume index at the end of each volume's last issue, the index pages should continue with the numerical sequence of the text pages. If instead a journal places its index just before the main text in a bound volume, the pagination should continue that of the preliminary pages in the volume's first issue. Alternatively, a system of alphanumeric designators (eg, I1, I2, and I3) may be used regardless of where the index appears.

Journals with both print and online editions usually use the same pagination scheme in both editions. However, online-only journals may have no page numbers (see Section 27.3.1.2.1, "Advantages and Disadvantages"). Instead, online-only journals may assign an electronic identifier called an "elocation-id" to each article in addition to a digital object identifier.

27.5.4 Parts of Bound Print Volumes

Traditionally, bound volumes of print journals include at least 4 parts: title page, volume table of contents, text of the volume, and index. Although online archives are replacing bound volumes, some journals and publishers still prefer this method for collecting a volume's worth of content.

To facilitate prompt binding of the issues of a volume, a title page, the volume table of contents, the information pages, and the index are usually published as the end pages of the last issue of the volume. If these pages are instead routinely sent to subscribers later in a separate form or are sent only on request, this practice should be noted on the masthead of each issue.

27.5.4.1 Title Page

The title page for a journal's print volume should carry information needed to fully identify the journal, the volume, the issues included, the first and last dates of issues, the journal's International Standard Serial Number and CODEN, and the publisher's name

Table 27.1. Title Page for Bound Volumes: Recommended Information

Information	Notes
Complete title	The journal's title should be the dominant element on the title page of each bound volume. Preferably, it should be located near the top of the title page. If the complete title is a compound title or includes a subtitle, all elements should be included on the title page.
	If the title consists of a spelled-out name and an acronym or other initialism, the title page should be designed so as to prevent ambiguity as to which is the proper title. If the initialism is the proper title, it should be superior to the complete name in both its location and font size, and vice versa if the complete name is the proper title.
ISSN and CODEN	These identifying serial numbers can be in a subordinate location, such as the foot of a volume's title page.
Publisher's name and location	If the journal has a sponsoring organization (eg, a professional society) that is not the publisher, the organization's name and location should appear on the title page along with the publisher's name.
Editor's name	The editor's name need not appear on the title page if it is on the masthead page or other information pages among a bound volume's preliminary pages.
Volume number and issue numbers	The volume number and the numbers for the first and last issues in a bound volume should be listed in arabic numerals. The issue numbers should be separated by an en dash. If volumes are divided into parts, each part's title page should list the issue numbers for only the issues included in that part.
Dates of the first and last issues	The dates should include the year, as in "January–December 2022" and "6 January–28 December 2022."

and location (see Table 27.1). The title page should be a right-hand page (ie, recto) but need not carry a page number. The editor in chief's name may be included on the title page, the masthead, or both.

The reverse (ie, verso) of the title page is usually blank, but it can provide additional information about the journal, such as the information that usually appears in the masthead of issues (see Section 27.6.2.3.4, "Masthead").

27.5.4.2 Table of Contents

The volume table of contents is conveniently formed from the tables of contents of all the issues in the volume (see Section 27.6.2.3, "Table of Contents"). The tables are separated into issues sequentially by date rather than organized in an uninterrupted sequence. The table should begin on a recto (see Section 27.5.3, "Pagination").

Additional preliminary pages, such as ones for the masthead and the information for authors, are placed between the volume table of contents and the table of contents for the volume's first issue.

27.5.4.3 Information Pages

These pages can include information carried in each issue that would be excluded from bound volumes (eg, information for authors published within advertising pages and masthead information). See Sections 27.6.2.5, "Information for Authors," and 27.6.2.3.4, "Masthead," and Tables 27.5 and 27.6.

27.5.4.4 Index

A volume index should list subjects and authors of the various types of articles in the journal (eg, original research, meta-analyses, systematic reviews, editorials, letters to the editor, and book reviews). Separate subject and author indexes are easier to consult than single indexes that combine both categories.

The index is preferably the last element of the text pages of the last issue in a volume (see Section 30.3, "Indexes"). The pagination of an index should be continuous with that of the text pages that precede the index. Alternatively, an index may use an alphanumeric page-numbering system (eg, I1, I2, and I3), particularly if the index is bound separately from the rest of the volume. Index pages should not be interrupted by advertising, news, or other content.

27.6 JOURNAL ISSUES AND THEIR PARTS

Issues are collections of articles and other documents assembled and published with a frequency and in dimensions that are economical for the publisher and convenient for readers. For consistency, most journals that publish both print and online versions continue to organize articles into issues, as do many online-only journals. Some notable exceptions, such as *PLOS ONE*, publish articles online individually without collecting them in an "issue." Even so, PDF versions of such articles may comply with many of the text guidelines described in Sections 27.6.2.4, "Text Pages," and 27.7, "Journal Articles and Their Parts."

27.6.1 Frequency, Numbering, and Dates

Most scholarly journals are published quarterly, monthly, or weekly. Within each volume, issues are identified sequentially with arabic, not roman, numerals beginning with 1. The needs of publishers, readers, and librarians are best served with consistent frequencies in which journals are issued on fixed dates. (For recommendations on dates of issues, see Table 27.2, and for recommendations on where to place issue dates, see Sections 27.6.2.1, "Cover," to 27.6.2.3, "Table of Contents," and Tables 27.1 and 27.3.)

27.6.2 Parts of Issues

27.6.2.1 Cover

In print journals, the front cover must carry the information needed to unequivocally identify a journal and the issue. The information recommended in Table 27.3 should appear on the front cover of every issue, placed in the same location and set in the same typefaces. The title should be the most prominent element on the cover. The preferred location for the title is the upper half of the cover. If a title is rendered in more than one language, the title in the journal's primary language should be emphasized by its size and location.

Additional information may be needed on the covers of special issues, supplements, issues that include a volume index, and issues during the first year of a title change (see Table 27.4).

Table 27.2. Issue Dates

Frequency of issues	Recommended dating
Weekly	The same day each week.
Quasiweekly (for example, 50 issues per year)	The same day each week in the weeks in which issues are published.
Semimonthly (twice a month)	Dates representing approximately equal intervals between issues. Convenient dates are the 1st and 15th days of each month, even if the posting and mailing dates do not coincide with the issue dates.
Monthly	The month and year of issuance. A day of the month need not be specified. The dates of posting and mailing should be as constant as possible from month to month.
Bimonthly (2-month intervals)	Both months and the year should be listed, even if the posting and mailing dates are not in the first month (eg, "January–February 2024").
Quarterly or 3 times a year	The span of months represented by each issue should be listed with the year, such as "July–September 2024" for a quarterly journal and "January–April 2024" for a journal published 3 times a year. Designations such as "Winter 2023" should be avoided.

Barcodes or quick response (QR) codes are widely used by popular magazines to identify a publication's International Standard Serial Number (ISSN), its title, and the particular issue. If used on a scientific journal, a barcode or QR code should be located in the lower left-hand corner of the front cover, and it should be oriented horizontally.

The cover should not be included in the pagination of the text.

27.6.2.2 Spine

If a print journal's spine is wide enough to accommodate legible type, the spine should carry the following information, which is critical for identifying an issue:

(1) The title of the journal or its standard abbreviation (see Section 27.4, "Journal Titles")
(2) The volume and issue numbers
(3) The issue date
(4) The volume page numbers in the issue

This information should be printed so that it can be read when the issue stands upright on a shelf. If an issue's spine is too narrow to accommodate type, the spine information should be positioned so that it can be read from left to right on the front cover.

27.6.2.3 Table of Contents

The table of contents of each issue must identify fully all of the articles and the sections into which they are grouped. Material of minor and ephemeral value might be identified solely by a section title, such as "News Notes."

For online journals, the table of contents is typically an issue's first web page. The

Table 27.3. Front Covers of Print Issues: Needed Information and Location[a]

Element	Recommendations
Complete title	The title should be in a dominant position on the upper portion of the cover. The title should include the journal's subtitle, if the journal has one.
Volume number and issue number	These numbers should be placed near the title. They should be in arabic numerals, such as "Volume 68, Number 9." If issues of a journal are combined, the numbers of the combined issues and their dates should be used, such as "Volume 127, Numbers 9–10, September–October 2022."
Issue date	This date should be placed near the journal's title. Its format changes based on the journal's frequency.
Weekly and semimonthly	A specific date, not the day of the week, should be used. The date should follow the journal's style rule. This style manual recommends using either the day-month-year sequence (eg, "15 January 2023") or the year-month-day sequence (eg, "2023 January 15") (see Section 13.3.3, "Sequence of Date Elements").
Monthly	Use month and year (eg, "December 2022" or "2022 December").
Bimonthly	Use both months and the year but use only one number for the issue (eg, "Number 2, March–April 2021" or "Number 2, 2021 March–April").
Quarterly or less frequent	Use both the first and last months and the year, but use only one number for the issue (eg, "Number 3, July–September 2020" or "Number 3, 2020 July–September"). Seasonal designations such as "Summer 2020" should be avoided.
Names of publisher and sponsor	If the publisher and sponsor are different, both names should appear on the cover. If they are the same, use the name only once. If the sponsor is different from the publisher, the sponsor's name may also be placed on the masthead or elsewhere at the publisher's discretion.
URL for journal's website	Use the journal's complete uniform resource locator, including "https://" or "http://."
Location of the table of contents	If the table of contents is not on the front or back cover or within the first 3 pages of the issue, include the page numbers for the contents on the cover.
Issue contents	If space is available, describe the contents of the issue on the cover.
Bar code or QR code	Barcodes and quick response codes are more widely used with popular magazines than with scientific journals. They identify a publication's International Standard Serial Number, its title, and the particular issue.

[a] For more information, see Section 27.6.2.1, "Cover."

article titles listed in an electronic table of contents are usually hyperlinked to online articles.

In a print journal, the table of contents of each issue is usually placed on a recto near the front of the issue. Care should be taken not to bury the contents page among advertisements and other introductory material. A few scientific journals list on their front covers the titles of all articles, which used to be a common practice. When the table of contents is on the front cover, it should be reproduced within the first 5 pages as a precaution against the front cover being damaged or lost.

Table 27.4. Front Covers of Print Issues: Additional Information for Particular Issues

Element	Recommendations
Volume index	The page numbers of the volume index should be listed on the cover of the issue in which the index appears.
Supplements and special issues	Supplements and special issues should be identified as such on their front covers. Supplement covers should include topic or title and correct citation (eg, "Volume 32, Supplement 1," and "Volume 32, Number 3, Supplement B"). When a special issue is mailed with a regular issue, the regular issue should be cited as Part 1 and the special issue as Part 2 (eg, "Number 9, Part 1"). (See Section 27.6.2.9, "Supplements and Special Issues.")
Change of title	The journal's previous title should appear below the new title on covers for a year after the title change. The previous title should be placed within parentheses or after the term "Formerly." (See Section 27.4.5, "Change of Title.")
Change in frequency and other major changes	A change in frequency should be announced prominently on the covers and in the tables of contents of the last issue before the change and the first issue with the change (see Section 27.3.3.1, "Change in Frequency of Issues"). Other significant changes should be noted on the cover of the first affected issue, as well as at the head of the table of contents and on the masthead page of that issue (see Section 27.3.3.3, "Other Changes"). All significant changes to a journal should start with a new volume, not in the middle of a volume.

27.6.2.3.1 BIBLIOGRAPHIC IDENTIFICATION

All information needed to identify a journal and the issue should be included at the top of the table of contents in both print and online editions. That information should consist of the following:

(1) Complete title and subtitle
(2) Abbreviated title in standard form (see Section 27.4, "Journal Titles," and Appendix 29.1)
(3) The ISSNs for the journal's print and electronic editions (see Section 27.4.4, "ISSN and CODEN"), which need not be adjacent to the complete title but should be in an obvious place
(4) Volume and issue numbers
(5) Date of issue (ie, the exact date for a weekly, biweekly, or twice-monthly journal; the month and year for a monthly journal; the month or month range with the year for a bimonthly or quarterly journal) (see Table 27.2)
(6) Notice of a change in title or frequency

27.6.2.3.2 INFORMATION ON ARTICLES AND SECTIONS

All articles, except those of minor and ephemeral value, must be described in the table of contents. Article titles on the contents page must match the exact wording used on the articles' title pages. Similarly, if space permits, author names in the contents should match the author bylines on the articles (see Section 29.3.3.1, "Author"). If space is limited in the contents, author names can be reduced to surnames and initials for given names as they are in bibliographic references.

Most scholarly journals include brief descriptions of major articles below the titles and author names in the contents. In addition, articles are grouped under sections such as "Original Research," "Case Reports," and "Editorials." Sections with short articles (eg, "Book Reviews," "Letters to the Editor," and "News Notices") may be referenced in the contents by their section headings only without listing the article titles. Only initial page numbers need to be given for articles listed in the table of contents.

Any regular feature that serves important needs of readers and authors (eg, a journal's policy page) should be listed with its page number in the contents, even if the feature appears in an advertising section.

Special features that do not appear in every issue (eg, information for authors, an erratum or retraction notice, and a volume index) must be listed in the table of contents by title and initial page number. When information for authors is not published in an issue, the table of contents should identify the issues in which that information does appear.

27.6.2.3.3 FORMAT AND TYPOGRAPHY

Bibliographic data, including the journal title, should appear above the heading "Table of Contents" (see Section 27.6.2.3.1, "Bibliographic Identification").

Section and article titles should generally follow the same sequence in the table of contents as in the issue. For those journals whose articles are published only online, however, articles may be ordered by date or by subject in the journals' electronic tables of contents.

One common practice for formatting print tables of contents is to list article titles on the left with author names directly below the titles and page numbers in a column to the far right of the article titles. Online journals tend to omit page numbers and instead hyperlink the titles in the contents to the full articles. Different typefaces may be used to distinguish the elements in tables of contents, such as boldface for titles, italics for author names, and roman for descriptions.

27.6.2.3.4 MASTHEAD

Each issue of a print journal should include a masthead, which describes the journal, its availability, its ownership, and any additional information typically needed by authors, readers, subscribers, librarians, dealers, indexers, bibliographers, archivists, and advertisers. (See Table 27.5 for details on the information to include on the masthead page.)

For print journals, the masthead should be placed within the first 5 pages, preferably beginning on the same page as or immediately adjacent to the table of contents. If the masthead begins on the same page as the table of contents, the layout and typography should make clear the information most likely needed by most users of the journal. The masthead's location should be constant from print issue to print issue. If the masthead is on the inside front cover or within pages that are not likely to be bound in volumes, it should be reproduced with the preliminary pages that are published as the end pages of the last issue of a volume (see Section 27.5.4, "Parts of Bound Print Volumes").

Table 27.5. Masthead

Type of information	Notes
Minimally needed information	
Title, subtitle, International Standard Serial Number, and CODEN	If a journal is published with titles in more than one language or with an acronymic title, this information should be included. In the case of a journal with both a complete title and an initialism, the one that serves as the proper title should be listed first.
Publisher	The publisher's name should be listed in the masthead with the publisher's mailing address, email address, telephone number, and fax number.
Volume number, issue number, and issue frequency	Frequency should be indicated as "weekly"; "biweekly," not "fortnightly"; "semimonthly" or "twice a month"; "monthly"; "bimonthly"; "quarterly"; or "semiannually" or "twice a year."
Required postal data[a]	This is the information required by the postal service of the country in which the journal is published or distributed.
Copyright information	This should consist of the legally required notice of copyright with the copyright symbol "©," the copyright year, and the name of the copyright holder. Also include the publisher's policy on fair-use copying and its mechanism for charging royalties for copying.
URL for the journal's website	Provide the journal's complete uniform resource locator, including "https://" or "http://." Also include the journal's website for submitting manuscripts.
Subscription rates, single-issue price, and ordering information	The address and phone number for placing orders should be listed, even if they are the same as those for the publisher. Ordering information should include terms of payment; accepted currency; payment methods, such as credit card, check, money order, and cabled payments; and availability of back issues in print and online.
Address-change procedure	Provide the mailing address and email address to which address changes should be sent.
Sponsorship	Provide the name and address of the journal's sponsor, if not the same as the publisher.
Advertising management	List names, addresses, and telephone numbers of staff or agencies that can provide information on online advertising, print display advertising, and classified advertising.
Logo of circulation auditor	Include the name of the circulation auditor along with its logo.
Statement and logo indicating publication on acid-free paper[a]	The National Information Standards Organization recommends using the following statement: "This paper meets the requirements of ANSI/NISO Z39.48-1992 (R2009) (Permanence of Paper)." The logo is an infinity symbol within a circle (ie, "♾").
Additional useful information	
Editor in chief, staff editors, and other staff members	Include with this information the names, addresses, email addresses, and telephone numbers of those who receive correspondence for the editor in chief.
Editorial board or committee	List the additional editors and associate editors who assist the editor in chief with peer review and acceptance decisions.
Publication committee	List the members of the sponsoring or publishing organization's publications committee.
Peer-review policy	Provide the details of this policy that are useful to authors.
Indexing and abstracting	Identify the online and print services providing indexing data and abstracts from the journal.
Statement on use of recycled paper[a]	Include this statement in addition to the statement on acid-free paper.

[a] These items should be excluded from the online masthead; they may appear in any position in the printed version.

For online journals, the masthead should be maintained as a separate web page rather than included with each issue.

27.6.2.4 Text Pages

27.6.2.4.1 FORMAT

Although text can be formatted in solid text blocks for both print and online journals, such a format is more appropriate for online content designed as responsive web pages, which adjust to different screens and resolutions. Print pages with a single column of text typically have lines of type that are too long for convenient reading. Print pages are easier to read if they are divided into 2 or 3 columns, separated from each other by a gutter of white space. Pages divided into columns require less space between lines (ie, "leading"), which can reduce the length of formatted articles. Multiple columns also accommodate variation in placement of figures and tables. Because the choice among formats hinges on various considerations, the decision should be made in consultation with a typographic designer.

When organized in columns on printed pages, text must begin at the upper left-hand corner of the text block, proceed down the first column, and continue at the top of the next column. Figures and tables should be aligned within the margins of the text block and be readable in the same orientation as the text. If a table is too large for this orientation, it can be rotated counterclockwise 90° so that its bottom is against the right margin of the page. Alternatively, a large table can be split between 2 facing pages, but the row headings may need to be repeated on the right-hand side of the table's second page. Larger tables may run across more than 2 pages. If this is done to accommodate additional column headings, the row headings must be repeated at least on the left-hand pages. If this is done to add rows instead, the column headings must be repeated.

In general, tables and figures are best placed in the upper part of print pages and the text in the lower part (see also Chapter 30, "Tables, Figures, and Indexes"). In contrast, in online journals, small images of tables and figures can be inserted to the right or below the first paragraph in which tables and figures are mentioned. Those images, in turn, can be hyperlinked to full-sized versions of the tables and figures.

27.6.2.4.2 INFORMATION CARRIED ON PRINT PAGES

Facing print pages of text should together carry the information that identifies the journal and issue in the "running head" on the top margin; in the "running foot," or "footline," in the bottom margin; or in both. The information should include the journal title or, if necessary, its abbreviation; the volume and issue numbers; and the issue date. If space permits, the running head or foot also should carry either a short version of the article title or the name of the journal section (eg, "Book Reviews" or "Letters to the Editor"). A widely used convention is to place article or section information in the running head and place the journal title, volume and issue numbers, and date in the running foot.

All print pages must be numbered sequentially, with odd numbers used for right-hand pages and even numbers for left-hand pages. The recommended position for page numbers, or "folios," is the outside edge of the running foot or running head. (See Section 27.5.2, "Numbering," for sequential numbering through each volume.)

Online articles posted to a single web page are not divided into pages, so they do not have running heads or feet. However, online articles created as PDFs from print articles will retain running heads and feet.

27.6.2.4.3 TRIM SIZE

The pages of all issues of a print journal should have the same trim size (ie, height and width). Many scientific journals use the trim size of 8.5 inches by 11 inches (ie, 21.5 cm by 28 cm). This size is desirable for a number of reasons:

- It accommodates more text per weight of paper than smaller trim sizes, which reduces both paper and postal costs.
- It facilitates advertising space.
- It conveniently allows for 2- or 3-column page formats.
- Bound volumes of journals with this trim size fit on most library shelving.

27.6.2.5 Information for Authors

To help authors prepare and submit manuscripts, scientific journals should publish both online and in print one or more pages describing journal policies and and providing instructions for manuscript requirements (see Table 27.6 for recommended content). These pages, often called "Information for Authors," should appear in the same place in each print issue, and they should be listed in the table of contents (see Section 27.5.4.2, "Table of Contents"). If a journal does not print its information for authors in each print issue, the URL to the online version should be part of the table of contents, the masthead page, or the cover of each print issue.

For assistance in preparing author information, editors can refer to other journals. The University of Toledo hosts a web page titled "Instructions to Authors in the Health Sciences"[31] that provides hyperlinks to the author instructions for numerous scientific journals.

27.6.2.6 Advertisements

The pagination for advertising pages in print editions should be in the same sequence as that for other pages that are likely to be omitted from online editions or from bound print versions for library use. A system of pagination distinguishing these pages from the text pages helps librarians and binderies identify pages to be excluded from bound volumes (see Section 27.5.3, "Pagination").

Advertising that might be confused with the scientific text of a journal should be refused, especially if it is formatted to resemble text pages (eg, "advertorials").

In addition, advertisements can sometimes raise ethical issues. For example, advertisements can be problematic if placed adjacent to articles discussing the same topics, such as an advertisement for an antidepressant next to an article on drug treatment for patients with depression. To avoid interspersing advertisements with editorial content, many science journals place advertisements only in their front and back matter.

Table 27.6. Information for Authors: Potential Content

Potential content	Specific items
Distribution	Circulation: domestic and foreign for print and online Categories of subscribers: personal, institutional, and student
Impact factors	Journal metrics such as those from Clarivate's Journal Citation Reports, Elsevier's Scopus Journal Analyzer, and Altmetric's Explorer for Publishers
Indexing and abstracting services	Names and contact information for services covering the journal such as the US National Library of Medicine, the Web of Science, Crossref, and SciFinder
Content	Categories of manuscripts considered Specifications for each category of manuscript, including appropriate content, expected format, and allowed length
Manuscript style and format	Details on manuscript format, including specifications for title page; limit on title length; and structure and length of abstract, text, and headings Details of scientific style, including units of measurement and nomenclature and recommended style manual, specialized dictionary, and standard dictionary Format for references Requirements for figures and tables
Submission of manuscripts	Journal's website for submitting manuscripts Recommended content of submission letter, including the name, address, email address, and telephone number of the author responsible for further communication Authorship statements Copyright-transfer requirements Other required information, such as research funding and recommendations for which peer reviewers to use or avoid Page charges and charges for color figures
Ethical considerations	Criteria for authorship, including basis for sequence of author names Identification of conflict of interest Definition of prior and duplicate publication, whether identical or not
Editorial and review process	Peer-reviewing policy, including indicating whether peer review is single-blind, double-blind, or open Procedures and time frames followed by the editors and peer reviewers Information provided to authors
Publication	Scheduling Handling of proofs Prepublication release of information, including author preprints Reprint orders

27.6.2.7 Corrections and Retractions

Statements on authors', editors', or publishers' errors; omissions' retractions; or other matters that merit being called to the attention of readers should be published in the same place in any print issue carrying such statements. The location should be prominent, such as on the table of contents, pages with editorials or letters to the editor, or in a section titled "Corrections." If these statements are not in the table of contents, their location in both the print and online editions should be identified in the print and online tables of contents. In addition, the corrections should be made to the online editions of the affected articles, and a statement describing the corrections should be included with each corrected online article.

Each statement should make clear its function. For example, the statement's title should begin with a term such as "Correction," "Notice of Error," or "Errata," and the title should include a term or brief phrase that identifies the subject or title of the article being corrected. The statement should fully identify the corrected item, and it should provide the bibliographic data for the original version of the article. The material corrected and the correction should be made clear in the notice.

Retraction notices should be published with attention to the same considerations for other notices. However, retractions must also identify the persons or organizations responsible for the errors, misinformation, or violations that led to the articles being retracted.

Both correction and retraction notices must be indexed in volume indexes. In addition, the index entries for the articles that are the subject of notices should include in parentheses cross-references to those notices. In scientific fields with indexing databases that flag indexed articles that have been corrected or retracted (eg, PubMed of the US National Library of Medicine), journals should call the notices to the attention of the relevant databases.

For additional guidance on contractions and retractions, see "CSE's Recommendations for Promoting Integrity in Scientific Journal Publications."[32]

27.6.2.8 Articles in Installments or in Series

Some articles may be too long to publish in a single issue, so they are published in 2 or more installments. If possible, article installments should be scheduled so that all parts appear in consecutive issues of a single volume.

Each installment of such an article should carry the article's title along with a designation for which part it is (eg, "Part 1" and "Part 2"). The part number should be in arabic, not roman, numerals. The concluding installment should note that it is the last in the series. Preferably, the title should include a notice such as "Part 2 (Conclusion)," or the article itself should state that it concludes the series.

The title pages of the second and subsequent parts of the article should include notices with the titles and page numbers of the preceding parts. If the subjects of the installments permit, the title of each part can reference the installment's content.

> Recent Developments in the Geomorphology of the United States: Part 1, The East
> Recent Developments in the Geomorphology of the United States: Part 2, The Midwest
> Recent Developments in the Geomorphology of the United States: Part 3, The West (Conclusion)

Like an article in installments, individual articles in a series should be published within a single volume if possible, and their titles should suggest that the articles are in a series.

> The Economics of Health Services: Canada
> The Economics of Health Services: France
> The Economics of Health Services: Great Britain

27.6.2.9 Supplements and Special Issues

To serve special editorial objectives, journals may occasionally publish supplements or special issues. These are usually collections of manuscripts on a specific topic or lectures and abstracts presented at a conference.

The print versions of supplements and other special issues should have the same design, format, and trim size as regular issues to avoid confusing readers and creating difficulties with binding (see Table 27.7 for details). Similarly, the online versions should mirror the specifications of the online editions of the regular issues.

Pagination for supplements and other special issues differs based on how and when they are published. If a supplement is designated as Part 2 of a specific issue and is to be bound immediately after Part 1, the pagination of Part 2 should be consecutive to that of Part 1. In contrast, a volume supplement can be paginated at the end of the usual set of issues and before the volume index. When a supplement is published in the middle of a volume or when it is published with an issue that is unpaginated, separate pagination is required (eg, the page numbers could be preceded by an alphabetical modifier, such as "S" for "supplement," as in "S1," "S2," and "S3").

The content of supplements and other special issues should be indexed in the volume's index, whether the supplement follows the volume's pagination or has its own.

Table 27.7. Supplements and Special Issues

Element	Notes
Supplement	
Cover	Separate cover but same design, format, and identification as the regular issues Indication of supplement, including title or topic Volume and issue numbers with designation that publication is a supplement, such as "Volume 31, Supplement 2," "Volume 31, Number 3, Supplement B," or "Volume 31, Number 3, Part 2"
Title page and table of contents	These should be specific to the supplement
Bibliographic identification	Include abbreviation "Suppl" for "supplement" or "Pt" for "part" Indicate the volume and issue numbers of regular issue with which the supplement was published, as in "31(3 Suppl B):2024" or "31(3 Pt 2):2024"
Title	If a supplement consists of a single article, such as a long review article or dissertation, the supplement title is the same as the article title For a supplement with multiple articles, the supplement title is preferably one that ties the articles together under which the individual article titles appear
Pagination	Preferably continuous with the regular issue the supplement accompanies If a supplement is paginated separately from the regular issue, an alphabetical designation such as "S" precedes the page numbers, which begin with the arabic number 1 (eg, "S1," "S16," and "S27")
Trim size	For print supplements, same trim size as print version of regular issues
Indexing	Indexed with issues and other supplements in the same volume
Special issue	
Cover and table of contents	Cover and table of contents may carry a title identifying the theme or title of the special issue
Format	Uses same design, format, and pagination as regular issues
Trim size	Uses same trim size as regular print issues

27.7 JOURNAL ARTICLES AND THEIR PARTS

Most of this section applies to aspects of format and style common to the principal kind of articles in scientific journals, while the unique aspects of specific types of journal articles are addressed in Sections 27.7.5.1, "Research Articles," through 27.7.5.6, "Book Reviews." In addition, this section is organized based on common practices for print articles and PDF versions of those print articles. Articles published in online formats typically contain the same elements, but the arrangement of those elements can vary widely.

27.7.1 Title Page

The first page of a print article and the top of an online article should carry all the information needed to enable readers to rapidly identify the contents, authors, and origin of the article, as well as its location in the journal. These elements are summarized in Table 27.8.

27.7.1.1 Article Title

The title should be as informative as possible within the length limit stipulated by the journal in its instructions to authors. A title should be straightforward and avoid hyperbolic rhetoric. A title may be a declaratory statement as long as it does not generalize the reported findings beyond what is supported by the described evidence. An interrogatory title is acceptable if it indicates the question considered (eg, the subject of the research reported).

Table 27.8. Title Page of a Scientific Article: Necessary Elements[a]

Element	Notes
Title	Should describe the subject Should include terms used in recognized thesauri or other vocabularies to increase likelihood that the article will be found in searches
Byline and "Author Contributions" statement	All author names in the sequence preferred by the authors or stipulated by the journal Each author's surname and at least one given name If the journal's policies permit, initialisms for academic degrees and other credentials after author names Each author's roles in the research reported in the article and their roles in preparing the manuscript The institutions at which the authors carried out their research
Abstract of the article	Abstract in the format and within the length specified by the journal
Bibliographic reference	The reference by which the journal would like the article identified in end references of other works
Beginning of text	Text begins after abstract, depending on availability of space
Footnotes	In print articles, footnotes may be needed if authorship is an extensive group, or they may be used to call attention to authors' current affiliations and addresses
Footline elements	Journal title, volume and issue numbers, copyright notice, initial page number, and digital object identifier

[a] See Section 27.7.1, "Title Page," for more details.

> Large-Scale Eradication of Rabies in Southern Belgium with Recombinant Vaccinia-Rabies
> Vaccine
> *not* Large-Scale Eradication of Rabies with Recombinant Vaccine
> (Without indicating the region and the specific vaccine, readers could generalize the title to
> indicate that any recombinant vaccine can eradicate rabies anywhere.)

> Does Astronomy Really Need the Hubble Space Telescope?
> *not* What Does Astronomy Need?
> (The second question does not focus on the specific issue.)

If possible, the title should start with a word or term representing the most important aspect of the article, with the remaining terms in descending order of importance. Terms subordinate to the subject of the research (eg, a description of the research design) can be relegated to a subtitle.

> Hypertension in Young Black Men Treated with Hydrochlorothiazide or Weight Reduction:
> A Randomized Controlled Trial
> *not* A Randomized Controlled Trial of Treatment of Young Black Men for Hypertension with
> Hydrochlorothiazide or Weight Reduction

Another factor to consider when writing article titles is whether they are easily searchable on online databases. When evaluating the results of online searches, most people consider titles to be the most important factor in deciding which articles to consider. This is particularly true for search results of databases and articles that do not have abstracts. In addition, many searchers restrict their searches to words in titles to increase the relevance of the results.

The following are additional guidelines for writing titles in general and increasing their searchability in databases:

(1) Do not write "erudite" or "cutesy" titles that appear to be clever but do not convey useful information to readers. For example, the title "Whither Oncology?" would not convey that an article discusses the future of surgical oncology, nor would the title appear in the results of a search for that topic.

(2) Be as specific as possible. Do not use broad terms when narrower concepts are discussed. For example, an article that discusses the use of fluoxetine to treat patients for ulcers should not be titled "Tranquilizers for Gastrointestinal Distress." A title search for "fluoxetine" and "ulcers" would not find that article.

(3) Do not leave readers in doubt about what is being discussed. Titles beginning with such wording as "A New Use for . . ." are not helpful because they do not specify what the new use is. Titles that note concepts but do not state the relation between the concepts are also not useful (eg, "Growth Hormone Therapy and Thyrotropin-Releasing Hormone").

(4) Use standard terms from formal scientific nomenclatures, such as the BIOSIS vocabulary guides[33] and Medical Subject Headings,[34] instead of common or nonstandard terms.

(5) Avoid using acronyms and other initialisms whenever possible because many initialisms have more than one meaning. For example, "MCV" stands for both "mean corpuscular volume" and "motor conduction velocity." In addition, because database searches are not case sensitive, searches cannot distinguish between initialisms with the same letters in different cases (eg, "Ca" for calcium and "CA" for cancer).

(6) Whenever possible, use the common or generic names for drugs and other chemicals instead of formulas and numbers, which require complicated searches. For example, in many databases, searching for the drug 2-chloro-10-(3-dimethylaminopropyl)-phen othiazine requires combining the constituents into the following search: "chloro AND

dimethylaminopropyl AND phenothiazine." Nevertheless, the results of such a search could include a number of irrelevant articles. Using the term "chlorpromazine" instead would prevent the confusion. Number designations for drugs and other chemicals also can produce confusing results. If, for example, the number for a chemical includes a hyphen, many online systems replace the hyphen with a space or eliminate the hyphen. Thus, "STI-571" would produce results for "STI 571" or "STI571." Use "imatinib" instead.

(7) Write out scientific names because many online databases do not permit searching for single letters, such as those used to abbreviate genera. Thus, if "*S. typhi*" is used in a title instead of "*Salmonella typhi*," the title might be found only in a search for the species "typhi," which could produce results such as *Rickettsia typhi* in addition to *Salmonella typhi*. In contrast, in databases that do permit searching for single letters (eg, "S" for "sulfur"), such searches produce innumerable irrelevant results.

(8) Be aware of guidelines for article titles issued by authoritative bodies. For example, one checklist item in the Consolidated Standards of Reporting Trials (CONSORT) statement[35] details what must be included in the titles of articles reporting on randomized clinical trials.

(9) Add alternative or synonymous terms in titles where appropriate. For example, titles with common names for plants or animals might include the proper taxonomic names, and titles with generic drug names might include the trade names if the research used specific products.

> The North American Distribution of the House Fly (*Musca domestica*)
> The Incidence of Adverse Effects from Phenytoin (Dilantin)

(10) Use simple word order and common word combinations. Online searchers often use adjacency (ie, one word directly in front of another) to retrieve more relevant results. For example, using "juvenile delinquency" in a title rather than "delinquency in juveniles" will increase the likelihood that the article will show up in searches.

(11) Do not use roman numerals in titles to designate parts of articles published in 2 or more installments. This will prevent false results, such as "Part III" of articles showing up in a search for "blood coagulation factor III."

(12) Do not assume that journals' titles assist in searches. Most databases do not search for individual words in journal titles. For example, researchers searching for articles about radiation injuries in dentistry may not find an article titled "Natural History and Prevention of Radiation Injury" that appears in *Advances in Dental Research*. A better title would be "Natural History and Prevention of Radiation Injury in Dentistry."

(13) Become familiar with database "stopwords." Every database has a list of words that cannot be used in searching. "Stopwords" are commonly used words that would overburden the system if they were searchable. Examples are "a," "the," "really," and "very." However, some meaningful words appear on stopword lists. "Being" is one example, so terms such as "well-being" and "human being" are not searchable. Consequently, synonyms such as "welfare" and "human," respectively, should be used in titles instead.

The conventions for typefaces and other standards appropriate to scientific nomenclature apply to titles. For example, genus and species should be in italics in a title, and the species name should be lowercased even though the other major words in the title are capitalized. (For more on capitalization in titles, see Section 9.3, "Titles and Headings within Text.")

Punctuation in titles should be kept to a minimum. Commas are acceptable for separating the elements of a series of related terms (see Section 5.3.3.1, "Serial Comma"), and colons are acceptable for coupling main and subordinate elements in a title (see Section 5.3.1.1, "General Uses"). In general, semicolons and dashes should be avoided.

27.7.1.2 Author Statement and Byline

Only those persons who qualify for authorship by standards accepted in the scientific community[27,36–39] should be listed as authors (see Section 1.3.1, "Authorship Criteria"). In general, these standards ensure that persons credited as authors can take public responsibility for an article because they participated in designing the research, carrying it out, and writing or revising drafts of the article. Journals can help ensure that these standards are met by requiring presumed authors to sign author statements before the journals accept the articles. In signing these statements, authors vouch that they participated in conducting and reporting the research and that the research is authentic and valid.

Some journals require that authors clearly document in this statement their roles in the research reported in the article and their roles in preparing the manuscript. This information usually appears between the end of the article and the reference list under the heading of "Contributors" or "Author Contributions." (See Chapter 2, "Publication Policies and Practices," for more information on the responsibilities of authors.)

In the "Author Contributions" section, as well as in bylines at the beginning of articles, authors should be identified by their surnames and at least one given name, not just initials. Given names, also known as "forenames," reduce ambiguity in author names, which helps ensure that the correct authors are credited for the research and that the research is indexed under the correct author names. Some journals make an exception to this convention by using 2 or more initials in place of given names. Exceptions also exist for names from cultures in which only surnames are used.

Provided that space and journals' policies permit, initialisms for academic degrees and other credentials may be listed after author names (see Section 8.1.1.1, "Names in Bylines"). However, a number of journals omit such credentials based on the scientific standard that the validity and importance of research should stand on the evidence, not on the "authority" of authors.

Some scientific journals allow authors to include in "Author Contributions" statements their ORCID iDs, which stands for "Open Researcher and Contributor ID identifiers." These are unique identification numbers that tie researchers' names and nonsensitive identification information to their published works, thus eliminating confusion in identifying authors with similar names.[40] Authors can register for ORCID iDs at www.orcid.org.

Journals should stipulate in their instructions to authors the sequence in which author names are listed in author statements and bylines. Journals should consistently apply this sequence to articles. Typically, journals sequence names in order of decreasing responsibility for the research and the manuscript. Other sequences include listing the principal investigator last and listing authors' names alphabetically by surname. (Also see Section 1.3.2, "Author Order.")

Using authors' correct surnames is extremely important because references, indexes, and databases use surnames for entries, and the users of such tools depend on surnames to search for articles. Surnames are not always last names. In some cultures, for example, the first name is the surname (eg, "Wu" is the surname of Chinese mathematician Wu Wenjun, PhD). In addition, not all compound surnames are hyphenated (eg, British physician Elizabeth Garrett Anderson, LSA, used "Garrett Anderson" as her surname). Some publishers use typography to identify surnames (eg, by placing surnames in all capital letters or by using an initial capital followed by small capitals). For more on identifying authors' surnames, see Section 8.1.2, "Conventions for Authors' Names," and Table 8.1.

For large-scale research involving many investigators, authorship can be indicated by a collective or corporate title.

CERN   Genomics and Randomized Trials Network
Chinese Research Group of Gallbladder Cancer   Scientific Coalition for UAP Studies

For articles with collective or corporate authors, additional information on participants is usually desirable. At a minimum, such articles should list the names, addresses, and email addresses of the investigators who will respond to inquiries about or criticism of the research. In addition, a note or footnote can list the names and affiliations of those who were responsible for writing the manuscripts and who signed the author statements. Additional listings can identify participants by function (eg, detector design and statistical analysis).

27.7.1.3 Author Affiliation and Site of Research

The institutions at which the authors carried out their work should be identified in the "Author Contributions" statement. In their print editions, many journals publish this information on the title page in a separate box or in a margin or footer. For a large group of authors from multiple institutions, this information may have to be provided in a footnote or in a sidebar on the title page. Online editions of journals often embed a hyperlink in the byline that takes readers to a separate web page with the authors' affiliations.

Author affiliations not directly related to research should not be included (eg, research articles by quantum chemist Angela Merkel, Dr rer nat, should not note that she was Germany's chancellor from 2005 to 2021). In addition, institutions involved in the research are often identified in a line beginning with "From the . . ." when the authors' affiliations do not make the institutions' connection to the research clear.

To enable others in the scientific community to communicate with the authors, the authors' mailing addresses, telephone numbers, and email addresses may be included with their affiliations (see Section 14.3, "Addresses"). If this information would require too much space on the title page of an article's print version, it may be provided in a note at the end of the article. In such cases, a footnote should be placed on the title page to indicate where the contact information is located. Alternatively, a journal may choose to limit the contact information to that for a single author, usually the "corresponding author," who agrees to respond to all communications.

27.7.1.4 Abstract

The National Information Standards Organization[3] (NISO) of the United States and the International Organization for Standardization[8] (ISO) mandate that an abstract be included with every journal article, essay, and discussion. Most journals also require abstracts.

An abstract should be placed between an article's "Author Contributions" statement and the article's text. The abstract should be no longer than the length limit stipulated in the journal's instructions to authors. In some disciplines, the maximum length is dictated by major indexing services, which stipulate the length of abstracts they will allow in their databases or the length at which they will truncate abstracts.

Because abstracts are reproduced in online databases and abstract journals, abstracts should be complete and understandable on their own.

Both NISO and ISO state that an abstract should contain the following elements: purpose, methodology, results, and conclusions. Many journals require that structured abstracts[41,42] be used for articles of clinical interest. In a structured abstract, each paragraph has its own heading. Typical headings are "Objective," "Design," "Setting," "Participants," "Interventions," "Measurements and Results," and "Conclusions." For articles that do not require structured abstracts, abstracts can consist of single paragraphs without headings.

Abstracts for research manuscripts should be informative, providing specific summaries of all elements of the content. For review articles and other similarly long articles of wide scope, abstracts need only sketch out the topics without summarizing evidence and conclusions.

Regardless of the nature of an abstract, abbreviations should not be used unless they are understood when standing alone (eg, "DNA," "pH," and "WHO") or after the terms have been spelled out on first mention. In addition, abstracts should not include bibliographic references or tabulated data.

As an aid to database searchers, abstracts may include synonyms for terms used in article titles. For example, if the term "renal" appears in a title, the word "kidney" may be used in the article's abstract.

See NISO standard Z39.14[3] and ISO standard 214[8] for additional information.

27.7.1.5 Bibliographic Reference

The title page of an article should include a bibliographic reference that other authors can use to reference the article in other works. The format of the bibliographic reference should be the same as that specified by the journal for end references (see Chapter 29, "References," for the formats recommended by this style manual).

In the print version of an article, the bibliographic reference is placed at the end of the abstract. In an article's electronic version, this reference may be placed at the top of the first display screen above the title of the article.

27.7.1.6 Keywords

Journals often ask authors to provide a brief list of "keywords," which are terms that characterize the main topics of an article. Such terms assist indexers and readers. When selecting keywords, use recognized vocabularies related to the discipline discussed, such as the BIOSIS Search Guide[33] and the MeSH thesaurus[34] of the US National Library of Medicine. Keywords are placed after the abstract.

27.7.1.7 Financial and Other Support

Information on financial support, such as funding from grants and donations of equipment and medications, is often placed at the end of articles with the acknowledgments (see Section 27.7.3.3, "Acknowledgments"). Alternatively, this information can be placed in a footnote on the title page of print articles. The International Committee of Medical Journal Editors[27] (ICMJE), for example, recommends that financial support be disclosed on the title page.

If grants and contracts are disclosed, the relevant grant and contract numbers should be provided, along with the names of the supporting institutions. In addition, authors should disclose potential conflicts of interest, such as receiving honoraria from and holding stock in companies whose products were studied or whose products compete with the studied products.

27.7.1.8 Text on the Title Page

In print journals, text usually begins on the title page unless the title, author and affiliation statements, and abstract are very long. In both print and online journals, the design, layout, and typefaces should enable readers to readily distinguish the text from the other elements on the title page. Generally, figures and tables should not be placed on the title page of print articles lest they distract from clearly presenting the other elements on the title page.

27.7.1.9 Footline

The title page of a print article should carry in its running foot the title of the journal, the volume and issue numbers, the issue date, and the copyright notice. Many journals also include in the footline the uniform resource locator (URL) for the journal's website and the article's digital object identifier (DOI) number.

27.7.2 Text

27.7.2.1 Text Headings

To help readers who scan articles, headings are used to break the text into sections with specific content. Additionally, headings break up what would otherwise be long uninterrupted blocks of type that give pages a "gray look," which is intimidating to readers. Headings can indicate structural elements (eg, Methods and Results) or subjects (eg, "Antisepsis Antibodies" and "Artificial Satellites").

Headings can have hierarchical levels. First-level headings indicate the main divisions of text within an article. Second-level headings, or "subheadings," divide subject matter within the main divisions. Each additional lower-level heading divides the section under the heading one level above. Each section should be divided by at least 2 lower-level headings, not a single one.

The different levels of headings must be readily apparent to readers. This can be accomplished by varying the style, size, and weight of typefaces. In general, the headings for all levels should be either in the same or a closely related typeface as the one used for the text or in a contrasting, but consistent, typeface (eg, a sans serif typeface for headings and a serif typeface for text). A typical hierarchy is illustrated below:

First level: Bold serif typeface in larger size than text with one line of space between heading and text. Flush left directly above first sentence of next section.

Second level: Bold sans serif typeface in same size as text in headline-style capitalization. Flush left directly above first sentence of next section.

Third level: Bold italic small capitals in same size and typeface as text. Flush left directly above first sentence of next section.

Fourth level: Bold italic in same size and typeface as text. Flush left but running into first sentence of new section.

Methods

Text text text text text text text text text

Chemical Analyses
Text text text text text text text text text

CALCIUM
Text text text text text text text text text

Ionized Calcium Text text text text text

The number of levels of headings should hinge on the length of sections. Many journals find 2 levels to be adequate for most articles.

27.7.2.2 In-Text References

In text, references are handled in 2 basic ways. The most common is to designate references with numbered superscripts in the text. The other is to place author surnames and date of publication within parentheses in the text. With both methods, the in-text references refer readers to full citations at the end of the articles. Journals use various names for the list of full citations, such as "References," "Reference List," "Literature Cited," and "Bibliography." (For recommendations on the location and spacing of in-text references, see Section 29.2.1, "Systems of In-Text References.")

27.7.2.3 Footnotes and Endnotes

Footnotes are occasionally used for references, for additional information about manufacturers and suppliers, and for other parenthetical information. Because they appear at the bottom of the same print pages as their related text, footnotes are convenient for readers. However, this practice is generally not used in the print editions of scientific journals, and it is not feasible in online editions because articles are not divided into pages. In contrast, footnotes are still used in some books. (For recommendations on footnotes in tables and recommendations on footnote signs, see Section 30.1.1.8, "Footnotes.")

In journal articles, the functions of footnotes can often be served with parenthetical statements in the text, provided that the statements are relatively brief.

> The patients in Group 1 were treated with metoprolol tartrate (supplied by Pfizer, Inc). Group 2 patients were treated with metoprolol succinate (supplied by AstraZeneca, PCL).

Endnotes serve the same function as footnotes. Like footnotes, endnotes are useful for comments that are too extensive or frequent for parenthetical statements, and they interrupt the text far less than parenthetical statements. Unlike footnotes, endnotes are placed at the end of the text.

Superscript lowercase letters can be used in text to reference both endnotes and footnotes. Using letters instead of arabic numbers prevents confusion with in-text references (see Section 29.2.4, "Placement of In-Text References"). While footnotes are listed at the bottom of print pages without a heading, endnotes appear under a heading such as "Notes."

> The patients in Group 1 were treated with metoprolol tartrate.[a] Group 2 patients were treated with metoprolol succinate.[b]

Notes
[a] Pfizer, Inc. Note that . . .
[b] AstraZeneca, PCL. This form is also . . .

The distinction between endnotes and footnotes has diminished in online journals because hypertext links take readers from in-text references to the citations. In addition, some journals, such as *Science*, combine references and endnotes in a single section and cite both in the text with the same sequence of superscript numbers.

27.7.2.4 Tables and Figures

Most journals place tables and figures just after they are first mentioned in the text. This is easier to accommodate online than in print because tables and figures can be inserted between paragraphs in online articles and because their in-text references can be hyperlinked to the tables and figures. In addition, if tables and figures are large, they can be scaled down to fit the online format and embedded with hyperlinks to full-sized versions.

For print articles, a useful general principle is to place tables and figures at the top of pages and place the text below tables and figures. The goal is to ensure a mix of tables, figures, and text. It is important to avoid placing too many tables and figures on individual print pages, which could visually confuse readers. When a print article has many figures, some may be combined into multipart figures that can be treated as single figures.

For additional information, see Sections 30.1, "Tables," and 30.2, "Figures: Graphs, Images, and Illustrations."

27.7.3 Appendixes, Addenda, and Acknowledgments

In addition to endnotes (see Section 27.7.2.3, "Footnotes and Endnotes"), the sections between an article's text and its end references can include appendixes, addenda, and acknowledgments.

27.7.3.1 Appendixes

Appendix sections can be used for aspects of an article's subject that are too long and detailed to be put in the text and that would interrupt the flow of information. Appendixes should be capable of standing alone without reference to the article they accompany. Content of this kind includes long descriptions of unusual methods, long quotations from cited documents, questionnaires, guidelines, and patient handouts.

If an article has more than one appendix, they should be numbered with arabic, not roman, numerals. In addition, each appendix should be headed with an explanatory title, such as "Appendix 1: Ethical Standards of 6 Major Scientific Societies," not simply "Appendix 1."

27.7.3.2 Addenda

An addendum is usually used for information that authors need to add to their article late in the publication process, such as after the article has been edited by a manuscript editor and the proofs have been approved by the authors. Addenda can provide such information as evidence from a just-published paper by other authors, additional validating evidence from the authors' own work, and overlooked references.

27.7.3.3 Acknowledgments

An "Acknowledgments" section can specify permissions obtained to cite unpublished work, identify grants and other kinds of financial support, and credit those whose contributions to the work did not justify authorship. If grants and contracts are mentioned in this section, provide the relevant grant and contract numbers, as well as the names of the supporting institutions. Note that the ICMJE's uniform requirements[27] call for biomedical journals to provide financial information on the title page instead of in the acknowledgments section (see Section 27.7.1.7, "Financial and Other Support").

27.7.4 References and Bibliography

At the end of a document, such as a journal article, book chapter, or book, all of the references cited in the work are presented in a list variously called "References," "Reference List," "Literature Cited," or "Bibliography."

If the list includes both works used to write the article and other works on the topic that might be of interest to readers, the list is often called a "Bibliography." Another option is to divide the references into 2 sections: "Cited References" and "Additional References."

For more on end references, see Section 29.3, "End References."

27.7.5 Types of Scientific Articles

Scientific research can be communicated in various types of articles. These articles fall into 2 main categories: primary and secondary. Primary, or original, research articles report the discoveries of authors who conducted studies.[43] In short, primary literature reports results obtained firsthand. Examples are articles based on randomized clinical trials, cohort studies, observational and experimental research, case-control studies, before-and-after studies, cross-sectional surveys, and diagnostic test assessments.[43] Theoretical articles, original theses and dissertations, case reports, and technical notes are also considered original research articles.

Secondary, or review, articles expand the knowledge communicated in primary literature by "revisiting, reviewing, analyzing, or synthesizing existing research, and presenting it in a new light."[43] In short, secondary literature is based on secondhand information. Examples are systematic reviews, meta-analyses, narrative review articles,[43] and book reviews.

Perspective, commentary, and opinion articles can be either primary or secondary literature, depending on the source of their content.[43] Written more quickly than original research, these articles communicate the authors' personal opinions on research topics.[44]

No single format can serve all possible needs of either authors or journals. A journal covering a clearly defined field of science should stipulate the specific formats most useful for that field. These formats, in turn, should be summarized in the journal's instructions to authors (see Table 27.6).

Basic to most scientific articles is that they have the structure of critical argument.[17] That structure consists of posing a question or hypothesis; presenting and assessing the evidence, both pro and con; and reaching an answer. In addition to following the structure of critical argument, articles on scientific research generally include the following elements. First, the means by which the investigators gathered their evidence needs to be described in detail sufficient to enable other investigators to replicate the research. Second, the evidence gathered by the investigators must be presented separately from evidence already in the scientific literature.

27.7.5.1 Research Articles

Research articles, also known as "original research," "original articles," and "research papers,"[45] are "the most important type of scientific paper."[43] These articles are usually divided into sections titled Introduction, Methods, Results, and Discussion, known by the acronym "IMRAD" (see Section 1.2.2, "Text"). However, depending on the nature of the article or the research, these headings may be modified or even omitted. For example, a report can open with a paragraph that serves the functions of an introduction but is not titled Introduction. If the materials used in the research need to be described in detail, the methods section can be titled Materials and Methods.

A research article is more likely to be clearly understood if the research is described in the sequence in which it was conceived, designed, and carried out; its results analyzed; and its conclusions reached. Whichever headings are assigned to sections, they should be selected with this principle in mind.

See Table 27.9 for more on IMRAD and the functions each article section serves. Another useful resource for research articles is the CONSORT statement.[35]

27.7.5.2 Review Articles

Properly conceived review articles have the same basic structure of critical argument as research articles. The headings in review articles should make clear the topics considered sequentially. Explicit description of the methods and standards applied in selecting the studies that were compared and analyzed helps review articles meet the same intellectual standards[46] as research papers. That description can be provided in Methods near the beginning of the review article or in the article's appendix.

Additional guidance on review articles can be found in the Preferred Reporting Items for Systematic Reviews and Meta-analyses (PRISMA) statement.[47]

27.7.5.3 Editorials

Editorials may comment on articles published in a journal's issue or on "recent innovations or topics of interest to the journal readers."[43] Because of their brevity, editorials

Table 27.9. Sections of a Research Report: Typical Headings and Functions

Headings for sections	Functions
Introduction	Describes the state of knowledge that gave rise to the question examined by or the hypothesis posed for the research States the question (not necessarily as an explicit question) or hypothesis May include a review of existing literature
Methods	Describes the research design; the methods and materials used in the research, such as study participants, their selection, equipment, laboratory procedures, and field procedures; and methods for analyzing findings Adheres to the highly specific needs for descriptions dictated by the particular science discipline, which should be specified by the journal
Results	Findings from the research Tables and figures supporting the text
Discussion	Brief summary of the decisive findings and tentative conclusions Examination of other evidence supporting or contradicting the tentative conclusions Authors' final conclusions Consideration of how the conclusions could be generalized Implications for further research
References	Citation of sources relevant to elements of the argument and the descriptions of methods and materials used

usually do not need text headings, although their intellectual structure should still have the elements of critical argument. Some journals, such as *Research Involvement and Engagement*, publish editorials that have abstracts, are divided into sections under headings, and have references.

Editorials are often written by the journal's editor in chief[43] or, in the case of special issues, by guest editors.[46] Some journals invite experts to write editorials on research topics covered in particular issues. Although not all journals indicate who wrote their editorials, those that do either place the "Author Contribution" statement under the editorial's title to emphasize the authority of the author, or theyplace the statement at the end of the text to lessen the emphasis on the author (see Section 27.7.1.2, "Author Statement and Byline").

27.7.5.4 Letters to the Editor

Like editorials, letters to the editor can focus on articles recently published in a journal, on recent innovations in the journal's scientific discipline, or on other topics of interest to the journal's readers. In some journals, reader-submitted scientific comment on articles is called "Discussion" or "Correspondence" instead of "Letters."

Letters are usually published largely as submitted, so they typically have the structure of a letter. Each letter or group of letters on the same topic should be headed with a short title to clarify for readers what topics the letters cover. Many journals edit the salutations of letters to a standard greeting, such as "To the Editor."

Journal editors should decide to what extent the authors of letters are identified. Options include name only, name and academic degrees, and name and postal address.

27.7.5.5 Commentaries

Commentaries are short articles that communicate the point of view of authors about controversial topics.[43] Authors of commentaries can support their arguments with a few tables, figures, and references.

The structure of commentaries varies from one journal to another. So it is critical for authors to consult a journal's instructions to authors before writing a commentary for the journal.

27.7.5.6 Book Reviews

The structure of a book review section and its contents should be designed for the convenience of the journal's readers. For example, readers might prefer book reviews that begin with each book's bibliographic information, price, and International Standard Book Number. The bibliographic information, in turn, may be more useful to readers if it begins with the book title followed by the author's name rather than adhere to the author-first format of end references. In addition, reviews might be grouped together by topics with appropriate headings.

Book reviews should do more than just summarize the content of books. For example, reviews can assess the value of books for the discipline covered by the journal. Reviews can also point out the revelations and deficiencies in the books. In addition, reviews can compare the reviewed books with other books on the same topic. Ultimately, reviews should recommend whether the books are worth reading and by whom.

27.8 REPRINTS

Reprints, also known as "offprints," are individual articles or groups of articles reproduced as a separate publication. Authors may order reprints to promote their work to other audiences, or institutions may use reprints for educational purposes.

Reprints of single articles should retain the format, pagination, and trim size of the original article. However, 2 or more articles combined in a single reprint may carry new, additional pagination that begins on the first right-hand text page with the arabic numeral 1. In such cases, the new page numbers should be accompanied by the original page numbers within parentheses.

The original bibliographic identifiers on the title pages of articles should appear on the reprints. If a reprint of one or more articles is published under a new cover, that cover should carry the articles' full titles, author designations, the original bibliographic identifiers, and a copyright statement.

27.9 POSTAL REQUIREMENTS

National postal services typically have requirements for notices that should be included in print journals concerning ownership, circulation, and other characteristics of journals. The postal service in the country of publication or distribution should be consulted for detailed information on its requirements.

CITED REFERENCES

1. National Information Standards Organization. Bibliographic references. NISO Press; 2005 [reaffirmed 2010]. (ANSI/NISO Z39.29-2005 [R2010]) https://groups.niso.org/higherlogic/ws/public/download/12969 /Z39_29_2005_R2010.pdf

2. National Information Standards Organization. The Dublin Core metadata element set. NISO Press; 2013. (ANSI/NISO Z39.85-2012). https://www.niso.org/publications/ansiniso-z3985-2012-dublin-core -metadata-element-set

3. National Information Standards Organization. Guidelines for abstracts. NISO Press; 2015. (ANSI/NISO Z39.14-1997 [R2015]). https://www.niso.org/publications/ansiniso-z3914-1997-r2015-guidelines-abstracts

4. National Information Standards Organization. International Standard Serial Numbering (ISSN). NISO Press; 2001. (ANSI/NISO Z39.9-1992 [R2001]). Also available at https://www.niso.org/

5. National Information Standards Organization. Permanence of paper for publications and documents in libraries and archives. NISO Press; 2010. (ANSI/NISO Z39.48-1992 [R2009]). https://www.niso.org /publications/z3948-1992-r2009-permanence-paper

6. National Information Standards Organization. Placement guidelines for information on spines. NISO Press; 2015. (ANSI/NISO Z39.41-1997 [S2015]). https://www.niso.org/publications/ansiniso-z3941-1997 -s2015

7. National Information Standards Organization. Syntax for the digital object identifier. NISO Press; 2005. (ANSI/NISO Z39.84-2005). Also available at https://www.niso.org/

8. International Organization for Standardization. Documentation—abstracts for publications and documentation. International Organization for Standardization; 1976 [reviewed and confirmed 2020]. (ISO 214:1976). https://www.iso.org/standard/4084.html

9. International Organization for Standardization. Information and documentation—electronic manuscript preparation and markup. International Organization for Standardization; 1994 [reviewed and confirmed 2021]. (ISO 12083:1994). https://www.iso.org/standard/20866.html

10. International Organization for Standardization. Information and documentation—International Standard Serial Number (ISSN). International Organization for Standardization; 2022. (ISO 3297:2022). https://www.iso.org/standard/84536.html

11. International Organization for Standardization. Documentation—presentation of contributions to periodicals and other serials. International Organization for Standardization; 1986 [reviewed and confirmed 2019]. (ISO 215:1986). https://www.iso.org/standard/4086.html

12. International Organization for Standardization. Information and documentation—presentation and identification of periodicals. International Organization for Standardization; 2019. (ISO 8:2019). https:// www.iso.org/standard/67723.html

13. International Organization for Standardization. Documentation—presentation of title information of series. International Organization for Standardization; 1985 [reviewed and confirmed 2019]. (ISO 7275:1985). https://www.iso.org/standard/13933.html

14. International Organization for Standardization. Information and documentation—the Dublin Core metadata element set—part 1: core elements. International Organization for Standardization; 2017 [reviewed and confirmed 2020]. (ISO 5836-1:2017). https://www.iso.org/standard/71339.html

15. International Organization for Standardization. Information and documentation—rules for the abbreviation of title words and titles of publications. International Organization for Standardization; 1997 [reviewed and confirmed 2020]. (ISO 4:1997). https://www.iso.org/standard/3569.html

16. Hall GM, editor. How to write a paper. 5th ed. BMJ Books; 2012.

17. Huth EJ. Writing and publishing in medicine. 3rd ed. Williams & Wilkins; 1999.

18. Day RA, Gastel B. How to write and publish a scientific paper. 9th ed. Greenwood; 2022.

19. Zeiger M. Essentials of writing biomedical research papers. 2nd ed. McGraw-Hill, Health Professions Division; 2000.

20. Kasdorf WE, editor. The Columbia guide to digital publishing. Columbia University Press; 2003.

21. National Academies of Sciences, Engineering, and Medicine, Committee on Electronic Scientific, Technical, and Medical Journal Publishing and Committee on Science, Engineering, and Public Policy, Policy and Global Affairs Division. Electronic scientific, technical, and medical journal publishing and its implications: report of a symposium. The National Academies Press; 2004.

22. Goldstein H. Acrobat/PDF resources. Goldray Consulting Group; 2010. https://goldray.com/webdesign/acrobat_resources.htm

23. MyMorph by the National Library of Medicine (US). Ver 2.0. [accessed 2023 Feb 3]. https://en.freedownloadmanager.org/Windows-PC/MyMorph-FREE.html

24. Riley J. National Information Standards Organization. Understanding metadata: What is metadata, and what is it for?: A primer. NISO Press; 2017. https://www.niso.org/publications/understanding-metadata-2017

25. Brand A, Daly F, Meyers B. Metadata demystified: a guide for publishers. Sheridan Press; 2003.

26. Bailey CW Jr. Scholarly electronic publishing bibliography 2010. Digital Scholarship; 2010. https://www.digital-scholarship.org/sepb/annual/sepb2010.pdf

27. International Committee of Medical Journal Editors. Recommendations for the conduct, reporting, editing, and publication of scholarly work in medical journals. International Committee of Medical Journal Editors [updated May 2022]. https://www.icmje.org/recommendations/

28. List of serials indexed for online users, 2022. National Library of Medicine (US); 2022. https://www.nlm.nih.gov/tsd/serials/lsiou.html

29. BIOSIS serial sources. Thomson; 2001.

30. Ulrich's periodicals directory. 50th ed. Bowker; 2011.

31. Instructions to authors in the health sciences. The University of Toledo, Mulford Health Science Library. http://mulford.utoledo.edu/instr/

32. Council of Science Editors, Editorial Policy Committee. CSE's recommendations for promoting integrity in scientific journal publications. Council of Science Editors. https://www.councilscienceeditors.org/recommendations-for-promoting-integrity-in-scientific-journal-publications

33. BIOSIS search guide 2001–2002. Biological Abstracts, Inc; 2001.

34. Medical Subject Headings 2023. National Library of Medicine (US); [accessed 2023 Feb 3]. https://www.nlm.nih.gov/mesh/MBrowser.html

35. Schulz KF, Altman DG, Moher D, for the CONSORT Group. CONSORT 2010 statement: updated guidelines for reporting parallel group randomized trials. Ann Intern Med. 2010;152(11):726–732. https://doi.org/10.7326/0003-4819-152-11-201006010-00232

36. Banik G, Baysinger G, Kamat P, Pient N. ACS guide to scholarly communication. American Chemical Society; 2020. https://doi.org/10.1021/acsguide

37. Huth EJ. Guidelines on authorship of medical papers. Ann Intern Med. 1986;104(2):269–274. https://doi.org/10.7326/0003-4819-104-2-269

38. Huth EJ. Scientific authorship and publication: process, standards, problems, suggestions. National Academies of Sciences, Engineering, and Medicine; Institute of Medicine; 1988.

39. Publication manual of the American Psychological Association. 7th ed. American Psychological Association; 2020.

40. ORCID [Open Researcher and Contributor ID]. ORCID. https://www.orcid.org

41. Ad Hoc Working Group for Critical Appraisal of the Medical Literature. A proposal for more informative abstracts of clinical articles. Ann Intern Med. 1987;106(4):598–604. https://doi.org/10.7326/0003-4819-106-4-598

42. Haynes RB, Mulrow CD, Huth EJ, Altman DG, Gardner MJ. More informative abstracts revisited. Ann Intern Med. 1990;113(1):69–76. https://doi.org/10.7326/0003-4819-113-1-69

43. Shoja MM et al. A guide to the scientific career: virtues, communication, research, and academic writing. John Wiley & Sons; 2020.

44. Majumder K. A young researcher's guide to perspective, commentary, and opinion articles. In:

Planning to write. Cactus. https://www.editage.com/insights/a-young-researchers-guide-to-perspective
-commentary-and-opinion-articles

45. Öchsner A. Introduction to scientific publishing: backgrounds, concepts, strategies. Springer-Verlag;
2013.

46. Mulrow C, Thacker S, Pugh J. A proposal for more informative abstracts of review articles. Ann Intern
Med. 1988;108(4):613–615. https://doi.org/10.7326/0003-4819-108-4-613

47. Page MJ. The PRISMA 2020 statement: an updated guideline for reporting systematic reviews. BMJ.
2021;372:n71. https://doi.org/10.1136/bmj.n71

ADDITIONAL REFERENCE

International Organization for Standardization. Information and documentation—the Dublin Core metadata
   element set—part 2: DCMI properties and classes. International Organization for Standardization; 2019.
   (ISO 15836-2:2019). https://www.iso.org/standard/71341.html

28 Books and Other Monographs

Editor: Kristen Swendsrud

28.1 DEFINITIONS

In scientific publishing, the nature of published media has drastically changed during the past few decades, primarily because electronic formats have surpassed print as the principal media for disseminating peer-reviewed materials published by and for academic and other scholarly institutions. Online versions and mobile-optimized options are now ubiquitous among academic publications, and many journals have moved to online-only formats. Today, writing is considered to be published simply by being made available online, regardless of whether the work appears in an electronic monograph, an e-journal, a database, an online archive, a blog post, a podcast, a video, or some other digital format.

The term "monograph" is typically defined as a scholarly treatise on a specific subject, usually limited in scope and written by a specialist on the topic. In this style manual, however, the term is used in its broader sense to mean a publication that is complete in a single volume or in a limited number of volumes, such as a book that is published in 3 physical pieces. In addition to treatises, monographs include textbooks, other books, technical reports, conference proceedings, dissertations, theses, bibliographies, collections of scholarly works like Festschrifts, and patents. Most of this chapter focuses on standard books. Technical reports and other monographs with special features are addressed in Section 28.6, "Specific Types of Monographs," and digital monographs are covered in Section 28.6.3, "Electronic Monographs."

Some books are published in what is known as a "monographic series." Each book in the series can stand on its own as a separate publication, but the series brings together titles on a specific subject or by a specific organization. For example, the book *Selenoprotein Structure and Function*, published by Elsevier Academic Press in 2022, is also volume 662 of the series *Methods in Enzymology*, and *The Sea: Nature and Culture*, published by Reaktion Books in 2021, is part of the series *Earth*.

In contrast, "journals" and "newspapers," also known as "serials," are publications that are intended to be published indefinitely at regular intervals, each issue of which normally contains separate articles or papers[1] (see Chapter 27, "Journal Style and Format").

Book publishers and printers use many technical terms to refer to parts within publications. Among the most widely used for print publications are "recto," meaning a right-hand page, and "verso," meaning a left-hand page. "Verso" is also used to refer to the back, or reverse, side of a page, as in the verso of the title page. Recto pages are always odd numbered, and verso pages are even numbered. A printed page number may be called a "folio" (although this term has additional meanings in publishing), and a page number at the bottom of the page is a "drop folio."

More specific terms are defined in their relevant sections within this chapter. Definitions of additional technical terms may be found in the glossaries in the standards of the National Information Standards Organization,[1-10] *The Chicago Manual of Style*,[11] and *Butcher's Copy-editing: The Cambridge Handbook for Editors, Copy-editors and Proofreaders*.[12]

28.2 FORMAT

Most monographs share the same main characteristics. They typically have 3 sections: the pages or screens of the opening section, usually called "front matter," "preliminary pages," or "prelims"; the main text of the publication, known as "subject matter," "main text," or "body text"; and the accessory closing pages or screens, usually called "back matter" or "end matter." See Table 28.1 for an alphabetical list of components and their location.

Monographs published in print are generally static in nature, meaning their content is fixed at the time of publication and does not change until a new or revised edition is published. In addition, the content in print monographs is in linear form, from front to back. In contrast, electronic monographs are more likely to be dynamic in their functionality rather than static. For example, electronic publications can be updated frequently, and they can offer such capabilities as hyperlinking. Links are commonly embedded in in-text references to take readers to the corresponding citations in the reference list, and the citations, in turn, have links to the reference sources themselves. In addition, electronic monographs may provide access to large data sets and other types of ancillary material, either within the text or through hyperlinks.

When a book is issued in more than one format, its publisher should inform readers as to which format has been designated as the "version of record," the authoritative version in which errata and other changes are recorded. The publisher may pick the online version as the version of record because it can be updated relatively easily and quickly compared with the print version.

Although it may no longer be necessary to indicate in end references whether authors consulted the online or print version of a monograph, the version cited should contain the information the authors refer to in their manuscript. If the print version of a monograph was used, for example, authors should not provide the online version's uniform resource locator in its end reference without ensuring that no significant differences exist between how the cited material was used in the print and online versions. (For more details on referencing electronic and print books, see Chapter 29, "References.")

There is no universal standard for the structure and format of books. In the United States and Canada, *The Chicago Manual of Style*[11] has become the de facto standard because of its detailed descriptions of conventional practices in book layout and formats. The legalities and customs of publishing in Canada are described on the website of the Library and Archives Canada.[13] Additionally, British practices are well reflected in Chapters 7–9 of *Butcher's Copy-editing: The Cambridge Handbook for Editors, Copy-editors and Proofreaders*.[12]

Other resources for book formatting include standards published by the International Organization for Standardization and the National Information Standards Organization in the United States. When producing official government documents, authors and publishers should adhere to government standards. In the United States, for example, publishing standards are issued through such agencies as the US Government Publish-

Table 28.1. Components of Books

Component	Location[a]	Section reference
Abbreviation list	Front matter	28.4.2.11.3, "List of Abbreviations"
Abstract	Front matter (verso)	28.4.2.4.9, "Abstract"
	Text	28.4.3.5, "Parts, Sections, and Chapters"
Acknowledgments	Front matter (verso)	28.4.2.4.8, "Copyright Acknowledgments, Permissions, and Credits"
	Front matter (recto)	28.4.2.10, "Preface"
	Back or front matter	28.4.4.5, "Acknowledgments of Contributions"
Addenda	Back matter (preferably begin on recto)	28.4.4.1, "Addenda"
Appendixes	Back matter (preferably begin on recto)	28.4.4.2, "Appendixes"
Barcode	Cover and jacket	28.4.1, "Cover, Jacket, and Endpapers"
Bibliography	Back matter (preferably begin on recto)	28.4.3.5.4, "Reference List"
Cataloging in Publication	Front matter (verso)	28.4.2.4.4, "Cataloging in Publication (CIP)"
Chapters' first page	Text (preferably on recto)	28.4.3.5, "Parts, Sections, and Chapters"
Colophon	Cover and jacket	28.4.1, "Cover, Jacket, and Endpapers"
	Front matter	28.4.2.3, "Title Page"
Colophon page	Back matter (verso or page after index or last page of book)	28.4.4.7, "Colophon Page"
Contributors	Front matter	28.4.2.11.2, "Chapter Contributors"
Copyright notice	Front matter (verso)	28.4.2.4.1, "Copyright Notice"
Copyright page	Front matter (verso)	28.4.2.4, "Verso of Title Page (Copyright Page)"
Country where printed	Front matter (verso)	28.4.2.4.6, "Location of Printing"
Cover	Cover and jacket	28.4.1, "Cover, Jacket, and Endpapers"
Dedication	Front matter (recto)	28.4.2.5, "Dedication"
Endpapers	Cover and jacket	28.4.1, "Cover, Jacket, and Endpapers"
Epigraph	Text or front matter	28.4.3.4, "Epigraph"
Errata	Front matter	28.4.2.7, "Errata"
Footnotes	Text	28.4.3.5.3, "Footnotes"
Foreword	Front matter (recto)	28.4.2.9, "Foreword"
Frontispiece	Front matter (preferably on verso)	28.4.2.2, "Frontispiece and Other Uses for Half Title's Verso"
Glossary	Back matter (preferably begin on recto)	28.4.4.4, "Glossary"
Half title page	Front matter (recto)	28.4.2.1, "Half Title"
Half title page, second	Text (preferably on recto)	28.4.3.3, "Second Half Title"
Half title page's verso	Front matter (verso)	28.4.2.2, "Frontispiece and Other Uses for Half Title's Verso"

Table 28.1. Components of Books (*continued*)

Component	Location[a]	Section reference
History of publication	Front matter (verso)	28.4.2.4.2, "History of Publication"
Illustration list	Front matter	28.4.2.8, "Lists of Illustrations and Tables"
Indexes	Back matter (preferably begin on recto)	28.4.4.6, "Indexes"
International Standard Book Number	Front matter (verso)	28.4.2.4.5, "International Standard Book Number (ISBN)"
Introduction	Front matter	28.4.2.11.1, "Introduction"
	Text	28.4.3.2, "Introduction"
Jacket	Cover and jacket	28.4.1, "Cover, Jacket, and Endpapers"
Notes	Back matter (preferably begin on recto)	28.4.4.3, "Notes"
Parts	Text	28.4.3.5, "Parts, Sections, and Chapters"
Permanence of paper statement	Front matter (verso)	28.4.2.4.7, "Permanence of Paper Statement"
Preface	Front matter (recto)	28.4.2.10, "Preface"
Printer	Front matter (verso)	28.4.2.4.10, "Printer"
	Back matter (verso or page after index or last page of book)	28.4.4.7, "Colophon Page"
Publisher's address	Front matter (verso)	28.4.2.4.3, "Publisher's Address"
Reference list	Text	28.4.3.5.4, "Reference List"
Running heads and feet	Text	28.5.2, "Running Heads and Running Feet"
Second half title page	Text (preferably on recto)	28.4.3.3, "Second Half Title"
Section numbering	Text	28.4.3.5.2, "Section Numbering"
Sections	Text	28.4.3.5, "Parts, Sections, and Chapters"
Series	Cover and jacket	28.4.1, "Cover, Jacket, and Endpapers"
	Front matter (recto)	28.4.2.1, "Half Title"
	Front matter (verso)	28.4.2.2, "Frontispiece and Other Uses for Half Title's Verso"
	Text (preferably on recto)	28.4.3.3, "Second Half Title"
Spine	Cover and jacket	28.4.1, "Cover, Jacket, and Endpapers"
Table of contents	Front matter (recto for first page)	28.4.2.6, "Table of Contents"
Table list	Front matter	28.4.2.8, "Lists of Illustrations and Tables"
Title page of book	Front matter (recto)	28.4.2.3, "Title Page"
Title page's verso	Front matter (verso)	28.4.2.4, "Verso of Title Page (Copyright Page)"
Title pages of chapters	Text (preferably on recto)	28.4.3.5, "Parts, Sections, and Chapters"

[a] For components that should begin on a recto or verso page, the preferred page type is indicated in the "Location" column.

ing Office,[14] and in the United Kingdom, this is done through the Office of Public Sector Information.[15]

28.3 PUBLICATION

"Publication" is defined as "the reproduction in tangible form . . . of copies of a work from which it can be read or otherwise visually perceived."[16] In addition, "[publication] covers any general distribution to the public of works in print and of non-print works, including visual, audiovisual, and electronic works."[1]

28.3.1 Publishers

Publishers of scientific monographs fall into 4 major categories: commercial scholarly publishers, professional organizations, university presses, and trade publishers. For-profit firms specializing in books for specific disciplines, such as the physical and life sciences, are the largest class of scientific publishers. They market their publications extensively and are widely known to their specific audiences. The publishing arms of professional organizations, such as the British Medical Association and the American Chemical Society, are the next largest group of scholarly publishers. The members of professional organizations are built-in audiences. These organizations often publish monographs that commercial publishers would reject because the potential markets are small. University presses produce high-quality publications, but their output is likely to be limited to monographs of regional or limited interest. Finally, trade publishers are large commercial organizations that produce publications of mass appeal. They publish scientific monographs if they anticipate that the general reading public will have enough interest in them.[17]

At both professional organizations and university presses, final publication decisions are made by editorial boards composed of association members and faculty, respectively. In contrast, at commercial and trade publishers, such decisions are usually made by the staff editors.[18]

Publishers negotiate contracts with authors and their agents, edit the authors' manuscripts, design the print and electronic monographs, produce the finished products, market the works, and arrange to distribute the works through various market channels.[19] (For more details, see Section 28.3.2, "Publication Process.")

28.3.2 Publication Process

The publication process for monographs begins when an author submits a book proposal to a publisher or when an acquisitions editor solicits a manuscript from an author on behalf of a publisher. The former is more common.

Before submitting a proposal, also called a "prospectus," an author should obtain the submission requirements of the publisher to which the author plans to submit the proposal. These requirements will specify such elements as text length and technical details like word-processing and illustration formats.

A proposal should usually contain the following information[12]:

- The rationale for the book (ie, its purpose)
- The book's intended audience
- Information on competing books and reasons why the proposed book would be better
- The proposed content and structure of the book
- A completed chapter
- Estimated completion date
- Marketing suggestions

If the proposal is accepted, an acquisitions editor or commissioning editor will work with the author to develop a contract that defines the relationship between the author and the publisher. See Chapter 2, "Publication Policies and Practices," for authors' and editors' responsibilities and ethical considerations.

After the author submits the manuscript, which is typically done through a submission website or email, a technical or substantive editor looks for such items as errors in fact, incoherent or ambiguous sentences, misleading language, and agreement between the level of complexity of the text and the anticipated audience's comprehension level. In addition, a copy editor ensures that the text and illustrations are clear, correct, and consistent; that the spelling, grammar, and punctuation are in accord with the publisher's house style; and that agreement exists among all of the parts, such as headings appropriately describing the text and in-text references matching the sources in the reference list. At some publishers, the copy editor and the technical or substantive editor are the same person.

In the editing process, the manuscript is marked up, usually using tracked changes in a word-processing program or annotations in a PDF file. Once substantive editing is complete, the manuscript is returned to the author for comment, after which the author and editors negotiate on any changes with which they disagree.

As part of the publication process, the copy editor prepares the manuscript for a designer by coding the text so that design software automatically generates the correct typefaces for chapter and section headings, body copy, illustration captions, and other elements. Once the book is electronically generated into pages, a typeset copy, often called the "galley proof," is returned to the author and editors for a last review, including checking the copy against the manuscript's last marked-up version.

More details of the entire publication process may be found in *The Chicago Manual of Style*[11] and other sources.[12,20] In addition, see Section 28.5, "Design Elements."

28.4 PARTS OF MONOGRAPHS

28.4.1 Cover, Jacket, and Endpapers

A book cover is composed of a front, a back, and a spine. A hardcover book typically has a jacket that wraps around the book cover, and the jacket's flaps tuck inside the front and back of the book. However, "lithocase" hardcover books and paperback books do not have jackets.

Displayed on a hardcover book's jacket are the book's title and the name of the author (or the editor if the book is a collection). This information appears on the front cover of lithocase hardcovers and paperbacks. It is important that the title on the jacket or cover match what is on the title page inside the book.

The jacket or cover also may include a subtitle, the title under which the book was previously published, an edition number, a volume number and series title if the book is part of a multipart collection, the publisher's name or its colophon (the publisher's trademark or a graphic device used as an embellishment), and the date of publication. The author's name may be shortened to the surname, and the publisher's name may be reduced to a single word or replaced by its colophon. For a book with a jacket, the cover may be left blank.

Standards for the information placed on book spines have been established by both the National Information Standards Organization (NISO) in the United States and the International Organization for Standardization (ISO). NISO standard Z39.41[6] reserves the top two-thirds of a spine for the author's or editor's name; the title; the edition number; and any "numerical, alphabetical, or chronological" data, such as the year of publication and the number or name of a volume, part, or series. The bottom third is divided into a blank area for library information, below which is printed the publisher's information. ISO standard 6357[21] reserves the top two-thirds of a spine for the title and, if sufficient space remains, the author's and publisher's names. At least the bottom 30 mm of the spine is left blank for library identification.

The back cover of a jacket, lithocase hardcover, or paperback usually carries an International Standard Book Number (ISBN) and a barcode.[11] Assigned to a book by its publisher, the ISBN should appear in a readable font above the barcode preceded by the letters "ISBN." The first 3 digits of USBNs for books are always "978." For additional information, see Section 28.4.2.4.5, "International Standard Book Number (ISBN)."

The barcode, or the "Bookland European Article Number" (Bookland EAN), begins with the first 12 digits of the ISBN with hyphens deleted. The 13th number of a book's barcode is a check digit, which is calculated from the other 12 digits to ensure that the barcode is correct. A separate 5-digit number is often added to the right of the barcode to indicate the retail price.[22] The Bookland EAN symbol should be printed at the bottom of the back of the jacket or cover, and it must be printed in a color that can be read by a scanner.[11] See *ISBN Book Industry Barcode*[22] for more details.

Quick response (QR) codes and other 2-dimensional matrix-style codes may also be used on jackets, book covers, and inside pages.[23] When used, 2-dimensional codes may replace or appear alongside standard barcodes. The advantage of such codes is that they allow readers to quickly access a great deal of information about books before purchasing them.[21]

In addition, the back of a jacket, lithocase hardcover, or paperback is often used for promotional purposes by the publisher. In addition, if a book has a jacket, the jacket's left inside flap may contain a summary of the book, and the right inside flap may contain a brief biography and a photograph of the author.

Some hardcover books have endpapers, or folded papers, which can be used for maps, chronologies, pronunciation tables, and other information that readers may wish to consult frequently. To make an endpaper convenient to consult, its first or last fold is pasted to the inside front or back cover, respectively, and the rest of the folded endpaper sits between the inside cover and the first or last leaf of the book. However, for books in library collections and other archives, endpapers may be partially obscured by bookplates and other acquisition information placed inside the covers.

28.4.2 Front Matter

The usual components of front matter (ie, the preliminary pages of a book) are listed in Table 28.1, along with their locations.

Many components of front matter typically start on right-hand pages (ie, rectos), but this rule can be ignored for some components if economical use of pages is a necessity. If the front matter begins with a half title page before a full title page, the half title page is counted as the first page, but if the front matter does not have a half title page, the first page is the full title page. The pages of the front matter are usually given lowercase roman-numeral page numbers. However, page numbers are usually not printed on the half title page, the title page, the left-hand pages (ie, versos) on the back side of the half and full title pages, and blank pages, even though these pages are counted for pagination purposes. For example, if the first page is the title page and it is followed by a verso, the first numbered page would be page iii.

The front matter components in the remainder of this section are listed in the order in which they usually appear in monographs. Depending on the nature of a book, other components may be incorporated into the front matter, and other sequences may be suitable.

28.4.2.1 Half Title

The right-hand page preceding the title page is generally used for the half title, also called the "false title." The half title page is a remnant from when books were printed without covers, so this recto is sometimes omitted to reduce the number of pages in a book. When included in a book, the half title page usually carries only the main title without the subtitle or author's name. This page sometimes carries a series name and, on rare occasions, an epigraph (ie, a brief quotation or motto that encapsulates the theme of the book).

Some books have additional half title pages. For example, a duplicate of the half title page (ie, second half title page) may be placed after the front matter at the front of the text. In addition, if a book is divided into parts or units, each may be preceded by a half title page bearing the title of the part.

28.4.2.2 Frontispiece and Other Uses for Half Title's Verso

The left-hand page between the half title page and the full title page is usually blank. Sometimes, this verso is used for a "frontispiece," which is an illustration that sets the tone by depicting an important aspect of the book (eg, a portrait of the subject of a biography).

Another use of this verso is to carry an advertisement, also known as an "ad card," "book card," "card page," and "face title." This advertisement lists other books by the author, usually with phrasing such as "Rita Levi-Montalcini, MD, is also the author of *Abbi il Coraggio di Conoscere* . . ." or "Other books by Rachel Carson, MS, are *Silent Spring*, *Under the Sea-Wind* . . ." If a book is part of a series, this verso may be used instead to list the series title, its editor, and the other titles in the series.

28.4.2.3 Title Page

The title page provides the full title of the book, including any subtitle. The title on this page is considered to be the authoritative title (ie, the title used in reference lists and library catalogs). The title on the half title page, cover, jacket, and other locations should agree with the title on the title page. (Note that the title cannot be copyrighted, even though the book contents can. See Table 3.1.)

The title page should be designed before the rest of the front matter, and it and the rest of the front matter should be in harmony with the main body of the book. The full title should follow the conventions of headline-style capitalization, as well as scientific conventions such as those for symbolization and italicization (see Chapters 4, 9, and 15–26).

The title and subtitle should be clearly delineated by type style and size, punctuation, and page layout. For example, the subtitle should appear below the main title and in a smaller type size.

Title:	**THE CSE MANUAL**
Subtitle:	Scientific Style and Format for Authors, Editors, and Publishers

The title page should include not only the book's full title but also the names of all authors and the publisher's name and location. When an organization serves as the author, the organizational name and its constituent parts should be listed in hierarchical order, according to ISO standard 1086.[24]

Organizational author in hierarchical order:	Council of Science Editors
	Style Manual Task Force

The title page may also include the names of such other contributors as the editor, the illustrator, the translator, and the author of the foreword or introduction. The roles of each contributor should be clearly stated, as in "Edited by Hany F. Shehata, PhD, and Braja Das, PhD" or "Petr Hliněný, PhD, and Antonín Kučera, PhD, editors." If each chapter or part of a book has its own author, only the names of the book's editors appear on the title page, while the chapter authors' names appear on a separate recto page in the front matter under the heading "Contributors" and with the chapter titles in the table of contents (see Section 28.4.2.11.2, "Chapter Contributors").

If a book has been revised or if it is a new edition, this information should be noted on the title page. Similarly, if the book is published in 2 or more volumes, the title page of each volume should include the volume number, the volume title, and the name of the volume editor. Alternatively, this information may be placed on a separate title page in each volume after the title page for the entire publication.

If a book provides a uniform resource locator (URL) to additional material, such as data sets and illustrations, that should be acknowledged on the title page as follows: "See [URL address] for additional materials."

ISO standard 1086[24] mandates that the publication date appear on the title page, which also is permitted by *Butcher's Copy-editing: The Cambridge Handbook for Editors, Copy-editors and Proofreaders*.[12] If the publication date is not given on the title page, it must appear on its verso.

The arrangement of elements on the title page depends on the book. Usually, the title is first, but the author's name may be placed above the title if the author is a widely known public figure or a top expert on the topic. In general, the proper sequence of elements from top down on the title page is as follows:

- Title
- Subtitle
- Author or editor information, including name and academic or research affiliation
- Edition number
- Information on accompanying materials
- Imprint information, namely the publisher's name; the publisher's location; and the year of publication, the copyright information, or both

In addition, the title page or its verso may include a colophon, which is a symbol or emblem representing the publisher's imprint. A colophon is required by British law but not by US law.

Colophon for the Council of Science Editors:

Furthermore, the title page or its verso may include details on the manufacture and design of the book, including such information as the name of the printer, the typeface used, the paper and other materials used, and the style of binding. The names of those responsible for the book's design and production are also sometimes included on the title page or its verso.

Occasionally, a book's designer may design the title page as a "spread" that spans pages ii and iii (ie, the pages normally used for the half title page's verso and the title page). In that case, the imprint information should be on page iii.

28.4.2.4 Verso of Title Page (Copyright Page)

The verso on the back side of the title page is often called the "copyright page" because it is usually where the copyright notice is located. In the United Kingdom, this verso is termed the "biblio page" because it also carries detailed information that identifies the book more fully than the information on the title page.

28.4.2.4.1 COPYRIGHT NOTICE

Although the copyright notice is customarily placed on the verso of the title page, it may appear on the title page instead. Until 1976, US copyright law specified that the copyright notice had to appear on the title page or its verso. Although this is no longer required, the custom still prevails.

For books published in the United States, the copyright notice must consist of the word "Copyright," the abbreviation "Copr," or the symbol "©," accompanied by the name of the copyright holder and the year of copyright for the current edition. The copyright information may include copyright dates for previous editions. In addition, the copyright notice may be accompanied by the words "All rights reserved" and the date of first publication of the current edition, which is usually but not always the copyright date.

For more details on copyright, including regulations in various countries, consult Chapter 3, "The Basics of Copyright."

28.4.2.4.2 HISTORY OF PUBLICATION

The copyright notice is often followed by statements noting the year of first publication, the years of subsequent editions, and the year of each impression. For further details, consult *The Chicago Manual of Style*.[11]

28.4.2.4.3 PUBLISHER'S ADDRESS

The publisher's address and that of its agents in other countries are often provided on the verso of the title page. The addresses may be complete addresses or abbreviated addresses with only the city, a URL to the publisher's website, or both.

28.4.2.4.4 CATALOGING IN PUBLICATION (CIP)

Cataloging in Publication[25] (CIP) is a prepublication cataloging program established in 1971 by the US Library of Congress to enable libraries to catalog books immediately upon their arrival. The US program is available only for books carrying a US place of publication. The Library of Congress creates a bibliographic record, known as "CIP data," for each book and transmits it to the publisher, which in turn prints the CIP data on the verso of the title page.

The Library of Congress distributes new CIP data to large libraries, bibliographic services, and book vendors around the world. Many of these organizations use the data to alert the library community to forthcoming publications and to facilitate acquisitions.[25] For details, search for "About CIP" at https://www.loc.gov.

Other countries have similar programs. For example, the British Library administers the British National Bibliography for the United Kingdom and Ireland (search for "Guide to the British National Bibliography" at https://www.bl.uk), the Library and Archives Canada runs Canada's Cataloguing in Publication program (search for "Cataloguing in Publication" at https://library-archives.canada.ca/), and the National Library of Australia oversees the Prepublication Data Service (search for "Prepublication Data Service" at https://www.nla.gov.au). For other countries, such as China and Malaysia, contact the appropriate national library for details.

28.4.2.4.5 INTERNATIONAL STANDARD BOOK NUMBER (ISBN)

The International Standard Book Number (ISBN) uniquely identifies a book. ISBNs are used for such purposes as keeping track of inventory by computer and handling orders.

Each 13-digit ISBN is unique to one title or one edition of a title from one publisher. If a book is published in hardcover, paperback, and online, each version will have a separate ISBN. Similarly, when a work is published in 2 or more volumes, a separate ISBN is assigned to each volume, unless the work is sold only as a set.

The ISBN should be printed on a book's jacket, the title page's verso, and the foot of the outside back cover. When volumes are sold only as a set, the same ISBN should be printed on the title page's verso of each volume.

The International ISBN Agency is located in London, United Kingdom (see https://www.isbn-international.org). However, publishers obtain ISBNs from the national ISBN agencies in the countries in which the publishers maintain their headquarters. National ISBN agencies include the following:

Country	ISBN agency	Website
United States	R.R. Bowker, LLC	https://www.isbn.org
United Kingdom	Nielsen BookData	https://nielsenbook.co.uk/isbn-agency/
Canada	Library and Archives Canada for English-language books and the Bibliothèque et Archives nationales du Québec for French-language books	https://www.bac-lac.gc.ca/eng/services/isbn-canada/Pages/isbn-canada.aspx and https://www.banq.qc.ca
Australia	Thorpe-Bowker	https://www.myidentifiers.com.au/identify-protect-your-book/isbn/buy-isbn
New Zealand	National Library of New Zealand	https://natlib.govt.nz/publishers-and-authors/isbns-issns-and-ismns
Ireland	Nielsen BookData	https://nielsenbook.co.uk/isbn-agency/
India	Raja Rammohun Roy National Agency for ISBN	https://isbnnew.inflibnet.ac.in/
Hong Kong	Hong Kong Public Libraries	https://www.hkpl.gov.hk/en/about-us/services/book-registration/isbn.html
South Korea	National Library of Korea	https://seoji.nl.go.kr
Kenya	Kenya National Library Service	https://www.knls.ac.ke
South Africa	National Library of South Africa	https://www.nlsa.ac.za
Brazil	Câmara Brasileira do Livro	https://www.cblservicos.org.br/isbn/

28.4.2.4.6 LOCATION OF PRINTING

A statement naming the country in which a book is printed is typically placed on the copyright page.

Printed in the United States of America   Printed in Canada

Such a statement also must appear on the cover or jacket of a book printed in a country other than the country of publication.[11]

28.4.2.4.7 PERMANENCE OF PAPER STATEMENT

NISO has established criteria to ensure that coated and uncoated paper will not deteriorate significantly for several hundred years under normal use and storage in libraries and archives. NISO set forth these standards in a document titled *Permanence of Paper for Publications and Documents in Libraries and Archives*,[5] which was subsequently approved by the American National Standards Institute (ANSI). When a book is printed on paper that meets this standard, the following symbol and statement should be printed on the verso of the title page: "⊖ This paper meets the requirements of ANSI/NISO Z39.48-1992 (R2009) (Permanence of Paper)."

28.4.2.4.8 COPYRIGHT ACKNOWLEDGMENTS, PERMISSIONS, AND CREDITS

Acknowledgments related to previously published or otherwise copyrighted material should be included on the copyright page. Such acknowledgments are necessary for anthologies. If copyright acknowledgments are lengthy, they may appear on a separate recto page in the front matter. In contrast, acknowledgments that recognize contributions from persons, groups, and others who were essential to producing a book appear in the preface (see Section 28.4.2.10, "Preface") or on a separate page in the back matter or front matter (see Section 28.4.4.5, "Acknowledgments of Contributions").

"Permission" is clearance from a copyright owner to quote passages or reproduce illustrations from another publication. Permissions may appear on the title page's verso, as can photo credits if they are not provided with the photos. When permissions and credits are extensive, it is sometimes desirable to list them on a separate recto page (see Section 28.4.2.11, "Miscellaneous Front Matter").

28.4.2.4.9 ABSTRACT

Some nonfiction books on narrow topics carry abstracts if the topics are not adequately reflected in the books' titles. A book's abstract may appear on the title page's verso or on the right-hand page following the verso. If a book's chapters have separate abstracts, each abstract should appear on the page preceding the start of the chapter or on the chapter's opening page.

For standards for writing abstracts, see NISO Z39.14[4] and ISO 214,[26] as well as the style manuals of such organizations as the American Medical Association and American Chemical Society.

28.4.2.4.10 PRINTER

The name of the compositor, printer, binder, or any combination thereof may be included on the title page's verso or on the colophon page (see Section 28.4.4.7, "Colophon Page").

28.4.2.4.11 ADDITIONAL INFORMATION FOR THE VERSO OF THE TITLE PAGE

Additional information on appropriate content for the title page's verso, such as credit for cover artwork, may be found in *The Chicago Manual of Style*.[11] Content specific for books published in the United Kingdom may be found in *Butcher's Copy-editing: The Cambridge Handbook for Editors, Copy-editors and Proofreaders.*[12]

28.4.2.5 Dedication

A "dedication" is an inscription that honors or compliments persons or entities important to the authors. Whether a book will carry a dedication and what it will say are up to the authors. However, brevity is encouraged. Although a book's dedication sometimes appears on the title page's verso, it usually appears on the right-hand page opposite the title page's verso. If this recto page is used for the dedication, the following verso should be blank.

28.4.2.6 Table of Contents

The "table of contents" lists the content of a book, usually by providing chapter titles and page numbers. The table of contents simplifies finding specific information in both the text and back matter, and the table clarifies the organization of the book. The table should be titled simply "Contents."

In cases in which the number of subheadings in chapters is limited and the subheadings would assist readers in finding content rapidly, subheadings can be listed under chapter titles. This is often done with reference books, for example. In such situations, the table of contents should be designed so that readers can readily distinguish chapter titles from subheadings.

In a book's table of contents, the beginning page numbers of chapters are listed immediately after the chapter titles. However, titles of parts or sections of the book are not followed by beginning page numbers in the table of contents.

For books published in the United States, the tables of contents traditionally follow the dedication. The table should exclude the front matter that precedes the table but include all front matter that follows the table (eg, errata, list of illustrations, and preface).

In contrast, European practice often places tables of contents at the end of books.

28.4.2.7 Errata

In rare instances, an "Errata" section is published in the front matter to list substantive errors discovered after a book is printed but before it is bound. This section is not used for typographic and other minor errors. This list can be placed right after the table of contents to increase the probability that readers will see it. If a book has already been bound when the errors are discovered, a separate leaf with the errata may be tipped into the book after the table of contents. Each item in the errata should list the page number, line number, error, and correction.

> page 85, line 8: For "Einstein's theory," read "Einstein's wife's theory."

28.4.2.8 Lists of Illustrations and Tables

When illustrations (eg, figures) and tables are major sources of information in a book, lists of each may be included in the front matter so that readers can be directed to them without consulting the text. Reference books frequently have such lists, especially for tables. In contrast, heavily illustrated books might be too burdened by long lists to use them.

If both illustrations and tables are listed, the list of illustrations is placed before the list of tables. Both kinds of lists should begin either at the end of the table of contents or on the right-hand page immediately after the table of contents.

For information on the structure and format of illustrations and tables, see Chapter 30, "Tables, Figures, and Indexes."

28.4.2.9 Foreword

A "foreword" (not to be confused with the homophones "foreward" and "forward") is often similar to an author's preface, but it is written by someone other than the author. The foreword often places the book in the context of related literature and points out merits of the book from the perspective of an authority other than the author or editor. The foreword is typically no more than 2 pages, and it ends with the name of the person who wrote it and, if desired, the person's title and affiliation.

28.4.2.10 Preface

A "preface" offers the author's perspective about a book as a whole, and it describes the book's purpose, sources, and extent. The preface may provide background information on why the author wrote the book, or it may describe the process of writing it. The preface differs from the introduction in that the introduction discusses the text and prepares the reader for understanding the content.

The end of the preface is one of the places where an author customarily acknowledges indebtedness to other authorities, to persons who contributed information, and to those who provided feedback or proofed the book (see also Section 28.4.4.5, "Acknowledgments of Contributions").

If a new preface is added when a new edition is published, it is placed before the prefaces to previous editions. Similarly, if a book has an editor's preface, it usually precedes the author's preface.

28.4.2.11 Miscellaneous Front Matter
28.4.2.11.1 INTRODUCTION

Although the introduction is usually part of a book's text (see Section 28.4.3.2, "Introduction"), it is sometimes included on a separate recto page in the front matter.

28.4.2.11.2 CHAPTER CONTRIBUTORS

For books with multiple authors in which only the editor's name appears on the title page, it is appropriate to list the authors' names in a contributors section in addition to listing the authors' names with the chapter titles in the table of contents. The contributors section may include the authors' affiliations and other relevant information. This section is among the front matter that usually follows the table of contents.

An alternative to a contributors list is to note each author's information at the beginning of the author's chapter, either below the author's name or in a footnote. However, if each author has an extensive biographical note, those notes can be placed at the end of the book.

28.4.2.11.3 LIST OF ABBREVIATIONS
When many abbreviations are used throughout the text, the front matter may include a separate list of abbreviations and the terms or names they stand for. Alternatively, abbreviations may be incorporated into a glossary at the back of the book (see Section 28.4.4.4, "Glossary").

28.4.3 Text
28.4.3.1 Pagination
The first page of the text, or body, of a book begins the sequence of pagination with arabic numerals. From this point, all text pages are counted consecutively whether the page numbers are printed on the pages. For example, the page number is often omitted on the first page of a each part and chapter if that page does not contain substantive text. In addition, page numbers are often omitted on pages containing only illustrations or tables. Nevertheless, both types of pages are counted in a book's pagination.

In many books, the first chapter starts on page 1 of the text section. However, in some books, the text section starts with other types of content, such as an introduction, a second half title page, and an epigraph (see the next 3 sections). Regardless of its use, the first page is customarily a recto page.

28.4.3.2 Introduction
An "introduction" defines the purpose, scope, and organization of the work, and it describes the author's approach. The introduction may be signed by the person who wrote it, but if the book's author wrote the introduction, a signature is not necessary. Note that the introduction is sometimes part of the front matter, not the text (see Section 28.4.2.11.1, "Introduction").

28.4.3.3 Second Half Title
A second half title page is usually used to indicate the start of a book's text when the book has a large amount of front matter. As the first page of the text, this recto page is counted as page 1, but the page number is usually omitted from the page.

Like the first half title page, a second half title page may carry only the main title. Sometimes, this page carries a series name or an epigraph (see Section 28.4.3.4, "Epigraph"). The verso of this page is normally blank, although the epigraph can be placed on this page instead of on the second half title page.

If the text of a book is divided into parts, a half title page may be used at the start of each part. In this function, the page is also called a "part title page."

28.4.3.4 Epigraph
An "epigraph" is a quotation intended to convey a theme or tone that resounds in the text. Its source need not be documented beyond the name of the quoted author and, perhaps, the title of the quoted work.

"To invent, you need a good imagination and a pile of junk."—Thomas A. Edison

"Those who are not shocked when they first come across quantum theory cannot possibly have understood it."—Niels Bohr, PhD, *Essays 1932–1957 on Atomic Physics and Human Knowledge*

A number of options exist for where to place an epigraph. It may replace the second half title page, becoming the first page of the text (see Section 28.4.3.3, "Second Half Title"). It may be placed on the verso or recto after the second half title page, making it the second or third page of the text, respectively. Alternatively, the epigraph can be placed in the front matter instead of the text. If that is the case, the epigraph is likely to be on the same page as the dedication or on the following page (see Section 28.4.2.5, "Dedication"). In rare situations, the epigraph is on the title page or half title page (see Sections 28.4.2.3, "Title Page" and 28.4.2.1, "Half Title").

Epigraphs may be used for the same purpose for individual chapters. These epigraphs should appear on chapter title pages between the title and the chapter's text.

28.4.3.5 Parts, Sections, and Chapters

Chapters of a book may be logically grouped into parts or sections. The number and title of each part appear on a right-hand page preceding the first page of the part's first chapter. This recto is counted in the pagination, but its page number is omitted. In addition, its verso is usually blank.

Each part or section may have its own introduction. A part's introduction may appear on a new right-hand page following the part's title page, on the verso of the title page, or on the title page itself.

Part numbers should be in arabic numerals. Chapters within parts are numbered consecutively throughout the book instead of each part beginning with chapter 1.

Most prose books are divided into chapters, regardless of whether the chapters are grouped into parts. Chapters in the same book may be of varying lengths. Like parts, each chapter usually starts on a new page. While a recto page is preferred for this purpose, a verso can be used if economical use of pages is desired.

A chapter should begin with a chapter number and title, also known as a "chapter head." Directly below those elements, the chapter's text begins. Unlike a part page, the first page of a chapter carries a "drop folio," a page number at the bottom of the page. However, this page does not have a "running head," a short title printed at the top of the page.

For books in which the chapters are written by different authors, the name of the author of each chapter is printed below the chapter number and title. If the author's affiliation and other identification are not provided in the front matter or back of the book (see Section 28.4.2.11.2, "Chapter Contributors"), they may be provided in a footnote on the first page of the chapter[11] or immediately below the chapter author's name on this page.[12]

Abstracts are not commonly used for book chapters. If chapter abstracts are needed, they should appear at the top of the chapters beneath the chapter titles or, if present, the author information.

28.4.3.5.1 SUBHEADINGS

As an aid to readers, chapters may be divided with subheadings, secondary subheadings, and tertiary subheadings. These subheadings should be succinct and meaningful, and the hierarchy of these subheadings should be accurately observed. When a chapter is divided, it should have at least 2 subheadings. Likewise, if the text between subheadings is divided, it should have at least 2 secondary subheadings, and text divided under any secondary subheading should have at least 2 tertiary subheadings.

A subheading is usually set on a separate line from the text above and below it, but if space is limited, the lowest-level subheading may be used at the beginning of the first line of the text in the new subsection.

Preferred Placement of Subheadings

Place each subheading on a separate line above the new section's first sentence.

Alternate Placement of Lowest-Level Subheadings—Set subheading on the same line as the new section's first sentence.

The levels of subheadings should use different typefaces and point sizes so that readers can visually distinguish them.

Subheadings Should Be a Larger Size and Different Typeface than Secondary Subheadings

Secondary Subheadings Should Be a Larger Size and Different Typeface than Tertiary Subheadings

Tertiary Subheadings Should Be a Smaller Size and Different Typeface than Secondary Subheadings

Subheadings should not contain reference citations or notes, and they should not be used to define abbreviations or terms.

The first sentence after a subheading should be complete unto itself and not assume that readers have read the subheading.

SUBHEADINGS

Subheadings should not contain reference citations.

not

SUBHEADINGS

They should not contain reference citations.

28.4.3.5.2 SECTION NUMBERING

Sections and subsections of chapters may be numbered to help guide readers quickly to parts of the text. This device is especially useful for reference books, which are unlikely to be read sequentially because they are used mainly to obtain answers to specific questions. This style manual is an example of a reference book that uses this device.

A common method for section numbering is "double numeration," in which each section begins with the chapter number, a demarcation point such as a period, and a section number. For example, the section number "4.8" would designate the 8th section of chapter 4.

An alternative method for instruction manuals and similar books is to follow the chapter number with a paragraph number. Under that method, every paragraph would have a section number.

28.4.3.5.3 FOOTNOTES

Footnotes should generally not be used in text pages (see Section 28.4.4.3, "Notes"). Exceptions include using footnotes in text as an alternative way to provide the authors' affiliations and other relevant information in books in which chapters are written by different authors (for other options, see Section 28.4.2.11.2, "Chapter Contributors").

If footnotes must be used, consult such sources as Chapter 14, "Notes and Bibliography," in *The Chicago Manual of Style*[11] and Chapter 9, "Other Parts of a Book," in *Butcher's Copy-editing: The Cambridge Handbook for Editors, Copy-editors and Proofreaders.*[12]

28.4.3.5.4 REFERENCE LIST

A "reference list" consists of works cited in a book, as well as uncited works on the same subject as the book's. References should preferably be listed at the end of the chapter in which they are cited (as is the case throughout this style manual). This practice is particularly desirable when chapters have numerous references. Readers will find references more readily at the end of the relevant chapters than in a long compilation in the back matter.

If the list consists only of works cited in the text, the citations are usually designated as "Cited References," "References," "Reference List," or "Literature Cited." However, if the list consists of both works used to write the book and other works that might be of interest to readers, it may be divided into "Cited References" and "Additional Resources." Alternatively, if cited and uncited works are combined, the list is often called a "Bibliography" and placed in the back matter. See Chapter 29, "References," for style guidelines for reference lists.

28.4.4 Back Matter

The "back matter," also called "reference matter" and "end matter," constitutes the pages after the main text of a book. Back matter includes any of the components described in the next 7 subsections, and the sequence is generally the one shown here. Pagination of this section continues with arabic numerals.

28.4.4.1 Addenda

Brief supplemental data that became available too late to include in the text may be added immediately after the text in an "addendum." In general, addenda should be avoided because their content is likely to be overlooked by readers, even though addenda are usually listed in tables of contents.

Addenda can also be provided as online-only material. However, this option is more valuable for mobile-optimized and other electronic versions of books than for print books.

28.4.4.2 Appendixes

Appendixes carry supplemental material that illustrates, expands on, or otherwise supports the text. Appendixes are separated from the text so that they do not distract readers from the text. Appendixes provide such supporting documents as lists, large tables, correspondence, speeches, forms, detailed protocols, and questionnaires.

If appendixes are in a book's back matter, the first appendix begins on a recto page. If space permits, this recto page is preceded by a half title page labeled "Appendix" or "Appendixes," depending on the number of supporting documents. If the work has a single appendix, it is not numbered, while if the book has 2 or more, the appendixes are numbered or lettered consecutively.

Alternatively, appendixes may be placed at the end of their respective chapters, especially if the book has different chapter authors and each appendix is needed for just a single chapter. When an appendix follows its chapter, it may begin on either a left- or right-hand page, or it may begin immediately after the text on the last page of the chapter.

Appendixes can also be posted online in an archived database. Just like online addenda, online appendixes are more useful in electronic books because readers can reach these additional resources nearly effortlessly through hyperlinks.

28.4.4.3 Notes

"Notes" provide information of interest to researchers and other scholars. Notes can follow the chapter to which they refer, which is common in books in which chapters are written by different authors. Alternatively, notes can be placed at the end of the book, in which case they should be grouped by chapter. For each chapter, the numbering of notes usually begins with the numeral "1." Notes are set in a slightly smaller type than the text.

Notes are generally preferred to footnotes in books and other monographs. Footnotes can interrupt the flow of reading, reduce text areas on pages, disarrange the orderly appearance of pages, and increase the difficulty and cost of page makeup.

28.4.4.4 Glossary

A "glossary" is a brief dictionary of terms and concepts used in a book. The glossary usually precedes the reference list in the back matter. Entries should be lowercased except for proper nouns.

28.4.4.5 Acknowledgments of Contributions

If too lengthy to appear in the preface (see Section 28.4.2.10, "Preface"), acknowledgments that recognize contributions from others who were essential to producing a book can be placed on a separate page in the back matter or sometimes the front matter. This page is used for such purposes as acknowledging indebtedness to authorities, to persons who contributed information, and to those who provided feedback or proofed the book.

28.4.4.6 Indexes

An "index" is an alphabetical list in the back matter that cites names, places, and topics discussed in the book and gives the page numbers on which these topics are discussed.

If section numbering is used in a book (see Section 28.4.3.5.2, "Section Numbering"), these section numbers may be used in the index rather than page numbers, as is the case in this style manual's Index.

In lengthy books that cite many authors and topics, it may be preferable to have separate indexes for authors and subjects. In such cases, the author index should precede the subject index.

An index can be set in smaller type than the text, and it can be laid out in 2 or more columns per page. Main headings should be clearly distinguished from subheadings in indexes. It is important to capitalize only the entry terms that are proper nouns. See Section 30.3, "Indexes," for additional details on formatting indexes.

28.4.4.7 Colophon Page

For the purposes of the back matter, a "colophon" is more extensive than a colophon used in the front matter. While a colophon in the front matter is just a trade emblem or other device used by the printer or publisher, the back matter version consists of details on the manufacture and design of the book, including the name of the printer, the typeface used, the paper and other materials used, and the style of binding. The names of those responsible for the book's design and production are also sometimes included in this colophon.

Over time, it has become more common for all of this information to be listed on the title page or its verso than in the back matter (see Section 28.4.2.3, "Title Page"). When printed in the back matter of a scholarly book, the colophon page usually falls right after the index's last page. In other books that include the colophon page as back matter, the page usually appears on the verso or recto page following the last page of the book.

28.5 DESIGN ELEMENTS

Good book design should complement the text "noiselessly" rather than ostentatiously demonstrate the skills of the designer.[27] Book design is a complex task for which many reference texts are available (see the selection of design texts included in the Additional Resources section of this style manual). The sections below cover 2 areas of design of particular interest to authors and editors.

28.5.1 Pagination

All text pages are numbered consecutively regardless of whether page numbers are printed on the pages. On the opening page of each chapter, the page number appears at the bottom of the page and is called a "drop folio." The most common location for page numbers on other pages is at the top of the page. Whether located at the top or bottom of pages, page numbers are generally flush with the left margin of the verso pages and the right margin of the recto pages. The outside margins are preferred to the center of pages because the page numbers are more readily seen by readers while skimming through a book.

In contrast to the text section of a book, the front matter is numbered with lowercase roman numerals, preferably in drop folios.

Page numbers are generally not printed on pages that contain only illustrations or tables.

In multivolume works, pagination may be consecutive throughout the volumes or begin anew with each volume. The former is preferred.

28.5.2 Running Heads and Running Feet

"Running heads" are headings that appear across the top of pages, while "running feet" appear below the text. Both are intended to indicate content. Often, the title of the book appears in the running head of left-hand pages, and shortened versions of chapter titles appear in the running head of right-hand pages. An alternative approach uses the running heads of verso pages for chapter titles and the running heads of recto pages for titles of chapter sections.

Both running heads and running feet are discretionary. Running heads are usually omitted on display pages such as chapter title pages and illustration pages. However, both running heads and running feet can be useful to readers in skimming the text, as well as in identifying the source of pages copied from a book.

28.6 SPECIFIC TYPES OF MONOGRAPHS

28.6.1 Technical Reports

According to the National Information Standards Organization[1] (NISO) in the United States, a "technical report," also known as a "scientific and technical report," is "a separately issued record of research results, research in progress, or other technical studies." The purpose of a technical report is to disseminate research results.[7] Most technical reports are issued by government agencies, usually at the national, state, or provincial level, but technical reports also originate from universities and other research institutions.

Most of the components of technical reports are akin to those of books, but technical reports have some special features. The major differences between a technical report and a standard book lie in their authorship and in information provided about sponsorship, including contract, grant, and report numbers.

A technical report needs to identify both the organization that funded the research and the organization that conducted the research, and the report needs to indicate which of those organizations published the report. In some cases, the same organization sponsors and conducts the research. For example, the National Institutes of Health (NIH) in the United States has intramural scientists who conduct some of the research funded by NIH. Often, however, the sponsoring organization provides funds to another organization to perform the research. These funds are awarded through grants and contracts. When this occurs, either the sponsoring organization or the performing organization may publish the report. Thus, there are 3 basic scenarios for technical reports:

- Written and published by the sponsoring organization
- Written by the performing organization and published by the sponsoring organization
- Written and published by the performing organization

The following sections address significant differences between technical reports and standard books. These recommendations are based on standard Z39.18 of the American National Standards Institute (ANSI) and NISO, which is titled *Scientific and Technical Reports—Preparation, Presentation, and Preservation.*[7] Consult this ANSI/NISO standard for further details. In addition, for information on citing technical reports, see Section 29.4.4, "Technical Reports."

28.6.1.1 Front Matter

28.6.1.1.1 COVER

For a technical report, a cover is optional. When used, a cover identifies the report's subject and indicates whether the report contains classified or proprietary information. The information that is relevant to the cover includes the following:

(1) Report number (alternatively, this number can be printed on the back cover or on both covers)
(2) Report title and subtitle
(3) Title and number of series
(4) Authors, principal investigators, editors, and compilers
(5) Publisher (the organization assuming responsibility for the publication, which may or may not be the sponsoring organization)
(6) Publication date
(7) Limitations on distribution of the report (see Section 28.6.1.1.3, "Notices")
(8) Sponsoring organization
(9) International Standard Book Number (ISBN) or International Standard Serial Number (ISSN) (also included on the back cover)
(10) Technical format of the report, such as e-publication, video recording, or audio recording
(11) Subject of the report

Legal or other requirements may mandate additional information.

28.6.1.1.2 TITLE SECTION

Unlike a cover, a title section is required for a technical report. The title section indicates the subject and content of the report, and it provides the information needed for bibliographic description and access. The title section also clearly identifies the organizations that conducted and sponsored the report. The information that is relevant for the title section includes the following:

(1) Report number (see standard Z39.23[8] from ANSI and NISO and standard 10444[28] from the International Organization for Standardization [ISO])
(2) Report title and subtitle
(3) Title and number of series
(4) Authors, principal investigators, editors, and compilers
(5) Performing organization, including the authors' affiliation with the organization
(6) Publication data, including place of publication, publisher, and date of publication

(7) Type of report and the period covered by the report
(8) Contract and grant numbers
(9) Sponsoring or issuing organization if different from the performing organization
(10) Subject descriptors

Legal or other requirements may mandate additional information be included in the title section.

28.6.1.1.3 NOTICES

Special notices are included on the verso of the title page or inside the front cover. Among the common notices is a copyright section, which includes the following elements:

(1) Copyright symbol
(2) Year of publication
(3) Copyright holder
(4) Full name of the sponsoring organization and any other organization providing funding for the report
(5) A statement providing authority to copy the contents or requiring permission to copy
(6) The complete name, address, and telephone number of the producer of the report
(7) Instructions for obtaining additional copies of the report

Other notices may detail limitations to the report, such as security classification, distribution restrictions, and proprietary status. Such notices may state the nature of the report, such as that the data are preliminary, the report is a draft or a preprint, the document is a working paper intended to elicit comments and ideas, or the paper will be presented at a professional meeting. Additionally, the limitations detailed in a notice can include legal conditions connected with the report, such as its use of brand, or trade, names. Although disclaimers generally should be avoided, they should be placed in a special notice if they are necessary.

28.6.1.1.4 REPORT DOCUMENTATION PAGE

Some agencies of the US government require that reports include a documentation page, such as the bibliographic data sheet of the National Technical Information Service or standard form SF298 of the US General Services Administration. US government agencies also specify where to place the documentation in reports.

For technical reports produced by academic and industrial institutions, a report documentation page is optional. When used by such institutions, the documentation page is typically placed in the back matter of reports.

28.6.1.1.5 ABSTRACT

An abstract is required for technical reports. The abstract summarizes in approximately 200 words the purpose, scope, and major findings of a report. It contains no references or illustrations. The abstract should be understood on its own without referring to the text.

The abstract usually appears in its own section between the title section and the table of contents. However, if a report documentation page is required, the abstract is included on that page. See ANSI/NISO standard Z39.14 (R2002), titled *Guidelines for Abstracts*,[4] and ISO standard 214, titled *Documentation—Abstracts for Publications and Documentation*.[26]

28.6.1.2 Text

28.6.1.2.1 SUMMARY

The text of a technical report begins with a 500- to 1,000-word summary that states the problem investigated, principal results, conclusions, and recommendations. The summary also defines all symbols, acronyms and other abbreviations, and unusual terms used in the report. A summary includes only material that is covered in the report.

Some organizations prefer the term "executive summary" to "summary," and some organizations place the summary at the end of the front matter rather than at the beginning of the text.

28.6.1.2.2 REPORT TEXT

The text of a technical report closely resembles a journal article in format and style with an introduction, description of methods and procedures, presentation of results, discussion, conclusions, recommendations, and references.

28.6.1.3 Back Matter

The back matter clarifies and supplements the text of a technical report. Back matter contains appendixes; a bibliography; lists of symbols, acronyms, and other abbreviations; a glossary; an index; and the distribution list. The last is a complete list of the names and addresses of the persons and organizations that will receive a copy of the report, thus creating a permanent record of the initial distribution.

28.6.2 Conference Proceedings

Conference proceedings consist primarily of the full papers or abstracts of papers presented at a conference or other type of meeting. Proceedings also may include such material as posters presented at the conference, lists of sessions and conference sponsors, information on the conference organizers and attendees, advertisements for supporting organizations, and information on forthcoming conferences by the same organizers.

Conference proceedings share many characteristics with books. The major differences lie in their titles. Conference proceedings often have 3 titles: the title of the book of proceedings, the title of the conference, and the title of the conference series.

Book title:	Health and Wellbeing e-Networks for All
Conference title:	MEDINFO 2019
Conference series title:	Proceedings of the 17th World Congress on Medical and Health Informatics

However, proceedings may have fewer titles if they do not have a book title, conference title, or both.

Conference title:	Intelligent Data Engineering and Automated Learning—IDEAL 2019
Conference series title:	Proceedings of the 20th International Conference on Intelligent Data Engineering and Automated Learning
Conference series title:	Proceedings of the American Society of Mechanical Engineers' Power 2021 Conference (POWER2021)

Besides listing all titles, the title page of conference proceedings should include the location, the inclusive dates, and the sponsoring organizations of the conference. The location information should consist of the venue, such as a university, as well as the city and, if not readily discernible from the city, the state or province and country. For further information, see ANSI/NISO standard Z39.82, *Title Pages for Conference Publications*.[10] For information on citing proceedings, see Section 29.4.3, "Conference Proceedings and Conference Papers."

28.6.3 Electronic Monographs
28.6.3.1 Advantages of Electronic Format

Books may be published in a variety of electronic media, including online, print and graphics file downloads, and audio or video file downloads. Likewise, electronic books can be read on a variety of devices, including computers and e-readers like mobile-optimized tablets and smartphones. The features of electronic books provide many advantages over print books, such as the following[29]:

- Although editorial and production costs will increase with the size of both electronic and print books, the restrictions on size are not the same with electronic books. For example, electronic books do not incur paper and printing costs. However, if publishers impose page charges on authors, those charges will increase with each additional page even for electronic books. Another size restriction is that excessively large electronic files may require a long time to download or large amounts of bandwidth to stream.
- The scope and usability of electronic books can be broadened by providing access to large data files, multimedia, and a range of other supplementary content that would be prohibited in print by cost, physical limitations, or both. For example, videos of surgical procedures and animations of complex molecular activities greatly enhance written descriptions. Such supplementary content can be stored online separately from an electronic book, and readers can reach that content via hyperlinks in the digital text.
- Hyperlinks further enhance readers' experiences by allowing readers to navigate to other places in a book. In addition, hyperlinks provide readers with immediate access to other documents, to websites, and to numerous other resources.
- The searchability of electronic books allows readers to look for words and phrases anywhere within the text.
- Specialized dictionaries can be built into digital texts to allow readers to check the meaning of terms without having to leave their place in the text.
- Color and typefaces are almost limitless online, dependent only on the instrument used for reading. Not only can publishers use colors and typefaces to distinguish elements of a book such as levels and sections, but readers can use those features to mark text and add notes.
- Editorial changes and updates can be made easily before publication without costly page changes, and errors can be corrected or text updated with similar ease after initial publication.
- Multiple versions of a book can be produced from the same text base.
- Electronic media can be optimized for readers who are blind or otherwise visually impaired and for readers with learning disabilities such as dyslexia.
- With no physical inventory to maintain, electronic books have minimal storage and distribution costs, and the danger of producing more copies than can be sold is eliminated.
- An electronic book need never go out of print, and printed copies can be generated through print-on-demand options in numbers as few as a single copy at a time.

The capabilities and versatility of electronic books increased as formatting evolved from Portable Document Format (PDF) to markup languages, such as HyperText Markup Language (HTML), Extensible HyperText Markup Language (XHTML), and Extensible Markup Language (XML). In some cases, the electronic formats used for electronic books are proprietary, and documents created in these formats cannot be read using platforms that support competing formats.

28.6.3.2 Metadata and Digital Object Identifiers (DOIs)

Although electronic books differ radically in physical form from standard print books, the content of electronic books should still include all the elements discussed in Section 28.4, "Parts of Monographs." In addition to those standard elements, electronic books have metadata and digital object identifiers (DOIs). Metadata provide standardized, descriptive, machine-readable information used by online search engines, such as Google, to locate documents based on search criteria. A DOI is an identification number for a publication that is persistent across networks and across time. As a consequence, many reference citation styles require including DOIs for resources that have them.

Two major metadata vocabularies are used for electronic books: the Dublin Core Metadata Initiative[30,31] (DCMI) and the Online Information Exchange for Books (ONIX for Books).[32] DCMI, in turn, has 2 options for metadata: Simple DCMI describes 15 metadata elements, including title, subject, creator, publisher, and language, while qualified DCMI has 3 additional elements, including audience and rights holder.[31] In contrast, ONIX for Books has more than 200 metadata elements.[32]

Metadata are designed to assist with the following functions[33]:

(1) Identifying, locating, and organizing resources
(2) Promoting interoperability across systems
(3) Providing for digital identification using such identifiers as ISBNs and uniform resource locators (URLs)
(4) Archiving and preservation

Two NISO publications are useful introductions to metadata: *Understanding Metadata*[33] and *Metadata Demystified: A Guide for Publishers*.[34]

Developed by the International DOI Foundation (https://www.doi.org), DOIs can be assigned to entire books or to individual chapters. Some publishers assign one DOI to a book regardless of changes in edition or format, whereas other publishers give a unique DOI to each version of a book.

The naming convention for DOIs consists of a prefix and a suffix. The prefix contains the name of a specific DOI directory and the content owner's identifier, while the suffix is a numerical or alphanumeric string supplied by the publisher. Each publisher registers with the International DOI Foundation to obtain its owner's identifier. Then for each document, the publisher submits a suffix to a DOI registration agency along with the document's URL and appropriate metadata.[9] For additional information, see Section 27.3.1.2.2, "Production Issues," and the *Encyclopedia of Library and Information Sciences'* chapter titled "Digital Object Identifier (DOI) System."[35]

CITED REFERENCES

1. National Information Standards Organization. Bibliographic references. NISO Press; 2010. (ANSI/NISO Z39.29-2005 [R2010]). https://www.niso.org/publications/ansiniso-z3929-2005-r2010

2. National Information Standards Organization. The Dublin Core Metadata Element Set. NISO Press; 2013. (ANSI/NISO Z39.85-2012). https://www.niso.org/publications/ansiniso-z3985-2012-dublin-core-metadata-element-set

3. National Information Standards Organization. Electronic manuscript preparation and markup. NISO Press; 2002. (NISO/ANSI/ISO 12083-1995 [R2002]).

4. National Information Standards Organization. Guidelines for abstracts. NISO Press; 2015. (ANSI/NISO Z39.14-1997 [R2015]). https://www.niso.org/publications/ansiniso-z3914-1997-r2015-guidelines-abstracts

5. National Information Standards Organization. Permanence of paper for publications and documents in libraries and archives. NISO Press; 2010. (ANSI/NISO Z39.48-1992 [R2009]). https://groups.niso.org/apps/group_public/download.php/13464/Z39-48-1992_r2009.pdf

6. National Information Standards Organization. Placement guidelines for information on spines. NISO Press; 2015. (ANSI/NISO Z39.41-1997 [S2015]). https://www.niso.org/publications/ansiniso-z3941-1997-s2015

7. National Information Standards Organization. Scientific and technical reports—preparation, presentation, and preservation. NISO Press; 2010. (ANSI/NISO Z39.18-2005 [R2010]). https://www.niso.org/publications/z39.18-2005-r2010/

8. National Information Standards Organization. Standard technical report number format and creation. NISO Press; 2015. (ANSI/NISO Z39.23-1997 [S2015]). https://www.niso.org/publications/z39.23-1997-s2015

9. National Information Standards Organization. Syntax for the digital object identifier. NISO Press; 2000. (ANSI/NISO Z39.84-2000).

10. National Information Standards Organization. Title pages for conference publications. NISO Press; 2001. (ANSI/NISO Z39.82-2001).

11. The Chicago manual of style: the essential guide for writers, editors, and publishers. 17th ed. The University of Chicago Press; 2017. Also available at https://www.chicagomanualofstyle.org

12. Butcher J, Drake C, Leach M. Butcher's copy-editing: the Cambridge handbook for editors, copy-editors and proofreaders. 4th ed. Cambridge University Press; 2006.

13. Library and Archives Canada. Government of Canada; [accessed 2023 Feb 6]. https://library-archives.canada.ca/

14. US Government Publishing Office. Style manual: an official guide to the form and style of federal government publishing. US Government Publishing Office; 2016. https://www.govinfo.gov/collection/gpo-style-manual

15. Office of Public Sector Information (UK). The National Archives (UK), Office of Public Sector Information; [accessed 2023 Feb 6]. https://discovery.nationalarchives.gov.uk/details/r/C16765

16. UNESCO Universal Copyright Convention. United Nations Educational, Scientific and Cultural Organization; 1952. (Universal Copyright Convention, Article VI. Copyright Office [US]); 1974. (International copyright conventions circular; 38c). http://portal.unesco.org/en/ev.php-URL_ID=15381&URL_DO=DO_TOPIC&URL_SECTION=201.html

17. Huth EJ. Writing and publishing in medicine. 3rd ed. Williams & Wilkins; 1999.

18. MLA handbook. 9th ed. Modern Language Association of America; 2021.

19. Reitz JM. ODLIS: online dictionary of library and information science. ABC-CLIO; [accessed 2023 Feb 6]. https://www.abc-clio.com/ODLIS/odlis_A.aspx

20. Clark G, Phillips A. Inside book publishing. 5th ed. Routledge; 2014. http://www.insidebookpublishing.com/

21. International Organization for Standardization. Documentation—spine titles on books and other publications. ISO; 1985. (ISO 6357). https://www.iso.org/standard/12665.html

22. ISBN book industry barcode. Bar Code Graphics, Inc; [accessed 2023 Feb 6]. https://www.barcode .graphics/isbn-book-industry-barcode/

23. Furht B, editor. Handbook of augmented reality. Springer; 2011. Also available at https://link.springer .com/book/10.1007/978-1-4614-0064-6

24. International Organization for Standardization. Information and documentation—title leaves of books. International Organization for Standardization; 1991. (ISO 1086). https://www.iso.org/obp/ui/#iso: std:5589:en

25. About CIP. Library of Congress (US), Cataloging in Publication Division; [accessed 2023 Feb 6]. https://www.loc.gov/publish/cip/about/index.html

26. International Organization for Standardization. Documentation—abstracts for publications and documentation. International Organization for Standardization; 1976. (ISO 214). https://inen.isolutions .iso.org/obp/ui#iso:std:iso:214:ed-1:v1:en

27. Peacock J. Book production. 2nd ed. Blueprint; 1995.

28. International Organization for Standardization. Information and documentation—international standard technical report number (ISRN). International Organization for Standardization; 1994. (ISO 10444).

29. Kasdorf WE, editor. The Columbia guide to digital publishing. Columbia University Press; 2003.

30. Dublin Core. c1995–2023; [accessed 2023 Feb 6]. https://www.dublincore.org/

31. Hillman D, creator. Using Dublin Core—the elements. Dublin Core Metadata Initiative; (revise 1995). https://dublincore.org/documents/usageguide/elements.shtml

32. ONIX for books. EDItEUR; [accessed 2023 Feb 6]. https://www.editeur.org/11/Books/

33. National Information Standards Organization. Understanding metadata. NISO Press; c2004.

34. Brand A, Daly F, Meyers B. Metadata demystified: a guide for publishers. NISO Press and Sheridan Press; c2003. https://www.niso.org/sites/default/files/2017-08/Metadata_Demystified.pdf

35. Paskin N. Digital Object Identifier (DOI) System. In Encyclopedia of library and information sciences. Taylor & Francis; 2008.

29 References

Editors: Peter J. Olson, ELS; Iris Y. Lo; Jessica LaPointe; and Kelly Newton

29.1 OVERVIEW

The bibliographic description of a published work is called a "reference." References fulfill 2 essential roles in ensuring intellectual integrity in the research and publishing process[1]:

- Giving credit to individuals and organizations whose published works have contributed to the research being reported
- Providing sufficient information to identify and locate published works

This chapter explains the components of a reference, describes how these components should be determined and formatted, and indicates how they should be combined to form references to the various types of works found in scientific literature.

In scientific publications, a reference consists of 2 main components:

- a number, letter, or abbreviated description that appears in the text of a document
- a detailed description of the cited source that is part of a list of sources at the end of the document

The first component, which is called an "in-text reference," refers readers to the corresponding second component at the end of the document. This list of end references has various headings, such as "Cited References," "References," "Reference List," and "Literature Cited." The term "Cited References" is used for the heading for end references in the chapters of this style manual.

This chapter provides information on the 3 main reference systems used in scientific literature to create in-text references. Although many methods exist for formatting end references (see the Additional Resources at the end of this guide for a representative list), this manual prefers the method outlined in *Citing Medicine: The NLM Style Guide for Authors, Editors, and Publishers*.[2] The recommendations in that guide, published by the National Library of Medicine in the United States, are based on principles set forth by the US National Information Standards Organization[1] and the International Organization for Standardization (ISO).[3]

Another critical recommendation of this chapter is to romanize names and other words in references written in non-Roman alphabets, such as Arabic, Chinese, Cyrillic, Greek, Hebrew, Japanese, and Korean. A form of transliteration, "romanization" means to represent in the Roman alphabet the letters or characters of another alphabet. Romanization does not translate references into English. Many systems of romanization exist. The ISO has a series of standards[4-11] covering various languages. Another frequently used authority is the *ALA-LC Romanization Tables*,[12] produced by a joint effort of the US Library of Congress and the American Library Association. In addition to recommending romanization, this style manual recommends translating non-English titles and other reference components wherever possible and placing those translations in square brackets.

29.2 IN-TEXT REFERENCES

The 3 major methods for referencing published works are called the "citation–sequence system," the "name–year system," and the "citation–name system." Each of these systems is described in the following sections. Although all 3 systems are widely used and accepted, this style manual recommends the citation–sequence system, and that system is used predominantly in the examples throughout this manual.

29.2.1 Systems of In-Text References

29.2.1.1 Citation–Sequence

In the citation–sequence system, numbers—or occasionally letters—are used within the text to refer to the end references. Unless a publication's instructions to authors require otherwise, format these numbers in superscript instead of enclosing them in parentheses. This will prevent confusion between in-text references and parenthetical numbers. End references, in turn, are numbered in the sequence in which they are first cited within the text. When the same references are used later in the text, the same numbers are repeated for the in-text references. For example, if a document by microbiologist Zinaida Vissarionovna Yermolyeva is the first one mentioned in the text, its corresponding end reference will be No. 1 in the list of references at the end of the document. Wherever else that document is cited in the text, "1" will be used in a superscript to reference it.

In-text reference:
Black women have worse outcomes across reproductive health measures compared with White women in the United States. These health inequities can be attributed to structural racism.[1]

End reference:
1. Alson JG, Robinson WR, Pittman L, Doll KM. Incorporating measures of structural racism into population studies of reproductive health in the United States: a narrative review. Health Equity. 2021;5(1):49–58. https://doi.org/10.1089/heq.2020.0081

If a statement in a document is supported by more than one reference, the numbers are listed together in the same in-text reference. For only 2 cited references, separate their numbers with a comma with no space. For 3 or more numbers in a continuous sequence, use only the first and last numbers, and connect them with an en dash, which can be inserted into text using Unicode 2013. If the numbers are not in a continuous sequence, separate them in the superscript using commas with no spaces. If some of the numbers are in a continuous sequence and some are not, use an en dash to connect the first and last numbers of each continuous sequence of 3 or more, and use commas to separate all the other reference numbers.

> Modern scientific nomenclature really began with Carl Linnaeus in botany,[1] but other disciplines[2,3] were not many years behind in developing various systems[4-7] for nomenclature and symbolization.
> . . . have been shown[1,2,5,7,11-15] to abrogate the requirements of T cells . . .

29.2.1.2 Name–Year

In the name–year system, in-text references consist of the surnames of up to 2 authors and the year of publication of the document. Enclose the surnames and year in parentheses.

> **In-text reference:**
> Randomized clinical trials of HIV vaccine candidates are ongoing (Fauci and Lane 2020).

> **End reference:**
> Fauci AS, Lane HC. 2020. Four decades of HIV/AIDS—much accomplished, much to do. N Engl J Med. 383(1):1–4. https://doi.org/10.1056/NEJMp1916753

The following sections explain how to vary the name–year system to accommodate such scenarios as multiple works written by the same author, authors with identical surnames, multiple authors, corporations and other organizations serving as authors, works without identifiable authors, and works whose citations contain multiple dates.

29.2.1.2.1 MULTIPLE WORKS PUBLISHED BY THE SAME AUTHOR IN DIFFERENT YEARS

Distinguish works by the same author published in different years by placing the years after the author's surname in chronological sequence in the in-text reference.

> **In-text reference:**
> Although HIV prevention measures continue to evolve (Saag 2015, 2021) . . .

> **End references:**
> Saag MS. 2021. HIV infection—screening, diagnosis, and treatment. N Engl J Med. 384(22):2131–2143. https://doi.org/10.1056/NEJMcp1915826
> Saag MS. 2015. Preventing HIV in women—still trying to find their VOICE [editorial]. N Engl J Med. 372(6):564–566. https://doi.org/10.1056/NEJMe1415750

29.2.1.2.2 MULTIPLE WORKS PUBLISHED BY THE SAME AUTHOR IN THE SAME YEAR

For 2 or more works published by the same author or authors in the same year, add alphabetic designators to the year in both the in-text reference and the end references. This method can be used for works with up to the same 5 authors listed in the end references in the same order. It can also be used for works by 6 or more authors with the same first-listed author (followed by "et al"), whether or not all the authors are the same (see also Section 29.2.1.2.4, "Two or More Authors").

In-text reference:
Changes in the body during long-duration spaceflight (Lee et al 2020a, 2020b) can lead to . . .

End references:
Lee SMC et al. 2020a. Arterial structure and function during and after long-duration spaceflight.
   J Appl Physiol. 129(1):108–123. https://doi.org/10.1152/japplphysiol.00550.2019
Lee SMC et al. 2020b. Efficacy of gradient compression garments in the hours after long-duration
   spaceflight. Front Physiol. 11:784. https://doi.org/10.3389/fphys.2020.00784

The sequence of the alphabetic designations is preferably determined by the sequence of publication of the works, running from earliest to latest. For example, an article published in January 2022 would be designated "2022a," and one published in February 2022 would be designated "2022b." If a temporal sequence cannot be determined, list the references in alphabetical order by article title.

29.2.1.2.3 AUTHORS WITH IDENTICAL SURNAMES

When different authors of 2 or more works published in the same year have identical surnames, include their initials in all in-text references to those works to differentiate the authors. Note that in-text references to multiple works by different authors are separated by a semicolon rather than a comma.

In-text reference:
Previous research on improving evapotranspiration measurement (Wang D et al 2019; Wang T
   et al 2019) has shown . . .

End references:
Wang D et al. 2019. Improving meteorological input for surface energy balance system utilizing
   mesoscale Weather Research and Forecasting Model for estimating daily actual evapotranspiration. Water. 12(1):9. https://doi.org/10.3390/w12010009
Wang T et al. 2019. An improved spatio-temporal adaptive data fusion algorithm for evapotranspiration mapping. Remote Sens. 11(7):761. https://doi.org/10.3390/rs11070761

29.2.1.2.4 TWO OR MORE AUTHORS

If a work has 2 authors, include both names in the in-text reference, separated by "and." For the end reference, however, separate the names with a comma.

In-text reference:
. . . a recent study of environmental and cultural factors affecting tool use among chimpanzees
   (Pascual-Garrido and Almeida-Warren 2021) has shown . . .

End reference:
Pascual-Garrido A, Almeida-Warren K. 2021. Archaeology of the perishable: ecological constraints and cultural variants in chimpanzee termite fishing. Curr Anthropol. 62(3):333–362.
   https://doi.org/10.1086/713766

If both authors have the same surname, add their initials to the in-text reference.

(Smith T and Smith UV 2019)

If a reference has 3 or more authors, the in-text reference should give only the first author's name followed by "et al" and the publication year. In contrast, the corresponding end reference should continue to list up to 5 authors' names. For references with 6 or more authors, use the first author's name and "et al."

. . . but later studies (Kawaguchi et al 2021) established that . . .

If the first authors' names and the years of publication are identical for several references, include enough coauthor names in the in-text references to eliminate ambiguity (see also Section 29.2.1.2.2, "Multiple Works Published by the Same Author in the Same Year").

> (Bleeker E, de Groot J, et al 2019) *to distinguish from* (Bleeker E, Dillen W, et al 2019)
> (Kawali A, Mahendradas P, Khanum A, et al 2021) *to distinguish from* (Kawali A, Mahendradas P, Sanjay S, et al 2021)

29.2.1.2.5 ORGANIZATIONS AS AUTHORS

If the author of a reference is a university, committee, corporation, group, or other organization, a shortened version of the name should be created for the in-text reference to avoid interrupting the text with a long string of words. If the organization does not have an established abbreviation to use for this purpose, create an initialism from the first letter of each major word in the name. To clearly connect the abbreviation in the in-text reference with the corresponding end reference, place the abbreviation within square brackets before the full name of the organization in the end reference. See Section 29.3.3.1.2, "Organizations as Authors," for more details.

> **In-text reference:**
> . . . a report on the development and adoption of nonradioisotopic technologies (NASEM 2021) recommended . . .

> **End reference:**
> [NASEM] National Academies of Sciences, Engineering, and Medicine (US). 2021. Radioactive sources: applications and alternative technologies. The National Academies Press. https://doi .org/10.17226/26121

However, if the document has few in-text references, full organizational names are acceptable. In such cases, the corresponding end references omit the organizations' abbreviations.

> **In-text reference:**
> . . . a report on the development and adoption of nonradioisotopic technologies (National Academies of Sciences, Engineering, and Medicine 2021) recommended . . .

> **End reference:**
> National Academies of Sciences, Engineering, and Medicine (US). 2021. Radioactive sources: applications and alternative technologies. The National Academies Press. https://doi.org/10 .17226/26121

29.2.1.2.6 WORKS WITHOUT IDENTIFIABLE AUTHORS

Do not use the term "anonymous" for works whose authors cannot be determined (see Section 29.3.3.1, "Author," for where to find author information). Instead, begin the in-text reference with the first word or first few words of the title, followed by an ellipsis and the year of publication. Use only as many words of the title as are needed to distinguish it from other titles being used as references.

> **In-text reference:**
> Drug dosage recommendations for elderly patients (Handbook . . . c2020) depart from . . .

> **End reference:**
> Handbook of geriatric drug therapy. c2020. Springhouse.

29.2.1.2.7 WORKS WHOSE CITATIONS CONTAIN MULTIPLE DATES

Some works have more than one date. This occurs with journals whose volumes span calendar years; with books published in several volumes over time; and with electronic documents that have a date of publication, a date of copyright, and a date of modification.

For in-text references referring to publications with a range of dates, list the first and last years of publication, separated by an en dash, as in "2019–2022." For publications with more than one date, include only one of these dates in the in-text reference in the following order of preference: (1) date of publication; (2) date of copyright; and (3) date of modification, update, or revision. In the end reference, however, include the date of modification, update, or revision in addition to the publication or the copyright date.

> **In-text references:**
> (Allen 1874–1879)
> (Hansson 2008)
> (Morris c1992)
>
> **End references:**
> Allen TF, editor. 1874–1879. The encyclopedia of pure materia medica. Boericke & Tafel. 10 vol.
> Hansson SO. 2008. Science and pseudo-science. The Stanford Encyclopedia of Philosophy; [revised 2021 May 20]. https://plato.stanford.edu/archives/fall2021/entries/pseudo-science/
> Morris C, editor. c1992. Academic Press dictionary of science and technology. Academic Press.

Because dates of modification, update, and revision are within square brackets in end references (see Sections 29.3.3.6, "Date," and 29.4.11, "Websites"), those dates also appear within square brackets if they are used in the in-text references. For example, if the encyclopedia article by Sven Ove, PhD, in the second example above were to include only a date of revision, it would be cited in the text as "(Hansson [revised 2021 May 20])."

29.2.1.2.8 WORKS WHOSE DATES CANNOT BE DETERMINED

On rare occasions, publications do not have any date. This occurs most frequently with older works. For in-text references for such resources, place the words "date unknown" within square brackets. Alternatively, an estimated date may be used, followed by a question mark (see also Sections 29.3.3.6, "Date," and 29.3.3.5, "Publisher").

> **In-text reference:**
> An early study by a surgeon in the US Army (Tripler [date unknown]) . . .
> *or*
> An early study by a surgeon in the US Army (Tripler [1857?]) . . .
>
> **End reference:**
> Tripler CS. [date unknown]. Delirium tremens: its nature and treatment. [publisher unknown].
> *or*
> Tripler CS. [1857?]. Delirium tremens: its nature and treatment. [publisher unknown].

29.2.1.2.9 ACCOMMODATIONS FOR LOCATION OF IN-TEXT REFERENCES

Additional variations in the name–year system are needed when an in-text reference is cited near an author's name and when multiple in-text references are cited at the same point in the text.

If the work of an author or author group is cited close to a mention of the author's name in the text and there is no uncertainty as to the author's identity, the in-text reference may be limited to the publication year, and the author's name can be omitted.

Licciardone and Aryal's research (2021) assessed longitudinal changes in 3 common measures . . .

When Chen's studies (2017, 2018a, 2018b, 2021) are examined closely . . .

When several in-text references are cited at the same point, list the references in chronological sequence from earliest to latest, separated by semicolons. Works published in the same year should be sequenced by month, but the months should not be included in the references. If the month information is not available, order the references alphabetically by author names.

The main contributors (Iaboni and Fischer 2017; Iaboni and Rapoport 2017; Beaumont et al 2018; Maust et al 2018; Zhang et al 2018; Hsu 2020; Ailabouni et al 2021) established beyond a doubt that . . .

29.2.1.3 Citation–Name

In the citation–name system, the in-text references are numbered, but the end references are sequenced first alphabetically by author and then by title (see Section 29.3.2, "Sequence of End References," for details). The numbers assigned to the alphabetically ordered end references are used for the in-text references regardless of the sequence in which references appear in the text of the work. For example, although the first work cited in the end references might be a work written by physicist John Bertram Adams, FRS, and the 56th, a work by oncologist Christoph Zielinski, MD, reference No. 56 could appear in the text before reference No. 1 if Dr Zielinski's work is cited before Adams's.

In-text references:

One study reported that osteopathic manipulative treatment was a significant mediator for patients with low back pain,[2] while another report suggested that osteopathic manipulation has the potential to benefit patients who are lactating.[1]

End references:

1. Conaway E, O'Donnell A. Osteopathic considerations for breastfeeding women. J Osteopath Med. 2021;121(10):805–811. https://doi.org/10.1515/jom-2021-0069
2. Licciardone J, Aryal S. Patient-centered care or osteopathic manipulative treatment as mediators of clinical outcomes in patients with chronic low back pain. J Osteopath Med. 2021;121(10):795–804. https://doi.org/10.1515/jom-2021-0113

When several in-text references occur at the same point, place their corresponding end reference numbers in numeric order. Separate in-text reference numbers that are not in a continuous numeric sequence by commas with no spaces. Similarly, if there are only 2 consecutive numbers, separate them with a comma. However, for 3 or more numbers in a continuous sequence, list only the first and last numbers, and connect them with an en dash by using Unicode 2013.

In-text references:

. . . have been shown[3,8–10,17,33] to abrogate the requirements of T cells . . .

End references:

3. Czyminski et al. . . .
8. Elling et al. . . .
9. Jiang et al. . . .
10. Lissert et al. . . .
17. Mizuki et al. . . .
33. Peterson et al. . . .

As with reference numbers in the citation–sequence system, in-text reference numbers in the citation–name system should be placed in superscript if possible, instead of enclosing them in parentheses.

29.2.2 Advantages and Disadvantages of the Systems

29.2.2.1 Citation–Sequence System

There are 2 main advantages to the citation–sequence system. First, because in-text references consist of numbers only, they save space and only minimally interrupt the flow of the text. This advantage is especially significant when long sequences of in-text references are provided, such as in review articles. Second, in-text and end references in this system follow standard bibliographic format, so little decision-making is required. In contrast, in the name–year system, a series of rules must be applied.

The citation–sequence system has 3 disadvantages. First, readers are unable to tell from the in-text reference numbers whether they are familiar with the subjects and authors of the cited works, so readers must refer to end references if they want to obtain this information. Second, because in-text references are numbered in the order in which they first appear within the work, the reference numbers change as the manuscript changes, which may result in misnumbering. Third, because the end references are ordered by their appearance within the text, the works of a specific author are not grouped together, so they cannot be located easily.

29.2.2.2 Name–Year System

The first advantage of the name–year system is that it is easier to add and delete references from a manuscript because doing so does not necessitate any renumbering. Second, readers may be able to recognize authors in the text without having to refer to the end references. Third, readers may be able to ascertain some useful context from the dates provided with the authors' surnames. Finally, because the end references are arranged alphabetically by the first authors, it is easy to locate works by specific authors.

The most significant disadvantage of the name–year system relates to the numerous rules that must be followed to properly format in-text references. Applying rules for multiple authors, works published by the same author in the same year, and works written by organizations can become burdensome. Finally, long in-text references in the name–year format can interrupt the flow of text.

29.2.2.3 Citation–Name System

The citation–name system has 2 advantages. First, it shares with the citation–sequence system the advantage that in-text superscript references minimally interrupt the flow of the text. Second, it shares with the name–year system the advantage that end references are listed in alphabetical order by the names of first authors, making identifying the works by specific authors easier because those works are grouped together.

The citation–name system has 2 disadvantages, both of which are shared with the citation–sequence system: Authors cannot be easily identified from the in-text references, and references may require renumbering as the manuscript is revised.

29.2.3 Rationale for Using the Citation–Sequence System within This Manual

Although all 3 of the systems for generating in-text references are accepted, the citation–sequence system is used within this manual. For the purposes of this manual, the advantages of using superscript numbers for in-text references so that the text is not interrupted and showing end references in standard bibliographic format outweigh the disadvantages.

Most of the examples of end references in the remainder of this chapter are formatted in the citation–sequence system, as are the end references in all the chapters of this style manual. These references can be converted easily to either of the other systems.

- To convert in-text references from the citation–sequence system to the name–year system, follow the rules described in Section 29.2.1.2, "Name–Year."
- To convert end references from the citation–sequence system to the name–year system, move the date information to immediately follow the authors' names. In addition, for books and similarly cited resources, change the semicolon after the publisher to a period. For details, see Section 29.2.1.2, "Name–Year."
- To convert to the citation–name system, reorder the end references alphabetically by the first authors' surnames. The format for end references remains the same. See Section 29.2.1.3, "Citation–Name."

29.2.4 Placement of In-Text References

This section discusses the location of in-text references. The recommendations in this chapter do not specify type styles for in-text references but simply indicate what information they should include. Some publications may choose to use italic or bold type to help distinguish in-text references for readers. However, too many variations in type styles may confuse more than help.

To avoid ambiguity about what is being referenced, an in-text reference should immediately follow the title, word, or phrase to which it is directly relevant, rather than appearing at the end of long clauses or sentences. As a consequence, many in-text references will be located at the beginning or in the middle of sentences. With regard to punctuation adjacent to in-text references, place any superscript numbers immediately after commas and ending punctuation but before colons, semicolons, and dashes.

For additional information about handling in-text references, such as multiple in-text references at one point and multiple in-text references by the same author, see Sections 29.2.1.1, "Citation–Sequence," 29.2.1.2, "Name–Year," and 29.2.1.3, "Citation–Name."

In-text references in a table are usually most appropriately put in the table's footnotes rather than in the table's field (see Section 30.1.1.8, "Footnotes"). If either the citation–sequence or citation–name system is used, superscript reference numbers in the field of a table could be misinterpreted as exponents. Consequently, superscript letters should be used in the table's field to refer to footnotes, which in turn cite in-text reference numbers.

Table footnotes with in-text references:
[a] Based on Canada's Copyright Act,[6] RSC 1985, c C-42, as amended, ss 6-12.
[b] Based on the United Kingdom's Copyright, Designs and Patents Act 1988[7] (c 48).
[c] Based on Australia's Copyright Act 1968, Act No. 63 of 1968[9] as amended, s 34, s 95, and s 96.

Similarly, do not insert in-text references within graphs, illustrations, and other figures. If such references are needed to support the data or methods, put them in figure captions.

In the citation–sequence system, an in-text reference that appears in a table footnote or figure caption but not in the text should be numbered according to the physical location of the table or figure in relation to the text (eg, in-text references before a figure would have lower superscript numbers than those in the figure's caption, and those after the figure would have higher numbers). In the name–year and citation–name systems, in-text references follow the same conventions in table footnotes and figure captions as they follow in the text.

29.3 END REFERENCES

At the end of each journal article, book chapter, or other document, list the references to sources that have been cited within the text, including those found only in tables and figures. These references should be under the heading "Cited References," "References," "Reference List," or "Literature Cited." References not cited in the text that are provided for additional reading or other purposes should be listed alphabetically by first-listed author under a separate heading (eg, "Additional References," "Additional Reading," or "Supplemental References"). In this style manual, end references are titled "Cited References," and "Additional References" is used for recommended reading.

29.3.1 General Principles

Authors should never place in an end reference list a document that they have not seen. Reading a summary of a cited document somewhere else is not sufficient grounds for citing the reference in a new work. The practice of citing documents only on the basis of information from other documents has led to perpetuating many erroneous references in the literature. For each document cited, authors should be able to point to a public archive that contains the document, or they should have in their possession a copy of the cited document. When it is not possible for authors to read an original document, the source of the information that the authors relied on should be cited, not the original document.

Virtually all publishers accept references to journal articles, books, book chapters, conference proceedings, conference papers, technical reports, dissertations, theses, and websites. Because other types of publications may be less accessible, whether they can be included in an end reference list is left to individual publishers and their editors. The types of references that may not be approved by publishers include the following:

- documents that have been accepted for publication but have not yet been published
- papers or abstracts of papers presented at meetings for which full articles were never published
- written personal communications
- documented personal conversations
- in-house documents, such as memoranda
- "trade" documents, such as manufacturers' catalogs and instruction manuals

Sections 29.4.1, "Journals," and 29.4.2, "Books," provide detailed information on the standard reference formats for journal articles and books. The rest of Section 29.4, "Examples of Citation–Sequence End References for Specific Types of Sources," provides information on the components of end references for other types of documents.

Some end reference components are required (eg, the title), whereas other components are optional (eg, the total number of pages in a book and the month of publication of a journal article). Although optional components may provide additional useful information to a reader, they are not necessary to identify or find a document.

Square brackets may be used in end references for such purposes as delineating material that is not included in a document but has been added to the reference.

29.3.2 Sequence of End References

In the citation–sequence system (see Section 29.2.1.1, "Citation–Sequence"), list and number end references in the order in which each referenced document is first cited in the text. In the citation–name system (see Section 29.2.1.3, "Citation–Name"), place end references in alphabetical order by author, and number the references in that same sequence. These numbers should be used for the in-text references regardless of where in the text they are first cited. For the name–year system (see Section 29.2.1.2, "Name–Year"), place end references in alphabetical order by author.

When using the citation–name and name–year systems, alphabetical sequencing is determined by the first author's surname (ie, family name). Should the surname of the first author be the same for 2 or more documents, the ordering is determined by the initials of the first author and, if necessary, the beginning letters of the surnames of the documents' subsequent authors. The following additional principles apply to alphabetizing author names.

(1) Determine whether to treat particles such as "de," "della," "la," "Van," and "zur" as part of surnames based on the authors' preferences (see Section 8.1.2, "Conventions for Authors' Names"). When the authors' preferences cannot be determined, follow the recommendations in Table 8.1. When particles are part of surnames, alphabetize according to the particles.

Al-Ghafiqi MIA	de Luzzi M	ten Asbroek ALMA
Carter RM	Kapany NS	van de Kamp P

(2) Alphabetize initial elements of surnames as they are written, not in terms of the full name that they represent. For example, do not alphabetize "St Louis" as if it were "Saint Louis" or "McGinnis" as if it were "MacGinnis." For an exception in British convention, see Section 30.3.1.3, " 'Mac,' 'Mc,' and 'M'.' "

(3) Ignore apostrophes within surnames. For example, order "M'Veagh" as if it were "Mveagh" and "A'Amar" as if it were "Aamar."

(4) Treat marked characters such as diacritics, accented letters, and ligated letters as though they were unmarked English letters.

æ treated as ae	ü treated as u	Ø treated as O	Ç treated as C

(5) When an organization serves as the author, do not use the word "the" at the beginning of the name unless it is part of the organization's official name. If the official name starts with capitalized "The," ignore the definite article in alphabetizing the name.

> American Chemical Society
> The King's College
> New Zealand Society of Soil Science
> The Royal Society of London for Improving Natural Knowledge
> Society of Interventional Radiology

(6) When a component of an organization is an author, place components in descending hierarchical order:

> University of North Carolina at Chapel Hill, Adams School of Dentistry, Division of Diagnostic Sciences

(7) In the name–year system, when an acronym or another initialism has been used for an organization as the in-text reference, list the initialism in square brackets at the beginning of the end reference but order the reference alphabetically according to the full name, not the initialism (see example in Section 29.2.1.2.5, "Organizations as Authors").

> [ACS] American Chemical Society
> [AACB] Australasian Association of Clinical Biochemists and Laboratory Medicine

In the citation–name and name–year systems, if no authors are associated with references, order the references by title based on the following principles:

(1) Ignore the articles "a," "an," and "the" at the beginning of a title when alphabetizing (eg, alphabetize "The Science Book" under "S").
(2) Alphabetize acronyms and other initialisms as if they were words, rather than the full name for which they stand. For example, "CAS Registry," in which "CAS" stands for "Chemical Abstracts Service," would be alphabetized as "Ca," not "Ch."
(3) Alphabetize a title beginning with a numeral as if the numeral were spelled out. For example, the title "10 Essential Rules for Healthy Living" would be ordered as if "10" were "Ten."

Sequence end references beginning with the same author or the same title words according to the following rules:

(1) For multiple sources by the same author, the same team of authors, or the same first author followed by "et al," order the references by title in the citation–name system but by year in ascending order in the name–year system.

> **Citation–name system:**
> Baba S et al. Multiple collisions in the East African–Antarctica Orogen: constraints from timing of metamorphism in the Filchnerfjella and Hochlinfjellet terranes in central Dronning Maud Land. J Geol. 2015;123(1):55–78. https://doi.org/10.1086/679468
> Baba S et al. SHRIMP zircon U-Pb dating of sapphirine-bearing granulite and biotite-hornblende gneiss in the Schirmacher Hills, East Antarctica: implications for neoproterozoic ultrahigh-temperature metamorphism predating the assembly of Gondwana. J Geol. 2010;118(6):621–639. https://doi.org/10.1086/656384

> **Name–year system:**
> Baba S et al. 2010. SHRIMP zircon U-Pb dating of sapphirine-bearing granulite and biotite-hornblende gneiss in the Schirmacher Hills, East Antarctica: implications for neoproterozoic ultrahigh-temperature metamorphism predating the assembly of Gondwana. J Geol. 118(6):621–639. https://doi.org/10.1086/656384
> Baba S et al. 2015. Multiple collisions in the East African–Antarctica Orogen: constraints from timing of metamorphism in the Filchnerfjella and Hochlinfjellet terranes in central Dronning Maud Land. J Geol. 123(1):55–78. https://doi.org/10.1086/679468

(2) If there are several references with the same first author, list sources written by a single author before items with coauthors, regardless of title. In such a grouping, list the multi-

author publications in alphabetical order by the second authors' surnames, regardless of the number of authors.

Citation–name system:

Kletz T. An engineer's view of human error. 3rd ed. CRC Press; 2001.

Kletz T. HAZOP and HAZAN: identifying and assessing process industry hazards. 4th ed. IChemE; 1999.

Kletz T, Amyotte P. Process plants: a handbook for inherently safer design. 2nd ed. CRC Press; 2010.

Kletz T, Amyotte P. What went wrong? case histories of process plant disasters and how they could have been avoided. 6th ed. Butterworth-Heinemann; 2019.

Kletz T, Chung P, Broomfield E, Shen-Orr C. Computer control and human error. Gulf Publishing Company; 1995.

Name–year system:

Kletz T. 1999. HAZOP and HAZAN: identifying and assessing process industry hazards. 4th ed. IChemE.

Kletz T. 2001. An engineer's view of human error. 3rd ed. CRC Press.

Kletz T, Amyotte P. 2010. Process plants: a handbook for inherently safer design. 2nd ed. CRC Press.

Kletz T, Amyotte P. 2019. What went wrong? case histories of process plant disasters and how they could have been avoided. 6th ed. Butterworth-Heinemann.

Kletz T, Chung P, Broomfield E, Shen-Orr C. 1995. Computer control and human error. Gulf Publishing Company.

(3) When organizations are listed as authors, group all items written by the same organization. Each overall organization listed without any indication of subsidiary components is considered one author. Each component of the same organization is another author. References written by an overall organization should be listed first followed by references written by the overall organization with a second organization. Next in order is the organization with parts. The more parts after the organization name, the lower the author is listed. Farther down the hierarchy are references whose authors consist of one organization with parts and another without parts. These references are followed by those in which both author organizations have parts. Use a semicolon to separate 2 organizations as authors and a comma to separate an organization from its parts.

Six sets of authors from the same 2 overall organizations:

International Union of Pure and Applied Chemistry.

International Union of Pure and Applied Chemistry; International Union of Biochemistry and Molecular Biology.

International Union of Pure and Applied Chemistry, Organic and Biomolecular Chemistry Division.

International Union of Pure and Applied Chemistry, Chemistry and the Environment Division, Subcommittee on Chemical and Biophysical Processes in the Environment.

International Union of Pure and Applied Chemistry, Chemistry and the Environment Division, Subcommittee on Chemical and Biophysical Processes in the Environment; International Union of Biochemistry and Molecular Biology.

International Union of Pure and Applied Chemistry, Chemistry and the Environment Division, Subcommittee on Chemical and Biophysical Processes in the Environment; International Union of Biochemistry and Molecular Biology, Nomenclature Committee.

(4) Apply the following hierarchy of delimiters in titles and author names when sequencing end references: period, colon, comma, space.

Titles beginning with the same word:

Aging.	Aging, society, and the life course.
Aging: mental aspects, social welfare and health.	Aging and ethnicity.

Authors:

De Forest L	Debye P	Le Pichon X	Leigh A

29.3.3 Components of End References

Journals and newspapers, also known as serials, are publications that are published at regular or stated intervals and that normally contain separate articles or papers.[1] In contrast, monographs are scholarly works on a specific subject, usually limited in scope and usually written by specialists in the topics. Monographs include textbooks, other books, treatises, technical reports, conference proceedings, dissertations, theses, bibliographies, festschrifts, and patents (see Section 28.1, "Definitions"). A major difference in the way end references are constructed for serials and monographs is that end references for serial articles do not list the publishers but end references for monographs do.

Consistent ordering of the components of end references is necessary to avoid confusion and to enable readers to locate needed information easily. The order in the chart below is prescribed by the National Information Standards Organization[1] (NISO) in the United States.

Journal and newspaper articles	Books and other monographs
Authors	Authors
Article title	Title
Content designator	Content designator
Journal or newspaper title	Medium designator
Edition	Edition
Medium designator	Secondary author
Date	Publisher
Volume	Date
Issue	Extent (total pages)
Location (article pages)	Physical description
Physical description	Series
Notes	Notes
Digital object identifier or other web page location	

Any component may be removed from this order for an entire list of end references to meet a particular need as long as all the other components remain in order. In addition, individual end references may not possess all the components in the table above. For example, not all monographs have editions, and not all are part of a series.

The information on authors, titles, and other components of end references provided in the rest of this section pertains primarily to journal articles and books. Conference publications, technical reports, patents, audiovisual materials, online documents, and other specialized types of publications have more complex rules (for details, see Section 29.4, "Examples of Citation–Sequence End References for Specific Types of Sources").

29.3.3.1 Author

A commonality among the citation–sequence, name–year, and citation–name systems is that end references begin with author information. NISO defines "author" as "a person, committee, organization, or other party responsible for the creation of the intellectual or artistic content of a work."[1]

If no author can be determined for a document, omit authorship from the end reference. Avoid using the term "anonymous" as a substitute for the author's name.[1] When the name–year system is being used and no author can be identified, begin the in-text reference with the first word or first few words of the title, followed by an ellipsis. Use only as many words as are necessary to distinguish the document from other cited documents without authors (see also Section 29.2.1.2, "Name–Year").

In-text reference:
Drug dosage recommendations for older adults (Handbook . . . c2000) depart from . . .

End reference:
Handbook of geriatric drug therapy. c2000. Springhouse Publishing.

29.3.3.1.1 PERSONAL AUTHORS

List the names of the authors in the order in which they appear in the original text. Begin with the first author's surname (ie, family name, or last name), followed by the initial letters of the author's given, or first, name and middle name. Although the number of names to include in end references varies among style manuals and other sources, this style manual recommends the following: When there are 5 or fewer authors, list all of them, including the 5th author. If there are 6 or more authors, list only the first author followed by "et al." When "et al" is used after only one author's name, do not place a comma between the name and "et al."

Separate the surname and initials of each author with a space. Separate successive author names with a comma and a space. The author list should end with a period. If the reference has editors or other secondary authors instead of primary authors (see Section 29.3.3.1.3, "Secondary Authors"), their names are followed by a comma and the type of secondary author.

Takagi Y, Harada J, Chiarugi A, Moskowitz MA.
Osunkoya AO et al.
McKee PH, Calonje E, Granter SR, editors.

Some publishers' house styles may prefer providing full first names. In this case, separate the surname from the first name with a comma, and separate successive names with semicolons.

Takagi, Yasushi; Harada, Jun; Chiarugi, Alberto; Moskowitz, Michael A.
Osunkoya, Adeboye O, et al.
McKee, Phillip H; Calonje, Eduardo; Granter, Scott R, editors.

FORMAT OF PERSONAL AUTHOR NAMES

The conventions for designating surnames vary by country and culture. For example, whether prefixes such as "von," "de," and "los" are part of surnames depends on an author's country or region of origin or the author's preference. In addition, in many Asian cultures, surnames may precede given names. If the list of authors does not indicate which part of the name is the surname, the publication's table of contents or, if applicable, its author index may. If the publication itself is not helpful, see Section 8.1, "Personal Names," and Table 8.1 for guidance on the format to use.

Other conventions for formatting author names for end references are as follows.

- Omit degrees, titles, and honors that may follow a personal name, such as "MD" and "PhD." Similarly, omit rank and honors that may precede a name, such as "Colonel" and "Sir."

- Place patronymics (eg, affixes like "Sr" and "II") after initials without punctuation. Convert patronymics in roman numerals to arabic ordinals.

  > Vincent T. DeVita Jr, MD, *becomes* DeVita VT Jr
  > John T. Evans IV, PhD, *becomes* Evans JT 4th

- Capitalize names and place spaces within names as they appear in the document being cited on the assumption that the author has approved the conventions used in the document. This is especially crucial for surnames with prefixes, which could appear in various forms.

  > Van Der Horn *or* van der Horn
  > De Wolf *or* de Wolf *or* DeWolf *or* Dewolf
  > Le Sage *or* LeSage *or* Lesage

- Omit punctuation such as periods within surnames except for hyphens and for apostrophes after particles (eg, O', D', and L').

  > Stephen J. O'Brien, PhD, *becomes* O'Brien SJ
  > Cecilia Payne-Gaposchkin, PhD, *becomes* Payne-Gaposchkin C
  > Kathryn St James, PsyD, *becomes* St James K
  > Alison St Paul, DMD, *becomes* St Paul A

- Convert given, or first, names and middle names to initials. The following rules apply:

  (1) Designate given names and middle names containing a prefix, a preposition, or other particle by the first letter only.

  > Philip D'Arcy Hart, MD, *becomes* Hart PD
  > De la Broquerie Fortier, MD, *becomes* Fortier D
  > Sir Frank Macfarlane Burnet, MD, *becomes* Burnet FM
  > W St John Patterson *becomes* Patterson WS

  (2) Disregard hyphens or en dashes joining given names.

  > May-Britt Moser, PhD, *becomes* Moser MB
  > Augustin-Louis Cauchy *becomes* Cauchy AL

  (3) Disregard traditional abbreviations of given names. Instead, use only the first letter of the abbreviation.

  > Charles Babbage, FRS, *becomes* Babbage C *not* Babbage Ch
  > Frank J. Fenner, MBBS, FRS, *becomes* Fenner FJ *not* Fenner FrJ

  (4) For transliterated names in which the original initial is represented by more than one letter, include the required letters, but capitalize only the first one.

  > Iu. O. Iakontov *becomes* Iakontov IuO
  > Astros Th. Skuladottir *becomes* Skuladottir ATh

AUTHOR ROLES IN SOME MONOGRAPHS

In references for some types of monographs, the authors' roles are noted after the authors' names. For example, authors of bibliographies are termed "compilers"; of patents, "inventors"; and of maps, "cartographers" (see Sections 29.4.6, "Bibliographies," 29.4.7, "Patents," and 29.4.9, "Maps"). In these references, authors' names are separated from their roles by a comma.

> Selden C, Kelliher R, compilers. Distance education in public health: January 1998 through October 2003: 471 citations. (bibliography).
> McClellan E, cartographer. Cholera epidemic of 1873 in the United States. (map).

AUTHORS OF PARTS OF MONOGRAPHS

When citing parts of a monograph or other book, such as chapters that have different authors, begin the end references with the authors of the parts and the titles of the parts, followed by the connective word "In" and a colon (see Section 29.4.2.10, "Parts and Contributions to Books"). Complete the references with the names of the books' editors and the other citation information for the books.

> Patel SM, Libby P. Atherosclerosis. In: Lilly LS, editor. Pathophysiology of heart disease: an introduction to cardiovascular medicine. 7th ed. Wolters Kluwer; 2020. p 118–141.

29.3.3.1.2 ORGANIZATIONS AS AUTHORS

An organization, such as a university, society, association, corporation, or governmental body, may serve as an author. However, if both a personal author and an organizational author appear on the title page of a document, use the personal author only and place the organization name in a note. An exception may be made to this rule when the organization has specific significance to the intended audience.

If a division or other part of an organization is the author of a publication, list the parts of the name in descending hierarchical order, separate the parts by commas, and place a period at the end of the complete name. This requirement may call for breaking and reordering the elements of an organizational author name as it appears on the referenced document.

> The Guidelines Subcommittee of the Endocrine Society *becomes* Endocrine Society, Guidelines Subcommittee.
> The Singular Perturbation Committee of the Chinese Mathematical Society *becomes* Chinese Mathematical Society, Singular Perturbation Committee.

The hierarchy need not include all possible components listed in a publication. Only the one most likely to be known by readers is needed. For example, although the National Cancer Institute (NCI) in the United States is hierarchically under the National Institutes of Health, which is part of the US Department of Health and Human Services, the NCI is well known in its own right. Thus, a book identified as coming from the NCI's Division of Cancer Prevention would be given the authorship designation "National Cancer Institute (US), Division of Cancer Prevention," not "Department of Health and Human Services (US), National Institutes of Health, National Cancer Institute, Division of Cancer Prevention."

A general rule is to include the word "the" at the beginning of an organizational name only when a capitalized "The" is part of the official name of the organization. For example, "the" would be omitted from "American Chemical Society" and "International Chiropractors Association" but used with "The Joint Commission" and "The University of British Columbia." Check organizations' websites to determine whether "The" is part of their official names.

In citing organizations that are national bodies, such as government agencies and professional societies, designate the country names after the organization names if the nations are not obvious from the organizations' formal names. Use the 2-letter country codes recommended by the International Organization for Standardization (ISO)[13] to indicate the nations, place the country codes within parentheses, and end the name with a period.

> National Academy of Sciences (US).
> KTH Royal Institute of Technology (SE).
> National Fire Service (GB).

Anglicize names of organizations in languages other than English. Whenever possible, follow a non-English name with a translation. Place all translations within square brackets.

> [China Ornithological Society].
> Universitatis Helsingiensis [Helsinki University].

If 2 or more organizations appear as authors, list them in the order listed in the publication, and separate them with a semicolon and a space.

> American Medical Association; The Federation of Medical Societies of Hong Kong.

When using the name–year system, use the organizations' initialisms or readily recognizable abbreviations to keep the in-text references short. Then in the end references, place the initialisms or other abbreviations within square brackets before the full names of the organizations (see Section 29.2.1.2, "Name–Year"). Alphabetize these abbreviations within the end references as if they were spelled out. For example, alphabetize "ASSA" as if it were spelled out as "Academy of Science of South Africa."

> **In-text reference in name–year system:**
> Decreasing agriculture's overall footprint is critical to achieving ecosystem balance (ASA, CSSA, SSSA 2021).
> The conference proceedings included action steps for building institutional cultures of fairness (Macy 2021).
>
> **End references in name–year system:**
> [ASA, CSSA, SSSA] American Society of Agronomy, Crop Science Society of America, Soil Science Society of America. 2021. Advancing resilient agriculture: recommendations to address climate change. Science Policy Office. https://www.agronomy.org/files/science-policy/issues/2021-acs-climate-solutions-statement.pdf
> [Macy] Josiah Macy Jr Foundation. 2021. Addressing harmful bias and eliminating discrimination in health professions learning environments: proceedings of a conference sponsored by Josiah Macy Jr Foundation in February 2020. Josiah Macy Jr Foundation. https://macyfoundation.org/assets/reports/publications/jmf_2020_monograph_web_4.19.2021.pdf

29.3.3.1.3 SECONDARY AUTHORS

A "secondary author" is "the person, committee, organization, or other party responsible for adopting, interpreting, or otherwise modifying the intellectual content of a preexisting work."[1] Editors, translators, illustrators, and producers are examples of secondary authors. The names of secondary authors are an optional component of an end reference, except when a reference does not have a personal or organizational author as its primary author.

When included in an end reference with a primary author, a secondary author's name should be placed after the title of the work or, if applicable, the edition statement. Place a comma after the secondary author's name, and then note the role or roles of the secondary author. Separate multiple secondary authors with commas if they have the same role or with semicolons if they have different roles. List secondary authors in the order in which they appear in the publication.

> Callard A. On anger. Chasman D, Cohen J, editors.
> Bécoulet A. Star power: ITER and the international quest for fusion energy. Butler E, translator.
> Alexander L. All in a drop: how Antony van Leeuwenhoek discovered an invisible world. Milden-berger V, illustrator.

If a work has no primary author but has an editor, translator, or illustrator, place the name of the editor, translator, or illustrator at the beginning of the end reference, followed by that secondary author's role.

> Gyamfi A, Williams I, editors. Big data and knowledge sharing in virtual organizations.
> Kasper D et al, editors. Harrison's principles of internal medicine. 20th ed.

If no primary or secondary author can be determined for a work, begin the end reference with the work's title. Using the term "anonymous" as a substitute for an author's name should be avoided.[1]

29.3.3.2 Title and Subtitle

Journals, monographs, and articles have titles, often referred to as "names." Specific rules for journal titles are discussed in detail in Section 29.4.1.1, "Journal Titles." Books and other monographs have one title for the entire work, and they may have individual volume, section, or chapter titles within the work. Some books may be part of a series with its own title (see Section 29.4.2.9, "Series of Books"). Just as chapter titles differentiate content within a book, article titles differentiate content within the same issue of a journal or other serial.

In both the citation–sequence and citation–name systems, the title for a journal article or a book follows the author list in the end reference. If no authors are listed in the work, the end reference begins with the title. In contrast, in the name–year system, the title follows the year of publication.

A title is a required component in all 3 reference systems. If no title is discernible, construct one from the first few words of the document, and place it within square brackets in the end reference.

29.3.3.2.1 FORMAT OF TITLES

In general, titles should appear in references just as they do in the original documents. For example, if an acronym in a title has periods, retain the periods in the end reference even though this manual generally recommends using acronyms without periods. However, there are 6 major exceptions to the general rule not to modify the words of a published title when creating an end reference.

(1) For the titles of monographs, articles, and most other works, use sentence-style capitalization regardless of how the original titles are capitalized. With sentence-style capitalization, capitalize the first word of a title and any proper nouns, proper adjectives, acronyms, and other initialisms (ie, as if the title were a sentence). When a title is divided into a title and subtitle by a colon, semicolon, or em dash, the first word in the subtitle should be lowercased unless that word is a proper noun, proper adjective, or an initialism. If the title and subtitle are separated by a period, question mark, or exclamation

point, the first word in the subtitle is capitalized as if the subtitle were a new sentence. If the title includes a Greek letter, chemical formula, or another special character that might lose its meaning if lowercased, retain the capitalization.

(2) Romanize titles that appear in a non-Roman alphabet. Regardless of what alphabet was used for the original non-English title, follow the conventions of the particular language regarding capitalization. Whenever possible, place the translated title within square brackets after the non-English version of the title.

(3) If a title contains a Greek letter or some other symbol that cannot be reproduced with the type fonts available, substitute the name for the symbol (eg, "Ω" becomes "omega").

(4) If a title is given in 2 languages because the text is bilingual (eg, as often occurs in Canadian and Brazilian scholarly works), give the English title first, followed by the non-English title in brackets.

(5) Treat the names of conferences in titles as proper nouns, applying headline-style capitalization (see Section 29.4.3, "Conference Proceedings and Conference Papers").

(6) When such terms as "final report" and "annual report" follow the titles of technical reports, treat these terms as edition statements, not as part of the titles (see Section 29.4.4, "Technical Reports").

Occasionally, a publication does not have a title. As noted in Section 29.3.3.2, "Title and Subtitle," under this unusual circumstance, construct a title from the first few words of the document, using enough words to make the title meaningful. Place the constructed title within square brackets in the end reference.

Some end references require content and medium designators, which follow titles to provide information on the nature and physical format of publications. For information on these designators, see Section 29.3.3.4, "Content and Medium Designators."

29.3.3.2.2 PUNCTUATION

Retain in the end reference whatever punctuation is in the original title. If no punctuation separates the title from a subtitle, use a colon followed by a space. End the title information with a period unless the original title ends in a question mark or exclamation point.

> Eight years to the moon: the history of the Apollo missions.
> Overlay journals, overlay reviews: has their time finally come?
> Novobiocin returns? But not as an antibiotic.
> Managing conflict of interest disclosure—where are we going?
> Welcome to science.org!

29.3.3.3 Edition

For books and other monographs, edition information "identifies a different form or version of a previously published work, such as 2nd edition or version 4.0."[1] Romanize statements in non-Roman languages, and if desired, place translations in square brackets after the non-English edition statements.

Express numbers representing editions in arabic ordinals.

> first *becomes* 1st   III *becomes* 3rd   quatrième *becomes* 4th

Place edition information after the title. Abbreviate the word "edition" and other descriptive words for print publications according to ISO standard 832.[14] As long as a reference has only one edition, it need not be identified as the "1st ed," but once later editions are published, the first edition should be identified as such when cited.

For an English-language edition statement, capitalize only the first word and any other terms that would normally be capitalized. For a non-English statement, abide by the conventions of the particular language regarding capitalization. In the end reference, follow the complete edition statement with a period.

> Eighth Edition *becomes* 8th ed   New revised edition *becomes* New rev ed
> Third American Edition *becomes* 3rd Am ed   Edizione Italiana *becomes* Ed Ital

For electronic publications, such terms as "version" are commonly used in place of "edition." While "version" can be abbreviated, less commonly used words such as "update" or "release" should not.

> Ver 4.0   Update 1.2

The edition statements for technical reports often use such terms as "final report" or "annual report" (see Section 29.4.4, "Technical Reports").

Edition statements for journals are affected by the long-held practice of treating journals differently in end references than books and other monographs (see Section 29.4.1, "Journals"). Unlike editions for monographs, which are published sequentially, journal editions are usually published simultaneously. Whereas editions for monographs may be designated as "1st," "2nd," and "3rd" editions, journal editions may be distinguished by such terms as "domestic," "international," "office," and "hospital" editions. In addition, journal editions are considered part of the journals' names.

Journal editions are abbreviated according to ISO standard 4.[15] Separate an edition abbreviation from the journal's name with a space, place the abbreviation within parentheses, and capitalize each word of the abbreviation. Do not use periods with the abbreviation, but place a period at the end of the title information in end references (see also Section 29.4.1.1, "Journal Titles").

> *The BMJ International Edition becomes* BMJ (Int Ed)
> The Chinese edition of the *European Heart Journal becomes* Eur Heart J (Chinese Ed)
> The English edition of the *International Journal of Chronic Obstructive Pulmonary Disease becomes* Int J Chron Obstruct Pulmon Dis (Engl Ed)

Like journals, newspapers may be published in different editions, and edition designations are considered part of the newspapers' names (see Section 29.4.10, "Newspapers").

29.3.3.4 Content and Medium Designators

Content and medium designators are used to provide optional information on the nature and format of a document. For journal articles, content designators indicate formats such as original research articles, meta-analyses, editorials, letters to the editor, news, and meeting abstracts. For books and other monographs, content designators include formats such as dissertations and bibliographies. For electronic documents, content designators identify formats such as databases and computer programs. Medium designators, in turn, are used to inform users that works have nonprint formats and that special equipment may be needed to access them, especially if the works are available only in older media (eg, "DVD," "videocassette," and "microfilm").

Because both content and medium designators clarify title information, place them within square brackets between the title and the period that ends the title information. Neither type of designator is capitalized, except for initialisms such as "CD-ROM" and "DVD."

Content designators:

Armann JP et al. SARS-CoV-2 IgG antibodies in adolescent students and their teachers in Saxony, Germany (SchoolCoviDD19): persistent low seroprevalence and transmission rates between May and October 2020 [preprint]. medRxiv 2020; 2020.07.16.20155143. https://doi.org/10.1101/2020.07.16.20155143

Nishimura E et al. Molecular and biological properties of insulin icodec, a new insulin analog designed to give a long half-life suitable for once-weekly dosing [abstract]. Diabetes 2020;69 (Suppl 1):236-OR. https://doi.org/10.2337/db20-236-OR

Ogola CA. The Sterkfontein western breccias: stratigraphy, fauna and artefacts [dissertation]. University of the Witwatersrand; 2009.

Medium designator:

Goldie S. A calendar of letters of Florence Nightingale [microfiche]. Oxford Microfilm Publications; 1977.

It is possible for an end reference to have both a content designator and a medium designator. In those cases, note the content designator first, and separate the 2 with a comma.

29.3.3.5 Publisher

NISO defines "publisher" as the "person, firm, or corporate body responsible for making a work available (issuing it) to the public."[1] The publisher is required in the end references for all publications except journal and newspaper articles, in which journal and newspaper titles are substituted for publisher names. If no publisher can be determined, use the words "publisher unknown" between square brackets.

In end references, the publisher information is generally placed after the title. End publisher information with a semicolon when using the citation–sequence and citation–name systems and a period when using the name–year system. When a journal or newspaper title is used in place of a publisher's name, the title is followed by a period in all 3 reference systems.

Record the publisher's name as it appears in the publication, using whatever capitalization and punctuation is found there. Omit an initial "The" if the definite article is not an official part of the publisher's name. For example, if the publisher is identified as "The Entomological Society of America," it would be listed in end references as "Entomological Society of America" because "The" is not part of the society's official name. For references to journal articles, abbreviate the journal titles using ISO rules[15] (see Section 29.4.1.1, "Journal Titles"). For newspaper articles, use the newspapers' entire titles.

Give names of publishers in their original language, provided the language uses the Roman alphabet. If desired, non-English names in the Roman alphabet may be followed with English translations of the names within square brackets. For publisher names in non-Roman alphabets, substitute romanized versions of the names, and place them inside square brackets.

Istituto Superiore di Sanità [Italian National Institute of Health]
[China Science Publishing & Media] *instead of* 科学出版社

If the name of a division or other part of an organization is included in the publisher information, list the names in hierarchical order from highest to lowest.

University of Pennsylvania, Museum of Archaeology and Anthropology
Wolters Kluwer Health, Lippincott Williams & Wilkins
EDP Sciences, Collection Rencontre

However, the hierarchy need not include all components listed in a publication if a subordinate component is as well known as or better known than the components higher in the hierarchical order. For example, although the National Weather Service (NWS) in the United States is hierarchically under the National Oceanic and Atmospheric Administration, which is under the US Department of Commerce, the NWS is well known in its own right. Consequently, the end reference for a publication published by the NWS's Climate Prediction Center could designate the publisher as "National Weather Service (US), Climate Prediction Center" instead of as "Department of Commerce (US), National Oceanic and Atmospheric Administration, National Weather Service, Climate Prediction Center."

When citing national organizations such as government agencies and professional societies with names that do not identify the nations, place the ISO 2-letter country codes[13] after the names in parentheses, just as would be done for national organizations that serve as authors (see Section 29.3.3.1.2, "Organizations as Authors").

Society for Industrial and Applied Mathematics (US)	Ministry of Earth Sciences (IN)
Royal Society of Medicine (GB)	

Do not list multiple publishers in end references. For a document with joint publishers or copublishers, use the first one listed or the one set in the largest type or in boldface. If desired, place the name of the other publisher in a note at the end of the reference (eg, "Jointly published by the Canadian Pharmacists Association").

29.3.3.6 Date

The date of publication is required for all end references. In end references for journal articles using the citation–sequence and citation–name systems, the date information follows the name of the journal. A semicolon is used to separate the date from the volume and issue numbers. If the journal does not have a volume number, a colon is used between the date and the issue number. For monographs, a period is used after the date information in end references in both the citation–sequence and citation–name systems. When using the name–year system, a period follows the date information in both journal and monograph references.

The following rules apply to determining the date to use in end references:

(1) If the document lists both publication and copyright dates, use only the date of publication, unless at least 3 years separate the 2 dates. In the latter situation, use both dates, beginning with the publication date, and separate them by a comma and a space, as in "2020, c2017." This convention alerts users that the information contained in the publication is older than the publication date implies.

(2) If the year of publication is questionable or if no year of publication can be determined, use the year of copyright, preceded by "c," as in "c2020."

(3) If neither a year of publication nor the copyright date can be found, place within square brackets the words "date unknown" or an estimated date, if one can be determined (but see item 5 below).

> al-Khwārizmī MM. Al-Kitāb al-mukhtaṣar fī ḥisāb al-jabr waʾl-muqābala [The compendious book on calculation by completion and balancing]. [date unknown]. *or* . . . [9th century].

(4) For a publication that was published over the course of multiple years, separate the first and last years with an en dash, as in "2020–2023." (The en dash can be created by using Unicode 2013.)

(5) When citing sources consulted online that include a "last modified" or similar date, use that date in addition to or instead of a date of publication or copyright. In the absence of any discernible date for an online source, use an access date instead. (Note that access dates are no longer required by this style manual for sources that list a date of publication, copyright, or modification.)

Although the year of publication is sufficient for most end references, the month and day of the month or the season are sometimes needed (eg, when citing newspaper articles and when listing a date of modification for an online source).

When the month is used, abbreviate it to the first 3 letters without a period, and place it after the year (eg, "September 2021" becomes "2021 Sep"). Separate multiple months of publication with an en dash, as in "2022 Jan–Feb" and "2022 Dec–2023 Jan." When including the day of month, place the day after the month abbreviation, as in "2022 Aug 3," and separate multiple days with an en dash, as in "2020 Mar 1–15."

When using seasons rather than months, capitalize the seasons, and do not abbreviate them (ie, "Summer," "Fall," "Winter," and "Spring"). Separate multiple seasons by an en dash, as in "2019 Fall–Winter."

29.3.3.7 Volume and Issue

Volume and issue information, if available, is required in end references for journals and monographs (see Sections 29.4.1, "Journals," and 29.4.2, "Books"). However, both volume and issue are usually omitted for newspaper articles (see Section 29.4.10, "Newspapers").

29.3.3.8 Location within a Work and Extent of a Work (Pagination)

Location is a required component of an end reference, while extent is an optional component.

The location within a work "indicates the specific point in a publication at which the item being referenced is located."[1] The pages on which a journal article resides and the pages of a book chapter are examples of location information.

The extent of a work is "a number that indicates the total physical extent or size of the work."[1] The total number of pages of a book and the length of a podcast are examples of extent information.

In this style manual, the terms "location" and "extent" are used in place of "pagination" to accommodate online and other nonprint publications that do not have traditional page numbers.

29.3.3.8.1 LOCATION WITHIN A WORK

For a journal article with continuous pagination, location consists of the first and last pages of the article, separated by an en dash. Use entire numbers rather than shortening the number for the last page by omitting the leading digits that the first and last pages have in common. If the pagination is discontinuous, as occurs when a print article is interrupted by advertisements, separate the groupings of page numbers with a comma and a space.

710–716 *not* 710–6   345–346, 348, 351–357 *not* 345–357

Place location information after the volume and issue numbers. If the reference does not have either a volume or issue number, place the location after the year of publication. Use a colon before the location, with no space between the colon and the location. A period follows the location.

> Quah BL. Overminus lens therapy for children with intermittent exotropia. JAMA Ophthalmol. 2021;139(4):476–477. https://doi.org/10.1001/jamaophthalmol.2021.0081

Some online-only journals may have a number or combination of letters and numbers called an "article number" or "citation number," which can be used for the location information.

> Okonofua JA, Saadatian K, Ocampo J, Ruiz M, Oxholm PD. A scalable empathic supervision intervention to mitigate recidivism from probation and parole. Proc Natl Acad Sci (US). 2021; 118(14):e2018036118. https://doi.org/10.1073/pnas.2018036118

Appendixes and supplements may also have locations with letters before or after page numbers. List these letters in uppercase or lowercase based on which style is used by the publications.

> Stefan N, Roden M. Diabetes and fatty liver. Exp Clin Endocrinol Diabetes. 2019;127(S 01): S93–S96. https://doi.org/10.1055/a-0984-5753

Unlike volume and issue numbers, locations that are expressed in roman numerals should be retained as roman numerals. Uppercase and lowercase the roman numerals as they appear in the publications.

> Arnaiz E, Lebrero R, de Godos I, Camargo-Valero MA, Muñoz R. Editorial: recent advances in pond and algal technologies for wastewater treatment and resource recovery. Water Sci Technol. 2020;82(6):iii. https://doi.org/10.2166/wst.2020.458
> Wardlaw JM et al. ESO guideline on covert cerebral small vessel disease. Eur Stroke J. 2021;6(2):CXI–CLXII. https://doi.org/10.1177/23969873211012132

Location information for a chapter or other part of a book or other monograph consists of the beginning and concluding pages of the chapter or part, separated by an en dash. If the chapter or part has different authors than the overall book, the part information is listed before the book information, and the location information is placed after the book's date information. The location is preceded by the letter "p" and a space, and the page numbers are followed by a period.

> Ferron G, Martinez A, Mezghani B. Complications and management of radical cytoreduction. In: Ramirez PT, Frumovitz M, Abu-Rustum NR, editors. Principles of gynecologic oncology surgery. Elsevier; 2018. p 182–191.

If the part being cited has the same authors as the entire book, place the title of the part and its location after the book information. Separate the part title from the location with a semicolon, followed by a space.

> Greenberg DA, Aminoff MJ, Simon RP. Lange clinical neurology. 11th ed. McGraw Hill; 2021. Table 1-2, Common pupillary abnormalities; p 14.

29.3.3.8.2 EXTENT OF A WORK

Although an optional component of an end reference, extent information may be useful to readers. For example, readers might consider a 10-page journal article to be substantive but not a 10-page book. For print books and other print monographs, express extent as the total number of pages, followed by a space and "p." If a book was published in more than one physical volume, provide the total number of volumes in place of the number of pages, as in "3 vol." The extent information is placed after the book's date information.

> Cowan MK, Smith H. Microbiology: a systems approach. 6th ed. McGraw Hill; 2021. 864 p.
> Wiesel SW, Albert TJ, editors. Operative techniques in orthopedic surgery. 3rd ed. Wolters Kluwer; 2022. 4 vol.

Because audiovisual materials such as podcasts, slides, DVDs, and CDs have no pagination, express their extent in terms of the total number of physical pieces and, if possible, run time. If the run time is available, units may be spelled out or abbreviated according to ISO standard 832[14] (eg, "3 CDs: 2 h, 23 min").

Many online publications lack fixed page numbers and are nonlinear, providing numerous hyperlinks. For that reason, the extent is usually omitted from end references for online publications.

29.3.3.9 Series

When a book or other monograph is part of a series or collection, it is likely to have a collective title in addition to its own title. Although series information is optional in end references, this information may aid in identifying and retrieving books in a series, especially if the series is numbered. In the citation–sequence system and the citation–name system, the series information follows the pagination information, or if pagination is omitted, it follows the date of publication. In the name–year system, the series information follows the publisher information. In all 3 systems, enclose the series information in parentheses. Provide the title of the series first, using sentence-style capitalization, and follow the title with any numeration for the series, such as volume and issue numbers. Separate the series title from the numeration by a semicolon. Conclude the series information with a period outside the closing parenthesis.

> Negm A, Elkhouly A, editors. Groundwater in Egypt's deserts. Springer; 2021. 442 p. (Springer water; vol 36). https://doi.org/10.1007/978-3-030-77622-0
> Asimellis G. Wave optics. SPIE; 2020. 396 p. (Lectures in optics; vol 3). https://doi.org/10.1117/3 .2506314

Some series have series editors who are different from the authors or editors of the volumes. If desired, the names of these authors can be listed at the beginning of the series information.

> Zhou XN, Engels D, Wang Y, editors. National Institute of Parasitic Diseases, China: 70 years and beyond. Academic Press; 2020. 458 p. (Rollinson D, Stothard JR, editors. Advances in parasitology; vol 110).

29.3.3.10 Connective Phrase

As the name implies, a "connective phrase" joins one component of an end reference to another. Connective phrases are followed by a colon. "In" is the most common connective phrase.

When citing a chapter that has different authors or editors than a book as a whole,

"In:" is used to connect information about the chapter to information about the book. "In:" is also used to connect information about papers in conference proceedings to information about the proceedings as a whole.

> Geisslitz S, Scherf K. The holy grail of ancient cereals. In: Boukid F, editor. Cereal-based foodstuffs: the backbone of Mediterranean cuisine. Springer International Publishing; 2021. p 269–301.

"Available from" and "Located at" are examples of other connective phrases. These 2 phrases precede information about the source or physical location of a publication, such as a clearinghouse or library collection.

> Available from: National Technical Information Service (US); PB91-182030.
> Located at: National Library of Medicine (US); WZ 345 P314n 1991.

29.3.3.11 Notes

"Notes" provide information about a publication that enhances an end reference or provides clarification. Although classified as "Notes," this information is not actually preceded by the term "Notes." There are only 3 circumstances in which notes are required:

- to provide a technical report's contract or report number (see Section 29.4.4, "Technical Reports")
- to provide an online document's digital object identifier or a uniform resource locator (as in many of the examples throughout this chapter)
- to indicate where to locate a hard-to-find manuscript (see Section 29.4.16.1, "Manuscripts Other than Journal Articles").

Other information contained in notes is not essential but may be of interest to readers. Examples of nonessential notes are the language of the publication if it is other than English; the name of the organization that sponsored the publication if it differs from the publisher; numbers that would facilitate acquiring the publication, such as its International Standard Serial Number (ISSN); and system requirements for electronic documents, including the computer operating system and the names and versions of any software required.

29.4 EXAMPLES OF CITATION–SEQUENCE END REFERENCES FOR SPECIFIC TYPES OF SOURCES

This section describes how to apply the citation–sequence system to end references for specific types of sources. Examples taken from actual sources illustrate the rules outlined elsewhere in this chapter. For details about a specific component in a reference, consult the applicable section in Section 29.3.3, "Components of End References."

29.4.1 Journals

This section describes many common variations for end references for journal articles. In most science disciplines, the standard is to keep journal references succinct while providing enough information to allow readers to identify and retrieve the original sources. This brevity originated in part from the need for print journals to conserve space. To this end, the name of the publisher is omitted, and while volume number and issue number are provided, the words "volume" and "number" are omitted. In addition, the date, volume, issue, and page numbers are separated by punctuation but not by spaces.

General format:
Authors. Article title. Journal title. Date;volume(issue):pagination.

Example:
Haasnoot M, Lawrence J, Magnan AK. Pathways to coastal retreat. Science. 2021;372(6548):1287–1290.

For any journal article consulted online, a uniform resource locator (URL) should be added to the end of the reference. No punctuation follows the URL. Note that this style manual no longer requires an access date in end references for online sources, provided the reference lists a date of publication, copyright, or modification.

Haasnoot M, Lawrence J, Magnan AK. Pathways to coastal retreat: the shrinking solution space for adaptation calls for long-term dynamic planning starting now. Science. 2021;372(6548): 1287–1290. https://doi.org/10.1126/science.abi6594

The URL in the example above is a digital object identifier (DOI). A DOI is a persistent URL consisting of 3 parts: "https://doi.org/," which is the home page of the International DOI Foundation; a prefix (eg, "10.1126" in the example above) assigned by a DOI registration agency, such as Crossref; and a suffix assigned by the publisher (eg, "science.abi6594" in the example above). If an article can be accessed by more than one URL, the DOI is preferred to any other URL for the end reference. If an article does not have a DOI, a URL that begins with the Hypertext Transfer Protocol Secure prefix "https://" is preferred to one that begins with the Hypertext Transfer Protocol prefix "http://."

As science journals continue to move largely or entirely online, many articles no longer have page numbers or even volume numbers. In many such journals, page numbers have been replaced with "citation numbers," also called "article numbers." These numbers often begin with the letter "e" for "electronic." Regardless of whether end references to online articles have citation numbers, DOIs or other URLs are sufficient for locating the articles.

Articles with citation numbers:
Le Merle E et al. Directional and frequency spread of surface ocean waves from SWIM measurements. J Geophys Res Oceans. 2021;126:e2021JC017220. https://doi.org/10.1029/2021JC017220
Chen Y, Zitello E, Guo R, Deng Y. The function of LncRNAs and their role in the prediction, diagnosis, and prognosis of lung cancer. Clin Transl Med. 2021;11(4):e367. https://doi.org/10.1002/ctm2.367

29.4.1.1 Journal Titles

In end references, journal titles typically consist of abbreviations of the significant words in the titles. However, do not abbreviate journal titles that consist of a single word (eg, "Nature"). To confirm the recognized abbreviations for journal titles, consult the International Organization for Standardization (ISO) at https://www.issn.org, the NLM Catalog at https://www.ncbi.nlm.nih.gov/nlmcatalog/journals/, or one of the other sources listed in Appendix 29.1. The abbreviated words should be capitalized, and periods should be omitted from the abbreviated terms. Place the journal titles in roman, not italic, type, and place a period at the end of the titles.

> *The Journal of the Anthropological Survey of India becomes* J Anthropol Surv India
> *International Journal of Cell Biology and Physiology becomes* Int J Cell Biol Physiol
> *Science remains* Science

Generally, journal titles in character-based languages (eg, Chinese and Japanese) are translated into English and abbreviated. However, never abbreviate titles that are retained in their original character-based languages.

> 日本航空宇宙学会論文集 *becomes* J Jpn Soc Aeronaut Space Sci

Journal subtitles are not considered part of the title proper and, therefore, are not included in journal abbreviations.

> *JAMA: The Journal of the American Medical Association becomes* JAMA

It is important to cite a journal name as it appeared when the cited article was published. For example, the *British Medical Journal* officially changed its title to *BMJ* in 1988 and to *The BMJ* in 2014, so articles published before 1988 should be cited under "Br Med J," not "BMJ."

Punctuation marks should be omitted from journal titles, except for parentheses to enclose an edition statement and square brackets to enclose a medium designator. If an edition statement appears in a journal, include the edition information in parentheses following the journal's title. Abbreviate the words in the edition statement according to ISO standard 4.[15]

> *The BMJ International Edition becomes* BMJ (Int Ed)
> *Indice de Especialidades Farmacéuticas Intercon. Edición para Farmacias becomes* Indice Espec
>   Farm Intercon (Ed Farm)
> *Enfermedades Infecciosas y Microbiología Clínica* (English Edition) *becomes* Enferm Infecc Micro-
>   biol Clin (Engl Ed)

A medium designator tells readers that a journal is published in a form other than print or online, such as microfilm, microfiche, or DVD.

> Aesthetic Reconstr Plast Surg [microfiche]

Some bibliographies and online databases show a place of publication after certain journal titles, such as "Oncology (Williston Park)." This is a library convention to indicate that the organization indexing the journal has 2 or more journal titles with the same name in its collection. The name of the city where the journal is published is used to distinguish the titles. Like the journal name, the place of publication is usually shortened, as in "J Surg Oncol (Tallinn)" for "*Journal of Surgical Oncology* (Tallinn, Estonia)."

In some circumstances, it may be desirable to list a journal as a whole in end references. Unlike references for individual articles, a reference for an entire journal lists the full name of the journal in unabbreviated format and includes any subtitle the journal's title might have. In addition, the journal's publisher is included if the publisher's name is not part of the journal's name. When the reference is to multiple volumes, the word "volume" is abbreviated as "Vol," and when citing a specific edition of the journal, the edition statement follows the title and is enclosed in parentheses.

Format for entire journal:
Title (Edition). Publisher. Notes.

Examples:
American Journal of Medical Quality: Official Journal of the American College of Medical Quality.
British Medical Journal (Clinical Research Edition). Published 1981 to 1988.
International Journal of Numerical Analysis and Modeling. University of Alberta.

Format for multiple volumes of same journal:
Title (Edition). Publisher. Beginning volume's number, year of first volume – ending volume's number, date of ending volume. Notes.

Examples:
American Journal of Roentgenology. American Roentgen Ray Society. Vol 1, 1913–Vol 9, 1922.
Astronomy Quarterly. Elsevier. Vol 1, 1977–Vol 8, 1991. Continues in part: Vistas in Astronomy.

29.4.1.2 Authors

29.4.1.2.1 NUMBER OF AUTHORS

List the names of up to 5 authors in end references, but if there are more than 5, list the first author only followed by "et al" (see Section 29.3.3.1, "Author").

Single author:
Wascher CAF. Heart rate as a measure of emotional arousal in evolutionary biology. Phil Trans R Soc B Biol Sci. 2021;376(1831):20200479. https://doi.org/10.1098/rstb.2020.0479

2 to 5 authors:
Dia A, Cifu AS, Shah AP. Management of patients with a patent foramen ovale with history of stroke or TIA. JAMA. 2021;325(1):81–82. https://doi.org/10.1001/jama.2020.22176
Wolf RM, Channa R, Abramoff MD, Lehmann HP. Cost-effectiveness of autonomous point-of-care diabetic retinopathy screening for pediatric patients with diabetes. JAMA Ophthalmol. 2020;138(10):1063–1069. https://doi.org/10.1001/jamaophthalmol.2020.3190

6 or more authors:
Lamprecht AL et al. Towards FAIR principles for research software. Data Sci. 2020;3(1):37–59. https://doi.org/10.3233/DS-190026
Ashraf M et al. Interaction between the distribution of diabetic retinopathy lesions and the association of optical coherence tomography angiography scans with diabetic retinopathy severity. JAMA Ophthalmol. 2020;138(12):1291–1297. https://doi.org/10.1001/jamaophthalmol.2020.4516

29.4.1.2.2 ORGANIZATION OR GROUP AS AUTHOR

See Section 29.3.3.1.2, "Organizations as Authors," for details.

National Association of School Nurses. NASN position statement: immunizations. NASN Sch Nurse. 2011;26(2):121–122. https://doi.org/10.1177/1942602X10398003
French Radiology Society, Artificial Intelligence Group; French College of Radiology Teachers; French Radiology Community. Artificial intelligence and medical imaging 2018: French Radiology Community white paper. Diagn Interv Imaging. 2018;99(11):727–742. https://doi.org/10.1016/j.diii.2018.10.003

29.4.1.2.3 TITLE IN PLACE OF AUTHOR

If a document has neither a personal nor an organizational author, begin the end reference with the article title. Avoid using the term "anonymous" as a substitute for author names.

Covid-19: drug shortage and surgery. Anticipating the threat. Bull Acad Natl Med (FR). 2020; 204(9):e114–e115. https://doi.org/10.1016/j.banm.2020.07.045

In the name–year system, the in-text reference should consist of the first few words of the article title followed by an ellipsis and the date. In the end reference, the date would follow the article title.

> **In-text reference for name–year system:**
> In the event of a second wave of Covid-19, France's National Academy of Medicine and its National Academy of Surgery (Covid-19 . . . 2020) recommend . . .

> **End reference for name–year system:**
> Covid-19: drug shortage and surgery. Anticipating the threat. 2020. Bull Acad Natl Med (FR). 204(9):e114–e115. https://doi.org/10.1016/j.banm.2020.07.045

29.4.1.3 Article Titles

For the title of a journal article, use sentence-style capitalization. For details, see Section 29.3.3.2, "Title and Subtitle."

> Hettiarachchi IL, Meng S, Chahine M, Li X, Zhu J. Stereoselective β-mannosylation via anomeric O-alkylation with L-sugar-derived electrophiles. Eur J Org Chem. 2021;(48):6682–6687. https://doi.org/10.1002/ejoc.202100903
> Velis CA, Cook E, Cottom J. Waste management needs a data revolution—is plastic pollution an opportunity? Waste Manag Res. 2021;39(9):1113–1115. https://doi.org/10.1177/0734242X211051199

For journal articles with non-English titles, follow the conventions of the particular language regarding capitalization. Whenever possible, follow a non-English title with a translation. However, for a title in a non-Roman alphabet, use an anglicized translation in place of the original title. Place all translated titles within square brackets. List the article's original language between the date, volume, and issue notation and the DOI.

> Steffen HM. Wie beeinflussen Parasiten das Verhalten ihres Wirts? Die parasitäre Manipulationshypothese [Behavioral changes caused by parasites? The parasite manipulation hypothesis]. Dtsch Med Wochenschr. 2020;145(25):1848–1854. German. https://doi.org/10.1055/a-1220-8737
> Zhao C, Xu G, Huang S. [Preliminary experimental study on improving cloud computing process with satellite data]. Meteorol Mon. 2020;46(12):1585–1595. Chinese. https://doi.org/10.7519/j.issn.1000-0526.2020.12.006
> Fujibe F, Matsumoto J, Kamahori H. [A climatological study of heavy rainfalls due to Typhoon Hagibis in 2019]. Tenki. 2020;67(10):595–607. Japanese. https://doi.org/10.24761/tenki.67.10_595

For articles in 2 languages, give the title in English followed by the non-English title in brackets. List both languages toward the end of the reference before the DOI.

> Coignard L, Martinez C, Bonnefond H, Charles R. Rethinking the prescription's comprehension: an example of care centers for Deaf people [Repenser la compréhension de l'ordonnance: l'exemple des soins aux Sourds]. Therapie. 2015;70(6):501–513. English, French. https://doi.org/10.2515/therapie/2015034

29.4.1.4 Content Designators

A content designator may be placed at the end of an article title in square brackets. For journal articles, content designators indicate such article formats as editorials, letters to the editor, news articles, and meeting abstracts. Although content designators are optional, they can provide useful information to readers.

Baldini L, Emery W. Data availability principles and practice [editorial]. J Atmos Oceanic Technol. 2020;37(12):2165. https://doi.org/10.1175/JTECH-D-20-0163.1

Sandgathe S et al. Exploring the need for reliable decadal prediction [meeting summary]. Bull Am Meteorol Soc. 2020;101(2):E141–E145. https://doi.org/10.1175/BAMS-D-19-0248.1

29.4.1.5 Dates of Journal Articles

The year of publication is required for all journal article references. In the citation–sequence and citation–name systems, the year follows the journal title, and semicolon separates the date from the volume and issue numbers. In the name–year system, the year follows the author names, and a period follows the date (see Section 29.2.1.2, "Name–Year").

End reference for citation–sequence and citation–name systems:
Gorthi RS, Kamel G, Dhindsa S, Nayak RP. COVID-19 presenting with diabetic ketoacidosis: a case series. AACE Clin Case Rep. 2021;7(1):6–9. https://doi.org/10.1016/j.aace.2020.11.010

End reference for name–year system:
Gorthi RS, Kamel G, Dhindsa S, Nayak RP. 2021. COVID-19 presenting with diabetic ketoacidosis: a case series. AACE Clin Case Rep. 7(1):6–9. https://doi.org/10.1016/j.aace.2020.11.010

The month and the day of month or the season are not usually included in the date. If a date consists of a range, use an en dash between the 2 dates (eg, "2020–2021)." For more details, see Section 29.3.3.6, "Date."

29.4.1.6 Volumes and Issues of Journal Articles

Volume and issue numbers must be included in journal references when they exist. Write all volume and issue numbers in arabic numerals (eg, change "VI" to "6"). Ignore words and abbreviations that may precede these numbers, such as "Volume," "Vol," "Number," and "No." or their German equivalents "Band," "Bd," "Heft," and "H." Separate multiple volumes or issues by an en dash, as in "1–2." Follow volume information with the issue information in parentheses, and follow the issue information with a colon. If no issue information exists, follow the volume number with a colon. Do not use spaces within the date, volume, and issue information.

Issue with volume:
Alsaffar D, Alzoman H. Efficacy of antioxidant mouthwash in the reduction of halitosis: a randomized, double blind, controlled crossover clinical trial. J Dent Sci. 2021;16(2):621–627. https://doi.org/10.1016/j.jds.2020.10.005

Volume with no issue or other subdivision:
Kim KY, Kim BS. The effect of regional warming on the East Asian summer monsoon. Clim Dyn. 2020;54:3259–3277. https://doi.org/10.1007/s00382-020-05169-7

Wen J et al. Antibody-dependent enhancement of coronavirus. Int J Infect Dis. 2020;100:483–489. https://doi.org/10.1016/j.ijid.2020.09.015

Issue with no volume:
Janer Torrens JD. La aplicación de la cláusula derogatoria del Convenio Europeo de Derechos Humanos con motivo de la crisis sanitaria derivada del COVID 19 [The application of the derogatory clause of the European Convention on Human Rights on the occasion of the sanitary crisis arising from COVID-19]. Rev Electron Estud Int. 2020;(40). French. https://doi.org/10.17103/reei.40.03

Multiple volume numbers:
Bibi S, Alam K, Chishtie F, Bibi H, Rahman S. Temporal variation of black carbon concentration using Aethalometer observations and its relationships with meteorological variables in Karachi, Pakistan. J Atmos Sol Terr Phys. 2017;157–158:67–77. https://doi.org/10.1016/j.jastp.2017.03.017

Some journals use the year of publication as the volume and, therefore, have no separate volume number. When such a journal also has no issue numbers or other subdivisions, follow the year of publication with a colon and the page, citation, or article numbers.

No volume or issue number:
Liu L, Bai P, Liu C, Tian W, Liang K. Changes in extreme precipitation in the Mekong Basin. Adv Meteorol. 2020:8874869. https://doi.org/10.1155/2020/8874869
Watts PJ et al. How do COVID-19 inpatients in the Denver metropolitan area measure up? Adv Med. 2020:8579738. https://doi.org/10.1155/2020/8579738

If an end reference is to a supplement, part, special number, or other division, add that information within the parentheses used for the journal's issue number. Use an abbreviation such as "Suppl," "Pt," or "Spec No."

Supplements, parts, and special numbers:
A supplement to issue No. 1 *becomes* (1 Suppl)
Supplement A to issue No. 4 *becomes* (4 Suppl A)
The first stand-alone supplement in a volume *becomes* (Suppl 1)
Part 1 of issue No. 5 *becomes* (5 Pt 1)
Special No. 2 to issue No. 6 *becomes* (6 Spec No. 2)

Note that when the abbreviation "No." is used for "number," a closing period is required to avoid ambiguity with the negative statement "no" (see Section 11.1.1, "Contraction Abbreviations").

Volume with numbered supplement:
Broccoli S et al. Understanding the association between mother's education level and effectiveness of a child obesity prevention intervention: a secondary analysis of an RCT [Influenza del titolo di studio materno sull'efficacia di un intervento di prevenzione dell'obesità infantile: analisi secondaria di un RCT]. Epidemiol Prev. 44(5–6 Suppl 1):153–162. English, Italian. https://doi.org/10.19191/EP20.5-6.S1.P153.085
Reddy PH. Lifestyle and risk factors of dementia in rural West Texas. J Alzheimer's Dis. 2019;72(Suppl 1):S1–S10. https://doi.org/10.3233/JAD-191280

29.4.1.7 Pagination of Journal Articles

Page numbers follow the volume and issue information. The page numbers are separated from the issue number by a colon without any spacing.

For continuous pages, list the first page number and last page number with an en dash between them. If an article skips pages, end the first set of page numbers with the page before the break, and start another group with the page on which the article resumes. Separate the groupings of page numbers with a comma and a space.

Use entire numbers rather than shortening the numbers for end pages (eg, "120–129," not "120–9"). Do not precede page numbers with the word "page" or the abbreviation "p." Retain any letters that a journal includes with page numbers. For example, "E" may be part of page numbers for an electronic article, and "S" may be added to page numbers for an article in a supplement. However, do not add letters if they are not used in the pagination of such articles. Retain roman numerals for page numbers written as such in the source, and use uppercase or lowercase depending on what was used in the original document, as in "ix–xx" or "IX–XX." For more details, see Section 29.3.3.8.1, "Location within a Work."

> Askarimarnani SS, Kiem AS, Twomey CR. Comparing the performance of drought indicators in Australia from 1900 to 2018. Int J Climatol. 2021;41(Suppl 1):E912–E934. https://doi.org/10.1002/joc.6737

29.4.1.8 Articles with No Page Numbers

Many online journal articles lack traditional page numbers, often because the journals are only available online or the articles have not yet been published in print. For online-only articles, some journals use article numbers or citation numbers, each of which consists of a string of numbers or a string of numbers and letters. In an end reference, the article or citation number takes the place of pagination after the volume and issue numbers. If an online article does not have an article or citation number, place a period after the issue number. For an article published ahead of print, replace the date, volume, issue, and page information with a statement in square brackets that notes the article was published ahead of print and provide the date the article was posted.

> **Online-only article with article or citation number:**
> Kubryakov AA, Kozlov IE, Manucharyan GE. Large mesoscale eddies in the western Arctic Ocean from satellite altimetry measurements. J Geophys Res Oceans. 2021;126(5):e2020JC016670. https://doi.org/10.1029/2020JC016670
> Lotz J et al. Transport processes in equine oocytes and ovarian tissue during loading with cryo-protective solutions. Biochim Biophys Acta Gen Subj. 2021;1865(2):129797. https://doi.org/10.1016/j.bbagen.2020.129797
> Anenberg SC et al. Using satellites to track indicators of global air pollution and climate change impacts: lessons learned from a NASA-supported science-stakeholder collaborative. GeoHealth. 2020;4(7):e2020GH000270. https://doi.org/10.1029/2020GH000270

> **Online-only article without article or citation number:**
> Jury MR, Gaviria Pabón AR. Dispersion of smoke plumes over South America. Earth Interact. 2021;25(1). https://doi.org/10.1175/EI-D-20-0004.1

> **Article published online ahead of print:**
> Corrigan TJ Jr, Businger S. The anatomy of a series of cloud bursts that eclipsed the US rainfall record. Mon Wea Rev. [published online ahead of print 2021 Dec 30]. https://doi.org/10.1175/MWR-D-21-0028.1

29.4.1.9 Portions of Journal Articles

Figures, tables, and other portions of articles can be referenced in other journal articles. The end reference to such an element begins with the reference to the entire document and ends with the title of the figure, table, or other component, along with its pagination. For details, see Section 29.3.3.8, "Location within a Work and Extent of a Work (Pagination)."

> Liu C, Gao M, Hu Q, Brasseur GP, Carmichael GR. Stereoscopic monitoring: a promising strategy to advance diagnostic and prediction of air pollution. Bull Am Meteorol Soc. 2021;102(4):E730–E737. Figure 1, Locations of MAX-DOAS network in China; E734. https://doi.org/10.1175/BAMS-D-20-0217.1

29.4.1.10 Journal Preprints

Preprints are draft versions of manuscripts that authors share with their research communities to elicit commentary and feedback, usually by posting the manuscripts to publicly accessible preprint servers prior to peer review. Popular preprint servers include arXiv for such fields as physics, mathematics, and computer science; medRxiv for medical and related fields; and bioRxiv for life sciences. References to a preprint should be updated if the manuscript advances to the ahead-of-print and published stages. A preprint should be identified as such in a content designator after the article title. List the name of the preprint server before the year, and include the article or citation number, if one exists, and the DOI.

> Cohen I, Berkov T, Gilboa G. Total-variation mode decomposition [preprint]. arXiv 2021;2105. 10044v2. https://doi.org/10.1007/978-3-030-75549-2_5
> Lopez Bernal J et al. Effectiveness of COVID-19 vaccines against the B.1.617.2 variant [preprint]. medRxiv 2021;2021.05.22.21257658. https://doi.org/10.1101/2021.05.22.21257658

Note that a journal article preprint is not the same as an article published by a journal online ahead of print. For an example of an end reference for an article published ahead of print, see Section 29.4.1.8, "Articles with No Page Numbers."

29.4.2 Books

End references to textbooks and other standard books are covered in this section. Special types of books, such as conference proceedings, technical reports, dissertations, theses, bibliographies, and patents, are treated in later sections in this chapter.

End references for books differ from journal article references in that book references list publishers instead of journal names. Note that because many book publishers have multiple locations and because most books can be identified from the publishers' names alone, this style manual now recommends omitting publishers' locations from end references.

> **General format:**
> Authors. Title. Edition. Publisher; year of publication.

> **Example:**
> Tilley RJD. Understanding solids: the science of materials. 3rd ed. Wiley; 2021.

29.4.2.1 Authors and Secondary Authors of Books

As with the authors of journal articles, book authors are listed by their last names followed by the initials of their first and middle names. For books with 5 or fewer authors, list all authors in end references, separating each author's name with a comma and a space. For books with 6 or more authors, list only the first author, followed by "et al." For additional details, see Sections 29.3.3.1.1, "Personal Authors," and 29.3.3.1.2, "Organizations as Authors."

> **Book with personal author:**
> Zhai L. Electromagnetic compatibility of electric vehicle. Springer; 2021.
> De Doncker RW, Duco WJP, Veltman A. Advanced electrical drives: analysis, modeling, control. Springer; 2020.

Book with organization as author:
Advanced Life Support Group. Acute psychiatric emergencies: a practical approach. Wiley-Blackwell; 2020.
National Academies of Sciences, Engineering, and Medicine. High and rising mortality rates among working-age adults. The National Academies Press; 2021.

Many scientific books are credited to editors rather than to authors. When that is the case, follow the editors' names with a comma and the word "editors."

Gupta HK, editor. Encyclopedia of solid earth geophysics. 2nd ed. Springer; 2021.
Jong EC, Stevens DL, editors. Netter's infectious diseases. 2nd ed. Elsevier; 2021.
Sorino C, Feller-Kopman D, Marchetti G, editors. Pleural diseases: clinical cases and real-world discussions. Elsevier; 2021.

If a book has editors, translators, illustrators, or other contributors in addition to authors, list the additional contributors after the title in the order provided in the book. Multiple contributor types are separated by a semicolon.

Nguyen-Kim MT. Komisch, alles chemisch! [Chemistry for breakfast: the amazing science of everyday life]. Pybus S, translator; Lenkova C, illustrator. Greystone Books, 2021. German, English.

If a book has neither a personal nor an organizational author or editor, begin the end reference with the title of the book. Avoid substituting the term "anonymous" for a book's authors (see Section 29.2.1.2.6, "Works without Identifiable Authors").

29.4.2.2 Titles of Books

Use sentence-style capitalization with book titles. Otherwise, do not change the words in book titles. Place book titles in roman type, not italics. For more details, see Section 29.3.3.2.1, "Format of Titles."

Dynamics, Geometry, Number Theory: The Impact of Margulis on Modern Mathematics becomes
Dynamics, geometry, number theory: the impact of Margulis on modern mathematics

For a non-English book title, follow the capitalization rules of the language in which the book is written. Include an English translation in sentence-style capitalization in brackets. List the book's language at the end of the reference.

Stein O. Grundzüge der Nichtlinearen Optimierung [Basics of nonlinear optimization]. 2nd ed. Springer Spektrum. 2021. German.

For titles in a non-Roman alphabet, romanize the title, and include an anglicized translation between square brackets.

Gartsman BI, Shamov VV, Gubareva TS. Rechnye sistemy Dal'nego Vostoka Rossii: chetvert' veka issledovanii [River systems of Pacific Russia: a quarter century of research]. Dal'nauka; 2015. Russian.

For titles provided in 2 languages, give both titles in the order in which they appear on the book with an equals sign between the 2 versions of the title.

Adoption of an opinion on ethical aspects of human stem cell research and use = Adoption d'un avis sur les aspects éthiques de la recherche sur les cellules souches humaines et leur utilisation. Rev ed. European Group on Ethics in Science and New Technologies to the European Commission; 2001. English, French.

29.4.2.3 Medium Designators for Books

A medium designator tells readers that the authors of a manuscript consulted a book in a nonprint format that might require special equipment or software to access, such as microfilm or CD-ROM. For contemporary electronic formats, including e-books and audiobooks, medium designators are optional. Place the medium designator in square brackets after the title.

> Yergin D. The new map: energy, climate, and the clash of nations [audiobook]. Penguin Books; 2020.

For online books, include a URL, preferably a DOI, instead of a medium designator (see also Section 29.4.1, "Journals").

> Agarwal A et al, editors. Oxford handbook of clinical surgery. 5th ed. Oxford University Press; 2022. https://doi.org/10.1093/med/9780198799481.001.0001

29.4.2.4 Editions of Books

If a book has been published in more than one edition, an edition statement is a required part of its end reference. For details, see Section 29.3.3.3, "Edition."

> Yang XX. Urban remote sensing. 2nd ed. Wiley-Blackwell; 2021.
> Gershenson, DM, Lentz GM, Valea FA, Lobo RA. Comprehensive gynecology. 8th ed. Elsevier Health Sciences; 2021.

29.4.2.5 Publishers of Books

The name of the publisher is a required component of an end reference to a book, but the publisher's location should be omitted. For details, see Section 29.3.3.5, "Publisher."

> Tarancón A, Esposito V, editors. 3D printing for energy applications. Wiley; 2021.

29.4.2.6 Dates of Books

The year of publication is a required component of an end reference to a book (see Section 29.3.3.6, "Date"). Separate the year from the publisher by a semicolon. If the date is a range of years, separate the beginning year from the ending year with en dash, as in "2022–2023." If the year of publication cannot be found on the book's title page or copyright page, provide the likely year of publication between brackets, as in "[2022]." If no year of publication can be determined or estimated, use the year of copyright, as in "c2022." In some older books, it may not be possible to find or estimate a publication date or copyright date. Under that circumstance, the end reference should use "[date unknown]."

> Louis ED, Mayer SA, Noble JM, editors. Merritt's neurology. 14th ed. Wolters Kluwer; 2021.

For a book that is in press, add "Forthcoming" before the date.

> Kasper D et al. Harrison's principles of internal medicine. 22nd ed. McGraw Hill Professional. Forthcoming 2026.

29.4.2.7 Pagination of Books (Extent)

Extent, or pagination, is an optional component of a book reference, but it can provide useful information to readers (see Section 29.3.3.8.2, "Extent of a Work"). When citing an entire book, record the total number of pages of the text of the publication. Many publishers and booksellers list total page counts in online book descriptions. When that is not the case, record the total number of pages for all text, including front and back matter. When books are published in more than one physical volume, cite the total number of volumes instead of the number of pages.

Pagination should follow the year of publication, and a period should be placed after both the year and pagination information. When the total number of pages are listed, follow that number with a space and the letter "p." If the number of volumes is provided instead, place an arabic numeral before the abbreviation "vol."

> Cormick C. The science of communicating science: the ultimate guide. CSIRO Publishing; 2019. 256 p.
>
> Mayers DL, Sobel JD, Ouellette M, Kaye KS, Marchaim D, editors. Antimicrobial drug resistance. 2nd ed. Springer; 2017. 2 vol.

For books in audiovisual formats, such as a DVD and CD-ROM, express extent as the number of physical pieces, as in "1 DVD," "3 CD-ROMs," and "56 slides." Online and other e-books usually do not have pagination unless they are in PDF format. Do not provide an extent for e-books without pages. For online books that are PDF files, the extent is determined in the same manner as it is for print books. See also Section 29.4.2.3, "Medium Designators for Books."

29.4.2.8 Volume Numbers

Citing one volume of a multivolume set is treated differently than citing all the volumes in a set (see Section 29.4.2.7, "Pagination of Books (Extent)"). When citing one volume, its number follows the title of the set, not the date of publication. If the individual volume has its own title, that title follows the volume number. Write these numbers in arabic numerals (eg, "III" becomes "3"). Unlike volume numbersfor multivolume sets and journal articles, volume numbers for books are preceded by the abbreviation "Vol," as in "Vol 3." Use "Vol" even for books written in non-English languages (eg, the Spanish "Tomo" and the Dutch "Deel" should be translated to "Vol").

> Rabinovich AB, Fritz HM, Tanioka Y, Geist EL, editors. Global tsunami science: past and future. Vol 3. Birkhäuser; 2019.
>
> McSweeney PLH, Fox PF, O'Mahony JA, editors. Advanced dairy chemistry. Vol 2. Lipids. 4th ed. Springer International Publishing; 2020.

29.4.2.9 Series of Books

When a book has a collective title in addition to its own title, it is part of a series. The collective title is an optional component of an end reference. When the series title is included, it should be listed within parentheses at the end of the reference, along with its series or volume number. The series title and number should be separated by a semicolon. For more details, see Section 29.3.3.9, "Series."

> Bora PK, Nandi S, Laskar S, editors. Emerging technologies for smart cities: select proceedings of EGTET 2020. Springer; 2021. (Lecture notes in electrical engineering; vol 765).
>
> Grumezescu AM, Holban AM, editors. Quality control in the beverage industry. Academic Press; 2019. (Science of beverages; vol 17).

Note that the abbreviation "vol" follows sentence-style capitalization, so it is lower-case when not at the beginning of a reference element.

29.4.2.10 Parts and Contributions to Books

At times, it may be appropriate to cite a portion of a book rather than the book as a whole. Chapters, sections, tables, graphs, photographs, appendixes, and the like are considered "parts" of books when they are written or compiled by all of the same authors of a book. In contrast, these components are treated as "contributions" when written by someone other than the authors of the book or by not all of the authors who wrote the book.

Because an end reference should start with the individuals or organizations responsible for the intellectual content of the entire document, authors of parts are treated differently in references than are authors of contributions. The information for a part of a book follows the information for the entire book. For a contribution, the end reference begins with the information for the contribution, followed by the word "In:" and the information for the entire book. For both parts and contributions, the references end with the letter "p" and the page numbers for the cited sections. In a reference for a part, the page information is separated from the name of the part by a semicolon, while in a reference for a contribution, the page information is separated from the date of publication by a period.

Parts:
Everard M. Ecosystem services: key issues. Routledge; 2017. Chapter 8, Regenerative landscapes: reversing the cycle; p 141–169.
Gupta RK. Numerical methods: fundamentals and applications. Cambridge University Press; 2019. Table 3.4, Properties and convergence of methods; p 116.

Contributions:
Geisslitz S, Scherf K. The holy grail of ancient cereals. In: Boukid F, editor. Cereal-based foodstuffs: the backbone of Mediterranean cuisine. Springer; 2021. p 269–301.
Voter AF. High-throughput screening to identify inhibitors of SSB-protein interactions. In: Oliveira MT, editor. Single stranded DNA binding proteins. Humana; 2021. p 117–133.

29.4.3 Conference Proceedings and Conference Papers

Although not all conference proceedings are published in books and many are now published in supplemental journal issues, they do share many characteristics with books (see Section 29.4.2, "Books"). The major difference in citing conference proceedings is that they often have 2 titles: the title of the published proceedings and the name of the conference.

When both titles are used, begin with the title of the proceedings, using sentence-style capitalization. Then list the conference information, which follows a specific order and uses specific punctuation. Begin with the number of the conference in an arabic ordinal number and the title of the conference in headline-style capitalization. Follow the conference title with a semicolon, the inclusive dates of the conference in the year-month-day format, another semicolon, the location of the conference, and a period.

The location information should consist of the city and country in which the conference was held. For a conference in the United States, Canada, or Australia, add the abbreviation for state, province, or territory after the city (see Table 14.1). For reference lists in publications with primarily US, Canadian, or Australian readers, the country name "United States," "Canada," or "Australia" may be omitted, respectively.

After the location information, list the publisher of the proceedings, followed by a semicolon, the date of publication, and a period. Add a URL, preferably a DOI, if the proceedings were consulted online.

General format for entire conference proceedings:
Editors. Title of proceedings. Number and name of conference; date of conference; place of conference. Publisher; date. URL, preferably DOI

Examples:
Chaudhuri K, Salakhutdinov R, editors. Volume 97: International Conference on Machine Learning. 36th International Conference on Machine Learning; 2019 Jun 9–15; Long Beach, CA, United States. Proceedings of Machine Learning Research; 2019. https://proceedings.mlr.press/v97/
Lechevallier Y, Saporta G, editors. Proceedings of COMPSTAT 2010. 19th International Conference on Computational Statistics; 2010 Aug 22–27; Paris, France. Physica-Verlag Heidelberg; 2010. https://doi.org/10.1007/978-3-7908-2604-3

Because conference proceedings are a collection of the papers presented at a conference, they typically have editors instead of authors. For listing editors and other secondary authors in end references for conference proceedings, follow the guidelines in Section 29.4.2.1, "Authors and Secondary Authors of Books."

An end reference for an individual conference paper is similar to that for a book contribution such as a chapter (see Section 29.4.2.10, "Parts and Contributions to Books"). Begin with the authors of the paper, followed by the title of the paper and a period. Then, use the connective phrase "In:" and provide the information for the proceedings. After the publisher information, add the letter "p" and the inclusive page numbers for the location of the paper within the proceedings.

General format for a conference paper included in a proceedings:
Authors of paper. Title of paper [content designator]. In: Editors. Title of proceedings. Number and name of conference; date of conference; place of conference. Publisher; date. Location within proceedings. URL, preferably a DOI

Example:
He K, Zhang X, Ren S, Sun J. Deep residual learning for image recognition. In: Proceedings of the 29th IEEE Computer Society Conference on Computer Vision and Pattern Recognition (CVPR); 2016 Jun 26–Jul 1; Las Vegas, NV, United States. IEEE Computer Society; 2019. p 770–778. https://doi.org/10.1109/CVPR.2016.90

In addition to DOIs or other URLs, useful notes for conference proceedings and conference papers include information on the sponsorship of the conference and the language of proceedings if other than English (see Section 29.3.3.11, "Notes").

Many paper presentations and poster sessions at conferences may never be published, or they may be published sometime in the unknown future. In such cases, follow the format for conference papers, but use the connective phrase "Paper presented at:" or "Poster session presented at:" instead of "In:" to connect author and title information with conference information.

Charles L, Gordner R. Analysis of MedlinePlus en Español customer service requests. Poster session presented at: ¡Futuro magnifico! Celebrating our diversity. MLA '05: Medical Library Association Annual Meeting; 2005 May 14–19; San Antonio, TX, United States.

29.4.4 Technical Reports

A technical report is "a separately issued record of research results, research in progress, or other technical studies."[1] Most technical reports are issued by governmental agencies, usually at the federal or state level, but some originate from universities and other research institutions. Technical reports share many characteristics with books as described in Section 29.4.2, "Books." The major differences in end references for technical reports are in the information provided about authorship and about sponsorship, which can include report, contract, and grant numbers.

In citing a technical report, it is necessary to identify both the sponsoring organization (ie, the organization that funded the research) and the performing organization (ie, the organization that conducted the research). It is also necessary to specify which of those organizations published the report. In some cases, the same organization performs more than one function. Thus, technical reports may be published under the following 3 scenarios:

- written and published by the sponsoring organization
- written by the performing organization and published by the sponsoring organization
- written and published by the performing organization

General format for a technical report written and published by the sponsoring organization:
Authors. Title of report. Sponsoring organization; year of publication. Report No.: number. URL, preferably DOI

Example:
Page E, Harney JM. Health hazard evaluation report. National Institute for Occupational Safety and Health; 2001. Report No.: HETA2000-0139-2824. https://www.cdc.gov/niosh/hhe/reports /pdfs/2000-0139-2824.pdf

General format for a technical report written by the performing organization and published by the sponsoring organization:
Authors (performing organization). Title of report. Sponsoring organization; year of publication. Report No.: number. Contract or grant No.: number. URL, preferably DOI

Example:
Behrend D, Armstrong KL, Baver KD (NVI, Inc). International VLBI Service for Geodesy and Astrometry 2019+2020 biennial report. National Aeronautics and Space Administration (US); 2021. Report No.: NASA/TP-20210021389. Contract No.: NNG1HSOOC. https://ntrs.nasa.gov /citations/20210021389

General format for a technical report written and published by the performing organization:
Authors. Title of report. Performing organization; year of publication. Report No.:number. Contract or grant No.: number. Sponsored by performing organization. URL, preferably DOI

Example:
Miller R, Murray J, Kennedy T. Goldstone radar measurements of the orbital debris environment: 2018. NASA Technical Reports Server (US); 2021. Report No.: NASA/TP20210015780. Contract No.: NASA 917091. Sponsored by the National Aeronautics and Space Administration (US). https://ntrs.nasa.gov/citations/20210015780

Unlike edition statements for books, edition statements for technical reports are expressed by such terms as "Annual report," "Interim report," and "Final report." Edition statements follow report titles, just as they follow book titles in end references. Use whatever wording is provided in the report, but use sentence-style capitalization.

Place report numbers, contract numbers, and grant numbers after the year of publication. List report numbers before the other 2 numbers.

Cite a part of a technical report, such as a chapter, table, or graph, in the same manner that a book part or contribution is referenced, varying the format based on whether the report part is written by the same set of authors as the report or a different set (see Section 29.4.2.10, "Parts and Contributions to Books").

29.4.5 Dissertations and Theses

Dissertations and theses are written in support of academic degrees above the baccalaureate level. Although some European and other countries use the term "thesis" to refer to material written for a doctorate, in this style manual, "thesis" refers to work at the master's level, and "dissertation" refers to work at the doctoral level.

Dissertations and theses are cited much like books, with the name of the academic institution serving as publisher (see Section 29.4.2, "Books"). For documents consulted online, include a URL.

> **General format for a dissertation or thesis:**
> Author. Title [content designator]. Publisher; year of publication. URL
>
> **Examples:**
> Danish F. Food insecurity among South Asian immigrant communities in the Inland Empire of Southern California [dissertation]. California State University; 2019. https://scholarworks.lib.csusb.edu/etd/891
> Fleury S. Le portage de repas à domicile: enjeu et impact de ce service sur le statut nutritionnel de la personne âgée dépendante [Home delivery meal: issue and impact of this service on the nutritional status of older people living at home] [dissertation]. Université Bourgogne Franche-Comté; 2021. French. https://tel.archives-ouvertes.fr/tel-03263444
> Kayembe Kashondo JJ. Infant feeding practices in Lubumbashi, Democratic Republic of Congo [master's thesis]. University of Washington; 2020. https://hdl.handle.net/1773/46328
> Liang S. Transport properties of topological semimetals and non-symmorphic topological insulator [dissertation]. Princeton University; 2020. https://arks.princeton.edu/ark:/88435/dsp01vm40xv52n

Cite a part of a dissertation or thesis in the same manner that a book part is referenced when it is written by the author of the entire book (see Section 29.4.2.10, "Parts and Contributions to Books"). Begin the reference with the information on the dissertation or thesis, and follow that with the information about the part.

> Hamilton DJ. The lived experience of homeless individuals with type 2 diabetes mellitus [dissertation]. Kent State University; 2020. Table 1, Demographic data; p 69–70.

29.4.6 Bibliographies

"Bibliographies" are stand-alone collections of references made for a specific purpose, such as bringing together references on a specific subject or by a particular author. With 2 exceptions, the citation format for a bibliography is identical to that of a book (see Section 29.4.2, "Books"):

- Bibliography citations describe authors as "compilers."
- The title of a bibliography is followed by the content designator "[bibliography]" if that term is not used in the title.

> **General format for a bibliography:**
> Compilers. Title of bibliography [content designator]. Publisher; year of publication. URL
>
> **Example:**
> Ritchie JC, Ritchie CA, compilers. Bibliography of publications of [137]Cesium studies related to erosion and sediment deposition. US Department of Agriculture, Agricultural Research Service; 2005. https://hrsl.ba.ars.usda.gov/cesium/Cesium137bib.htm

29.4.7 Patents

A "patent" is a title of legal protection of an invention issued by a government office.[1] Patents have 2 types of authors:

- the inventor of a device, process, or other entity being patented
- the assignee, which is the individual or organization holding legal title to a patent

List the names of multiple inventors in the order in which they appear on the patent. Follow the names of the inventors with a semicolon and the name of the assignee.

After the title of the patent, place the adjective form of the country or region granting it (eg, "United States" or "Japanese") before the word "patent." Then record the patent number as it appears on the patent document, followed by a period and the date of publication. The patent number will begin with a 2-digit ISO country code.

> **General format for a patent:**
> Inventors; assignee. Title of patent. Adjective form of the country or region issuing the patent followed by country- or region-coded patent number. Full date of publication.
>
> **Examples:**
> Engelhardt TP et al, inventors; Roche Diabetes Care, Inc, assignee. Diabetes manager for glucose testing and continuous glucose monitoring. United States patent US10568511B2. 2020 Feb 25.
> Braig JR, Keenan R, Rule P, Rivas G, Seetharaman M, inventors; Optiscan Biomedical Corp, assignee. System und Verfahren zur Analyse von Flüssigkeitsbestandteilen für Glucoseüberwachung und -kontrolle [Fluid component analysis system and method for glucose monitoring and control]. European patent EP2695573B1. 2021 Apr 28.

29.4.8 Databases and Data Sets

An end reference for a database usually begins with the database's title, while a reference for a data set usually begins with the names of the authors.

Databases may have version numbers, the equivalent of edition numbers. The dates of databases may need to be expressed as a range (eg, "2001–2023"). If a database is ongoing, leave the second date blank. As with website citations (see Section 29.4.11, "Websites"), database and data set citations require a URL, preferably a DOI.

> **General format for a database:**
> Title of database. Version. Publisher; beginning date–ending date [date updated]. URL
>
> **Example:**
> The Human Protein Atlas. Version 21.0. SciLifeLab; 2003– [updated 2021 Nov 18]. https://www.proteinatlas.org/
>
> **General format for a data set:**
> Authors. Title of data set. Publisher; date. URL
>
> **Example:**
> Cerezer F, Cáceres N, Dambros C. Effect of productivity on community size explains the latitudinal diversity gradient of South American small mammals [data set]. Dryad; 2021. https://doi.org/10.5061/dryad.6hdr7sr0f

29.4.9 Maps

In this section, "maps" are those that are published as stand-alone items. For citing maps in books and atlases, see Section 29.4.2.10, "Parts and Contributions to Books." The end reference for a map begins with the name of the map's cartographer, followed by a comma and the word "cartographer." If the cartographer is unknown, begin the end reference with the map's title or description. A content designator may be added in square brackets after the title to clarify the nature of the map. Maps consulted online may not include a date of publication or revision, in which case, the access date can be used (see also Section 29.4.11, "Websites").

General format for a map:
Author, cartographer. Title of map [content designator]. Publisher; date of publication. URL if map is online

Examples:
Doyon R, Donovan T, cartographers. AIDS in Massachusetts, 1985–1991 [demographic map]. University of Massachusetts, Department of Geology and Geography; 1992.
Montrose Point Bird Sanctuary, Chicago, Illinois. Google Maps; [accessed 2022 Jan 3]. https://goo .gl/maps/GJALYGBXFZ8TSGK5A

29.4.10 Newspapers

References to newspaper articles are similar to those for journal articles (see Sections 29.4.1, "Journals," and 29.3.3.3, "Edition"). The main differences are that the dates always include day and month, not just year, and the URLs are usually not DOIs.

General format for newspaper articles consulted online:
Author. Title of article. Newspaper name (edition). Year, month, and day of publication. URL

Example:
Zimmer C. To speed vaccination, some call for delaying second shots. The New York Times. 2021 Apr 9 [updated 2021 April 27]. https://www.nytimes.com/2021/04/09/health/covid-vaccine -second-dose-delay.html

General format for newspaper articles consulted in print:
Author. Title of article. Newspaper name (edition). Year, month, and day of publication;section:page number.

Example:
Zimmer C. To speed up, some call for delaying second shots. The New York Times (New York Ed). 2021 Apr 10;Sect A:4.

29.4.11 Websites

The rules for citing websites do not differ markedly from those for many of the other types of sources discussed in this chapter. The main difference is that a URL is a required component. Additionally, any date of modification, update, or revision must be recorded in addition to or instead of a publication date. If no date can be determined, record an access date instead (see Section 29.4.9, "Maps," for an example).

General format for an article on a website:
Author. Title of article. Publisher; date of publication [date updated]. URL

Examples:
Health equity considerations and racial and ethnic minority groups. US Centers for Disease Control and Prevention, National Center for Immunization and Respiratory Diseases, Division of Viral Diseases; 2020 Apr 30 [updated 2022 Jan 25]. https://www.cdc.gov/coronavirus/2019 -ncov/community/health-equity/race-ethnicity.html
Zooming into the Sun with solar orbiter. The European Space Agency; 2022 Mar 24. https://www .esa.int/Science_Exploration/Space_Science/Solar_Orbiter/Zooming_into_the_Sun_with_Solar _Orbiter

29.4.12 Blogs

Blog posts are cited much like newspaper articles (see Section 29.4.10, "Newspapers"). Add a content designator after the title of the post if the nature of the work is not evident from either the title of the post or the title of the blog. Use sentence-style capitalization for the title of posts and headline-style capitalization for the title of blogs (see Section 29.4.1.1, "Journal Titles").

General format for a blog post:
Author's name. Title of post [blog post]. Title of blog. Published date. URL

Example:
Tao T. The inverse theorem for the U^3 Gowers uniformity norm on arbitrary finite abelian groups: Fourier-analytic and ergodic approaches [blog post]. What's New. 28 Dec 2021. https://terrytao.wordpress.com/2021/12/28/the-inverse-theorem-for-the-u3-gowers-uniformity-norm-on-arbitrary-finite-abelian-groups-fourier-analytic-and-ergodic-approaches/

29.4.13 Social Media

Social media posts are typically not cited in end references of scientific manuscripts. More reliable sources are usually available. If a post needs to be cited (eg, for a post's thread of commentary), the end reference should be modeled on the examples that follow.

General format for a Facebook post:
Username or group page name. Title, text, or partial text of post. Facebook; date of posting. URL

Example:
National Institute of Mental Health (US). Clinical research is critical to understanding and treating mental illnesses. Learn about why people participate in clinical research and what to expect during a clinical trial. http://nimh.nih.gov/clinicaltrials. #CTD2021. Facebook; 2021 May 20. https://www.facebook.com/10160825448806978

General format for a LinkedIn post:
Username or group page name. Title, text, or partial text of post. LinkedIn; date of publication. URL

Example:
World Health Organization. Local action is how we can improve our health and our climate . . . LinkedIn; 2023 Nov 12. https://www.linkedin.com/feed/update/urn:li:activity:7129546272310472704/

General format for an X (formerly Twitter) post:
Display name (username). Title or text of post. X; date of posting. URL

Example:
NASA (@NASA). Docking confirmed. @SpaceX's cargo Dragon is now at the @SpaceStation with a delivery of supplies and research . . . X; 2023 Nov 11. https://twitter.com/NASA/status/1723283017299329534

Content posted before Twitter's 2023 rebranding as X need not be updated to refer to the new name. Note, however, that "twitter" should appear in the URL for functionality.

29.4.14 Audiovisual Sources
29.4.14.1 Podcasts

Add a medium designator after the title of a podcast if it is not clear from the rest of the reference that the source is a podcast. Use sentence-style capitalization for the title of the episode and headline-style capitalization for the title of the podcast (see Section 29.4.1.1, "Journal Titles"). Although it is optional, the length of the episode may be added after the date.

General format for a podcast:
Narrator's name. Title of podcast episode [medium designator]. Title of podcast program. Producer; date. Extent. URL

Example:
Petrie B. One in a million: what you need to know about the Johnson & Johnson vaccine pause [podcast]. Petrie Dish. Texas Public Radio; 2021 Apr 19. 27:09 min. https://www.tpr.org/podcast/petrie-dish/2021-04-19/one-in-a-million-what-you-need-to-know-about-the-johnson-johnson-vaccine-pause

29.4.14.2 Videos

The following examples show how to cite a YouTube video and a TED Talk. Other types of videos may be cited similarly. See also Section 29.4.14.3, "Movies and Television Shows."

General format for a YouTube video:
Title of video [medium designator]. Title of program. Producer or publisher. YouTube; date posted. Extent. URL

Example:
Coronavirus: How to teach kids about COVID-19 [YouTube video]. BrainPOP. YouTube; 2020 Mar 23. 4:32 min. https://www.youtube.com/watch?v=GoXxmzKdick

General format for a TED Talk:
Speaker. Title of talk [medium designator]. Title of conference; date of talk; location. Extent. URL

Example:
Bianco F. How we use astrophysics to study earthbound problems [TED Talk video]. TED2019; 2019 Apr; Vancouver, BC, Canada. 5:09 min. https://www.ted.com/talks/federica_bianco_how _we_use_astrophysics_to_study_earthbound_problems

29.4.14.3 Movies and Television Shows

To cite a motion picture or television show, start with the name of the director. Add a medium designator after the title. Though optional, a length may be added after the date (see also Section 29.3.3.8.2, "Extent of a Work").

General format for a movie:
Director's name, director. Title of motion picture [motion picture]. Production company or dis-tributor; year of release. Extent.

Example:
Orlowski J, director. The social dilemma [motion picture]. Netflix; 2020. 94 min.

Episodes from television shows are usually cited as a part of a named series. This is similar to the approach for citing a chapter or part of a book (see Section 29.4.2.10, "Parts and Contributions to Books").

General format for a television episode:
Director's name, director. Title of episode [medium designator]. In: Title of show. Network; original air date.

Example:
Cendrowski M, director. The Einstein approximation [TV episode]. In: Big bang theory. CBS; 2010 Feb 1.

29.4.15 Legislative Citations

The examples of common legislative citations in this section are based on the standards followed in the United States and may not be applicable to laws, bills, and other legisla-tive documents from other countries. References to legislative documents are described in more detail in *The Bluebook: A Uniform System of Citation*.[16]

General format for a public law:
Name of act including year introduced, public law No., statute information (date of enactment).

Example:
CARES Act of 2020, Pub L No. 116-136, 134 Stat 281 (2020 Mar 27).

General format for an unenacted bill:
Name of bill including year introduced, bill number, Congress. (year bill was introduced).

Example:
Anti-Racism in Public Health Act of 2021, HR 8178, 116th Cong. (2021).

General format for the Code of US Federal Regulations:
Name of the regulation, location in the Code of US Federal Regulations (year regulation was added to the code).

Example:
Regulations for the Protection of Human Subjects, 45 CFR Sect 46.104 (2018).

General format for a hearing:
Hearing name: name of the committee or subcommittee, Congress number, session number. (full date of hearing).

Example:
Coping with compound crises: extreme weather, social injustice, and a global pandemic: hearing before the Subcomm on Environment of the House Comm on Science, Space, and Technology, 116th Cong, 2nd Sess. (2020 Sep 30).

Note that the names of laws, bills, and regulations use headline-style capitalization, while the names of hearings use sentence-style capitalization.

Because these formats are prescribed by the legislative community, they are not modified to meet the guidelines of the citation–sequence, citation–name, or name–year system. For example, the date in the title of legislation is not taken out of sequence in end references to adhere to the name–year system. However, in-text references in the name–year system are handled the same way as are books without authors. Begin the in-text reference with the first word or first few words of the title, followed by an ellipsis and the year of publication.

(CARES Act . . . 2020)   (Anti-Racism . . . 2021)

In contrast, an item in the *Federal Register* is cited the same way as a standard journal article (see Section 29.4.1, "Journals").

Findings of research misconduct. Fed Regist. 2020;85(13):3394–3395.

29.4.16 Unpublished Sources

Unpublished sources include manuscripts not distributed to an audience and personal communication, such as letters, email, and conversations. Unpublished documents that are available to scholars in an archive or depository can usually be included in a reference list. However, some publishers do not permit end references to any form of unpublished material. Note that preprints are not considered unpublished material (see Section 29.4.1.10, "Journal Preprints").

29.4.16.1 Manuscripts Other than Journal Articles

"Manuscripts" are unpublished books or other documents and may be either typed or handwritten. In end references, cite only documents that reside in public archives, such as libraries and other repositories that permit public access. The location of an unpublished manuscript should be provided as a note at the end of a reference. Both collections of documents and individual manuscripts may be cited.

General format for an individual manuscript:
Author. Title. Date. Physical description. Notes with location; manuscript's call number.

Example:
Stearns AA. Armory Square Hospital nursing diary. 1864. 70 leaves. Located at: National Library of Medicine, History of Medicine Division, Bethesda, MD, US; MS B 372.

General format for a manuscript collection:
Author. Title. Date. Physical description. Notes with location; collection's call number.

Example:
Axelrod J. The Julius Axelrod papers. 1915–1998. 22 boxes. Located at: National Library of Medicine, History of Medicine Division, Modern Manuscripts Collection, Bethesda, MD, US; MS C 494.

If an unpublished manuscript has no discernible author, begin the reference with the title instead of substituting "anonymous" for the author's name.

If a manuscript does not have a formal title, construct a title from the first few words of the text, using enough words to make the constructed title meaningful. Place the constructed title within square brackets. In addition, by convention, manuscript collections have constructed titles, such as "Collected papers of . . ." or "Papers and collection of . . ." Unlike a constructed title for a single manuscript, the constructed title for a manuscript collection should not be placed within square brackets.

For an end reference for an individual manuscript, include either a single year or, if appropriate, a span of years. For a manuscript collection, use the first and last years of the items in the collection, separated by an en dash, as in "1978–2020." If there is no date on a manuscript or a manuscript collection but a date or dates can be estimated, place the year or years within square brackets, as in "[1858]" or "[1920–1930]." If no year can be determined, use "date unknown" within square brackets.

Because viewing manuscripts and manuscript collections often requires making special arrangements with the libraries or other archives where the manuscripts are housed, it is highly recommended that the extent of the item be included in the reference. Describe the extent of a manuscript collection in such terms as the number of items within the collection, the number of boxes holding the collection, and the number of linear feet of shelf space occupied by the collection.

1,500 items: 10 boxes   200 items: 3.5 feet   5 boxes

If an individual manuscript is paginated, provide the total numbered pages, as in "15 p." If a manuscript is unpaginated, describe the number of physical pages in terms of leaves, as in "5 leaves."

In noting the location where a manuscript or manuscript collection can be found, begin with the words "Located at:" and provide the names of the institution and the department, as well as the city and country. If the institution is in the United States, Canada, or Australia, also include the state, province, or territory. Follow the location information with the call number or other finding aid that the institute has assigned to the manuscript or collection. Use commas to separate segments of the location and a semicolon to separate institution information from the institute's call number or other finding aid. End location information with a period.

Additional optional information that may be included in notes are the language of the manuscript if other than English, further descriptions of the material such as its history, and any restrictions on the use of the material by the library or other archive.

29.4.16.2 Personal Communications

This style manual recommends placing references to personal communications, such as letters and conversations, within a manuscript's running text instead of using formal end references. Describe the nature and source of the cited information within parentheses, and clearly indicate that the citation is not included in the reference list.

> . . . local and systemic reactions were expected to be stronger after the second dose (2022 letter from TD Perez to author; unreferenced, see "Notes") . . .

Unless the personal communication is accessible to the public, the authors citing the communication must provide the publisher with written permission from the cited person, if living, or from the cited organization. The permission should be acknowledged in an "Acknowledgments" or a "Notes" section at the end of a journal article or the end of a book's main text. Such statements may have additional details, such as the reason for the communication.

APPENDIX 29.1 AUTHORITATIVE SOURCES FOR JOURNAL TITLE ABBREVIATIONS

Standard ISO 4:1997[15] of the International Organization for Standardization (ISO) is the recognized standard for abbreviating words in journal titles. The database for this standard, which is maintained by the ISSN International Centre in Paris, France, is available by subscription at https://www.iso.org/standard/3569.html.

Alternative sources for journal title abbreviations are listed below. Each of these sources is considered authoritative within its specific field or discipline. Although many of these sources use the ISO standard as their basis, some do not. For example, ISO does not permit abbreviations of the anglicized versions of titles in character-based languages (eg, Chinese, Japanese, and Korean), but some of these sources do permit the translated titles to be abbreviated.

- Abbreviations of names of serials. American Mathematical Society. https://www.ams.org /msnhtml/serials.pdf
- CAS Source Index (CASSI) search tool. American Chemical Society. https://cassi.cas.org
- Embase content: list of journal titles in Embase. Elsevier. https://www.elsevier.com /solutions/embase-biomedical-research/embase-coverage-and-content
- Inspec: list of journals indexed. The Institution of Engineering and Technology. https:// www.theiet.org/publishing/inspec/
- Journal Citation Reports. Clarivate. https://jcr.clarivate.com/jcr/home
- Journals indexed in AGRICOLA. National Agricultural Library (US). https://agricola.nal .usda.gov/jia/
- NLM catalog: journals referenced in the NCBI databases. National Library of Medicine (US), National Center for Biotechnology Information. https://www.ncbi.nlm.nih.gov /nlmcatalog/journals

CITED REFERENCES

1. National Information Standards Organization (US). Bibliographic references. NISO Press; 2005 [revised 2010 May 22]. (ANSI/NISO Z39.29-2005 [R2010]). https://www.niso.org/publications/ansiniso-z3929 -2005-r2010

2. Patrias K. Citing medicine: the NLM style guide for authors, editors, and publishers. 2nd ed. National Library of Medicine (US); 2007 [updated 2015 Oct 2]. https://www.ncbi.nlm.nih.gov/books/NBK7256/

3. Information and documentation—guidelines for bibliographic references and citations to information resources. International Organization for Standardization; 2021. (ISO 690:2021). https://www.iso.org/standard/72642.html

4. Information and documentation—transliteration of Arabic characters into Latin characters. International Organization for Standardization; 1984. (ISO 233:1984). https://www.iso.org/standard/4117.html

5. Information and documentation—transliteration of Arabic characters into Latin characters—part 2: Arabic language—simplified transliteration. International Organization for Standardization; 1993. (ISO 233-2:1993). https://www.iso.org/standard/4118.html

6. Information and documentation—transliteration of Armenian characters into Latin characters. International Organization for Standardization; 1996. (ISO 9985:1996). https://www.iso.org/standard/17893.html

7. Information and documentation—transliteration of Cyrillic characters into Latin characters—Slavic and non-Slavic languages. International Organization for Standardization; 1995. (ISO 9:1995). https://www.iso.org/standard/3589.html

8. Information and documentation—transliteration of Devanagari and related Indic scripts into Latin characters. International Organization for Standardization; 2001. (ISO 15919:2001). https://www.iso.org/standard/28333.html

9. Information and documentation—transliteration of Georgian characters into Latin characters. International Organization for Standardization; 1996. (ISO 9984:1996). https://www.iso.org/standard/17892.html

10. Information and documentation—transliteration of Hebrew characters into Latin characters—part 2: simplified transliteration. International Organization for Standardization; 1994. (ISO 259-2:1994). https://www.iso.org/standard/4162.html

11. Information and documentation—transliteration of Thai. International Organization for Standardization; 1998. (ISO 11940:1998). https://www.iso.org/standard/20574.html

12. Library of Congress (US), Cataloging Policy and Support Office. ALA-LC romanization tables. Library of Congress (US); 1997 [updated 2017 Nov 28]. https://www.loc.gov/catdir/cpso/roman.html

13. The international standard for country codes and codes for their subdivisions. International Organization for Standardization; 2021. (ISO 3166). https://www.iso.org/iso-3166-country-codes.html

14. Information and documentation—bibliographic description and references—rules for the abbreviation of bibliographic terms. International Organization for Standardization; 1994. (ISO 832:1994). https://www.iso.org/standard/5195.html

15. Information and documentation—rules for the abbreviation of title words and titles of publications. International Organization for Standardization; 1997. (ISO 4:1997). https://www.iso.org/standard/3569.html

16. Columbia Law Review, Harvard Law Review, University of Pennsylvania Law Review, Yale Law Review, compilers. The bluebook: a uniform system of citation. 21st ed. Harvard Law Review Association; 2020. https://www.legalbluebook.com/

ADDITIONAL REFERENCES

Banik GM, Baysinger G, Kamat P, Pient N. The ACS guide to scholarly communication. American Chemical Society; 2020. https://doi.org/10.1021/acsguide

The Chicago manual of style: the essential guide for authors, editors, and publishers. 17th ed. The University of Chicago Press; 2017. https://www.chicagomanualofstyle.org

NLM Catalog: journals referenced in the NCBI databases. National Library of Medicine (US); 2005. https://www.ncbi.nlm.nih.gov/nlmcatalog/journals

Publication manual of the American Psychological Association. 7th ed. American Psychological Association; 2020.

Turabian KL. A manual for writers of term papers, theses, and dissertations: Chicago style for students and researchers. 9th ed. The University of Chicago Press; 2018.

30 Tables, Figures, and Indexes

Editors: Thomas A. Lang, MA, and Jessica S. Ancker, PhD, MPH

30.1 TABLES

In scientific publications, data need to be presented as accurately, completely, clearly, and concisely as the written accounts of how the data were collected, analyzed, and interpreted. Thus, tables and figures are as important as text and should be prepared with the same care and attention. As with the text of scientific publications, tables, graphs, and images should help readers understand, find, remember, and use information as quickly and as easily as possible.

Tables are to be distinguished from series of items listed in running text, short lists, and tabulations. Lists and tabulations are embedded in the text, and they rely on the preceding text to explain their meaning. Although lists and tabulations have headings and 1 or 2 columns, they are not separated from the text with rules, numbers, or titles. In contrast, a table has a number, a title, and both row and column headings. A table is an independent unit set off from the running text and placed near where it is first cited.

Series in running text:
The drug had unexpected side effects on patients: nausea, vomiting, headache, fever, fatigue, and body aches. Treatment consisted of . . .

List:
The drug had unexpected side effects on patients:

Nausea
Vomiting
Headache
Fever
Fatigue
Body aches

Treatment consisted of . . .

Tabulation:
The drug had unexpected side effects on patients:

Nausea   12%
Vomiting   10%
Headache   10%
Fever   6%
Fatigue   4%
Body aches   3%

Treatment consisted of . . .

Preparing effective tables often calls for creativity rather than rigidly applied rules. The following guidelines may be helpful, but they do not apply in every case. For the principles of table construction, see Chapter 4 of *How to Write, Publish, and Present in the Health Sciences: A Guide for Clinicians and Laboratory Researchers*.[1]

(1) Tables, including their titles, headnotes, and footnotes, should be complete enough to understand without undue reference to the text.
(2) Tables should be orderly, logical, and relatively simple.
(3) The data, units of measurement, abbreviations, and symbols in tables need to be consistent with those in the text.
(4) Entries in each cell of the data field must be consistent with both the row and column headings.

(5) Tables containing similar types of information should have parallel formats.
(6) Tabular data should only rarely be duplicated in text and figures.
(7) Tables should not be divided into parts (eg, "Table 1a" and "Table 1b").

In designing tables, follow the instructions for authors of the publication to which the manuscript will be submitted. Ideally, tables are created with the table function in a word-processing program, and each heading or data element is placed in its own cell. Tables created in regular text have to be spaced and aligned with manual tab stops and line breaks, which are easily lost in file transfers and conversions.

30.1.1 Parts of a Table

Tables have up to 9 parts: a table number, a title (not a caption, which is formatted differently), a headnote, column headings, row headings, the data field (ie, the individual data cells), expanded abbreviations, footnotes, and a credit line. Although not every table will have all of these parts, every table should have at least a title, column and row headings, and a data field (see Figure 30.1).

Table number. Table Title (Boldface with Headline-Style Capitalization). Headnote (presents information more important than that included in a footnote; written in roman type with sentence-style capitalization) (major horizontal rule)

	Spanner heading[a] (minor rule)			
	Decked heading (minor rule)			
Stub (major rule)	Column heading	Column heading	Column heading	*P* value
Stub heading[b]				
Row heading	Data field			0.XX
Row heading	(minor rule)			0.XX
"Cut-in" heading[c] (alternative to a stub heading)				
Row heading	Data field			0.XX
Row heading	(minor rule)			0.XX
Total (major rule)				

Expanded abbreviations (eg, EMA, European Medicines Agency)
Footnotes
[a] Spanner headings are sometimes called "straddle headings."
[b] Stub headings organize row headings into groups.
[c] Cut-in headings break a table into 2 or more subtables by grouping related rows. Cut-in headings are centered in the data field to emphasize the start of each subtable. Because cut-in headings and stub headings serve the same purpose, they normally do not appear in the same table.
Credit line (eg, "Data are from Brett et al[39]")

Figure 30.1. Table components and terms. The table number precedes the table title. Additional qualifying information can be included after the title in a headnote or below the data field in a footnote. Stub headings and cut-in headings organize row headings into groups. Row headings and column headings identify the variables whose values are reported in the data field. Major horizontal rules, also called "horizontal lines," should be placed below the title, below the column headings, and below the data field. Minor horizonal rules are placed directly below spanner and decked headings, as well as across the entire table above cut-in headings.

30.1.1.1 Table Number

If a document has only one table, the table is not numbered and is cited in the text as "Table." This practice is necessary to avoid confusion because citing "Table 1" implies that the document has at least 2 tables.

If a document has 2 or more tables, label them with the word "Table," and number them with consecutive arabic numerals in order of mention in the text (eg, "Table 1" and "Table 2"). In books and other long documents that are divided into chapters or sections, tables are more easily found if their numbers combine the chapter or section number with numbers indicating the order in which the tables appear in the section or chapter. For example, "Table 1.1" and "Table 1.2" would be the first 2 tables in a document's first chapter or section, whereas "Table 3.4" would be the fourth table in the third chapter or section.

The word "table" and the number should be in boldface type. The table number is followed by a period, a space, and a table title.

30.1.1.2 Table Title

Table titles must be clear and detailed enough to allow readers to understand the fundamental aspects of the tables without undue reference to the text.

Table titles begin one space after the period after table numbers, and the titles are set in boldface type.

Titles consist of a single phrase in headline-style capitalization (see Section 9.3, "Titles and Headings within Text"). Unlike table numbers, titles do not have a closing period, unless followed by a headnote. Titles should identify the data in the data field (see Section 30.1.1.6, "Data Field"). They should not simply repeat column and row headings, nor should they state the authors' interpretation of the data or the authors' conclusion.

Besides identifying the data in tables, titles may include such information as interventions and outcomes; the source of the data (eg, subjects, species, chemicals, or physical elements); dates and other key details of data collection, provided these apply to all data in the tables; and sample sizes. Alternatively, this information may be placed in headnotes, footnotes, or headings.

> **Table 12. Values of Water Quality Variables for Samples Taken from Lake Tanganyika in Africa between June and August 2022**
>
> *not*
>
> **Table 12. Mean Dissolved Oxygen, pH, Hardness, Chlorophyll a, and Temperature for 16 Samples from Lake Tanganyika in Africa from June through August 2022**

Abbreviations may be included in titles (see Section 30.1.3, "Preparing and Designing Tables"). However, unless a unit of measure applies to every cell in the data field, unit abbreviations are best introduced in column or row headings (see Sections 30.1.1.4, "Column and Spanner Headings," and 30.1.1.5, "Row and Stub Headings," respectively).

In large documents with a series of similar tables, the titles should allow readers to easily distinguish among the tables. To show that tables are related, each title should begin with the same phrase. The remainder of each title should use terms that distinguish the tables from one another.

Table 1. Infectious Diseases in China: Incidence by Socioeconomic Class
Table 2. Infectious Diseases in China: Incidence by Region
Table 3. Infectious Diseases in Japan: Incidence by Socioeconomic Class
Table 4. Infectious Diseases in Japan: Incidence by Region

30.1.1.3 Headnote

A headnote is useful for expanding or qualifying the information in a title. The headnote follows the title, and these 2 elements are separated by a period and a space. Because the headnote appears above the data field, the information in it is emphasized more than it would be in a footnote, which appears below the data field. Unlike the title, the headnote uses sentence-style capitalization, and it is set in roman type, not bold-face. In addition, the headnote does not have a closing period unless it consists of full sentences.

Table 5. Number of Prescriptions Written between 1 January and 20 June 2022 by Medical Specialty. Data are from a 2022 survey of health care professionals at 23 clinics with more than 5 physicians.

In some disciplines, especially those in the basic life sciences, table titles and their accompanying headnotes can summarize the experimental conditions and, sometimes, interpret the data. Such titles and headnotes can be paragraphs long. Information presented in these titles and headnotes should rarely be duplicated in the running text.

30.1.1.4 Column and Spanner Headings

Column headings, also called "heads" or "boxheads," identify the data in the columns of a table. A heading is required for each column of a table, including the column on the far left that is above the row headings. Headings that span 2 or more columns are called "spanner headings" (see Figure 30.1). Minor spanner headings under major spanner headings are called "decked headings."

In tables reporting the results of studies involving explanatory (ie, independent) and response (ie, dependent) variables, the explanatory variables (eg, group names) are usually—but not always—identified in the column heads. This arrangement allows values for the explanatory variables to be placed side by side, which generally makes groups easier to compare. Tables in this more common arrangement report P values or estimates and confidence intervals in a column (see Figure 30.2).[2] However, other factors, such as the number of columns or rows, the available space on a page, and the orientation of the table on the page, may affect how a table is organized[3] (see Section 30.1.3, "Preparing and Designing Tables").

Each column heading should consist of a word or short phrase using sentence-style capitalization. Text in a column heading should be flush left or centered over the column. Other elements in the column can be aligned by decimal points, opening or closing parentheses, or another element common to all cells in the column (see Figure 30.3).

A common heading format for a variable consists of the name of the variable, a descriptive statistic, and the unit of measurement. Headings that identify groups often give or repeat the sample size.

(a)

Table 1. Achievement Scores for Children Taught the New Curriculum, by Grade Level

Grade	Median test scores[a]		
	Reading	Writing	Math
3rd, $n = 53$	23	19	16
5th, $n = 67$	60	55	49
7th, $n = 54$	79	81	63
P value	<0.001	<0.001	<0.001

[a] Scores could range from a low of 0 to a high of 100.

(b)

Table 1. Achievement Scores for Children Taught the New Curriculum, by Grade Level

Skill	Median test scores[a]			P value
	3rd grade, $n = 53$	5th grade, $n = 67$	7th grade, $n = 54$	
Reading	23	60	79	<0.001
Writing	19	55	81	<0.001
Math	16	49	63	<0.001

[a] Scores could range from a low of 0 to a high of 100.

Figure 30.2. Groups are more easily compared side to side, which means treatment groups and explanatory variables are best placed in column headings. (a) However, numbers are more easily compared vertically in columns, and for this reason, some basic science disciplines place explanatory variables in row headings so that the values of response variables are in columns.[2] (b) Comparing groups side by side means that P values and estimates and confidence intervals generally appear in the far right column, which is a more familiar location than along the bottom of the data field. In practice, however, orientation can be dictated by the space available for the table. In addition, the footnote is important because it reports both the range of scores and how to interpret them. However, such information is often missing from published articles.

(a) Undesirable alignment			(b) Desirable alignment		
23.4	726 (78)	↑	23.4	726 (78)	↑
0.234	0.02 (0.002)	↓	0.234	0.02 (0.002)	↓
234	4,819 (7)	↓	234	4,819 (7)	↓
2.34	57 (32)	↑	2.34	57 (32)	↑

Figure 30.3. Elements in the cells of a table should be organized sensibly. (a) Placing numbers flush left or flush right makes their relative values difficult to see. Unaligned symbols create the possibility that their lateral position has meaning, which is not the case here. (b) The elements are positioned more sensibly. The relative values of the numbers in the left column are easily seen because the numbers are aligned on their decimal points. Numbers in the middle column could not be aligned on decimal points but are aligned on the opening parentheses of the values for standard deviation. In the right column, centering the arrows clearly indicates that the arrows' only meaning is an increase or a decrease in quantity.

Television watching, mean (SD), h/d   Live births ($n = 34$)
Salmon caught, No. (%): 35 (83)   Culture media: bile salts No. 3
Bone mineral density, range, g/cm^2

Avoid labeling groups with numbers and letters (eg, "Group 1" and "Group 2"), which are "arbitrary codes" that readers have to remember as they read a document. Instead, give groups descriptive labels (eg, "low-dose group" and "high-dose group").

A heading should not consist solely of a unit designation (eg, "%" or "g") unless the variable is identified in a spanner heading or is clear from the table title and any accompanying text. Providing additional descriptive information is generally better (eg, "Positive, %" or "Mass, g"). In particular, avoid using "No." or "n" alone in a column heading. Instead, use explicit wording (eg, "No. of participants" or "Patients, n"), or explain the meaning in a footnote.

To conserve space, use abbreviations, symbols, and other short forms in column and row headings. If necessary, define these elements in footnotes. The same abbreviations and symbols may not need to be redefined in later tables if the tables are placed close together.

To save space and to organize complex material, the common elements of 2 or more adjacent column headings can sometimes be consolidated under a spanner heading, also known as a "straddle heading" (see Figure 30.1). All information in a spanner heading must apply to every column encompassed by the spanner. Every column under the spanner must either have its own column heading or share a decked heading. In addition, a horizontal rule, called a "spanner rule," should run above all the columns to which the spanner applies, as well as to all the colums to which decked headings apply.

Nesting of spanners (ie, placing a spanner heading above 2 or more decked headings) is sometimes appropriate. However, nesting that creates more than 2 levels above the initial column headings (ie, "complex nesting") can be confusing. Similarly, a spanner should not encompass all the column headings. Instead, incorporate the information into the table's title.

In addition to using abbreviations and spanners to save space, headings can be broken into multiple lines at appropriate places (see Section 30.1.5, "Horizontal Rules and Spacing").

30.1.1.5 Row and Stub Headings

Each row heading describes the type of information that appears in the cells to the right of the row heading. Guidelines for row headings match those for column headings (see Section 30.1.1.4, "Column and Spanner Headings").

Sitting above the row headings in the left-most column of a table is a cell called the "stub." It serves as the column heading for the row headings, and like other column headings, the stub describes the type of information contained in the cells directly beneath it. A table may also have 2 or more "stub headings" that differentiate between sets of row headings (see Figure 30.1).

The row headings under the stub can be organized into groups by using either "stub

(a)

Table 1. Physiological Effects of Stress in Lake Trout. Stress was the result of handling and transport.

Effect	Nonstressed fish, % ($n = 78$)	Stressed fish, % ($n = 93$)
Elevated corticosteroids	16	45
Metabolic effects	14	68

(b)

Table 1. Physiological Effects of Stress in Lake Trout. Stress was the result of handling and transport.

Effect	Nonstressed fish, $n = 20$		Stressed fish, $n = 23$	
	Juvenile, %	Spawning, %	Juvenile, %	Spawning, %
		Elevated corticosteroids		
Lower lymphocyte counts	3	2	12	9
Lower sex hormone concentrations	5	6	14	10
		Metabolic effects		
Retarded growth	2	4	19	7
Reduced oxygen uptake	3	5	26	16

Figure 30.4. (a) A table displaying 2 variables, each of which has 2 values. (b) Adding 2 spanner headings and 2 cut-in headings allows 2 additional variables to be presented, which adds useful information to the table. Adding more than 2 levels of spanner headings is unwise, however. The number of cut-in headings or flush-left stub headings depends on the data presented, but too many may justify breaking the table into 2 or more tables.

headings" or "cut-in headings," just as spanner and decked headings organize columns into groups.

Each stub heading describes all the row headings between it and the next stub heading. Stub headings are left-justified and use sentence-style capitalization. The only information in a row with a stub heading is the wording of the stub heading itself. That information may include the units for the data if the units apply to all the row headings under the stub heading

As an alternative to stub headings, cut-in headings break a table into 2 or more subtables by grouping related rows (see Figure 30.4). Whereas stub headings act as qualifiers for row headings, cut-in headings often identify different groups, each of which has the same set of row headings.

Each cut-in heading should be centered, and it should be separated from the rows above it by a horizontal rule across the entire width of the data field with one line space above and one below the rule.

The decision to use left-justified stub headings or centered cut-in headings is a matter of style and judgment (see Figure 30.1). In general, use left-justified stub headings when they are brief (eg, no more than 4 words and numbers), the number of row headings under each stub heading is limited, and the total number of stub headings is high. Cut-in headings can distinguish different sets of row headings when the row descriptions are long, the number of row headings between cut-in headings is extensive, or the total number of cut-in headings is limited.

When placed under cut-in headings, row headings should be left-justified. However, when placed under stub headings, row headings should be indented. If row headings are longer than one line, the additional lines should align under the first line, regardless of whether the first line is left-justified or indented. Data entries in the columns to the right of multiple-line row headings should align with the top line of the row headings, not the bottom line.

30.1.1.6 Data Field

The body of a table, or "data field," may contain numbers, text, or symbols. Each entry appears in a cell formed by the intersection of a column and a row. The data in each cell must be consistent with the respective column and row headings (see Figure 30.5).

The quantities or values in the text should be identical to those in the accompanying tables. Thus, if a number in the text is rounded, then the value in the accompanying table should be rounded to the same extent (see Sections 12.1.3.3, "Significant Digits," and 12.1.3.4, "Rounding Numbers"). Numbers should be rounded to the nearest mean-

(a)

Trait	Holstein cows ($n = 63$) No. (%)	Hereford cows ($n = 71$) No. (%)
Female	36 (57)	35 (49)
BVDV vaccinated	60 (95)	71 (100)
Age	22 (9)	25 (11)

BVDV, bovine viral diarrhea virus

(b)

Trait	Holstein cows ($n = 63$)	Hereford cows ($n = 71$)
Female, No. (%)	36 (57)	35 (49)
BVDV vaccinated, No. (%)	60 (95)	71 (100)
Age, mean (SD), weeks	22 (9)	25 (11)

BVDV, bovine viral diarrhea virus

Figure 30.5. The data in each cell in the data field should be consistent with the row and column headings. (a) Age is not a number and a percentage, so the variable "age" introduces an inconsistency to the table. (b) Putting the unit of measurement in the row headings creates consistency for the variable "age" in the data field. The problem with (a) is common because researchers tend to report age and sex in the same column with other baseline variables that have different units of measurement. As an alternative to the solution in (b), the out-of-place unit (ie, "age" in this example) might instead be reported in a headnote or footnote.

ingful digit. For example, although a patient's weight might be measured accurately to a hundredth of a kilogram, rounding to a tenth of a kilogram may be sufficient for many applications (eg, 68.3 kg instead of 68.31 kg).

The way in which a data field is organized may depend on whether the intent is to help readers see patterns in the data or to help them find specific values (ie, the "look-up" function of a table). If the purpose is to aid in seeing patterns in the data, row or column headings should be ordered by some characteristic of the data. For example, in a table showing the prevalence of a disease in several countries, the prevalence data could be listed from high to low in one column, allowing readers to focus on the pattern of the data before considering the corresponding countries in the row headings. Alternatively, the countries could be listed in a column in alphabetical order, which allows readers to quickly find the prevalence data for a given country.

30.1.1.6.1 NUMBERS IN DATA FIELD

When possible, align numbers within a column on the decimal point, whether the decimal point is actual or implied. If the entries do not have a decimal point, align columns on another common element, such as parentheses or a multiplier (eg, "×10") (see Figure 30.3).

Alignment can be challenging if entries lack a common element. In such situations, it may be preferable to left-justify or center all entries.

Often, values should be summed across columns, down rows, or both. If values in a table have been rounded and the sums of the columns or rows are different than the actual totals, explain in a footnote that the totals do not sum to 100% because of rounding.

30.1.1.6.2 TEXT IN DATA FIELD

Single words or short phrases in cells in the data field may be either left-justified or centered. However, if any entries require more than one line, all entries in a data field should be left-justified. If needed to aid comprehension, place a blank line between rows in a table containing only text.

Entries consisting of complete sentences are seldom effective or necessary, with the exception of quotations. When complete sentences are necessary, punctuate in sentence style.

Similar entries in a column or row should use the same or parallel phrasing (eg, all nouns, all noun phrases, all verbs, or all verb phrases).

> Treated species, Untreated species *not* Treated species, Species not treated
> Cut, Sutured, Ablated *not* Cut, Suturing, Ablation

Conventional table design calls for an initial capital for each cell entry and each heading (see Figures 30.1 and 30.2). However, technical tables may contain words, abbreviations, or symbols for which lowercase initial letters are essential for recognition (eg, pH, cDNA, c-Jun, sin, log, and mtDNA). If readers might be misled by the use of capitals for words that should not be capitalized in running text, use lowercase and uppercase as dictated by the conventions for those terms. For other exceptions to using sentence-style capitalization in tables, see Section 30.1.3, "Preparing and Designing Tables."

To conserve space in a data field, use where possible the standard symbols and abbreviations that the intended publisher allows without definition or expansion. Other abbreviations and symbols should be defined in the table's footnotes, even if they are defined in the text. Defining abbreviations and symbols in the table's footnotes allows the table to stand alone.

30.1.1.6.3 SYMBOLS IN DATA FIELD

Symbols (eg, arrows, check marks, and bullet points) should usually be centered in cells. Variation in the horizontal or vertical placement of symbols can be confusing because such variation may not be meaningful (see Figure 30.3). Each symbol should be defined in a footnote, unless the symbol is among those that the intended publisher does not require to be defined or expanded.

30.1.1.6.4 EMPTY CELLS

A table may have cells for which no information is available or possible. An empty cell should never be used to represent a value of zero or indicate no difference. Empty cells can raise concerns that data were unintentionally omitted, especially if a data field has only one empty cell. Unless it is obvious that some cells should be empty, avoid ambiguity by placing an ellipsis (ie, ". . .") or an appropriate abbreviation (eg, "nd" for "no data") in each empty cell to indicate that it is not supposed to contain data. Specify the meaning of the abbreviation in a footnote, and avoid using similar abbreviations with different meanings in the same table (eg, "NA" for "not applicable," "not assessed," or "not available").

Cases in which it is obvious that some cells should be empty are ones in which entries would not make sense. For example, in a table showing the harvest results of 3 plots of land, 2 of which were treated with different fertilizers and 1 of which was untreated, the cell giving the contents of the fertilizer for the untreated plot will obviously be empty.

Although empty cells are less likely to be interpreted as omissions in a large table with many empty cells, the empty cells can make tracking across rows difficult. Tracking across rows can also be difficult if a table is wide. In such situations, shading every other row or otherwise dividing the data field horizontally can make tracking easier. When a row or column has many empty cells or identical entries (eg, a "+" symbol in numerous rows), consider redesigning the table.

30.1.1.7 Expansion of Abbreviations

Define abbreviations immediately below the data field by stacking the definitions vertically. Separate each abbreviation from its expansion by a comma. Alternatively, list the definitions on one line by separating them with semicolons.

> TEE, transesophageal echocardiography
> CVD, cardiovascular disease
> OS, overall survival
> *or*
> TEE, transesophageal echocardiography; CVD, cardiovascular disease; OS, overall survival

30.1.1.8 Footnotes

Footnotes explain special aspects of a table's title, column headings, row headings, and data. Superscript lowercase letters are used to direct readers to the related footnotes. Superscript letters are preferred to numbers and symbols for the following reasons:

(1) Alphabetical order is readily recognizable.
(2) Letters are unlikely to be misunderstood as reference citations or typographic errors.
(3) Because most scientific tables contain numbers, using letters reduces the likelihood that footnote symbols will be mistaken for exponents.
(4) More letters are available to serve as footnote signs than are symbols such as asterisks and daggers.
(5) In some word-processing programs, finding the appropriate symbols is not straightforward.

Assign footnote letters in alphabetical order from left to right beginning with the title and proceeding line by line to the bottom of the table. However, if a table contains spanners (see Section 30.1.1.4, "Column and Spanner Headings"), the assignment order for column headings is from left to right as follows (with footnote letters in parentheses corresponding to the example table below):

- the stub column heading (footnote a)
- spanner headings, or first-level spanners (footnote b)
- unspanned column headings, or column headings with no spanner headings above them (footnotes c and d)
- decked headings, or minor spanners below first-level spanners (footnote e)
- column headings under spanner headings but not decked headings (footnote f)
- column headings under decked headings (footnotes g and h)
- row headings under stub headings (footnote i)
- data cells (footnotes j through n)

| | | Spanner heading[b] | | | |
		Decked heading[e]			
Stub heading[a]	Column heading[c]	Column heading[g]	Column heading[h]	Column heading[f]	Column heading[d]
Row heading[i]	Data[j]	Data[k]	Data[l]	Data[m]	Data[n]

The footnotes themselves are placed below the expanded abbreviations. The footnotes are listed in alphabetical order based on their designated footnote letters. Just as they are in the table, footnote letters should be in lowercase and superscript in the footnotes. Each superscript letter should be flush with the left margin of the table and separated from its footnote by a regular space, not a superscript space. The footnotes should use sentence-style capitalization, and full sentences should end with a period. Carryover lines should be aligned with the first word of the footnote so that the footnote letter stands out.

30.1.1.9 Credit Lines

A credit line, also called a "source line," identifies the source of the information in a table or the table itself. A credit line may be placed as a headnote, a footnote, or the lowest element of the table below the data field. If a credit line is needed, the form depends on the publication's style for reference citations and the required specificity. The credit line should take one of the following forms.

(1) "Reprinted from . . ." (for an exact reproduction)
(2) "Adapted from . . ." (when either the information or the form of the table has been changed from the original)
(3) "Based on . . ." (when information comes from a particular source but the table is not reproduced or adapted)

When a table is reproduced or adapted from a published source, it is not sufficient to cite the source and list it in the reference list. Permission to reuse the table must be obtained from the copyright holder (see Section 3.3.3, "Permission to Republish Scientific Works"). In addition, the credit line should adhere to the form and content requested by the copyright holder, which may vary from the publication's style for reference citations.

When a few cells contain data from a source other than the main source, the overall source can be identified with a credit line, and the data from the minor source can be credited in footnotes. If the data come from several sources, none of which provides most of the data, the sources may be credited with footnotes for the column or row headings or with in-text references to the manuscript's end references, such as a superscript "[10]" or "(Widder 2021)."

30.1.2 Text References to Tables

In general, interpret the data in the manuscript's text, and cite the table with the supporting data in parentheses at the end of a sentence.

> The tensile strength of the new alloy was double that of the old one (see Table 3).
> *not*
> Table 3 shows the results of tensile strength testing. [This information should be apparent in the title of the table, so it need not be mentioned in the text.]

In some cases, however, clarity and simplicity are achieved by referring to the table in the subject of the sentence.

> Table 3 shows that tensile strength of the new alloy was double that of the old one.

30.1.3 Preparing and Designing Tables

Before preparing or editing a table, determine whether the data might be more effectively presented in the text or in figures.[4] Data that would require only 1 or 2 columns and only 2 or 3 rows should most likely be converted to text, unless the data are numbers with many digits, which would be easier to read and compare in a table. In some cases, however, even long tables should be converted to text. *How to Write and Publish a Scientific Paper*[5] offers an excellent discussion of data that are more appropriate for text than for tables (eg, reporting in text data that would require many empty cells or zeros if placed in tables).

The question of which data should be in rows and which in columns has no simple answer. As discussed in Section 30.1.1.4, "Column and Spanner Headings," groups are more easily compared side by side, which is easier to accomplish if the response (dependent) variables are identified in row headings and the explanatory (independent) variables, such as group names, are identified in column headings. The final choice should be based on the constraints of size (eg, average screen settings for readers of online publications and page size for print publications), the width of cell entries, and the number of rows and columns. The number of rows is virtually unlimited because they can extend down numerous screens or printed pages, whereas the number of columns is limited, at least in print publications, by page width (or page height for tables turned sideways). In addition, more space is required between columns than between single-spaced rows.

Authors and editors can use several techniques to reduce table size:

- Edit tightly.
- Remove columns and rows that do not contain enough variation to justify inclusion by moving that information to the text or a footnote.
- Combine columns under spanner heads to avoid duplicating common elements in column headings.
- Use abbreviations and symbols.

Designers can contribute to reducing table size by using such techniques as the following:

- Use a narrow or condensed font.
- Reduce the "kerning" value to move letters closer together.
- Optimize the horizontal and vertical dimensions of the cells (eg, reduce the height of the rows and the width of the columns).
- Expand the left and right text margins in each column as much as possible. Setting the margins at the limits of the width of cells frees a small amount of space.

30.1.4 Designating Units of Measurement

The units of measurement in a table should correspond to those in the text. If all the numbers in the data field are in the same unit of measurement, designate that unit in the table title or headnote. If the units vary across columns or rows, designate the relevant units in the column or row headings. In the rare circumstances in which cells have units that differ from those specified in the column or row heads and in which those discrepancies cannot be resolved, explain the exceptions in individual footnotes.

In most cases, the most appropriate units of measurement will be those from the Système International d'Unités, or SI (the International System of Units). Some of the most notable exceptions to this guideline are found in medicine (eg, blood pressure is almost always reported as millimeters of mercury rather than in the SI unit pascal, and unit of concentration is almost always reported as nanograms per deciliter instead of in the SI unit picomole per liter).

In tables, SI units should not be converted with SI prefixes to smaller or larger units if some of the cells contain numbers smaller than 0.1 or greater than 1,000, even though this violates standard recommendations for SI units (see Section 12.2.1.2, "Other Recommendations for Writing SI Units"). For example, if all of the numbers in a row or column use the SI unit meter, do not convert to kilometers the one entry that is greater than 1,000 m. Instead, keep the unit of measurement consistent for the entire row or column to aid readers in comparing the magnitude of differences among the values. This practice also eliminates both the need for a footnote to explain an exception and the risk that readers will miss the footnote.

When the data in a row or column are not expressed in SI units, very large or very small numbers should be expressed with multipliers in the row or column headings, preferably using scientific notation (eg, "Study participants $\times 10^3$") (see Section 12.1.5, "Scientific Notation"). Avoid alternative formats such as words (eg, "thousands" or "millions") and numerals (eg, "1,000s" or "1,000,000s"). Even in scientific notation, multipliers can cause uncertainty. In such cases, explanatory footnotes may be needed to ensure that the scientific notations are not misconstrued.

When the range of values is extremely wide, it may be necessary for each cell to be in full scientific notation (eg, "1.596×10^{-4}" in one cell and "1.732×10^6" in another). In such tables, the column or row heading would provide only the common unit of measurement. Because of the space problems created by the width of such entries, this option should be avoided if possible.

When a table reports nonscientific information, such as budget data, traditional headings are acceptable (eg, "€, millions").

30.1.5 Horizontal Rules and Spacing

Most tables have only 3 table-width horizontal lines, called "rules." The first of these rules is below the title, the second is below the column headings, and the third is below the data field (see Figure 30.1). Additional rules may include a line below spanner and decked headings, one above a cut-in heading, and one below a column of numbers to indicate that the total is in the next row down. These latter horizontal rules usually are set in a smaller point size than the first 3.

Vertical lines are unnecessary if columns are properly separated. Vertical lines generally clutter a table without adding clarity. Observing appropriate spacing is especially important when a table's columns are numerous and the data values are similar to each other. In tightly spaced columns, shading every other column may help define columns.

To make large tables of numbers easier to read, break the rows by placing a blank row after every 3 to 5 rows. Alternatively, as long as none of the table's columns is shaded, shading alternate rows can provide visual markers without adding rows.

30.1.6 Common Errors in Table Design

Avoid the following errors when designing tables:

- The amount of information does not justify a table (see Figure 30.6).
- The title unnecessarily repeats column or row headings.
- The column or row headings are unclear.
- The column or row headings are not sensibly organized.
- Elements in columns are poorly aligned.
- Data are inconsistent with their row or column headings.

<table>
<tr><td colspan="4">(a)</td></tr>
<tr><td colspan="4">Table 1. Effect of Temperature on the Growth of 2 Shrimp</td></tr>
</table>

	Length, mean (SD), mm		
Species	23 °C	25 °C	27 °C
P. borealis	11.1 (1.4)	13.6 (2.0)	14.7 (1.5)

(b)

The length (SD) of *P. borealis* was 11.1 (1.4) mm at 23 °C, 13.6 (2.0) mm at 25 °C, and 14.7 (1.5) mm at 27 °C.

Figure 30.6. (a) The amount of data is not sufficient to justify a table. (b) With so few values, the same data can be more efficiently reported in the text.

- The data field is too crowded to determine any patterns in the data.
- Numbers are reported with unnecessary or "spurious" precision, which takes space and adds complexity without increasing clarity.
- The denominators for percentages are missing, unclear, or incorrect.
- The data contain mathematical errors, especially in calculated values such as totals and percentages.
- Cells are left blank without explanation.
- Useful row or column totals are missing.
- Information is missing, such as group sizes or units of measurement.
- Abbreviations are not defined.

30.2 FIGURES: GRAPHS, IMAGES, AND ILLUSTRATIONS

30.2.1 General Information on Figures

Figures are displays of data and other information. Figures include graphs, such as charts; photographs and biomedical images, such as radiographs, blots, and gels; and illustrations, such as diagrams, drawings, and maps.

When preparing a figure, strive for the characteristics listed below:

- The figure should have a purpose, and it should support claims made in the text.
- The figure should be as simple as possible. It should minimize elements that distract from seeing and understanding the data.
- The figure should emphasize data over other elements, such as axis and scale marks and labels in the data field.
- The data and other information displayed should be accurate and not misleading.
- In general, data presented in a figure should not be duplicated in the text or in a table. Exceptions include reporting the most important data in the text.
- Similar elements within a figure and in related figures should be presented consistently.
- Illustrations should be visually appealing and professionally drawn.
- Photographs and medical images should be sharp and in focus.

For all figures in a document, use consistent styles and formats (eg, capitalization of labels, typeface, type size, symbol use, and caption format). However, consistency may be relaxed for figures reproduced from other published works.

Use the same terms, symbols, and abbreviations in figures as in the text. Abbreviations introduced in the text should usually be redefined in the figures to allow the figures to stand alone.

More detail on the characteristics of various kinds of graphs, images, and illustrations and their preparation can be found in Chapters 4, 9, and 10 of *How to Write, Publish, and Present in the Health Sciences: A Guide for Clinicians and Laboratory Researchers*[1] and in Chapters 16, 17, and 18 of *How to Write and Publish a Scientific Paper.*[5]

30.2.1.1 Figure Number

If a manuscript has only one figure, the figure is not numbered, and it is cited in the text simply as "Figure."

If a manuscript has 2 or more figures, label each with the word "Figure" or the abbreviation "Fig," depending on the preference of the intended publication or publisher. Number the figures with consecutive arabic numerals in order of their mention in the text. Like the table number, the word "Figure" or the abbreviation "Fig" and its number are in boldface, and the number is followed by a period.

For long documents, reports, or books that are divided into sections or chapters, figures may be found more easily by using designations that combine section or chapter numbers with consecutive numbers based on the order in which the figures are mentioned in a section or chapter (eg, "Figure 47.1" and "Figure 47.2" or "Fig 64.5" and "Fig 64.9"). In each section or chapter, numbering should begin with 1, and the section or chapter number should be separated from the figure number by a period.

Unlike tables, figures can have parts (eg, "Figure 1a" or "Fig 6b"). Figure parts are 2 or more adjacent data displays or images that are related in some way and that share a common caption. For example, figure parts can show several measurements from the same sample, the same measurement from several samples, pre- and posttest values, and time-series photographs.

Figure parts need not be cited in the text. The reference to the entire figure is adequate. Figure parts are best introduced in figure captions.

30.2.1.2 Figure Caption

A figure caption must allow readers to identify and understand the fundamental aspects of the figure without undue reference to the text. Information in a figure caption or its headnote rarely need to be duplicated in the text.

Unlike table titles, which are placed above tables, figure captions are typically placed below figures. A figure caption consists of a single phrase in sentence case and roman type. A figure caption does not have a closing period, unless the figure has parts or a headnote (see Section 30.2.1.3, "Headnote").

Captions for figures with parts typically explain the information that applies to all parts first. The information specific to each part is then described after identifying the part by a lowercase letter enclosed in parentheses (see Section 5.3.6.1, "General Uses").

> **Figure 3.** Rainfall in the greater Seattle area, 2021. (a) March, (b) June, (c) September, and (d) December.

Figure captions should identify the data in the data fields or the images in the image areas, as well as explain any relevant relationships or comparisons. Captions should not simply repeat the axis labels or note the type of figure (eg, "A bar chart showing . . ." or "A Kaplan-Meier curve indicating . . ."). In general, captions should describe the data and not interpret them or state a conclusion based on them.

> **Figure 5.** Association between air temperature and survival of Tanner crabs released as illegal bycatch in the Bering Sea bottom trawl fishery
> *not*
> **Figure 5.** A plot of air temperature and survival of Tanner crabs released as illegal bycatch in the Bering Sea bottom trawl fishery

Some disciplines, especially those in the basic sciences, make exceptions to the guidelines above. They may allow a figure's caption or headnote to summarize the experimental conditions and even interpret the information in the figure.

30.2.1.3 Headnote
Just as a table headnote follows the table's title, a figure headnote follows the figure's caption. Also like a table headnote, a figure headnote contains information that qualifies the description in the caption. Unlike a table headnote, a figure headnote may contain expanded abbreviations, keys that identify symbols in the data field or image area, and credit information.

30.2.1.4 Text References to Figures
As with in-text references to tables, most references to figures should be placed in parentheses in the text. However, in some cases, clarity and simplicity are achieved by integrating figure references into sentences. If a reference to a figure begins a sentence, spell out the word "Figure." Elsewhere, use either "Figure" or the abbreviation "Fig," depending on the preference of the intended publication or publisher.

> Figure 5 shows the result of this combination.
> The incubator (see Figure 2) was custom made.
> The incubator shown in Fig 2 was custom made. The one shown in Fig 3 is a commercial model.

Phrasing such as that in the third example above is reserved for figures that show drawings or photographs of something physical, as opposed to numerical data. In addition, avoid suggesting that a graph is evidence. Rather, state that a graph "presents" evidence. In contrast, photographic, diagnostic, and laboratory images are indeed evidence.

> Separation was complete after 20 h (see Figure 5).
> *not* Figure 5 shows that separation was complete after 20 h.
>
> Additive X was the most effective (see Fig 13).
> *not* Figure 13 proves that additive X was the most effective.
>
> The echocardiogram in Figure 3 clearly shows mitral regurgitation.

30.2.2 Graphs

Graphs are visual presentations of data and the relations among those data. "Graphs" are sometimes defined as a subset of "charts," and other times, "charts" and "graphs" are defined as 2 different types of figures. More commonly, the 2 terms are synonymous. Throughout this chapter, the term "graph" is treated as encompassing items commonly called "charts," "plots," and "curves" (eg, column and bar charts, box plots, and Kaplan-Meier curves). This section of the style manual focuses on Cartesian graphs, also known as "coordinate graphs." These graphs use 2 or 3 axes to plot data values. As is apparent from Figures 30.7 to 30.18, many of the guidelines in this section are applicable to the other types of graphs mentioned above.

In preparing graphs, authors should adhere to the instructions to authors of the publication to which they intend to submit their manuscript. Doing so increases the likelihood that the graphs can be published as submitted or replicated by the publication's designers. In general, graphs should present data clearly, and they should be of professional quality, with legible symbols and type, complete labeling, appropriate resolution and line thicknesses, and easily distinguished shadings.[6]

Graphs and other figures prepared as posters or as slides are seldom suitable for print and electronic media without modification. The visual strengths and weaknesses of posters and slides differ from those of print and online media (see Figure 30.7). In addition, because the default formats of graphing and statistical software programs are design to assist in analyzing data, they rarely meet the specifications of publishers. Accordingly, most graphs should be prepared by professional graphic artists.

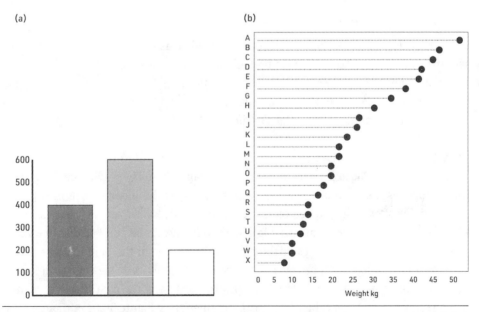

Figure 30.7. (a) A simple column chart is suitable for a slide or a poster, but it does not contain enough data to justify using it in a publication. (b) A dot chart can display far more data[8] and can easily be understood on a printed page. However, a dot chart is difficult to read from a distance when presented on a slide or poster.

(a)   (b)

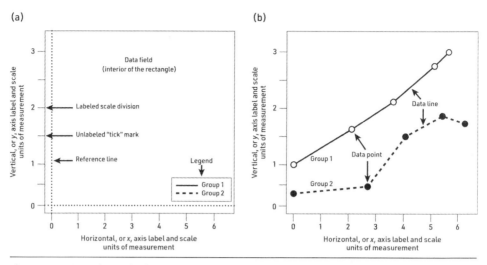

Figure 30.8. Parts of a graph. The zero-zero point has been moved away from the axis lines so that the values of zero can be seen without the axis lines interfering.[8] The tick and scale marks are outside the data field to reduce clutter in the data field. Data lines and key points should be labeled in the data field. A legend is necessary only when the data lines cannot be labeled directly.

At a minimum, a graph should have a data field, axes, axis labels, scales, plotting, symbols, and a caption (see Figure 30.8). Depending on its format, a graph may also have a legend, as well as a figure number, a headnote, and a credit line (see Sections 30.2.1.1, "Figure Number," 30.2.1.3, "Headnote," and 30.1.1.9, "Credit Lines," respectively).

30.2.2.1 Data Field

Data fields should be well defined, and they should be as free from distracting characteristics as possible.[7–9] One common problem with data fields is that the values for zero may not be noticed if they are graphed on the thick axis lines. To avoid this problem, move the zero-zero point of the grid a little up and away from the axis lines, and graph the values of zero along thinner lines in the data field (see Figure 30.8).[8]

30.2.2.2 Axes and Scales

For most purposes, the dependent, or response, variable should be graphed on the vertical axis, known as the "ordinate axis" or "y axis," whereas the independent, or explanatory, variable should be graphed on the horizontal axis, known as the "abscissa axis" or "x axis." As an independent variable, time should always be graphed on the x axis.

Extend the scale on the axes slightly beyond the range of the plotted values. If an extremely large range must be covered and cannot be practically shown with a single, uninterrupted scale, indicate a break in both the scale and the data field with paired diagonal lines, which are called "discontinuity lines" (see Figure 30.9). Although most graphs begin at the zero-zero point, beginning one or both scales at zero may not be practical if either scale shows large numbers. For example, a value of zero might be physically or physiologically impossible and adding discontinuity lines might be distracting. In

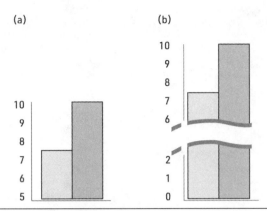

Figure 30.9. Most graphs begin on a baseline of zero, and most readers assume that all graphs will have that baseline. Comparing 2 sets of data with unequal baseline values can be misleading. Here, the graph on the right looks larger than that on the left, even though they show the same values. (a) This graph has a baseline of 5, not zero, so only the tops of the columns are shown. As a result, the relative lengths of the columns do not create a fair comparison, although this image is easy to remember. (b) A better way to graph the data is to break both the scale and the baseline to highlight the fact that the scale is not continuous.

such a case, the scales can begin with a value slightly smaller than the minimum value of the graphed data.

Graphs are not meant to communicate exact values, so grid lines are not usually included in data fields. Indicate values on the scales with labeled scale marks and unlabeled "tick marks" at intervals. Place scale marks, tick marks, and numerical values outside the data field, just left of the y axis and just below the x axis (see Figure 30.8). For numerical values, appropriate multiples are typically 2, 5, 10, 20, and 25. For numbers less than 1, include zero before the decimal.

In addition to using a left-side scale, some graphs use a right-side scale outside the data field, either to graph a third variable or to give the values of the data in another unit (eg, kilograms are graphed on the left scale and pounds on the right). The right-side vertical axis should be scaled, numbered, and labeled in the same manner as the left-side scale. Each data line in such a graph must be clearly identified, preferably in the data field, and linked to its respective scale. If lines cannot be identified in the data field, they should be identified in the graph's caption or in a legend.

When values graphed on the right-hand scale are not mathematically related to those on the left-hand scale, the visual relationship between the respective data lines can be distorted or manipulated by varying the ratio of one vertical scale to the other (see Figure 30.10). No rule governs the use of a third scale in this situation, and the space available for the graph may be a limiting factor.

The variable plotted on each axis should be identified with an axis label consisting of a word or short phrase. Use sentence-style capitalization for the labels, and note the unit of measurement for each axis after a comma at the end of the axis label. Each label should be centered on its corresponding axis and run parallel to that axis (ie, the y axis label runs upward, and the x axis label runs from left to right). When a multiplier is

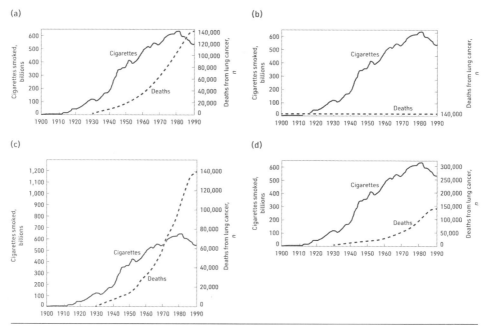

Figure 30.10. Graphs in which a third variable is plotted on the right vertical axis can be deceptive. There is no rule about how to scale this third axis, other than to keep in mind that it can bias the interpretation. (a) The data seem to show that the number of people who died from lung cancer increased in proportion to the number of cigarettes smoked, implying an association and possibly causation. (b) Graphing the number of patients who died from lung cancer on a right-hand scale that is proportional to the left-hand scale produces a flat line essentially on zero because 140,000 deaths is so much smaller than 1 billion cigarettes smoked. The visual message is that smoking is not related to death from lung cancer. (c) The same data with the expanded right scale seem to indicate that the number of people who died of lung cancer increased more rapidly than cigarette consumption, suggesting the 2 variables are perhaps unrelated. (d) The same data with a compressed right scale imply a weaker relationship between smoking and death from lung cancer.

needed for the axis labels, the multiplier should multiply the variable, not the unit of measurement.

> Median recovery time, d *not* Median days until recovery

Make axis labels as unambiguous and descriptive as possible (eg, "Red blood cells, n"). Occasionally with complex variables, suitable axis labels may consist only of units of measurement. In such cases, the units should be spelled out instead of represented by symbols, especially if units of the Système International d'Unités (SI) are combined with non-SI units (see Section 12.2.1.2, "Other Recommendations for Writing SI Units").

> **An axis label with one unit of measurement for a graph with complex variables:**
> Nanograms per cell per day *not* ng·cell^{-1}·d^{-1} *and not* ng·cell^{-1}·d^{-1}

If an axis is labeled with only a simple unit, the appropriate variable should be added. For example, when the unit for an axis is kilometers, the axis label usually should be "Distance, km."

> **Standard axis label for a unit of measurement:**
> Calcium uptake, ng/cell·d

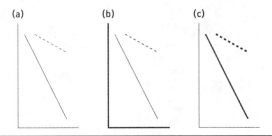

Figure 30.11. Emphasize the data.[10,11] (a) Data lines and axis lines of the same weight do not guide the eye immediately to the data. The eye does not know where to focus. (b) If the axis lines are thicker than the data lines, the eye goes to the axis lines, not to the data. (c) If the data lines are thicker than the axis lines, the eye instantly goes to the data, which is the objective.

30.2.2.3 Plotting Symbols

Plotting symbols need to be distinct and legible and provide good contrast when overlapping other symbols denoting equal or similar values. Open and closed circles provide the best contrast, and they are more effective than the combination of open circles and open squares. A good symbol system also allows overlapping values to be distinguished. The symbols ○, ●, ◉, ⊙, ⊗, and ⊘ work well, but there are many other options. (For examples, see *The Elements of Graphing Data*.[8])

Different data lines can be shown as solid lines (eg, ⸺ and ▬▬) or dashed lines (eg, ‒ ‒ ‒, ▬▬ ▬▬ ▬▬, ‒ ‒ ‒, and ====).[8] In keeping with the principle of emphasizing the data, make the data lines thicker and darker than the scale lines (see Figure 30.11).

Like other figures, graphs should be simple and contain no more information than is needed to make the point. For example, the dot charts described by William S. Cleveland, PhD,[8] are more efficient for presenting data and focusing attention on the data than are column and bar charts. In column and bar charts, the thickness and height of columns and bars focus attention away from the data, the values of which are indicated only at the end of the columns and bars (see Figure 30.12).

Several data lines can be displayed on a single graph, provided they do not obscure other values. When the number of lines is great or when overlap hinders following any of the lines, consider using "small multiples." As described by Edward R. Tufte, PhD,[10-12] small multiples graph each line separately in its own figure part (see Figure 30.13). Such multiple graphs should have identical X and Y axes and identical dimensions.

Make symbols, shadings, and line patterns consistent among all graphs in a manuscript. For example, the symbols representing control and treatment groups in a manuscript's first graph should be the same for these groups in later graphs. However, if the series of graphs represent 2 or more experiments, different symbols can help distinguish the results of the experiments in their respective graphs.

30.2.2.4 Legends

Labeling lines and symbols directly in the data field is preferable but not always possible. If direct labeling is not feasible, a "legend" can be added in a graph's data field to define the data points or lines (see Figure 30.8). Legends are most effective in graphs

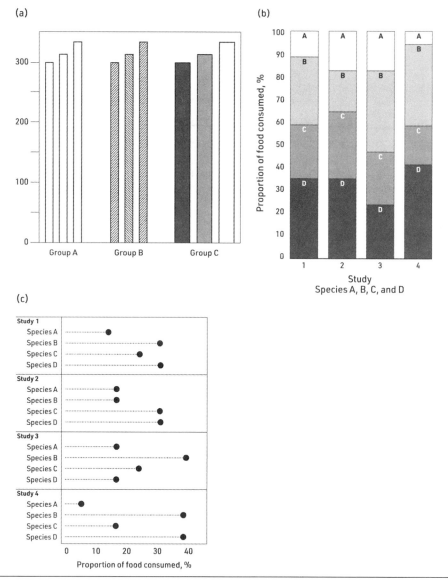

Figure 30.12. Graphs should present data as efficiently as possible. (a) Column and bar charts are prone to common errors, such as confusing spacing, optical illusions, and unintended emphasis created by high contrasts in shading.[1] (b) In this divided column chart, the body of each column draws more attention than the top of each segment within each column, which gives the value of the data. In addition, comparing the same segments across columns is difficult because they are not graphed on a common baseline. (c) In contrast, a dot chart graphing the same data emphasizes the data and allows the value of each segment to be compared with every other value.[8] In addition, a dot chart can present more data in the same space because it does not use wide columns. Emphasize the last dot, which represents the value; the rest of the dots should be smaller and in a lighter color.

(a)

(b)

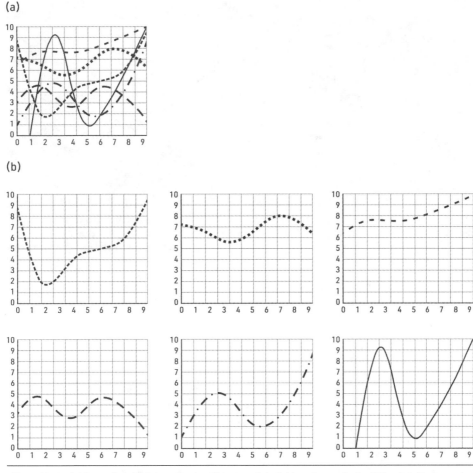

Figure 30.13. (a) Several variables graphed on the same data field can be confusing if the data lines overlap or cross other lines. (b) In such cases, "small multiples" can be an elegant solution. All of the multiples must have the same scales and dimensions.[10–12]

with several different variables, provided the graphs have sufficient space to accommodate them without obscuring the data. If a legend would obscure the data, a caption or headnote may be more desirable for identifying elements in the data field.

30.2.2.5 Common Errors in Designing Graphs
Avoid the following errors when designing graphs:

- Beginning one or both scales with a number other than zero (see Figure 30.9). Readers often assume that the baseline is zero and may not notice when it is not. To indicate unequivocally that the baseline has been modified, both scales should begin at zero and the axis lines and the data field should be broken to indicate omitted values.
- Graphing 2-dimensional data on a 3-dimensional graph (see Figure 30.14), which can distort the relationships in the data. The third dimension is rarely necessary, and its purpose is usually only to make the graph more interesting.

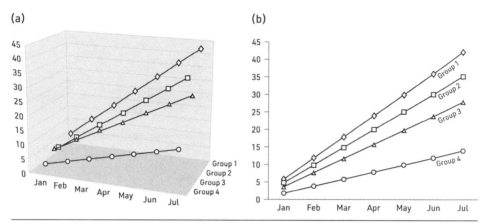

Figure 30.14. Some commercial spreadsheet programs allow data to be graphed in several ways to generate more interesting graphs. (a) Graphing 2-dimensional data on a 3-dimensional graph is neither necessary nor wise because readers will be unable to accurately compare the data. (b) The same data presented on a 2-dimensional graph are easily read.

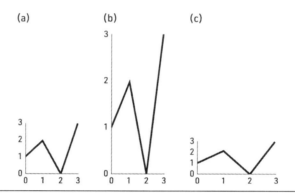

Figure 30.15. "Aspect ratio" is the ratio of the height to the width of the graph. Varying the aspect ratio can change how the graph is interpreted. (a) The x and y scales are the same. (b) Lengthening the y axis emphasizes the increasing values. (c) Shortening the y axis minimizes the decreasing values. No rule covers setting the aspect ratio. The only guidance is to be aware that the ratio can bias interpretation.

- Extending or compressing the length of one or both axes to change the visual impression of the data (see Figure 30.15). This will change the "aspect ratio," even if the graph remains technically accurate.
- Including unnecessary data (see Figure 30.16). For example, if the pattern of the data is more important than individual values, the data can be graphed as a smooth curve. In addition, error bars may not be necessary.
- Plotting paired data (2 or more data points taken from the same subject) on individual graphs that do not indicate the pairing can distort the analysis and mislead its interpretation (see Figure 30.17).
- Not graphing specific differences or comparisons, forcing readers to draw their own conclusions, which may be incorrect (see Figure 30.18).

30.2.3 Images

Images in scientific publishing vary widely in nature and format (eg, standard photographs, echocardiograms, micrographs, diagrams, and illustrations). Generally, image areas should be well defined, contain little distracting information, and provide good

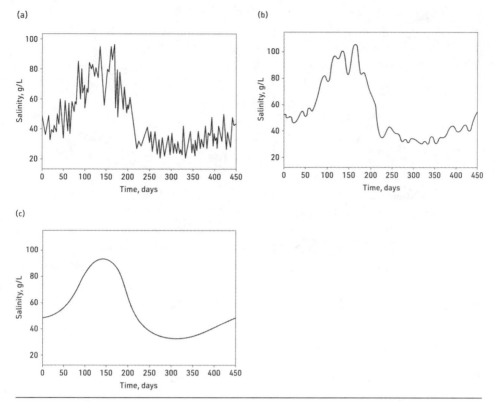

Figure 30.16. (a) Often, graphing all the data is unnecessary and even undesirable because the details can distract from the main message of the graph. (b) and (c) Smoothing the curves reduces the amount of detail, allowing readers to focus attention on the pattern of the data. How much detail to add depends on the intended readers and the purpose of the graph.

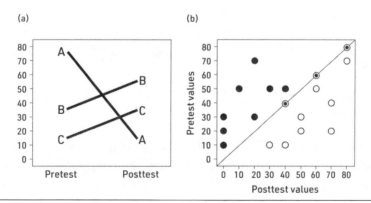

Figure 30.17. Paired data should remain paired when graphed and analyzed. (a) The pretest mean is 11.6, and the posttest mean is 10. However, the 1.6-point drop is misleading when the pairing is considered because the scores rose in 2 of the 3 subjects. (b) A standard way to plot paired data is to graph one variable on the *y* axis and one on the *x* axis. The diagonal "line of unity" indicates no difference between pretest and posttest values.

(a)

(b)

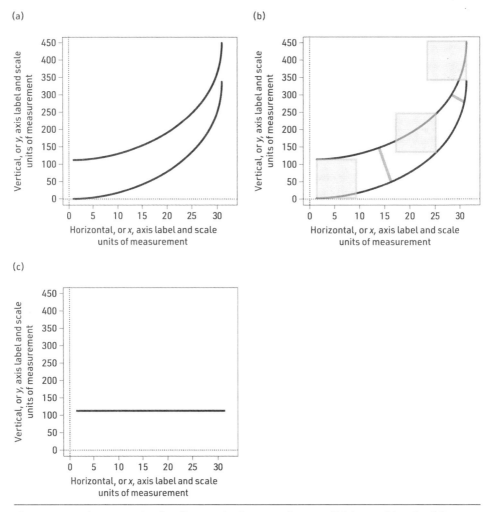

(c)

Figure 30.18. When comparing data lines, make the comparisons explicit by graphing the differences. (a) Readers are unlikely to correctly perceive the differences between these 2 lines. (b) This image demonstrates the optical illusion of (a) in which the eye compares the closest points on the 2 lines, not the vertical difference, which conveys the actual mathematical comparison. (c) A graph of the differences between the 2 lines in (a) and (b) shows that the lines are actually parallel.

contrast between foreground and background. In addition, images should be composed or cropped to direct readers' attention to the details of interest. Images should meet the specifications in the instructions to authors of the intended publisher, including file size and format (eg, jpg or tiff) and naming conventions (eg, "Colins Fig 5").[1]

When images are submitted for publication, supportive information may need to be provided separately, especially for micrographs and biomedical images.[1,13] Adequately documenting an image requires providing the following information[13]:

- The subject of the image (ie, explain what is being shown)
- The implications of the image in the context of the research
- The circumstances under which the image was acquired (eg, the equipment and settings, the position of the subject, the stains, and information on whether the image is from the current research or another source)

- The reason for selecting the image for publication over similar images (eg, the image illustrates typical or atypical conditions)
- Any modifications made to the image. Not disclosing modifications to digital images (eg, micrographs and images of blots and gels) is scientific misconduct, especially if the modifications intentionally falsify or otherwise misrepresent the data.[14]
- Superimposed circles, lines, arrows, or other graphic elements on the image that draw viewers' attention to important features. The image should be submitted electronically with the graphic devices in editable form so that they can be modified to meet the publisher's specifications.

30.2.3.1 Standard Photographs

Standard, or "normal lens," photographs are images that correspond roughly to what the human eye sees.[15] These images are preferred because they depict a natural relationship between the various elements in a scene. In contrast, images obtained through telephoto and wide-angle lenses are somewhat distorted.

30.2.3.1.1 SCALE

To make normal-lens photographs meaningful, indicators of scale are often needed. Such an indicator may be as simple as an object included in a photograph if the size of the object is widely known (eg, a geologist's hammer could serve as the scale in a photo of a rock formation). In many cases, however, scale indicators have to be added to images.

Although a statement of magnification (eg, "×1,000") should be included in the caption for a photograph, visual scales are more useful. They are also more useful than other mathematical scales (eg, a bar denoting a given length on a micrograph).

30.2.3.1.2 CALLOUTS

The font for callouts that identify structures or parts in an image should be consistent with that used in other figures in the manuscript. Before adding callouts to photographs, the photos should be reduced, enlarged, or cropped to the size requested by the publisher so that the type size is unaffected when the photos are printed. Authors should submit photos electronically with the callouts in editable form so that they can be modified by the publisher. Many journals also require that the original, unaltered image be submitted as well.

30.2.3.1.3 COMMON ERRORS IN PREPARING PHOTOGRAPHS

Errors in normal-lens photographs can be related to technique, composition, and the physical characteristics of photos. Most of the errors listed in this section can also occur in other types of photographs, such as photomicrographs and clinical photographs obtained through fiberoptic scopes.

PROBLEMS RELATED TO THE PHOTOGRAPHIC TECHNIQUE:
- Out-of-focus or hazy images
- Poor image resolution (For print publication, photographs should have a resolution of at least 300 dots per inch [dpi]. For online publication, photographs can have a resolution as low as 72 dpi.)
- The presence of shadows and reflections in images
- Inadequate illumination, which can darken parts of an image

- Poor color registration
- Artifacts, such as spots on photographs caused by dust on the camera lens
- Shallow depth of field (ie, the distance between the closest and farthest objects in focus is narrow, leaving objects outside that field blurry)
- "Motion blur," typically caused by movements of either the subject or the camera at the moment the photograph is taken
- Excessive "visual noise," such as graininess caused by taking photographs in low light

PROBLEMS WITH THE COMPOSITION OF PHOTOGRAPHS:
- The subject is not emphasized
- The subject is not distinct from the background
- The contrast between elements in an image is poor
- Important details are too small to see

PROBLEMS WITH THE PHYSICAL CHARACTERISTICS OF PHOTOGRAPHS:
- Not large enough to capture the relevant surrounding context of the subject of the image
- Poorly cropped, resulting in irrelevant content included in the image
- Detail lost during the printing process, which occurs when images are excessively enlarged or reduced
- Not indicating the top of photographs, potentially resulting in publishing the photographs in the wrong orientation

30.2.3.2 Clinical and Laboratory Images

Clinical and laboratory images, such as radiographs and gels, are unique among graphic elements because they do not report, compare, or summarize information. Instead, they are the information.[13] This information must be interpreted, which can be affected by such factors as biological variation, image acquisition and quality, artifacts, diagnostic standards, and the training and skill of the interpreter.

The guidelines below are intended to help authors report what readers need to assess the accuracy, completeness, validity, and credibility of the interpretation and implications of clinical and laboratory images.[13,14]

30.2.3.2.1 NATURE OF IMAGES

Tell readers what they are looking at and what they should look for. The reason for acquiring the images should be made clear. If applicable, tell why the specific images were chosen for publication, as opposed to other, related images.

Note any biological, historical, or physical characteristic of the subject or sample that readers need to know to understand the images. Indicate whether the images show a typical, atypical, extreme, or selected case. Identify any differences in the source or acquisition of images that have been grouped for comparison.

It is rarely necessary to identify the medium in image captions (eg, "A radiograph showing . . ." or "This micrograph . . .").

30.2.3.2.2 ACQUIRING IMAGES

If applicable, name or reference the equipment, software, and protocol used to acquire the images. Describe how the subject or sample was prepared for imaging and the conditions under which the images were acquired.

30.2.3.2.3 MODIFYING IMAGES

Disclose any processing, modifications, or enhancements of the original images, including cropping and digital manipulation. Be prepared to submit both the modified and the original images to the publisher for comparison. Before submitting the images, remove from them any patient identifiers (eg, names and patient numbers). Obtain permission from photographed subjects to publish any full or partial images in which their identities might be determined. Covering the eyes with a black bar is not adequate to protect the identity of people in images.

30.2.3.2.4 DETAILS OF IMAGES

In each image's caption or on the image itself, indicate what the image shows and where it shows it, including areas where something is missing. If relevant, report the resolution and magnification of the images, and indicate the orientation of the subject or sample in the images (eg, "sagittal plane," "overhead view," or "the northwest corner of the property").

30.2.3.2.5 ANALYSIS, INTERPRETATION, AND IMPLICATIONS OF IMAGES

Report any quantitative or qualitative data needed to interpret the images. If necessary, describe the circumstances under which the images were analyzed or evaluated (eg, inter-rater agreement or blind evaluation). Interpret the meaning and implications of the images in light of that information.

30.2.3.3 Illustrations: Line Drawings

The term "illustration" sometimes refers to any visual element in a text: drawings, paintings, photographs, graphs, charts, and medical images (eg, retinal scans and electrocardiograms). In this chapter, "illustration" refers only to "line drawings," which include anatomical drawings, flowcharts, schematic drawings of processes and systems, and decision trees. Line drawings consist only of lines or dots that form images through line length, width, direction, configuration, texture, shading, and related text. They differ from continuous-tone art, which has a range of colors, including grays. Traditionally, the 2 types of drawings are handled differently in the publication process, with continuous-tone images being processed in the same way as photographs (see Section 30.2.3.1, "Standard Photographs").

The uses of line drawings can range from depicting a realistic image (eg, the details of an experimental apparatus) to illustrating a purely conceptual one (eg, the details of a chemical reaction). A major advantage of a line drawing is that it can focus attention on a subject without including extraneous or detracting details. The choice of how realistic or conceptual to make a drawing depends on the audience and the purpose of the drawing.

Drawings for scientific publications are best prepared by professional illustrators, graphic artists, or technical drafters who specialize in the relevant scientific disciplines. Good line drawings are not only illustrative and informative but also visually pleasing.

As with all other images, line drawings should emphasize the subject. Only details important to understanding the subject should be included, key aspects should be identified, and relationships in the drawings should be clear. The lines should be clean, sharp, and contrast well with the background. In addition, drawings should be sized to be legible if they have to be reduced for publication. Many journals specify in their instructions for authors the desired dimensions of published images, such as 1- or 2-column width or maximum image areas in inches or centimeters.

If line length, width, direction, configuration, or texture denote specific information, a legend should define each element.

All elements of line drawings must be readable in published documents. Consequently, lines should be at least 1-point thick (ie, 1/72 inch, or 0.35 mm). Labels should be set in at least 8-point type, and superscripts and subscripts should be set in at least 6-point type. A sans serif font, such as Helvetica or Eurostyle, is usually preferred for the small type in line drawings. Lines connecting labels to the parts of the drawing they identify should be a lighter weight than the lines that form the image itself.

30.2.3.4 Maps

Maps are an integral part of publishing in geoscience and some other disciplines. Maps combine elements of cartography, statistics, graphic design, and fine art. Maps also have a variety of formats (eg, line drawings, pictorial maps, and arial photos). When a map is included with a scientific manuscript, its purpose should be clear.

Maps are primarily visual, so they must be legible and carefully rendered. To create a map with a clear message, care must be taken to balance the size, number, and density of elements; the contrast between different elements; the number and use of colors; and the relationship between background and foreground images.

By their very nature, maps distort the spatial relationships of areas, distances, directions, angles, and gross shapes. In particular, be careful of distortions created by emphasizing certain elements, de-emphasizing contradictory elements, using provocative symbols, and overloading the image with details.

In addition, authors, data specialists, and editors should make sure that the names of attributes plotted on maps match the names of those attributes in geographic metadata files (see Section 14.1.2, "Authorities for Names").

The guidelines in this section are for printed maps, although many of these guidelines also apply to maps viewed online. The most common parts of maps are described below. Not all maps will have all of these parts.

More information on designing and editing maps can be found in *Suggestions to Authors of the Reports of the United States Geological Survey*.[16]

30.2.3.4.1 TITLE

A map's title should identify the locations or data depicted. The title should be a dominant element on the map, but it should not obscure important details. The title may appear anywhere on the map, or it may be placed above or below the map. The title should use headline-style capitalization.

30.2.3.4.2 SOURCES

The sources of the data used to create a map are typically placed in small type at the bottom of the map, often outside the map's border.

30.2.3.4.3 PREPARER

The preparer of the map, whether a person or group, is usually acknowledged along with the source of the data.

30.2.3.4.4 COPYRIGHT AND PUBLICATION DATE

The copyright holder and the date of publication should be listed on the map. Also, the map's publisher should be named if it differs from the copyright holder. This information is typically placed in small type at the bottom of the map, often outside the border.

30.2.3.4.5 PROJECTION

To compensate for differences caused by Earth being spherical and maps being flat, map designers rely on projections (eg, Mercator, Goode homolosine, and sinusoidal). Projections distort the shape of the planet in different ways, so each projection has strengths and weaknesses. For example, the Mercator projection of a world map is square, and latitude and longitude lines intersect perpendicularly, resulting in distortions that increase the relative size of geographic areas the farther they are from the equator. The world depicted by the sinusoidal projection is shaped like a squat sphere in which each pole comes to a point, and distortions narrow geographic areas the farther they are from the equator or the prime meridian, now officially the "International Earth Rotation and Reference Systems Service Meridian."[17]

Map designers should choose the projection that best illustrates the points being made in the manuscript. In addition, the projection should be noted in small type at the bottom of the map.

30.2.3.4.6 COMPASS ROSE

A compass rose indicates directions on a map. The term "compass rose" takes its name from the tradition of indicating 8 or even 16 points on an artistic compass drawn on maps, but a single arrow pointing north is now sufficient.

A compass rose is required if the top of a map is not oriented north. The compass rose can be placed anywhere on the map as long as it does not obscure important information.

30.2.3.4.7 SCALE

On a map, the scale is the ratio of the map dimensions to actual geographic distances. It is useful to provide the scale both as a ratio (eg, 1:50) and as a conversion unit (eg, 1 cm = 750 km). The scale is often placed at the bottom of the map close to the projection.

30.2.3.4.8 COORDINATE SYSTEM

To assist readers in finding specific sites, many maps feature a coordinate system, such as lines of latitude and longitude (also known as "parallels" and "meridians," respectively), numerical scales along the map sides (eg, Cartesian coordinates), or a grid pattern with alphabetized and numbered rows and columns. Associated with a coordinate system is a map index that lists the coordinates of locations on the map, usually in alphabetical order.

The details of the chosen coordinate system should be obvious, and they should be consistent with the map's index. A different coordinate system may be needed for an inset map (see 30.2.3.4.12, "Inset Map").

30.2.3.4.9 IDENTIFICATION SYMBOLS

Maps use a variety of symbols to identify various features of geography, demography, economic activity, and other types of information. These symbols can include pictographs, icons, colors, shading, cross-hatching, stippling, and even typefaces and type sizes. The symbols need to be legible and distinct from each other, and they should be defined in a legend (see Section 30.2.3.4.11, "Legend"). Standard symbol systems are available for several applications and should be used when applicable.

30.2.3.4.10 QUANTITATIVE SYMBOLS

Many maps overlay statistical graphics on locations to indicate quantity, intensity, density, gradation, and other measures. The symbols should be distinct enough to make meaningful comparisons.

30.2.3.4.11 LEGEND

On a map, the legend defines the meaning of graphic conventions, identification symbols (eg, pictographs, icons, and cross-hatching), and quantitative symbols (eg, dotted lines for national boundaries, orange for high temperatures, and stippling for forests). The legend is usually placed on the edge of the map in a position that does not obscure any of the map's details.

30.2.3.4.12 INSET MAP

An inset map is a small map inside the primary map. Drawn on a larger scale than the primary map, the inset map expands a section of the primary map to display more details in that section (eg, a small map of a major city on a large state highway map). An inset map may be placed anywhere on the primary map. An inset map usually has a different coordinate system for locating places than does the primary map.

30.2.3.4.13 LOCATOR MAP

A locator map is a small map drawn at a small scale to indicate the location and orientation of the primary map in relation to a larger geographical area (eg, a drawing of a country with a shaded area indicating the region depicted on the primary map). The indicator map may be placed anywhere on the primary map.

30.3 INDEXES

Because the purpose of an index is to enable readers to find information easily, the needs of likely readers should guide which terms are indexed, which cross-references are included, and what depth of indexing is provided. Guidance on these issues can be found in texts on indexing.[19–21] Another helpful source is *The Chicago Manual of Style*.[22]

After these initial decisions are made, the structure of the index needs to be deter-

Table 30.1. Checklist for Decisions about Content and Formatting of Indexes[a]

What unique aspects of the topic and the audience should the index consider? For example, if the text focuses on geography, does each place name need to be indexed? If the text is medical, does each disease and each treatment need to be indexed? If the text is historical, do dates need to be indexed?
Is only a subject index needed, or should authors of cited references also be indexed?
Does the index need to observe scientific conventions, such as italicizing the genus and species names for organisms and using small capitals for some prefixes of chemical names?
Should entries be alphabetized using the letter-by-letter or word-by-word system?
To how many levels should terms be indexed? Should the index have one level only for entries, top-level entries with second-level subterms, or top-level entries with second- and third-level subterms? For example, if a solvent is indexed, should each of its chemical components also be indexed?
Should subterms be stacked or set in the "run-in" format?
How far should subterms be indented?
When entries extend beyond one line (ie, "turnovers"), how far should the subsequent lines be indented?
Should the index entries reference page numbers, section numbers, or both?
How much spacing should separate the entry terms from the page or section numbers?
When listed in subterms, should cross-references be listed at the top or bottom of the subterms under the same top-level entry?

[a] These decisions should be made by the indexer and publisher and discussed with the designer to ensure the layout of the index follows the indexer's and publisher's intentions.

mined (see Table 30.1). Among these decisions is determining how many levels the index should have. Indexes typically have either a single level, with all terms treated equally, or 2 levels, with top-level terms and subterms (eg, indexing a river as a top-level term and its primary tributaries as subterms). Under rare circumstances, topics benefit from indexes with more than 2 levels.

In general, indexes with 2 levels are organized either with the subterms stacked and indented below their respective top-level terms or in a "run-in" format with each top-level term and its respective subterms on the same line. With the run-in format, each top-level term is separated from its subterms by a colon, and the subterms are separated from each other with semicolons. In both the stacked format and the run-in format, the subterms are alphabetized.

> **Stacked format:**
> eponymic terms
>   capitalization 90
>   clinical genetic syndromes 237
>   rules for use 132–133
>
> **Run-in format:**
> eponymic terms: capitalization 90; clinical genetic syndromes 237; rules for use 132–133

For top-level terms and subterms, use only nouns, and do not include articles before the nouns.

> "equator" *not* "the equator"

Retain all the words in compound terms. Reducing such entries to their root nouns could render the terms unhelpful or meaningless in an index.

> ground truth   North Pole   red blood cell
> intensive care   polar bear   somatic dysfunction

Because readers will infer that a subterm is subordinate to the term at the level above it, a preposition indicating the relation of the subterm to the entry above is seldom needed. When a preposition is needed for clarity, follow logical grammatical order and alphabetization.

> metastases
>   from liver 45–92
>   to liver 1–5

Initialisms and other abbreviations are not good index terms because nearly all abbreviations have more than one meaning, even in the same field. In cardiology, for example, "TVP" can mean "tricuspid valve plasty," "tricuspid valve prolapse," "temporary transvenous pacing," and "transplant vasculopathy."[18]

If initialisms and other abbreviations are chosen as index headings, intended readers should know those abbreviations as well as or better than the full terms. For example, in indexes intended for most audiences, "AIDS" and "DNA" may be preferred to "acquired immunodeficiency syndrome" and "deoxyribonucleic acid," respectively.

30.3.1 Alphabetization

30.3.1.1 Letter-by-Letter or Word-by-Word System

Index entries can be alphabetized in 2 different ways, depending on how terms with 2 or more words are treated. The generally preferred system is letter-by-letter alphabetization. In this system, all letters in terms with 2 or more words are alphabetized as if all the terms run together as one word. As a consequence, terms with one word and terms with 2 or more are intermixed in this system because the spaces between the multiword terms is ignored, as is any punctuation between the words.

The second method is the word-by-word system. This system groups terms with the same first word and then alphabetizes them based on the sequence of letters in the second word. Because alphabetizing stops at the end of the first word and restarts at the beginning of the second, 2-word terms are placed above one-word terms that begin with all the same letters as the first word in 2-word terms.

In both systems, alphabetizing restarts after a comma or a parenthesis.

Letter-by-letter alphabetization:	Word-by-word alphabetization
null hypothesis	null hypothesis
nullification	null results
null results	nullification
solar cell	solar cell
solarization	solar panel
solar panel	solarization
wattage	Watt, James
Watt, James	wattage

Computer programs designed specifically for indexes can sort letter by letter. However, general word-processing and desktop publishing programs may not, which would require indexers to correct alphabetization.

30.3.1.2 Prefixes

For chemical names that have numerals as prefixes, entries are alphabetized by the letters in the names, not the numerals. However, numerical order is preferred for sorting entries of chemical names with the same root but different numbers. Other descriptive prefixes (eg, chloro-, α-, β-, D-, and L-) are ignored in alphabetizing entries.

| butanol | 2-butanol | 1-butanone | β-carotene |
| 1-butanol | tert-butanolysis | 2-butanone, 3-hydroxy- | α_1-globulin |

Computer programs may sort such terms incorrectly, placing those that begin with numbers or symbols at the end of the alphabetical sequence and those that begin with letters alphabetically according to the prefix. If so, the terms should be rearranged in the proper sequence.

30.3.1.3 "Mac," "Mc," and "M'"

Letter-by-letter sequence is recommended for names that begin with "Mac" and "Mc," as well as for names that begin with "M" followed by an apostrophe. British practice, however, is to alphabetize names that begin with all 3 prefixes as if they were all spelled "Mac."

Recommended sequence:	British sequence:
Macadam, J	Macadam, J
MacGillivray, W	McCarthy, D
Marconi, G	M'Clintock, J
McCarthy, D	McGaugh, S
McGaugh, S	MacGillivray, W
M'Clintock, J	Marconi, G

Letter-by-letter sequence is preferred for surnames with other prefixes (eg, "D'Marco" and "O'Brian").

30.3.1.4 Abbreviations

Letter-by-letter sequence is recommended for alphabetizing abbreviations. Some indexes, however, place abbreviations at the beginning of their respective alphabetic sections.

Recommended sequence:	Alternative sequence:
CBC	CBC
Chemical bond	CT
Conduction	Chemical bond
CT	Conduction

30.3.1.5 Non-English Roman Characters

When alphabetizing non-English words beginning with a ligature or a letter with a diacritic, treat those letters as though they were unmarked English letters, just as ligatures and diacritics are ignored with author names in reference lists (see Section 29.3.2, "Sequence of End References").

In some European languages, words beginning with ligatures or diacritics may be placed in a different order. For example, in a Danish dictionary, words beginning with the ligatures "æ," "ø," and "å" follow words beginning with "z."

30.3.2 Capitalization

Indexes of scientific publications should not follow arbitrary capitalization, such as capitalizing main-entry terms and lowercasing subterms. Instead, the rules on capitalizing and lowercasing initial letters specified in this style manual should be (see Chapter 9,

"Capitalization"). In the sciences, capitalized and lowercased letters differentiate many classes of terms. For example, proprietary names (ie, trademark names) of drugs are capitalized, and nonproprietary names (ie, generic names) are not (see Section 9.4.8, "Trademarks"). In addition, genus and species are differentiated by capitalizing the genus and lowercasing the species (see Section 9.4.5, "Taxonomic Names").

30.3.3 Punctuation

When indexing personal names, alphabetize by surnames, and separate surnames from given names with a comma.

> Burns, Louisa 112
> Downs, Marion 17
> Verbeek, Rogier 89

In contrast, punctuation is generally not needed to separate terms from page numbers in either stacked or run-in indexes. Spacing suffices.

> anatomy 57
> geology 162
> zoology 98

However, commas are needed to separate multiple page references for the same entry.

> black hole 15, 69, 101
> light-year 26, 64–67, 89
> satellite 5–8, 76–84

If an index contains terms that incorporate numbers, it may be visually helpful to separate the terms from the page or section numbers with a comma.

> history
>   14th century, 12–13
>   1799–1825, 40, 44–47
>   1850–1890, 52–57, 66, 70
>   Freemasons, 99, 102

If the index contains entries in which punctuation is meaningful, as in entries for chemical names, extra space may be preferred to a comma between the terms and the page numbers, even for entries without punctuation. For this purpose, use the em space, created by Unicode 2003.

> 1-butanol, 2-cyclohexyl-   83, 123, 267
> $C_5H_{10}O_2$ 272–274, 410
> dextrin-1,6-glucosidase   5, 7, 10
> 1,1-dichloroethylene   42, 117
> vitamin B   11

30.3.4 Page Numbers

When an entry is for consecutive pages, list the numbers for both the initial page and the final page, not just the initial page or each page separately. Separate the initial page number from the final page number with an en dash, which is slightly longer than a hyphen and which is created with Unicode 2013 (see Section 5.3.5.1, "En Dash").

> Refraction 17–21 *not* Refraction 17 *and not* Refraction 17, 18, 19, 20, 21

30.3.5 Section Numbers

Texts that are organized by section numbers can be indexed by section numbers, page numbers, or both. For entries with both section numbers and page numbers, the section numbers should appear first, and the page numbers should be placed within parentheses.

Both section and page numbers:	Section numbers only:	Page numbers only:
age 203.1 (4–8)	age 203.1	age 4–8
AIDS	AIDS	AIDS
certification 203.5 (17–20)	certification 203.5	certification 17–20
confidentiality 203.6 (21)	confidentiality 203.6	confidentiality 21
anemia 400.3 (23)	anemia 400.3	anemia 23

30.3.6 Cross-referencing

To make an index more helpful, references can be included to related entries or to the terms under which the information being sought is indexed. Such cross-references are indicated with the words "see also" and "see," respectively, which are often italicized to make clear that they are not the entry terms.

"See" is used only to direct readers to another entry, so it is placed between the 2 related terms on the same line. "See also" precedes a subterm to direct readers to information in addition to that provided by an entry's other subterms. The phrase "see also" can appear either before or after the other subterms under the top-level term. The preferred method is to place the cross-reference entry below the other subterms. Which method to use should be decided at the outset of the indexing process, as should the placement, spacing, and typeface for cross-references.

It is inconsiderate to refer readers to a cross-reference that provides only one page number. Instead, cite the page number in both entries, and do not cross-reference them. In addition, some information warrants having both top-level entries and subterm entries under other top-level entries.

> barns
>   Amish 25
>   New England 5–10
>   Pennsylvania 11, 20–22, 25, 60–68
>   tertiary use 40–45
>   *see also* houses
> BIA *see* Bureau of Indian Affairs
>
> houses 16–22
>   *see also* radon
> Pennsylvania
>   Amish 25
>   cities 60–68
>   Lake Erie 20–22
>   rivers 11
>   *see also* barns
> rivers 11

30.3.7 Clarifying Homonyms

Homonymic personal names (ie, the names of 2 or more persons spelled exactly the same) and other words with the same spelling but with different referents should be differentiated to avoid misdirecting index users. Options include identifying position, birthplace, or birth date for persons and specifying type for objects. The information that differentiates the entries in the index does not necessarily have to be mentioned in the text.

CITED REFERENCES

1. Lang TA. How to write, publish, and present in the health sciences: a guide for clinicians and laboratory researchers. American College of Physicians; 2010.

2. Lang TA. Up and down or side by side: structuring comparisons in data tables. AMWA J. 2018;33(3):104–110. https://cdn.ymaws.com/www.amwa.org/resource/resmgr/journal/issues/2018/AMWA_Journal_33.3_online_OCT.pdf

3. Harris RL. Information graphics. A comprehensive illustrated reference. Oxford University Press; 1999.

4. Gelman A, Pasarica C, Dodhia R. Let's practice what we preach: turning tables into graphs. Am Stat. 2002;56(1):121–130. https://www.jstor.org/stable/3087382

5. Day RA, Gastel B. How to write and publish a scientific paper. 9th ed. Greenwood; 2022.

6. Loos EM. Evaluating scientific illustrations: basics for editors. Sci Ed. 2000;23(4):124–125.

7. Cleveland WS, McGill R. Graphical perception and graphical methods for analyzing scientific data. Science. 1985;229(4716):828–833. https://doi.org/10.1126/science.229.4716.828

8. Cleveland WS. The elements of graphing data. 2nd ed. Hobart Press; 1994.

9. Cleveland WS. Visualizing data. Hobart Press; 1993.

10. Tufte ER. Envisioning information. Graphics Press; 1990.

11. Tufte ER. Visual explanations: images and quantities, evidence and narrative. Graphics Press; 1997.

12. Tufte ER. The cognitive style of PowerPoint. Graphics Press; 2004.

13. Lang TA, Talerico C, Siontis GCM. Documenting clinical and laboratory images in publications: the CLIP principles. CHEST. 2012;141(6):1626–1632. https://www.doi.org/10.1378/chest.11-1800

14. Rossner M, Yamada KM. What's in a picture? The temptation of image manipulation. J Cellular Bio. 2004;166:11–15. http://doi.org/10.1083/jcb.200406019

15. Harris G. The standard lens. Learning with Experts. 15 May 2013. https://www.learningwithexperts.com/photography/blog/the-standard-lens

16. Hansen WR, editor. Suggestions to authors of the reports of the United States Geological Survey. 7th ed. United States Geological Survey; 1991. https://doi.org/10.3133/7000088

17. International Earth Rotation Service. Federal Agency for Cartography and Geodesy; 2013. https://www.iers.org/IERS/EN/Organization/About/History/history.html

18. Lang TA. Abbreviations: expectations, permutations, revelations, reservations, and applications of shortened words and phrases. AMWA J. 2019;34(4):152–157. https://cdn.ymaws.com/www.amwa.org/resource/resmgr/journal/issues/2019/amwa_journal_34.4_web.pdf

19. Lancaster FW. Indexing and abstracting in theory and practice. 3rd ed. University of Illinois, Graduate School of Library and Information Science; 2003.

20. Mulvaney NC. Indexing books. 2nd ed. The University of Chicago Press; 2005.

21. Wellisch HH. Indexing from A to Z. 2nd ed. HW Wilson; 1996.

22. The Chicago manual of style: the essential guide for writers, editors, and publishers. 17th ed. The University of Chicago Press; 2017. Also available at https://www.chicagomanualofstyle.org

ADDITIONAL REFERENCES

American Society for Indexing. Indexing evaluation checklist. American Society for Indexing. c2023. http://www.asindexing.org/about-indexing/index-evaluation-checklist/

Briscoe MH. Preparing scientific illustrations: a guide to better posters, presentations, and publications. 2nd ed. Springer-Verlag; 1996.

Few S. Now you see it: an introduction to visual data sensemaking. Analytics Press; 2021.

Few S. Show me the numbers. Designing tables and graphs to enlighten. Analytics Press; 2004.

Frankel F. Envisioning science: the design and craft of the science image. MIT Press; 2002.

Hodges ERS, editor. The Guild handbook of scientific illustration. 2nd ed. John Wiley & Sons; 2003.

International Standards Organization. Information and documentation—guidelines for the content, organization and presentation of indexes. 2nd ed. International Standards Organization; 1996. (ISO 999).

Kasdorf WE. The Columbia guide to digital publishing. Columbia University Press; 2003.

Monmonier M. How to lie with maps. 3rd ed. The University of Chicago Press; 2018.

Schofield EK. Quality of graphs in scientific journals: an exploratory study. Sci Ed. 2002;25(2):39–41.

Tufte ER. The visual display of quantitative information. 2nd ed. Graphics Press; 2001.

Wainer H. Graphic discovery. A trout in the milk and other visual adventures. Princeton University Press; 2005.

Wainer H. How to display data badly. Am Stat. 1984;38(2):137–147. https://doi.org/10.1080/00031305.1984.10483186

Wright P. Presenting technical information: a survey of research findings. Instruct Sci. 1977;6(1):93–134. https://doi.org/10.1007/BF00121082

31 Typography and Manuscript Preparation

Editors: Audrey Daniel Lusher and Simona Fernandes, MSc, ELS

31.1 TYPE SPECIFICATIONS

Text attributes (eg, roman, italic, and boldface), letter case (eg, capital letters and small capitals), and position of characters (eg, superscript and subscript numbers and letters) convey special meanings in scientific publications. These conventions are summarized in this manual in Chapter 10, "Type Styles, Excerpts, Quotations, and Ellipses," and throughout Part 3, "Special Scientific Conventions." Although authors should be familiar with the basics of these conventions, they need not be expert in applying them. Authors rely on copy editors to ensure that manuscripts adhere to such conventions, especially those that are specific to scientific fields and publishers.

The design specifications and detailed typographic characteristics used by publications are usually established by publishers and editors in consultation with graphic designers. Specifications are typically developed when launching a new journal or redesigning an existing journal. The elements that need to be specified are summarized in Table 31.1.

Publishers also establish requirements for formatting manuscripts for submission. For example, publishers may require that authors use a specific software for word processing the text and for rendering chemical and mathematical formulas. Publishers may also stipulate formats for figures, photographs, and other graphic elements. Before drafting a manuscript for a publication, authors should review the publication's instructions for authors, specifically the section on manuscript style and format (see Table 27.6).

In the electronic processes overwhelmingly used today, copy editors modify accepted manuscripts by using style functions in word-processing software and coding to apply the publication's style and format, such as the typefaces for titles, headings, text, and tables; the spacing between type lines, known as "leading"; special characters; and other typesetting aspects of manuscripts. In traditional paper-based publication processes,

Table 31.1. Specifications for a Print Journal: Dimensions, Paper, Design, and Typographic Details

Characteristic	Notes
Trim size	Width and height of the cover and pages.[a]
Paper: kind and weight	Selection determined by the kind and quality of paper needed and the cost. The kind influences the quality of color and halftone illustrations (eg, photographs and radiographs).
Layout of pages[b]	
Type-page margins	Space around the type page (ie, the area surrounding the text).
Columns	Pages typically have a 1-, 2-, or 3-column layout for the text.
Typefaces	Determined in part by aesthetic and readability considerations. The need for typefaces that have special characters for chemistry, mathematics, and physics may limit the options.
Text sizes	Sizes differ for titles, headings, body text, references, figure captions, and the various elements of tables.
Text colors[c]	Color is typically limited to titles, text heads, in-text callouts, tables, figures, and appendixes.
Leading	Space between lines, specified for each size of type.
Title page for articles	The layout for this page typically consists of the article title, the author byline, the authors' affiliations, the corresponding author's contact information, an abstract, the article's bibliographic reference, keywords, information on financial and other support, the beginning of the text, and the running foot used at the bottom of all pages of the article.[d] Any conflicts of interest that the editor in chief believes should be shared with readers can also be listed on this page (eg, authors serving as employees of an organization that funded the research or authors serving on the journal's advisory board).
Text format	Formatting includes indentations, spacing between paragraphs, and spacing around figures and tables.
Text heads and subheads	Typefaces and sizes differ based on the level of the headings.[e]
Table characteristics	Characteristics include column and row headings, rules, spacing, and footnote signs.[f]
Design of special pages	Special pages, such as the table of contents, have designs that differ from that of articles.[g]
Cover design	Includes such essential elements as the journal's title, volume number, and issue number.[h,i]

[a] See Section 27.6.2.4.3, "Trim Size."
[b] See Section 27.6.2.4, "Text Pages."
[c] Color may be expensive and impractical for some printed publications, but it is usually an option for online and other digital publications. For digital formats, choose colors that are aesthetically pleasing on all monitors, regardless of the monitors' color calibration.
[d] See Section 27.7.1, "Title Page."
[e] See Section 27.7.2.1, "Text Headings."
[f] See Section 30.1, "Tables."
[g] See Sections 27.6.2.3, "Table of Contents," and 27.6.2.5, "Information for Authors."
[h] See Section 27.6.2.1, "Cover."
[i] With a book instead of a journal, the cover may include such elements as the book's title, authors, editors, and publisher. If a book has a jacket, these elements may be on the jacket, and the book cover may be blank. See Section 28.4.1, "Cover, Jacket, and Endpapers."

copy editors mark up manuscripts to indicate these specifications, which typesetters or designers apply when creating galleys and page proofs.

The remainder of this chapter summarizes the basics of preparing journal and book manuscripts for publication. This chapter does not attempt to describe all of the decisions, steps, and processes involved. Detailed guidance on manuscript preparation can be found in publications such as *The Chicago Manual of Style*[1] and *The Copyeditor's Handbook*.[2]

31.2 CHARACTERISTICS OF TYPE

Manuscript editors should be familiar with the chief characteristics of type: typeface styles and weights, type sizes, and character and line spacing. These characteristics must be specified for both print and electronic publications. Although authors typically are not involved with setting these specifications, having basic knowledge of typographic characteristics may help authors understand how their manuscripts have been prepared for publication.

Further guidance on the elements of typography can be found in *The Chicago Manual of Style*.[1] In addition, *The Elements of Typographic Style*[3] has clear and concise descriptions of the practical and aesthetic aspects of typographic design and typefaces.

31.2.1 Typeface Styles

A "typeface," or simply a "face," is a coherently designed group of letters, numerals, punctuation, symbols, and additional characters used together (eg, Helvetica or Times New Roman). In comparison, a "font" is a complete set of characters of a specific typeface in a given size, weight, and style (eg, 12-point bold Helvetica or 10-point italic Times New Roman). The meanings of "typeface" and "font" have been blurred in recent years: "Font" is now often applied where "typeface" has traditionally been used in the printing industry and among typographers.[4]

The classifications of typefaces are complex. One classification is that of Western typography, which has 3 major typefaces. The most widely used typeface in printing English-language texts is the "roman typeface," which is based on the style of capital, or uppercase, letters that were used in classical Roman culture and related lowercase letters that were developed later.[5] The second major typeface, known as "italic typeface," is based on a style of lowercase letters developed in Italy early in the 16th century. The third is the "Gothic/Old English typeface," which is based on the style used in European manuscripts, particularly those from German-speaking countries, between the end of the 12th century and the 20th century.

> Times New Roman is a common roman typeface.
> *Italic typefaces like Freestyle Script resemble cursive writing.*
> Old English Text MT is an example of a Gothic/Old English typeface.

Roman typefaces and Gothic/Old English typefaces can be modified to mimic italic typefaces by slanting the characters. This style manual follows the convention of referring to all "upright" typefaces as "roman" and all "slanted" ones as "italic." Section 10.1.1, "Italic Type," summarizes the general and scientific uses of the italic typeface.

Typefaces are also classified as being either serif or sans serif. Serifs are short straight or curved lines at the ends of the main strokes of letters. Sans serif typefaces lack serifs.

Circled are the capital T's 4 serifs and the lowercase t's 2 serifs.

Tt

Sans serif typefaces terminate with straight edges.

Council of Science Editors (a roman serif typeface)
Council of Science Editors (a roman sans serif typeface)
Council of Science Editors (an italic version of a serif typeface)
Council of Science Editors (an italic version of a sans serif typeface)

Each typeface has an identifying name, which may be the designer's surname, a reference to the typeface's geographic or historical origin, or a term that evokes the aesthetic character of the design.

The roman serif typeface Times New Roman was designed for *The Times* of London.
Developed by Swiss designers, the roman sans serif typeface Helvetica is named after the Latin word for "Swiss."

There are many other varieties of typefaces for the Roman alphabet. Among the most notable are Fraktur typefaces, or blackletter script, which were developed in the early days of print media and evolved with the German language to convey its unique features.

Some typefaces work better than others for unique purposes, such as algebras like Lie[6] and other symbol systems. Some typefaces are not used in scientific publication at all, such as script typefaces designed to simulate handwritten letters.

Greek letters, which are widely used in scientific and mathematic publications, are available in many typefaces. However, uppercase and lowercase Greek letters are represented as symbols rather than alphabetical letters in typefaces. For example, Unicode 03A8 produces the Greek capital letter psi in a number of typefaces, including Times New Roman as Ψ, Helvetica as Ψ, and Franklin Gothic Demi as Ψ.

Additional variations within typefaces are useful in scientific publishing. For example, some publications use small capitals in combination with initial capitals as a means of distinguishing surnames from given names in bylines (see Section 8.1.2, "Conventions for Authors' Names"). Small capitals are letters with the basic structure of capital letters but at the height of lowercase letters.

Ōsumi Yoshinori, DSc   Maria Goeppert Mayer, PhD   Aubrey de Grey, PhD

31.2.2 Typeface Weights

Many typefaces have closely related typefaces that differ in weight, a characteristic determined mainly by the width of letter strokes. For example, a derived typeface with heavy or wide strokes is known as the "boldface" version of the typeface. Other common weights are "italic" and "boldface italic."

Archaeologists discovered a 3,400-year-old city on the west bank of the Nile. (in boldface type)
A star's corona is far hotter than its surface. (in italic type)
A clone is a type of plant cultivar. (in boldface italic type)

The general and scientific uses of boldface type are summarized in Section 10.1.2, "Boldface Type," while those of italic type are summarized in Section 10.1.1, "Italic Type."

31.2.3 Type Sizes and Spacing

In the publishing industry, linear dimensions are measured in 4 units: the "point," the "pica," the "em," and the "en." Points and picas are related to each other, as are ems and ens. Both points and picas can be defined as fractions of an inch.[4]

> 1 point = 1/12 pica = 1/72 inch (approximately 0.35 mm)
> 12 points = 1 pica = 1/6 inch (approximately 4.2 mm)
> 72 points = 6 picas = 1 inch (approximately 25.4 mm)

Points are used to measure the height of characters in a font; the spacing between characters, words, and lines of type; and the thickness of ruled lines. The size of a font is defined as the number of points in the "maximal vertical dimension" of its characters. A font's maximal vertical dimension is measured from the lowest point of the characters' descenders (eg, the downward stroke of the lowercase letter "j") to the highest point of the characters' ascenders (eg, the upward stroke of the letter "h") (see Figure 31.1). Another essential dimension of fonts is the "x-height," also known as the "corpus height." This is the height of lowercase characters without ascenders and descenders, such as x and m.

While the vertical linear dimensions of characters are measured in points, horizontal linear dimensions are measured in units of em and en. The name "em" is derived from the width of the capital letter M. The "en" has traditionally been defined as half the length of the em. However, these 2 units are less rigidly defined in today's electronic composition systems. Ems and ens are used to specify horizontal distances for such

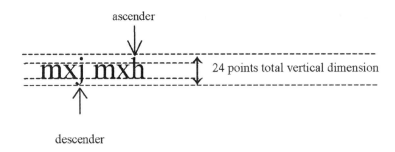

Enzymes are proteins.   [in 8-point Times]

Enzymes are proteins.   [in 10-point Times]

Enzymes are proteins.   [in 12-point Times]

Enzymes are proteins.   [in 14-point Times]

Enzymes are proteins.   [in 18-point Times]

Figure 31.1. Visual representations of ascender, descender, and sample typeface sizes.

design elements as word and letter spacing, indentations, and dashes (see Sections 5.3.5.2, "Em Dash (—)" and 5.3.5.1, "En Dash (–)").

Picas, in turn, are used to measure the vertical and horizontal dimensions of a type page (ie, the portion of a page reserved for text, figures, tables, and other content), the spacing between columns of type (sometimes given in ems instead), and other elements in the page design. However, modern design software may use inches or metric units instead of picas for these purposes. In addition, the outside dimensions of an entire page (ie, the margins surrounding the type page) are often stated in inches or metric units, even when the type page and the elements within it are measured in picas.

31.3 COPYEDITING

Before an accepted manuscript is sent to a typesetter or graphic designer for typesetting, a copy editor (known as a "subeditor" in British parlance and sometimes called a "manuscript editor" in American parlance) checks the manuscript to ensure that it adheres to the publisher's style specifications (eg, the style details outlined in this manual) and scientific conventions. This editor also corrects errors found in references, spelling, punctuation, and other elements of the manuscript. Some aspects of this process are frequently automated by editing software that can correct spelling and grammar or correctly format references. The changes are typically tracked by the editing software so that the copy editor can review and, if necessary, revise the changes.

If the authors' intent pertaining to any of the details of the manuscript is not clear, the copy editor should clarify the items with the manuscript's corresponding author, preferably before the manuscript is typeset. Alternatively, if the editor believes that proposed edits clarify the authors' intended meaning, the editor can accept the changes within the manuscript's electronic file and ask the corresponding author to confirm that the changes more clearly convey the intended meaning.

Publishers establish the specifics of author approval of edited manuscripts, including at what stage it occurs, how and in what form the document is sent to the corresponding author, and the requested turnaround time. The timing and process of each stage are designed to fit with other steps of the production process and schedule.

See *The Chicago Manual of Style*[1] for detailed information on copyediting.

31.4 MANUSCRIPT MARKING FOR TYPESETTING

31.4.1 Electronic Markup

Most publishers insist that authors use word-processing software to prepare manuscripts and, in so doing, indicate type characteristics such as italic face, boldface, small capitals, and special characters. After copyediting these electronic manuscripts, copy editors return the manuscripts with queries to the corresponding authors. Depending on the publishers' protocols, the queries may be embedded in the text, outlined in a series of electronic comments in the text, listed at the beginning or end of the manuscripts, or detailed in emails accompanying the copyedited manuscripts.

After queries have been addressed and the manuscripts have been revised to the satisfaction of the corresponding authors and the editors, the manuscripts are typeset. Electronic typesetting and layout programs automatically apply the specific typefaces, weights, sizes, styles, and formats used by the manuscripts' publishers. For example, many systems convert styles applied in word-processing programs into type characteristics in layout software or into markup languages. Alternatively, manuscripts may be coded manually to designate the functional elements of text, equations, tables, and other elements. This coding may take the form of generic codes, style tags, or markup languages.[1] The systems that process the manuscripts for print and electronic publication are programmed to recognize the codes and tags and automatically apply stipulated specifications to galleys, page layouts, and electronic screens.

31.4.2 Hard-Copy Markup

Traditional hard-copy publishing processes have all but disappeared in scientific publishing. Compared with electronic markup, preparing paper manuscripts for publication is more labor intensive and time consuming for copy editors, typesetters, and even authors.

In copyediting paper manuscripts, editors use handwritten copyediting marks to correct misspellings, awkwardly worded sentences, incorrectly formatted references, and other problems. Queries to authors must be clearly distinguished from copyediting marks and instructions to typesetters and graphic designers. For example, author queries may be placed within brackets and marked specifically for the authors' attention (eg, "A: Please confirm that this change is accurate" or "Au: Is the unit of measure centimeters or millimeters?"). When such copyediting changes and queries are extensive, parts of manuscripts or accompanying documents may have to be retyped.

Copy editors also mark up the text for typesetters and graphic designers to designate such elements as boldface type, superscripts, subscripts, special characters, and different levels of headings. In addition, printing instructions should be placed adjacent to the element in question, and the instructions should be circled to indicate that they are not editing-related revisions to the manuscript. However, editors need not mark up elements for which specifications have been defined and that are readily identified by typesetters (eg, running text, single-level headings, article titles, and abstracts).

For examples of copyediting and proofreading marks and markup symbols, see Tables 31.2 and 32.1.

31.5 MARKUP LANGUAGES

For many scholarly journals and books, content is published in both print and digital formats. In fact, a growing number of journals and books are published only electronically. Digital content, in turn, is rendered in different formats to accommodate reading on computers, tablets, cell phones, and other devices. Various tagging systems and tools are used to prepare manuscripts in full-text format and Portable Document Format (PDF) for these different media.

Table 31.2. Typographic Conventions and Markings[a]

Convention	Mark	Example	Printed appearance
Italic type[b]	Single underline	Escherichia coli	*Escherichia coli*
Boldface type[c]	Wavy underline	the Paramyxoviridae	**the Paramyxoviridae**
Capital letter[d]	Triple underline	professor of geology	Professor of Geology
Small capital[e]	Double underline	d-cystathionine	D-cystathionine

[a] Because the overwhelming majority of journals now import word-processed documents into typesetting software, these marks have a limited role in publishing today. Additional marks can be found in Table 32.1.
[b] For additional discussion, see Section 10.1.1, "Italic Type."
[c] For additional discussion, see Section 10.1.2, "Boldface Type."
[d] For additional discussion, see Section 10.1.3, "Capital Letters."
[e] For additional discussion, see Section 10.1.4, "Small Capital Letters."

Although some tagging systems are proprietary, standards have been developed for the uniformity and interchangeability of electronic manuscripts. For example, the US National Information Standards Organization's standard for preparing and tagging electronic manuscripts[7] draws on the International Organization for Standardization's Standard Generalized Markup Language[8] (SGML). In addition, Extensible Markup Language[9] (XML), which is now widely used for tagging documents for electronic delivery, is derived from SGML but is less complex. Both XML and SGML are metalanguages that make up and define markup systems.[10] As such, they specify the structure of documents rather than their appearance.

In contrast, the following tagging systems define appearance but not structure: HyperText Markup Language (HTML), which is derived from SGML; HTML5, an improved and more advanced version of HTML; Extensible HyperText Markup Language (XHTML); and Adobe Acrobat's PDF.[10]

XML and SGML do not specify tags. Rather, tags are defined by the users according to their need (eg, a user may define the tag "<heading>" for a heading). Once documents are tagged, different processing systems are used to prepare the tagged documents for various media (eg, print, online applications, e-books, and mobile devices). Each system has a style sheet for its medium that converts the tags into the system's own codes, which in turn determine the final appearance of manuscripts in that medium.[7] These tags also allow content to be manipulated to extract material for additional utilities (eg, online abstracting and indexing in databases to increase searchability and reusability).

A required component for manuscripts marked up in SGML is document type definition (DTD). While optional with XML, DTD helps establish consistency within a set of documents.[10] By specifying the tags to be used and the rules for using them, DTD defines the overall structure of a document before it is coded in SGML or XML.

For further discussion of document structure and XML, see Chapter 1, "Elements of Scientific Publication."

CITED REFERENCES

1. The Chicago manual of style: the essential guide for writers, editors, and publishers. 17th ed. The University of Chicago Press; 2017. Also available at https://www.chicagomanualofstyle.org

2. Einsohn A, Schwartz M. The copyeditor's handbook: a guide for book publishing and corporate communications. 4th ed. University of California Press; 2019.

3. Bringhurst R. The elements of typographic style. 4th ed. Hartley & Marks; 2013.

4. Luna P. Typography: a very short introduction. Oxford University Press; 2018.

5. roman. Encyclopedia Britannica; [accessed 2023 Jan 20]. https://www.britannica.com/topic/roman -typeface

6. Hall BC. Lie algebras. In: Lie groups, lie algebras, and representations. Springer; 2015. p 49–75.

7. National Information Standards Organization (US). Electronic manuscript preparation and markup. NISO Press; 1995. (NISO/ANSI/ISO 12083-1995 [R2002]).

8. International Organization for Standardization. Information processing—text and office systems—Standard Generalized Markup Language (SGML). International Organization for Standardization; 1989. (ISO 8879). ISO 8879:1986/Amd 1:1988; ISO 8879:1986/Cor 1:1996; ISO 8879:1986/Cor 2:1999.

9. W3C XML Core Working Group. Extensible Markup Language (XML) 1.0. 5th ed. W3C recommendation. W3C Consortium; 2008. http://www.w3.org/TR/xml/

10. Kasdorf B. XML and PDF: why we need both. An introduction to the two core technologies for publishing. JP News J Publ. 2003;(2):1,3–14.

ADDITIONAL REFERENCES

Hall F. The business of digital publishing: an introduction to the digital book and journal industries. Routledge; 2013.

Mrva-Montoya A. Editing skills in the era of digital revolution. In The 6th IPEd National Editors Conference, Perth. 2013.

Saller CF. The subversive copyeditor: advice from Chicago. 2nd ed. The University of Chicago Press; 2016.

32 Proof Correction

Editors: Mary Warner, CAE; Kevin R. Brown; and Michelle Cathers, MLS

32.1 PROOFS

Most publishers provide authors with prepress proofs of articles, chapters, books, and other documents so that authors can correct inaccuracies and rectify other problems in the documents. Proofs show the text, tables, figures, and supplementary material in their final locations on pages. Most commonly, proofs are provided in a Portable Document Format (PDF), or they are posted on online-only digital interfaces (eg, Proof Central and e.Proofing). Publishers often email PDF proofs as attachments or send authors hyperlinks to downloadable PDFs or digital interfaces.

Authors are responsible for checking the proofs against the original text and other sources, indicating any corrections that need to be made, and replying to queries that copy editors made on the proofs (see Section 32.3, "Annotating PDF Proofs"). Usually, this is the last time authors see their work before publication, so thoroughness is critical. Proofs must be returned to the copy editors, printers, or publishers by a specific deadline. This is often done by email in the form of an attached annotated PDF or some other electronic method.

Because timing is critical, copy editors should inform corresponding authors when to expect proofs so that authors can make arrangements to review the proofs in a timely manner or take appropriate steps if they will not be available at that time, such as asking a coauthor or other contributor to the work to check the proofs.

32.2 CHECKING PROOFS

Proofreading encompasses a wide variety of tasks for authors, copy editors, and proofreaders, such as ensuring that all components of the document are present, determining whether the changes made in copyediting were appropriate, verifying that document specifications were followed, checking end-of-line hyphenation, and checking page numbers and running heads. Several important aspects of correcting proofs are described briefly in the rest of this chapter. Detailed guidelines for copy editors and proofreaders are available in *The Chicago Manual of Style*[1] and *Mark My Words*.[2]

32.2.1 Text and Composed Tables

32.2.1.1 Authors' Responsibilities

When asked to check proofs, corresponding authors should be instructed to do the following:

(1) Authors should read proofs at least twice. During the first reading, authors should compare the proofs against the original manuscripts to ensure that all components have been included and rendered correctly. For this reading, an author may ask another person to read the manuscript aloud while the author follows the text of the proof. During the second reading, authors should read the proof for sense.

(2) The authors' highest priority should be to carefully confirm the accuracy of scientific, technical, and mathematical content; measurements and dosages; and other data in the proofs. Given their scientific expertise, authors are far more likely to identify these types of errors than are copy editors and proofreaders.

(3) Authors should check the correctness of any changes made by copy editors and respond to all of the copy editors' queries. Rather than making changes or deletions, authors should only comment on queries, suggestions for changes, and corrections made by copy editors or proofreaders at an earlier proofing stage.

(4) At the proof stage, authors should limit their changes to correcting errors. The proof stage is not the time to make trivial changes, improve prose style, delete material, or add material (unless absolutely essential for scientific or scholarly reasons). Making such changes at this stage can introduce errors, cause production delays, and increase composition costs.

(5) Should authors want to add some text on important observations made since submitting their manuscripts, those additions need to be approved by the journal's or book's scientific editor, not by copy editors. This also applies to changing author order in bylines and to adding or deleting authors. In the case of a peer-reviewed article, it is generally unethical to add content after the manuscript has been accepted for publication unless the additional content is judged acceptable by the manuscript's peer reviewers or the scientific editor. As an alternative, the scientific editor may decide to provide the new material in a dated addendum or "note added in proof," which will obviate the need for changing the text.

(6) Authors need not concern themselves with a publisher's style points because the copy editors and proofreaders will pay close attention to those.

(7) Authors should not delete any identifying or other information on proofs that printers will need. These marks will be deleted by graphic designers or printers before the documents are published.

32.2.1.2 Accuracy

Authors, copy editors, and proofreaders should carefully check proofs against the original text to ensure that meaning has not been changed and to confirm that equations and numerical data are accurate, especially data in tables. They should also check for proper spelling, punctuation, separation of paragraphs, order of headings, and citation of references, figures, and tables. In addition, they should confirm that tables and figures are appropriately located in relation to their first mention in the text.

32.2.1.3 Typography

The publisher's copy editors, proofreaders, and typesetters are responsible for correcting problems with typography and alignment. Authors, in turn, should verify the proper alignment of statistical data, chemical formulas and equations, and mathematical expressions. In addition, authors should make sure that symbols agree with recognized conventions.

32.2.1.4 Word Division

Although most computer systems for composition have rules for hyphenating words at the ends of lines, the extent to which those systems are programmed to handle exceptions to general rules differs. So copy editors, proofreaders, and authors should take some care in checking proofs for word divisions at line endings. In the American system, word division is based on pronunciation (eg, "En-gland"), whereas in the British system, word division is based on the etymological derivation of words (eg, "Eng-land").

In checking divided words, copy editors, proofreaders, and authors should pay particular attention to the division of specialized scientific terms and Latin words. When the division of such words is in question, give preference to the word divisions recommended in the publisher's scientific dictionary of choice as opposed to the publisher's general dictionary of choice.

For terms that are usually spelled with hyphens, the break should be at an existing hyphen if possible. Chemical names, for example, frequently use hyphenated terms (see Chapter 17, "Chemical Formulas and Names"). When breaking scientific terms at the end of a line, the hyphen should fall at the end of the first line, not at the beginning of the next line.

> L-glutamine amido-   *not*   L-glutamine ami-   *not*   L-glutamine amido
> ligase   do-ligase   -ligase

If possible, avoid a break that produces part of a word that can be read as a word in its own right with a meaning that might be confusing, amusing, or offensive in the context (eg, avoid breaking "readjust" as "read-just"). If breaking such a word cannot be avoided, break the word elsewhere even if a standard rule for breaks must be violated. Alternatively, the sentence can be rearranged to avoid an inappropriate break.

For an equation or other mathematical expression in running text that must be broken at the end of a line, try to break the expression after an operator symbol, such as the equals symbol, not at the beginning of the following line (see also Section 12.4, "Mathematics in Text and Display").

> . . . and if $x =$ *not*   . . . and if x
> $3.51y + 75.89$, then . . . $= 3.51y + 75.89$, then . . .

More detailed guidance on word breaks can be found in *New Hart's Rules*.[3] In addition, *The Chicago Manual of Style*[1] includes recommendations for word breaks in several non-English languages.

32.2.2 Figures

When checking proofs, authors, copy editors, and proofreaders should ensure that all figures are incorporated into the proofs and that they have been organized in numerical order and oriented correctly. If figure proofs are supplied separately from the text proofs and there is a potential for incorrect orientation in the layout, mark the top of each figure proof.

Examine the proof of each figure to ensure that all elements (eg, lettering, symbols, and lines) are included and legible. Indicate clearly any corrections that are needed. Depending on the process used for figures, the publisher may be able to make the corrections or the authors may be required to correct the originals.

In checking proofs for halftones (eg, photographs and radiographs), ensure that all necessary labels are included and that areas of interest are visible. Also make sure that the contrast is not too light or dark.

Review each figure's caption to ensure that the numbering on the caption and figure match and to confirm that the caption adequately describes the figure.

32.2.3 Supplementary Materials for Online Publication

In proofing supplementary material that is intended exclusively for online publication or as an appendix in print or online, the same considerations should be given for accuracy and completeness as are given in proofing text, tables, and figures. However, additional components must be checked. In particular, copy editors, proofreaders, and authors must confirm that electronic documents look and function as intended (eg, special characters appear as intended on screen and hyperlinks to parts of the document and to external resources work). Supplementary material provided exclusively online need not be typeset. It may be formatted as a Word, Excel, or PDF file.

32.3 ANNOTATING PDF PROOFS

Since publishers, printers, and journal offices moved to paperless workflows, it has been standard practice to send authors proofs in Portable Document Format (PDF) via email instead of mailing hard-copy galley proofs. Publishers generally prefer that authors annotate, or mark, the PDF proofs rather than print them out and indicate changes in handwriting.

Many best practices exist for annotating PDFs, but much depends on which version of Adobe Acrobat or other PDF software authors use. Authors who do not have PDF software on their computers can download the most recent version of Adobe Acrobat Reader at no cost at https://get.adobe.com/reader. Adobe Acrobat Reader provides limited but sufficient annotation functions on its "Comment" ribbon. Adobe Acrobat Professional has additional editing tools.

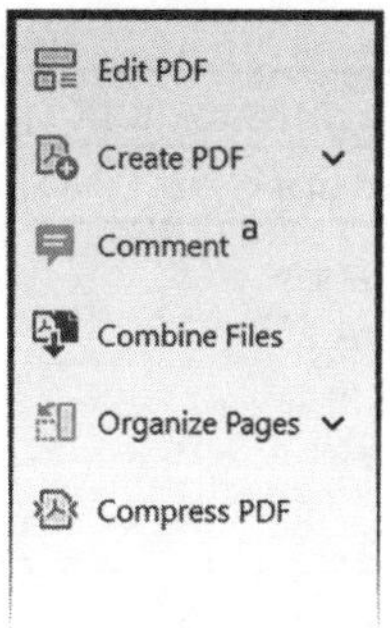

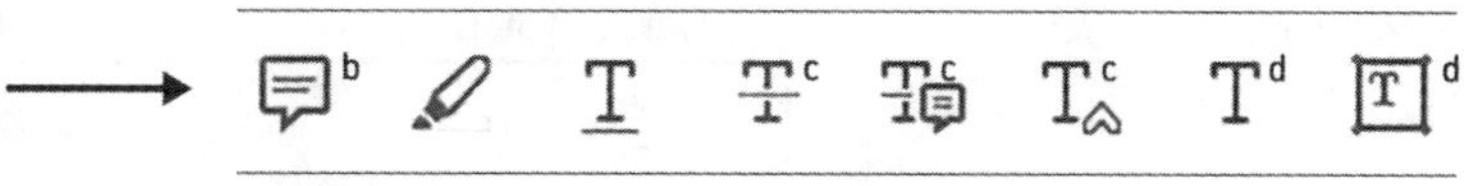

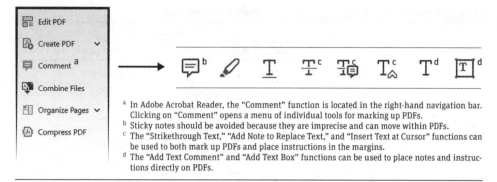

[a] In Adobe Acrobat Reader, the "Comment" function is located in the right-hand navigation bar. Clicking on "Comment" opens a menu of individual tools for marking up PDFs.

[b] Sticky notes should be avoided because they are imprecise and can move within PDFs.

[c] The "Strikethrough Text," "Add Note to Replace Text," and "Insert Text at Cursor" functions can be used to both mark up PDFs and place instructions in the margins.

[d] The "Add Text Comment" and "Add Text Box" functions can be used to place notes and instructions directly on PDFs.

Figure 32.1. Adobe Acrobat Reader offers tools in its "Comment" ribbon for marking corrections on page proofs that are in Portable Document Format.

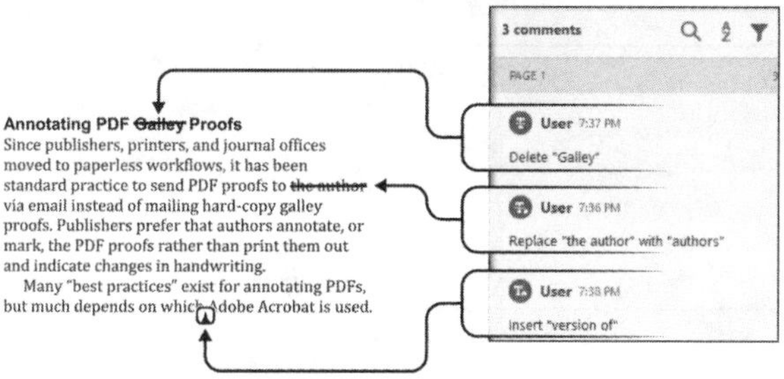

Figure 32.2. Examples of annotated corrections on a proof page in Portable Document Format.

Authors should use only the editing tools that allow recommended changes to stand out on PDFs so that publishers can find them quickly. With Adobe Acrobat Reader, for example, this can be accomplished with the "Comment" ribbon's functions for "Add Note to Replace Text," "Insert Text at Cursor," "Add Text Box," and "Strikethrough Text" (see Figures 32.1 and 32.2). Authors should avoid using editing tools that modify PDF proofs without drawing attention to the changes, such as Adobe Acrobat Professional's "Replace Text" and "Insert Text" functions. Publishers use PDF proofs to update manuscripts' digital master files, not as the finalized documents to print or post online. Consequently, changes integrated into the text of PDF proofs will be missed in updating the master files. In addition, authors should not use sticky notes because they are imprecise and can move within the PDFs, creating confusion as to what the intended corrections are.

32.4 MARKING PAPER PROOFS

Although hard-copy proofs have decreased significantly in use, some publishers continue to use them, and some others allow authors to return hand-marked printed copies of PDF proofs. For marking corrections on paper proofs or printed-out proofs, authors should use standard proofreaders' marks and symbols (see Table 32.1 for American marks[4] and British marks[5,6]). Although no standardized system of European proofreaders' marks exists, most European printers recognize American and British marks.

Table 32.1. American and British Proofreaders' Marks and Symbols

Instruction	American marks[a]	
	Marginal mark	In-text mark
Delete	*(symbol)*	the red book
Close up	*(symbol)*	the bo ok
Delete and close up	*(symbol)*	the b ook
Restore deletion	stet	the red book
Insert in line	red	the book
Substitute in line	red	the black book
	e	the book
Insert space in line	#	the book
Equalize spacing	eq #	the yellow book
Lead (space between lines)	# *or* ld	The red book was lost.
Remove leads (space) between lines	*(symbol)* # *or* *(symbol)* ld	The red book ___ was found.
Insert hair space or thin space	hr # *or* thin #	100/000
Begin new paragraph	*(symbol)* *or* L	The red book was lost. The black book was found.
Run paragraphs together	no *(symbol)*	The black book was lost. The red book was found.
Insert 1-em quad (indent)	*(symbol)*	The red book
Insert 2-em quad (indent)	*(symbol)* *or* [2]	was found
Insert 3-em quad (indent)	*(symbol)* *or* [3]	at night.
Move to left	[	the book
Move to right	]	the book
Center	ctr	the book
Move up	*(symbol)*	the book
Move down	*(symbol)*	the book

Table 32.1. American and British Proofreaders' Marks and Symbols (*continued*)

British marks[b]

Marginal mark	In-text mark	Corrected text
	the red book	the book
	the bo ok	the book
	the bbook	the book
	the red book	the red book
⋀ red	the book	the red book
red	the black book	the red book
e	the book	the book
Y	thebook	the book
X	the/yellow / book	the yellow book
extend text mark (specify the space if necessary)	The red book was lost.	The red book was lost.
extend text mark (specify the space if necessary)	The red was found.	The red book was found.
thin Y	100/000	100 000
	The red book was lost. The black book was found.	The red book was lost. The black book was found.
	The black book was lost. The red book was found.	The black book was lost. The red book was found.
⌐1	The red book	The red book
⌐2	was found	was found
⌐3	at night.	at night.
	the book	the book
	the book	the book
[ ]	the book	the book
	the book	the book
	the book the book	the book
		the book

Table 32.1. American and British Proofreaders' Marks and Symbols (*continued*)

Instruction	American marks[a]	
	Marginal mark	In-text mark
Align vertically	‖ *or* align	‖The book / was lost / in the fog.
Align horizontally	═ *or* straighten	The book w̲a̲s found.
Transpose	tr.	teh book was found. / The found book was.
Spell out	sp	He arrived 1st.
Push down quad (spacing material)	↓	the book
Reset broken letter	✗	the book
Turn right side up	⊚	the green book
Lowercase letter	lc	the green book
Capitalize as marked	cap	The good Book
Set as small capital	sc	D-glucose
Set in italic type	ital	The Good Book
Set in roman type (upright type)	rom	the book
Set in boldface type	bf	The Good Book
Set in lightface type	lf	The book
Set in capitals and small capitals	c+sc	Dong geng
Set in boldface italics in capitals and lowercase	bf ital c+lc	a style manual
Wrong font; reset	wf	body type
Reset as superscript (superior)	2	$e = mc2$
Reset as subscript (inferior)	2	H2S
Insert as superscript (superior)	v	1203
Insert as subscript (inferior)	∧2	HO
Period (full stop)	⊙	Read the book
Comma	⌃	leaves, buds and branches
Semicolon	⁏	Think then decide.
Colon	⁝	Read these books
Hyphen	=/=	graft versus host disease
Apostrophe	⌄	Donovans Demise
Double quotation marks	❛❛/❜❜	He said book.

Table 32.1. American and British Proofreaders' Marks and Symbols (*continued*)

British marks[b]		
Marginal mark	In-text mark	Corrected text
‖	‖The book was lost in the fog.	The book was lost in the fog.
=	The book $\overline{\text{was}}$ found.	The book was found.
⎿⏋	The book was found. The found book was.	The book was found. The book was found.
first	He arrived ⁄st.	He arrived first.
⊥	the book	the book
✕	the book	the book
↻	the green book	the green book
≢	the green book	the green book
≡	The good Book	The Good Book
=	D-glucose	ᴅ-glucose
⊔⊔⊔	The Good Book	*The Good Book*
⊥	the book	the book
∼∼∼	The Good Book	**The Good Book**
∼∿∼	The book	The book
= ≡	Dong geng	Dᴏɴɢ Geng
⊔⊔⊔ ∼∼∼	a style manual	***A Style Manual***
⊗	body type	body type
²⁄	$e = mc$	$e = mc^2$
₂⁄	H₂S	H_2S
ᵇ⁄	1203	1203^b
₂⁄	H₂O	H_2O
⊙	Read the book	Read the book.
⹁	leaves, buds and branches	leaves, buds, and branches
⹊	Think then decide.	Think; then decide.
⦂	Read these books	Read these books:
⊨∣ ∣=∣	graft versus host disease	graft-versus-host disease
⸲	*Donovans Demise*	*Donovan's Demise*
⁶⁶ ²²	He said book.	He said "book."

Table 32.1. American and British Proofreaders' Marks and Symbols (*continued*)

Instruction	American marks[a]	
	Marginal mark	In-text mark
Single quotation marks	↜/↝	"Don't cry, Fire!"
Question mark	?	Is this your book/
En-dash (rule)	⊥/N	pages 10 15
Em-dash	⊥/M	His book find it!
3-em dash	3/M	His reply, "nuts"
Parenthesis marks	(/)	report Smith 1992 was
Brackets (square brackets)	[/]	the book a manual
Slash (slant line, oblique)	/	5 ms

Authors need to indicate corrections to hard-copy galleys or page proofs both in the text and in the margin nearest each in-text proofreaders' mark (see Figure 32.3). In examining proofs, copy editors and proofreaders will scan the margins for corrections, so in-text proofreaders' marks may be missed if they are not accompanied by marginal marks. Authors should also use the margins to indicate instructions. Instructions should be circled so that they are not accidentally set in type.

In-text proofreaders' marks include the caret (ie, "^") to show where an addition is to be inserted and a line drawn through a character or word that is to be deleted. The accompanying marginal marks specify the requested changes. For example, marginal marks may indicate which characters or words should be inserted at an in-text caret, or these marks may be proofreaders' symbols, such as the space sign (ie, "#") and the deletion sign (ie, " ℘ "). Place each marginal mark next to the line of type to which it applies, preferably in the right margin. To indicate more than one correction to a single line, arrange the corrections in sequence from left to right, and separate the corrections with forward slashes (ie, "/").

Copy editors and proofreaders, in turn, may place queries to the authors in the margins of proofs. Corresponding authors should respond to all such queries. If, for example, a query recommends making a change to the typeset copy that is not acceptable, the corresponding author should draw a line through the suggested change in the query, add a marginal note (eg, "OK as set" or "stet"), and place dots under the material that must be retained. Like other instructions, the marginal note should be circled so that it is not accidentally set in type.

Handwritten corrections should be neat and legible. If proofs are to be returned by fax, use a pen with dark ink, not a pencil, so that the markings are suitable for transmission.

If a correction or insertion consists of more than one line, type it into a Word document or an email. Before each correction, place a letter and the page number of the proof where the copy should be inserted (eg, "Insert A for page 2"). On the proof, clearly indicate the

Table 32.1. American and British Proofreaders' Marks and Symbols (*continued*)

Marginal mark	British marks[b]	Corrected text
	In-text mark	
⸲ ⸲	"Don't cry, 'Fire!'"	"Don't cry, 'Fire!'"
?	Is this your book	Is this your book?
⊢1en⊣	pages 10/15	pages 10–15
⊢1em⊣	His book/find it!	His book—find it!
⊢3em⊣	His reply, /////!	His reply, ———!
( )	report Smith 1992 was	report (Smith 1992) was
[ ]	the book/a manual	the book [a manual]
⊘	5 m/s	5 m/s

[a] National Information Standards Organization.[4]
[b] British Standards Institution.[5]

corresponding location for the new copy. Do not provide the original manuscript or a revised manuscript because the publisher will not use it to update the proof, given that the typesetting stage of production has passed.

32.5 RETURNING PROOFS

Authors must return proofs to whomever the publisher stipulates in the instructions sent with the proofs. If figures have to be revised, authors must provide updated images to the publisher.

Proof Marked for Corrections

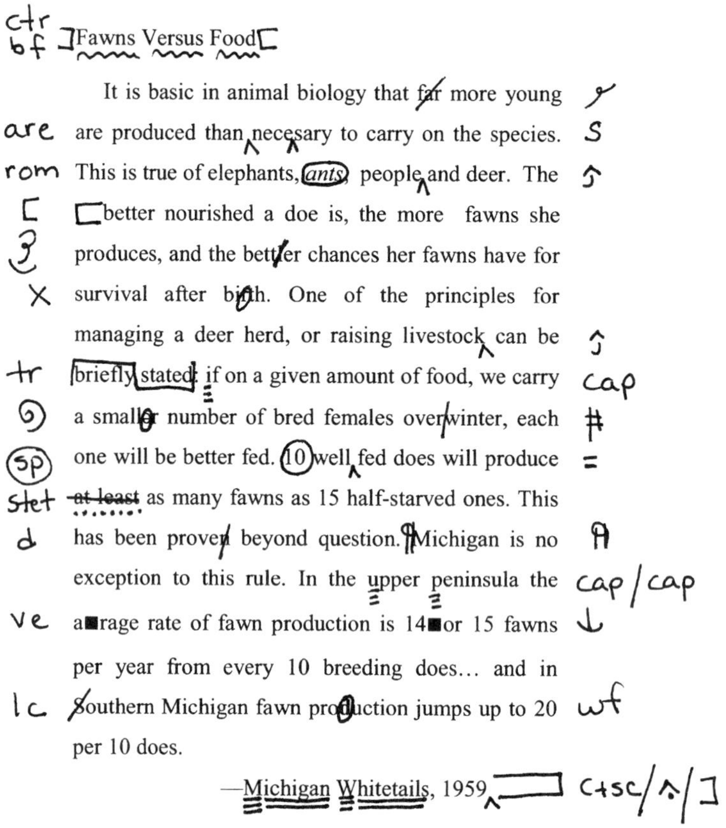

ctr
bf   ]Fawns Versus Food[

        It is basic in animal biology that far more young   ⸮

are   are produced than necesary to carry on the species.   S

rom   This is true of elephants, (ants) people and deer. The   ⸜

[[better nourished a doe is, the more fawns she

⸑   produces, and the better chances her fawns have for

✗   survival after birth. One of the principles for

      managing a deer herd, or raising livestock can be   ⸜

tr   briefly stated: if on a given amount of food, we carry   cap

⑤   a smaller number of bred females over winter, each   #

(sp) one will be better fed. ⑩well fed does will produce   =

stet at least as many fawns as 15 half-starved ones. This

d   has been proved beyond question. ¶Michigan is no   ⁋

      exception to this rule. In the upper peninsula the   cap/cap

ve   a■rage rate of fawn production is 14■or 15 fawns   ⤓

      per year from every 10 breeding does… and in

lc   Southern Michigan fawn production jumps up to 20   wf

      per 10 does.

                    —Michigan Whitetails, 1959 ⌐ ⌐   c+sc/⌃/]

Proof after Corrections

Fawns Versus Food

It is basic in animal biology that more young are produced than are
necessary to carry on the species. This is true of elephants, ants, people,
and deer. The better nourished a doe is, the more fawns she produces,
and the better chances her fawns have for survival after birth. One of
the principles for managing a deer herd, or raising livestock, can be
stated briefly: If on a given amount of food, we carry a smaller number
of bred females over winter, each one will be better fed. Ten well-fed
does will produce at least as many fawns as 15 half-starved ones. This
has been proved beyond question.

Michigan is no exception to this rule. In the Upper Peninsula the
average rate of fawn production is 14 or 15 fawns per year from every
10 breeding does … and in southern Michigan fawn production jumps
up to 20 per 10 does.

                    — MICHIGAN WHITETAILS, 1959.

Figure 32.3. Marked paper proof and corrected proof.

CITED REFERENCES

1. The Chicago manual of style: the essential guide for writers, editors, and publishers. 17th ed. The University of Chicago Press; 2017. Also available at https://www.chicagomanualofstyle.org

2. Smith P. Mark my words: instruction and practice in proofreading. 3rd ed. EEI Press; 1997.

3. New Hart's rules: the Oxford style guide. 2nd ed. Oxford University Press; 2014.

4. National Information Standards Organization. Proof corrections: American National Standard proof corrections. National Information Standards Organization; 1991. (ANSI/NISO Z39.22-1989). No longer in print.

5. British Standards Institution. Copy preparation and proof correction: specification for typographic requirements, marks for copy preparation and proof correction, proofing procedure. British Standards Institution; 2005. (BS 5261-2:2005).

6. Butcher J, Drake C, Leach M. Butcher's copy-editing: the Cambridge handbook for editors, copy-editors and proofreaders. 4th ed. Cambridge University Press; 2006.

ADDITIONAL REFERENCE

Lomangino K, Kaufman CS, Wills AJ. Digital art in scholarly periodical publishing. Sheridan Press; 2002.

33 Digital Standards of Scholarly Journal Publishing

Editor: Sun Huh, MD, PhD

33.1 INTRODUCTION

Brand improvement strategies for scholarly journals are mainly based on enhancing the quality of articles and increasing accessibility.[1] Improvements in both these domains can be achieved through internationalization and meticulous compliance with digital standards. With access to high-speed internet connections increasing worldwide, journal websites have become more important than print journals to the point that some scholarly journals now publish only online.

In the scholarly journal market, digital technology has been led mainly by large commercial publishers and online-only journal-publishing companies since early in this century. A good example is BioMed Central,[2] which emerged shortly after the PubMed Central database[3] was launched in the United States in 2000. Although some academic society journals were among the first to offer their content online in the late 1990s, these and other nonprofit journals often lack sufficient budgetary and technological resources. Consequently, many society journals have been kept busy replicating the digital achievements that commercial and online-only publishers have been continually introducing and quickly applying to their journals.

This chapter summarizes the digital standards and technologies related to online publishing that scholarly journals must implement, regardless of the size of their publishers. The content of this chapter is relevant mostly for professional staff who specialize in digital standards and technologies, as opposed to scientist editors, whose primary roles are to determine which articles to publish and to set journal policies.

The accessibility of a journal depends mainly on the digital technology used by the journal's website. It is impossible to be competitive in the journal market without adopting up-to-date digital technology. In fact, to become competitive, nonprofit organizations outside North America and Western Europe have begun to publish online-only journals. For example, in 2006, the *Journal of Educational Evaluation for Health Professions*[4] became the first online-only journal published in South Korea.

33.1.1 Definition of Digital Standards in Journal Publishing

A digital standard outlines product testing criteria for software and smart devices. Consumers and manufacturers use digital standards to evaluate whether products meet specific performance criteria, such as protecting consumer privacy, information security, freedom of speech, and product ownership.[5] For example, Journal Article Tag Suite Extensible Markup Language,[6] better known as "JATS XML," was developed by the National Information Standards Organization in the United States as a standard for journal article production. In addition, the Dublin Core Metadata Initiative[7] was developed by the International Organization for Standardization (ISO) as a standard for metadata. Digital object identifiers[8] (DOIs) have become the standard uniform resource identifiers for scholarly journals and books, and EPUB[9] is the ISO standard for book browsers.

33.1.2 Progress of Online and Digital Technologies

After the internet was introduced in journal publishing, the National Institutes of Health's National Center for Biotechnology Information (NCBI) in the United States launched PubMed[10] in 1996. PubMed emerged as the most useful early-generation web database in the life sciences. Additional database tools were subsequently introduced, primarily by the NCBI, global commercial companies, and nonprofit organizations.

PubMed and similar databases quickly became essential to most researchers for searching the literature, especially in the fields of medicine, the physical sciences, technology, and engineering.

Digital technologies are constantly evolving, with some gradually disappearing and being replaced by more innovative technologies. For example, metaverse-based platforms,[11] which are popular for entertainment and daily-life–related activities, have the potential to become integral to journal publishing in the future.[12] Because many members of younger generations are accustomed to interacting in the metaverse as avatars, journal editors and publishers may find the metaverse helpful in recruiting younger authors, reviewers, and readers.

33.2 JOURNAL HOME PAGE

A journal's home page is a gateway to basic information about the journal, providing hyperlinks to the journal's aims and scope, editorial board, instructions to authors, and publishing policies. Especially critical, the home page is the primary avenue through which readers access the journal's full-text articles.

A journal's home page should be well organized so that readers can easily find, view, and understand its content and hyperlinks. Most of the content provided on or hyperlinked to a journal's home page is the same as that found in the table of contents and other front matter of a print journal. However, home pages have features unique to online publications, such as search functions for articles and metric-driven functions like most-read articles.

33.2.1 Value of Using English on the Home Page

Making English the primary language of a journal's home page may make it more accessible and intelligible for readers throughout the world. For example, researchers in the Czech Republic may be better able to access, read, and comprehend the home page of a Japanese journal if its primary language is English. Furthermore, it is difficult to find journals when they do not have English-language websites. Even if they locate non-English journals, researchers may not fully understand the content if they have to rely on computer-based translation programs, which have limitations.

Consequently, online journals whose primary language is not English should provide the content of their home pages in both their primary language and English. In addition, these journals should provide abstracts, if not entire articles, in both languages. For example, the home page and abstracts of the *Brazilian Journal of Science* are available in Brazilian Portuguese, English, and Spanish, and articles are published in whichever of those 3 languages they were written.

33.2.2 Elements of Journal Home Pages

Journal home pages should include content, menus, and submenus that link to journal information, instructions to authors, and current and previous issues.

The following sections outline the information to post or link to on a journal's home page.

33.2.2.1 Basic Journal Information

Official journal title: This should be the official, or primary, journal title that is registered with the International Standard Serial Number (ISSN) International Centre.[13]

ISO-and NLM-abbreviated titles: Both the International Organization for Standardization (ISO) and the National Library of Medicine (NLM) in the United States assign abbreviated titles to journals. ISO abbreviations are based on ISO standard 4:1997, which is titled *Information and Documentation—Rules for the Abbreviation of Title Words and Titles of Publications*,[14] and NLM abbreviations are listed online in the *NLM Catalog: Journals Referenced in the NCBI Databases*.[15] The ISO's and NLM's abbreviations are almost the same. They differ mainly in that the ISO's abbreviations use periods, and NLM's do not. For example, the ISO abbreviation for the *Australian Journal of Zoology* is "Aust. J. Zool.," while the NLM abbreviation is "Aust J Zool." This style manual recommends using most abbreviations without periods (see Section 11.2, "Punctuation and Typography").

Publication frequency: There are many common publication frequencies: weekly, biweekly, monthly, bimonthly, quarterly, triennially, biannually, and annually. Less common are such frequencies such as 7 times a year, 8 times a year, and no fixed frequency. However, some online-only journals do not announce their frequency because they publish articles both in ahead-of-print versions immediately after acceptance and in versions of record once manuscript edits are approved. The *Journal of Educational Evaluation for Health Professions* is an example of an online-only journal that has no issue numbers.

Publication start year: This is the year in which the journal began publishing.

Journal history: A brief description suffices for the history of the journal, including the start year, title changes, changes of ownership, changes of publishing or printing companies, language changes, and significant policy changes (eg, converting to an open-access publication).

ISSN: International Standard Serial Numbers (ISSNs) are 8-digit codes used to identify journals, magazines, newspapers, and other periodicals in print and electronic media. Each ISSN takes the form of the acronym "ISSN," followed by a colon and 2 groups of 4 digits separated by a hyphen. The eighth digit is a check digit based on the 7 preceding digits. This check digit, which can be used to confirm that the first 7 digits are correct, may be an "X" if the computation of the first 7 digits equals 10. Although ISSNs are maintained by the ISSN

International Centre, publishers should contact the ISSN centers in their countries to obtain ISSNs[13] (for a list of some national ISSN centers, see Section 27.4.4, "ISSN and CODEN"). Publishers should obtain separate ISSNs for the print and electronic versions of their journals. Publishers can also obtain linked ISSNs, which serve as umbrella numbers for journals in all their media. Those different types of ISSNs are designated as follows[13]:

- "ISSN (print)" or "pISSN" is for print media.
- "ISSN (online)" or "eISSN" is for online or other electronic media.
- "ISSN (linking)" or "ISSN-L" groups all media of the same serial publication, irrespective of how many components are in the series.

  Neurointervention's ISSNs:
  Printed version: pISSN: 2093-9043
  Electronic version: eISSN: 2233-6273
  Linked version: ISSN-L 2093-9043

Copyright: Usually, the copyright holder is the publisher. For example, the copyright holder of *The Lancet* is its publisher, Elsevier. The year of publication can be added to the copyright statement, but it is not mandatory.

  © 2023 Elsevier

If the owner is different from the publishing company, the copyright holder is often the owner. For example, *The British Journal for the Philosophy of Science* is published by The University of Chicago Press Journals for the journal's copyright holder, the British Society for the Philosophy of Science.

  © 2023 British Society for the Philosophy of Science

However, some journals published on behalf of professional associations are owned by the publisher. The *Journal of Quantitative Analysis in Sports* is an official publication of the American Statistical Association, but its publisher and copyright holder is Walter de Gruyter, GmbH.

  © 2023 Walter de Gruyter, GmbH

With many open-access journals, authors retain copyright to their articles. For example, authors hold the copyright to articles published in *Discover Materials*, which is published by Springer Nature Switzerland, AG.

License: A license is a grant by the holder of a copyright or patent that defines how others can use the copyrighted material or patent.[16] Although there are many kinds of licenses, Creative Commons licenses[17] are commonly used by scholarly journals. Copyright holders can select from 6 different Creative Commons copyright licenses, which grant various degrees of permission and impose various degrees of restrictions (see Section 3.3.4, "Creative Commons Licenses"). A Creative Commons license protects those who use or redistribute works from copyright infringement as long as they abide by the conditions specified in the license.

Language information: The language used for the main text of articles and for abstracts should be specified.

Abstracting and indexing status: Databases in which a journal is indexed or was previously indexed should be listed in the journal information. Some international literature databases are described below:

- Agricola[18] (ie, AGRICultural OnLine Access) is a database produced by the National Agricultural Library in the United States. Its more than 5.2 million records are divided into 2 sets: journal article citations, including abstracts, and bibliographic records describing monographs, serials, audiovisual materials, and online content. Agricola collects resources encompassing all aspects of agriculture and allied disciplines, including animal and veterinary sciences, entomology, plant sciences, forestry, aquaculture and fisheries, farming and farming systems, agricultural economics, food and human nutrition, and Earth and environmental sciences.

- BIOSISPreviews[19] covers life sciences and biomedical research from journals, books, meetings, and patents. Maintained by Clarivate, BIOSIS indexes more than 5,300 journals. As of February 2022, this database held approximately 28 million records dating back to 1926.
- Chemical Abstracts Service[20] (CAS) is the world's leading source of chemical resources, including journals and reference materials on chemicals. CAS is maintained by the American Chemical Society.
- CINAHL Database[21] focuses on authoritative nursing and allied health literature. As of March 2022, CINAHL contained more than 3,000 journals. It is maintained by EBSCO.
- The Directory of Open Access Journals[22] (DOAJ) lists open-access journals. To be added to DOAJ, journals undergo a rigorous review process. As of March 2022, DOAJ includes more than 17,500 journals.
- Embase[23] is a comprehensive biomedical literature database that includes 2,900 journals with some 41 million records as of March 2022. It is maintained by Elsevier.
- Engineering Village[24] provides access to 12 engineering document databases, including Ei Compendex and Inspec. It is maintained by Elsevier.
- PubMed, MEDLINE, and PubMed Central (PMC) are literature database services provided by the NLM's National Center for Biotechnology Information in the United States. PubMed is an abstract database that includes the journals indexed in both MEDLINE and PMC. As of March 2022, PubMed contained more than 33 million citations,[10] and PMC archived approximately 7.8 million articles.[3]
- Clarivate[25] maintains the Science Citation Index Expanded (SCIE), the Emerging Sources Citation Index (ESCI), the Social Sciences Citation Index (SSCI), and the Arts & Humanities Citation Index (A&HCI). SCIE is a literature database for the sciences. ESCI is a collection of journals in fields that publish scholarly work but do not generate enough citations to be listed elsewhere. SSCI focuses on the social sciences, and A&HCI is dedicated to the arts and humanities. Altogether, SCIE, ESCI, SSCI, and A&HCI had approximately 70 million records from nearly 21,000 journals as of March 2022.
- Scopus[26] is a literature database with journals in all fields. Maintained by Elsevier, Scopus had more than 43,000 source titles as of March 2022.

Journals are also indexed by national journal databases. Examples include J-J-STAGE[27] in Japan, the Korea Citation Index[28] in South Korea, the Portal of Croatian Scientific and Professional Journals,[29] the Russian Science Citation Index,[30] the SciELO Citation Index[31] in Latin America, and Sinta[32] (ie, the Science and Technology Index) in Indonesia. The number of journals indexed by each of these databases ranges from 770 to nearly 7,000.

Aims and scope: The aims and scope are essential for all scholarly journals. Aims typically focus on journals' commitment to advancing science in their disciplines. Scope usually defines the topics on which journals are and are not interested in publishing articles.

Readership: This section clarifies a journal's target audience. Although anyone may read the journal, clarifying the target audience will help researchers determine whether the journal is relevant to their fields of study.

Independent domain: A journal's home page should have an independent domain that identifies the journal's website. The main difference between a uniform resource locator (URL) and a domain name is that the URL is a complete address to a specific web page while a domain name is a subsection of the URL.[33] For example, the domain name of *Science and Technology Indonesia* is sciencetechindonesia.com, and the URL to its archives page is https://sciencetechindonesia

.com/index.php/jsti/issue/archive. An independent domain name is usually required to obtain an eISSN.

Publishers that have multiple online journals may include the same company domain within each journal's independent domain. For example, the independent domains for the journals *GeoBios* and *Microelectronic Engineering*, which are both part of Elsevier's ScienceDirect platform, are sciencedirect.com/journal/geobios and sciencedirect.com/journal/microelectronic-engineering, respectively. Likewise, the independent domain for the American Medical Association's flagship publication, *JAMA*, is jamanetwork.com/journals/jama, while the independent domain for *JAMA Ophthalmology* is jamanetwork.com/journals/jamaophthalmology.

Although most journals' domain names end in ".org" or ".com," some journals use different endings, such as country codes. For example, parasitol.kr is the domain name for *The Korean Journal of Parasitology*, and scientiamarina.revistas.csic.es is the domain for *Scientia Marina*, which is published in Spain.

QR codes: Quick response codes,[34] better known as "QR codes," can instantly take readers to websites or specific content, such as photographs and videos.

Contact information: This section indicates to whom authors, reviewers, and readers should send inquiries. The journal office's address, telephone number, and email address should be provided. Alternatively, the journal can use a web-based contact form.

33.2.2.2 Scientist Editors and Professional Editorial Staff

Editorial board: A science journal's editorial board consists mainly of scientists with expertise in the journal's scientific discipline. Their roles in reviewing and approving articles and editorial policy should be described in detail in this section of a journal's home page. The following titles represent positions typically held by editorial board members:

- The editor in chief presides over the editorial board, has the right to make the final decisions, and has overall responsibility for the journal. This position is sometimes called simply "editor."
- The deputy editor or associate editor supports the editor in chief and assumes duties that the editor in chief is unable to perform (eg, when the editor in chief has a competing interest related to a submission).
- Assistant editors also support the editor in chief.
- Section editors handle manuscripts in specific fields.
- The senior editor supervises editorial work.
- The editorial consultant advises the editor in chief and other editors. The editorial consultant does not need to be a specialist in the journal's discipline.
- Editorial board members are scientist editors who lend their subject expertise to the journal, providing support to the editor in chief. Sometimes, they are referred to as "editors." A broad range of duties can be assigned to editorial board members. These duties are typically specified by the editor in chief or publisher.
- The ethics editor deals with ethical issues, assisting the editor in chief in deciding those issues.
- The statistical editor provides the editor in chief with technical expertise regarding statistics.
- The legal consultant is a lawyer or other legal professional who resolves legal issues raised during the publishing process.

Professional editorial staff: Unlike most editorial board members, the editorial staff typically consists of publishing professionals, not scientists. Usually, a journal's professional editorial staff is not listed on a journal's home page, especially if staff members work for a commercial publishing company. However, society journals may list their editorial staffs separate from the editorial board, using such terms as "managing editor," "manuscript editor," "language editor," "layout editor," "illustrator," "website editor," and "administrative staff."

33.2.2.3 Publisher Information

Publisher's name, address, and website URL: This information should be clearly presented on a journal's home page.

Business status: It should be specified whether the journal is published for commercial or nonprofit purposes. Sources of financial support can also be described (eg, subscriptions, advertising, article fees, and academic society sponsorship).

Management team: This can consist of the editorial staff and business staff of the publishing company.

33.2.2.4 Publishing Policies

Editorial policies: Although a hyperlink to editorial policies can be posted on a journal's home page, these policies are usually included in the instructions to authors.

Publication ethics policies: Policies related to publication ethics cover such issues as ethical oversight (eg, research ethics, informed consent, and approval by an institutional review board), authorship criteria, complaint process, and appeals.

Competing-interest statement: This statement should explain that competing interests exist when the private interests and the official responsibilities of a person in a position of trust diverge.[16] For example, researchers may have a competing interest if they would benefit financially from their research results. Similarly, editors directly involved in reviewing or processing a manuscript may have a competing interest if their financial or personal relationships would interfere with judging a manuscript appropriately.

Plagiarism policy: This section explains what the journal considers to be plagiarism, as well as the journal's plagiarism-checking process and its penalties for plagiarism.

Peer-review policy: The peer-review process is described, including whether the journal uses single-blind, double-blind, or open peer review.

Advertising policy: This section should clearly state the journal's policy regarding advertisements for products related to the journal's content.

Data sharing and reproducibility policy: The journal's policy on data sharing should be described, even if the journal does not require data sharing.

Preprint policy: This policy should explain whether the journal will accept as new submissions manuscripts that have been posted to preprint servers. The policy should also indicate whether the journal will allow preprints to be cited in new manuscripts.

Repository policy: This section should indicate whether authors are permitted to deposit submitted manuscripts in institutional repositories before final publication.

Corrections: This section should explain the processes for handling corrections, editorial expressions of concern, and retractions. The section should define the consequences for egregious behavior, such as violations regarding competing interests, human and animal rights, and informed consent.

Principles of transparency: Journals should specify whether they adhere to the most recent edition of the "Principles of Transparency and Best Practice in Scholarly Publishing,"[35] which was developed by the Committee on Publication Ethics, the Directory of Open Access Journals, the Open Access Scholarly Publishers Association, and the World Association of Medical Editors. Several journals have announced their compliance with these principles,[36,37] which are summarized in Table 33.1.

Table 33.1. Principles of Transparency and Best Practice in Scholarly Publishing[a]

Category	Item	Subitems	Description
Basic journal information	Website	Aims and scope	Purpose and range of academic fields that a journal covers should be stated on its website.
		Readership	Target readers of a journal should be stated on its website.
		Authorship criteria	Criteria that authors should follow, including not simultaneously submitting manuscripts to multiple journals and creating redundant publications.
		Electronic ISSN	Electronic ISSN should be displayed with print ISSN.
		Print ISSN	Print ISSN should be displayed with electronic ISSN.
	Name of journal	Uniqueness of name	A journal's name should be unique so that it cannot be confused with another journal or mislead potential authors and readers.
	Peer-review process	Statement of the review process	Any policies related to a journal's peer-review procedures should be clearly described on the journal's website.
		Methods of peer review	A journal should specify whether it uses single-blind, double-blind, or open peer review.
		No guarantee of manuscript acceptance	Neither a journal's main website nor its manuscript-submission website should guarantee manuscript acceptance or very short time for peer review.
	Ownership and management		Ownership and management information should be clearly indicated on a journal's website.
	Governing body	Editorial board	The full names and affiliations of the members of a journal's editorial board and other governing bodies should be listed. The members of these bodies should be recognized experts in subject areas within the journal's scope.
	Editorial team and contact information		A journal should provide the full names and affiliations of its scientist editors on its website, as well as the contact information for its editorial office, including its full mailing address.
	Author fees		Any fees or charges that are required for manuscript processing and for publishing materials should be stated clearly.
	Publishing schedule		The frequency at which a journal publishes should be clearly indicated on its website.
Publication ethics	Process for identifying and dealing with allegations of research misconduct	Ethical and professional standards	In no case should a journal or its editors encourage such misconduct or knowingly allow it to take place.
		Steps to prevent research misconduct	Publishers and editors should take reasonable steps to identify and prevent the publication of papers wherein research misconduct has occurred, including plagiarism, citation manipulation, data falsification, and data fabrication.
		COPE's guideline	If a journal's editor or publisher is aware of allegations of research misconduct relating to a published article, the editor or publisher should follow COPE's guidelines or similar guidelines to address the allegations.

Table 33.1. Principles of Transparency and Best Practice in Scholarly Publishing[a] (*continued*)

Category	Item	Subitems	Description
Publication ethics (*cont.*)	Publication ethics	Authorship and contributorship	A journal should adhere to publishing ethics and policies on authorship and contributorship.
		Complaints and appeal	A journal should indicate in its publishing ethics and policies how the journal will handle complaints and appeals.
		Competing interests	A journal should outline on its website its publishing ethics and policies on competing interests.
		Data sharing and reproducibility	A journal should clearly state its policies on data sharing and reproducibility.
		Ethical oversight	A journal should detail ethical oversight in its publishing ethics and policies.
		Intellectual property	A journal should state its publishing ethics and policies on intellectual property.
		Postpublication discussion	A journal should outline its options for postpublication discussions and corrections.
Copyright and archiving	Copyright and licensing	Licensing information	Licensing information should be clearly described in the guidelines on a journal's website, and licensing terms should be indicated on all published articles, in both HTML and PDF formats.
		Creative Commons	If a journal allows authors to publish under a Creative Commons license, the specific license requirements should be noted.
		Policies on posting accepted articles with third parties	Any policies on the posting of final accepted versions and published articles on third-party repositories should be clearly stated.
	Access	Open access	A journal should state the ways in which the entire journal and individual articles are available to readers, such as through open access, subscriptions, and pay-per-view fees.
		Subscription	
	Archiving		A journal should clearly indicate its plan for backing up and preserving the journal's electronic content.
Profit model	Revenue sources		A journal's business model and its revenue sources should be clearly stated on the journal's website.
	Advertising		If a journal accepts advertising, it should indicate its advertising policy, including what types of advertisements it considers, who makes decisions regarding accepting ads, and whether ads are displayed at random or whether they can be linked to reader behavior or content.
	Direct marketing		Any direct marketing activities that are conducted on behalf of a journal, including solicitation of manuscripts, should be appropriate, well-targeted, and unobtrusive.

ISSN, International Standard Serial Number; COPE, Committee on Publication Ethics; HTML, Hypertext Markup Language; PDF, Portable Document Format

[a] Adapted from Choi HW, Choi YJ, Kim S.[36]

33.2.3 For Contributors

Submission information: The mandatory information for authors should be stated on a journal's home page.

Research and publication ethics: The journal should indicate whether it expects research to adhere to internationally acceptable policies.

Author checklist: The author checklist should detail all the elements required for manuscript submission.

Manuscript submission: The journal should specify how authors are required to submit manuscripts. Authors should be instructed in how to use the journal's manuscript management system or other manuscript-submission methods, including email and surface mail.

Submission fee: This fee is usually used to offset publication costs or to discourage inappropriate submissions. This fee should be detailed. If the journal does not charge a submission fee, that should be explained.

Article-processing charges: Processing charges are usually imposed on articles to make them open access so that they are not password protected or placed behind a paywall. Some journals offer open access as an option to authors, and some publish only open-access articles. Like a submission fee, article-processing fees should be outlined in detail in the information for contributors, and if a journal does not impose processing fees, that should be explained.

Instructions to peer reviewers: Clear instructions should be posted so that reviewers understand the journal's peer-review policy and process.

33.2.4 Current and Past Articles

Access to issues: A journal's home page should provide easy access to a journal's most recent issue, as well as to the journal's most-cited or most-read articles. In addition, the home page should have a menu item for past issues.

Article search: The journal should have a search function that is easy to find on its home page and convenient to use.

Journal metrics: The journal can provide such metrics as its average number of weeks between article submission and first decision and the frequency with which its articles are cited elsewhere.

33.3 PRESENTATION OF JOURNALS ONLINE

33.3.1 HTTPS URLs

Hypertext Transfer Protocol Secure (HTTPS) is a protocol in which encrypted data are transferred over secure connections, such as Transport Layer Security and Secure Sockets Layer. HTTPS encrypts data between a client and a server, protecting against eavesdropping, forging of information, and tampering with data. HTTPS-encrypted uniform resource locators (URLs) have "https://" as a preface. Generally, journals use such URLs to redirect users to secured connections to online articles and other content.[38] Table 33.2 shows the difference between HTTPS and Hypertext Transfer Protocol, or HTTP.[39]

33.3.2 Portable Document Format (PDF)

The most basic method of providing journal articles online is to convert them to the Portable Document Format (PDF). Based on PostScript language, each PDF file replicates exactly the print version of an article, encapsulating text, typefaces, vector graphics,

Table 33.2. Differences between HTTPS and HTTP

	HTTPS	HTTP
Full name	Hypertext Transfer Protocol Secure	Hypertext Transport Protocol
Security	secured	unsecured
Data delivery port	sends data over port 443	sends data over port 80
Operation layer	operates at the transport layer	operates at the application layer
SSL certificates	SSL certificate signed by a certificate authority	no SSL certificate required
Domain validation	requires at least domain validation, and certain certificates require legal document validation	does not require domain validation
Encryption	data is encrypted before sending	no encryption
Speed	somewhat slower than HTTP	

SSL, secure sockets layer

raster images, and other information. Most desktop publishing programs can convert files to PDFs.

Although many journal websites offer PDFs, journals usually treat these files as an alternative option for accessing articles. PDFs can be difficult to read online because they look just like print articles. PDF files are broken into pages, which in turn are divided into columns of type. So readers have to constantly reposition PDFs on their screens to move from one column and one page to the next.

33.3.3 Hypertext Markup Language (HTML)

Hypertext Markup Language (HTML) is the standard markup language for documents displayed in web browsers. HTML displays can be assisted by technologies such as Cascading Style Sheets (CSS), which determine how content is rendered on screens, and scripting languages, which are computer languages such as JavaScript.[40]

Compared with PDF files, HTML files are more conducive to reading online because articles are rendered in one long column that readers can scroll through without repositioning the text.

33.3.4 Journal Article Tag Suite Extensible Markup Language (JATS XML)

Another basic component of presenting the content of online journals, Journal Article Tag Suite Extensible Markup Language[41] (JATS XML) defines a set of elements and attributes in Extensible Markup Language (XML) for tagging journal articles.

When an article posted on a website is converted from an HTML file to a JATS XML file, the article becomes compatible with various other formats, such as PubReader and EPUB 3.0. Furthermore, JATS XML files can be converted to PDFs,[42] as well as to files for Crossref XML, PubMed XML, and the Directory of Open Access Journals' XML format.[43,44]

In addition, JATS XML files are preferred by some organizations that use automated technologies to perform quality assurance, content enrichment, taxonomy/indexing, and compliance checking.

JATS XML is also used in presenting content in non-English languages on the web. This requires that the languages have been declared in the International Organization for Standardization's language code ISO 639 and that the content is encoded in Unicode Transformation Format–8-bit, better known as UTF-8.

33.3.5 PubReader

PubReader[45] is a web tool provided by PubMed Central and Bookshelf that makes reading easy in environments other than the desktop, such as tablets, mobile phones, and other small-screen devices. When using PubReader to read content on such devices, the typeface size appears uniform, and the experience feels like reading a book. PubReader can also be used on desktops, laptops, and multiple web browsers.

PubReader is free and open to the public, and users can download it from GitHub, Inc.[46] Moreover, if PubReader is installed on a publication's server, users do not need to install the program on their devices.

33.3.6 EPUB

The EPUB[47] format provides a means of representing, packaging, and encoding structured and semantically enhanced web content for distribution in a single-file container. Unlike the environment for viewing journals on the web, EPUB configures the screen and displays content like a printed document. It is a standard format of e-books, and it is widely supported by hardware readers. However, unlike PubReader, EPUB must be installed as an add-on for each browser.

33.3.7 Journal Smart Apps

Because readers use smartphones and tablets more than desktop computers to review online content, publishers have created and disseminated journal apps for smartphones. Although journals can be viewed through smartphones and tablets through the internet without journal-specific apps, such apps can offer readers a more straightforward user interface, and they may be able to provide content even when there is no internet connection.

33.4 CROSSREF SERVICES

Crossref was launched in 2000 by the Publishers International Linking Association, Inc, which is one of the registration agencies for digital object identifiers (DOIs) recognized by the International DOI Foundation[48] (see Table 33.3).

Crossref[49] is a not-for-profit membership organization focused on making research outputs easy to find, cite, link to, access, and reuse. From issuing DOIs to scholarly journal articles, books, grants, and other documents, Crossref expanded its offerings to such services as Check for Update (Crossmark), Cited-By, Funder Registry, Metadata Retrieval, Metadata Plus, Event Data, Participation Report, and Text and Data Mining (TDM). The services provided by Crossref create networks among online journals, and they have become the digital standards of online publications.

Table 33.3. Select DOI Registration Agencies

Agency	Coverage
Airiti, Inc.	Traditional Chinese materials.
BSI Identify	Provides a unique digital identifier for manufacturers to verify product safety by ensuring access to accurate product identification and information.
China National Knowledge Infrastructure (CNKI)	China-based information resources, including Chinese science, technology, economics, humanities, social science, and politics. CNKI publishes databases containing e-journals, newspapers, dissertations, proceedings, reference works, yearbooks, and more.
Crossref	Scholarly and professional research content in journal articles, books, conference proceedings, searchable metadata databases, and more.
DataCite	Research data and other research outputs. Additional services make it easy to connect, share, and assess research outputs with the broader research ecosystem.
The Institute of Scientific and Technical Information of China (ISTIC)	Linking services for Chinese journals, dissertations, books, conference proceedings, and other literature resources and management services for Chinese scientific data sets and multimedia resources, such as audiovisual assets.
Japan Link Center (JaLC)	Scientific and academic metadata and content from various sources that promote science and technology, such as national institutes, universities, and commercial publishers. Works with Crossref to register English-language journals in Japan.
Korea Institute of Science and Technology Information (KISTI)	Scientific data project for depositing, accessing, and sharing scientific data in Korea, including journal articles, proceedings, official government documents, patent information, research and development reports, and traditional Korean knowledge.
Multilingual European DOI Registration Agency (mEDRA)	Relation tracking between intellectual property entities and certification of voluntary deposits, including time stamping and digital signatures.
Publications Office of the European Union (OP)	Assigns DOIs on the behalf of institutions, offices, agencies, and other bodies in the European Union.

DOI, digital object identifiers

33.4.1 Digital Object Identifier (DOI)

Digital object identifiers[50] (DOIs), which are a type of uniform resource name (URN), are hyperlinks that direct readers to such digital objects as full-text articles and books on the web. Unlike uniform resource locators (URLs), which are frequently deleted or modified, DOIs and other URNs are permanent, ensuring uninterrupted access to documents.

A DOI is a unique alphanumeric string. It begins with the DOI resolver "https://doi .org/," which serves as the domain name. The resolver is followed by a DOI directory prefix and a suffix (see Figure 33.1). The "prefix" is a unique numerical string assigned to a publisher by a DOI registration agency, such as Crossref, the Publications Office of the European Union, or the China National Knowledge Infrastructure. The "suffix" is a unique alphanumeric string or series of strings assigned by the publisher to identify the digital object. For example, in the DOI https://doi.org/10.6087/kcse.70, "10.6087" is the prefix Crossref assigned to the Korean Council of Science Editors, and "kcse.70" is the suffix that the Korean Council of Science Editors assigned to an original article in *Science Editing* titled "Increasing Number of Authors per Paper in Korean Science and Technology Papers."

Figure 33.1. Components of digital object identifiers

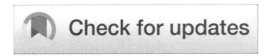

Figure 33.2. By clicking on the Crossmark icon on an online article, readers can learn about changes to the article, such as revisions and corrections.

Publishers list DOIs on the title pages of journal articles and other digital objects. When these documents are cited elsewhere, their DOIs can be included in end references so that readers can quickly access the cited documentsm. Upon clicking on the DOIs in the end references, readers are referred via DOI registration agencies to the cited documents at their correct locations on the web. If content moves, the publisher updates the DOI registration agency's database, and DOIs will redirect readers to the content at its new location.[50] However, when users copy DOIs into browsers instead of clicking on the DOIs in end references, the users may sometimes obtain search results that describe the documents rather than the documents themselves.

To be able to assign DOIs to articles and other digital objects, publishers need to become members of an appropriate DOI registration agency (see https://www.doi.org for a list of DOI registration agencies). The DOI agency, in turn, provides publishers with DOI prefixes and log-in information for the agency's deposit system. To maintain their DOIs and metadata, publishers need to adhere to the criteria that the DOI agency recommends.

33.4.2 Check for Update (Crossmark)

Check for Update (Crossmark) is a service that informs users of changes to a document, such as revisions, corrections, and withdrawal of the content. Crossmark also allows users to identify the final version of a document. In addition, Crossmark provides information on publications, such as their history, their access policies for original texts, the location of their supplements, their peer-review process, and their licensing policies, as well as information on research funding agencies.[50] A Crossmark[51] can be attached to any article for which a Crossref DOI has been deposited (see Figure 33.2).

33.4.3 Cited-By Function

If a publisher implements Crossref's Cited-By function when assigning a DOI to an article, readers of that target article can trace where it has been cited (see Figure 33.3). Users can navigate to the full text of the citing articles to see which authors and journals cited the target article.

Crossref member publishers can retrieve citation counts through a public application programming interface. These publishers can then display the citing works on their

Figure 33.3. Crossref's Cited-By results tell readers where an article they are reading has been cited. This 2009 article in the *Korean Journal of Women Health Nursing* has been cited 41 times in other articles. By clicking on the Crossref icon, users can navigate to the full text of those articles (in pop-up screen) to see which authors and journals cited the target article.

websites. However, only citations in articles deposited in Crossref can be seen. Therefore, Crossref's Cited-By results are different from those of other literature databases, such as Scopus and the Web of Science Core Collection.

For more information, search for "Cited-By" on Crossref's website at https://www .crossref.org.

33.4.4 Funder Registry

The Crossref Funder Registry,[52] which was previously called FundRef, was established to standardize the reporting of research grant support to facilitate text mining and analysis. Funders that are registered on this registry designate a standard name, and they are assigned a standard identifier. Authors, in turn, provide the funders' names, identifiers, and grant numbers when they submit their manuscripts. Publishers, in turn, supply this information to Crossref when registering articles through the e-submission process.

Funding information for individual articles can be checked by clicking the Crossmark logo on the articles. In addition, the funders of published research can be searched at https://search.crossref.org/funding.

Through Funder Registry, funders can track which of their grants resulted in research that was published. Funders also can obtain the DOIs and metadata of funded articles. Government funders can use Funder Registry data to be more transparent in informing the public of the results of government-funded research. Universities and other research institutions can quickly evaluate the academic achievements of their faculties and identify major funders. Researchers and the publishers of scholarly journals can analyze the research funders for various types of research.

33.4.5 Similarity Check

Similarity Check[53] is a plagiarism-detection service provided by Crossref and powered by iThenticate. Similarity Check provides immediate feedback on similarities to other published content. Editors can use Similarity Check to screen submitted manuscripts at the editorial desk, or Similarity Check can be configured to automatically screen all submissions as part of the e-submission system.

33.4.6 Metadata Retrieval and Metadata Plus

The services provided by Crossref's Metadata Retrieval[54] include validating records; filling metadata gaps; collecting related research objects; matching and linking citations; aggregating and integrating content; searching and discovering; and measuring, reporting, and applying metrics.

Metadata Search and Simple Text Query are used to look up a small number of DOIs. To look up a substantial number of DOIs or to look up metadata records, the following application programming interfaces (APIs) can be used: REST API, XML API, OAI-PMH, and Open URL.

Metadata Plus is a paid premium service, unlike Metadata Retrieval, Metadata Search, and Simple Text Query. The advantages of the premium service include more rapid, larger-scale bulk data retrieval and more queries.

33.4.7 Event Data

The Event Data service,[55] jointly developed by Crossref and DataCite, collects raw data and related context on citations to DOIs or other URLs of Crossref- or DataCite-registered works. These citations are in both traditional and nontraditional venues. The nontraditional venues include news articles, Wikipedia pages, and social media. Among the advantages of this API service is that it allows users to find more comments on published works and to assess the impact of published research beyond citations in research works.

33.4.8 Participation Report

Crossref Participation Reports show member publishers the percentage of their deposits that were registered with 10 key metadata elements. These key elements add context and richness, and they allow for easier discovery and broader and more varied use[56] (see Table 33.4[57]). In addition, publishers can use the reports to measure the amount of metadata they provide to Crossref and to compare themselves with other publishers.

33.4.9 Text and Data Mining (TDM)

Text and Data Mining (TDM) uses Crossref's REST API to show researchers where the full text of a work is located on the web and provides a license that explains whether the work can be text mined and under what permissions. Consequently, this service facilitates data mining of research.

Both researchers and the public can use TDM[58] for free. Such use is legally protected in some countries. For example, the United Kingdom revised its Intellectual Property Act to allow TDM to be used for nonprofit purposes.

Journal publishers can use TDM to help readers search for articles related to those in the journal. To use TDM, researchers must have access to the content through open access, an institutional subscription, or another subscription.

Table 33.4. Items Tracked in Crossref Participation Report[a]

Items	Explanation
References	Percentage of content registered with Crossref that includes reference lists in its metadata
Open references	Percentage of registered references that is openly available
ORCID iDs	Percentage of content containing ORCID iDs
Funder Registry IDs	Percentage of registered content that contains the name and Funder Registry ID of at least one of the organizations that funded the research
Funding award numbers	Percentage of registered content that contains at least one funding award number
Crossmark-enabled	Percentage of content using the Crossmark service, which identifies the final versions of documents, as well as changes made to documents, such as revisions, corrections, and withdrawal of the content
Text-mining URLs	Percentage of registered content containing full-text URLs in the metadata so that researchers can easily locate content for the purposes of text mining and data mining
License URLs	Percentage of registered content that contains URLs that point to licenses that explain the terms and conditions under which readers can access content
Similarity Check URLs	Percentage of registered content that includes full-text links for the Similarity Check service
Abstracts	Percentage of content that includes abstracts in the metadata, providing further insights into the content of the work

ORCID iD, Open Researcher and Contributor Identifier ID; URL, uniform resource locator
[a] Based on Tolwinska A.[57]

33.4.10 Research Organization Registry (ROR)

Crossref and DataCite are among the 4 founding organizations of the Research Organization Registry (ROR),[59] which was launched in January 2019 to develop open, sustainable, and unique identifiers for research organizations around the world. By adopting ROR identifiers as part of their academic infrastructure, institutions and grantees can more efficiently track publications generated by specific institutions and funded by specific funding bodies,

33.5 UNIFORM IDENTIFIERS

33.5.1 Uniform Resource Locator (URL) and Uniform Resource Name (URN)

Uniform resource locators (URLs) and uniform resource names (URNs) are different kinds of uniform indicators. Both define where resources can be located and retrieved on the internet. They differ from each other in that URNs are persistent, even after resources cease to exist or become unavailable. URNs are typically registered with authorities so that their registrations can be revised to redirect users when the locations of the resources change.[60]

URNs, such as digital object identifiers (see Section 33.4.1, "Digital Object Identifier (DOI)") and Archival Resource Keys, are used to locate journal articles and other digital objects on the web. URNs can also be used for author identification, as are International

Standard Name Identifiers, Open Researcher and Contributor Identifier IDs, Scopus Author Profile identifiers, and Researcher IDs.

33.5.2 Archival Resource Keys (ARKs)

Archival Resource Keys (ARKs) are multipurpose URNs widely used by libraries, archives, museums, data centers, publishers, and government agencies to provide reliable references to scientific, cultural, and other scholarly objects.

33.5.3 International Standard Name Identifier (ISNI)

The International Organization for Standardization (ISO) established the International Standard Name Identifier[61] (ISNI) as its certified global standard for identifying contributors who creative works and those active in their distribution, including researchers, inventors, writers, artists, visual creators, performers, producers, publishers, and aggregators. As of March 2020, approximately 13 million individuals and 1.6 million organizations had ISNIs. ISNIs are used by libraries, publishers, databases, and rights management organizations.

As with DOIs, ISNIs can be obtained from a number of registering agencies (select the "Get an ISNI" menu item at https://isni.org), and they are generally requested by the organizations that produce the work, not individuals. Consequently, an author may have more than one ISNI if more than one organization registers the author with registering agencies. In addition, an ISNI record may not have the complete list of an author's works because the information was provided by one or more organization, not the author.

Although the ISNI is the ISO standard number for authors, it is not widely used in scholarly journals and books as an author identifier.

33.5.4 Open Researcher and Contributor ID Identifier (ORCID iD)

The Open Researcher and Contributor ID identifier[62] (ORCID iD) is a researcher's identification developed by ORCID, a global, nonprofit organization. Launched in 2012, this ORCID service is free to researchers. Unlike most ISNIs, ORCID iDs are created by individuals, not institutions. In registering for ORCID iDs, researchers list their published research, their funding, and their biographical information, including education, employment, positions, distinctions, memberships, and services. After creating unique ORCID iDs, researchers update their records over time.

ORCID iDs can be used to retrieve researchers' research from such citation databases as Crossref Metadata Search, Europe PubMed Central, and Scopus.[62] Although not an ISO standard, ORCID is the most widely used service for researchers' identifiers. In addition, some journals have mandated that authors include their ORCID iDs when they submit manuscripts so that those identifiers can be listed with the "Author Contributions" statements (see Section 27.7.1.2, "Author Statement and Byline").

33.5.5 Scopus Author Profile

Scopus Author Profile is a free service that Elsevier, BV, provides at https://www.scopus.com. Researchers whose works are published in Scopus journals are provided with

Scopus Author Profile identifiers. Researchers, in turn, can edit their Scopus Author Profile records to include works published in other journals. This service provides citation frequencies for the researchers' works in Scopus journals, as well as document and citation trends and data from the Hirsch index.

33.5.6 Researcher ID

Web of Science's Researcher ID is a free service that has been integrated into Publons. Researcher ID records include authors' publications, their total citations, their Hirsch index, their verified reviews, and verified editor records. Researchers can select works to list in their records from the Web of Science Core Collection records, and they can import other works by synchronizing with ORCID, by providing DOIs, and by uploading files.

33.6 AUXILIARY SERVICES

The auxiliary services that publishers may offer with their scholarly journals and books include Quick Response codes, podcasts, journal metrics, and Altmetrics.

33.6.1 Quick Response (QR) Codes

When used by scholarly journals, Quick Response[34] (QR) codes enable users to navigate directly to journal home pages and to specific resources such as figures and videos.

33.6.2 Podcasts

Scientific, technical, and medical journals frequently use podcasts for interviews in which researchers explain the significance of their work.

33.6.3 Journal Metrics
33.6.3.1 Definitions and Types

When researchers are selecting a target journal for their manuscripts, journal metrics are an important consideration in addition to such other factors as the journal's scope, its publisher, and its article-processing charges (see Figure 33.4). Although journal metrics are quantitative tools, each metric only tells a part of the story of a journal's quality and impact, and each has its limitations.[63] Consequently, authors should consider more than one type of metric and other characteristics of journals before deciding to which journal to submit their work. Editors and publishers can promote their journals and help authors by listing various journal metrics clearly on their journals' home pages.

Although publishers can generate metrics based on their own data, the term "journal metrics" usually refers to metrics aggregated from a network of journals. It is convenient to import and analyze data from stable and extensive literature databases, such as the Web of Science Core Collection, Journal Citation Reports, Scopus, SCImagoJR, Crossref metadata, and the Eigenfactor Project. Each database's application programming interface (API) may be used to obtain data on a single journal or even an individual article.

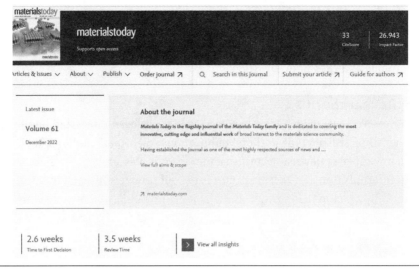

Figure 33.4. The upper right-hand side of *Materials Today*'s home page features the external metrics of CiteScore and the Journal Impact Factor, while the lower left-hand side of the page provides internal metrics on the journal's time to first decision and its review time.

33.6.3.2 Journal Impact Factor

The most well-known journal metric is Clarivate's Journal Impact Factor, which is based on the Web of Science Core Collection. The Journal Impact Factor is one of the metrics contained in the Journal Citation Reports.[64]

The Journal Impact Factor compares the number of citations that indexed journals make in the current year to the citable articles that a specific journal published during the previous 2 years. For example, a journal's 2024 Impact Factor would be a calculation based on the number of citable articles the journal published between 2022 and 2023 compared with the number of times those articles were cited in indexed journals in 2024.

Authors should assess a journal's Impact Factor relative to other journals in the same category, not to the highest possible Impact Factor. In 2022, the highest Impact Factor was 503.702, which was more than 400 points higher than the next 2 highest Impact Factors. Typical Impact Factors are significantly more modest. For example, the median Impact Factor of journals in the 178 categories of Clarivate's Science Citation Index Expanded (SCIE) ranged from 0.625 to 4.752 in 2020.

33.6.3.3 Journal Citation Indicator (JCI)

Launched in 2021, the Journal Citation Indicator (JCI) is part of Journal Citation Reports, as is the Journal Impact Factor. JCI differs from the Impact Factor in 2 ways: JCI analyzes the number of citations to articles published in the previous 3 years, and it compares that data to the average citation count of journals in the same subject category. With JCI's field-normalized measures, a value of 1.0 indicates that a journal's published articles received citations equal to the average citation count in the journal's subject category.[65]

33.6.3.4 SCImago Journal Rank (SJR)

A well-known alternative metric to the Journal Impact Factor is the SCImago Journal Rank (SJR), which is based on Scopus.[66] Unlike the Journal Impact Factor, SJR is not simply a calculation of the number of citations. Instead, SJR is based on a complex mathematical calculation that takes into account the prestige of the citing journals and those journals' similarity to the cited journal.[67]

33.6.3.5 CiteScore

Similar to the Journal Impact Factor, the CiteScore is based on the number of citations made to a journal's articles in indexed journals. However, the CiteScore compares journals indexed in Scopus, not SCIE or any of Clarivate's other indexes. In addition, the citations are divided by the number of the same document types indexed in Scopus instead of the number of citable articles in the journal being measured. Another significant difference is that CiteScore averages both sets of numbers over the same 4 years.

33.6.3.6 Other Metrics Based on the Journal Citation Reports and Scopus

Other journal metrics based on the Journal Citation Reports and Scopus measure 5-year impact, immediacy, half-lives, and citations, excluding "self-cites" in which articles are cited in other articles in the same journal. An additional Clarivate measure worth considering is the Article Influence Score.

33.6.3.7 Metrics Based on Crossref Metadata

Publishers can use an API to access real-time Crossref metadata on the publications they have deposited with Crossref. This free service generates metrics for Crossref citations and for source titles that cite a journal.

33.6.3.8 Altmetrics

Altmetrics[68] are metrics and qualitative data that complement traditional, citation-based metrics. Whereas traditional citation metrics are based on literature databases, Altmetrics measure the impact of research based on its mention in social media, media reports, and reference management tools. Those data are then analyzed multidimensionally. A journal's Altmetrics will break down the mentions for a journal article into categories such as news outlets, tweets, and Facebook postings.

The Altmetric service monitors such sources as the following for mentions of research:

- Public policy documents
- Mainstream media
- Online reference managers
- Postpublication peer-review platforms
- Wikipedia
- Patents, using data from IFI CLAIMS
- Open Syllabus Project
- Blogs
- Dimensions citation data
- Research highlights

- Facebook
- LinkedIn
- Google+
- Pinterest
- X (formerly Twitter)
- YouTube
- Reddit
- Stack Overflow

33.6.3.9 Journal-Generated Metrics

Journals can generate some interesting metrics on their own. These include the average time from submission to first decision, the average time from submission to publication, the average time from acceptance to publication, and the acceptance rate. Journals can retrieve those metrics from their manuscript-submission systems. Especially important for authors is the time between submission and first decision because if their manuscript is rejected, they will want to revise it and submit it to another journal as quickly as possible.

33.7 RESEARCH DATA AND DATA SHARING

Many journals require authors to include research data with their submissions for the peer-review process, for transparency, for reuse, or for helping to ensure reproducibility.[69] Research data can range from raw data (eg, DNA sequences, measurement results, and survey results) to data produced during analyses. Data sharing is especially advantageous for meta-analyses and mega-analyses because raw data are preferable to article results for these analyses. In addition, data sharing can obviate the need for researchers to repeatedly carry out the same work to generate data.

Online public databases have been created to which researchers can upload large data sets so that they can be shared with others. Examples include ClinicalTrials.gov, which is maintained by the National Library of Medicine in the United States. Research data also can be deposited to data repository sites and journal sites.

33.7.1 Various Degrees of Data Sharing

Data sharing is mandatory for some journals but optional for others. Policies with varying degrees of data sharing are summarized in Table 33.5.[70] Journal editors should adopt the data-sharing policy that would be most appropriate for their journals. Even if a journal has no data-sharing policy, it is recommended that the journal explain that on its home page.

33.7.2 Repositories for Data Deposits

More than 2,000 data repositories exist, and lists are available at the Registry of Research Data Repositories[71] and the Repository Finder.[72] The most widely used repositories for general data sets are Dryad, figshare, Harvard Dataverse, Open Science Framework, Zenodo, and Mendeley Data.[73]

Table 33.5. Various Degrees of the Data-Sharing Policy[a]

	Data availability statement is published	Data have been shared	Data have been peer reviewed	Example journals
Encourages data sharing	Optional	Optional	Optional	
Expects data sharing	Required	Optional	Optional	*British Journal of Social Psychology*
Mandates data sharing	Required	Required	Optional	*Ecology and Evolution*
Mandates data sharing and peer reviews data	Required	Required	Required	*Geoscience Data Journal*

[a] Based on Wiley's Data Sharing Policies.[70]

33.8 PREPRINTS

According to the Committee on Publication Ethics,[74] a preprint is a "scholarly manuscript posted by the author(s) in an openly accessible platform, usually before or in parallel with the peer review process." Launched in 1991, ArXiv.org was the first official preprint server. For a directory of preprint servers, search https://asapbio.org using the phrase "list of preprint servers."

Preprints account for approximately 4% of the published literature,[75] but the majority of preprints are not published by journals.

Journals should announce their preprint policies, stating both whether they accept submissions of preprint manuscripts and whether preprint articles can be included as references.[76]

33.9 INNOVATIONS IN ABSTRACTS

33.9.1 Graphic Abstracts

A graphic abstract is a visual equivalent of a written abstract.[77] Some journals have introduced graphic abstracts to help their readers quickly understand the content. A graphic abstract should be easy, summational, and intuitive (see Figure 33.5). Graphic abstracts are usually produced by professional designers who understand the content of the articles.

33.9.2 Audio Abstracts

Some journals provide audio recordings of abstracts with their online articles.[78] These audio recordings are usually made by article authors.

33.9.3 Video Abstracts

Video abstracts are peer-to-peer video summaries of journal articles that are 3 to 5 min long.[79] These videos may include graphic presentations. A video abstract enables experts in the same field to understand an article's content quickly so that they can decide whether they are interested in reading the full text.

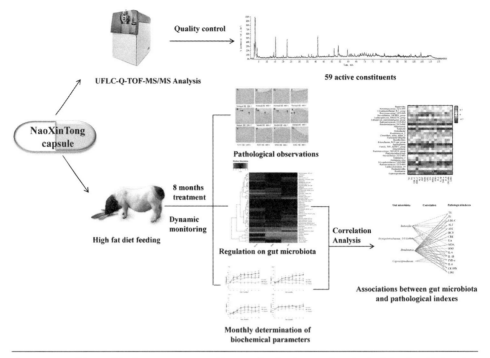

Figure 33.5. This graphic abstract was created by *Frontiers in Pharmacology* for its 3 October 2019 article titled "Naoxintong Capsule Inhibits the Development of Cardiovascular Pathological Changes in Bama Minipig through Improving Gut Microbiota."

33.10 DIGITAL ARCHIVING

From the point of view of researchers, continuous access to the full text of journal articles is essential. This infrastructure concern is especially critical to journals that do not publish print versions and those that have ceased publication or consolidated with other journals. In addition to preserving their home pages and full-text articles, journals should deposit their content in digital archives or institutional repositories.

A digital archive "stores collections of digital information such as documents, videos, and pictures in a digital format to provide long-term access to the information."[80] Archiving journals ensures that future generations can access the research published in those journals.

Journals should establish digital archiving policies that outline where they will deposit their past, present, and future content so that the content can be accessed in the event that the journals cease publication. In addition, journals should announce their digital archiving policies so that researchers are aware of them.

33.10.1 Mandatory Metadata for Archiving

The General International Standard Archival Description[81] provides a guideline for describing archival materials. It defines 26 data elements, the following 6 of which are mandatory:

(1) Reference code, which is a unique identifier that links to the description
(2) Title, which is the name of the record
(3) Name of creator, which is the entity or individual who created or cultivated the record
(4) Dates of creation, which are the dates when the record was created
(5) Extent of the unit of description, which is the number of physical or logical units in arabic numerals and the unit of measurement
(6) Level of description, which is the level of arrangement of the unit of description (eg, fonds, subfonds, series, subseries, files, and items)

33.10.2 Hyperlinks for Archives

Most archives continue to use uniform resource locators (URLs) when referencing the locations of digital objects. However, some archives are changing this practice. For example, OCLC, formerly called the Online Computer Library Center, uses persistent uniform resource locators (PURLs) to which changeable URLs are mapped.[82] The American Chemical Society relies on journal articles' digital object identifiers (DOIs), and it retains the original manuscript numbers assigned to articles at the beginning of the publication process. The Defense Technical Information Center of the US Department of Defense uses the Handle system developed by the Corporation for National Research Initiatives.[83]

33.10.3 International Archives

Some archiving agencies that generally accept e-journals from all over the world are listed below.

33.10.3.1 PubMed Central (PMC)

PubMed Central[3] (PMC) is an archive that the National Library of Medicine in the United States established in 2000. Originally limited to the full text of English biomedical journals, PMC started accepting non-English MEDLINE journals in 2019. Furthermore, non-MEDLINE, non-English journals are now eligible to apply to PMC if they carry a considerable number of English articles. As of March 2022, PMC has archived 7.8 million articles from 2,578 full-participation journals, 41 deposit programs, and 307 "portfolio" journals that publish articles funded by the National Institutes of Health in the United States.

33.10.3.2 LOCKSS

LOCKSS,[84] which stands for "Lots Of Copies Keep Stuff Safe," is a low-cost preservation system for digital literature based at the Stanford Libraries in the United States. One of the system's programs, the Global LOCKSS Network, archives journals. While publishers do not pay fees to deposit materials in this network, only libraries that pay membership fees to LOCKSS can access the archives. As of March 2022, the Global LOCKSS Network has archived approximately 15,500 journals.

33.10.3.3 CLOCKSS

CLOCKSS is an abbreviation for "Controlled LOCKSS." It is a collaboration of the world's leading academic publishers and research libraries.[85] Discontinued journals deposited at CLOCKSS are made available under open access. As of March 2022, CLOCKSS has preserved 48.5 million journal articles and 360,000 books from 435 participating publishers.

33.10.3.4 Portico

Portico[86] is an electronic archive operated by ITHAKA, a nonprofit organization that preserves more than 350,000 electronic journals and e-books.

33.10.3.5 Public Knowledge Project Preservation Network (PKP PN)

Public Knowledge Project Preservation Network (PKP PN) is a LOCKSS-based network limited to journals hosted on the Open Journal Systems (OJS) platform.

33.10.3.6 Fatcat

The Fatcat service is a versioned, user-editable catalog of research publications, including journal articles, conference proceedings, and data sets. Fatcat is hosted on the Internet Archive, which is a nonprofit digital library. As of April 2022, Fatcat includes more than 32 million manuscripts from more than 185,000 journals.

33.10.4 Notification of Discontinuation of a Journal to Relevant Organizations

When a publication is discontinued, it must inform relevant organizations such as indexing databases, libraries that catalog the journal, and the International Standard Serial Number center for the publisher's country. When appropriate notification is provided, the journal's discontinuation is recorded in WorldCat,[87] the world's largest network of library content and services. It is recommended that discontinued archived journals specify which type of Creative Commons license covers their content (see Section 3.3.4, "Creative Commons Licenses"). Journals should also issue a disclaimer stating that they do not take any legal responsibility for previously published journal content.

33.11 MANUSCRIPT MANAGEMENT SYSTEMS

A manuscript management system is an online system that handles a journal's entire process of submission, peer review, and final decision. It is also called an "online submission system" and an "e-submission system."

33.11.1 Widely Used Systems

All large international commercial publishers, as well as most society journals, use manuscript management systems. Two of the commercial systems that are widely used internationally are ScholarOne by Clarivate and Editorial Manager by Elsevier. A well-known free system is Open Journal Systems by the Public Knowledge Project.[88] Many lesser-known e-submission systems exist throughout the world. For example, some countries have specific systems that are used by society journals in those countries.

Any manuscript management system can be implemented smoothly as long as the system is well suited to the journal's needs. To achieve that, editors and publishers should understand their existing workflow well, and they should engage in close communication with the system's engineers during initial construction.

If journal editors and publishers are interested in implementing a new manuscript management system, they should consult publishing colleagues to elicit information about their experiences before contracting companies to install manuscript management systems.

33.11.2 What to Consider When Selecting a Manuscript Management System

Before selecting a manuscript management system, editors and publishers should ask questions such as the following[89]:

(1) How many publishers are customers of the system?
(2) What are the costs for initial implementation, annual maintenance, and processing of the journal's historical average number of annual submissions?
(3) Is the processing fee discounted for manuscripts rejected before peer review?
(4) Are there training expenses for users, such as editors, reviewers, and staff?
(5) What file formats does the system accept?
(6) Is it possible to upload supplementary files or data sets in addition to manuscript files and figure files?
(7) Does the system have analysis and report functions for submissions?
(8) Does the system provide editors with customized service and support?
(9) Are the following functions available?

- Version management and control
- Hosting in the cloud
- Workflow chart
- Automatic emails
- Automatic delivery of files to the editor or publisher
- Reference hyperlinks
- Bibliography check
- Automatic searching of external databases to find similar articles
- Support for author taxonomy
- Support for various formats of submitted picture files
- Automatic retrieval from ORCID of Open Researcher and Contributor ID identifiers and association of those identifiers with submitted manuscripts
- Online signature for copyright transfer
- E-commerce options that allow authors to pay publication fees and article-processing charges
- Manuscript submission by an author other than the corresponding author
- Ability to provide Similarity Check or similar results to peer reviewers
- PDF conversion
- Ability to transfer all existing data should systems be changed later

33.11.3 Manuscript Exchange Common Approach (MECA)

To avert compatibility problems when manuscripts or data are transferred from one system to another, the National Information Standards Organization (NISO) in the United

States established the Manuscript Exchange Common Approach[90] (MECA) project in 2020. NISO's intention is for MECA to define a standard protocol for formatting manuscript metadata and assembling manuscript files so that they can be sent between electronic systems, such as from preprint servers to submission systems and from submission systems to compliance checkers.

33.12 ARTIFICIAL INTELLIGENCE PROGRAMS IN JOURNAL PUBLISHING

Defined as using computers to imitate intelligent human behavior,[16] artificial intelligence (AI) is rapidly increasing in writing, processing, and publishing manuscripts. These programs allow authors and editors to work more efficiently.

Similarity Check[53] and Plagiarism Checker X[91] are among the many AI programs that authors and peer reviewers can use to identify content that would be considered plagiarized from other sources. Programs like Proofig[92] can be used to look for image manipulation, and others like Ripeta[93] can be used to run quality checks and assess manuscripts for reproducibility. Google Translate is a frequently used AI program that translates manuscripts in 81 languages, and Papago[94] translates manuscripts in 15 languages. There are even AI programs that authors can use for selecting target journals.[95]

33.12.1 Target Journal Selection Tools

OA Journal Finder[96] follows a search algorithm and uses a "validated journal index" to assist authors in selecting target journals. Clarivate's Manuscript Matcher[97] applies complex algorithms, Web of Science information, and statistics from Journal Citation Reports to provide a "match score" that suggests impactful journals to submit specific manuscripts to. Elsevier's JournalFinder[98] leverages its in-house "fingerprint engine" and subject-specific vocabularies in identifying journal choices.

33.12.2 Grammar-Checking Tools

AI-backed, grammar-checking tools, such as Grammarly[99] and Wordvice AI,[100] can be used to review manuscripts to determine whether they are ready for submission.

33.12.3 Bibliography and Citation Management Tools

Bibliography and citation management tools are valuable to authors, manuscript editors, and copy editors. To find references that are relevant to a manuscript's topic, tools like CoCites, Citation Gecko, and Connected Papers use 2 innovative, unconventional methods: "Co-citation" determines the frequency with which 2 or more research articles appear together in the same reference lists, and "bibliographic coupling" is a measure of the interdependence between 2 articles, based on the presence of shared references.[101] Other reference management programs include EndNote[102] and Mendeley,[103] both of which are widely used.

33.12.4 Peer-Review and Quality-Assessment Tools

AIRA[104] from Frontiers is one of the first AI-supported tools used for peer review of scholarly manuscripts. In reviewing manuscripts, AIRA makes recommendations regarding grammar, style, figures, and legends; identifies potentially plagiarized content; and issues warnings about potential competing interests. PubSURE Report,[105] an AI-backed assessment tool trained with millions of published articles, examines "reporting hygiene" related to readability, adherence, and comprehensiveness. StatCheck[106] helps psychology journal editors detect statistical errors in submitted manuscripts.

CITED REFERENCES

1. Chi Y. Scientific publishing in the Asian century: an international perspective. Sci Ed. 2016;3(2):112–115. https://doi.org/10.6087/kcse.76

2. BioMed Central. Wikipedia; [updated 2022 Oct 13]. https://en.wikipedia.org/wiki/BioMed_Central

3. PubMed Central. National Library of Medicine (US); [accessed 2023 Feb 11]. https://www.ncbi.nlm.nih.gov/pmc/

4. Huh S. To broaden the horizon of the Journal of Educational Evaluation for Health Professions. J Educ Eval Health Prof. 2006;3(2):112–115. https://doi.org/10.3352/jeehp.2006.3.1

5. The Digital Standard. Wikipedia; [updated 2022]. https://en.wikipedia.org/wiki/The_Digital_Standard

6. Journal Article Tag Suite. National Library of Medicine (US), National Center for Biotechnology Information; [accessed 2023 Feb 11]. http://jats.nlm.nih.gov/

7. Dublin Core Metadata Innovation. Dublin Core Metadata Initiative; c1995–2022. https://www.dublincore.org/

8. International Organization for Standardization. Information and documentation—digital object identifier system. International Organization for Standardization; 2012. (ISO 26324:2012). https://www.iso.org/standard/43506.html

9. International Organization for Standardization. Digital publishing—EPUB accessibility—conformance and discoverability requirements for EPUB publications. International Organization for Standardization; 2021. (ISO/IEC 23761:2021). https://www.iso.org/standard/76860.html

10. PubMed.gov. National Library of Medicine (US); [accessed 2023 Feb 10]. https://pubmed.ncbi.nlm.nih.gov/

11. Metaverse. Wikipedia; [accessed 2023 Feb 10]. https://en.wikipedia.org/wiki/Metaverse

12. Kim K. Metaverse in journal publishing. Sci Ed. 2022;9(1):1–2. https://doi.org/10.6087/kcse.256

13. International Standard Serial Number International Centre. [updated 2022 Jul 15]. https://www.issn.org

14. International Organization for Standardization. Information and documentation—rules for the abbreviation of title words and titles of publications. International Organization for Standardization; 1997 [confirmed 2020]. (ISO 4:1997). https://www.iso.org/standard/3569.html

15. NLM catalog: journals referenced in the NCBI databases. National Library of Medicine (US); [accessed 2023 Feb 11]. https://www.ncbi.nlm.nih.gov/nlmcatalog/journals/

16. License. Merriam-Webster dictionary. Merriam-Webster; [accessed 2023 Feb 10]. https://www.merriam-webster.com/

17. Creative Commons license. Wikipedia; [accessed 2023 Feb 10]. https://en.wikipedia.org/wiki/Creative_Commons_license

18. Welcome to AGRICOLA. National Agricultural Library (US). https://agricola.nal.usda.gov/

19. BIOSIS™ Previews on Web of Science. Clarivate. c2022. https://clarivate.com/webofsciencegroup/solutions/webofscience-biosis-previews/

20. CAS. American Chemical Society' c2022. https://www.cas.org

21. CINAHL Database. EBSCO; [accessed 2023 Feb 10]. https://www.ebsco.com/products/research
-databases/cinahl-database

22. Directory of Open Access Journals: find open access journals & articles. DOAJ; [accessed 2023 Feb 10].
https://doaj.org

23. Embase content coverage. Elsevier; c2022. https://www.elsevier.com/solutions/embase-biomedical
-research/coverage-and-content

24. Engineering Village. Elsevier; c2022. https://www.engineeringvillage.com

25. Web of Science Group: master journal list. Clarivate; c2022. https://mjl.clarivate.com/search-results

26. Welcome to Scopus Preview. Elsevier, BV; [accessed 2023 Feb 10]. https://scopus.com

27. J-J-STAGE. Japan Science and Technology Agency; [accessed 2023 Feb 10]. https://www.jstage.jst.go.jp

28. Korea Citation Index. National Research Foundation of Korea; [accessed 2023 Feb 10]. https://www
.kci.go.kr

29. SRCE. Croatian Scientific and Professional Journals; [accessed 2023 Feb 10]. https://www.srce.unizg
.hr/en/

30. Russian Science Citation Index. Clarivate; c2022. http://webofscience.help.clarivate.com/Content
/russian-science/russian-science-citation-index.htm

31. SciELO Citation Index. Clarivate; c2020. http://wokinfo.com/products_tools/multidisciplinary/scielo/

32. Sinta is evolving. Republic of Indonesia, Ministry of Education, Culture, Research, and Technology;
[accessed 2023 Feb 10]. https://sinta.kemdikbud.go.id

33. Difference between URL and domain name. Pediaa.COM; [2018 Jun 15]. https://pediaa.com/difference
-between-url-and-domain-name

34. Chang JH. An introduction to using QR codes in scholarly journals. Sci Ed. 2014;1(2):113–117. https://
doi.org/10.6087/kcse.2014.1.113

35. Principles of transparency and best practice in scholarly publishing. 4th ed. Committee on Publication Ethics, Directory of Open Access Journals, Open Access Scholarly Publishers Association, and World
Association of Medical Editors; 2022. https://doi.org/10.24318/cope.2019.1.12

36. Choi HW, Choi YJ, Kim S. Compliance of "principles of transparency and best practice in scholarly
publishing" in academic society published journals. Sci Ed. 2019;6(2):112–121. https://doi.org/10.6087
/kcse.171

37. Choi YJ, Choi HW, Kim S. Compliance of "principles of transparency and best practice in scholarly publishing" in Korean academic society-published journals listed in Journal Citation Reports. Sci Ed.
2020;7(1):24–33. https://doi.org/10.6087/kcse.186

38. Definition of "Https." The Economic Times; [accessed 2023 Feb 10]. https://economictimes
.indiatimes.com/definition/https

39. Jackson B. HTTP vs HTTPS: what is the difference between HTTP and HTTPS? KeyCDN; c2016 [updated
2022 Jun 1]. https://www.keycdn.com/blog/difference-between-http-and-https

40. HTML. Wikipedia; [updated 2022 Dec 29]. https://en.wikipedia.org/wiki/HTML

41. Huh S. Journal Article Tag Suite 1.0: National Information Standards Organization standard of
journal extensible markup language. Sci Ed. 2014;1(2):99–104. https://doi.org/10.6087/kcse.2014.1.99

42. jats-xml-to-pdf. GitHub, Inc; c2021. https://github.com/elifesciences/jats-xml-to-pdf

43. NLM/JATS to CrossRef Deposit XML XSLT. GitHub, Inc; c2022. https://github.com/CrossRef/jats
-crossref-xslt

44. PeerJ/jats-conversion. GitHub, Inc; c2020. https://github.com/PeerJ/jats-conversion

45. PubReader. National Library of Medicine (US); [updated 2022 Apr 28]. https://www.ncbi.nlm.nih
.gov/labs/pmc/about/pubreader/

46. NCBI/PubReader. GitHub, Inc; c2022. https://github.com/ncbi/PubReader

47. Garrish M, Cramer D, editors. EPUB 3.2. W3C Community Group; c1999–2019. https://www.w3.org
/publishing/epub32/epub-spec.html

48. DOI Registration Agencies. International DOI Foundation; c2022. https://doi.org/registration
_agencies.html

49. Pentz E. Role of Crossref in journal publishing over the next decade. Sci Ed. 2022;9(1):53–57. https://doi.org/10.6087/kcse.263

50. Lammey R. CrossRef developments and initiatives: an update on services for the scholarly publishing community from CrossRef. Sci Ed. 2014;1(1):13–18. https://doi.org/10.6087/kcse.2014.1.13

51. Lammey R. How to apply CrossMark and FundRef via CrossRef extensible markup language. Sci Ed. 2014;1(2):84–90. https://doi.org/10.6087/kcse.2014.1.84

52. Lammey R. Funders: how Crossref fits with funders. Crossref; [updated 2020 Dec 1]. https://www.crossref.org/community/funders/

53. Luschek K. Similarity Check: a service provided by Crossref and powered by iThenticate—Similarity Check provides editors with a user-friendly tool to help detect plagiarism. Crossref; [updated 2020 Apr 08]. https://www.crossref.org/services/similarity-check/

54. Kemp J. Metadata retrieval: the collective power of our members' metadata is available to use through a variety of tools and APIs—allowing anyone to search and reuse the metadata in sophisticated ways. Crossref; [updated 2020 Apr 08]. https://www.crossref.org/services/metadata-retrieval/

55. Rittman M. Event data. Crossref; [updated 2020 Sep 8]. https://www.crossref.org/services/event-data/

56. Tolwinska A. Participation reports help Crossref members drive research further. Sci Ed. 2021; 8(2):180–185. https://doi.org/10.6087/kcse.253

57. Tolwinska A. Documentation: participation reports. Crossref; [updated 2020 Apr 8]. https://www.crossref.org/documentation/reports/participation-reports/

58. Lammey R. CrossRef text and data mining services. Sci Ed. 2015;2(1):22–27. https://doi.org/10.6087/kcse.32

59. Lammey R. Solutions for identification problems: a look at the Research Organization Registry. Sci Ed. 2020;7(1):65–69. https://doi.org/10.6087/kcse.192

60. Uniform Resource Name. Wikipedia; [updated 2022 Jun 12]. https://en.wikipedia.org/wiki/Uniform_Resource_Name

61. ISNI. ISNI International Agency, Ltd; [accessed 2023 Feb 11]. https://isni.org/

62. ORCID.Open Researcher and Contributor ID; [accessed 2023 Feb 11]. https://orcid.org/

63. Understanding journal metrics: using data to choose a journal for submission. Taylor & Francis; c2022. https://authorservices.taylorandfrancis.com/publishing-your-research/choosing-a-journal/journal-metrics/

64. Kim K, Chung Y. Overview of journal metrics. Sci Ed. 2018;5(1):16–20. https://doi.org/10.6087/kcse.112

65. Szomszor M. Introducing the Journal Citation Indicator: a new, field-normalized measurement of journal citation impact. Clarivate; 2021 May 20. https://clarivate.com/blog/introducing-the-journal-citation-indicator-a-new-field-normalized-measurement-of-journal-citation-impact/

66. SJR: Scimago Journal & Country Rank. Scimago Lab; c2007–2022. http://www.scimagojr.com

67. Guerrero-Bote VP, Moya-Anegón F. A further step forward in measuring journals' scientific prestige: the SJR2 indicator. J Informetr. 2012;6(4):674–688. https://doi.org/10.1016/j.joi.2012.07.001

68. Who's talking about your research? Altmetric; [accessed 2023 Feb 10]. https://www.altmetric.com/

69. Principles and obstacles for sharing data from environmental health research: workshop summary. National Academies Press (US); 2016. https://www.ncbi.nlm.nih.gov/books/NBK362433/

70. Wiley's data sharing policies. Wiley; c2000–2022. https://authorservices.wiley.com/author-resources/Journal-Authors/open-access/data-sharing-citation/data-sharing-policy.html

71. re3data.org—Registry of Research Data Repositories. [accessed 2023 Feb 11]. https://doi.org/10.17616/R3D

72. Repository Finder. DataCite; [accessed 2023 Feb 11]. https://repositoryfinder.datacite.org/

73. Generalist repositories. Scientific Data; [accessed 2023 Feb 11]. https://www.nature.com/sdata/policies/repositories#general

74. Preprints. Committee on Publication Ethics; [accessed 2023 Feb 11]. https://publicationethics.org/node/38176

75. Xie B, Shen Z, Wang K. Is preprint the future of science? A thirty year journey of online preprint services. arXiv; 2021 Feb 17. https://arxiv.org/abs/2102.09066

76. Choi YJ, Choi HW, Kim S. Preprint acceptance policies of Asian academic society journals in 2020. Sci Ed. 2021;8(1):10–17. https://doi.org/10.6087/kcse.224

77. Graphical abstract. Wikipedia; [updated 2022 Dec 12]. https://en.wikipedia.org/wiki/Graphical_abstract

78. Huh S. Revision of the instructions to authors to require a structured abstract, digital object identifier of each reference, and author's voice recording may increase journal access [editorial]. J Educ Eval Health Prof. 2013;10:3. https://doi.org/10.3352/jeehp.2013.10.3

79. Video abstract. Wikipedia; [updated 2022 Dec 23]. https://en.wikipedia.org/wiki/Video_abstract

80. What is digital archiving? Falcon Document Solutions; 2019 Mar 1. http://falcondocs.com/blogs/2019/03/01/what-is-digital-archiving

81. International Council on Archives, Sub-Committee on Descriptive Standards. ISAD(G): General International Standard Archival Description—second edition. International Council on Archives; 2011 Sep 1. https://www.ica.org/en/isadg-general-international-standard-archival-description-second-edition

82. PURL administration. Internet Archive. [accessed 2023 Feb 11]. http://purl.oclc.org/

83. HDL.NET information services. Handle.Net Registry. Corporation for National Research Initiatives; 2023 Feb 2. http://www.handle.net/

84. LOCKSS. Stanford University; [accessed 2023 Feb 11]. https://www.lockss.org

85. CLOCKSS. CLOCKSS; [accessed 2023 Feb 11]. https://clockss.org

86. Portico. ITHAKA; [accessed 2023 Feb 11]. https://www.portico.org

87. WorldCat. WorldCat; [accessed 2023 Feb 11]. https://www.worldcat.org

88. Open Journal Systems. Public Knowledge Project; [accessed 2023 Feb 11]. https://pkp.sfu.ca/ojs/

89. Kim S, Choi H, Kim N, Chung EK, Lee JY. Comparative analysis of manuscript management systems for scholarly publishing. Sci Ed. 2018;5(2):124–134. https://doi.org/10.6087/kcse.137

90. Randall L, Ubnoske S. MECA and JATS compatibility: a case study utilizing the JATS Compatibility Meta Model. In: Journal Article Tag Suite Conference (JATS-Con) Proceedings 2020 [Internet]. National Library of Medicine (US), National Center for Biotechnology Information; 2020. https://www.ncbi.nlm.nih.gov/books/NBK555480/

91. #1 plagiarism checker. Plagiarism Checker X, LLC; [accessed 2023 Feb 11]. https://plagiarismcheckerx.com/

92. Proofig. Proofiger, Ltd; [accessed 2023 Feb 11]. https://www.proofig.com/

93. Ripeta. Ripeta, c2012–2023. https://ripeta.com/

94. Papago web translator. NAVER Corp; [accessed 2023 Feb 11]. https://papago.naver.com

95. Abdul Razack HI, Mathew ST, Ahmad Saad FF, Alqahtani SA. Artificial intelligence-assisted tools for redefining the communication landscape of the scholarly world. Sci Ed. 2021;8(2):134–144. https://doi.org/10.6087/kcse.244

96. Open Access Journal Finder: find the best suited English journals for your paper. Enago; c2022. https://www.enago.com/academy/journal-finder/

97. Introducing: Manuscript Matcher. Clarivate EndNote. 2022. https://endnote.com/product-details/manuscript-matcher/

98. Find the perfect journal for your article. Elsevier, BV, JournalFinder; [accessed 2023 Feb 11]. https://journalfinder.elsevier.com/about

99. Grammarly: great writing, simplified. Grammarly, Inc; [accessed 2023 Feb 11]. https://www.grammarly.com/1

100. Heintz K, Roh Y, Lee J. Comparing the accuracy and effectiveness of Wordvice AI Proofreader to two automated editing tools and human editors. Sci Ed. 2022;9(1):37–45. https://doi.org/10.6087/kcse.261

101. Gadd E. AI-based citation evaluation tools: good, bad or ugly? The Bibliomagician; 2020 Jul 23. https://thebibliomagician.wordpress.com/2020/07/23/ai-based-citation-evaluation-tools-good-bad-or-ugly/

102. Accelerate your research with EndNote 20. Clarivate EndNote; [accessed 2023 Feb 11]. https://endnote.com/product-details/compare-previous-versions

103. Mendeley Reference Manager. Mendeley, Ltd. c2022. https://www.mendeley.com/reference-management/reference-manager

104. Frontiers Communications. Artificial intelligence to help meet global demand for high-quality, objective peer-review in publishing. Frontiers Science News. 2020 Jul 1. https://blog.frontiersin.org/2020/07/01/artificial-intelligence-to-help-meet-global-demand-for-high-quality-objective-peer-review-in-publishing/

105. The first AI-powered manuscript submission marketplace connecting authors and journals. Editage Insights. 2019 Sep 5. https://www.editage.com/insights/the-first-ai-powered-manuscript-submission-marketplace-connecting-authors-and-journals

106. Epskamp S, Nuijten MB. StatCheck: extract statistics from articles and recompute p values (version 1.3.1). Package StatCheck; [accessed 2018 May 4]. https://cran.r-project.org/web/packages/statcheck/index.html

ADDITIONAL REFERENCES

ARK overview: what ARKs are and why you would use them. ARK Alliance; [updated 2021 Mar 30]. https://arks.org/about/ark-overview/

Data citation. Dataverse Project; [accessed 2023 Feb 11]. https://dataverse.org/best-practices/data-citation

Data Citation Synthesis Group. Joint declaration of data citation principles. Martone M, editor. FORCE11; 2014. https://doi.org/10.25490/a97f-egyk

Hennessey J, Ge SX. A cross disciplinary study of link decay and the effectiveness of mitigation techniques. BMC Bioinformatics. 2013;14(Suppl 14):S5. https://doi.org/10.1186/1471-2105-14-S14-S5

Huh S. Coding practice of the Journal Article Tag Suite extensible markup language. Sci Ed. 2014;1(2):105–112. https://doi.org/10.6087/kcse.2014.1.105

International Organization for Standardization. Information technology—automatic identification and data capture techniques —QR code bar code symbology specification. International Organization for Standardization; 2015. (ISO/IEC 18004:2015). https://www.iso.org/standard/62021.html

Kelly L. JATS to EPUB: Unraveling the mystery. Journal Article Tag Suite Conference (JATS-Con) Proceedings 2010. National Library of Medicine (US), National Center for Biotechnology Information; 2010. https://www.ncbi.nlm.nih.gov/books/NBK47314/

Kemp J. Documentation: REST API. Crossref; [updated 2020 Apr 8]. https://www.crossref.org/documentation/retrieve-metadata/rest-api/

Kim K, Chung Y. Overview of journal metrics. Sci Ed. 2018;5(1):16–20. https://doi.org/10.6087/kcse.112

Kim SY, Yi HJ, Huh S. Current and planned adoption of data sharing policies by editors of Korean scholarly journals. Sci Ed. 2019;6(1):19–24. https://doi.org/10.6087/kcse.151

Kim YC. Korean Institute of Medical Education and Evaluation presidential address: the role of KIMEE as a medical education accreditation agency during the coronavirus disease 2019 pandemic [editorial]. J Educ Eval Health Prof. 2021;18:2. https://doi.org/10.3352/jeehp.2021.18.2

Lammey R. The basics of CrossRef extensible markup language. Sci Ed. 2014;1(2):76–83. https://doi.org/10.6087/kcse.2014.1.76

Lammey R. Crossref at 20 years: what do the community need? Sci Ed. 2020;7(2):125–129. https://doi.org/10.6087/kcse.206

Lammey R. How publishers can work with Crossref on data citation. Sci Ed. 2019;6(2):166–170. https://doi.org/10.6087/kcse.165

National Information Standards Organization. The Dublin Core Metdata element set. NISO Press; 2013. (ANSI/NISO Z39.85-2012). http://www.niso.org/publications/ansiniso-z3985-2012-dublin-core-metadata-element-set

Park HJ. How to share data through Harvard Dataverse, a repository site: a case of the World Journal of Men's Health. Sci Ed. 2022;9(1):85–90. https://doi.org/10.6087/kcse.270

Registry of open access repositories. University of Southampton, School of Electronics and Computer Science; [accessed 2023 Feb 11]. http://roar.eprints.org

Root Zone Database. Internet Assigned Numbers Authority. https://www.iana.org/domains/root/db

ROR search. Research Organization Registry; [accessed 2023 Feb 11]. https://ror.org/search

Schmidt HE. Scientific, technical, and medical podcasting in Korea. Sci Ed. 2016;3(1):43–48. https://doi
.org/10.6087/kcse.62

Smart P. The evolution, benefits, and challenges of preprints and their interaction with journals. Sci Ed.
2022;9(1):79–84. https://doi.org/10.6087/kcse.269

Types of QR code. Denso Wave, Inc; [accessed 2023 Feb 11]. https://www.qrcode.com/en/codes/index.html

Universal Numerical Fingerprint (UNF). Dataverse Project; [updated 2021 Dec 13]. https://guides.dataverse
.org/en/5.9/developers/unf/index.html?highlight=unf

Vitaliy-1/docxToJats. GitHub, Inc. c2023. https://github.com/Vitaliy-1/docxToJats

Welcome to S-Space. Seoul National University Library; [accessed 2023 Feb 11]. https://s-space.snu.ac.kr

Yoon JW, Chung EK, Schalk J, Kim J. Examination of data citation guidelines in style manuals and data
repositories. Learn Publ. 2020;34(2):198–215. https://doi.org/10.1002/leap.1349

ADDITIONAL RESOURCES

Style Manuals
    General
    Scientific

Dictionaries
    General
    Scientific

Usage and Prose Style

Editing

Publishing

Graphics and Design

Standards for Editing and Publishing

The works listed here are books and other sources useful in scientific writing, editing, and publishing. Additional sources, many of them with a narrower scope, are listed under "References" or "Additional References" at the end of each chapter.

STYLE MANUALS
General
Australian Public Service Commission. Style manual. Commonwealth of Australia; c2022. https://www.stylemanual.gov.au/

The Chicago manual of style: the essential guide for writers, editors, and publishers. 17th ed. The University of Chicago Press; 2017. Also available at https://www.chicagomanualofstyle.org/

Columbia Law Review, Harvard Law Review, University of Pennsylvania Law Review, Yale Law Journal, compilers. The bluebook: a uniform system of citation. 21st ed. The Harvard Law Review Association; 2020. Also available at https://www.legalbluebook.com/

Hacker D, Sommers N. A pocket style manual. 9th ed. Bedford/St. Martin's; c2021.

Kanigel R, editor. The diversity style guide. https://www.diversitystyleguide.com/

Lunsford AA. Easy writer: a pocket guide. 8th ed. Bedford/St. Martin's; c2022.

Modern Language Association. MLA handbook. 9th ed. Modern Language Association of America; 2021.

New Hart's rules: the Oxford style guide. 2nd ed. Oxford University Press; 2014.

Public Services and Procurement Canada, Translation Bureau. Writing tips plus; [modified 5 May 2022]. https://www.noslangues-ourlanguages.gc.ca/en/writing-tips-plus/index-eng

Skillin ME, Gay RM. Words into type. 3rd ed. Prentice Hall; 1974.

Strunk W, White EB. The elements of style. 4th ed. Simon and Schuster Trade; 1999.

Turabian KL. A manual for writers of term papers, theses, and dissertations: Chicago style for students and researchers. 9th ed. The University of Chicago Press; 2018.

US Government Publishing Office. Style manual: an official guide to the form and style of federal government publishing. US Government Publishing Office; 2016. https://www.govinfo.gov/collection/gpo-style-manual

Virag K, editor. Editing Canadian English: a guide for editors, writers, and everyone who works with words. 3rd ed. Editors' Association of Canada; 2015.

Waddingham A. New Oxford style manual. Oxford University Press; 2016.

Yin K. Conscious style guide. https://consciousstyleguide.com/

Younging G. Elements of Indigenous style. Brush; 2018.

Scientific

AIP style manual. 4th ed. American Institute of Physics; 1990. https://publishing.aip.org/wp-content /uploads/2021/03/AIP_Style_4thed.pdf

American Society of Agronomy. Publications handbook and style manual. American Society of Agronomy; [updated November 2021]. Jointly published with the Crop Science Society of America and the Soil Science Society of America. https://www.agronomy.org/publications/journals/author-resources/style-manual/

ASA style guide. 6th ed. American Sociological Association; 2019.

ASM style manual for journals and books. American Society for Microbiology; 1991.

Banik GM, Baysinger G, Kamat PV, Pienta NJ, editors. ACS guide to scholarly communications. American Chemical Society; c2022. https://pubs.acs.org/page/acsguide

Bates RL, Adkins-Heljeson MD, Buchanan RC, editors. Geowriting: a guide to writing, editing, and printing in earth science. 5th rev. ed. American Geological Institute; 2004.

Browner WS. Publishing and presenting clinical research. 3rd ed. Lippincott Williams & Wilkins; c2012.

Day RA, Gastel B. How to write and publish a scientific paper. 9th ed. Greenwood; 2022.

Day RA, Sakaduski N. Scientific English: a guide for scientists and other professionals. 3rd ed. Greenwood; 2011.

Gustavii B. How to write and illustrate a scientific paper. 2nd ed. Cambridge University Press; 2008. https:// doi.org/10.1017/CBO9780511808272

Hall GM, editor. How to write a paper. 5th ed. BMJ Books; 2012.

Hansen WR, editor. Suggestions to authors of the reports of the United States Geological Survey. 7th ed. United States Geological Survey; 1991. https://doi.org/10.3133/7000088

Huth EJ. Writing and publishing in medicine. 3rd ed. Williams & Wilkins; 1999.

International Committee of Medical Journal Editors. Recommendations for the conduct, reporting, editing, and publication of scholarly work in medical journals. International Committee of Medical Journal Editors; [updated 2022 May] https://www.icmje.org/recommendations/

JAMA Network Editors. AMA manual of style: a guide for authors and editors. 11th ed. Oxford University Press; 2020. Also available at https://www.amamanualofstyle.com

Lang TA, Secic M. How to report statistics in medicine: annotated guidelines for authors, editors, and reviewers. 2nd ed. American College of Physicians; 2006.

Matthews JR, Matthews RW. Successful scientific writing: a step-by-step guide for the biological and medical sciences. 4th ed. Cambridge University Press; 2014. https://doi.org/10.1017/CBO9781107587915

McMillan VE. Writing papers in the biological sciences. 7th ed. Bedford/St. Martin's; 2021.

Montgomery SL. The Chicago guide to communicating science. 2nd ed. The University of Chicago Press; 2017.

Patrias K. Citing medicine: the NLM style guide for authors, editors, and publishers. 2nd ed. National Library of Medicine; 2007 [updated 2015 Oct 2]. https://www.ncbi.nlm.nih.gov/books/NBK7256/

Publication manual of the American Psychological Association. 7th ed. American Psychological Association; 2020.

Swanson E, O'Sean A, Schleyer A. Mathematics into type. Updated ed. American Mathematical Society; 1999.

Terryberry KJ. Writing for the health professions. 2nd ed. XanEdu; 2017.

Waldron A, Judd P, Miller V, editors. Physical Review style and notation guide. American Physical Society; 1983 [revised 1993 Feb; minor revision 2005 Jun; minor revision 2011 Jun]. https://cdn.journals.aps .org/files/styleguide-pr.pdf

Wheatley D. Scientific writing and publishing: a comprehensive manual for authors. Cambridge University Press; 2021. https://doi.org/10.1017/9781108891899

WHO style guide. 2nd ed. World Health Organization; 2013.

Wilkins GA. IAU style manual. International Astronomical Union; 1989. https://www.iau.org/static /publications/stylemanual1989.pdf

Zeiger M. Essentials of writing biomedical research papers. 2nd ed. McGraw-Hill, Health Professions Division; 2000.

DICTIONARIES

General

The American heritage dictionary. 5th ed, updated. Houghton Mifflin Harcourt; 2015. https://ahdictionary .com/

Cambridge essential British English dictionary. Cambridge University Press; [continually updated]. https:// dictionary.cambridge.org/us/dictionary/essential-british-english/

The Chambers dictionary. 13th ed. Chambers; 2014.

Lindberg CA, compiler. The Oxford American writer's thesaurus. 3rd ed. Oxford University Press; 2012.

Longman dictionary of American English. 5th ed. Pearson; 2014. https://www.ldoceonline.com/
Merriam-Webster dictionary. Merriam-Webster; [continually updated]. https://www.merriam-webster.com/
New Oxford dictionary for writers and editors. 2nd rev ed. Oxford University Press; 2014.
Oxford English dictionary. 2nd ed. Oxford University Press; [continually updated]. https://www.oed.com/

Scientific

Allaby M. A dictionary of geology and earth sciences. 5th ed. Oxford; 2020. https://doi.org/10.1093/acref
/9780198839033.001.0001
ASTM dictionary of engineering science and technology. 10th ed. ASTM International; 2005.
Atkins T, Escudier M. A dictionary of mechanical engineering. Oxford University Press; 2013. https://doi
.org/10.1093/acref/9780199587438.001.0001
Dictionary of physics. Palgrave Macmillan; 2004. An updated translation of the Lexicon der Physik, published in German.
Dorland's medical dictionary online. Elsevier; c2022. Includes Dorland's illustrated medical dictionary.
https://www.dorlandsonline.com/dorland/dictionary
Earl R, Nicholson J. The concise Oxford dictionary of mathematics. 6th ed. Oxford University Press; 2021.
https://doi.org/10.1093/acref/9780198845355.001.0001
Hine R, editor. A dictionary of biology. 8th ed. Oxford University Press; 2019. Also available at https://doi
.org/10.1093/acref/9780198821489.001.0001
Jerrard HG, McNeil DB. A dictionary of scientific units: including dimensionless numbers and scales. 6th
ed. Springer; 1992. Also available at https://link.springer.com/book/10.1007/978-94-011-2294-8
Larrañaga MD, Lewis RJ Sr, Lewis RA. Hawley's condensed chemical dictionary. 16th ed. Wiley; 2016. https://
doi.org/10.1002/9781119312468
Law J, editor. A dictionary of science. 7th ed. Oxford, 2017.
Law J, Rennie R, editors. A dictionary of chemistry. 8th ed. Oxford University Press; 2020. https://doi.org
/10.1093/acref/9780198841227.001.0001
Lawrence E, editor. Henderson's dictionary of biology. 16th ed. Pearson; 2016.
McGraw-Hill dictionary of mathematics. 2nd ed. McGraw-Hill; c2003.
Nelson D, editor. The Penguin dictionary of mathematics. 4th ed. Penguin Books; 2008.
Rennie R, Law J, editors. A dictionary of physics. 8th ed. Oxford University Press; 2019. https://doi.org/10
.1093/acref/9780198821472.001.0001
Stedman's medical abbreviations, acronyms & symbols. 5th ed. Lippincott Williams & Wilkins; 2012.
Stedman's online. Wolters Kluwer Health; c2022. https://stedmansonline.com/

USAGE AND PROSE STYLE

Many books offer guidance on clear writing and appropriate usage. This list includes
primarily guides with broad scope.

Aarts B. The Oxford dictionary of English grammar. 2nd ed. Oxford University Press; 2014. https://doi.org
/10.1093/acref/9780199658237.001.0001
Baron D. What's your pronoun: beyond he and she. Liveright; 2020.
Bernstein TM. The careful writer: a modern guide to English usage. Atheneum; 1973.
Brown RW. Composition of scientific words. Rev. ed. Smithsonian Institution Press; 1956.
Butterfield J, editor. Fowler's dictionary of modern English usage. 4th ed. Oxford University Press; 2015.
Casagrande J. The joy of syntax: a simple guide to all the grammar you know you should know. Ten Speed
Press; 2018.
Dreyer B. Dreyer's English. Random House; 2019.
Fogarty M. Grammar Girl's quick and dirty tips for better writing. Holt Paperbacks; 2008. https://www.quick
anddirtytips.com/
Garner BA. The Chicago guide to grammar, usage, and punctuation. The University of Chicago Press; 2016.
Garner BA. Garner's modern English usage. 5th ed. Oxford; forthcoming.
Gower E. The complete plain words. Gowers R, reviser. Penguin Books; 2015.
Seely J. Oxford A–Z of grammar and punctuation. 3rd ed. Oxford University Press; 2020.
Truss L. Eats, shoots & leaves: the zero tolerance approach to punctuation. Gotham Books; 2004.
Walsh B. The elephants of style: a trunkload of tips on the big issues and gray areas of contemporary
American English. McGraw-Hill; c2004.
Williams JM, Bizup J. Style: lessons in clarity and grace. 12th ed. Pearson; 2016.

EDITING

ACES: The Society for Editing (US). https://aceseditors.org/

Bűky E, Schwartz M, Einsohn A. The copyeditor's workbook: exercises and tips for honing your editorial judgment. University of California Press; 2019.

Butcher J, Drake C, Leach M. Butcher's copy-editing: the Cambridge handbook for editors, copy-editors and proofreaders. 4th ed. Cambridge University Press; 2006.

Chartered Institute of Editing and Proofreading (UK). https://www.ciep.uk/

Editor & Publisher (E&P). Curated Experiences Group. Vol. 1, 1901–. https://www.editorandpublisher.com/

Editorial Freelancers Association (US). https://www.the-efa.org/

Editors of Color Database. https://editorsofcolor.com/database/

Einsohn A, Schwartz M. The copyeditor's handbook: a guide for book publishing and corporate communications. 4th ed. University of California Press; 2019.

Flann E, Hill B, Wang L. The Australian editing handbook. 3rd ed. Wiley; 2014.

Germano W. On revision: the only writing that counts. The University of Chicago Press; 2021.

Ginna, Peter. What editors do: the art, craft, and business of book editing. The University of Chicago Press; 2017.

Professional editorial standards. Editors Canada; 2016. https://www.editors.ca/publications/professional-editorial-standards-2016

Rude C, Eaton A. Technical editing. 5th ed. Longman; 2010.

Saller CF. The subversive copy editor. 2nd ed. The University of Chicago Press; 2016.

The Scholarly Kitchen. Society for Scholarly Publishing. https://scholarlykitchen.sspnet.org/

Science Editor. Council of Science Editors; 2000–. Continues: CBE Views. https://www.csescienceeditor.org/

Smart P, Maisonneuve H, Polderman A, editors. Science editors' handbook. 2nd ed. European Association of Science Editors; 2013. https://ease.org.uk/publications/science-editors-handbook/

White JV. Editing by design: for designers, art directors, and editors. Allworth Press; 2003.

PUBLISHING

Abel R, Newlin L, editors. Scholarly publishing: books, journals, publishers, and libraries in the twentieth century. John Wiley & Sons; c2002.

Bailey CW Jr. Scholarly electronic publishing bibliography 2010. Digital Scholarship; 2010. https://www.digital-scholarship.org/sepb/annual/sepb2010.pdf

Belcher WL. Writing your journal article in twelve weeks: a guide to academic publishing success. 2nd ed. The University of Chicago Press; 2019.

Clark GN, Phillips A. Inside book publishing. 6th ed. Routledge; 2019.

Hames I. Peer review and manuscript management in scientific journals: guidelines for good practice. Wiley; 2007. https://doi.org/10.1002/9780470750803

Kasdorf WE, editor. The Columbia guide to digital publishing. Columbia University Press; 2003.

Lee M. Bookmaking: the illustrated guide to editing, design and production. 3rd ed. W.W. Norton & Company; 2009.

Page G, Campbell R, Meadows J. Journal publishing. Rev ed. Cambridge University Press; 1997.

GRAPHICS AND DESIGN

Bringhurst R. The elements of typographic style. 4th ed. Hartley & Marks; 2013.

Cleveland WS. Visualizing data. Hobart Press; 1993.

Felici J. The complete manual of typography: a guide to setting perfect type. 2nd ed. Adobe Press; 2011.

Hill W. The complete typographer: a manual for designing with type. 3rd ed. Prentice Hall; 2010.

Hodges ERS, editor. The Guild handbook of scientific illustration. 2nd ed. John Wiley & Sons; 2003.

Lupton E. Thinking with type: a critical guide for designers, writers, editors, and students. Princeton Architectural Press; 2010.

Rosen S. Describing visual resources toolkit: describing visual resources for accessibility in arts & humanities publications. https://describingvisualresources.org/

Tufte ER. The visual display of quantitative information. 2nd ed. Graphics Press; 2001.

STANDARDS FOR EDITING AND PUBLISHING

In the United States, standards issued by the National Information Standards Organization (NISO) are available from NISO Press (https://www.niso.org/publications/press). A NISO-developed and approved standard becomes an American National Standard after the American National Standards Institute (ANSI) verifies that the process for approval has met the ANSI criteria.

Standards of the International Organization for Standardization (ISO) can be purchased from ISO (https://www.iso.org/) or the American National Standards Institute (https://www.ansi.org/).

In Canada, the standards issued by the Canadian Standards Association and by ANSI, ISO, and other standards organization can be purchased from the Standards Council of Canada (https://www.scc.ca/).

British standards can be purchased from BSI (the British Standards Institution, https://www.bsigroup.com/); in the United States, they can be purchased from the American National Standards Institute.

See the individual chapters of this manual for specific standards.

INDEX

Page numbers followed by t indicate a table. Page numbers followed by f indicate a figure. Page numbers followed by App indicate an appendix.